**Human and Animal Filariases**

## *Titles of the Series "Drug Discovery in Infectious Diseases"*

Selzer, P.M. (ed.)

**Antiparasitic and Antibacterial Drug Discovery**

**From Molecular Targets to Drug Candidates**

2009
Print ISBN: 978-3-527-32327-2, also available in digital formats

Becker, K. (ed.)

**Apicomplexan Parasites**

**Molecular Approaches toward Targeted Drug Development**

2011
Print ISBN: 978-3-527-32731-7, also available in digital formats

Caffrey, C.R. (ed.)

**Parasitic Helminths**

**Targets, Screens, Drugs and Vaccines**

2012
Print ISBN: 978-3-527-33059-1, also available in digital formats

Jäger, T., Koch, O., Flohé, L. (eds.)

**Trypanosomatid Diseases**

**Molecular Routes to Drug Discovery**

2013
Print ISBN: 978-3-527-33255-7, also available in digital formats

Doerig, C., Späth, G., Wiese, M.

**Protein Phosphorylation in Parasites**

**Novel Targets for Antiparasitic Intervention**

2013
Print-ISBN: 978-3-527-33235-9, also available in digital formats

Unden, G., Thines, E., Schüffler, A. (eds)

**Host – Pathogen Interaction**

**Microbial Metabolism, Pathogenicity and Antiinfectives**

2016
Print-ISBN: 978-3-527-33745-3, also available in digital formats

Müller, S., Cerdan, R., Radulescu, O. (eds.)

**Comprehensive Analysis of Parasite Biology**

**From Metabolism to Drug Discovery**

2016
Print-ISBN: 978-3-527-33904-4, also available in digital formats

Meng, C. Q., Sluder, A. E. (eds)

**Ectoparasites**

**Drug Discovery Against Moving Targets**

2018
Print-ISBN: 978-3-527-34168-9, also available in digital formats

# Human and Animal Filariases

## Landscape, Challenges, and Control

*Edited by Ronald Kaminsky and Timothy G. Geary*

*Series Editor*
*Paul M. Selzer*

WILEY-VCH

**Editors**

***Ronald Kaminsky***
ParaConsulting
Altenstein 13
79685 Häg-Ehrsberg
Germany

***Timothy G. Geary***
McGill University
Institute of Parasitology
21111 Lakeshore Road
Sainte-Anne-de-Bellevue, H9X 3V9 QC
Canada

*and*

Queen's University-Belfast
School of Biological Sciences
Microbes & Pathogen Biology
19 Chlorine Gardens, Belfast BT9 5DL
Northern Ireland

**Series Editor**

***Paul M. Selzer***
Boehringer Ingelheim Animal Health
Binger Straße 173
55216 Ingelheim am Rhein
Germany

**Cover**
The human filariae Wuchereria bancrofti, Brugia timori and Brugia malayi are the causative agents of the disease elephantiasis which is illustrated by a wooden figurine of the Basonge people from the Democratic Republic of the Congo (courtesy of P. Mäser, for details see chapter 2). Photo credits Science Museum London, Wellcome Trust collection (wellcomecollection.org; CC BY 4.0); CDC/R.S. Craig (phil.cdc.gov). The graphic sketches the life cycle of the animal pathogenic Dirofilaria immitis and potential chemical intervention periods. D. immitis causes heartworm disease in dogs and other canids. The inner circle represents the life cycle of D. immitis within the mammalian host (dog) and the vector (mosquito). The length of the arrows approximately reflects the development time of each stage. The outer circle shows prevention and treatment options depending on the stage of development of the parasite. Macrocyclic lactones are used as preventive treatment up to 60 days post-infection against D. immitis L3 and L4 larval stages. Melarsomine, the only registered heartworm adulticide, is efficacious against adult D. immitis, which can be diagnosed around 180 days post-infection. Ectoparasiticides or repellents can be used to prevent mosquitos from feeding on dogs and cats, reducing the potential for infection (courtesy of S. Noack et al., for details see chapter 9).



**Library of Congress Card No.:** applied for

**British Library Cataloguing-in-Publication Data**
A catalogue record for this book is available from the British Library.

**Bibliographic information published by the Deutsche Nationalbibliothek**
The Deutsche Nationalbibliothek lists this publication in the Deutsche Nationalbibliografie; detailed bibliographic data are available on the Internet at <http://dnb.d-nb.de>.



**Print ISBN:** 978-3-527-34659-2
**ePDF ISBN:** 978-3-527-82343-7
**ePub ISBN:** 978-3-527-82342-0
**oBook ISBN:** 978-3-527-82341-3

**Cover Design** Adam Design, Weinheim, Germany
**Typesetting** Straive, Chennai, India
**Printing and Binding** CPI Group (UK) Ltd, Croydon, CR0 4YY

C096372_061022

# Contents

## List of Contributors

***David Abraham***
Thomas Jefferson University
Sidney Kimmel Medical College
Department of Microbiology and Immunology
Philadelphia, PA 19107
USA

***Lyric Bartholomay****
University of Wisconsin-Madison
School of Veterinary Medicine
Department of Pathobiological Sciences
1656 Linden Dr, Madison WI 53706
USA
lbartholomay@wisc.edu

***Boakye A. Boatin****
McGill University
Institute of Parasitology
Montreal, Quebec H2X 3V9
Canada
University of Ghana Legon
Noguchi Memorial Institute for Medical Research
Lymphatic Filariasis Support Centre for Africa Accra Ghana
5569 Oakwood Drive Stone Mountain, GA 30087
USA

***Dwight D. Bowman****
Cornell University
College of Veterinary Medicine
Department of Microbiology and Immunology
Ithaca, NY 14853-6401
USA

***Collette Britton****
University of Glasgow
Institute of Biodiversity Animal Health and Comparative Medicine, College of Medical, Veterinary and Life Sciences
Bearsden Road, Glasgow G61 1QH
Scotland
UK
Collette.Britton@glasgow.ac.uk

***Douglas S. Carithers***
Boehringer Ingelheim Animal Health
3239 Satellite Blvd, Duluth, GA 30096
USA

***Eileen Devaney****
University of Glasgow, Institute of Biodiversity, Animal Health and Comparative Medicine
College of Medical, Veterinary and Life Sciences
Bearsden Road, Glasgow G61 1QH, Scotland
UK
Eileen.Devaney@glasgow.ac.uk

***Jason Drake***
Global Technical Marketer – Pet Health Parasiticides
Elanco Animal Health
2500 Innovation Way
Greenfield, IN 46140
USA

***Thomas Duguet***
INVENesis Sàrl
Rue de Neuchâtel 15A
CH-2072 St-Blaise
Switzerland

***Christian Epe****
Boehringer Ingelheim Animal Health
Binger Str. 173, 55216 Ingelheim am Rhein
Germany
christian.epe@boehringer-ingelheim.com

***Christopher Evans***
University of Georgia
College of Veterinary Medicine
Department of Infectious Diseases
Athens, GA 30602
USA

***Timothy G. Geary****
McGill University
Institute of Parasitology
21111 Lakeshore Road
Sainte-Anne-de-Bellevue, H9X 3V9 QC
Canada
Queen's University-Belfast
School of Biological Sciences, Microbes & Pathogen Biology
19 Chlorine Gardens, Belfast BT9 5DL
Northern Ireland
timothy.g.geary@mcgill.ca

***John O. Gyapong***
Vice Chancellor's Office
University of Health & Allied Sciences AS
PMB 31 Ho, VH-0194-8222 Volta Region
Ghana

***John Harrington***
Boehringer Ingelheim Animal Health
1730 Olympic Drive, Athens GA 30601
USA

***Natalie A. Hawryluk****
Bristol Myers Squibb
Global Health Research & Early Development, 10300 Campus Point Drive, Suite 100
San Diego, CA 92121
USA
nataliehawryluk@gmail.com

***Achim Hoerauf***
University Hospital Bonn, Institute for Medical Microbiology, Immunology and Parasitology
Venusberg-Campus 1
Building 63, 53127 Bonn
Germany
Cluster of Excellence of the University of Bonn
Bonn
Germany
German Center for Infection Research (DZIF)
Partner site Bonn-Cologne
Bonn
Germany

***Marc P. Hübner****
University Hospital Bonn, Institute for Medical Microbiology
Immunology and Parasitology
Venusberg-Campus 1, Building 63, 53127 Bonn

Germany
Cluster of Excellence of the University of Bonn
Bonn
Germany
German Center for Infection Research (DZIF) Partner site Bonn-Cologne
Bonn
Germany
huebner@uni-bonn.de

***Joseph Kamgno***
Epidemiology and Biostatistics
Centre for Research on Filariasis and other Tropical Diseases (CRFilMT)
Street 1.411, Fouda Quarter Yaounde
Cameroon
University of Yaoundé I
Department of Public Health, Faculty of Medicine and Biomedical Sciences
P.O. Box 1364, Yaoundé
Cameroon

***Ronald Kaminsky****
ParaConsulting
Altenstein 13, 79685 Häg-Ehrsberg
Germany
para.C@gmx.de

***Phillip Kaufman***
Texas A&M University
Department of Entomology
College Station, TX 77843-2475
USA

***Jennifer Ketzis***
Ross University School of Veterinary Medicine, Biomedical Sciences
Basseterre, St. Kitts
West Indies

***Darrell Klug****
Darrell Klug Consulting
5 Waldron Ct., Greensboro, NC 27408
USA
Darrell.Klug1@gmail.com

***Regina Lizundia***
Elanco Tiergesundheit AG
Mattenstrasse 24A, 4058 Basel
Switzerland

***Alan Long***
Boehringer-Ingelheim
3239 Satellite Blvd NW
Duluth, GA 30096
USA

***Sara Lustigman****
Lindsley F Kimball Research Institute
New York Blood Center, Laboratory of Molecular Parasitology,
New York, NY 10065
USA
slustigman@nybc.org

***Charles D. Mackenzie****
Neglected Tropical Disease Support Center
Task Force Global Health
330 West Ponce de Leon Avenue,
Decatur, GA 30030
USA
cmackenzie@taskforce.org

***Ben Makepeace***
University of Liverpool, Institute of Infection & Global Health
Department of Infection Biology
Liverpool L3 5RF
UK

***Pascal Mäser****
Swiss Tropical and Public Health Institute
Department of Medical Parasitology and Infection Biology
Kreuzstrasse 2, 4123 Allschwil
Switzerland
University of Basel
Petersplatz 1
Basel
Switzerland
pascal.maeser@swisstph.ch

***Tanja McKay****
Arkansas State University
Department of Biological Sciences,
Jonesboro, AR 72467
USA
tmckay@astate.edu

***Andrew Moorhead****
University of Georgia, College of Veterinary Medicine
Department of Infectious Diseases
Athens, GA 30602
USA

***Sandra Noack***
Boehringer Ingelheim Animal Health
Binger Str. 173
Ingelheim am Rhein 55216
Germany

***Kenneth Pfarr***
University Hospital Bonn
Institute for Medical Microbiology, Immunology and Parasitology
Venusberg-Campus 1, Building 63, 53127 Bonn
Germany
German Center for Infection Research (DZIF)
Partner site Bonn-Cologne
Bonn
Germany

***Nils Pilotte***
Smith College
Department of Biological Science
Northampton, MA 01063
USA

***Roger Prichard****
McGill University, Institute of Parasitology
21111 Lakeshore Road
Sainte Anne-de-Bellevue, H9X3V9
Canada
roger.prichard@mcgill.ca

***Kapa D. Ramaiah***
12, Bhaktavatsalam Street, Tagore Nagar
Lawspet, Puducherry 605008
India

***Frank O. Richards Jr***
453 Freedom Parkway
Atlanta, GA 30307
USA

***Lucien Rufener****
INVENesis Sàrl
Rue de Neuchâtel 15A
CH-2072 St-Blaise
Switzerland

***Heinz Sager****
Elanco Tiergesundheit AG
Mattenstrasse 24A
4058 Basel
Switzerland
heinz.sager@elancoah.com

***Sandra Schorderet-Weber****
Consultant Parasitology
Neuchâtel
Switzerland
sandra.schorderet@hotmail.com

***Paul M. Selzer****
Boehringer Ingelheim Animal Health
Binger Str. 173, Ingelheim am
Rhein 55216
Germany
paul.selzer@boehringer-ingelheim.com

***Ashley Souza***
Neglected Tropical Diseases Support
Center, The Task Force for Global
Health
330 West Ponce de Leon Avenue,
Decatur, GA 30030
USA

***Sabine Specht****
Drugs for Neglected Diseases initiative
15 Chemin Camille-Vidart, 1202
Geneva
Switzerland

***Sofija Todorovic***
Arkansas State University
Department of Biological Sciences
Jonesboro, AR 72467
USA

***Lucienne Tritten****
University of Zurich
Institute of Parasitology
Winterthurerstrasse 266a
CH-8057 Zurich
Switzerland
lucienne.tritten@swisstph.ch
Swiss Tropical and Public Health
Institute
Kreuzstrasse 2
CH-4123 Allschwil
Switzerland
University of Basel
CH-4000 Basel
Switzerland

***Alexandre Vernudachi***
INVENesis France
Bâtiment 311, Route de Crotelles
37380 Nouzilly
France

***William H. White***
Elanco Animal Health
Alfred-Nobel-Str. 50, 40789 Monheim
Germany

***Steven Williams***
Smith College
Department of Biological Science
Northampton, MA 01063
USA

***Timothy K. Wu***
Cornell University, College of
Veterinary Medicine
Department of Microbiology and
Immunology
Ithaca, NY 14853-6401
USA

# Foreword

While recent global attention has been rightfully focused on viruses (coronaviruses, Ebola, etc.) and bacteria (MSRA, tuberculosis, etc.) as sources of infectious diseases, one should not overlook the continued importance of parasites in human and animal health. The World Health Organization (WHO) reported for 2018 about 228 million cases of malaria worldwide with associated 405 000 deaths [1]. Less known is the impact of human and animal pathogenic filariae which are causing severe disease in both humans and animals. An estimated 180 million humans are infected with filarial parasites resulting in considerable suffering and disability. Filariasis is considered to the second leading cause of disability with DALYs (disability-adjusted life years) estimated to be 5.549 million [2].

The economic and health impacts of diseases like "river blindness" (onchocerciasis) continue to be dramatic both for the individual [3] and society as a whole [4]. Lymphatic filariasis (LF) is considered to be a "Neglected Tropical Disease" in humans and causes illness and suffering in more than 125 million individuals. The main causative agents of lymphatic filariasis include the mosquito-borne filarial nematodes *Wuchereria bancrofti* and *Brugia malayi*. An estimated 90% of LF cases are caused by *W. bancrofti* (Bancroftian filariasis). Neglected Tropical Diseases like these still cause severe disease, suffering, and economic loss in affected countries [5].

Treatments and prevention of onchocerciasis and LF generally rely on community-based approaches using donated drugs such as ivermectin, a compound originally developed in veterinary medicine [6], or diethylcarbamazine [7], an anthelmintic discovered in 1947 that due to side effects in humans can't be used in onchocerciasis-endemic regions.

In animals, filariae cause heartworm disease in dogs and cats, a widespread and often fatal parasitic infection of pet and feral animals, with canine and feline heartworm being the economically most important filarial infections. The global animal health heartworm market is exceeding US$ 2 billion per year in pet owner spend [8]. With that, it is the most important single disease/parasitic infection market in all of animal health. Prevention and treatment of *Dirofilaria immitis*, the parasite causing heartworm disease, is the focus of intense research in all major animal health corporations. Given the necessary investment in research, compound libraries, and whole organism-screening systems, etc., it can be assumed that currently only the top four animal health companies (Zoetis, Boehringer Ingelheim Animal Health,

Elanco, and MSD Animal Health) have the resources and financial stamina to truly bring innovation to market. With the cost of biotechnology dropping dramatically, it might be possible, however, that smaller animal health startup companies become active in this field.

Filariae are a prime example for the concept of One Medicine. Different species of these parasites cause illness and often fatal disease in humans and animals and have a grave economic impact for both. Antiparasitic and particularly anthelmintic treatments for human health are often based on animal health compounds and as filariases are "Neglected Tropical Diseases" in humans while of strong economic importance in animal health, research in animal health is often the driving force for new interventions. It is noteworthy that compounds widely used in human health like ivermectin and related macrocyclic lactones, emodepside (which is currently in a clinical trial against onchocerciasis in human health [9, 10]) and others, were discovered in animal health and subsequently tested and used in human health. This is contrary to the usual pattern of active ingredients proven in human health being tested and utilized in animal health and, again, a good example of the benefits of a One Health approach.

One Health is an approach that recognizes that the health of people is deeply connected to the health of animals and our shared environment. One Health is not new, but it has become more important in recent years. This is because many factors have changed interactions between people, animals, plants, and our environment [11]. With growing human populations that expand into wildlife areas previously undisturbed by human settlement and humans living in close contact with domestic and wild animals, opportunities for diseases and parasites to pass between animals and humans increase. With ever-accelerating climate change and land use, disruptions in environmental conditions can provide new habitats for diseases and parasites and allow them to more easily pass between animals and humans.

Research into filariases in animals and humans as presented in this book is a hallmark of the One Health approach. Parasitic diseases research and treatment in animals have a direct effect on the available treatment and prevention options in humans and with that a large impact on the economic wellbeing of millions of people. That reasoning behind One Health is why organizations like the Bill and Melinda Gates Foundation, the Drugs for neglected Disease Initiative (DnD*i*), GALVmed, and others support research in parasitic and other infectious diseases.

Last but not least, human and animal health is important not just for tropical areas of the world where human filariae are endemic, but also effectively for the entire world. Economic hardship, inability to generate incomes, or live in certain parts of the world due to parasite populations or endemic diseases lead to suffering and mass migrations which increase economic burdens both for countries where citizens leave and those where they arrive.

Research in infectious diseases and parasites like the comprehensive material presented in this book is paramount for the future of our global society. Without continued pioneering work to understand the prevalence, pathogenesis, economic impact, and treatment and prevention of filariases, the economic impact will only increase and could make entire normally fertile regions around river deltas uninhabitable.

The work done for the discovery and development of new heartworm drugs for dogs and cats has a direct positive effect and relationship with the work done on human filarial diseases providing the benefit of One Health for both humans and animals.

The detail and quality of the work in this book from the description of the parasites, detailed chapters on the diseases caused by filariae in humans and animals all the way to current and future chemotherapy followed by an outlook on drug discovery for novel antifilarials and even approaches including genetics, vector control, and potential vaccines will contribute greatly to the understanding of these important parasites and thus will help with treatment, control, and possibly eradication in both animals and humans.

May 2022

*Dr. med. vet. Fabian M. Kausche*
*Trustee at GALVmed;*
*Chairman of the board at PetMedix, Ltd.;*
*Member of the board at Pet Flavors, LLC;*
*Member of the board at Sequent Scientific, Pvt Pty;*
*Member of the Scientific Advisory Committee at Rejuvenate Bio, Inc.*

## References

**1** WHO (2019). *World Malaria Report 2019*. Geneva: World Health Organization https://www.who.int/publications/i/item/world-malaria-report-2019 (accessed 12 March 2021).

**2** Fenwick, A. (2012). The global burden of neglected tropical diseases. *Public Health* 126: 233–236.

**3** Ubachukwu, P.O. (2006). Socio-economic impact of onchocerciasis eith particular reference to females and children: a review. *Animal Res. Int.* 3 (2): 494–504.

**4** https://www.who.int/apoc/onchocerciasis/disease/en/ (accessed 12 March 2021).

**5** Taylor, M.J., Hoerauf, A., and Bockarie, M. (2010). Lymphatic filariasis and onchocerciasis. *Lancet* 376 (9747): 1175–1185.

**6** Ōmura, S. and Crump, A. (2014). Ivermectin: panacea for resource-poor communities? *Trends Parasitol.* 30: 445–455.

**7** Hawking, F. (1962). A review of progress in the chemotherapy and control of filariasis since 1955. *Bull. World Health Org.* 27: 555–568.

**8** Klug and Drake, (2022) Chapter 13: Global economics of heartworm disease, in *Human and Animal Filariases*, (eds. R. Kaminsky and T.G. Geary), Wiley-VCH, Weinheim, Germany.

**9** https://dndi.org/research-development/portfolio/emodepside/ (accessed 12 March 2021).

**10** Krücken, J., Holden-Dye, L., Keiser, J. et al. (2021). Development of emodepside as the first safe, short-course adulticidal treatment for human onchocerciasis – the fruit of a successful industrial–academic collaboration. *PLOS Pathog.* 17: e100968.

**11** https://onehealthinitiative.com/ (accessed 12 March 2021).

# Preface

Pathogenic filariae affect the wellbeing of hundreds of millions of people and animals. The vector-borne human filarial parasites cause onchocerciasis (river blindness), lymphatic filariasis (elephantiasis), and loiasis (eyeworm). More than 200 million people live in onchocerciasis-endemic areas, and about 1.39 billion people are at risk in lymphatic filariasis areas in more than 72 countries. It is estimated that about 380 million dogs and 350 million cats are at risk of being infected with filariae, with canine heartworm as the most prominent filarial disease in dogs. This book aims to provide insights regarding the current landscape, the gaps and challenges, and current and future approaches for control of both human and animal filariases.

The first section of this volume is titled "Human and Animal Filariae and Their Diseases," providing a comprehensive overview of human and animal filariae and the diseases they cause. Firstly, arguments are presented which foster a "One Health" approach to review human and animal filariases and explore mutual benefits. Furthermore, a strong foundation is laid, based on the biological background, the description of the various diseases and the current gaps, diagnostic possibilities, and treatment options. A thorough assessment of current chemotherapeutic interventions (which are still the mainstay of control) is outlined as well as the importance of drug resistance. A consideration of current elimination and eradication programs for human filariases, and finally, an economic overview particularly of canine heartworm, closes this section.

The section on "Drug Discovery for Novel Antifilarials" starts with a discussion of the similarities and discrepancies in requirements (product profiles) for new antifilarials. Subsequently, various authors outline the state-of-the-art discovery processes for identifying new antifilarial lead compounds. They focus on the current status of *in vitro* discovery approaches, advantages, and handicaps of available *in vivo* rodent models, and finally on *in vivo* assays to explore and monitor the activity of active compounds on target parasites. Finally, the antifilarial drug pipeline, as much as is publicly available, is highlighted. As an area for discovery of new drugs, the host–filariae interface is advanced in particular.

The section on "New Frontiers for Control of Antifilarial Diseases" closes this volume. These contributions show the potential of exploring improved technologies for genomic, transcriptomic, and metabolic approaches for the discovery of novel points of intervention. Furthermore, they outline the major advances and obstacles in vaccine research against the background that an effective antifilarial vaccine has the potential of a breakthrough in control of filariases, particularly of canine heartworm. Alternatively, opportunities and current gaps and challenges of vector control methods are presented. Finally, the potential to therapeutically intervene with the rickettsia-like endosymbionts *Wolbachia* as a particular target in most filarial species is presented.

We thank Prof. Paul M. Selzer, the series editor, and many representatives of Wiley for the opportunity to embark on this volume and for their continued guidance and support. We also thank the authors who have generously contributed their time and expertise. The result of all these efforts is a volume that provides a comprehensive view on human and animal filariases for physicians, veterinarians, biologists, public health decision makers, and other interested people in academia and industry.

May 2022

*Ronald Kaminsky*
*Timothy G. Geary*

# Part I

# Human and Animal Filariae and Their Diseases

# 1

# Breaking the Silos – Obstacles and Opportunities for One Health in Filariases

*Ronald Kaminsky*[1,*] *and Timothy G. Geary*[2,3,*]

[1] *ParaConsulting, Altenstein 13, Häg-Ehrsberg 79685, Germany*
[2] *Institute of Parasitology, McGill University, 21111 Lakeshore Road, Sainte-Anne-de-Bellevue, QC H9X 3V9 Canada*
[3] *School of Biological Sciences, Queen's University-Belfast, 19 Chlorine Gardens, Belfast BT9 5DL, Northern Ireland*

**Abstract**

Despite major similarities in biology and transmission, human and animal filarial parasites exhibit a number of species-specific characteristics that prompt the question if a One Health approach is sui for filariases. We elucidate that applying the One Health concept to filariases is not motivated by the pathology of these diseases nor their geographic overlap and only to a minor extent by the zoonotic potential of animal filariases. Instead, the benefits of adopting a One Health view on this disease complex are evident in the areas of drug resistance, the well-being of humans and their pets, and even more importantly for the discovery of new anthelmintics and research on the basic biology of the host–parasite interface that may lead to entirely novel treatment strategies.

## 1.1 Introduction

Why should one combine chapters on scientific research and reviews into human and animal filariases in a single book? An obvious reason is that these parasites exhibit a number of biological similarities; the pathogenic filariae belong within the superfamily of Filarioidea and the same family of Onchocercidae [1], and they all cause vector-borne diseases (meaning that all are adapted to live in two very distinct kinds of hosts, arthropods, and mammals). However, the preferred sites of infection and thus the pathologies they cause are quite different, even within the same host [2, 3], and their respective competent vectors also differ a great deal in biology [4, 5]. In a more pragmatic approach, the present control methods are

*Corresponding authors.

*Human and Animal Filariases: Landscape, Challenges, and Control*, First Edition.
Edited by Ronald Kaminsky and Timothy G. Geary.

quite different for human and animal filariases and product profiles differ substantially [6]; however, the currently applied control methods rely to a large extent on the same chemical class, the macrocyclic lactones [7–10]. The common history of chemical control of filariases relates back to the discovery and development of ivermectin, firstly for veterinary purposes but subsequently applied for control of human onchocerciasis and lymphatic filariasis. In addition, it is for good reasons that Satoshi Ōmura and William C. Campbell were awarded the 2015 Nobel Prize in Medicine for that breakthrough innovation. Even now, control programs for human filariases [7] rely on ivermectin (among other drugs), and many veterinary products [9] contain ivermectin or subsequently developed macrocyclic lactones as the active pharmaceutical ingredients.

The One Health approach is currently endorsed by many authorities and has become popular in the scientific public health community [11, 12]. The term "One Health" was first used in 2003 for the valuable consideration of a combined perspective on the emerging severe acute respiratory disease (SARS) [12]. Subsequently, the correlation and deep connections between human and animal health, including wildlife health, and the need for an interdisciplinary and collaborative approach to respond to emerging diseases, were clearly outlined (Wildlife Conservation Society One World-One Health www.oneworldonehealth.org Sept 2004) [13], although the principles of the One Health concept originated several decades ago as "One Medicine, One World" [11]. The concept has not been applied to the study of parasites as frequently or intensively as might be desired, and, in our experience, veterinarians, physicians, and parasitologists do not always work together to the extent that they could or should, despite the excellent chances for mutual benefit.

## 1.2 Indicators for "One Health" Diseases

The obvious indicators for a link between research in human and animal diseases are (i) the origin of the pathogen, (ii) shared geographic or microhabitats, and (iii) a zoonotic characteristic of the disease. A number of emerging infections can be traced to animals, including wildlife, such as the pathogenic avian influenza H5N1 or SIV/HIV, associated with changes in human activities [14–16]. More recently, it has been hypothesized that SARS-CoV-2 originated from a β-coronavirus in the sarbecovirus (SARS-like virus) group that naturally infects bats and pangolins [17–19]. The risk of exposure may rise when the hosts of the same pathogen share common close habitats, such as the distribution of *Escherichia coli* in cattle grazing next to a lettuce field. Furthermore, at least 60% of human diseases are multi-host zoonoses [20], including parasitic infections such as leishmaniasis, human African trypanosomiasis, schistosomiasis, soil-transmitted helminthiasis, and lymphatic filariasis. Many of these diseases have been grouped as "Neglected Zoonotic Diseases" [21].

**Table 1.1** Differences and potential synergies for human and animal filariases

| Differences | Common/ different | Human health | Animal health |
|---|---|---|---|
| Geography | Different | Predominantly tropical, subtropical countries | Heartworm endemic areas in North America, etc. |
| Financial resources of involved communities | Different | Resource-limited; most drugs are donated | Heartworm control products are a major component of AH revenue, as well as for veterinary clinics |
| Treatment schedules | Different | Ideally once/year | Monthly to yearly |
| Zoonotic potential different for some species | Different | *D. repens*; *D. immitis* in human only anecdotal | — |
| | Different | — | No known animal reservoir for *W. bancrofti* |
| | Different | — | No animal host confirmed for *O. volvulus*, but related cattle species exist (*O. ochengi*) |
| Primary life stages targeted for chemotherapy | Different | L1, adult fertility | L3/L4, L1 |
| Vectors | ±: overlapping mosquito species, but flies not relevant for heartworm | Mosquitoes/ black flies | Mosquitoes |
| **Possible synergies** | | | |
| Zoonotic potential for some species | + | *Brugia malayi, Brugia pahangi* | Cats |
| | + | *Onchocerca lupi* | Dogs, cats |
| | + | *Dirofilaria repens* | Dog |
| Current drugs | + | Ivermectin, moxidectin, doxycycline, and diethylcarbamazine | Macrocyclic lactones, arsenicals, and doxycycline |
| Drug targets | + | Table 1.3 | Table 1.3 |
| Vaccine targets | + Common epitopes | *O. volvulus* | *D. immitis* |
| Vector control | ± for mosquitoes | For LF | For *D. immitis* |
| Costs | + low cost of goods | Affordable for public health resources of local communities | Competitive margins for animal health industries |
| Diagnostics | + Common protein or nucleic acid technologies | All human filariae | *D. immitis* |

## 1.3 Zoonotic Characteristics of Human and Animal Filariases

Although eight filariae species have been reported to infect humans [22, 23], the zoonotic potential of filarial parasites appears to be limited. They all rely on insect vectors for transmission, but most of them express a more or less strict host specificity such that each species is confined to a single or few specific definitive and intermediate hosts [12]. The human pathogenic species *Brugia malayi* and *Brugia pahangi* can also infect cats, but the epidemiological significance of this alternative host is not known. Nevertheless, they are grouped as lymphatic filariases in the Neglected Zoonotic Diseases list [21], and cats can serve as competent hosts for *B. malayi*, with reported prevalence reaching as high as 20% in endemic feline populations [24]. Other than *Onchocerca volvulus*, the cause of onchocerciasis, only one other species in this genus, *Onchocerca lupi*, can use humans as host, although it is far more commonly found in dogs and cats (Table 1.1). The medical significance of this parasite has only recently been appreciated. *O. lupi* infection is now also proposed as an emerging zoonosis [25, 26]. Infections of humans with the canine pathogen *Dirofilaria immitis* occur, but the parasites almost never mature into adult stages and are described mostly as anecdotal, single case reports. However, the usually non-pathogenic species *Dirofilaria repens*, with a primary canine host, has higher zoonotic potential than *D. immitis*. Human infection is usually characterized by subcutaneous nodules, but larva migrans-like symptoms may also occur and, notably, larvae may reach the eye, becoming visible in the conjunctiva. Some reports have described the presence of microfilariae in humans [27].

## 1.4 Are Human and Animal Filariases Suitable for a "One Health" Approach?

Applying the One Health concept to filariases is not motivated by common pathological manifestations of these diseases nor their geographic overlap and only to a minor extent by the zoonotic potential of animal filariases (Table 1.1). Instead, the benefits of adopting a One Health view on this disease complex are evident in the areas of pharmacology of antifilarial drugs (including drug discovery and drug resistance), the use of common technology platforms for diagnosis and vaccine control, aspects of vector biology, and implications for the well-being of humans and their pets. Research on the basic biology of the host–parasite interface that may lead to entirely novel treatment strategies also illustrates the great potential of a One Health approach to filariases.

### 1.4.1 Pharmacology of Antifilarial Drugs

As reviewed in this volume [7–10], chemotherapy of human and veterinary filariases relies to a significant extent on the use of macrocyclic lactones, in particular the prototype of this class, ivermectin. Although ivermectin has some filariid species- and host-specific effects [28, 29], the drug has microfilaricidal and temporary sterilization effects against human and veterinary filariae. Although microfilaricidal activity may be due to inhibition of secretion of parasite-derived immunomodulatory factors, a mechanistic explanation of the prolonged but reversible inhibition of fertility caused by the drug remains elusive. In contrast, the activity of ivermectin against L3 and L4 larvae of *D. immitis*, the basis for its use as a heartworm disease preventative, is not fully duplicated in *O. volvulus* or LF parasites, for unknown reasons. Although the microfilaricidal effects of diethylcarbamazine are evident against veterinary and human filariae, macrofilaricidal effects are only pronounced in LF parasites. The basis for the discrepancy between the profound pathology associated with killing of microfilariae in onchocerciasis and heartworm infections, but not in LF, is yet unresolved. Thus, although many commonalities are observed for antifilarial chemotherapy in human and veterinary medicine, the differences could provide a basis for comparative studies that may illuminate strategies for safer and more effective interventions.

### 1.4.2 Drug Resistance

Drug resistance is a well-known and urgently considered obstacle in animal health, particularly for livestock but also more recently for companion animals. Producers of small ruminants and cattle have experienced the disastrous effects of drug-resistant gastrointestinal nematodes, even to the point of forced abandonment of sheep farming in some areas with high-level resistance to all available anthelmintics. This major stressor has resulted in considerable investment in research to understand, monitor, and combat the issue of drug resistance in livestock animals [30]. These methods are now being applied to supplement human STH control programs, as concerns about the development of resistance to albendazole and mebendazole are heightened by the expansion and intensification of mass drug administration programs. In this case, extensive molecular biology work has clearly identified three alleles in a nematode beta-tubulin gene that cause benzimidazole resistance, and it is possible to monitor for the presence and spread of these alleles in human STH species [31]. Recently, one of these alleles (a change from phenylalanine to tyrosine at residue 167 of the beta-tubulin gene) has been reported to be present in *Ancylostoma caninum* (hookworms) in dogs in the United States [32], proving that benzimidazole resistance is a threat in hookworms and encouraging intensified monitoring for this mutation in areas that receive intensive treatment with these drugs for human STH infections.

A similar situation has developed in canine heartworms; recent experiments have proven that macrocyclic lactone-resistant *D. immitis* populations have appeared in the United States [33]. These resistant populations can break through previously effective macrocyclic lactone regimens, and microfilariae of these parasites are unaffected by these normally effective drugs. A mixture of genomic and phenotypic assays has conclusively demonstrated that resistant populations are genetically distinct from wild-type parasites and support the hypothesis that the phenotype of macrocyclic lactone resistance is multigenic. Although genomic analyses have not yet been able to conclusively identify the genes that cause this phenotype, single nucleotide polymorphisms (SNPs) have been found that can identify resistant parasites with high confidence. The phenotype extends to all members of this drug class, but further work is needed to define the quantitative shift in sensitivity and to determine if the extent of resistance is the same for all macrocyclic lactones. At this time, new drugs or drug regiments that are fully effective against resistant parasites have not been identified or confirmed.

Although resistance to macrocyclic lactones has been suspected in human filariases (particularly in *O. volvulus*; [33]), the lack of a convenient laboratory host for these parasites has greatly limited the opportunity for experimental validation. It is to be hoped that, once the genes responsible for resistance to macrocyclic lactones in *D. immitis* are identified, research can be initiated to characterize and monitor them in populations of *O. volvulus* that have been intensively treated with ivermectin.

### 1.4.3 Antifilarial Drug Discovery

Almost all medicines used in veterinary practice were originally developed for human use, with the notable exception of antiparasitic drugs, many of which were developed for use in animals (Table 1.2). The examples include the majority of drugs used to treat coccidian infections of poultry and, particularly, anthelmintics. Indeed, only one drug used as an anthelmintic in animals was originally discovered in a human-use screening operation: diethylcarbamazine [10], which was discovered

**Table 1.2** Anthelmintics discovered for AH, which were repurposed for HH

| Active ingredient | Indication for animal health | Year of market entry | Indication for human health |
|---|---|---|---|
| Thiabendazole | GI nematodes | 1964 | Derivatives in use (mebendazole, albendazole, and flubendazole) |
| Albendazole | GI nematodes | 1981 | Lymphatic filariases<br>GI nematodes<br>Tapeworms (*Taenia* and *Echinococcus*) |
| Pyrantel | GI nematodes | 1970s | GI nematodes |
| Oxantel | GI nematodes | 1970s | GI nematodes |
| Ivermectin | GI nematodes, heartworm, arthropods | 1981 | Filariases, mites, lice |
| Moxidectin | As for ivermectin | 1990 | Onchocerciasis |
| Praziquantel | Tapeworms | 1975 | Schistosomiasis, other trematodes |
| Triclabendazole | *Fasciola* spp. | 1983 | *Fasciola hepatica* |

All but diethylcarbamazine and doxycycline.

in a program looking for drugs for the treatment of lymphatic filariasis and only later transitioned for use as a heartworm preventative in dogs (now replaced by macrocyclic lactones for this indication).

However, relatively little investment was made in the animal health industry to discover new drugs for heartworm infections over the past 20 years. The most important reason for this status was the excellent record of efficacy and safety of the macrocyclic lactones, which greatly reduced the opportunity for new medicines to penetrate an already well-satisfied market. Furthermore, the necessity to maintain the long heartworm life cycle in dogs to detect efficacy endpoints requires much longer discovery programs than for GI nematodes, for example, and greatly limits the ability of academic researchers to operate in this area (along with animal use regulations that restrict the use of dogs for exploratory research). Finally, the marked consolidation of the animal health industry has led to a significant overall decline in the amount of private resources that can be devoted to the discovery of new drugs for prevention of heartworm disease.

Instead, significant investment has more recently been targeted for the discovery of new anthelmintics with macrofilaricidal activity for human use, especially for onchocerciasis, for which control programs that rely solely on the microfilaricidal action of ivermectin (and now moxidectin) may not achieve the goals of control programs in a cost- and time-effective manner. These efforts have led to the identification of several compounds that are in clinical trials or are candidates for such trials, including the veterinary anthelmintic emodepside, which has antifilarial activity in many animal models, auranofin, imatinib, and several antibiotics with anti-*Wolbachia* activity [10, 34, 35]. Although these compounds have known mechanisms of action, their antifilarial activity was discovered in phenotypic and infected animal models. Among them, only emodepside has been reported to have activity against *D. immitis* [36]. It is also important to recognize that other veterinary anthelmintics, such as monepantel [37] or derquantel [38], may have utility for filariases; further research is needed to support or reject this possibility.

Until recently, it has not been possible to maintain *Wuchereria bancrofti*, *O. volvulus*, or *D. immitis* in convenient laboratory rodent hosts to permit transition from *in vitro* to *in vivo* assays before testing promising compounds in dogs, a major limitation in the ability of academic or small industrial labs to participate in heartworm drug discovery programs. In the absence of such models, scientists commonly rely on surrogate filariid species maintained in permissive rodent hosts (e.g. *Litomosoides sigmodontis* in mice or *Brugia* spp. in jirds; see Ref. [39]) to identify compounds with promising antifilarial activity. These models can identify compounds with preventative activity, as well as microfilaricides and macrofilaricides, and represent a significant synergy in the One Health context. As reviewed in Ref. [39], novel immunosuppressed rodent models now permit more facile drug screening studies for *D. immitis* (mice and rats) and *O. volvulus* (mice). It remains somewhat challenging to procure infective larvae of *O. volvulus* for routine use, but this is simple for *D. immitis*, and it is possible that the heartworm screens could be used to generate and characterize new compounds with high likelihood of activity against the relevant stages of human filariid species.

It is important to emphasize in this regard that there is a disconnect between the life stages targeted for antifilarial chemotherapy in human and veterinary medicine. The overwhelming emphasis in veterinary medicine is to discover compounds that prevent maturation to adult parasites by targeting L3 and L4 stages as they develop in the host. Microfilaricidal activity is permissible but is not generally a therapeutic priority and can be a drawback (as seen with diethylcarbamazine). Adulticidal activity is clearly a drawback, as killing adult heartworms can lead to significant pathology in the host. In contrast, available human antifilarial drugs primarily target microfilariae, both in the host and developing in the adult female parasite, and preventative chemotherapy is not practiced or practical for these infections. In the absence of proven resistance to ivermectin, the emphasis has been on finding drugs that safely kill adult parasites. Thus, much research on potential drug targets in human filarial parasites may be applicable to heartworms, but it remains to be seen if the macrofilaricidal compounds now under evaluation for human filariases will find ready applications in veterinary medicine for heartworm prevention.

### 1.4.4 Discovery of Common Drug Targets

Because of their parasitic nature and their close phylogenetic relationship, human- and animal pathogenic filarial share some common drug targets (Table 1.3). The list includes targets for which activity against both human and animal filariids has been demonstrated, at least *in vitro*. There are chances that interference or inhibition of other targets in one filarial species, e.g. *D. immitis*, may be evident and relevant in other species. Activity in a particular mechanism-based screen cannot guarantee that the compound would be suitable for use against all filarial species as other parameters must also be met, such as stage of the life cycle, proper pharmacodynamics and pharmacokinetic characteristics, and safe toxicological profile in the respective host. Nevertheless, activity against a specific target could offer a valuable starting point for a drug discovery program with therapeutic implications for all filariases.

**Table 1.3** Selected drug targets shared by animal and human pathogenic filariae.

| Target | Function | References |
|---|---|---|
| Glutamate-gated chloride channels | Secretion, fertility | [10] |
| Intestinal proteases | Specific digestive enzymes of parasitic nematodes involved in feeding process | [40] |
| Peptide GPCRs | Motility, development, and feeding | [41] |
| sloK channel | Motility | [36, 42, 43] |
| *Wolbachia* | Viability and development | [35] |
| Kinases | Viability | [10, 34] |

## 1.5 Insights into Host–Parasite Interactions: New Therapeutic and Diagnostic Opportunities

Current control of filariases in both human and veterinary medicine relies on drugs discovered empirically; except for diethylcarbamazine, all were discovered for use in other indications (trypanosomes, gastrointestinal nematodes, and bacteria). As noted, our understanding of the molecular pharmacology underlying their therapeutic benefit remains incomplete, and indeed, our understanding of the basic biochemistry and physiology of filariid parasites has been little advanced over the past decades. In part, this reflects the difficulty of maintaining sufficiently large numbers of parasites at all stages of the life cycle in laboratories, as culture systems that can replicate the life cycle in the absence of hosts have not been developed. The use of surrogate (non-target) filariid species is necessary even now as it is quite challenging if not essentially impossible to obtain living specimens of, for instance, adult *O. volvulus*, *W. bancrofti*, and *D. immitis*. The situation is made more complex by the fact that we do not know if parasites removed from the host and placed in culture accurately reflect their biology *in situ* and for how long they are useful surrogates *in vitro* (see, for example, Ref. [44]).

However, the development of sensitive and highly quantitative technology platforms for genomic, proteomic, metabolomic, transcriptomic, and microRNA (miRNA) analyses is revolutionizing our ability to interrogate the host–filariid parasite interface and the molecular language that serves to maintain or prevent the establishment of a chronic infection. Comparative studies may eventually provide insights into the basis for host–parasite specificity, focusing on the species-specific molecules that are essential for enabling a chronic infection (possibly including proteins, metabolites, and/or non-coding RNAs). Work in model or surrogate filariid species may allow rapid extrapolation to medically important species, an important benefit of a One Health paradigm. Some advances have been made in our ability to perform functional genomics experiments in filariae [45], but more intensive investment in this area has the potential to radically transform our understanding of the host–parasite interface and to reveal new targets and novel strategies for prevention and control of these infections in humans and animals.

Clinically important advances may be expected from these studies, not only in terms of new targets for chemotherapy [46]. Identification of critically important immunomodulatory proteins can lead to the rational selection of vaccine antigens; by neutralizing those proteins, we may be able to convert permissive into non-permissive hosts for pathogenic filariid species. Similarly, obtaining the menu of abundantly secreted parasite-derived proteins and nucleic acids can be expected to offer new strategies for stage- and species-specific diagnosis of pathogenic species in field-friendly, cost-effective platforms. New vaccines and diagnostics can rationally be evaluated in lab animal models before development for use in people and/or dogs and cats.

## 1.6 Health Benefits

Although the direct health benefits of chemotherapy for human filariases are obvious and profound, the indirect human health benefits of chemotherapy for prevention of heartworm disease should be included in a consideration of the One Health landscape around filariid parasites. In many parts of the world, companion animals, particularly dogs and cats, have been integrated into family life, sometimes to the extent that pets are considered to be family members. The pet–owner bond, particularly as it relates to the well-being of these animals, contributes to a large extent to the overall life experience of the involved people, and as such, healthy pets can contribute to human health by providing many positive psychological and physical benefits for their owners [47–49]. Thus, although treatment of human filariases leads to direct improvements in the well-being of communities, families, and individuals, significant health benefits are also apparent in companion animal owners who are free from worry over possible heartworm infections and pet ill health.

## 1.7 Conclusions

Despite significant differences in vectors, tissue location, pathology, and strategies for control, the phylogenetic and pharmacological similarities among the important filarial species that parasitize humans and animals merit the application of a One Health approach to their study. Much can be learned about their diagnosis, physiology, biochemistry, and host manipulation strategies in comparative analyses that will benefit researchers, physicians, and veterinarians, as well as scientists who strive to develop better tools to control them. Research on filarial parasites that cause neglected tropical diseases and heartworm has for too long been focused on empirical discovery of diagnostics and treatments; very little emphasis has been placed on understanding the complex biology of the host–parasite interface from which novel approaches may emerge. The research highlighted in this book identifies areas of work that will benefit scientists in both sectors and can encourage joint efforts to enhance our ability to eliminate these parasites as significant health burdens.

## Acknowledgment

We are thankful to JakobZinsstag for critical reviewing of the manuscript.

## References

1 Mäser, P. (2022). Filariae as organisms. In: *Human and Animal Filariases* (ed. R. Kaminsky and T.G. Geary), Chapter 2. Weinheim, Germany: Wiley-VCH.

2 Mackenzie, C.D. (2022). Human filarial infections: reflections on the current understanding of their importance, pathobiology and management. In: *Human and Animal Filariases* (ed. R. Kaminsky and T.G. Geary), Chapter 3. Weinheim, Germany: Wiley-VCH.
3 Bowman, D.D. and Wu, T.K. (2022). Canine filariasis (heartworm) – disease and current gaps. In: *Human and Animal Filariases* (ed. R. Kaminsky and T.G. Geary), Chapter 4. Weinheim, Germany: Wiley-VCH.
4 Bartholomay, L. (2022). Vector control approaches to interrupt transmission of human filarial parasites. In: *Human and Animal Filariases* (ed. R. Kaminsky and T.G. Geary), Chapter 22. Weinheim, Germany: Wiley-VCH.
5 Todorovic, S., McKay, T., and Kaufman, P. (2022). Vector control approaches for canine filariasis. In: *Human and Animal Filariases* (ed. R. Kaminsky and T.G. Geary), Chapter 23. Weinheim, Germany: Wiley-VCH.
6 Specht, S. and Kaminsky, R. (2022). Product profiles for new drugs against human and animal filariasis. In: *Human and Animal Filariases* (ed. R. Kaminsky and T.G. Geary), Chapter 14. Weinheim, Germany: Wiley-VCH.
7 Specht, S., Kamgno, J., and Geary, T.G. (2022). Antifilarial chemotherapy: current options for humans. In: *Human and Animal Filariases* (ed. R. Kaminsky and T.G. Geary), Chapter 7. Weinheim, Germany: Wiley-VCH.
8 Ketzis, J. and Epe, C. (2022). Antifilarial chemotherapy – current options in veterinary medicine. In: *Human and Animal Filariases* (ed. R. Kaminsky and T.G. Geary), Chapter 8. Weinheim, Germany: Wiley-VCH.
9 Noack, S., Harrington, J., Carithers, D.S. et al. Heartworm disease – intervention and industry perspective. In: *Human and Animal Filariases* (ed. R. Kaminsky and T.G. Geary), Chapter 9. Weinheim, Germany: Wiley-VCH.
10 Geary, T.G., Long, A., and Tritten, L. (2022). The antifilarial drug pipeline. In: *Advances in Control of Heartworm and Human Filariases* (ed. R. Kaminsky and T.G. Geary), Chapter 10. Weinheim, Germany: Wiley-VCH.
11 Atlas, R.M. (2013). One Health: its origins and future. *Curr. Top. Microbiol. Immunol.* 365: 1–13.
12 Mackenzie, J.S. and Jeggo, M. (2019). The one health approach – why is it so important? *Trop. Med. Infect. Dis.* 4: 88.
13 Mackenzie, J.S., McKinnon, M., and Jeggo, M. (2014). One health: from concept to practice. In: *Confronting Emerging Zoonoses* (ed. A. Yamada, L. Kahn, B. Kaplan, et al.), 163–189. Tokyo: Springer.
14 Taylor, L.H., Latham, S.M., and Woolhouse, M.E. (2001). Risk factors for human disease emergence. *Philos. Trans. R. Soc. Lond. B Biol. Sci.* 356 (1411): 983–989.
15 Daszak, P., Cunningham, A.A., and Hyatt, A.D. (2001). Anthropogenic environmental change and the emergence of infectious diseases in wildlife. *Acta Trop.* 78: 103–116.
16 Fenton, A. and Pedersen, A.B. (2005). Community epidemiology in theory and practice: a conceptual framework for describing transmission dynamics in multiple hosts. *Emerg. Infect. Dis.* 11: 1815–1821.
17 Morens, D.M., Daszak, P., Markel, H., and Taubenberger, J.K. (2020). Pandemic COVID-19 joins history's pandemic legion. *mBio* 11: e00812–e00820.

**18** Tiwari, R., Dhama, K., Sharun, K. et al. (2020). COVID-19: animals, veterinary and zoonotic links. *Vet. Quart.* 40: 169–182.

**19** Sharun, K., Dhama, K., Pawde, A.M. et al. (2021). SARS-CoV-2 in animals: potential for unknown reservoir hosts and public health implications. *Vet. Quart.* 41: 181–201.

**20** Cleaveland, S., Laurenson, M.K., and Tylor, L.H. (2001). Diseases of humans and their domestic mammals: pathogen characteristics, host range and the risk of emergence. *Philos. Trans. R. Soc. Lond. B-Biol. Sci.* 356: 991–999.

**21** Webster, J.P., Gower, C.M., Knowles, S.C. et al. (2016). One health – an ecological and evolutionary framework for tackling Neglected Zoonotic Diseases. *Evol. Appl.* 9: 313–333.

**22** Kausar, S. (2020). Filariasis. In: *Helminthiasis* (ed. O.O. Okwa). Intechopen.com, London, UK https://doi.org/10.5772/intechopen.80144.

**23** Centers for Disease Control and Prevention (2021). Parasites. https://www.cdc.gov/parasites

**24** Palmieri, J.R., Masbar, S., Marwoto, H.A. et al. (1985). The domestic cat as a host for Brugian filariasis in South Kalimantan (Borneo), Indonesia. *J. Helminthol.* 59: 277–281.

**25** Otranto, D., Dantas-Torres, F., Cebeci, Z. et al. (2012). Human ocular filariasis: further evidence on the zoonotic role of *Onchocerca lupi*. *Parasites Vectors* 5: 84.

**26** Cantey, P.T., Weeks, J., Edwards, M. et al. The emergence of zoonotic *Onchocerca lupi* infection in the United States–a case-series. *Clin. Infect. Dis.* 62: 778–783.

**27** Genchi, C. and Kramer, L. (2017). Subcutaneous dirofilariosis (*Dirofilaria repens*): an infection spreading throughout the old world. *Parasites Vectors* 10: 517.

**28** Campbell, W.C. (1982). Efficacy of the avermectins against filarial parasites: a short review. *Vet. Res. Commun.* 5: 251–262.

**29** Zahner, H. and Schares, G. (1993). Experimental chemotherapy of filariasis: comparative evaluation of the efficacy of filaricidal compounds in *Mastomys coucha* infected with *Litomosoides carinii, Acanthocheilonema viteae, Brugia malayi* and *B. pahangi*. *Acta Trop.* 52: 221–266.

**30** Kotze, A.C., Gilleard, J.S., Doyle, S.R., and Prichard, R.K. (2020). Challenges and opportunities for the adoption of molecular diagnostics for anthelmintic resistance. *Int. J. Parasitol. Drugs Drug Resist.* 14: 264–273.

**31** Vlaminck, J., Cools, P., Albonico, M. et al. (2020). Piloting a surveillance system to monitor the global patterns of drug efficacy and the emergence of anthelmintic resistance in soil-transmitted helminth control programs: a Starworms study protocol. *Gates Open Res.* 4: 28.

**32** Jimenez Castro, P.D., Howell, S.B., Schaefer, J.J. et al. (2019). Multiple drug resistance in the canine hookworm *Ancylostoma caninum*: an emerging threat? *Parasites Vectors* 12: 576.

**33** Prichard, R.K. (2022). Drug resistance in filariae. In: *Advances in Control of Heartworm and Human Filariases* (ed. R. Kaminsky and T.G. Geary), Chapter 11. Weinheim, Germany: Wiley-VCH.

**34** Hawryluk, N. (2022). The antifilarial drug pipeline. In: *Human and Animal Filariases* (ed. R. Kaminsky and T.G. Geary), Chapter 18. Weinheim, Germany: Wiley-VCH.

35 Hübner, M.P., Pfarr, K., and Hoerauf, A. (2022). *Wolbachia* endosymbionts as treatment targets for filarial diseases. In: *Human and Animal Filariases* (ed. R. Kaminsky and T.G. Geary), Chapter 24. Weinheim, Germany: Wiley-VCH.

36 Hübner, M.P., Townson, S., Gokool, S. et al. (2021). Evaluation of the *in vitro* susceptibility of various filarial nematodes to emodepside. *Int. J. Parasitol. Drugs Drug Resist.* 17: 27–35.

37 Godel, C. (2012). Drug targets of the heartworm, "*Dirofilaria immitis*". PhD-thesis. Journal: University of Basel. https://doi.org/10.5451/UNIBAS-006021331Corpus ID: 82476422.

38 Verma, S., Kashyap, S.S., Robertson, A.P., and Martin, R.J. (2017). Functional genomics in *Brugia malayi* reveal diverse muscle nAChRs and differences between cholinergic anthelmintics. *Proc. Natl. Acad. Sci. U.S.A.* 114: 5539–5544.

39 Schorderet-Weber, S. and Specht, S. (2022). In vivo models for the discovery of new antifilarial drugs, In: *Human and Animal Filariases* (ed. R. Kaminsky and T.G. Geary), Chapter 16. Weinheim, Germany: Wiley-VCH.

40 Wang, Q., Rosa, B.A., Jasmer, D.P., and Mitreva, M. (2015). Pan-Nematoda transcriptomic elucidation of essential intestinal functions and therapeutic targets with broad potential. *EBioMedicine* 2: 1079–1089.

41 Atkinson, L.E., McCoy, C.J., Crooks, B.A. et al. (2021). Phylum-spanning neuropeptide GPCR identification and prioritization: Shaping drug target discovery pipelines for nematode parasite control. *Front. Endocrinol.* 12: 718363.

42 Harder, A., Schmitt-Wrede, H.P., Krücken, J. et al. (2003). Cyclooctadepsipeptides--an anthelmintically active class of compounds exhibiting a novel mode of action. *Int. J. Antimicrob. Agents* 22: 318–331.

43 Bah, G.S., Schneckener, S., Hahnel, S.R. et al. (2021). Emodepside targets SLO-1 channels of *Onchocerca ochengi* and induces broad anthelmintic effects in a bovine model of onchocerciasis. *PLoS Pathog.* 17: e1009601.

44 Ballesteros, C., Tritten, L., O'Neill, M. et al. (2016). The effect of *in vitro* cultivation on the transcriptome of adult *Brugia malayi*. *PLoS Negl. Trop. Dis.* 10: e0004311.

45 Devaney, E. and Britton, C. (2022). Functional genomics of filariae. In: *Human and Animal Filariases* (ed. R. Kaminsky and T.G. Geary), Chapter 20. Weinheim, Germany: Wiley-VCH.

46 Duguet, T.B. and Tritten, L. (2022). The host-helminth interface as a rich resource for novel drug targets. In: *Human and Animal Filariases* (ed. R. Kaminsky and T.G. Geary), Chapter 19. Weinheim, Germany: Wiley-VCH.

47 McConnell, A.R., Brown, C.M., Shoda, T.M. et al. (2011). Friends with benefits: on the positive consequences of pet ownership. *J. Per. Social Psychol.* 101: 1239–1252.

48 Cordaro, M. (2013). Pet loss and disenfranchised grief: implications for mental health counseling practice. *J. Mental Health Couns.* 29: 283–294.

49 Mubanga, M., Byberg, L., Nowak, C. et al. (2017). Dog ownership and the risk of cardiovascular disease and death – a nationwide cohort study. *Sci. Rep.* 7: 15821.

# 2

# Filariae as Organisms

*Pascal Mäser*[1,2,*]

[1] Swiss Tropical and Public Health Institute, Department of Medical Parasitology and Infection Biology, Kreuzstrasse 2, 4123 Allschwil, Switzerland
[2] University of Basel, Petersplatz 1, Basel, Switzerland

**Abstract**

The filariae are a clade of nematodes that consists of vector-borne parasites with a unique life cycle. Adults live and reproduce inside a mammal or other vertebrate host (other than fish). They are viviparous: the female worm does not release eggs but first-stage larvae, the microfilariae. These are picked up by a blood-feeding insect or tick, which serves as an intermediate host in which the microfilariae develop further to third-stage larvae. When, during the next blood meal, such a larva infects a vertebrate host, it will develop into an adult parasite, reproduce, and the cycle will be closed. The filariae are a burden to human and animal health. In particular, the subfamily Onchocercidae, which comprises important pathogens such as *Onchocerca volvulus* (the causative agent of river blindness), *Wuchereria bancrofti* and *Brugia malayi* (the causative agents of elephantiasis), and *Dirofilaria immitis* (the dog heartworm), causes considerable health and economic burdens. Thus, the filariae are of high interest to physicians as well as veterinarians. At the same time, they are fascinating study subjects in basic science due to their many biological peculiarities. The evolutionary biologist is captivated by the mutualistic symbiosis between onchocercid filariae and the intracellular bacterium *Wolbachia*; the geneticist is puzzled by the fact that the (male) filaria is the only nematode that possesses a Y chromosome; and the immunologist has to concede that the filariae have gained the higher degree of mastery in controlling the human immune system. This introductory chapter attempts to do justice to both aspects of the filariae: their importance as pathogens and their intriguing biology.

## 2.1 What's So Special about the Filariae?

The vast and immensely diverse phylum Nematoda (the roundworms) includes a defined group of highly specialized, vector-borne parasites: the Filarioidea,

*Corresponding author.

*Human and Animal Filariases: Landscape, Challenges, and Control*, First Edition.
Edited by Ronald Kaminsky and Timothy G. Geary.

commonly called filariae. The filariae have no free-living stages in their life cycles; adult worms as well as all larval stages are obligate endoparasites. The adults live and reproduce inside a mammal, bird, reptile, or amphibian as the definite host, while the larvae are disseminated by a blood-sucking insect or another hematophagous arthropod that serves as an intermediate host. The conserved reproduction and transmission cycles indicate that the filariae are monophyletic and of terrestrial origin, whereas the nematodes on the whole probably have marine ancestry [1].

The term filaria derives from *filum*, which means thread in Latin, while nematode comes from *nema*, which means thread in Greek. This somewhat awkward duplication is attributable to the fact that the filariae were discovered and named (Müller 1787) before the Nematoda were defined as an animal phylum [2]. At least the repetition reminds us that the filariae are the most thread-like of animals. An adult female *Onchocerca volvulus* measures up to 700 mm in length but only 0.4 mm in diameter!

Taxonomically, the Filarioidea form a superfamily, of which the family Onchocercidae is the best studied and the subject of the present book. The Onchocercidae comprises all the human-pathogenic species of the Filarioidea, most notably the causative agents of the neglected tropical diseases, river blindness (onchocerciasis) and elephantiasis (lymphatic filariasis or disfiguring lymphatic edema; Figure 2.1). The Onchocercidae also comprises several pathogens of veterinary importance, such as the dog heartworm. Table 2.1 lists the human-pathogenic Filarioidea and a selection of animal-pathogenic species.

### 2.1.1 Microfilariae and Macrofilariae

A distinctive feature of the filariae is the fact that they are ovoviviparous, i.e. the female adult gives birth to larvae that have hatched from eggs inside her body.

**Figure 2.1** The disease elephantiasis illustrated by a wooden figurine of the Basonge people from the Democratic Republic of the Congo, and a patient suffering from edema of the lower legs and feet. Source: Wellcome Trust Collection/CC BY 4.0; R.S. Craig/Center for Disease Control (CDC) – PHIL/Public Domain.

**Table 2.1** Selected species of Filarioidea.

| Family and species | Final host | Vector | Disease | Distribution |
|---|---|---|---|---|
| **Filariidae** | | | | |
| *Parafilaria multipapillosa* | Horse | *Haematobia* spp. biting flies | Summer bleeding | Global |
| *Stephanofilaria* spp. | Bovines | *Haematobia* spp. | Summer wounds | Europe |
| **Setariidae** | | | | |
| *Setaria* spp. | Ungulates | *Haematobia* spp., mosquitoes | Cavity filariasis | Global |
| **Onchocercidae** | | | | |
| *Onchocerca volvulus* | Human | *Simulium* spp. (blackflies) | Subcutaneous filariasis (river blindness) | Tropical Africa |
| *Loa loa* | Human | *Chrysops* spp. (deer flies) | Subcutaneous filariasis (loiasis) | West Africa |
| *Wuchereria bancrofti* | Human | Mosquitoes | Lymphatic filariasis | Pantropical |
| *Brugia malayi* | Human | Mosquitoes | Lymphatic filariasis | South and southeast Asia |
| *Brugia timori* | Human | Mosquitoes | Lymphatic filariasis (Timor filariasis) | Lesser Sunda Islands |
| *Mansonella streptocerca* | Human, Chimpanzee | *Culicoides* spp. (midges) | Subcutaneous filariasis | Tropical Africa |
| *Mansonella perstans* | Human | *Culicoides* spp. | Cavity filariasis | Sub-Saharan Africa, Central and South America |
| *Mansonella ozzardi* | Human | *Culicoides* spp., *Simulium* spp. | Cavity filariasis | Central and South America |
| *Dirofilaria repens* | Dog | Mosquitoes | | Old world |
| *Dirofilaria immitis* | Dog | Mosquitoes | Dirofilariasis (canine heartworm) | Global |
| *Brugia pahangi* | Cat | Mosquitoes | Feline filariasis | Southeast Asia |
| *Onchocerca ochengi* | Bovines | *Simulium* spp. | Intradermal onchocerciasis | Africa |

*(Continued)*

**Table 2.1** (Continued)

| Family and species | Final host | Vector | Disease | Distribution |
|---|---|---|---|---|
| *Onchocerca lienalis* | Bovines | *Simulium* spp. | | Global |
| *Pelecitus* spp. | Birds | Mallophaga (bird lice) | Cutaneous | |
| *Cardiofilaria* spp. | Birds | *Culex* spp. | Blood | |
| *Acanthocheilonema viteae* | Rodents | Argasidae (soft ticks) | Subcutaneous filariasis | Deserts |
| *Litomosoides sigmodontis* | Cotton rats | *Ornithonyssus bacoti* (rat mite) | | Central and South America |

Source: Compiled mainly from [3, 4].

The released larvae are called microfilariae, and the adults are called macrofilariae. Figure 2.2 depicts four different examples of microfilariae.

The microfilariae circulate in the blood (or upper dermis in the case of *O. volvulus*) and infect the intermediate host when it takes a blood meal. Interestingly, in some species of filariae, the presence of microfilariae in the peripheral blood of the

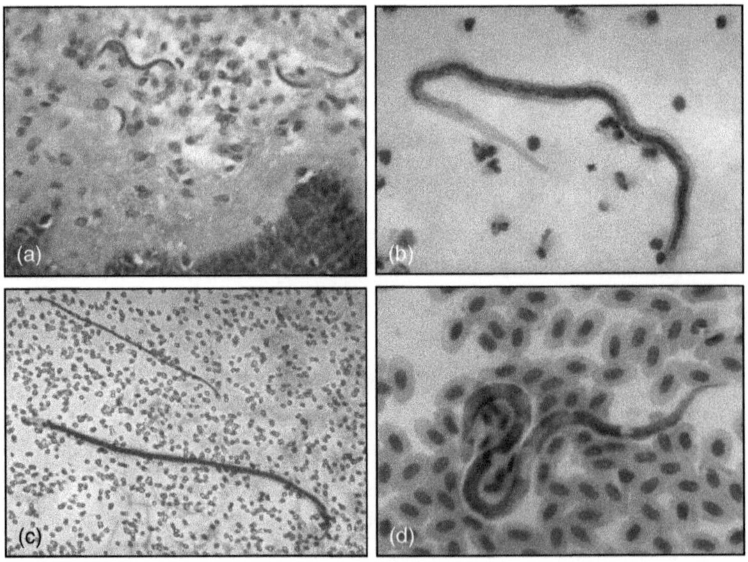


**Figure 2.2** Examples of microfilariae. (a) Histological preparation of *O. volvulus* microfilariae in the subcutaneous tissue of a patient. They are unsheathed and measure about 300 μm. (b) *Loa loa* microfilaria stained with hematoxylin. It is sheathed and measures about 250 μm. (c) Microfilaria of *W. bancrofti* (sheathed, about 280 μm) and *M. ozzardi* (unsheathed, about 180 μm) on a membrane filter stained with Giemsa. (d) Microfilariae of unknown species in the blood of a mistle thrush (*Turdus viscivorus*; note the bird's nucleated erythrocytes). Source: H. Zaiman (a, b); R. Müller (c); H.P. Striebel (d).

mammalian host is synchronized with the feeding habit of the vector. Thus, microfilariae of *O. volvulus*, which are transmitted by *Simulium* spp. (biting black flies; Table 2.1), have diurnal periodicity, whereas microfilariae of *Wuchereria bancrofti*, which are transmitted by night-feeding *Culex* or *Anopheles* mosquito spp. (Table 2.1), have nocturnal periodicity. However, in regions where *W. bancrofti* is transmitted by day-feeding *Aedes* spp., the microfilariae have diurnal periodicity. Other microfilariae, such as those of *Mansonella* spp., are aperiodic. The clues perceived by the microfilariae are blood oxygen tension and body temperature, both of which are lower during resting times of the mammalian host [5]. However, the molecular nature of the mechanisms governing the periodicity of microfilariae remains to be elucidated.

Microfilariae are the diagnostic stages for most filariases, since the macrofilariae usually are inaccessible. Obviously, the periodicity needs to be taken into account when examining blood for the presence of microfilariae. The microfilariae are similar in size, 250–300 μm (with the exception of the somewhat shorter *Mansonella*; see Figure 2.2c), but they can be distinguished microscopically based on the presence or absence of a sheath and by the arrangement of the nuclei in the tip of the microfilarial tail (Table 2.2). The sheath is the modified remnant of the egg shell and serves as a protective layer that is impermeable to antibodies [6].

### 2.1.2 Filariae at the Dawn of Tropical Medicine

Microfilariae were described for the first time in 1843, when Gruby and Delafond observed a high number of them in the blood of an infected, but apparently healthy dog [7]. The species was termed *Filaria immitis* (Leidy, 1956), later renamed to *Dirofilaria immitis* (lat. *dirus*, dreadful; *immitis*, relentless). Similar microfilariae were subsequently found in human body fluids such as urine, testicular hydrocele, and blood. By 1878, Patrick Manson had established the epidemiological link between elephantiasis and the presence of microfilariae in the blood [8]. The species was originally named *Filaria sanguinis hominis* but ultimately renamed to *Wuchereria bancrofti*, in honor of Otto Wucherer, who had found microfilariae in the patients' urine, and Joseph Bancroft, who had discovered the adult macrofilariae [9].

**Table 2.2** Distinctive characteristics of the microfilariae of human pathogens.

| Species | Periodicity | Sheath | Posterior nuclei |
|---|---|---|---|
| *W. bancrofti* | Mainly nocturnal | Present | No nuclei in tip of tail |
| *B. malayi* | Nocturnal | Present | Two distinct nuclei in tip |
| *L. loa* | Diurnal | Present | Nuclei extend to tip |
| *O. volvulus* | Diurnal | Absent | Nuclei extend to tip |
| *M. perstans* | Aperiodic | Absent | Large terminal nucleus |
| *M. ozzardi* | Aperiodic | Absent | No nuclei in tip of tail |

For details see (http://www.parasite-diagnosis.ch/microfilariaedk).

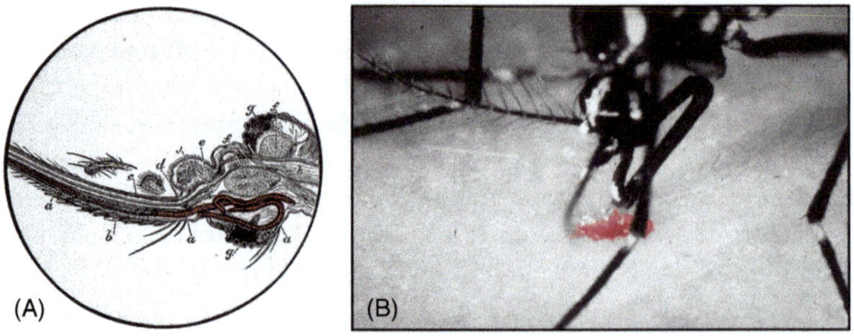


**Figure 2.3** (A) Sir Patrick Manson's drawing of nocturnal *W. bancrofti* microfilariae in the head of a mosquito that had taken a blood meal from a lymphatic filariasis patient (*a*, microfilariae; *b*, labium; *c*, labrum; *d*, base of hypopharynx; *e*, duct of venenosalivary gland; *f*, cephalic ganglia; *g*, eye; *h*, oesophagus; *j*, pharyngeal muscle). (B) An infected mosquito takes a blood meal. Microfilariae have escaped from the mouthparts and are present in the drop of fluid on the skin. They will enter the skin through the puncture wound. Source: P. Manson/Wellcome Trust Collection/CC BY 4.0 (A); R. Müller, Medical Helminthology, London (B).

Microfilaria can reach peak levels of several thousand per milliliter of blood. Manson reasoned that if the millions of circulating microfilariae in a dog infected with *D. immitis*, or in a patient infected with *W. bancrofti*, all matured to macrofilariae, the host would immediately be killed. Therefore, microfilariae had to complete their development outside the mammalian host's body. Considering all kind of escape routes, Manson favored the hypothesis that the microfilariae hijack mosquitoes to leave the bloodstream. He dissected female mosquitoes that had fed on elephantiasis patients and indeed found live microfilariae (Figure 2.3). Manson even described the nocturnal rhythm of the circulating microfilariae in the patient [8]. However, he rejected the – now obvious – hypothesis that the mosquito itself would transmit the parasites to a new host, since the dogma at the time was that female mosquitoes have a blood meal only once, whereupon they find water, lay eggs, and die [10, 11]. Nevertheless, the discoveries by Manson in 1877 mark the beginning of Tropical Medicine as a research discipline. It was Manson's work and his advice to Ronald Ross that paved the way for the discovery that malaria is also transmitted by mosquitoes [9].

*Onchocerca volvulus* (gr. *onkos*, hook; *kerkos*, tail; lat. *volvulus*, roll, or small clew) microfilariae were first described in 1875 [12], but it was not until 1931 that the causality was understood between the parasite *O. volvulus*, the vector *Simulium* spp., and the disease river blindness [13–16]. *Brugia malayi* is named after the Dutch parasitologist Steffen Lambert Brug, who discovered the parasite in Indonesia in 1927 and originally named it *Filaria malayi* [9]. *Mansonella perstans* was discovered 1890 in London and was originally named *Filaria sanguinis hominis minor* due to the microfilaria's small size (Figure 2.2). On a historical note, it is of interest that at first, *M. perstans* was thought to be the causative agent of sleeping sickness, since it was discovered in a West African patient hospitalized with this dreaded disease [17, 18].

A more recent milestone in filarial research was the isolation of avermectins from *Streptomyces avermitilis* in 1975 and the subsequent development of ivermectin (trade name Mectizan for human use), for which Satoshi Omura and William C. Campbell were awarded the 2015 Nobel Prize in Medicine (along with Tu Youyou for the discovery of the antimalarial drug artemisinin). Developed for the veterinary sector, ivermectin was repurposed for human medicine due to its high activity against onchocercid microfilariae [19]. Merck & Co. launched the Mectizan Donation Program in 1987, which aims to eliminate onchocerciasis and lymphatic filariasis. In 2007, Colombia was the first country to announce that it had eliminated onchocerciasis [20]. See Chapter 12 for more information on the success and challenges of filariasis elimination programs.

### 2.1.3 Genomic Insights

The first genome sequence of a parasitic nematode was that of *B. malayi* in 2007 [21]. As is frequently the case with obligate parasites, *B. malayi* has undergone a genomic reduction in the course of evolution. While *Caenorhabditis elegans* has 19 000 protein-coding genes [22], *B. malayi* only has about 12 000 [21], and the heartworm *D. immitis* has even fewer [23]. The onchocercid pathogens have lost anabolic pathways such as purine and pyrimidine de novo synthesis, whose end products they salvage either from their hosts or from endosymbiont bacteria (see below). On the other hand, filarial genome sequences have also revealed expansions of particular gene families that are important for immune evasion, tissue adherence and penetration, or for nutrient salvage [24–26]. A surprising finding was the presence of a Y chromosome in male *B. malayi* and *O. volvulus*, which had been suspected for a long time [27] and confirmed by genomics [24]. This indicates that these onchocercid parasites have a XX/XY-based genetic sex determination system, which further distinguishes them from other nematodes such as the free-living species *C. elegans*, which has a XX/X0-based system (yet other nematodes such as *Strongyloides* have disposed of their sex chromosomes altogether and rely on environmental cues for sex determination [28]).

### 2.1.4 Mutualism with Endosymbiont Bacteria

The Onchocercidae are not only parasites but also hosts: the majority of onchocercid species harbors an intracellular symbiont, the bacterium *Wolbachia* (also called *Wolbachia pipientis* since it was first discovered in the mosquito *Culex pipiens* [29]). Those onchocercid species that lack *Wolbachia* probably lost it in the course of evolution [30]. Estimated to be present in the majority of insects, *Wolbachia* is one of the most widespread bacteria on earth [31]. Interestingly, the *Wolbachia* of onchocercid nematodes has smaller genomes than those of arthropods [31]. In arthropods, *Wolbachia* is a parasite that cunningly manipulates the reproduction of its hosts in order to maximize the likelihood of vertical transmission to the next generation via the eggs [31]. In the Onchocercidae, this has culminated in a mutualistic relationship in which the worm has become fully dependent on *Wolbachia* for reproduction. When the endosymbiont is killed by antibiotic treatment, the worms become infertile and ultimately die as well [32, 33]. What exactly the bacteria contribute is unknown. They are thought to provide nutrients for which the filariae are auxotrophic, such as nucleotides or porphyrins. However, genome sequencing has revealed that these anabolic pathways are also absent in *Loa loa*, one of the few onchocercid species that does not possess *Wolbachia* [26]. Whatever the molecular nature of the mutualistic relationship, the fact that filariae such as *O. volvulus*, *W. bancrofti*, or *D. immitis* cannot survive without *Wolbachia* may turn out to be their Achilles' heel; see Chapter 24 for approaches to cure filariasis with antibiotics. The endosymbiont bacteria also contribute to the pathology of the filariases. *Wolbachia* released from dead worms activate toll-like receptors and thereby trigger inflammation [34].

## 2.2 Life Cycles of the Filariae

Like all nematodes, the filaria have a developmental cycle that comprises an egg, four larval stages (L1–L4), and adults. Unlike other nematodes, the filariae are ovoviviparous: the larvae hatch from the egg inside the uterus and are released from their mother as microfilariae. By definition, the progression from one life cycle stage to the next is accompanied by molting (Figure 2.4). However, sexually immature adults are sometimes called L5 larvae, and some authors also distinguish between a microfilaria and a matured L1 larva in the vector.

The L3 larva is the infective stage for the mammalian host (Figure 2.4). The so-called dauer-hypothesis [35] proposes that the L3 infective stage of parasitic nematodes is analogous to the L3 dauer stage of free-living nematodes such as *C. elegans*. The dauer larva is an arrested stage that enables the worms to survive periods of unfavorable conditions. Thus, the ability to make dauer forms would predispose free-living nematodes to evolve the ability to infect animals and become parasites [36].

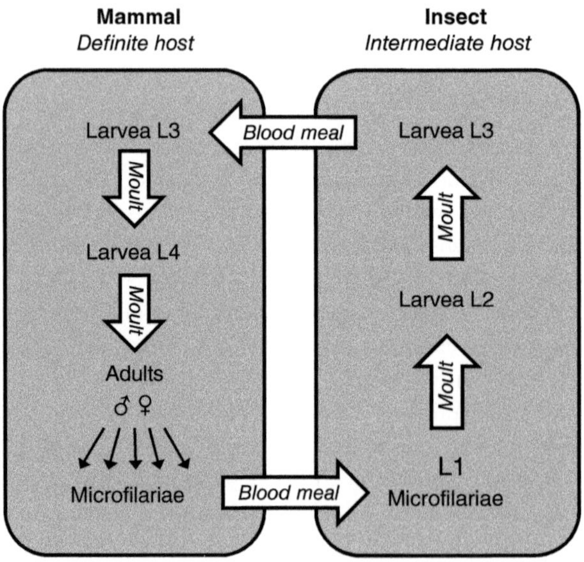


**Figure 2.4** Generalization of the life cycle of an onchocercid pathogen. The development in the mammalian host (left) takes from 6 to 12 months, depending on the species of filaria. The development in the arthropod vector (right) takes between 6 and 30 days. This also depends on the species of filaria, but even more so on the ambient temperature.

Once in the mammalian host, an L3 molts to an L4 larva. The further development to the adult stage often involves tissue migration of the larva and may take several months. Adult filariae come in two sexes only; there are no hermaphrodites. The males are smaller than the females. Since sexual reproduction takes place in the vertebrate, this is the definite host (in contrast to other insect-borne parasites such as *Plasmodium*, *Theileria*, or *Trypanosoma*, for which the definite host is the arthropod). Figure 2.4 depicts the general life cycle of an onchocercid parasite. In the following section, three individual examples are considered in more detail: *O. volvulus* for subcutaneous filariasis (Figure 2.5), *W. bancrofti* for lymphatic filariasis (Figure 2.6), and *D. immitis* for heartworm disease (Figure 2.7).

### 2.2.1 *Onchocerca volvulus*

*Onchocerca volvulus* is transmitted by blackflies (*Simulium* spp.), tiny biting flies that need oxygen-rich – i.e. moving – waters to breed. They are of the dipteran suborder Nematocera and hence more closely related to mosquitoes than to tabanids or muscids (which belong to the Brachycera). Only the female blackfly takes blood meals. When an infected fly has a meal from a human host, L3 larvae are deposited on the skin and penetrate through the bite wound into the subcutaneous tissue. They molt first to L4 and then to adults, which develop over several months. Mature macrofilariae measure 35–70 cm (♀) and 2–5 cm (♂). They reside in subcutaneous nodules (onchocercomata), characteristic granuloma that are formed in reaction to the worms. An onchocercoma will attract newly invading worms. Female macrofilariae produce unsheathed microfilariae (Table 2.2), up to a thousand per day for up

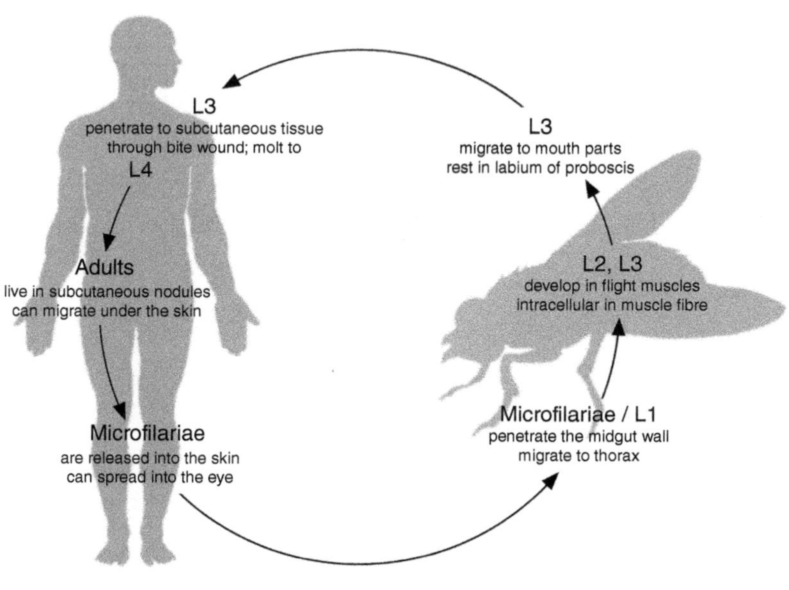


**Figure 2.5** Life cycle of *Onchocerca volvulus*, a causative agent of subcutaneous filariasis. Source: After Simonsen et al. [3] and the DPDx resource of the CDC (http://www.cdc.gov/dpdx/az.html).

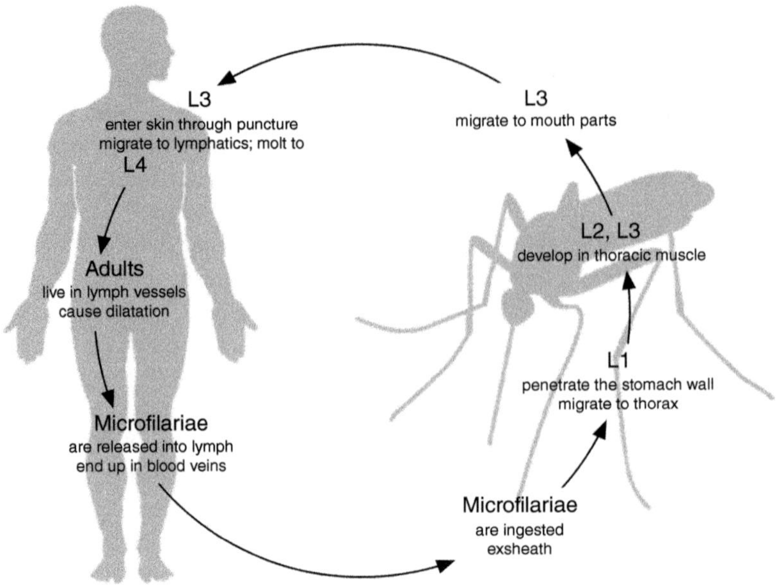


**Figure 2.6** Life cycle of *Wuchereria bancrofti*, a causative agent of lymphatic filariasis. Source: After Simonsen et al. [3] and the DPDx resource of the CDC (http://www.cdc.gov/dpdx/az.html).

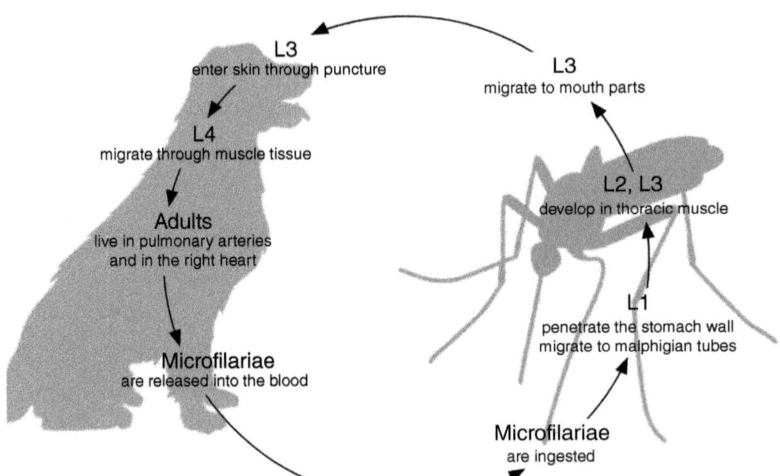


**Figure 2.7** Life cycle of *Dirofilaria immitis*, the causative agents of heartworm disease. Source: After Eckert et al. [4] and the DPDx resource of the CDC (http://www.cdc.gov/dpdx/az.html).

to 11 years. Microfilariae are first detectable in the skin 10–15 months after infection and can survive in the skin for up to two years. Thus, the total burden of microfilariae in a heavily infected person can be over 100 million. When microfilariae are ingested by another blackfly, they penetrate the midgut wall and migrate to the musculature of the thorax. After two molts, the infective L3 larvae progress to the proboscis, ready to infect the next person. The development in the vector takes 6–12 days [3, 37].

### 2.2.2 *Wuchereria bancrofti*

Of the three species that cause lymphatic filariasis (Table 2.1), the most common is *W. bancrofti*. It is transmitted by a variety of mosquitoes, e.g. *Culex*, *Anopheles*, *Mansonia*, and *Aedes* spp. During a blood meal of an infected female mosquito, L3 larvae enter the skin of the human host via the puncture wound. The larvae migrate to a lymph node, molt to L4, and develop to adults over a period of several months. Adult worms measure about 9 cm (♀) and 4 cm (♂), and they preferentially reside in the inguinal lymph nodes. They have a life span of about 10 years and can produce thousands of microfilariae per day. The microfilariae are sheathed (Table 2.2). They migrate from the lymphatic vessel into the bloodstream and end up in the capillary system of the lung. Triggered by the low alveolar oxygen tension during the night, microfilariae leave the capillaries of the lung and circulate in the peripheral blood. Once ingested by a female mosquito, the microfilariae exsheath, penetrate the stomach wall, and migrate to the thorax muscles. After two molts, the emerged L3 larvae migrate further to the proboscis. The development of *W. bancrofti* in the mosquito depends on the temperature and takes at least 10 days [3, 37].

### 2.2.3 *Dirofilaria immitis*

The dog heartworm *D. immitis* has become a global threat to canine welfare [38]. It can be transmitted by many different species of mosquitoes. The definitive hosts are dogs and other canids, such as coyotes. When an infected mosquito takes a blood meal, infective L3 larvae migrate from the mosquito's proboscis through the bite wound into the subcutaneous tissue of the dog, where they molt to L4 within one to two weeks. The L4 larvae continue their journey along muscle fibers and, after the final molt, penetrate into the veins as young adults. About three months after infection, they can be found in the pulmonary artery and the right heart. Sexual maturity is not reached until 180 days post-infection. The adult worms measure up to 30 cm (♀) and 18 cm (♂). The females release sheathless microfilariae that measure about 260 μm. The density of microfilariae in the peripheral blood peaks in the evening. Once ingested by a mosquito, the microfilariae invade the Malpighian tubules and develop further to "sausage-like" L1, L2, and finally L3 larvae, which then migrate via the thoracic muscle to the proboscis. Development in the mosquito takes between 8 and 30 days, depending on the temperature [4].

## 2.3 Pathology of the Filariases

Given their size, life span, and fecundity, filariae are remarkably harmless. Imagine that a female *W. bancrofti* macrofilaria is up to 10 cm long, can live for many years in the human body, and produces some 1000 microfilariae per day – and yet, a majority of infected people have no or only mild symptoms. In fact, the parasites are more harmful dead than alive! Dead worms, and in particular the *Wolbachia* released by them, trigger inflammatory responses that exacerbate pathology.

The pathogenic potential of dead worms is illustrated by the fact that the antifilarial drugs have more severe adverse effects in infected patients than in uninfected subjects [3]. Thus, the pathology emanating from dead parasites must be taken into account in the chemotherapy of filarial infections. This is particularly critical in the treatment of heartworm disease in dogs. For a dog with a high adult worm burden, treatment with a macrofilaricidal drug can be life-threatening, as the dying worms in the right pulmonary artery tend to accumulate in the right heart. Because of such complications, the American Heartworm Society recommends year-round chemoprophylaxis for all dogs (www.heartwormsociety.org). Drug-induced pathology due to dying microfilariae is also of concern in humans. This is why mass administration of ivermectin, whether for river blindness or *Anopheles* mosquitoes, is problematic in tropical African regions where loaisis is co-endemic. At high microfilaraemia of *L. loa*, ivermectin treatment can cause severe, potentially fatal neurological adverse events [39]. This is not due to the release of *Wolbachia* (*L. loa* does not possess *Wolbachia*), but is thought to be due to the presence of the *L. loa* microfilariae in the cerebrospinal fluid after treatment [39].

In general, most filarial infections are initially mild. However, symptoms can become increasingly severe over time, and the infection can progress from asymptomatic to acute inflammatory attacks to chronic pathology [40]. In animals as well, the pathology of filarial infections is usually not life-threatening. An exception is *D. immitis*, the canine heartworm, which can cause congestive caval syndrome and heart failure in dogs. In humans, the most severe pathologies are river blindness and elephantiasis.

River blindness is caused by migrating microfilariae of *O. volvulus*, which often end up in the eye. Over time, chronic inflammation caused by dead microfilariae leads to opacification of the cornea. Lymphatic filariasis is caused by *W. bancrofti* and *Brugia* spp. (Table 2.1). Elephantiasis is an extreme form of lymphedema of the legs (Figure 2.1) or scrotum (hydrocele), caused by adult macrofilariae that obstruct lymph flow. Tables 2.3 and 2.4 summarize the common pathologies caused by filariae in humans and animals, respectively.

The wide distribution of filariases and their high incidence in tropical regions of Africa and Asia result in a large cumulative burden to human health. According to the 2017 estimates of the Global Burden of Disease study [42], lymphatic filariasis (with 1 364 000 disability-adjusted life years) and onchocerciasis (1 343 000 DALY) impose a higher burden to human health than some of the other, more lethal insect-borne diseases, such as visceral leishmaniasis (511 000 DALY), African trypanosomiasis (79 000 DALY), or yellow fever (314 000 DALY) (http://ghdx.healthdata.org/gbd-results-tool). So overall, human infections with filariae cause little or no mortality but impose a very high global burden of morbidity, mainly in sub-Saharan Africa.

**Table 2.3** Filarial pathologies in human health.

| Pathology | Causative agent | Pathogenesis |
|---|---|---|
| River blindness | *O. volvulus* | Dying microfilariae in the eye cause chronic keratitis, resulting in opacification of the cornea |
| Elephantiasis, hydrocele | *W. bancrofti B. malayi* | Extreme form of lymphedema caused by macrofilaria that block lymph flow |
| Tropical pulmonary eosinophilia (TPE) | *W. bancrofti B. malayi* | Asthmatic attacks caused by a hypersensitive immune response to microfilariae |
| Leopard skin<br>Skin atrophy | *O. volvulus* | Spotty depigmentation of the skin, mostly on the legs, resulting from chronic inflammation |
| Chyluria<br>(milky urine) | *W. bancrofti* | Abdominal lymphatic vessels blocked by macrofilariae dilate and rupture, inflow of chyle to the urinary excretory system |
| Immunosuppression | All filariae | The general state of immunosuppression in chronic filariasis may predispose carriers to other infections or vaccination failure |
| Nodding syndrome | *O. volvulus* (?) | Unresolved etiology; a clinical trial might be able to clarify the potential role of *O. volvulus* [41] |

Source: Based on [3].

**Table 2.4** Filarial pathologies in animal health.

| Pathology | Causative agent | Pathogenesis |
|---|---|---|
| Caval syndrome in dogs: intravascular hemolysis and heart failure | *D. immitis* | At high worm burden, macrofilariae in the pulmonary artery may recede into the right heart and from there into the caval veins, obstructing blood flow |
| Equine summer bleeding | *P. multipapillosa* | Stimulated by sunlight, subcutaneous nodules burst and release a bloody exudate with microfilariae (which will be taken up by biting flies) |
| Cerebrospinal setariosis in horses and sheep | *S. digitata* | If transmitted to an accidental host, the L3 larvae of *Setaria* of bovines may end up in the CSF and cause neuropathology |

Source: Eckert et al. [4].

## 2.4 Conclusion

A search in NCBI's PubMed for publications with the term "*Caenorhabditis*" returned about 30 000 hits; "*Wuchereria*" only gave 3000, and "*Mansonella*" as few as 400 (August 2019). Thus, filariae are neglected parasites also in terms of research activities, in spite of their rich and fascinating biology. This research deficit is caused not only by a lack of investment. A major hurdle is the difficulty of maintaining the parasites in the laboratory. There are no *in vitro* cultivation systems for filarial nematodes and, with the exception of *B. malayi* rodent models, hardly any *in vivo* models. A further complication is the biohazard risk posed by infected vectors, which requires experiments to be performed in biosafety level 3 laboratories. The rodent-pathogenic filariae *Acanthocheilonema viteae* and *Litomosoides sigmodontis* (Table 2.1) may serve as substitutes.

The problematic and cumbersome nature of experimental models used to study filariae may be the reason why the older literature, at times, is more informative than recent studies. See, for instance, the studies by Frank Hawking on the circadian rhythm of microfilariae [5]. However, readdressing basic research questions of filarial biology with modern technology would certainly be fruitful, to illuminate the mechanisms of microfilarial periodicity or the molecular nature of immunomodulation. For applied research, the long-term goals are to close existing gaps in diagnosis and treatment of the filariases and, ideally, to develop antifilarial vaccines. Immediate questions to be answered include how to optimize control and elimination programs for lymphatic filariasis and river blindness; how to optimize prophylaxis for dog heartworm; and how – given the close phylogenetic relationship of human- and animal-pathogenic onchocercids – to draw synergies between human and animal health, in particular for drug discovery.

## Acknowledgment

I would like to thank Hanna Walter for the pictures from the Swiss TPH library collection.

## References

**1** Blaxter, M. and Koutsovoulos, G. (2015). The evolution of parasitism in Nematoda. *Parasitology* 142 (Suppl 1): S26–S39.

**2** Diesing, K.M. (1857). *Sechzehn Arten von Nematoideen*. Vienna: K.K. Hof-und Staatsdruckerei.

**3** Simonsen, P.E., Fischer, P.U., Hoerauf, A., and Weil, G.J. (2014). The filariases. In: *Manson's Tropical Diseases* (ed. J. Farrar, P.J. Hotez, T. Junghanss, et al.), 737–765. Elsevier Saunders.

**4** Eckert, J., Friedhoff, K.T., Zahner, H., and Deplazes, P. (2008). *Lehrbuch der Parasitologie für die Tiermedizin*. Stuttgart: Enke Verlag.

**5** Hawking, F. (1967). The 24-hour periodicity of microfilariae: biological mechanisms responsible for its production and control. *Proc. R. Soc.* 169: 59–76.

**6** Zahner, H., Hobom, G., and Stirm, S. (1995). The microfilarial sheath and its proteins. *Parasitol. Today* 11: 116–120.

**7** Gruby, D. and Delafond, H.M. (1843). Note sur une altération vermineuse de sang d'un chien determiné par un grand nombre d'hématozoaires du genre Filaire. In: *Annales de Chimie et de Physique* (ed. J.L. Gay-Lussac, F. Arago, Chevreul, et al.), 381–382. Paris: Fortin, Masson et Cie.

**8** Manson, P. (1878). On the development of *Filaria sanguinis hominis* and on the mosquito considered as a nurse. *Trans. Linn. Soc. Zool.* 14: 304–311.

**9** Nelson, G.S. (1996). Lymphatic filariasis. In: *The Wellcome Trust Illustrated History of Tropical Diseases* (ed. F.E.G. Cox), 294–303. London: The Wellcome Trust.

**10** Delaporte, F. (2008). Manson's triple error. *Parasite* 15: 495–500.

**11** Marti, H. (2019). The discovery of helminth life cycles. *Adv. Parasitol.* 103: 1–10.

**12** O'Neill, J. (1875). On the presence of a filaria in 'craw-craw'. *Lancet* 105: 265–266.

**13** Muller, R.L. (1996). Onchocerciasis. In: *The Wellcome Trust Illustrated History of Tropical Diseases* (ed. F.E.G. Cox), 304–309. London: The Wellcome Trust.

**14** Hissette, J. (1931). Sur l'existence d'affections oculaires importantes d'origine filarienne dans certains territoires du Congo. *Ann. Soc. Belg. Med. Trop.* 11: 45–46.

**15** Blacklock, B. (1926). The development of *Onchocerca volvulus* in *Simulium damnosum*. *Ann. Trop. Med. Parasitol.* 20: 1–48.

**16** Robles, R. (1917). Enfermedad nueva en Guatemala. *La Junventud Medica* 17: 97–115.

**17** Simonsen, P.E., Onapa, A.W., and Asio, S.M. (2011). *Mansonella perstans* filariasis in Africa. *Acta Trop.* 120 (Suppl 1): S109–S120.

**18** Manson, P. (1891). The *Filaria sanguinis hominis* major and minor, two new species of haematozoa. *Lancet* 137: 4–8.

**19** Van Voorhis, W.C., Hooft van Huijsduijnen, R., and Wells, T.N. (2015). Profile of William C. Campbell, Satoshi Omura, and Youyou Tu, 2015 Nobel laureates in physiology or medicine. *Proc. Natl. Acad. Sci. U.S.A.* 112: 15773–15776.

**20** Nicholls, R.S., Duque, S., Olaya, L.A. et al. (2018). Elimination of onchocerciasis from Colombia: first proof of concept of river blindness elimination in the world. *Parasites Vectors* 11: 237.

**21** Ghedin, E., Wang, S., Spiro, D. et al. (2007). Draft genome of the filarial nematode parasite *Brugia malayi*. *Science* 317: 1756–1760.

**22** Consortium CeS (1998). Genome sequence of the nematode *C. elegans*: a platform for investigating biology. *Science* 282: 2012–2018.

**23** Godel, C., Kumar, S., Koutsovoulos, G. et al. (2012). The genome of the heartworm, *Dirofilaria immitis*, reveals drug and vaccine targets. *FASEB J.* 26: 4650–4661.

**24** Grote, A., Lustigman, S., and Ghedin, E. (2017). Lessons from the genomes and transcriptomes of filarial nematodes. *Mol. Biochem. Parasitol.* 215: 23–29.

**25** Cotton, J.A., Bennuru, S., Grote, A. et al. (2016). The genome of *Onchocerca volvulus*, agent of river blindness. *Nat. Microbiol.* 2: 16216.

**26** Desjardins, C.A., Cerqueira, G.C., Goldberg, J.M. et al. (2013). Genomics of *Loa loa*, a *Wolbachia*-free filarial parasite of humans. *Nat. Genet.* 45: 495–500.

**27** Underwood, A.P. and Bianco, A.E. (1999). Identification of a molecular marker for the Y chromosome of *Brugia malayi*. *Mol. Biochem. Parasitol.* 99: 1–10.

**28** Streit, A. (2014). How to become a parasite without sex chromosomes: a hypothesis for the evolution of *Strongyloides* spp. and related nematodes. *Parasitology* 141: 1244–1254.

**29** Hertig, M. and Wolbach, S.B. (1924). Studies on rickettsia-like micro-organisms in insects. *J. Med. Res.* 44: 329–374.7.

**30** Keroack, C.D., Wurster, J.I., Decker, C.G. et al. (2016). Absence of the filarial endosymbiont *Wolbachia* in seal heartworm (*Acanthocheilonema spirocauda*) but evidence of ancient lateral gene transfer. *J. Parasitol.* 102: 312–318.

**31** Landmann, F. (2019). The *Wolbachia* endosymbionts. *Microbiol. Spectrum* 7: https://doi.org/10.1128/microbiolspec.BAI-0018-2019. PMID: 30953430.

**32** Bandi, C., McCall, J.W., Genchi, C. et al. (1999). Effects of tetracycline on the filarial worms *Brugia pahangi* and *Dirofilaria immitis* and their bacterial endosymbionts *Wolbachia*. *Int. J. Parasitol.* 29: 357–364.

**33** Hoerauf, A., Nissen-Pahle, K., Schmetz, C. et al. (1999). Tetracycline therapy targets intracellular bacteria in the filarial nematode *Litomosoides sigmodontis* and results in filarial infertility. *J. Clin. Invest.* 103: 11–18.

**34** Taylor, M.J., Cross, H.F., and Bilo, K. (2000). Inflammatory responses induced by the filarial nematode *Brugia malayi* are mediated by lipopolysaccharide-like activity from endosymbiotic *Wolbachia* bacteria. *J. Exp. Med.* 191: 1429–1436.

**35** Crook, M. (2014). The dauer hypothesis and the evolution of parasitism: 20 years on and still going strong. *Int. J. Parasitol.* 44: 1–8.

**36** Ogawa, A., Streit, A., Antebi, A., and Sommer, R.J. (2009). A conserved endocrine mechanism controls the formation of dauer and infective larvae in nematodes. *Curr. Biol.* 19: 67–71.

**37** Wenk, P. and Renz, A. (2003). *Parasitologie; Biologie der Humanparasiten*. Stuttgart: Thieme Verlag.

**38** Bowman, D.D. and Atkins, C.E. (2009). Heartworm biology, treatment, and control. *Vet. Clin. N. Am.: Small Anim. Pract.* 39: 1127–1158.

**39** Boussinesq, M., Gardon, J., Gardon-Wendel, N., and Chippaux, J.P. (2003). Clinical picture, epidemiology and outcome of *Loa*-associated serious adverse events related to mass ivermectin treatment of onchocerciasis in Cameroon. *Filaria J.* 2 (Suppl 1): S4.

**40** Babu, S. and Nutman, T.B. (2012). Immunopathogenesis of lymphatic filarial disease. *Semin. Immunopathol.* 34: 847–861.

**41** Idro, R., Anguzu, R., Ogwang, R. et al. (2019). Doxycycline for the treatment of nodding syndrome (DONS); the study protocol of a phase II randomised controlled trial. *BMC Neurol.* 19: 35.

**42** Murray, C.J.L. and Lopez, A.D. (2017). Measuring global health: motivation and evolution of the Global Burden of Disease Study. *Lancet* 390: 1460–1464.

# 3

# Human Filarial Infections: Reflections on the Current Understanding of Their Importance, Pathobiology, and Management

*Charles D. Mackenzie**

*Neglected Tropical Disease Support Center, Task Force Global Health, 330 West Ponce de Leon Avenue, Decatur, GA 30030, USA*

**Abstract**

Filarial infections of humans are still medically significant conditions in many of the tropical regions of the world, despite considerable success in recent years toward controlling and eliminating those filariae that cause the most significant disease, namely onchocerciasis and lymphatic filariasis. Filariae, parasites that invade the host and multiply totally within the host's internal organs, establish unique relationships with their specific hosts. The consequences of these filaria-host relationships characteristically present as a spectrum, both in terms of the clinical presentation and the associated immunological and pathological responses.

The different clinical presentations of human filarial infections are related to the particular tissues infected and on the type of pathogenic events that follow and induce damage to these tissues. Central to the pathological changes in filaria-infected tissues are inflammatory responses associated with the microfilarial stage of these nematodes; these potentially dangerous clinical responses often pose a medical challenge following the use of chemotherapeutic agents that specifically damage this stage of the parasite. Unfortunately, the current anthelmintic-based control and treatment protocols commonly target this larval stage. However, study of the clinical responses often seen when targeting this parasitic stage with drugs has both informed our knowledge of these diseases and their pathogenesis, as well as catalyzed efforts to find agents that directly affect other stages of filariae where there is less likelihood of adverse reactions. Despite the challenges associated with the use of anthelmintics, chemotherapy with ivermectin, albendazole and diethylcarbamazine have been the mainstay in both individual patient treatment and in global control and elimination programs directed at onchocerciasis and lymphatic filariasis for well over 20 years. The global efforts against these two infections have been one of the most successful public health initiatives in tropical medicine in the past century.

Many gaps remain in our understanding of these complex infections. A lack of detailed understanding of their pathogenesis remains, as with many tropical infections, in

*Corresponding author.

*Human and Animal Filariases: Landscape, Challenges, and Control*, First Edition.
Edited by Ronald Kaminsky and Timothy G. Geary.

part due to the lack of detailed autopsy and non-invasive imaging studies, and in part due to major research and funding initiatives being directed largely at optimizing the distribution of currently available drugs rather than what are regarded as more academic issues. Whether the comparatively minimal level of current knowledge and range of available drugs will be sufficient to achieve control and elimination of these important infections remains to be seen.

## 3.1 Introduction

Humans can be infected by filarial nematodes, sometimes with little apparent consequence to the infected individual and in other cases with serious clinical and physical consequences. This theme of "variability" underscores much of the pathobiology of human and animal filarial infections and is seen in tissue changes, in host immune responses, and in the clinical presentations resulting from these infections. Filariae are one of the relatively few pathogenic nematodes in which adult stages develop, live, and reproduce within internal tissues of the body rather than in an externally linked cavity such as the gut, as seen with the other nematodes commonly infecting humans. Consequently, the mechanisms for survival evolved by filariae are likely to be somewhat different, and perhaps more complicated, than those of the more commonly and better studied parasitic nematodes such as *Ascaris* sp. There are many valuable reviews of filarial infections and the diseases they cause [1–3], and it is not the aim of the present discussion to duplicate the catalogue of well-described biological information on human filariae; rather, this discussion will highlight some of the less clear issues in human filarial infections and identify some important questions remaining about their pathobiology and their impact on human health that still need to be addressed.

The clinical images of filariasis – blind elders being led by children, grossly swollen and disfigured limbs, and grossly swollen male genitalia – have been well known across the world for centuries, although many who see these images in all likelihood do not fully understand what caused these disfigurements. In reality, prior to the establishment of the programs to control and eliminate onchocerciasis in 1974 and lymphatic filariasis in 2000, the general understanding of these diseases, and their effects on patients and their families, has remained for much of the past century one of rumor rather than of fact. Loiasis and mansonellosis, two other human filarial infections, in general have not received much attention as, at least in their untreated state, they have been regarded as being relatively unimportant and are often recognized as clinical conditions, only of interest to a few dedicated physicians and investigators. An exception to this generalization has been the recognition that clinically significant, and often lethal, outcomes can occur following anthelmintic treatment of patients with high microfilarial loads of these two infections [4], especially with loiasis. Indeed, post-chemotherapy reactions seen in many filaria-infected individuals constitute an important phenomenon that has driven much research into understanding the pathogenesis and treatment of these diseases.

## 3.2 Historical Aspects

Filarial infections were recorded in humans in early Egyptian times. Swollen limbs are depicted on Egyptian hieroglyphics from 2000 BCE, and drawings from the Edo period in Japan (600 CE) show swollen male genitals and limbs [5]. For most of the latter half of the twentieth century, few dedicated care centers and investigators around the world focused on lymphatic filariasis; namely, clinical aspects in India and Sri Lanka [6, 7], Haiti [8], Ghana [9], and Brazil [10], and with immunology, entomology, chemotherapy, and pathogenesis studies in the United Kingdom [11] and United States [12, 13]. The initiation of control and elimination programs for both onchocerciasis and lymphatic filariasis brought a welcome renewed focus on these devastating causes of disease and disability of people living in areas where healthcare support has generally been minimal at best.

Onchocerciasis, at least as a dermal condition, was likely first recorded by Naval Surgeon O'Neill in captured Africans being freed in West Africa during British anti-slavery activities in 1782; he used the local term "craw-craw" or "kru kru" in his original description. It should be noted, however, that these are common names used for severe skin conditions in coastal West Africa and likely include other dermal conditions, such as scabies, in addition to onchocerciasis. Although onchocerciasis is primary a dermal condition, the more dramatic presentation of loss of vision and the development of blindness, seen in heavily endemic areas of Africa, dominated the early clinical descriptions of this condition. It is not entirely clear where the common name given to onchocerciasis, "River Blindness," originated. However, in Sudan and now South Sudan, the disease has been known for almost a hundred years as "Jur River Blindness," this name being recorded in official reports submitted in the 1920s by the colonial medical staff in the township of Wau along the Jur River in what is now western South Sudan [14]; this is likely the first use of the term "river blindness" for this infection.

The blindness caused by onchocerciasis catalyzed the first major multi-country effort to control a human parasitic disease. In 1974, a major vector elimination intervention began in West Africa, the onchocerciasis control project (OCP), with the ultimate goal of reducing and preventing blindness [15]. A major reason for starting the OCP was the economic effect that onchocercal blindness in these West African countries caused through the abandonment of fertile river-associated areas of farm land. The target approach in this initiative, which focused on 11 West African countries, was to protect some 30 million people against infection by aerial spraying larvicidal agents to destroy larvae of the black fly vector in the rivers of endemic areas. Thus, the goal of reducing blinding eye disease in this large area of Africa was based solely on a single entomologically-based approach. Although this had an important effect on the disease, it was not totally successful and a chemotherapeutic intervention with ivermectin was added in 1987 to move more rapidly toward the goal of controlling this disease [16]. It is important to recognize that the introduction of ivermectin, and its distribution to the majority of all inhabitants of endemic areas in a mass drug administration (MDA) campaign, was a historically transformative event in providing health care to people living in underserved areas of the developing world.

In the period before the introduction of ivermectin, it should be noted that other chemotherapeutic approaches were in use, albeit most often with significant adverse consequences. In Eastern Africa in 1970s and 1980s, physicians were either already using the "Sudan regimen" – employing the very toxic drug suramin [17]. More widely used across Africa was the filaricidal drug diethylcarbamazine (DEC) that had been described in Mexico in 1949 [18]; some countries outside the OCP areas at this time were also including vector control activities [19], usually with chemotherapy. In Latin America, other than the use of DEC, the main approach to control of the infection in Mexico, Guatemala, and Ecuador was removal of the adult worm-containing subcutaneous nodules though extensive national surgical campaigns; these were largely unsuccessful in curtailing the infection and were almost universally unpopular. These pre-ivermectin interventions across the endemic areas of the world had variable and usually unsuccessful effects in reducing onchocerciasis, and it was not until ivermectin was included as a safe and suitable treatment for onchocerciasis that substantial movement toward control of this disease began to be seen across Africa, America, and the Middle East, recognized by award of the 2015 Noble Prize for Medicine [20, 21].

It is important to reiterate that chemotherapy has a central role in any historical reflection concerning the understanding of the pathogenesis of filarial infections. Three chemotherapeutic agents loom large in the history of control and treatment of both lymphatic filariasis and onchocerciasis: DEC [22], albendazole [23], and ivermectin [16]. DEC treatment was described for onchocerciasis in Latin America by Luis Mazzotti in 1948 [22, 24], and for lymphatic filariasis in 1947 [25]. The clinically evident adverse side effects that this drug causes in onchocerciasis, known as "Mazzotti reactions," stimulated the search for more tolerable drugs and the eventual selection of ivermectin for use in humans. The donation of this drug in 1987 by Merck & Co. for onchocerciasis control, as mentioned above a ground breaking event in public health, reinvigorated control programmes and led to the concept that onchocerciasis may be eliminated as public health problem from at least 10 countries by 2030 [26]. This drug company's inspiring donation was followed 10 years later by GSK's donation of albendazole, and then Eisai's donation of DEC in 2013, both for the elimination of lymphatic filariasis as a public health problem.

In contrast to the two major filariases, *Loa* sp. and the three *Mansonella* sp., feature little historically other than in basic morphological descriptions and other general parasitological aspects. Loiasis has featured more prominently in recent years as it has been associated with serious adverse reactions to chemotherapy that have restricted to some degree the implementation of global MDA-based elimination programs for the two major filariae. Very high loads of *Loa loa* circulating microfilariae are strongly associated with the majority of fatalities related to the distribution of the generally very safe ivermectin in *Loa* endemic areas in Africa [27]; severe responses to treatment of this infection have also been reported with DEC [28]. The first of the *Mansonella* parasites was first described by Patrick Manson in a sleeping sickness patient from West Africa he was treating in London in 1870 [29, 30]. In general, the clinical diseases due to loiasis and mansonellosis have only been the interest of filarial specialists and those treating expatriate cases, and in general, the parasites are not believed to induce major clinical outcomes. However, whether it is true that no significant clinical effects are caused by these filariae is now being investigated, perhaps as an additional peripheral benefit from the global focus on the two major filarial infections.

## 3.3 The Parasites

Eight filariae are known to infect humans, affecting lives in mild to significant ways [1, 3]. They all enter the human host through blood-feeding insect vectors (Table 3.1). Host specificity is an important phenomenon in these nematodes, and as a general rule, each filarial species is confined to a single, or at least very few, specific host(s) and vectors; their life cycle being able to be fully completed only in its respective primary host, with very few, often experimentally induced, exceptions. Although there are many different filariae across the animal kingdom, this host specificity likely involves intimate and complex adaptive processes that include a range of different biological protective mechanisms, including both innate and adaptive processes.

**Table 3.1** Human filarial infections.

| Filarial parasite | Common names[a)] | Vector | General global location |
|---|---|---|---|
| *Onchocerca volvulus* | "River blindness"<br>Craw craw<br>"Sowdah" | *Simulium* sp. ("black flies") | Latin America, Africa, and Yemen |
| *Wuchereria bancrofti* | Bancroftian filariasis<br>"Elephantiasis" | Mosquitoes | India, Su-saharan Africa, Western Pacific, South East Asia, and Americas |
| *Brugia malayi* | Filariasis | Mosquitoes | South East Asia |
| *Brugia timori* | Filariasis | Mosquitoes | Indonesia |
| *Mansonella ozzardi* | "Ozzardi filariasis" | Midges (Culicoides) | Americas |
| *Mansonella perstans* | "Perstans filariasis" | Midges (Culicoides) | Africa, Central and South America |
| *Mansonella streptocerca* | "Streptocerca filarisis" | Midges (Culicoides) | Africa |
| *Loa loa* | Eye worm, Calibar swelling | *Chrysops* sp. | Africa |
| *Dirofilaria* sp. | "Heartworm" | Mosquitoes | Uncommon zoonotic infection |

a) Names used generally, or locally in some cases, and with some forms of the disease.

### 3.3.1 Onchocerciasis

Human onchocerciasis, caused by *Onchocerca volvulus*, is still present in 31 countries in sub-Saharan Africa, although some of those in West Africa (e.g., Senegal, Niger, and Mali) are approaching very low levels as a result of MDA-based control programs [31]. Latin America now has only two (of the original seven) endemic countries, Venezuela and Brazil, that have ongoing transmission; the remaining endemic populations here are the very mobile Amerindian tribes (e.g., Yanomami) who live in the cross-border jungle area between these two countries [32]. The only other country outside of Africa to have endemic onchocerciasis is Yemen, where transmission is occurring in valleys facing the Red Sea [33, 34]; here, the vector is a member of the *Simulium* complex *Simulium rasyani* [34].

In all filariae, three parasitic stages of this nematode are generally the most important for understanding the pathobiology, clinical disease, and the interventions for individual treatment and infection control: microfilariae (L1), infective larvae (L3), and adult form (L5/adult). The parasite enters the vector from the human host skin to begin its development during the blood-seeking bites of *Simulium* sp. black flies [34], and these bites are also mechanism for infection, or reinfection, of the host when infective larvae (L3) move from the vector to continue development to adults and reproduction in humans. The bites of these flies create small pools of blood in the upper dermis; the insect's saliva is likely to be an attractant to dermal microfilariae in small local lymphatics [35, 36]. The adult female and male worms are detectable in fibrous, palpably firm, nodules commonly found in subcutaneous tissues or intramuscular fascial planes. These subcutaneous fibro-inflammatory masses are commonly associated with bony prominences, namely in the iliac crest area, in the skin overlying chest ribs and the base of the spine (the area above the cauda equina), as well as bony areas of the heads of infected children. Why these palpable nodules are most commonly associated with bony prominences is unclear. In addition, little is known as to whether additional adult worms and/or nodules are present in deeper body tissues; rare autopsy studies have shown that adult worms can be found in deeper fascial tissues of the upper leg close to the major long bones [37, 38]. It is often assumed by epidemiologists that most nodules are palpable in dermal tissues; the common lack of nodules in microfilariae- or antibody-positive individuals questions this premise. How newborn microfilariae migrate from nodules to the dermis, and to ocular tissues, is also not clear. Recent studies suggest that *O. volvulus* are generally more closely associated with the lymphatic system than was previously thought [37] and the suggestion that they can travel via this vascular system is plausible. However, their physical distribution in the body suggests that microfilariae probably remain relatively close to their originating adult worm. The observation that microfilariae are not present in diagnostic skin biopsies ("skin snips") following chloroquine administration [38] supports the idea that microfilariae are mobile and can at least move away from their location in the upper dermis relatively quickly. It is also possible that the movement of microfilariae contributes to the persistent and often intense pruritus associated with this infection.

The vectors in all endemic locations belong to the *Simuliid* family (Figure 3.2), although the capacity of the different species in this genus to transmit this parasite varies considerably depending on the geographic location. For example, the main vector in Ecuador, *Simulium exiguum,* is a voracious human feeder and effective transmitter of the worm, whereas *Simulium ochraceum* in Southern Mexico has poor host capacity due the destructive effects of their cibarial armature through which the parasites must past during their uptake into the fly. It also noted that humans bitten by black flies can, and often do, recognize antigens in the saliva of these vectors and can become hypersensitive to bites [39]; whether these immune responses to vector antigens affect transmission of onchocerciasis is not known.

### 3.3.2 Lymphatic Filariasis

Three parasites comprise the group causing lymphatic filariasis in humans: *Wuchereria bancrofti* (inducing "bancroftian filariasis") and two "Brugian filariases" caused by *Brugia malayi* and *Brugia timori.* In 2000, some 120–140 million people were estimated to be infected or exposed to infection. Adult worms of this group, as their common name implies, predominantly reside in lymphatic complexes of the pelvic girdle or the axillae of the upper torso and thus induce clinical consequences in the respective draining anatomical segment (e.g., a limb). Why these worms choose to lie in a supposedly immunologically active site, such as afferent lymphatic vessels and the sinuses of lymph nodes, remains a fascinating biological conundrum.

The vectors of lymphatic filariasis come from four genera of mosquitos: *Culex* sp. (common in Latin America but also Pacific, Asia and East Africa), *Aedes* sp. (common in the Pacific and Asia), *Mansonia* sp. (in Latin America, Pacific, and Asia), and *Anopheles* sp. (the commonest vector in Africa). Of increasing interest is urban transmission of this infection [40] and here a better understanding of the associated entomological aspects is needed, as is the development of optimal approaches to carrying out MDA in high population density locations, procedures for which will differ somewhat from those employed in more usual rural settings for MDA.

There are many similarities between the two major forms of lymphatic filariasis, Bancroftian and Brugian, although there are important differences [41]; chronic lymphedema and acute systemic attacks are seen in both. However, hydrocele is not seen in Brugian filariasis. Why this is so is not clear, but it may relate to differences in location of the adult worm between the two forms of filariasis.

*Brugia malayi,* unlike other human filariae, can infect cats naturally, and thus in programmatic efforts to eliminate this form of filariasis from endemic areas, such as Malaysia, additional surveying of the feline population was essential [42]. *Brugia malayi* is also endemic in southern areas of Thailand where domestic cats serve as the major reservoir host [41]. Another, non-human, filarial nematode, *Brugia pahangi* also infects cats and is a useful experimental model for the human disease.

### 3.3.3 Loiasis

*Loa loa* is transmitted by blood-feeding *Chrysops* sp., insects of the group often known as deer or horse flies. Untreated infections with *L. loa* are generally regarded as having minor clinical significance, with the most significant presentations being sub-cutaneous angio-edematous swellings and migration of the adult worm across the visible conjunctival tissues of the eye ("eye worm"). The dermal swellings, "Calabar swellings," were first seen in, and named after, the coastal Nigerian town of Calabar in 1895; these lesions are associated with degeneration of the adult worms in deep dermal tissues [43]. The migration of adult *Loa* across the eye is a directly observable example of the mobility of filarial parasites. It is now recognized that a wide variety of atypical clinical presentations (e.g., arthralgia, urticaria, etc.) can occur with *L. loa* infection and the frequency of these sign and symptoms increases with increasing parasite microfilarial loads [44, 45].

The main concern in patients with heavy loads of *Loa* microfilariae is the possibility of developing severe post-treatment responses, that are sometimes fatal, after microfilaricidal chemotherapy [46]. This is of particular concern where loiasis is present in endemic areas being treated for onchocerciasis and lymphatic filariasis. Primates can also carry a form of *Loa* sp., but it is not clear whether these parasites are also transmitted to, and can survive in, humans.

### 3.3.4 Mansonellosis

Three filarial species of the genus *Mansonella* can infect humans [30]: *Mansonella perstans* (found in Africa and Latin America), *Mansonella ozzardi* (only in the New World), and *Mansonella streptocerca* (found only in Central Africa). All are transmitted by *Culicoides* sp. flies (biting midges), with *M. ozzardi* transmission also occurring through the bites of black flies (*Simulium amazonicum*). The location of the adult worms in their host differs among the *Mansonella* species, *M. perstans* being more commonly found in body cavities such as the peritoneal cavity or the pleural cavity, and occasionally in the pericardial sac. On the other hand, *M. streptocerca* and *M. ozzardi* are found in sub-cutaneous tissues. However, very little is known about the clinical effect of these infections, nor is much known about their geographic prevalence.

### 3.3.5 Incidental Filarial Conditions

Zoonotic filarial infections occur and are occasionally identified, but the true incidence is not clear [47, 48]. One of the commonest examples is infection with the canine/feline parasite *Dirofilaria* sp., but these and virtually all zoonotic filarial infections, except for *Brugia* in cats (as mentioned above) rarely if ever fully develop and complete the parasitic cycle in humans, with the parasites usually dying causing focal, usually chronic, histopathological lesions [49]. These lesions usually serve more as oddities in clinical diagnosis and only when they are misdiagnosed for important diseases, such as lung cancer, do they have real clinical significance.

Given that there are at least 11 different filarial nematodes of non-human primates (including primate *Loa* sp. and *Mansonella* sp.), it is likely that in tropical environments, there is a significant chance that primate parasites do, on occasion, enter humans through the bites of blood-feeding insects; although given the typical host specificity of filariae, these most likely do not fully develop in humans to any great degree.

## 3.4 The Pathogenesis and Presentation of Human Filarial Infections

Before discussing the respective pathologies and clinical outcomes of filarial infections, it is worth considering the events and phenomena that occur in the interactions between filarial worms and their hosts. These are aspects that influence the various pathologies seen, the consequent disease presentations and responses to chemotherapy. These important factors can also guide research into these infections and diseases.

The clinical presentation of each of these human infections varies considerably, partially due to different levels of adaptation to the host's natural and developed immunity in different parasite life stages. The mechanisms that allow these niche host–parasite relationships to exist are complex and are not clearly understood. This complexity underscores a plasticity in the biochemistry of filarial worms, their interactions with host immune system, and ultimately their ability to survive and multiply in arguably "hostile" host tissues such as the lymphatic system. Filarial nematodes have the ability to adapt to different biochemical environments and, for example, to vary their energy sources [50]. Despite this ability to adapt to seemingly hostile environments, host specificity is an important concept for these nematodes. There are many different filariae across the animal kingdom, and most have the ability to survive in their particular host almost undetected in an immunological sense; the mechanisms by which they survive, which may or may not be similar in each of these host–parasite situations, are likely keys to developing strategies for treatments (e.g., vaccines, new drug targets, etc.).

A clear factor in this intimate host–filariae relationship is the ability of these parasites to suppress the host's immune system, a phenomenon made obvious by clinically notable cases in which this suppression is in fact absent and severe clinical responses occur – often situations associated with death or degeneration of certain stages of the worms. The immunologically active condition of reactive oncho-dermatitis ("sowdah") is a clear example of this in onchocerciasis where there appears to be heightened specific cellular immunity [51]. This concept of filarial immunological spectrum, paralleled by a clinical spectrum, has been described in lymphatic filariasis and onchocerciasis [39]. The presence, and increasing loads, of living healthy parasites in these infections is likely to be paralleled with suppression of Th1 immune responses induced by the worms; correspondingly, the development of clinically evident disease is likely the result of a failure of this immunomodulation. Which specific worm-derived antigens or molecules are involved in modulating the array of immune responses that influence clinical outcomes remains largely unresolved, but it most likely the case that microfilariae or microfilarial-related antigens (e.g., uterine products associated with the development of microfilariae) are important contributors; many of the inflammatory cellular tissue changes seen involve the presence of microfilariae or are linked to active productive female parasite's uterus and are often seen in the tissues adjacent to the worm's vaginal opening.

In clinicopathological terms, there are three different types of host response to filarial parasites, tissue responses that translate into specific clinical outcomes. First, is the form characterized by lack of a cellular tissue response by the host, a quiescent phase in which apparently healthy parasites are present in tissues and fluids; however, immunological indicators, such as specific antibodies, indicate that the host is "aware" of the presence of the worm in this phase. Second, there can be an active inflammatory response to obviously degenerating worms, and in many filarial infections, this is often most obvious clinically with regard to degeneration of microfilariae. These responses involve the activation of specific cellular components such as eosinophils and specific immunological factors and are often clinically severe, and often acute, reactions. In the third form, cellular responses to dead and primarily biologically inactive worms and worm components are evident; this type of tissue reaction involves a strong macrophage component, essentially a "foreign-body" type response, and probably involves much less of an immunologically specific inflammatory response. These different tissue responses, and their different immunological components, underscore the spectrum of clinical presentations seen in filarial infections and reflect the extent of adaptation of each filariae to their specific hosts.

The second of these types of tissue responses, the active responses targeted to microfilariae or reproductive components (e.g., uterus-derived components), are central to the development of major pathologies in most, if not all, filarial infections. This is well demonstrated by reactions that are associated with death of microfilariae as a result of chemotherapy. Histopathological observations indicate that inflammation associated with active degeneration of microfilariae reflects initially a more acute cellular reaction (e.g., eosinophil infiltration) than is seen with tissue responses to other stages; thus, microfilariae are arguably the most "pathogenic" stage of filarial infection. The ability of a tissue in which microfilariae are degenerating and dying – often in large numbers – to handle the associated inflammation is central to the external clinical responses that result. For example, inflammatory responses to microfilariae in sensitive and poorly recovering tissues, such as the retina of the eye or dermal tissues – as happens in onchocerciasis – results in significant clinical presentations and most often permanent damage to these tissues. In contrast, destruction of microfilariae in organs that are better able to handle the inflammatory consequences, such as the spleen – which occurs in lymphatic filariasis and loiasis – does not result in such serious clinical outcomes, at least as far as we know.

The signature clinical responses to microfilaricidal chemotherapy in filarial infections are known as Mazzotti reactions. First described in 1947, they involve both local and systemic signs and symptoms that vary in severity from mild local dermal reactions to serious and severe systemic reactions of an anaphylactoid character, including papular eruptions, dermal edema, arthralgia, and swollen and tender lymph nodes [24]. The severity is related to the microfilarial load and the tissue in which the parasites are located; for example, microfilariae in the skin, as in onchocerciasis, induce killing-related responses and severe dermal edema occurs. Differences in severity are noted with different microfilaricidal drugs, ivermectin inducing less severe Mazzotti reactions than does DEC. DEC, in addition to damaging the parasite, directly influences inflammatory biochemical pathways [22], e.g., the arachidonic acid pathway, and it is very likely that this agent exaggerates the inflammation initiated by its degenerative effect on the microfilariae. Ivermectin, on the other hand, likely does not compound the host's response to the dying parasite in this manner, although it does induce relatively mild Mazzotti reactions.

One common component of filariae that has support for being the key contributor to the development of pathology is the bacterial endosymbiont, *Wolbachia* sp. found in most human filariae with the exception of *L. loa* [52] and some *Mansonella* sp. *Wolbachia* are clearly an active and important component of filariae, and thus it is not surprising that the host recognizes *Wolbachia* antigens [53] when the worm releases them, e.g., when worms degenerate and die. Whether this endosymbiont is the only, or indeed the major, pathology-inducing filarial component remains unanswered.

### 3.4.1 Onchocerciasis

The common picture of onchocerciasis is a chronic condition with dermatological or blinding clinical presentations (Tables 3.2 and 3.3; Figure 3.3). However, it should be recognized that both the dermal and ocular stages of this infection have acute phases, seen characteristically in terms of the responses to microfilariae in the first periods of infection, for example in children in endemic areas, or expatriate cases [54]. Nevertheless, the most obvious and signatory presentation of the disease is the presence of persistent fibrous subcutaneous nodules that contain adult worms; these lesions have often been the presentation that has signaled the likely presence of *O. volvulus* in an individual, and indeed often in a geographical area. The presence of a nodule in a patient was the first sign that onchocerciasis was present in two Latin American countries, Guatemala and Ecuador [55, 56].

These nodules are classically seen in the area of the iliac crest as single entities or, in case with heavy parasite loads, as clusters containing up to 8–9 nodules, each usually between 0.5 and 1 cm in diameter. These lesions are painless unless they are lying on sensitive areas of the anatomy, such as nerves or bony prominences. In children, they can be often be found on the head and in all ages can also be found on the rib cage, or at the base of the spine in the *cauda equina* area. Differential diagnosis of these lesions includes dermal and sebaceous cysts, and cysts induced by cysticercosis infections. Whether palpably detectable onchocercal nodules are a true reflection of the load of adult worms is unclear; limited autopsy studies have shown

**Table 3.2** Clinical presentations of dermal onchocerciasis.

| Presentation | Pathogenesis | Characteristics | Associated features |
|---|---|---|---|
| Intolerant pruritus | Irritation due to the presence, movement, and degeneration of mf | Persistent itching and the presence of scratch marks on the skin | In severe cases trauma to the skin from overactive scratching |
| Acute papular dermatitis | Local reaction to degenerating mf | Discreet raise papule | |
| Dermal atrophy | Wearing out of the skin through years of microfilaria-induced damage | Thin dermis, dry skin | "Hanging groin" |
| Hyperpigmentation | Irritation of epidermal basal layer and the pigment cells | Increases pigmentation associated with focal area of inflammation | Mottled skin |
| Depigmentation | Damage to the basal layer of the epidermis | Variable amounts of pigment loss – minor to complete loss (e.g., "oncho shins") | Damage due to blackly bites, and scratching finger nails |
| Subcutaneous nodules | Fibrous granulomas surrounding nest of adult parasites | Firm, variably movable masses | Sometimes in multiple clusters, especially in aged individuals |
| Lymphadenopathy: atrophy | Worn out lymphoid tissue in chronic onchocerciasis | Shrunken, hard, lymph nodes (e.g., in groin and axillae) | Seen in very chronic cases with high parasite loads |
| Lymphadenopathy: enlargement | Hyper-responsive lymphoid tissue in Reactive Onchodermatitis (ROD) | Swollen, sometime painful, lymph nodes drain the affected localized are of ROD | A very active and pruritic condition with extensive dermal pathology (also known as "sowdah") |

that onchocercal nodules (sometime called "onchocercomas") can be found deep in the fascia associated with muscles attached to the upper femur in humans [57] and in an experimentally-infected chimpanzee (Mackenzie, unpublished observations); in both situations, it was not possible to detect nodules by external palpation.

Cellular composition of a nodule includes a predominance of collagen and aggregates of lymphocytic cells and plasma cells, together with sheets of macrophages and giant cells associated with degenerating adult worms. The presence of additional cell types with lymphatic vessel immunoprofiles (phenotypes) has led to the hypothesis that the genesis of an onchocercal nodule involves blockage of a dermal lymphatic bed by mature or maturing parasites [37]. Nodules contain an extensive vascular network commonly served by one or two major vessels; ligation of the vascular stalk by surgeons carrying out nodulectomy contributes significantly to a clean surgical outcome. Onchocercal nodules often carry 6–10 females together with 1 or 2 male worm; the age of the worm often being judged by biochemical indicators such as the amount of hemosiderin in their intestines. It has been a common practice to examine

**Table 3.3** Major clinical presentations of ocular onchocerciasis.

| Presentation | Pathogenesis | Characteristics | Associated features |
|---|---|---|---|
| Punctate corneal keratitis | Reactions associated with the death of microfilariae | Known as "snow flake" opacities. Microfilariae also be seen without any associated visible reaction | Are often reversible lesions |
| Sclerosing keratitis | Replacement of the corneal with disoriented collagen tissues | Can result in complete oblation of the cornea | Irreversible Autoimmune mechanisms have been suggested |
| Limbal globules | Accumulation of cells at conjunctival-corneal border | Round globules at the limbus linked to gimbal vessels | Following anthelmintic treatment these lesions contain eosinophil leucocytes |
| Iritis | Inflammation of the uveal tact | Often induces a "tear drop" deformity | Associated with the development of second cataracts |
| Choroidore-tinopathy | Associated with the presence and degeneration of microfilariae. Autoimmune mechanism also postulated. | Visual loss, reduced field of vision Characteristic retinal pigment alterations | Visual loss, night blindness |
| Optic neuritis | Similar to choroidoretina | Swollen optic disc | Loss of vision |
| Retinal vascular leakage | Seen after the use of diethylcarbamazine | Active leakage of vessels in tracer studies | Reduction in vision |

adult worms after removal of host tissues via digestion, usually after incubation in collagenase and dispase. Adult worms are then broken up by vortex shaking to release embryonic forms to allow quantification of the different developing forms to produce an "embryo-gram," a numerical profile of the stages present. This approach is often used in chemotherapy studies, particularly those aimed at sterilizing adult worms.

Although the presence of nodules is the recognizable presentation of this infection, onchocerciasis is primarily a more generalized dermatological condition induced by microfilariae predominantly residing in the skin. In addition, there is the well-known extension of the habitat of these worms to ocular tissues in cases with high microfilarial loads. The outcomes of the presence of microfilariae in ocular tissues and the consequence of blindness have led to the commonly used name for the disease, river blindness. Dermal changes in onchocerciasis reflect a range of responses in the clinical-immunological spectrum of this disease; Figure 3.3 summarizes the various stages in the development of the three major categories of dermal disease in onchocerciasis, namely juvenile onchodermatitis, reactive onchodermatitis, and chronic onchodermatitis. For the first six to eight years after infection, the majority of lesions, seen as discrete papular dermatitis and punctate corneal lesions, appear almost exclusively to be reactions to dying microfilariae. These reflect an active immune response against microfilariae; indeed, children will present with papular or ocular responses even though standard iliac crest skin biopsies are negative for parasites. Following this initial ability of the host to destroy microfilariae and keep the numbers of this stage in the skin low, patients, as they age, appear to move into either a phase of heightened immunoactivity with subsequent skin damage (reactive onchodermatitis), or more commonly, specific immune-modulation begins and the microfilarial load begins to increase, reaching high levels in hyperendemic areas (>500 microfilariae/mg of skin in some cases). The more commonly recognized dermatological presentation, chronic onchocercal skin changes, is likely due to many years of repeated events of microfilarial destruction and local tissue damage, with a consequent wearing out of the tissues in response to these inflammatory events, resulting in the well-described changes in pigmentation and atrophic degeneration of the skin.

Clinically, acute responses to the death of individual microfilariae in the skin are seen as discrete papular lesions (Figure 3.1) which over time become more indurated, and in some cases edematous. An associated alteration in pigmentation of the epidermis usually occurs over time, initially with an increase in pigmentation, although over time and with repeated rounds of parasite death, together with dermal trauma from constant scratching, depigmentation can also occur in certain locations (over nodules, and on the lower legs) (Table 3.2). Thus, long-term changes in the majority of those infected are dermal atrophy with altered pigmentation resulting from an accumulation of damaging events related to the destruction of the microfilariae and self-induced trauma induced by the intense pruritus (damage from scratching) over many years – the often-presented picture of onchocercal dermatitis. Children younger than seven to eight years old living in endemic areas commonly present with acute papular reactions related to microfilarial death but often have low parasite loads (skin snip counts), likely indicating that they still have an active immune response that is largely killing new microfilariae in skin and those that have invaded the cornea. Gradually, if they do not develop reactive onchodermatitis, the dermal parasitic load increases as the ability of the immune system to destroy the parasites becomes ineffective. Nevertheless, microfilariae continued to die and degenerate either naturally as they age (a life span for *O. volvulus* microfilariae of 15–16 months is suggested), causing focal damage to the associated dermal tissues and contributing to the eventual debilitation of the skin seen in typical chronic onchocerciasis.

Reactive onchodermatitis differs from this long-term slow degeneration of dermal tissues. It is more clinically active with an intense, more localized presentation – often seen to affect only a single limb – first described in Yemen and assigned the general Arabic name "Sowdah" (Arabic for "black") due to the associated hyperpigmentation [33, 34]. This immune-active form appears to be more common in areas of interrupted or low transmission, likely where the host is presented with new parasites on an interrupted, episodic basis and the immune system is unable to move into a permanently suppressed state. This form of dermatological presentation differs from the more general degenerative chronic phase in

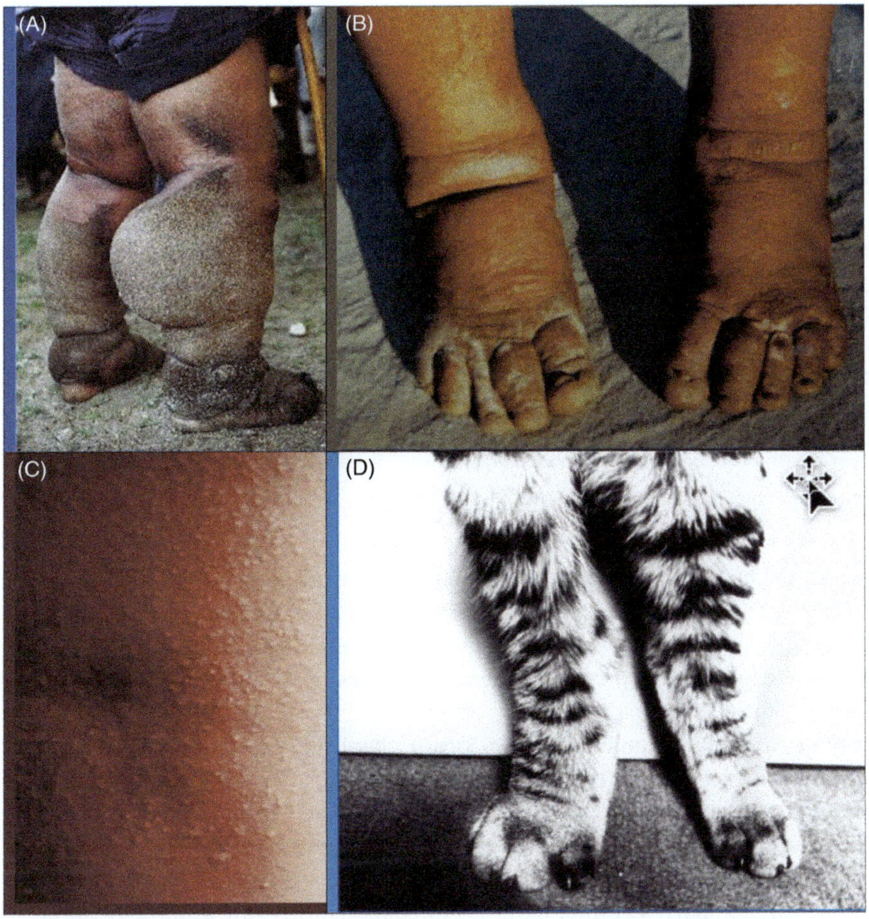


**Figure 3.1** Onchocerciasis and lymphatic filariasis. Lymphatic filariasis with severe dermal changes (A); severely affected toes in lymphatic filariasis (B); acute papular eruption in onchocerciasis (C); lymphoedematous left leg in a cat infected with Brugia (D).

that there is a very active cellular immune response [58]. This active dermatological presentation of onchocerciasis can be found in many endemic areas across Africa and Latin America where the history of Sowdah patients suggests that they have relatively recently entered an onchocerciasis endemic area and thus may have not yet reached a state of immunosuppression to this parasite. Hypo-endemic areas, or areas where transmission appears to be intermittent, often have more reactive onchodermatitis patients than locations with regular annual transmission.

A major contributor to degeneration and destruction of the skin is the constant scratching with finger nails; this condition is often labeled as "intolerable pruritus." Bites of the vector blackly *Simulium* (Figure 3.2) also cause damage; the creation of a blood pool by cutting mouthparts can resultant in damage to the upper dermis and the pigmented basal layer of the epidermis leading to depigmentation. Much weight has been placed on the loss of pigment in onchocerciasis cases, and depigmented

**Figure 3.2** A *Simulium* sp. blackly, the blood seeking vector of *Onchocerca volvulus*.

ankles ("oncho shins") are regarded by many as a signatory indicator of the presence of onchocercal infection. However, the presence of these areas of dermal depigmentation may also be a reflection of biting by the vector, and its presence, rather than being only caused by the presence of parasite, may be related to the intensity of vector feeding (Figure 3.3).

Although ocular changes (Table 3.3) are classically divided between anterior and posterior segments, with suggestions by investigators that changes in these tissues are due to fundamentally different pathogenic mechanisms. It is highly likely that the primary pathogenic event in both segments relates to degeneration of microfilariae and the accompanying inflammatory responses that damage the eye's sensitive, poorly healing, tissues. Punctate keratitis can be directly observed to be related to the death of microfilariae; however, the pathogenesis of sclerotic keratitis, a condition seen more commonly in dustier and dryer endemic regions, remains less clear [59, 60]. Microfilariae are reported to be present in many posterior segment anatomical structures, such as the optic nerve, retina, and vitreous [61], and it likely these parasites contribute to changes in the posterior segment. The suggestions that autoimmune phenomena, or perhaps immune-complex initiated pathology, are involved in both corneal and posterior segment lesions, still remain to be fully defined, as does the basis of the uveitis sometimes seen following chemotherapy. Certainly, damage to vascular components of the posterior segment occurs after treatment with microfilaricidal drugs; this was shown in a study on DEC-induced retinal lesions [62]. Ocular onchocerciasis has not been an active area of research in recent years, in part due to reduced prevalence of high loads of infection in many areas of Africa as a result of the implementation of ivermectin MDA. Arguably, this reduced prevalence of severe ocular pathology may also be due to elimination of the use of DEC for treatment of this disease, as ivermectin does not appear to produce the severe inflammatory sequelae often seen with DEC.

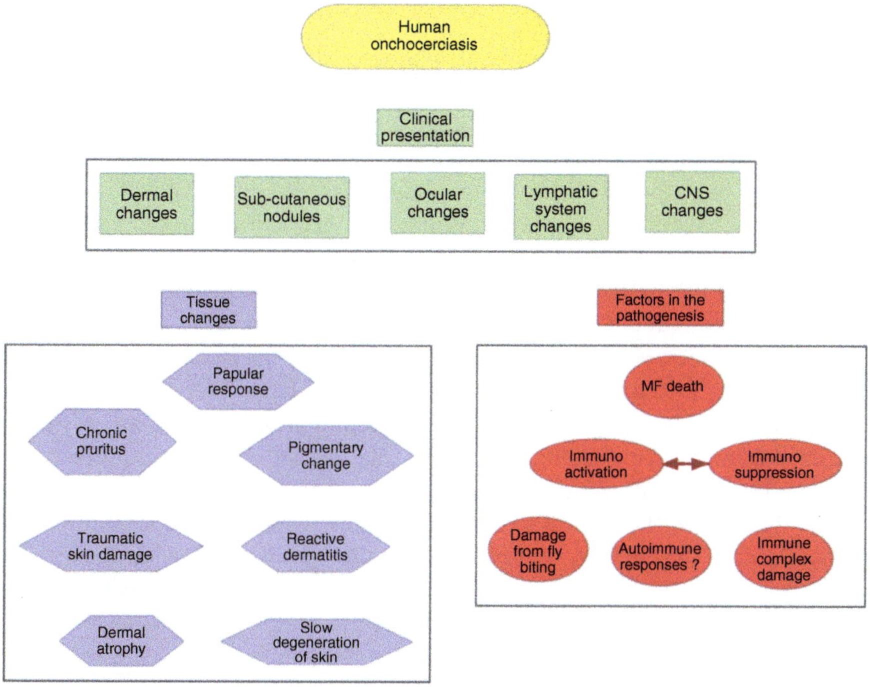


**Figure 3.3** The major components in the patho-biology of onchocerciasis. The range of clinical presentations (green), the tissue changes (mauve), and the major contributors to the pathology presented (red).

Causative association between onchocerciasis and a range of central nervous system (CNS) syndromes has been long been noted. Clinical presentations such as epilepsy, "Nodding disease," and the related Nakalanga syndrome have all been proposed as being caused by *O. volvulus* [63, 64]. These conditions, all seen in children, appear to have an epidemiological link to onchocerciasis. Taken together, these may represent a spectrum of CNS-related, onchocerciasis-induced, pathologies – developmental, mental, and other CNS-associated signs and symptoms that are caused by the parasite, or perhaps by immuno-cross reactivity between parasite components and CNS tissues. The suggestion that it is an auto-immune condition results from a study showing identity between parasite and CNS tissue leiomodin-1 [65]. The presence of onchocercal parasites in the CNS is not reported as a common finding in the few autopsies performed to date; however, microfilariae can be found in spinal fluid following treatment with DEC [66]. How microfilariae enter the CNS is not known, although via the extensive CNS blood vasculature is an obvious possibility, or perhaps through the recently described lymphatic system of the CNS [67]. Clearly, much more needs to be understood about the relationship between onchocerciasis (and other filariases) and the CNS. The now known association between microfilarial death and the lethal development of CNS pathology in loiasis after chemotherapeutic treatment underscores the need to refocus on the CNS aspects of filariasis [68].

### 3.4.2 Lymphatic Filariasis

As with onchocerciasis, the details of pathogenesis of lymphatic filariasis still remain relatively vague or incompletely understood at best. Autopsy reports on patients suffering from lymphatic filariasis have been in general confined to descriptions of various changes associated with the male reproductive organs [69] and unusual pathologies found in individual patients [70]. The two classical presentations of lymphatic filariasis are first lymphedema and associated dermal and systemic changes (Figure 3.1; Tables 3.4 and 3.5), and second the accumulation of fluid in the *tunica vaginalis* of the scrotum. Other pathologies occur, often related to either parasite-induced defects in the lymphatic system, such as the relatively rarely seen condition of chyluria [71], or damage due to the local effects of trapped microfilariae, e.g., renal disease in Brugian filariasis [72].

Lymphatic changes associated with the development of lymphedematous limbs are initiated by abnormal enlargement and dilation, contrary to the often-stated idea that the primary event is a physical blockage by adult worms. Dilation leads to inability of the vessels to pump and move lymph in a cranial direction. Lymphatic valve dysfunction occurs as the vessels enlarge, which leads to lymph stasis. Growth factors such as vascular endothelial growth factor (VEGF) have been shown to be involved in lymphedema due to lymphatic filariasis and other conditions in which this pathology occurs [73]. Experimental studies have shown that products from adult filarial parasites induce lymphatic endothelial cell proliferation [74], most likely in association with other inflammatory cells present in the vicinity of the infected vessels, that produce growth factors that induce lymphatic

**Table 3.4** Clinical presentation of lymphatic filariasis (Bancroftian).

| Presentation | Pathogenesis | Characteristics | Associated features |
|---|---|---|---|
| *Lymphoedema* | | | |
| "Elephantiasis"[a) ("Severe Skin Disfiguration" – SSD) | Exepidermal and dermal disfiguration | Skin folds, dermal nodules, "mossy foot," extensive skin cracks (especially on the heel and toes) | Secondary infections |
| Hydrocele | Fluid accumulation in the tunica albuginea of the scrotum | Slow enlargement of the sac on one or both sides of the scrotum | Can cause sharp pain in the lower abdomen during strenuous exercise |
| Chyluria | Impaired lymphatic system, obstruction to flow in the thoracic duct, lymphatic leak into the urinary system | Chyle (lymph fluid) in the urine | Can be self-limiting, but in severe cases lead to malnutrition |
| Systemic issues | | | |

a) The term "elephantiasis" should be discouraged as being stigma-inducing, i.e., having skin like an elephant is stigmatizing.

**Table 3.5** Genital changes in human Bancroftian filariasis.

| | Condition | Anatomical location | Pathogenesis | Associated features |
|---|---|---|---|---|
| Males | Hydrocele | Fluid accumulation in the tunica albuginea of the scrotum | Slow enlargement of the sac on one or both sides of the scrotum | Can cause sharp pain in the lower abdomen during strenuous exercise |
| | Scrotal lymphoedema | Scrotal connective tissue | Defective lymphatic draining of the scrotum | Distinct from hydrocele |
| | "Elephantiasis" of scrotal skin | Scrotal skin | Consequences of decreased lymphatic drainage | Treatment similar to that of limb "elephantiasis" |
| Females | Cystic swelling | Canal of Nuck | Cystic changes | Treatment by case to case assessment |
| | Edema and epidermal changes | Vulval tissues | Poor lymphatic drainage and local epidermal function | Local topical treatment |

proliferation. It is likely that, as lymphedema persists, fibrotic changes occur in the dermal tissues and reversibility of the condition becomes more and more difficult. This event underscores the importance of implementing effective treatment and the reversal of developing lymphedematous conditions as early as possible, especially in younger age groups.

The presence of static lymphatic fluid most likely leads to a general dysfunction of subcutaneous connective and the vitally active dermal tissues, with consequent inability of the skin to protect against secondary infections, to heal in the normal manner, and to maintain the homeostatic mechanisms it normally performs [75]. These progressive changes all contribute to the development of disfigurement and changes in the limbs, breast, and genital dermal tissues that are characteristic outcomes of this devastating infection. The term "elephantiasis" is often used for the severest dermal changes, but the author believes, as do others, that this term is socially inappropriate to both the patient and to the animal to which it refers and should therefore be avoided; it is unfortunate that this term has become probably the most used and globally recognized name for this condition, a term such as "severe filarial dermatitis" or similar terminology is much preferred.

Acute filarial attacks associated with lymphedema, also known as episodes of acute dermatolymphangioadenitis or ADLA [76], are commonly severe local and systemic events that involve lymphatic vessels and lymph nodes draining the affected lymphedematous limb. The majority opinion on the etiology of these debilitating attacks is the introduction of bacteria to the tissues and vessels of the affected limb [77]; it is also possible that additional or other pathogenic factors play a role in the genesis of these events, such as parasite-related inflammatory responses. Abscesses have also been reported to be associated with the death of adult worms lying in lymphatics; however, most of these reports are from patients in whom anthelmintics such as DEC were used, and it is probable that the nest of degenerating and dying adult worms induces a local inflammatory reaction that becomes clinically evident as an abscess. Infections with *B. timori* are thought to induce more abscesses than the other human filariae.

Hydrocoele, also known as "filaricele," which is seen only in Bancroftian filariasis, is the result of increased fluid in the sac between the *tunica vaginalis* and *tunica albuginea* covering the surface of the testis [78]. This cavity, which is lined with an extension of the abdominal mesothelium, is created at the time of the descension of the testis. It is a cavity, or "sac," that commonly becomes excessively filled with fluid in *W. bancrofti*-infected male patients; the primary origin of this fluid is most likely the secretory mesothelial cells lining this tunic, associated with a compromised natural drawing down of this fluid by mechanical impairment of lymphatic drainage. Adult filarial parasites residing in the lymphatic vessels of the spermatic cord are the likely cause for dysfunction of these lymphatics, reducing drainage of fluids from the tissues of the scrotum (including the fluid in the *tunica* sac), leading to the slow development of a hydrocoele. It should be noted that Brugian filariasis does not induce hydrocoele in infected males; the reason for this difference remains unclear. The question of whether an equivalent to male hydrocoele occurs in women is often raised. Despite obvious anatomical differences, there is indeed a somewhat similar condition, at least in its fundamental etiology, which can occur in women related to the Canal of Nuck, an abnormal extension of the parietal peritoneum; cysts can occur in this structure, but this is likely to be a rare event. There have unfortunately been very few investigations into the effects of lymphatic filariasis on female genitalia, and indeed of any consequences on female health and well-being in general.

Other conditions that can affect the scrotum in lymphatic filariasis include "lymph scrotum" (superficial scrotal lymphangiomatosis) with the development of lymphatic vesicles; it is not a common finding in LF, and the pathogenesis remains unclear. The various dermatological changes seen more often in the skin of affected lymphedematous limbs can also be seen on the male scrotum and on the external labia in females (Figure 3.4).

### 3.4.3 Loiasis

Loiasis is best known clinically for the adverse responses that patients with high loads of circulating microfilariae can suffer following microfilaricidal chemotherapy. These were first described with DEC and, importantly in recent times, with ivermectin in onchocerciasis and lymphatic filariasis elimination programs [27, 28]. Autopsies on these cases of encephalopathy in humans [79] and experimentally infected baboons [68], although few, indicate that the primary lesion in the CNS and other tissues is vascular blockage involving dying microfilariae (Figure 3.5), resulting in anoxia and degeneration of the surrounding parenchyma. In both species, hemorrhages are also found in various tissues along with intense tissue eosinophilia soon after treatment, supporting the concept that inflammation associated with parasite death is central to this phenomenon.

Untreated *L. loa* infection can also induce pathology. The archetypical clinical presentation characteristic of loiasis is the development of the classical "Calabar"

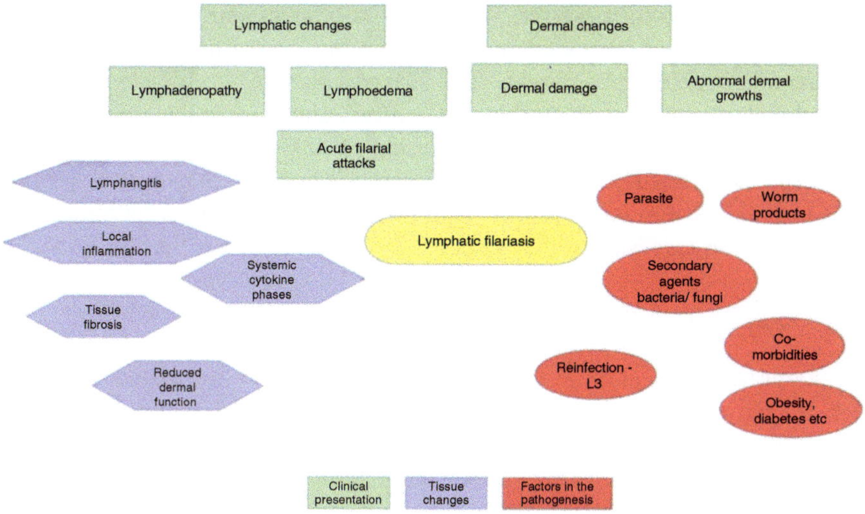


**Figure 3.4** Components in the pathogenesis of lymphatic filariasis. Source: [68], Figure 10, p 08 / PLOS / Public Domain.

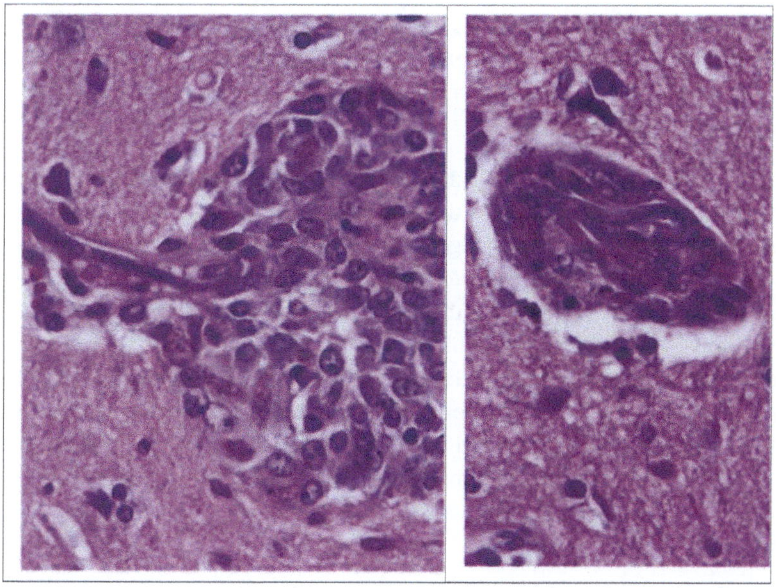

**Figure 3.5** Loiasis: Blocked vessels in the CNS of a hypermicrofilariaemic baboon [68]; blockage containing Loa microfilaria and chronic inflammatory cells with damage to vessel wall.

swelling, a subcutaneous response to dying adult worms [43]. In contrast to these reactions to degenerating worms, an important characteristic of this infection, as with many filariases, is lack of pathology or tissue reaction to healthy migrating adult worms. For example, adult worms can be seen migrating across the conjunctival membranes of the external ocular tissues, often causing great surprise and consternation to the observer. The frequency of observing these worms in the eye ("eye worm") has been used as a surrogate for estimating the endemicity of this infection in at-risk populations. The lack of a significant physical reaction in the conjunctival tissues to migrating adult worms underscores the ability of these worms whilst alive and healthy to avoid causing adverse tissue responses. However, it is not that an infected host does not in other ways recognize the parasites' presence in the body; blood eosinophilia, specific antibody responses, and other biomarkers are all present in *Loa*-infected individuals.

### 3.4.4 Mansonellosis

In general, *Mansonella* infections are thought to be well-tolerated and most often asymptomatic, on occasion inducing mild signs and symptoms such as lymphadenopathy, pruritus, vague abdominal symptoms, and joint pain; it is assumed that the pathogenesis of these presentations is similar to those occurring in lymphatic filariasis [80]. In Brazil, *M. ozzardi* infections are associated with corneal lesions. At present, there is no standardized therapy, although a combination of DEC and mebendazole is often used to treat individuals with *M. perstans* microfilaremia. Ivermectin has also been proven to be effective and safe against the microfilariae, as has doxycycline which kills the essential endosymbiotic *Wolbachia* bacteria present in some strains of *M. perstans* and *M. ozzardi.*

### 3.4.5 Incidental – Dirofilariasis

Filarial worms whose primary hosts are animal species can infect humans, although the development of these infections is usually quite limited as these parasites are not adapted to surviving in human beings. Human pulmonary dirofilariasis is a relatively rare condition caused by *Dirofilaria* sp. that is transmitted to humans by mosquitoes. This parasite enters the subcutaneous tissue and travels through the blood system, most often dying and embolizing the pulmonary vessels, causing a small infarction usually in the peripheral areas of the lung, which commonly appears as a solitary density on radiography, termed a "coin lesion." It is often misdiagnosed as an early neoplastic lesion and has on occasion led to surgery to remove the lesion – histopathology then confirms that it is a parasite and not cancer.

### 3.4.6 Comment

As seen from the discussion above, much remains to be learned about the pathogenesis and clinical presentation of filarial infections, even for those on which there is much global attention at present, onchocerciasis and lymphatic filariasis. Much of the available information comes from experimental models, and although this assists in informing the human condition, given the specific nature of each of the filarial complexes, confirmation of such information will always be needed in the natural infection. Acquisition of autopsy material is always likely to be a challenge in tropical countries for various understandable reasons (weak pathology services, religious prohibition, etc.), but such opportunities should be taken if available. It is also more likely that as medical imaging systems improve and become more common in endemic countries, non-invasive approaches to obtaining images and other information about filarial conditions will contribute significantly to understanding the pathogenesis and support more classical methods such as histopathology.

## 3.5 Treatment and Control of Human Filarial Diseases

The availability of a chemotherapeutic agent that can safely reduce infection levels in an individual has been central to treatment and control of filarial infections and disease. As described above, agents have been tested and used since the late 1940s. However, early treatments were associated with serious adverse events, and it was not until 1987, and the availability of the anti-microfilarial drug ivermectin, that large-scale MDA programs became safe and feasible. Nevertheless, the identification of a drug that specifically targets the adult worm (and perhaps the infective larval form) in a safe manner – for example, one that is lethal or permanently sterilizes the female – is still the most prized target in filarial treatment and control. Other treatment approaches that were once hoped for, such as vaccines, have not to date been successful.

### 3.5.1 Onchocerciasis

The major available treatment option in onchocerciasis is to kill microfilariae with avermectins (currently ivermectin and likely soon moxidectin) and thus reduce clinical disease in individuals, and to break the transmission cycle if used on an epidemiological scale in MDA programs.

#### 3.5.1.1 Individual Treatment

Reducing the microfilarial load with microfilaricidal drugs is effective in reducing symptoms, and together with supportive treatment to reduce the intense pruritus (e.g., an antihistamine), dermal support creams for damaged skin, and surgical removal of adult-containing nodules, are the current optimal approaches to treat individuals infected with *O. volvulus*. As noted, the primary microfilaricide used is ivermectin, with a second-generation macrocyclic lactone, moxidectin, possibly available in the near future. Ivermectin appears to work more slowly than the previously used microfilaricide DEC, and to produce only minimal adverse reactions. It is likely that ivermectin acts by blocking the parasite's ability to protect itself against the host's innate immune response, possibly by blocking the worms release of protective secretions [80]. Ivermectin reduces the live microfilarial load in the anterior segment of the eye, although damage already inflicted on poorly healing ocular tissues, such as the retina, that occurs with this infection is unfortunately usually permanent.

Removal of palpable nodules (nodulectomy) has always been regarded as an important approach to reducing the parasite load in an individual; nodulectomy as an approach has been used not only for individual treatment but also on a mass campaign basis for endemic communities in Latin America, although these unpopular campaigns were not effective in eliminating transmission of the parasite.

#### 3.5.1.2 Mass Treatment

As mentioned above, the major West African program to reduce disease by eliminating the vector using aerial larvaciding was greatly enhanced by the addition of mass distribution of ivermectin in 1987 for the West African countries under the OCP direction [15]. In 1995, the use of ivermectin for onchocerciasis was extended to cover all endemic countries in Africa with the establishment of the African Programme for Onchocerciasis Control (APOC) [81]. APOC was formed with the extended mandate of reducing the prevalence of both blinding and dermatological disease by distributing ivermectin on an annual basis to everybody (above four years of age, not infirm or pregnant) living in meso- and hyperendemic communities (Figure 3.6). A similar program was set up in Latin America (Onchocerciasis Elimination Program for the Americas – OEPA), although here the goal was the stricter "elimination of transmission"; ivermectin was used twice a year in all six OEPA endemic countries. The small Middle Eastern focus of onchocerciasis, confined to Yemen, initially took the approach of only providing symptomatic treatment for those individuals with severe clinical presentations of reactive onchodermatitis (Sowdah), i.e., a clinical, not an MDA approach; Yemen changed to the more extensive MDA approach in 2015 [34].

A major shift in the global program for onchocerciasis control has taken place in the last few years with the move from control of the disease as a "public health problem" to the goal of "elimination of transmission" [82]. One of the greatest challenges facing this new goal is the need to re-map specific endemic areas of onchocerciasis; the previous MDA program that began during the 1990s did not include hypoendemic areas (except in Sudan). At the time, it was believed that chronic skin disease and blindness were not a threat in areas of low endemicity; however, Sudan carried out MDA in areas defined as hypoendemic as the Sudanese team emphasized to

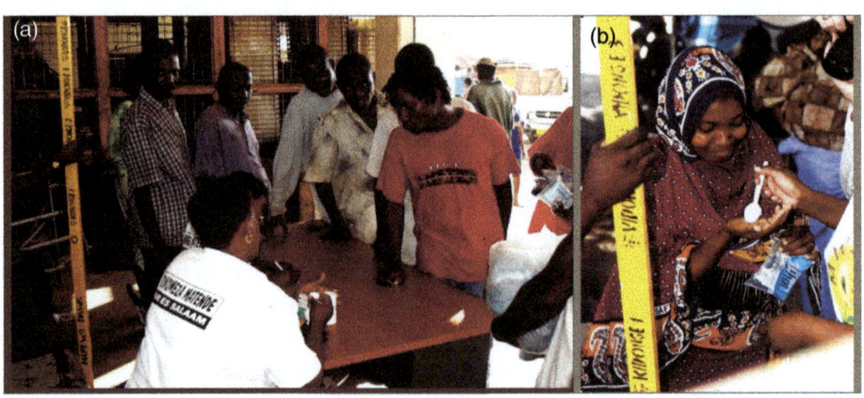


**Figure 3.6** Mass Drug Administration (MDA) in the Tanzanian Lymphatic Filariasis National Program. (a) A drug distributor dividing out the drugs for a villager. (b) A villager happily receiving the MDA drugs into her hand with a bag of water used to assist swallowing of the drugs in her other hand.

APOC the fact that some of the most severe reactive onchodermatitis cases are actually most commonly found in hypoendemic areas and on the periphery of meso- and hyperendemic areas.

The new global goal, elimination of transmission, likely requires MDA treatment for everyone living in endemic areas; thus, there is a need to redraw the map of onchocerciasis endemicity based on the presence of transmission. This substantial task, and the need for improvements in the recommended diagnostic detection tests and their usage for remapping, and for defining when to stop treatment when transmission is broken, are central to achieving success in the global program for onchocerciasis. However, elimination of transmission has been achieved in countries in Latin America [83–86], and in focal areas of Africa (e.g., Sudan, Uganda, and others) [87–89]. Four of the six OEPA countries have achieved breaking of transmission, leaving only Venezuela and Brazil with ongoing transmission. It is vital that such successes be monitored and surveillance continued to ensure that these important gains are maintained and there is no recrudescence in these sites [90] (Table 3.6).

Since OCP activities ceased in West Africa, vector control for onchocerciasis has not been a major activity in the remaining endemic areas of Africa, Latin America, or Yemen. The reduction of fly breeding sites by the "slash and clear" removal of vegetation has been proposed [91] and is now being applied in a few locations in Africa; whether this community approach to vector control is a major advance over

**Table 3.6** Important areas in need of investigation in three major aspects of filarial infections and diseases.

| General area | Items to better understand |
|---|---|
| Parasite biology | Better understanding of vulnerable target areas in filarial worm's biochemistry suitable for chemotherapeutic treatment – especially in adult worms and infective larvae |
| Infection assessment | Identification of factors released from parasites into the circulation, notably from fertile female worms |
| Elimination of transmission (onchocerciasis and lymphatic filariasis) | The development of point of use assessment tools and an efficient and reliable protocol for their use in endemic areas |
| | A clear confirmation of required reduction in parasite prevalence needed to break transmission permanently |
| Patient support | Ensuring lymphatic filariasis cases are treated early in the development of their lymphedematous conditions |
| | Improved treatment for the dermatological changes that occur in onchocerciasis and lymphatic filariasis |
| | Improved care and support for those already with onchocercal blindness and loss of vision |
| Chemotherapy | The development of safe chemotherapeutic agents that target adult and infective larvae and that can be used in all individuals two years old and above |

a fully implemented and maximized MDA strategy with a microfilaricide (and perhaps an adulticide if one is safely developed) where there is high coverage of the population remains an unanswered question. Arguably, the most important current needs for onchocerciasis programs are the development of a practical point-of-use diagnostic test, appropriate instructions for their use, and an increase in the local capacity in Africa to handle testing of the samples that country programs generate [92]. The use of twice a year ivermectin, and the incorporation of moxidectin into the menu of protocols available for achieving final success and handling difficult areas (e.g., cross-border situations, persistent hotspots, sites with highly mobile populations) are both important considerations for national programs as they look toward completing their elimination programs.

### 3.5.2 Lymphatic Filariasis

Lymphatic filariasis, transmitted by multiple species of mosquitoes, has always had a component of vector control, as with the other major tropical infection transmitted by this insect, malaria. It should be noted that antimalarial activities based on vector control in many locations assist in prevention of transmission of lymphatic filariasis. However, the major tool for controlling and eliminating this infection remains chemotherapy. Anthelmintic treatment for this parasitic infection in individual patients has been achieved using a combination of albendazole and either ivermectin or DEC. Unlike onchocerciasis, in which DEC caused problems, the use of this drug in lymphatic filariasis although on occasion causing significant adverse reactions, has been generally regarded as safe, with the severity of adverse reactions being related to the microfilariae load. In recent studies, the combination of these three drugs in lymphatic filariasis has been shown to be more effective than the dual combinations, at least in the short term; there appears to a more effective response when using this triple drug therapy in the first two years after the start of treatment [93].

#### 3.5.2.1 Individual Treatment

The major focus with individual treatment in lymphatic filariasis is patients who suffer from either lymphedema and its dermal consequences, or those affected with genital pathology, predominantly hydrocoele in males. Although extensive modern treatments are used in developed countries for lymphatic filariasis, many of these procedures, for various reasons, are often not appropriate for patients living in medically under-resourced environments; this restriction also applies to many patients suffering from lymphatic filariasis-induced lymphedema. This has led to recommendations on an essential package of care (EPC) for this infection by WHO developed from the experience of clinical field workers in this area [93]; this arm of the Program is termed Morbidity Management and Disability Prevention (MMDP) [94], or additionally Disease Management and Disability Inclusion (DMDI). The essential component in the EPC is to carry out skin hygiene and care (washing, antibiotic and anti-fungal creams, and basic physiotherapy techniques); this regimen has been remarkably effective in improving the condition of many patients with lymphedema [95, 96]

Acute filarial attacks – episodes of lymphadenopathy, lymphangitis, dermal exfoliation in the affected areas (legs, arms, breath) together with systemic malaise ("feelings of fever") – can greatly debilitate a patient for up to six or seven days in the severest of cases and are recommended to be given supportive treatment. This treatment is essentially anti-inflammatory agents and, as it is understood that secondary bacterial infections contribute to the initiation of these attacks, antibiotics. Doxycycline, with both antibiotic and anti-inflammatory properties, is currently being considered for treatment of lymphatic filariasis patients with lymphedema [97].

Patients who have very chronic and extensive forms of lymphedema are a special medical concern and require more medical oversight than is usually available from health workers in rural village settings. These patients commonly have extensive disfiguration of the lymphedematous limb, open wounds, and often suffer from multiple episodes of acute filarial episodes in relatively short period of time (e.g., every month). These patients also commonly have comorbidities such as hypertension, diabetes, and obesity, and often require extensive and long-term medical attention and thus need to be referred to specialist clinics.

It is important to recognize the two major components in the pathogenesis of lymphedema when addressing how best to improve care and develop new treatment regimens for patients suffering from filarial lymphedema and the associated acute filarial episodes. The first major component is dysfunction of the lymphatic system in the affected limb, resulting in lymph stasis and compromising the ability of associated tissues (muscles and connective tissue, dermis, and epidermis) to maintain homeostasis and health. Consequently, interventions that move static lymph fluid away from the affected side, preventing fluid stasis in the soft tissues and the recovery of function of these tissues, are likely to be successful, e.g., short stroke massage. The second major component in pathogenesis to consider is degradation of the ability of dermal and associated tissues to carry out homeostatic functions; this underscores the importance of using topical treatments to restore dermal tissues to their normal status, e.g., consistent washing, topical oils, and creams (restoration of dermal hydration and reduction in dryness), as well as assisting healing of open wounds. These procedures apply to all cases of lymphatic filariasis.

The major form of genital filariasis – hydrocoele in male in Bancroftian endemic areas – is treated by standard surgical intervention, and if carried out under regular "good surgical practice," is most often resolved. Simple aspiration of the fluid from the *tunica vaginalis/albuginea sac* is also carried out, but in general this is not as successful in resolving the issue, with recurrence common and more chance of secondary infection than with surgery. Two techniques for hydrocoelectomy are commonly used: "excision" with the removal of the hydrocoele sac (essentially the *tunica vaginalis*) – known as Lord's Technique, and second, "eversion" of the hydrocoele sac, known as Jaboulay's Technique. Both approaches are used, with the excision method preferred by most surgeons operating on thick-walled sacs.

An essential component in all hydrocoele surgeries in lymphatic filariasis patients, who often reside in medically underserved locations, is the need for good pre-operative steps (correct diagnosis, hygiene, and antibiotic coverage) and careful post-operative follow-up and monitoring [98, 99].

Tropical pulmonary eosinophilia (TPE) is a condition common in Tamil populations in India; it is often associated with lymphatic filariasis, although it is seen in less than 1% of those infected with filaria and is believed to be a hypersensitivity to microfilarial antigens [100]; the current recommendation for treatment of TPE is DEC for up to three weeks.

The expansion of the range of interventions that can be added to the EPC for lymphatic filariasis patients is the subject of many current discussions. Ways to address the mental health needs of lymphatic filariasis patients (stigma, mental depression) and their rehabilitation back to a more regular life and ability to support their families and contribute to community life are also currently the focus of many concerned investigators and care-givers; experiences with other tropical diseases such as leprosy are contributing to new approaches to care for lymphatic filariasis patients [101].

#### 3.5.2.2 Mass Treatment

In 1997, a global effort aimed at eliminating lymphatic filariasis as a public health problem was initiated (World Health Assembly adopted Resolution WHA 50.29). This resolution was to carry out MDA – i.e. providing medicines to everyone who is healthy and eligible for them in an endemic area – using two co-administered anthelmintics in all endemic areas: albendazole and DEC, or albendazole and ivermectin in countries where onchocerciasis is endemic in Africa. When the global program began in 2000, the number of people infected was in excess of 120 million, with over 40 million suffering clinically, and at risk of becoming infected likely over 200 million.

Many of the 73 lymphatic filariasis endemic countries have, as with onchocerciasis, been covered under an MDA program for 20 years, or in certain cases much longer (e.g., China and South Korea). This has had a marked effect in reduction of parasite loads, reflected in a correspondingly dramatic reduction in the development of new clinical cases. Such efficacy data are vital to demonstrate the value of the global program; after all, it would never have started had this parasite not induced such debilitating and life-destroying disease. It should be emphasized that it is important to link the two arms of the global LF elimination program (MDA and MMDP) together wherever possible. The provision of care for those symptomatically infected is a strong motivating factor to those without clinical symptoms to take the MDA drugs. A national program provides care that leads to improvement in their fellow villagers who unfortunately have the clinical disease is a powerful message that assists in improving MDA coverage [96].

By mid-2020, with some 7.7 billion treatments administered, 18 countries declared the elimination of LF as a public health problem, and another 22 had completed MDA in all their endemic areas, and the prevalence of the infection in many countries had reached very low levels, with some 573 million people no longer needing

MDA. However, the 2021–2030 period will be important, as almost 900 million people in 49 countries still remain threatened by this infection and require continuing MDA. The economic importance of lymphatic filariasis has always been an advocacy factor for developing an elimination program, and a study in 2016 suggested that >24 billion US dollars can be saved by elimination of the infection from the globe through reduction in loss in work potential and the health care costs this disease causes [102].

### 3.5.3 Loiasis

#### 3.5.3.1 Individual Treatment

Treatment of individuals has been generally with DEC together with supportive treatment (antihistamines, etc.), although it should be noted that patients with high parasite loads can suffer from serious CNS pathology. The treatment now often used more safely is albendazole [103].

#### 3.5.3.2 Control of Loiasis

There are no major programs in place at the field level to eliminate or control the transmission of loiasis. As mentioned above, problems do arise with this infection, and specifically with individuals harboring high loads of *L. loa* microfilariae residing in areas that are included in elimination programs for onchocerciasis and lymphatic filariasis. Thus, considerable efforts have been made to avoid the adverse effects by screening patients for high levels of *L. loa* microfilariae and excluding them from microfilaricidal treatment [104]. The use of the "loascope," a point of use tool that assists in the detection of dangerously high loads of circulating microfilariae has been crucial for screening those at risk.

### 3.5.4 Mansonellosis and Incidental Filarial Infections

Most of the unusual filarial infections are incidental findings and in most cases are treated with standard antifilarial drugs; in many cases, they are detected after biopsy in pathology reports. No control programs exist for mansonellosis or for incidental filarial infections.

### 3.5.5 Comment

It is not surprising that distributing a drug to millions of people in poorly served areas of developing countries comes with many challenges, and it is a tribute to the endemic countries participating in MDA activities that so much progress has been achieved. To reach elimination, the challenges that still exist must be addressed (Table 3.6); these include ensuring that there is routine drug coverage of >80% of the eligible population and that habitually non-compliant people are eventually reached and treated (as untreated individuals, they remain potential catalysts for resurgence of transmission). Specific endemic areas with persistent transmission (hotspots) have been seen in a number of areas and will likely need special attention and possibly additional interventions (such as multiple rounds of MDA, or the use of new anthelmintics) to eliminate them as transmission sites. The constant migration of people in endemic areas poses a special challenge, particularly in areas where countries share national borders; managing implementation of MDA in such areas requires an additional level of country-to-country communication and organization.

A major effort is ongoing to identify areas in onchocerciasis-endemic countries where MDA is still required; this has been termed "onchocerciasis elimination mapping" and consists of various steps, such as "exclusion mapping" (i.e., defining and eliminating from consideration all areas that are unlikely to support transmission), and "epidemiological mapping" of areas where there is the possibility of transmission but where MDA has not yet occurred (i.e., defining the presence of vector breeding sites and testing a sample of residents from the catchment area). This essential activity is needed if elimination of transmission is to be achieved in onchocerciasis, but in many cases, it is a huge undertaking for national programs, and ways of making this essential mapping activity more practical are needed.

An important additional challenge facing national elimination and control programs is the fact that, especially for onchocerciasis, the tests used to monitor progress of control to support decisions at major points (e.g., stopping MDA, confirming success) need to be improved, as do the field protocols to implement these tests [105, 106]. In both major global elimination programs, there is always a question of whether the criteria used to define elimination are sufficiently strong to prevent breakthrough resurgence; this remains a challenge for onchocerciasis in particular, compounded by the fact at present, the diagnostic criteria for stopping treatment and defining the cessation of transmission in the vector are not as robust and practical as they could be.

A programmatic issue commonly discussed in onchocerciasis is whether MDA twice a year would be more effective than the standard once a year protocol. In the OEPA program countries, twice a year dosing achieved the goal, even in endemic areas with intense transmission, such as Ecuador [84]. Two parasitological reasons support the use of twice a year treatment: First, repopulation of the skin with microfilariae after a single dose of ivermectin occurs after seven months, thus, retreating at six months would prevent this repopulation and limit the possibility of transmission occurring in the second half of the year. Second, it is likely that ivermectin has a direct effect on the health of adult worms, reducing their life span [107], and an additional dose of ivermectin could enhance this much wanted effect.

In a similar vein, there is the question of whether adding albendazole to the standard annual dose of ivermectin might speed progress to breaking transmission of *O. volvulus*. Various studies that have addressed this question have not produced a definitive answer. However, it is possible that the addition of a second anthelmintic to the protocol, one that damages the worm through different target mechanisms, could have at least some useful additive negative effect on parasite viability. Programmatically, adding albendazole to ivermectin is entirely feasible, as this already is used in MDA for lymphatic filariasis. This is the basis of the positive additive or synergistic effects seen in the triple therapy protocols now being rolled out for lymphatic filariasis [93]. However, the challenge of using a cocktail of anthelmintics in onchocerciasis-endemic regions that includes DEC should be emphasized, given the negative experience of ocular pathology in onchocerciasis patients administered DEC [62]. Use of the triple therapy as currently defined should be contraindicated in onchocerciasis-endemic countries, unless it can be guaranteed that individuals harboring *O. volvulus* microfilariae will not be treated.

The recent unexpected arrival of the corona virus pandemic has, as everywhere, had a major impact on filarial elimination programs, with an interruption to most, if not all, MDA activities, surveys, and hydrocoelectomy surgeries; the only component that some countries have been able to maintain is supporting direct care for patients with lymphatic filariasis-induced lymphedema. Interestingly, the experience and communication systems established by MDA programs over the years have been used by many countries to assist in emergency responses to the COVID-19 outbreak. The long-lasting impact of this pandemic on filarial elimination programs remains to be seen, but there is no doubt that the experience these countries gained in responding positively to the outbreak in the past few years indicates they are resilient and will move forward positively.

## 3.6 General Discussion

Filarial infections are characterized by an intimate relationship with the host, one that is complex and involves a number of biological systems – immunosuppression, immune-avoidance mechanisms, and other homeostatic phenomena. The complexity of the relationship between living filarial parasites and their host in all likelihood contributes to the difficulty in developing immune-based protection through vaccine approaches, although efforts are still being made to experimentally manipulate the host system and induce protection to filariae. Nevertheless, chemotherapeutic approaches, especially those that safely target the reproductive cycle of the adult female and perhaps the male worm, are arguably the currently most sensible approach to control. Macrofilaricidal drugs are available, although there have been concerns about the safety of the top candidates [108–110]. Breaking the life cycle at the pre-embryonic or embryonic stage, and thus preventing development of the pathogenic stage, the microfilariae, certainly seems a most logical research goal. This is not to say that other approaches will not also enhance the progress toward elimination of the two major filarial diseases – vector control being an obvious important adjunct mechanism in this regard.

As is obvious from the overall discussion above, there has been a global focus on the two dominant filarial infections of humans with relatively little attention to others (e.g., loiasis). Only where it has affected MDA programs against onchocerciasis and lymphatic filariasis has another human filarial parasite attracted research attention. This raises the question as to whether more effort should be devoted to the other filariae. There are good reasons to investigate these lesser-known filariae, as it is likely that they do induce tissue changes in at least some infected people. Second, understanding the mechanisms that allow them to live more pathologically peacefully in the host may provide important keys to developing control methods for filariae in general. It is likely that any chemotherapeutic agent that can be used safely against one type of filariae will also be useful for treating other filarial infections. The challenge in achieving such an agent is, as emphasized in the discussion above, the adverse reactions associated with destruction of large numbers of microfilariae, which has plagued the discovery pathway.

The control, and now elimination, of lymphatic filariasis and onchocerciasis in an increasing number of countries is amongst the major public health successes of all time. Nevertheless, eradication, a goal that many are now focusing on, is still a tremendous challenge. Where there has been apparently successful removal of the parasite – in Latin America with onchocerciasis, for example, and in China, the Pacific, south-east Asia (e.g., Vietnam), and countries in Africa (e.g., Togo, Malawi) with lymphatic filariasis, it is vital that surveillance is maintained to ensure that these parasites and the diseases they cause are truly eliminated. Long-term surveillance, although essential, is not easy to carry out and sustain; appropriate surveillance, whether via case detection or through more systematic entomological, serological, or parasitological surveys, must be carried out for a relatively long period with practical approaches, such as regular sampling of blood bank collections for parasite antigens or antibodies. Once success is reached in an MDA program, national government interest in the program, understandably, often wanes as other medical issues become a more important focus. Despite this, long-term surveillance needs to be carried out to ensure that the successes achieved are not lost.

It is also important that we continue to ask pertinent and practical questions about the biology of this group of diseases. The mechanisms that these parasites utilize to invade and live long-term in their hosts are still poorly understood but are an extraordinary example of adaptation in biology. The goal of elimination is highly desired, but is important to achieving this that we continue to better understand the pathobiological mechanisms involved in these infections, and the ways that these mechanisms might be manipulated to the disadvantage of the parasite and the advantage of the host. Many questions remain about these diseases, some perhaps more academic than practically useful to the overall goals of transmission elimination and provision of care for those already affected. For example, "is there a lower limit of prevalence at which the infection naturally dies out?" – many have speculated on this, but we really do not know enough about the factors involved, which differ from location to location. Another essential question is: Do we have the right criteria by which to measure success? Continuing refinement of such indicators is the most prudent approach here.

It should be noted that there is always a challenge in achieving the appropriate balance between scientific quest and public health, between research for more basic knowledge about each organism and the best way to eliminate them as global medical problems. However, without more information about the biological nature of these parasites (Table 3.6), there is an increased likelihood that incorrect assumptions are being made with potentially high cost in terms of the effect on field work.

The effort to control and now eliminate the two major human filariases has led to major changes in human health in developing countries. The mechanisms and systems built to distribute antifilarial drugs to many millions of people, at the necessarily high coverage levels essential to achieve elimination, have brought much needed health improvements to previously underserved populations. In addition, these methods and systems are now used more widely for the greater goal of universal health care [99]. This focus on efficient mass drug distribution, which has had major successes and great impact in reducing the prevalence of clinical onchocerciasis and lymphatic filariasis, has nevertheless had the effect of directing current research activities away from more basic research into filarial parasites and their diseases. Consequently, many major questions about filarial nematodes remain unsolved. Despite very significant advances in reducing clinical disease, the achievement of the current global goal for complete elimination of transmission of onchocerciasis, and the lesser goal of elimination of lymphatic filariasis as a public health problem, is likely to be challenging. Better understanding of these diseases and their complexity through more rather than less basic research, perhaps targeted to specific questions and areas, could assist in overcoming these challenges.

## Acknowledgments

Timothy Geary is thanked for his most valuable comments and suggestions regarding this manuscript.

## References

1 Kausar, S. (2020). Filariasis. *Helminthiasis*. (Omolade Olayinka Okwa). Intechopen.com. https://doi.org/10.5772/intechopen.80144 2020. eBook (PDF) ISBN: 978-1-83880-663-7.

2 Sungpradit, S. and Sanprasert, V. (2020). Lymphatic filariasis. *Molecular Advances in Tropical Disease Drug Discovery*. Elsevier Inc. pp. 64–94. https://doi.org/10.1016/B978-0-12-821202-8.00004-9.

3 Centers for Disease Control and Prevention (2021). Parasites. https://www.cdc.gov/parasites.

4 Mogoung-Wafo, A.E., Nana-Djeunga, H.C., Domche, A. et al. (2019). Prevalence and intensity of *Loa loa* infection over twenty-three years in three communities of the Mbalmayo health district (Central Cameroon). *BMC Infect. Dis.* 19: 146. https://doi.org/10.1186/s12879-019-3776-y.

5 Kimura, E. (2011). The Global Programme to eliminate lymphatic filariasis: history and achievements with special reference to annual single-dose treatment with diethylcarbamazine in Samoa and Fiji. *Trop. Med. Health.* 39: 17–30.

6 Shenoy, R.K., Suma, T.K., Kumaraswami, V. et al. (2009). Antifilarial drugs, in the doses employed in mass drug administrations by the Global Programme to eliminate lymphatic filariasis, reverse lymphatic pathology in children with *Brugia malayi* infection. *Ann. Trop. Med. Parasitol.* 103: 235–247.

7 Ismail, M.M., Jayakody, R.L., Weil, G.J. et al. (1998). Efficacy of single dose combinations of albendazole, ivermectin and diethylcarbamazine for the treatment of bancroftian filariasis. *Trans. R. Soc. Trop. Med. Hyg.* 92: 94–97.

8 Lammie, P.J., Eberhard, M.L., Addiss, D.G. et al. (2017). Translating research into reality: elimination of lymphatic filariasis from Haiti. *Am. J. Trop. Med. Hyg.* 97(Suppl 4): 71–75.

9 Mante, S.D. and Seim, A.R. (2007). *West African Lymphatic Morbidity Project – Surgical Handbook*, 2e. Norway: Health Development International. https://hdi.no/wp-content/uploads/2018/03/lymphatic-filariasis-guide.pdf.

10 Dreyer, G., Santos, A., Norøes, J. et al. (1998). Ultrasonographic detection of living adult *Wuchereria bancrofti* using a 3.5-MHz transducer. *Am. J. Trop. Med. Hyg.* 59: 399–403.

11 Grenfell, B., Michael, E., and Denham, D.A. (1991). A model for the dynamics of human lymphatic filariasis. *Parasitol. Today* 7: 318–323.

12 Ottesen, E.A. (1984). Immunological aspects of lymphatic filariasis and onchocerciasis in man. *Trans. R. Soc. Trop. Med. Hyg.* 78(Supplement): 9–18.

13 Chandrashekar, R., Curtis, K.C., Ramzy, R.M. et al. (1994). Molecular cloning of *Brugia malayi* antigens for diagnosis of lymphatic filariasis. *Mol. Biochem. Parasitol.* 64: 261–271.

14 Williams, J.F., Mackenzie, C.D., and Darwood, M. (1985). Current distribution of onchocerciasis in Sudan. *Sudan Med. J. (Suppl.)* 21: 9–17.

15 Boatin, B. (2008). The onchocerciasis control programme in West Africa (OCP). *Ann. Trop. Med. Parasitol.* 102(Suppl 1): 13–17.

16 Cupp, E.W., Mackenzie, C.D., and Unnasch, T. (2011). The importance of ivermectin to human onchocerciasis: past, present, and the future. *Res. Rep. Trop. Med.* 2: 81–92.

17 Abdel-Fattah, A., Youssif, N.M., Shenouda, O.A. and Morsy, T.A. (1985). On the treatment of onchocerciasis in Sudan. *J. Egypt Soc. Parasitol.* 15: 329–333.

18 Gustavson-Moringlane, I.L. and Bengtsson, E. (1981). Eosinophil leucocytic reactions from diethylcarbamazine in filariasis, *Ann. Trop. Med. Parasitol.* 75(6): 615–621. https://doi.org/10.1080/00034983.1981.11687492.

19 Brown, A.W. (1962). A survey of simulium control in Africa. *Bull. World Health Organ.* 27: 511–527.

20 Tambo, E., Khater, E.I., Chen, J. et al. (2015). Nobel prize for the artemisinin and ivermectin discoveries: a great boost towards elimination of the global infectious diseases of poverty. *Infect. Dis. Poverty* 4: 58. https://doi.org/10.1186/s40249-015-0091-8.

21 Campbell, W.C. (2016). Lessons from the history of ivermectin and other antiparasitic agents. *Annu. Rev. Anim. Biosci.* 4: 1–14. https://doi.org/10.1146/annurev-animal-021815-111209.

**22** Mackenzie, C.D. and Kron, M.A. (1985). Diethylcarbamazine: a review of its action in onchocerciasis, lymphatic filariasis and inflammation. *Trop. Dis. Bull.* 82: R1–R37.

**23** Addiss, D., Gamble, C.L., Garner, P. et al. (2005). Albendazole for lymphatic filariasis. *Cochrane Database Syst. Rev.* 4, CD003753. https://doi.org/10.1002/14651858.CD003753.pub3.

**24** Awadzi, K. and Gilles, H. (1992). Diethylcarbazine in the treatment of patients with onchocerciasis. *Br. J. Chin. Pharmacol.* 34: 281–288.

**25** Hewitt, R.I., Kushner, S., Stewart, H.W. et al. (1947). Experimental chemotherapy of filariasis III. Effect of 1-diethylcarbamyl-4-methylpiperazine hydro-chloride against naturally acquired filarial infections in cotton rats and dogs. *J. Lab. Clin. Med.* 32: 1314–1329.

**26** NTD Modelling Consortium Onchocerciasis Group (2019). The World Health Organization 2030 goals for onchocerciasis: insights and perspectives from mathematical modelling. *Gates Open Res.* 3: 1545. https://doi.org/10.12688/gatesopenres.13067.1.

**27** Gardon, J., Gardon-Wendel, N., Demanga-Ngangue, J. et al. (1997). Serious reactions after mass treatment of onchocerciasis with ivermectin in an area endemic for *Loa loa* infection. *Lancet* 350: 18–22. https://doi.org/10.1016/S0140-6736(96)11094-1.

**28** Carme, B., Boulesteix, J., Boutes, H., and Puruehnce, M. (1991). Five cases of encephalitis during treatment of loiasis with diethylcarbamazine. *Am. J. Trop. Med. Hyg.* 44: 684–90. https://doi.org/10.4269/ajtmh.1991.44.684.

**29** Manson, P. (1891). The filaria sanguinis hominis major and minor, two new species of haematozoa. *Lancet* 137: 4–8.

**30** Mediannikov, O. and Ranque, S. (2018). Mansonellosis, the most neglected human filariasis. *New Microbes New Infect.* 26: S19–S22. https://doi.org/10.1016j.nmni.2018.08.016.

**31** Boakye, D., Tallant, J., Adjami, A. et al. (2018). Refocusing vector assessment towards the elimination of onchocerciasis from Africa: a review of the current status in selected countries. *Int. Health.* 10(Suppl 1): i27–i32. https://doi.org/10.1093/inthealth/ihx066.

**32** Sauerbrey, M. (2008). The Onchocerciasis elimination program for the Americas (OEPA). *Ann. Trop. Med. Parasitol.* 012(Suppl 1): 25–29. https://doi.org/10.1179/136485908X337454.

**33** Fawdry, A.L. (1957). Onchocerciasis in South Arabia. *Trans. R. Soc. Trop. Med. Hyg.* 51: 253–25634.

**34** Al-Kubati, A.S., Mackenzie, C.D., Boakye, D. et al. (2018). Onchocerciasis in Yemen: moving forward towards an elimination program. *Int. Health* 10(Suppl 1): i89–i96. https://doi.org/10.1093/inthealth/ihx055.

**35** Mackenzie, C.D., Al-Kubati, A.S., Al-Qubati, Y. et al. (2018). Serological survey of human onchocerciasis in Yemen. *Am. J. Trop. Med. Hyg.* https://doi.org/10.4269/ajtmh.18-0051.

**36** Attout, T., Hoerauf, A., Dénécé, G. et al. (2009). Lymphatic vascularisation and involvement of Lyve-1+ macrophages in the human onchocerca nodule. *PLoS One* 4: e8234. https://doi.org/10.1371/journal.pone.0008234.

**37** Mackenzie, C.D., Huntington, M.K., Wanji, S. et al. (2010). The association of adult *Onchocerca volvulus* with lymphatic vessels. *J. Parasitol.* (96): 219–221. https://doi.org/10.1645/GE-2236.1. PMID: 19803543.

**38** Guderian, R.H., Anselmi, M., Beck, B.J. et al. (1991). The effect of antimalarial chloroquine therapy and prophylaxis on concurrent infection with *Onchocerca volvulus* in Ecuador. *Trans. R. Soc. Trop. Med. Hyg.* 85: 634–638. https://doi.org/10.1016/0035-9203(91)90372-6. PMID: 1780994.

**39** Piessens, W.F. and Mackenzie, C.D. (1982). Immunology of lymphatic filariasis and onchocerciasis, Chapter 18. In: *Immunology of Parasitic Infections*, 2e (ed. S. Cohen and K.S. Warren), 622–653. Oxford: Blackwell Scientific Publications.

**40** Koudou, B.G., de Souza, D.K., Biritwum, K.W. et al. (2018). Elimination of lymphatic filariasis in west African urban areas: is implementation of mass drug administration necessary?. *Lancet Infect. Dis.* 18: E214–E220. https://doi.org/10.1016/S1473-3099(18)30069-0.

**41** Simonsen, P.E., Fischer, P.U., Hoerauf, A., and Weil, G. (2014). The filariases. In: *Manson's Tropical Diseases*, 23e (ed. J. Farrar), 737–765. Elsevier.

**42** Mallawarachchi, C.H., Chandrasena, N., Wickramasinghe, S. et al. (2018). A preliminary survey of filarial parasites in dogs and cats in Sri Lanka. *PLoS One* 13: e0206633. https://doi.org/10.1371/journal.pone.0206633.

**43** Padgett, J.J. and Jacobsen, K.H. (2008). Loiasis: African eye worm. *Trans. R. Soc. Trop. Med. Hyg.* 102: 983–989.

**44** Saito, M., Armstrong, M., Boadi, S. et al. (2015). Clinical features of imported loiasis: a case series from the Hospital for Tropical Diseases, London, *Am. J. Trop. Med. Hyg.* 93: 607–611. https://doi.org/10.4269/ajtmh.15-0214.

**45** Buell, K.G., Whittaker, C., Chesnais, C. et al. (2019). Atypical clinical manifestations of loiasis and their relevance for endemic populations. *Open Forum Infect. Dis.* 6: ofz417. https://doi.org/10.1093/ofid/ofz417.

**46** Chesnais, C.B., Takougang, I., Paguélé, M. et al. (2017). Excess mortality associated with loiasis: a retrospective population-based cohort study. *Lancet Infect. Dis.* 17: 108–116. https://doi.org/10.1016/S1473-3099(16)30405-4.

**47** Oriel, T.C. and Eberhard, M.L. (1998). Zoonotic filariaisis. *Clin. Microbiol. Rev.* 11: 366–381.

**48** Tahir, D., Davoust, B., and Parola, P. (2019). Vector-borne nematode diseases in pets and humans in the Mediterranean Basin: an update. *Vet. World* 12(10): 1630–1643. https://doi.org/10.14202/vetworld.2019.1630-1643.

**49** González-Miguel, J., Rosarioa, L., Rota-Nodaria, E. et al. (2010). Identification of immunoreactive proteins of *Dirofilaria immitis* and *D. repens* recognized by sera from patients with pulmonary and subcutaneous dirofilariosis. *Parasitol. Int.* 59: 248–256.

**50** VandeWaa, E.A., Foster, L.A., DeRuiter, J. et al. (1993). Glutamine-supported motility of adult filarial parasites in vitro and the effect of glutamine antimetabolites. *J. Parasitol.* 79: 173–180.

**51** Bartlett, A., Turk, J., Ngu, J.L. et al. (1978). Variations in delayed hypersensitivity in dermal onchocerciasis. *Trans. R. Soc. Trop. Med. Hyg.* 72: 372–377.

**52** McGarry, H.F., Pfarr, K., Egerton, G. et al. (2003). Evidence against *Wolbachia* symbiosis in *Loa loa*. *Filaria J.* 2. 9. https://doi.org/10.1186/1475-2883-2-9.

**53** Punkosdy, G.A., Addiss, D.G., and Lammie, P.J. (2003). Characterization of antibody responses to *Wolbachia* surface protein in humans with lymphatic filariasis. *Infect. Immun.* 71: 5104–5114. https://doi.org/10.1128/iai.71.9.5104-5114.

**54** Glover, M., Murdoch, M., and Leigh, I. (1991). Subtle early features of onchocerciasis in a European. *J. R. Soc. Med.* 84: 435.

**55** Robles, R. (1919). Onchocercose humaine au Guatemala produisant la cécité et "l'erysipé le du littoral" (Erisipela de la costa). *Bull. Soc Pathol. Exot.* 12: 442–460.

**56** Shelley, A.J. and Arzube, M. (1985). Studies on the biology of Simuliidae (Diptera) at the Santiago onchocerciasis focus in Ecuador, with special reference to the vectors and disease transmission, *Trans. R. Soc. Trop. Med. Hyg.* 79: 328–338. https://doi.org/10.1016/0035-9203(85)90373-6.

**57** Duke, B.O. (1970). Onchocerciasis: deep worm bundles close to hip joints. *Trans. R. Soc. Trop. Med. Hyg.* 64: 791–792.

**58** Ali, M.M., Mukhtar, M., Baraka, O.Z. et al. (2002). Immunocompetence may be important in the effectiveness of Mectizan (ivermectin) in the treatment of human onchocerciasis. *Acta Trop.* 84: 49–53.

**59** O'Day, J. and Mackenzie, C.D. (1985). Ocular onchocerciasis: current approaches to diagnosis and treatment. *Trop. Doctor* 15: 87–94.

**60** Hall, L.R. and Pearlman, E. (1999). Pathogenesis of onchocercal keratitis (river blindness). *Clin. Microbiol. Rev.* 12: 445–453.

**61** Rodger, F.C. (1959). New observations on ocular onchocerciasis; related pathological methods and the pathogenesis of the various eye lesions. *Bull. World Health Organ.* 16: 495–508. PMID: 13472406; PMCID: PMC2538324.

**62** Bird, A.C., ElSheikh, H., Anderson, J., and Fuglsang, H. (1980). Changes in visual function and in the posterior segment of the eye during treatment of onchocerciasis with diethylcarbamazine citrate. *Br. J. Ophthalmol.* 64: 191–200.

**63** Colebunders, R., Njamnshi, A.K., van Oijen, M. et al. (2017). Onchocerciasis-associated epilepsy: from recent epidemiological and clinical findings to policy implications. *Epilepsia Open* 2: 145–152. https://doi.org/10.1002/epi4.12054.

**64** Föger, K., Gora-Stahlberg, G., Sejvar, J. et al. (2017). Nakalanga syndrome: clinical characteristics, potential causes, and its relationship with recently described nodding syndrome. *PLoS Negl.Trop. Dis.* 11: e0005201. https://doi.org/10.1371/journal.pntd.0005201.

**65** Johnson, T.P., Tyagi, R., Lee, P.R. et al. (2017). Nodding syndrome may be an autoimmune reaction to the parasitic worm *Onchocerca volvulus*. *Sci. Transl. Med.* 9: eaaf6953. https://doi.org/10.1126/scitranslmed.aaf6953.

**66** Duke, B.O., Vincelette, J., and Moore, P.J. (1976). Microfilariae in the cerebrospinal fluid, and neurological complications, during treatment of onchocerciasis with diethylcarbamazine. *Tropenmed. Parasitol.* 27: 123–132. PMID: 941247.

**67** Louveau, A., Smirnov, I., Keyes, T.J. et al. (2015). Structural and functional features of central nervous system lymphatic vessels. *Nature*. https://doi.org/10.1038/nature14432. PMID: 26030524.

**68** Wanji, S., Eyong, E.J., Tendongfor, N. et al. (2017). Ivermectin treatment of *Loa loa* hypermicrofilaraemic baboons (*Papio anubis*): assessment of microfilarial loads, haematological and biochemical parameters and histopathological changes following treatment. *PLoS Negl.Trop. Dis.* https://doi.org/10.1371/journal.pntd.0005576.

**69** Janssen, K.M., Willis, C.J., Anderson, M. et al. (2017). Filariasis orchitis-differential for acute scrotum pathology. *Urol. Case Rep.* 13: 117–119. https://doi.org/10.1016/j.eucr.2017.04.002.

**70** Mand, S., Debrah, A.Y., Klarmann, U. et al. (2011). The role of ultrasonography in the differentiation of the various types of filaricele due to bancroftian filariasis. *Acta Trop.* 120S: S23–S23.

**71** de Araújo, P.S.R., de Souza Junior, V.R., de Souza, A. et al. (2018). Chiluria in a lymphatic filariasis endemic area. *BMC Res. Notes* 11, 269. https://doi.org/10.1186/s13104-018-3357-y.

**72** Langhammer, J., Birk, H.W., and Zahner, H. (1997). Renal disease in lymphatic filariasis: evidence for tubular and glomerular disorders at various stages of the infection. *Trop. Med. Int. Health* 2(9): 875–884. https://doi.org/10.1046/j.1365-3156.1997.d01-404.x. PMID: 9315046.

**73** Bennuru, S., Maldarelli, G., Kumaraswami, V. et al. (2010). Elevated levels of plasma Angiogenic factors are associated with human lymphatic filarial infections. *Am. J. Trop. Med. Hyg.* 83(4): 884–890. https://doi.org/10.4269/ajtmh.2010.10-0039.

**74** Weinkopff, T., Mackenzie, C., Eversole, R. and Lammie, P.J. (2014). Filarial excretory-secretory products induce human monocytes to produce lymphangiogenic mediators. *PLoS Negl.Trop. Dis.* 8(7): e2893. https://doi.org/10.1371/journal.pntd.0002893.

**75** Sotiropoulou, P.A. and Blanpain, C. (2012). Development and homeostasis of the skin epidermis. *Cold Spring Harbor Perspect. Biol.* 4: a008383. https://doi.org/10.1101/cshperspect.a008383.

**76** El-Nahas, H.A., El-Shazly, A.M., Abulhassan, M. et al. (2011). Impact of basic lymphedema management and antifilarial treatment on acute dermatolymphangioadenitis episodes and filarial antigenaemia. *J. Global Infect. Dis.* 3: 227–232. https://doi.org/10.4103/0974-777X.83527. PMCID: PMC3162808.

**77** Shenoy, R.K. (2008). Clinical and pathological aspects of filarial lymphedema and its management. *Korean J. Parasitol.* 46: 119–125. https://doi.org/10.3347/kjp.2008.46.3.119.

**78** Otabil, K.B. and Tenkorang, S.B. (2015). Filarial hydrocele: a neglected condition of a neglected tropical disease. *J. Infect. Dev. Ctries* 9(5): 456–462. https://doi.org/10.3855/jidc.5346.

**79** Kamgno, J., Boussinesq, M., Labrousse, F. et al. (2008). Encephalopathy after ivermectin treatment in a patient infected with *Loa loa* and *Plasmodium* spp. *Am. J. Trop. Med. Hyg.* 78: 546–551.

**80** Ta-Tang, T., Crainey, J.L., Post, R.J. et al. (2018). Mansonellosis: current perspectives. *Res. Rep. Trop. Med.* 2018(9): 9–24.

**81** Amazigo, U. (2008). The African Programme for Onchocerciasis control (APOC). *Ann. Trop. Med. Parasitol.* 102(Suppl): 19–22.

**82** Mackenzie, C.D., Homeida, M.M., Hopkins, A. and Lawrence, J.C. (2012). Elimination of onchocerciasis from Africa: possible? *Trends Parasitol.* 28: 16–22.

**83** Nicholls, R.S., Duque, S., Olaya, L.A. et al. (2018). Elimination of onchocerciasis from Colombia: first proof of concept of river blindness elimination in the world. *Parasite Vectors.* 11: 237. https://doi.org/10.1186/s13071-018-2821-9. PMID: 29642939; PMCID: PMC5896109.

**84** Lovato, R., Guevara, A., Guderian, R. et al. (2014). Elimination of clinical disease from the Ecuadorian endemic focus of onchocerciasis. *PLoS Negl. Trop. Dis.* 8: e2821. https://doi.org/10.1371/journal.pntd. 0002821.

**85** Richards, F., Rizzo, N., Diaz-Espinoza, C.E. et al. (2015). One hundred years after its discovery in Guatemala by Rodolfo Robles, *Onchocerca volvulus* transmission has been eliminated from the central endemic zone, *Am. J. Trop. Med. Hyg.* 93: 1295–1304. https://doi.org/10.4269/ajtmh.15-0364.

**86** Rodríguez-Pérez, M.A., Fernández-Santos, N.A., Orozco-Algarra, M.E. et al. (2015). Elimination of onchocerciasis from Mexico. *PLoS Negl.Trop. Dis.* 9(7): e0003922. https://doi.org/10.1371/journal.pntd.0003922. PMID: 26161558; PMCID: PMC4498594.

**87** Zarroug, I.M., Hashim, K., ElMubark, W.A. et al. (2016). The first confirmed elimination of an onchocerciasis focus in Africa: Abu Hamed, Sudan, *Am. J. Trop. Med. Hyg.* 95: 1037–1040. https://doi.org/10.4269/ajtmh.16-0274.

**88** Katabawa, M.N., Habomugisha, P., Khainza, A. et al. (2020). Historical elimination of onchocerciasis from Victoria Nile focus in Central Uganda verified using WHO criteria. *Am. J. Trop. Med. Hyg.* 102: 411–1416. https://doi.org/10.4269/ajtmh.20-0064.

**89** Katabarwa, M.N., Zarroug, I.M.A., Negussu, N. et al. (2020). The Galabat-Metema cross-border onchocerciasis focus: the first coordinated interruption of onchocerciasis transmission in Africa. *PLoS Negl.Trop. Dis.* 14: e0007830. https://doi.org/10.1371/journal.pntd.0007830. PMID: 32027648; PMCID: PMC7004312.

**90** Guevara, A., Salazar, E., Vicuña, Y. et al. (2021). Use of Ov16-based serology for Post-elimination surveillance of onchocerciasis in Ecuador. *Am. J. Trop. Med. Hyg.* 103. 1569–1571. https://doi.org/10.4269/ajtmh.20-0082.

**91** Jacob, B., Loum, D., Munu, D. et al. (2021). Optimization of slash and clear community-directed control of *Simulium damnosum* Sensu Stricto in northern

Uganda. *Am. J. Trop. Med. Hyg.* 104: 1–10. https://doi.org/10.4269/ajtmh.20-1104.

**92** Shott, J., Ducker, C., Unnasch, T.R., and Mackenzie, C.D. (2018). Establishing quality assured (QA) laboratory support for onchocerciasis elimination in Africa. *Int. Health* 10(suppl): i33–i39. https://doi.org/10.1093/inthealth/ihx059. PMID: 29471345.

**93** Weil, G.J., Bogus, J., Christian, M., Dubray, C. et al. (2019). The safety of double- and triple-drug community mass drug administration for lymphatic filariasis: a multicenter, open-label, cluster-randomized study. *PLoS Med.* 16: e1002839. https://doi.org/10.1371/journal.pmed.1002839.

**94** World Health Organisation (2012). Morbidity management and disability prevention (MMDP). https://www.who.int/lymphatic_filariasis/managing-morbidity/en/89.

**95** Douglass, J., Graves, P., and Gordon, S. (2016). Self-care for management of secondary lymphedema: a systematic review. *PLoS Negl. Trop. Dis.* 10(6): e0004740.

**96** Mackenzie, C.D. and Mante, S. (2021). Caring for patients in the global programme to eliminate lymphatic filariasis. *Int. Health.* 13(Supplement 1): S48–S54. https://doi.org/10.1093/inthealth/ihaa080.

**97** Horton, J., Klarmann-Schulz, U., Stephens, M. et al. (2020). The design and development of a multicentric protocol to investigate the impact of adjunctive doxycycline on the management of peripheral lymphoedema caused by lymphatic filariasis and podoconiosis. *Parasites Vectors.* 13: 155. https://doi.org/10.1186/s13071-020-04024-2. PMID: 32228663.

**98** World Health Organisation (2019). Surgical approaches to the urogenital manifestations of lymphatic filariasis. Report from an informal consultation among experts, World Health Organization; WHO/CDS/NTD/PCT/2019.04. https://apps.who.int/iris/rest/bitstreams/1240105/retrieve.

**99** Krishnasastry, S. and Mackenzie, C.D. (2021). Alternative approaches to lymphoedema care in lymphatic filariasis. *PLoS Negl. Trop. Dis.* 15(4): e0009293. https://doi.org/10.1371/journal.pntd.0009293.

**100** Mullerpattan, J.B., Udwadia, Z.F., and Udwadia, F.E. (2013). Tropical pulmonary eosinophilia – a review. *Indian J. Med. Res.* 138: 295–302. PMCID: PMC3818591 PMID: 24135173.

**101** Ton, T.G., Mackenzie, C., and Molyneux, D.H. (2015). The burden of mental health in lymphatic filariasis. *Inf. Dis. Poverty* 4: 34. https://doi.org/10.1186/s40249-015-0068-7.

**102** Gedge, L.M., Bettis, A.A., Bradley, M. et al. (2018). Economic evaluations of lymphatic filariasis interventions: a systematic review and research needs. *Parasites Vectors.* 11: 75.

**103** Kamgno, J. and Boussinesq, M. (2002). Effect of a single dose (600 mg) of albendazole on *Loa loa* microfilaraemia. *Parasite* 9: 59–63. https://doi.org/10.1051/parasite/200209159. PMID: 11938697.

**104** Kamgno, J., Pion, S., Chesnais, C. et al. (2017). Test and treat strategy to combat onchocerciasis in *Loa loa* endemic regions. *N. Engl. J. Med.* https://doi.org/10.1056/NEJMoa1705026.

**105** Mackenzie, C.D. (2019). A much-needed advance in the diagnosis of river blindness. *J. Infect. Dis.* pii: jiz309. https://doi.org/10.1093/infdis/jiz309.

**106** Bennuru, S., Oduro-Boateng, G., Osigwe, C. et al. (2020). Integrating multiple biomarkers to increase sensitivity for the detection of *Onchocerca volvulus* infection. *J. Infect. Dis.* 221: 1805–1815.

**107** Geary, J., Lovato, R., Wanji, S. et al. (2015). A histochemical study of the Nras/let-60 activity in filarial nematodes. *Parasites Vectors.* 8: 353. https://doi.org/10.1186/s13071-015-0947-6.

**108** Geary, T.G. and Mackenzie, C.D. (2011). Progress and challenges in the discovery of macrofilaricidal drugs. *Expert Rev. Anti Infective Ther.* 9: 681–695.

**109** Mackenzie, C.D. and Geary, T.G. (2011). Flubendazole; a candidate for a field usable macrofilariacide. *Expert Rev. Infect. Agents.* 9: 497–501.

**110** Geary, T.G., Mackenzie, C.D. and Silber, S.A. (2019). Flubendazole as a macrofilaricide: history and background. *PLoS Negl. Trop. Dis.* 13(1): e0006436. https://doi.org/10.1371/journal.pntd.0006436.

# 4

# Canine Filariasis (Heartworm) – Disease and Current Gaps

*Dwight D. Bowman* and Timothy K. Wu*

*Cornell University, College of Veterinary Medicine, Department of Microbiology and Immunology, Ithaca, NY 14853-6401, USA*

**Abstract**

This review presents a brief introduction to heartworm infections and the associated disease. This is followed by a history of the market introductions of the various macrocyclic lactone containing heartworm preventive products that began in 1987 with the approval of HeartGard-30®. Discussion then follows the development and pre- and post-approval testing of the efficacy of the additional macrocyclic lactone molecules and different formulations of these molecules with some being designed with the dose-limiting parasites being third-stage and young fourth-stage heartworm larvae in the host within 30 days of their acquisition by mosquito bite (180 days for ProHeart® 6 and a year for ProHeart® 12), while others chose other parasites as the dose-limiting organism, e.g. hookworms, roundworms, and whipworm, and in one case, fleas. In 2011, as part of a product's approval (Trifexis, a combination of milbemycin oxime with added spinosad for flea control), it did not prevent the development of a heartworm field isolate. Also, we now know that ivermectin, milbemycin oxime, moxidectin, and selamectin in product formulations have all failed against at least one field isolate. The history over the last 35 years or so also shows that many of these products are now failing to control heartworm infections at the internal parasite control dosages that are much higher dosages than initially required for heartworm prevention. Finally, there is discussion of the gaps in prevention and therapy. One is whether or not treatment should be year-round or not year-round. Stopping of therapy each year may let the last few "resistant" fourth-stage larvae in a dog to mature and produce microfilariae. Thus, the seasonal withdrawal of prevention appears potentially problematic. Then, there is the transmission gap; the macrocyclic lactones do not stop transmission, but the isoxazolines might, and the killing of the few mosquitoes that bite dogs is unlikely to produce isoxazoline resistance in the mosquito population. The greatest gap would be caused by the total loss of macrocyclic lactones for heartworm control. Less than perfect backups exist, e.g. daily diethylcarbamazine or injections of an arsenical such as melarsomine every four months, but perhaps most hopefully, the

*Corresponding author.

*Human and Animal Filariases: Landscape, Challenges, and Control*, First Edition.
Edited by Ronald Kaminsky and Timothy G. Geary.

biggest breakthrough for heartworm control would be to develop a vaccine that would prevent infection. The most disconcerting gap that might have been the least or most important gap during the last 35 years was the unwillingness of many to consider that resistance would happen or was happening when dogs were getting infected when being given approved products that had elevated efficacy margins of at least 2–10 times the minimal required preventive dose at the time of product approval. This last gap held back cohesive teamwork on intervention plans for more than a decade.

## 4.1 Introduction: Heartworm, Pathogenesis, Pathology, and Disease Presentation

Infections of dogs and also cats with heartworms is clearly the cause of horrendous disease that will have life-changing effects and can be fatal. Worms that mature from infectious larvae inoculated by mosquitoes appear as young adult worms that begin to appear in the lungs 70–120 days after infection. There they mature, and adults begin to produce circulating microfilariae some six months after the infection was initiated [1]. Adult females are about 19.5–29.5 cm ($7\frac{5}{8}$–$11\frac{1}{2}$ in.) long, and the males are around 14.7–18.4 mm ($5\frac{3}{4}$–$7\frac{3}{8}$ in.) long. Unlike intestinal worms, when *Dirofilaria immitis* adults die or are killed with adulticides, the dead adult *D. immitis* have no way to leave the body. Instead, they are pushed further down into the lungs, where they very slowly decompose, and the cuticular collagen is the last portion of the worms persisting in tissues. In a 1989 publication on heartworm in shelter dogs and cats in Florida, 22 of 712 cats (3.1%) and 510 of 876 dogs (59.4%) were infected, and within these infected dogs, the mean number of worms was 23.4, the median 11, and they harbored from 1 to 317 adult worms. In total, 38% of the infected dogs and none of the infected cats had circulating microfilariae [2]. Cats typically are only infected with 1–3 worms, usually to a maximum of 9 [3].

The pathogenesis of lesions in dogs and cats affected with dirofilariasis can be directly related to the organism's life cycle. Following initial inoculation of third-stage larvae by mosquito vectors and subsequent maturation of these larvae in the subcutis, fascia, muscle tissue, abdomen, and thorax, immature adult larvae migrate via the jugular and other venous circulation to the pulmonary arteries where they mature to adulthood [4]. During this initial larval migration, some larvae occasionally become lost, resulting in aberrant migration, and adult nematodes may be found in unusual locations, including the eye, brain, peritoneal cavity, and systemic circulation in dogs [5–8]. Aberrant migration occurs more commonly in cats, reflecting the unsuitability of cats as hosts for *D. immitis*. In these cases, adults may be found in the body cavities, systemic arteries, and central nervous system in feline hosts [9]. In cases when immature worms do reach the pulmonary arteries in cats, they often die, resulting in a severe inflammatory response attributed to pulmonary intravascular macrophages within the capillary beds of the lung [8, 10]. The resulting disease, known as heartworm-associated respiratory disease (HARD), presents clinically as dyspnea, coughing, wheezing, and asthma-like attacks, and manifests histologically as eosinophilic pneumonia and occlusive medial hypertrophy of the pulmonary arteries with villous [8, 10]. As adults, it is believed that mature heartworms downregulate the activity of pulmonary intravascular macrophages, resulting in mostly asymptomatic infections in cats [10]. Nonetheless, infected cats may be affected by acute death syndrome (a disease separate from HARD), the pathogenesis of which remains unknown, although an acute anaphylactic reaction is suspected [11]. Additionally, given the smaller size of feline lungs relative to adult nematodes, even a small number of worms may result in cardiovascular changes including pulmonary artery obstruction, as well as pneumothorax and chylothorax [12].

In dogs with adult heartworms in the pulmonary arteries, clinical signs typically include coughing, exercise intolerance, decreased appetite, and weight loss [13]. Lesions are commonly seen in the pulmonary arteries and lung parenchyma, with severity correlated to the number of worms present (Figures 4.1 and 4.2) [14]. Histologic changes typically include villous fibromuscular proliferation in the tunica intima of pulmonary arteries, as well as eosinophilic endarteritis, smooth muscle hyperplasia in the tunica media, and fibrosis (Figure 4.3) [15]. The resultant increase in vascular permeability results in pulmonary edema and pneumonia. The changes in the pulmonary arteries result in reduced luminal diameter of the vessels and subsequent pulmonary hypertension. In response to the increased overload, the right heart may undergo hypertrophy and dilation. In severe cases, the right heart changes may not be adequate to handle the increased overload, and decompensated right heart failure develops, manifesting as ascites and ultimately hepatic cirrhosis.

In some cases, adult worms, which usually live in the pulmonary arteries, move against the blood flow through the heart and enter the vena cava. This displacement causes the medical condition known as caval syndrome, which requires immediate life-protecting surgical intervention to extract the offending worms [16]. Caval syndrome is typically marked by a sudden onset of labored breathing, pale gums, and dark bloody or coffee-colored urine due to hemoglobinuria [13]. Caval syndrome is the result of partial inflow obstruction to the right heart and interference

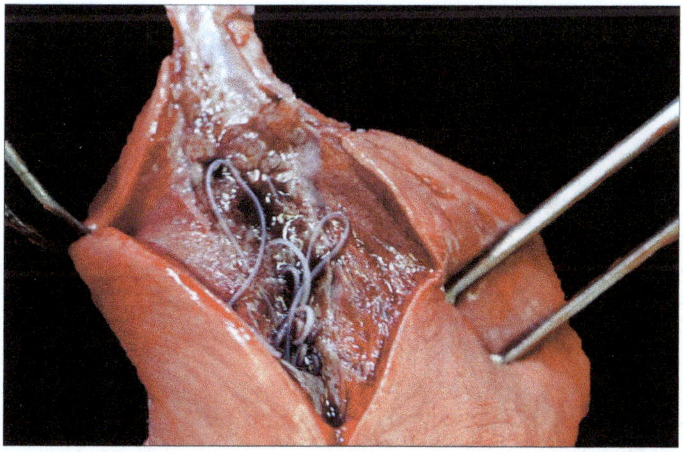

**Figure 4.1** *Dirofilaria immitis* infection in a dog, 150 days post-infection. Upon opening the pulmonary artery, the lumen is filled with numerous thin, long, white filarial nematodes.

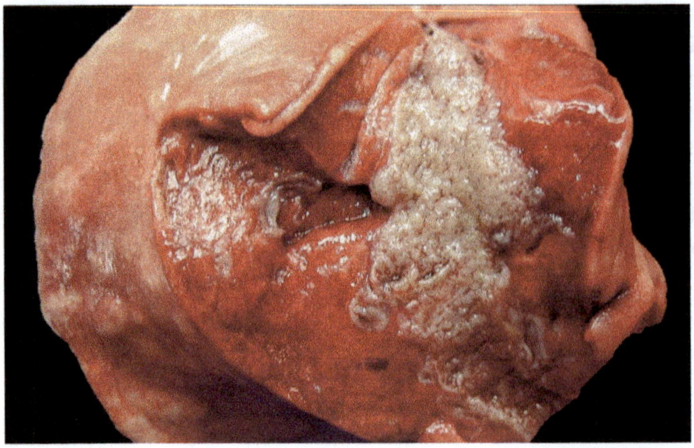

**Figure 4.2** *Dirofilaria immitis* infection in a dog, 150 days post-infection; the worms entered the pulmonary arteries only about 80 days before image was taken. The pulmonary artery has been incised and reflected to reveal the luminal surface of the vessel, with extensive, irregular villous proliferation of the intimal surface.

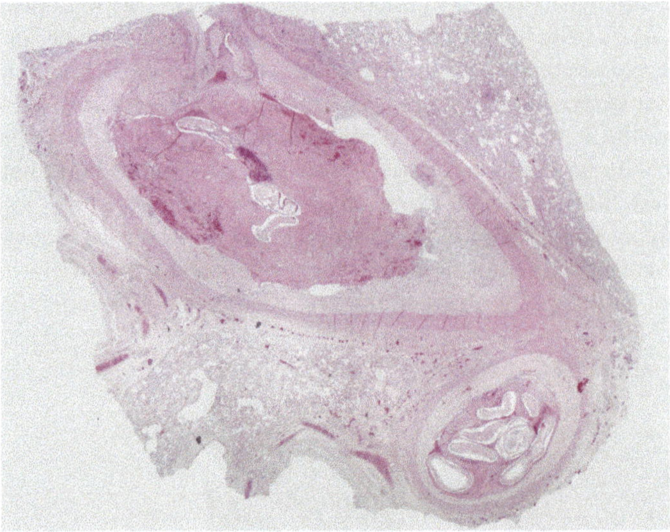

**Figure 4.3** Histology of pulmonary dirofilariasis. The pulmonary arteries are partially occluded by a mixture of fibrin and red blood cells, admixed with tangential and cross sections of adult *Dirofilaria immitis* nematodes. The tunica intima is circumferentially thickened and proliferative, with occasional villous projections of fibrointimal hyperplasia. Adjacent alveoli are collapsed, with loss of air space.

with the tricuspid valve [14]. Studies have revealed that caval syndrome is due to mechanical obstruction of the right atrium and right atrioventricular valve orifice, rather than any virulence factors produced by the nematodes [17–19]. Subsequent to this obstruction, red blood cells entering the right atrium and ventricle are subject to lysis as they are forced by vascular flow through a whisk of tangled and tumbling heartworms, leading to hemolytic anemia [14]. Disseminated intravascular coagulation (DIC) can also occur with caval syndrome as a result of intravascular hemolysis. Hepatic and renal dysfunction is also described with caval syndrome, though the exact cause is unclear [14]. Caval syndrome is less common in cats, but can occur even with low worm numbers [9].

Worms that have died or have been killed with treatment may induce thromboemboli or result in arterial obstruction. Thrombosis of pulmonary vessels occurs most often in the pulmonary arteries of the caudal lung lobes and in rare cases may lead to pulmonary infarction [14]. Other lesions related to the *D. immitis* infection include glomerulonephritis caused by antigen–antibody complex deposition, as well as nodular skin disease due to circulating microfilariae [14, 20].

## 4.2 Current Gaps

According to the Companion Animal Parasite Council (CAPC) map for heartworm testing for the year 2020 (at CAPCVET.org), 15 441 743 dogs were antigen tested and 201 888 dogs (1.31%) were positive. As there are 70 000 000 dogs in the United States, we can estimate that somewhere near 1 million dogs have a heartworm infection. Cases that go without diagnosis and treatment persist for years, causing chronic lung disease. Also, whether the worms are killed quickly with melarsomine or slowly with a macrocyclic lactone, dogs would still harbor worms undergoing slow degradation in granulomatous areas of the lungs. Also, it is likely that somewhere around 0.1–1% of infected animals would require life-saving surgical intervention for caval syndrome or to remove atopically located adults. It must be remembered that the CAPC data are being collected from dogs taken to veterinarians. Since these are the dogs seeing veterinarians, it must also be recognized that many dogs, both strays and pets, do not receive the protection of heartworm preventive therapy. There will also be a proportion of cats infected, and cats also develop disease. Treatment is expensive compared with prevention. As they say, "an ounce of prevention is worth a pound of cure," and in this case, the cure is likely to cost more than five times the cost of regular annual prevention. The following section discusses concerns that the current preventive molecules may not be working forever.

Nine years ago, one author of this chapter published a paper entitled "Heartworms, macrocyclic lactones (MLs), and the specter of resistance to prevention in the United States" [21]. Now, unfortunately, the specter has taken corporeal form and been made clearly manifest in two recent field studies of dogs from throughout the United States. These studies examined the safety and non-inferiority of two new products tested alongside the long-standing backbone of heartworm prevention, HeartGard® Plus [22, 23]. In these two studies, none of the dogs receiving the new products developed infections, but six dogs receiving HeartGard® Plus (two in one study and four in the other) developed heartworm infections. In one of the trials [22], the enrolled dogs were from throughout the United States, and 167 dogs (68% of total dogs) received a moxidectin, pyrantel, and sarolaner combination product (Simparica Trio™), while 82 dogs (32% of total dogs) received the combination of ivermectin and pyrantel, HeartGard® Plus. Of the 318 dogs receiving one or the other treatment, the only ones developing heartworm infections were two dogs from the lower Mississippi River valley that had received the HeartGard Plus. In the other trial [23], one group of dogs ($n = 236$) received extended release moxidectin, i.e. ProHeart® 12, while a second group of dogs ($n = 218$) received ivermectin and pyrantel, i.e. HeartGard® Plus. This study yielded results similar to the Simparica Trio™ study, with none of the dogs treated with ProHeart® 12 testing positive for heartworm infection, while four HeartGard® Plus-treated dogs, again also from the lower Mississippi River valley, developed heartworm infections, with three of the four having circulating microfilariae [23].

Within the work cited above [18] is the statement: "Every currently marketed compound (i.e., ivermectin, milbemycin oxime, moxidectin and selamectin) has been shown to be less than 100% efficacious against a resistant strain(s) in at least one controlled study. This includes Advantage Multi®, Bayer (imidacloprid + moxidectin) applied topically 30 days post-inoculation with the JYD-34 ML-resistant strain, in which one of the eight treated dogs had two adult heartworms at necropsy (McCall, unpublished data)." Thus, it would appear that all macrocyclic lactones that have been developed for heartworm prevention, including the macrocyclic lactone in these two new products, now appear to be also potentially at risk of failing, since the parasites seem to be able to develop even when the products are given at higher concentrations or on multiple occasions. Usually, resistance can be overcome by increasing the amount of drug delivered or by increasing the number of doses or dosing frequency, but at some point, the drug may fail, or it may require levels that are dangerous or deadly for the host. Overall, this is what has historically been occurring with heartworm preventives, and we may now be close to the maximum doses that are safe for all dogs.

If one looks at the history of the approval of the macrocyclic lactone products since first introduced as heartworm preventives, it is obvious that there has been a continued increase in the amount of product utilized in the preventives and that much higher doses are now required for prevention. If one looks at the original new animal drug application (NADA) summaries available through the Freedom of Information (FOI) act presented on the FDA's Center for Veterinary Medicine website, one can follow some of these changes. Ivermectin in early trials as HeartGard® was 100% effective when administered at only 3 μg/kg, and it was marketed at 6–12 μg/kg [24]. The NADA states: "The recommended minimum dose of 6 μg of ivermectin per kilogram of body weight was selected although a lower dose (3.3 μg/kg) may be as effective. In addition to the results of the dose titration and confirmation trials, factors concerning the practicality of incorporating minute amounts of drug in a reasonably sized tablet with good drug uniformity and the practical use of the product were considered. The target dose of 6 μg per kilogram of bodyweight was selected from titration study 10855 as the lowest dose providing 100% protection when the dosing interval was extended to 60 days to simulate a missed-dose circumstance." In the case of milbemycin oxime in the product Interceptor® [25]. "These studies established the minimal effective dose for heartworm prevention at 0.1 mg/kg [=100 μg/kg] and for hookworm control at 0.5 mg/kg [=500 μg/kg]. To support a dual claim for heartworm prevention and hookworm control, the finished pharmaceutical dosage forms (tablets) were formulated to provide a target dose of 0.5 mg/kg body weight. In target animal safety studies, the selected dose (0.5 mg/kg) was determined to have a wide margin of safety." After the launch of Interceptor, HeartGard-30®-Plus was launched with the inclusion of pyrantel pamoate for the control of adult *Toxocara canis*, *Toxascaris leonina*, *Ancylostoma caninum*, and *Uncinaria stenocephala*, with studies showing 100% efficacy in the prevention of heartworm infection [26]. In the case of Sentinel® [27], the product was shown to be equivalent to Interceptor® [11], and thus for heartworm, it was efficacious at 0.5 mg/kg, which, again, was 10 times the minimum effective dose required for killing month-old L3 heartworm larvae. Tablets of ProHeart® [28] containing moxidectin, rather than ivermectin, were 100% efficacious when marketed at the "recommended dose rate of 3 μg moxidectin/kg (1.36 μg/lb) body weight," and "100% effective in preventing the development of a one month-old heartworm infection of 50 L3 larvae of *D. immitis* in dogs when administered as a single oral treatment at 1.5, 3.0, and 6.0 μg/kg." Also, the "0.5 μg moxidectin/kg body weight dose level was 100% effective against the two-month heartworm infections." (Table 4.1)

The selamectin-containing product Revolution® was approved for the prevention of heartworm at a dose of 6 mg to be applied topically [30]. When Revolution® was topically dosed to dogs after experimental infections with 50 L3 larvae at 3 mg/kg on either 30 or 45 days after infection or with 6 mg/kg at 30, 45, or 60 days after infection, none of the dogs developed heartworm infections. The dose incorporated into this product was determined by its ability to kill fleas feeding on the infected pets, not based on heartworm efficacy.

The next product to appear was a second product containing moxidectin. This was ProHeart® 6 [31], a sustained-release injectable that was ultimately marketed to be dosed at 0.17 mg/kg; however, in the two trials with experimental infections where dogs were treated with 0.06, 0.17, or 0.5 mg/kg, none of the dogs developed patent infections. The dose of 0.17 mg/kg was chosen to provide efficacy against hookworms [31].

Then, three more heartworm preventives based on heartworm prevention with ivermectin entered the market. In 2002, a topical treatment for dogs, Advantage® Duo, containing imidacloprid with ivermectin for heartworm prevention, was examined at 40, 80, and 160 μg/kg, and all doses were 100% efficacious against dogs infected 30 days before product application. The product dose was set at 80 μg/kg topically applied ivermectin along with 10 mg/kg imidacloprid [32]. In 2003, the FOI for Tri-Heart® Plus [33], a generic version of HeartGard® Plus, showed that the new product was equally as efficacious (100%) in the prevention of heartworm infection when given 30 days after infection as the original product's oral ivermectin dose of 6 μg/kg. Work on Iverhart Max® (ivermectin, pyrantel pamoate, and praziquantel) [35] for approval employed the experimental infection of three groups of eight dogs, one untreated group, one group treated with praziquantel, and one group treated with the combination of all three ingredients; this oral formulation

**Table 4.1** This is a temporal list of the Freedom of Information Summaries of Different New Animal Drug Applications (NADAs) cited in the paper.

| (A)NADA | Product name | Drug(s) | Date | Route | Rate-limiting parasite(s) | References |
|---|---|---|---|---|---|---|
| 138–412 | HeartGard-30® | Ivermectin | 1987 | Oral | Heartworm | [24] |
| 140–915 | Interceptor® | Milbemycin | 1990 | Oral | Hookworm & Roundworm | [25] |
| 140–971 | HeartGard-30® Plus | Ivermectin + pyrantel | 1993 | Oral | Heartworm, Hookworm & Roundworm | [26] |
| 141–042 | Immiticide | Melarsomine dihydrochloride | 1995 | Injection | Heartworms (4-mo-old to adult) | [29] |
| 141–051 | ProHeart® for Dogs | Moxidectin | 1997 | Oral | Heartworm | [28] |
| 141–084 | Sentinel® Tablet | Milbemycin + lufenuron | 1997 | Oral | Hookworm and Roundworm + Fleas | [27] |
| 141–152 | Revolution® | Selamectin | 2000 | Topical | Fleas | [30] |
| 141–189 | ProHeart® 6 | Moxidectin – sustained release | 2000 | Injection | Hookworms | [31] |
| 141–208 | Advantage® Duo | Ivermectin + imidacloprid | 2002 | Topical | Heartworm + Fleas | [32] |
| 200–338 | Tri-Heart® Plus | Ivermectin + pyrantel | 2003 | Oral | Heartworm (HeartGard Plus) Control | [33] |
| 141–251 | Advantage Multi® | Moxidectin + imidacloprid | 2006 | Topical | Intestinal Helminths + Fleas | [34] |
| 141–257 | IverHart Max® | Ivermectin + pyrantel + praziquantel | 2006 | Oral | Heartworm + Roundworm + Tapeworm | [35] |
| 141–321 | Trifexis® | Milbemycin + spinosad | 2011 | Oral | Heartworm + Fleas | [36] |
| 141–333 | Sentinel® Spectrum® | Milbemycin + Lufenuron + praziquantel | 2011 | Oral | Heartworm + Fleas | [37] |
| 141–338 | Interceptor® Spectrum® | Milbemycin + praziquantel | 2012 | Oral | Hookworm and Roundworm + Tapeworm | [38] |
| 141–519 | ProHeart® 12 | Moxidectin – sustained-release | 2019 | Injection | Heartworm & Hookworm | [39] |
| 141–521 | Simparica Trio™ | Moxidectin + sarolaner + pyrantel | 2020 | Oral | Heartworm and Hookworm | [40] |

The list is not all inclusive since there are products that were approved, such as Coraxis® (a topical moxidectin without imidacloprid, NADA 141–417) that provided no additional testing data). This is true for most ANADAs that are "abbreviated" NADAs for the approval of generic copies of new animal drug products that have been previously approved and shown to be safe and effective. The column headed with "REF" refers to the appearance of the NADA-ANADA within the manuscript and Reference section of the test.

also had 100% efficacy against heartworm with the previously registered ivermectin dose of the "pioneer" product HeartGard® [24].

The next heartworm product containing moxidectin that entered the market was Advantage Multi® for Dogs [34]. Advantage Multi® for Dogs provides a dose of moxidectin of 2.5 mg/kg and has a claim for preventing heartworm infection along with adults and larvae of *A. caninum* and *U. stenocephala*, and adults *T. canis*, *T. leonina*, and *Trichuris vulpis*. The NADA states, see: "Dosage Characterization for the Treatment and Control of Intestinal Nematodes, which establishes a minimum effective dose for moxidectin." Relative to heartworm, Advantage Multi® for Dogs was fully efficacious when applied 30 or 45 days after dogs were inoculated with infective heartworm larvae [34].

The examination of the above work reveals a major difference in how two groups of products were approved relative to the determination of the dose required to prevent heartworm infections when dogs were experimentally infected with L3 larvae. In one group of products, such as HeartGard [24] and ProHeart® Tablets [28], the products were designed with heartworm as the dose-limiting parasite. However, in the second group of studies, the dose-limiting parasite used to set the amount of active ingredient in the product was not heartworm, but instead, was internal nematode parasites (Interceptor®, Sentinel®, ProHeart-6, Advantage® Duo, and Advantage Multi® [25, 27, 31, 32, 34] or fleas (Revolution®) [30]. Thus, when these products later failed to prevent heartworm infections, they were not working at the original drug concentration that was efficacious at a much lower dose level than the final prescribed doses in the marketed product. Thus, conversely, when products failed to prevent the disease, in most cases, the heartworms were surviving in the presence of much higher levels of drug than were protective when the product was originally approved.

The history of milbemycin oxime approvals shows the evolution of what occurred as the same molecule was submitted for approval in different formulations over a period of years at the same original preventive dose of 0.5 mg/kg. The argument could be made that the differences in efficacy were due to changes in product formulation for each newly developed product that affected the efficacy of the active ingredients, but it is also likely that it represents changes that had developed in the nematodes that resulted in reduced efficacy of the filaricide in the different formulations. It also needs to be remembered when discussing these products that the pharmacokinetics are such that oral ivermectin and milbemycin oxime only remain at therapeutic levels in the dog for one to a few days, i.e. they are not at protective levels all month long. Therefore, soon after treatment, the drug is not at levels that prevent heartworm infection; instead, when a dog is treated each month, newly acquired larvae that are <30 days old (or 45 or 60 days in some trials) at the time when the next treatment is administered, which for ProHeart6 is 180 days after the last injection of ProHeart 6, the sustained release product. In the following discussion, it is also important to remember that when Interceptor was first approved, it was 100% efficacious against month-old heartworm infections at 1/10 the approved dose that was chosen to kill adult roundworms and hookworms, i.e. in 1990, i.e. Interceptor had been shown to be 100% effective against month-old heartworm larvae at a minimum effective dose of 50 μg/kg. In 1997, Sentinel, the combination of milbemycin oxime and lufenuron, was shown to be 100% efficacious in similar studies, but at this time using a dose that was 10 times that initially shown to be effective at killing heartworm larvae that were < 30 days of age [27]. Confusion first arose with the approval of the next milbemycin oxime-containing product, Trifexis®, in 2011 [36]. In laboratory work as part of Trifexis development, it was found that "While one isolate was fully susceptible to this ML providing 100% prevention after a single dose administered 30 days after inoculation with heartworm (HW) L3, efficacy of a single treatment against the second isolate was less than 100%. A study was undertaken with this second field isolate to assess the effectiveness of currently marketed ML (Macrocyclic Lactones), in this case administering a single dose of ivermectin (IVM) or milbemycin oxime (MBO), in dogs challenged with HW L3 1 month before treatment." In further work that is described in the report [41], 42 dogs were divided into three groups of 14, and all received 50 infective larvae of an isolate of *D. immitis* (now known as the MP3 isolate). The three groups of dogs received either no treatment, oral ivermectin as HeartGard Plus, or oral milbemycin oxime as Interceptor. At necropsy, it was found that all 14 control dogs were infected with heartworms, but in both the ivermectin and milbemycin oxime-treated groups, one dog in each group had one worm [41]. This was followed by additional studies by Snyder et al. [42] that also appear within the NADA [36]. One of these studies used four groups of dogs (groups 1–4): (group 1) sham-treated controls, (group 2) treated 30 and 60 days after infection, (group 3) treated 30, 60, and 90 days after infection, and (group 4) treated 47 and 75 days after infection. All the control dogs (group 1) were infected. One heartworm was recovered from a dog in group 2 that had been treated on days 30 and 60 after infection (group 2). No worms were recovered from dogs in groups 3 and 4; i.e. when treated 30, 60, and 90 days after infection (group 3) or when treated 47 and 75 days after infection (group 4). In a separate study with a different isolate, no worms were recovered from dogs treated 30 or 45 days after infection. Based on these studies, the product was approved as "For heartworm prevention, give once monthly for at least 3 months after exposure to mosquitoes." The next approved product containing milbemycin oxime was Sentinel® Spectrum®, and this product examined only a regimen of six consecutive monthly treatments, which prevented the dogs in the study from being infected [37]. The same method of only examining six months of preventive therapy was also used in the approval of Interceptor™ Spectrum [38]. Approval for heartworm prevention was: "SENTINEL SPECTRUM should be administered at monthly intervals, beginning within 1 month of the dog's first seasonal exposure to mosquitoes and continuing until at least 6 months after the dog's last seasonal exposure." The NADA for the milbemycin oxime/afoxolaner product, i.e. NexGard SPECTRA®, did not receive approval in the United States, since it did not provide 100% efficacy against the heartworm isolate JYD-34, while was approved in the European Union by the Committee for Medicinal Products for Veterinary Use [43], with the following annotation: "In a further two studies using the USA isolate JYD-34 efficacy of circa 70% was achieved using two, three, or six treatments at monthly intervals. This may indicate reduced susceptibility of this isolate to the effects of milbemycin oxime. There is evidence of an increase of reports on 'product prevention failures' with respect to *D. immitis* ('lack of efficacy' in heartworm prevention) in the USA and it is now accepted that lack of efficacy related to resistance to macrocyclic lactones in *D. immitis* has developed in the USA."

The history of heartworm prevention with milbemycin oxime provides a significant amount of material to consider relative to the recognition of resistance and understanding how selection changes the ability of a population to live in higher concentrations of a given product or group of molecules. Again, when first approved in 1990, the milbemycin oxime component in heartworm preventives was 100% efficacious at 1/10 the marketed dose when given once 30 days after a dog was infected. We now have two labels that say "prescribe monthly," one that says "prescribe for three months after heartworm is no longer a threat," and another that says to "prescribe the product for six months after heartworm is no longer a threat." Unfortunately, the recent formulation, NexGard SPECTRA®, was not preventive even with six consecutive treatments against one isolate. These are basically all the same active ingredients delivered at the same dose (again, not all in the same formulations). So now, even though utilizing 10 times the originally approved dose, the worms survive. More intriguing therapeutically is the fact that it seems that multiple treatments have a real impact on the development of the heartworms already in the dogs. The work of Snyder et al. made it clear that these products had effects on worms after they were more than 30 or 45 days of age [37, 42]. This strongly suggests two things. First, it supports year-round prevention, i.e. not stopping treatment will be much more efficacious if a resistant isolate is present in the dog receiving treatment. Second, it suggests that repeated monthly treatments that maintain elevated drug concentrations in the tissues of a dog likely provide increased efficacy against resistant heartworms.

Selamectin, the active ingredient in Revolution® [30], has not appeared in another product for heartworm prevention in dogs, and again the therapeutic dose for this product was originally set for the control of fleas, which, at the time of approval, required higher doses than needed for heartworm prevention [30]. In work done soon after the approval of this product, of eight dogs that were given monthly treatments with Revolution® for 12 months beginning 90 days after each was inoculated with 50 infective heartworm larvae, three each had a single female worm [44]. In a study with one of the early resistant isolates, "MP3" (the isolate that caused the initial complications in the development of Trifexis® and Sentinel® Spectrum®), it was shown that when dosed at the heartworm preventive dose 30 days after infection with 100 infective heartworm larvae, that efficacy of Revolution® (selamectin) was 95.5% [45]. With the JYD-34 isolate (the isolate cited by the European Medicines Agency (EMA) relative to the lack of efficacy with NexGard Spectra), dogs were each given three consecutive monthly treatments with Revolution® 30, 61, and 90 days after receiving 50 L3 larvae of JYD-34 and the efficacy with Revolution was only 28.8% [38]. These studies clearly show that the original heartworm preventive dose defined at approval for selamectin was not as efficacious against this isolate as it had been with other isolates at the time of its development.

Moxidectin, like most other macrocyclic lactones, has different pharmacokinetics when applied topically versus orally, and the pharmacodynamics for moxidectin are such that the product remains at therapeutic levels much longer when delivered topically than the others used for heartworm prevention. Also, when moxidectin is applied topically to dogs on a monthly basis, the dog develops and maintains high serum levels of moxidectin [46]. These levels protect dogs against incoming hookworm infections [47]. These same levels were also characterized pharmacologically and shown to be protective against heartworm when dogs were treated 28 days after four consecutive monthly treatments [46]. Similarly, dogs treated just once 28 days prior to the inoculation of heartworm L3 larvae were also all protected against infection [48]. The pharmacokinetic data from the dogs infected 28 days after the fourth Advantage Multi® application show that steady-state levels of moxidectin in the dogs 28 days after four monthly treatments are higher than the level found after a single application [46]. Therefore, repeated applications of Advantage Multi® provide increased and continuously high drug levels that might provide protection similar to that which occurs with repeated monthly treatments with oral milbemycin oxime.

Two new canine products that contain moxidectin for heartworm prevention have entered the market in the United States. One is ProHeart® 12, a sustained-release injectable product that protects dogs from incoming infections for 12 months [39]. The other is an oral product, Simparica Trio™, that contains moxidectin for heartworm prevention, sarolaner for arthropod treatment and control, and pyrantel for adult roundworm and hookworm treatment [40]. These products have been shown in laboratory and in field trials to prevent 100% of dogs from being infected with heartworms under experimental and natural challenges [22, 23]. Unlike the topically applied moxidectin, neither product has been shown to reach steady-state pharmacokinetics after multiple applications. Thus, there is reason to suspect that dogs are again susceptible to infection when the drug is withdrawn. ProHeart® 6 was examined using a resistant isolate from Earle, Arkansas, USA (Jd2009), two control dogs and four treated dogs were infected with 80 L3 heartworm larvae 180 days after drug inoculation and examined 150 days later. All six dogs became infected, with no significant difference between the worm counts (15 and 43 for the controls and 19, 33, 24, and 15 for the treated dogs) [49]. This strongly suggests that, as for the other products discussed above, one must be concerned about dogs that are not provided with follow-up prevention. This is the case for both Proheart® 6 and ProHeart® 12, because the products are not designed to protect for more than the approved period, and there is no evidence that the Simparica Trio™ product protects for any given time after the last treatment application. It must be remembered that, for all of these products, because they are thought to kill only young heartworm stages, they are tested by infecting dogs and then waiting the scheduled treatment interval before a single dose is given. This interval is 30 days for the typical monthly products. For ProHeart® 6 and ProHeart® 12, testing is different; the dogs are expected to be protected for at least 180 or 365 days, respectively, and therefore, 6 months or 12 months after the respective product is delivered, the dogs are challenged with L3 heartworm larvae. Of course, as noted above, this simple scenario does not necessarily hold now for the monthlies, since to achieve 100% efficacy, some of the products require three or more monthly treatments.

One obvious gap in heartworm prevention that can be concluded from all the work with different product is that stopping and starting treatment every six or eight months is not a good idea. It is clear that currently, multiple, i.e. year-round, treatments are better than treating dogs only part of the year. The reason is simply that a single treatment with many of the products is no longer preventing heartworm infections unless there are two, three, or perhaps more, consecutive monthly treatments after the dog is infected. Year-round heartworm prevention should, therefore, now be considered as required for protection in much of the United States and the rest of the world if the treatments are must be given six months after the end of the area's mosquito season.

Data presented in a recent report of oral moxidectin for the prevention of heartworm infection support this conclusion; the report includes an excellent summary of all the work performed with different products against the JYD isolate [50]. This work and summary make it clear that repeated preventive treatments are important for the control of resistant isolates. Oral moxidectin, present in weighed portions of ground ProHeart® tablets to provide an exact dose, was given to dogs infected with 50 L3 larvae of the JYD-34 isolate. Thirty days later, dogs were treated with oral moxidectin in capsules at 3, 6, 12, or 24 μg/kg. Treatment efficacies compared with untreated controls were, respectively, 19.0%, 25.5%, 33.3%, and 53.2%, indicating that the magnitude of the dose mattered (it should be remembered that at approval, a single treatment of 3 μg/kg was 100% efficacious). Another group of dogs in this study was given repeated treatments of moxidectin at 3 μg/kg 30, 60, and 90 days after infection, with efficacy of 44.4% compared with the untreated controls. In an additional study, when moxidectin at 24, 40, or 60 μg/kg was given to groups of dogs on each of days 0, 28, and 56 after infection, efficacy was 98.8%, 100%, and 100%, respectively, clearly demonstrating that the efficacy of multiple monthly treatments with the newly increased doses warrants the use of continuing repeated treatments of dogs to prevent heartworm disease.

The original NADA for HeartGard® 30 [24] included two studies in which dogs were infected with 50 larvae and treated with 3.3, 6, 12, 25, or 50 μg/kg ivermectin 45 days after infection. In one study, all dogs receiving ivermectin were protected, while in a second, a single dog in the 3.3 μg/kg group and a single dog in the 12.0 μg/kg group each had one worm [24]. When the same dosing schedule was followed at 60 days after infection, three dogs receiving 3.3 μg/kg had 2, 2, or 16 worms, and one dog receiving 12.0 μg/kg had 5 worms [24]. These data suggested that ivermectin as formulated did not work as well against older larvae as against 30-day-old larvae, but this was not a direct comparison as to what occurs with multiple treatments. The recent field studies with Simparica Trio™ and ProHeart® 12 compared with HeartGard® Plus, however, suggest that multiple treatments with ivermectin in the HeartGard® Plus dose may not be effective in preventing infection [22, 23]. This is supported by work with the resistant MP3 isolate that treatment of 30-day-old larvae in dogs with single preventive doses of HeartGard® Plus Chewables for Dogs (ivermectin with pyrantel), Interceptor® Flavor Tabs (milbemycin oxime), Revolution® (selamectin), or Advantage Multi® for Dogs (moxidectin and imidacloprid), generated efficacies of 95.6%, 95.4%, 95.5%, and 100%, respectively [45]. For the JYD-34 isolate, when HeartGard® Plus (ivermectin & pyrantel), Revolution® (selamectin), or Trifexis® Chewable for Dogs (milbemycin oxime and spinosad) was given to dogs infected with L3 larvae on days 30, 61, and 90 after inoculation, efficacies were only 29%, 28.8%, and 52.2%, respectively, whereas Advantage Multi for Dogs (moxidectin & imidacloprid) given only once 30 days after infection was 100% efficacious [51]. In a similar study with a single treatment (presented as a late-breaking poster in New Orleans at American Heartworm Society in 2016), one of eight dogs treated with Advantage Multi for Dogs harbored two adult worms [22]. In a study in which six dogs were each treated twice with ProHeart 6 at the recommended 180 day interval, after which six untreated dogs and the six treated dogs were infected with 50 heartworm L3, only one treated dog developed an adult heartworm infection, a single male worm, while the controls had 21–37 heartworms [52]. Thus, in the case of moxidectin containing products, i.e. the monthly orally administered Simparica Trio; ProHeart 6 and ProHeart 12; and the monthly topically applied Advantage Multi for Dogs, the above work strongly suggests that the products be used year-round so that blood levels are maintained to be able to also treat any worms that are in the dog that may have already lived through a single treatment.

This annual gap in therapy could also have long-term impacts on the genetics of heartworms, i.e. abbreviating prevention with some of the products may play a role in selecting for resistance in incoming heartworms. When therapy stops in the case of the monthly oral products, e.g. HeartGard, Interceptor, Trifexis®, Sentinel, etc., there is essentially no drug left in the dog a few days into the month after the last treatment. Thus, we can consider these dogs as not having any significant role in the selection of resistant or nonresistant isolates; all heartworms in a macrocyclic-lactone-free environment should have an equal chance to survive to adulthood unless the dogs are restarted on therapy. In cases in which dogs have received regular treatment with Revolution, ProHeart 6, Advantage Multi, Simparica Trio, and ProHeart 12, residual product remains for varying periods. Therefore, when heartworms enter these dogs, drug levels may be lower than those needed for prevention, but potentially not at levels that are too low to exert selective pressure on incoming infective L3 and L4 developing larvae. This is what occurred in the case of antibiotic prophylaxis for traveler's diarrhea, for which travelers would take low-dose antibiotics to prevent the development of diarrhea [53]. One example of this usage was doxycycline. In the late 1970s, doxycycline was prescribed as preventive therapy because of its high efficacy against enterotoxigenic *Escherichia coli* (ETEC); however, within five years, this use was discontinued due to the high proportions of tetracycline resistant ETEC; and higher doses of doxycycline were associated with gastric upset. And "Today, because of the high risk of antimicrobial resistance and potential side effects from using doxycycline, no guidelines recommended its use." In fact, after looking at the potential of using other antibiotics, such as trimethoprim/sulfamethoxazole, erythromycin, mecillinam, and fluoroquinolones, it is now recommended "that prophylactic antibiotics should be restricted to some high-risk travelers or short-term critical trips" [53]. This is the same argument relative to the generation of resistance that has been made for the reduction of the amount of antibiotics in the feed of cattle [54]. In the case of nematodes, it is well recognized that routine ingestion of subtherapeutic doses of products can result in increased selection of resistance traits in gastrointestinal nematodes, such as *Haemonchus contortus* [55]. Also, one may always wonder what the impact of the ProHeart 6 recall might have had on the development of resistance by *D. immitis*. Twelve million doses of ProHeart 6 had been administered to dogs by the time of the withdrawal, and when it was withdrawn in September of 2004, how many of the millions of dogs treated within the last six months were placed on another heartworm prevention product at the time of their scheduled retreatment at six months? If they were not placed on a preventive, many dogs that would have harbored subtherapeutic concentrations of macrocyclic lactone in their bodies.

Another gap that now might be fixed by the isoxazolines is the protection of other dogs against transmission of heartworms from a dog with circulating microfilariae undergoing treatment for an existing heartworm infection. There has been a movement to add doxycycline to the treatment of all dogs receiving melarsomine therapy not just to kill *Wolbachia* living within the lateral cords and other tissues of the adult heartworms to minimize pathogenic effects as the worm die from treatment, but also to prevent transmission between dogs by interfering with larval development in mosquitoes. Concerns with the use of doxycycline in heartworm therapy are twofold. First is the potential of unexpected consequences associated with utilizing doxycycline without monitoring what else might be occurring with this treatment. As mentioned above, there should be concerns about dogs developing tetracycline-resistant ETEC that would provide a severe zoonotic risk to owners and others dealing with canine fecal matter. Also, in Grand Canary, Spain, a study examined *Staphylococcus aureus* from dogs treated for heartworms with doxycycline and seven dogs without doxycycline therapy. The samples were collected by swab from the nose and perineum. After treatment, 73.3% (11 of 15 isolates) were resistant or showed intermediate susceptibility to erythromycin, doxycycline, or chloramphenicol, while only two of nine (22.2%) of isolates were resistant before treatment, with most of the resistant isolates being from dogs that had been treated with doxycycline for a month [56]. In this same study, 48 isolates of *Enterococcus* were obtained before and after treatment; before treatment, 8 of 25 isolates (32%) were resistant to one antibiotic, and after treatment, 15 of 23 (65%) of isolates were resistant to at least one antibiotic. Also, two outbreaks of severe respiratory disease in kenneled dogs in the United Kingdom were shown to be associated with doxycycline-resistant *Streptococcus equi* subspecies *zooepidemicus*. In an earlier similar outbreak in shelter dogs in California, the *S. equi* subspecies *zooepidemicus* isolate was found not susceptible to doxycycline, the commonly used antibiotic within the shelter [57].

Again, the two arguments to use the doxycycline are to minimize pathology due to antigen release from dead worms and to stop the development of heartworms in mosquitoes. Dillon et al. [58] found no difference in the lungs of cats that received repeated subcutaneous or intravenous ground heartworms from either an untreated dog or a dog treated for a month with doxycycline; no immunopathogenic differences were noted whether the helminth extracts were from worms recovered from treated or untreated dogs. Relative to mosquitoes and microfilariae, the easiest means of preventing transmission would be to fully examine the ability of the isoxazolines to kill mosquitoes; afoxolaner has high efficacy against *Aedes aegypti* feeding on a treated dog [59]. Other work with *Anopheles*, *Aedes*, and *Culex* has shown that fluralaner and afoxolaner both have high efficacies against mosquitoes, and in the studies in which drug was added to human blood, fluralaner appeared somewhat more potent than afoxolaner [60]. If the isoxazolines prove to be excellent at killing fed mosquitoes, they provide the opportunity, until such time that they may be utilized for crop protection or directly for mosquito control, to selectively target only the small percentage of mosquitos that have taken a bloodmeal from a dog, which should minimize the chance of resistance developing in mosquito populations. Then, the microfilariae in the treated dogs will slowly die, i.e. microfilariae without treatment can survive circulating in a dog for up to two years [61], but nonresistant microfilariae should die sooner if the dog remains on preventive therapy [22, 62, 63]. Also, if doxycycline is not used, the potential unintended consequences of resistance may be avoided because bacteria in the dogs and the *Wolbachia* within *D. immitis* worms would not be targets of antibiotic therapeutic pressure.

The ultimate negative outcome causing the potential largest gap would be the realization or acceptance that the macrocyclic lactones as a class no longer remain useful for heartworm prevention in dogs. There is a maximum as to how much macrocyclic lactone can go into a dog without causing signs, and the MDR1 (Multi-Drug Resistance Gene) allele present in many collies and some other breeds, if homozygous, can allow the development of severe neurologic signs and sometimes fatal outcomes associated with macrocyclic lactone treatment. Thus, what if the class fails totally as the preventive of choice. Currently, no published studies show that drugs in two new anthelminthic drug classes – the depsipeptides [64] and the amino-acetonitrile derivatives [65] – have been tried against *D. immitis*. However, since they have existed for many years, it is likely that they have been tested perhaps without effect, likely in the form of the molecules emodepside, a depsipepside (in Profender®, with praziquantel in a spot on for cats and a tablet for dogs in the EU), and as monepantel, the amino-acetonitrile derivative drench for sheep (in Zolvix™ and Zolvix™ Plus with added abamectin). It is hard to create a world-changing class of molecules, and the macrocyclic lactones, exemplified by ivermectin, were just such a molecule. Hence, the awarding of the Nobel Prize in Physiology and Medicine to Drs. Campbell and Ōmura "for their discoveries concerning a novel therapy against infections caused by roundworm parasites" (https://www.nobelprize.org/prizes/lists/all-nobel-laureates-in-physiology-or-medicine). We can hope for a workaround through combinations, formulations, and application methods, which is where we are now with Advantage Multi®, ProHeart® 12, and Simparica Trio™. At this point though, there are only so many molecules to choose from, and until they go off patent, they are not really shared between the product developers.

If the macrocyclic lactones do fail, there are still fallback options. The old daily, diethylcarbamazine, went into retirement seemingly before resistance had developed to it. It has the complication it can cause dangerous toxicity in microfilaremic dogs. Thus, it will require that dogs, once again, be checked for microfilariae before beginning treatment, i.e. antigen testing cannot be the sole diagnostic method used. It needs to be remembered that melarsomine dihydrochloride was approved in 1995 as Immiticide for the treatment of adult heartworm infections [29]. HeartGard-30 was approved for heartworm prevention on 2 March 1987. Until HeartGard came along, the only other option was a daily, Filaribits (diethylcarbamazine). The initial plan for melarsomine dihydrochloride had been to provide a product that would be given every four months to dogs [66–68], or three treatments annually where heartworm was being transmitted year round, two injections where heartworm transmission stopped in the winter (one treatment in spring and one in late summer), and a single treatment in late spring where winters were long. With the development and launch of macrocyclic lactone preventives and company mergers and product ownership shifts, the new objective was to approve it as an adulticide. So, if the macrocyclic lactones do fail, there are backups, far from optimal, but heartworms will still be able to be controlled.

There is one other way that a gap caused by the failure of the macrocyclic lactones to prevent heartworms could be closed. Work was proceeding apace at about the same rate once for both a heartworm vaccine and preventive pharmaceutical development, and the preventive pharmaceuticals won the race. Thus, the pressure was removed from the need for a vaccine, and it became very difficult to receive private or public funds for vaccine research. We have the heartworm genome to suggest vaccine targets [69], but there has not been a great deal published with respect to a vaccine against heartworm since the mid 1990s. If the vaccine was "feline" safe, a vaccine would be a logical preventive for cats, because the worms do not typically do very well in them. Also, a vaccine for dogs to prevent heartworm would be more than novel – it would be an absolute game-changer that would likely also have implications for human health. It seems a logical time to again consider this alternative preventive measure, and the 2018 Nobel Prize in Physiology and Medicine that went to Drs. James P. Allison and Tasuku Honjo "for their discovery of cancer therapy by inhibition of negative immune regulation" clearly demonstrates the huge strides in made immunology since the early 1990s (https://www.nobelprize.org/prizes/lists/all-nobel-laureates-in-physiology-or-medicine).

There remains a philosophical and academic gap. It was fairly obvious to many when: microfilariae were no longer being cleared by monthly treatments after adulticide therapy; dogs were noted by the Center for Veterinary Medicine (CVM) of the FDA to be undergoing failures occurring mainly in the Lower Mississippi River Valley [70]; milbemycin oxime was reported by its manufacturer Elanco Animal Health [41, 42] not to be protecting dogs against an isolate when administered at the recommended minimum dose of 50 mg/kg (which, again, is 10 times that originally needed for heartworm prevention); when work was presented on studies funded by the Novartis Animal Health demonstrating that resistance was real and extended across the macrocyclic class at the Annual meeting of the American Society of Veterinary Parasitologists in 2013, it was repeatedly stated by many that an overall ineffectiveness of available heartworm preventives had not been demonstrated, with most failures still considered to be due to poor owner compliance and weather e.g. heavy hurricane activity, and increased mosquito numbers and newly introduced mosquito species. Then came two defining moments, first when the European Union in its document approving NexGard SPECTRA® on 6 November 2014 wrote: "There is evidence of an increase of reports on 'product prevention failures' with respect to *D. immitis* ('lack of efficacy' in heartworm prevention) in the USA and it is now accepted that lack of efficacy related to resistance to macrocyclic lactones in *D. immitis* has developed in the USA." This unfortunate news was followed by the work presented in the two field efficacy studies on Simparica Trio™ [22] and ProHeart® 12 [23], which demonstrated 2 of 138 HeartGard Plus treated dog positive in the first study and 4 of 218 dogs positive in the second. Thus, there were six heartworm antigen-positive dogs (1.6% of the 356 HeartGard Plus control dogs) that were required to be fully compliant and monitored for compliance with their HeartGard® Plus treatments. Also, three of these dogs had circulating microfilariae that developed from infections in the area of the country that has been the focus of the potential resistance issue since 2005 [70]. It should be remembered that JYD-34 came from western Illinois, just across the border from Missouri. This gap was the resistance to the possibility of and the use of the word "resistance." It 2005, folks could have done a much better job of working together for intervention and mitigation, but there was no interest in building the needed consensus to work at combatting what has become the current reality. And hopefully, we are not going to find ourselves in a few years in the position where the only options are routine arsenical therapy every four months or going back to daily diethylcarbamazine.

## References


1 Kotani, T. and Powers, K.G. (1982). Developmental stages of *Dirofilaria immitis* in the dog. *Am. J. Vet. Res.* 43 (12): 2199–2206.

2 Courtney, C.H. and Zeng, Q.-Y. (1989). The structure of heartworm populations in dogs and cats in Florida. In: *Proceedings of the Heartworm Symposium '89* (ed. G.F. Otto), 1–6. Washington, DC: American Heartworm Society.

3 Dillon, R. (1986). Feline heartworm disease. In: *Proceedings of the Heartworm Symposium '86* (ed. G.F. Otto), 149–154. Washington, DC: American Heartworm Society.

4 Abraham, D. (1988). Biology of *Dirofilaria immitis*, Chapter 2. In: *Dirofilariasis* (ed. P.F.L. Boreham and R.B. Atwell), 30–46. Boca Raton, FL: CRC Press Inc.

5 Carastro, S., Dugan, S., and Paul, A. (1992). Intraocular dirofilariasis in dogs. *Compend. Cont. Educ. Pract. Vet.* 14 (2): 209–217.

6 Patton, C.S. and Garner, F.M. (1970). Cerebral infarction caused by heartworms (*Dirofilaria immitis*) in a dog. *J. Am. Vet. Med. Assoc.* 156 (5): 600–605.

**7** Grimes, J.A., Scott, K.D., and Edwards, J.F. (2016). Aberrant heartworm migration to the abdominal aorta and systemic arteriolitis in a dog presenting with vomiting and hemorrhagic diarrhea. *Can. Vet. J.* 57 (1): 76–79.

**8** Simón, F., Siles-Lucas, M., Morchón, R. et al. (2012). Human and animal dirofilariasis: the emergence of a zoonotic mosaic. *Clin. Microbiol. Rev.* 25 (3): 507–544.

**9** American Heartworm Society (2014). Summary of the Current Feline Guidelines for the Prevention, Diagnosis, and Management of Heartworm (*Dirofilaria immitis*) Infection in Cats.

**10** Lee, A.C. and Atkins, C.E. (2010). Understanding feline heartworm infection: disease, diagnosis, and treatment. *Top Companion Anim. Med.* 25 (4): 224–230.

**11** Litster, A., Atkins, C., and Atwell, R. (2008). Acute death in heartworm-infected cats: unraveling the puzzle. *Vet. Parasitol.* 158 (3): 196–203.

**12** Bowman, D. (2002). *Dirofilaria immitis* (Leidy, 1856) Railliet and Henry, 1911. In: *Feline Clinical Parasitology* (ed. D.D. Bowman, C.M. Hendrix, D.S. Lindsay and S.C. Barr), 331–334. Ames, IA: Iowa State University Press.

**13** American Heartworm Society (2020). Heartworm in Dogs.

**14** Bowman, D. and Atkins, C. (2009). Heartworm biology, treatment, and control. *Vet. Clin. North Am. Small Anim. Prac.* 39 (6): 1127–1158.

**15** Miller, L.M. and Gal, A. (2006). Cardiovascular system. In: *Pathologic Basis of Veterinary Disease*, 6e (ed. J.F. Zachary), 614. St. Louis, MO: Mosby Elsevier.

**16** Jackson, R.R., Seymour, W.G., Growney, P.J., and Otto, G.F. (1977). Surgical treatment of the caval syndrome of canine heartworm disease. *J. Am. Vet. Med. Assoc.* 171 (10): 1065–1069.

**17** Kitagawa, H., Sasaki, Y., Ishihara, K., and Kuwahara, Y. (1990). Development of artificial model of caval syndrome in canine heartworm disease. *Jpn. J. Vet. Sci.* 52 (5): 1029–1035.

**18** Kuwahara, Y., Kitagawa, H., Sasaki, Y., and Ishihara, K. (1991). Cardiopulmonary values in dogs with artificial model of caval syndrome in heartworm disease. *Jpn. J. Vet. Sci.* 53 (1): 59–64.

**19** Buoro, I.B.J. and Atwell, R.B. (1984). Development of a model of caval syndrome in dogs infected with *Dirofilaria immitis*. *Aust. Vet. J.* 61 (8): 267–268.

**20** Scott, D.W. (1979). Nodular skin disease associated with *Dirofilaria immitis* infection in the dog. *Cornell Vet.* 69 (3): 233–240.

**21** Bowman, D.D. (2012). Heartworms, macrocyclic lactones, and the specter of resistance to prevention in the United States. *Parasites Vectors* 5: 138.

**22** Kryda, K., Six, R.H., Walsh, K.F. et al. (2019). Laboratory and field studies to investigate the efficacy of a novel, orally administered combination product containing moxidectin, sarolaner and pyrantel for the prevention of heartworm disease (*Dirofilaria immitis*) in dogs. *Parasites Vectors* 12: 445.

**23** McTier, T.L., Kryda, K., Wachowski, M. et al. (2019). ProHeart® 12, a moxidectin extended-release injectable formulation for prevention of heartworm (*Dirofilaria immitis*) disease in dogs in the USA for 12 months. *Parasites Vectors* 12: 369.

**24** NADA 138-412 (1987). HEARTGARD-30® – original approval.

**25** NADA 140-915 (1990). INTERCEPTOR® – original approval.

**26** NADA 141-971 (1993). HEARTGARD-30® – Plus – original approval.

**27** NADA 141-084 (1997). Sentinel® Tablets – original approval.

**28** NADA 141-051 (1997). ProHeart® for Dogs – original approval.

**29** NADA 141-042 (1995). IMMITICIDE® (melarsomine dihydrochloride) Injectable.

**30** NADA 141-152 (2000). REVOLUTION™ (selamectin) – original approval.

**31** NADA 141-189 (2000). ProHeart® 6 (moxidectin) Sustained Release Injectable for Dogs.

**32** NADA 141-208 (2002). Advantage® DUO (imidacloprid/ivermectin).

**33** NADA 200-338 (2003). TRI-HEART Plus (ivermectin/pyrantel) Chewable Tablets.

**34** NADA 141-251 (2000). Advantage Multi® for Dogs – original approval.

**35** NADA 141-257 (2006). IVERHARTMAX Chewable Tablets (ivermectin/pyrantel pamoate/praziquantel).

**36** NADA 141-321 (2011). TRIFEXIS – original new animal drug application.

**37** NADA 141-333 (2011). Sentinel® Spectrum® (milbemycin oxime/lufenuron/praziquantel) – original new animal drug application.

**38** NADA 141-338 (2012). Interceptor Spectrum™ (milbemycin oxime/praziquantel).

**39** NADA 141-519 (2019). ProHeart 12® (moxidectin) Dogs – original new animal drug application.

**40** NADA 141-521 (2020). Simparica Trio™ (sarolaner, moxidectin, and pyrantel chewable tablets) Dogs.

**41** Snyder, D.E., Wisemann, S., Cruthers, L.R., and Slone, R.L. (2011). Ivermectin and milbemycin oxime in experimental adult heartworm (*Dirofilaria immitis*) infection of dogs. *J. Vet. Int. Med.* 25 (1): 61–64.

**42** Snyder, D.E., Wisemann, S., Bowman, D.D. et al. (2011). Assessment of the effectiveness of a combination product of spinosad and milbemycin oxime on the prophylaxis of canine heartworm infection. *Vet. Parasitol.* 180: 253–266.

**43** European Medicines Agency. (2014). Center for Medicinal Products for Veterinary Use (CVMP) Assessment Report for NEXGARD SPECTRA (EMEA/V/D/003842/000) International non-proprietary name: afoxolaner/milbemycin oxime, 5 pages.

**44** McCall, J.W., Hack, R., McCall, S.D. et al. (2002). Evaluation of repeated monthly dosing of selamectin against *Dirofilaria immitis* beginning three months after experimental inoculation of larvae in dogs. In: *Recent Advances in Heartworm Disease: Symposium '01* (ed. R.L. Seward), 141–148. Batavia, IL, USA: American Heartworm Society.

**45** Blagburn, B.L., Dillon, A.R., Arther, R.G. et al. (2011). Comparative efficacy of four commercially available heartworm preventive products against the MP3 laboratory strain of *Dirofilaria immitis*. *Vet. Parasitol.* 176: 189–194.

**46** Bowman, D.D., Grazette, A.R., Basel, C. et al. (2016). Protection of dogs against canine heartworm infection 28 days after four monthly treatments with Advantage Multi® for dogs. *Parasites Vectors* 9: 12.

**47** Cruthers, L.R., Arther, R.G., Basel, C.L. et al. (2008). New developments in parasite prevention. Bayer selected proceedings. North American Veterinary Conference (NAVC). *Vet. Forum* 25 (Suppl 3B): 15–20.

**48** Bowman, D.D., Ohmes, C.M., Hostetler, J.A. et al. (2017). Efficacy of 10% imidacloprid + 2.5% moxidectin topical solution (Advantage Multi® for Dogs) for the prevention of heartworm disease and infection all month long. *Parasites Vectors* 10 (Supplement 2): 478.

**49** Bourguinat, C., Lee, A.C.Y., Lizundia, R. et al. (2015). Macrocyclic lactone resistance in *Dirofilaria immitis*: failure of heartworm preventives and investigation of genetic markers for resistance. *Vet. Parasitol.* 210: 167–178.

**50** McTier, T.L., Six, R.H., Pullins, A. et al. (2019). Preventive efficacy of oral moxidectin at various doses and dosage regimens against macrocyclic lactone-resistant heartworm (*Dirofilaria immitis*) strains in dogs. *Parasites Vectors* 12: 444.

**51** Blagburn, B.L., Arther, R.G., Dillon, R.D. et al. (2016). Efficacy of four commercially available heartworm preventive products against the JYD-34 laboratory strain of *Dirofilaria immitis*. *Parasites Vectors* 9: 191.

**52** Bowman, D.D., McTier, T.L., Adams, E.L. et al. (2017). Evaluation of the efficacy of ProHeart® 6 (moxidectin) against a resistant isolate of *Dirofilaria immitis* (JYD-34) in dogs. *Parasites Vectors* 10 (Suppl 2): 502.

**53** Diptyanusa, A., Ngamprasertchai, T., and Plyaphanee, W. (2018). A review of antibiotic prophylaxis for traveler's diarrhea: past to present. *Trop. Dis.Travel Med. Vaccines* 4: 14.

**54** FDA Center for Veterinary Medicine (2018). Supporting antimicrobial stewardship in veterinary settings - goals for fiscal years 2019 – 2023, https://www.fda.gov/files/animal%20&%20veterinary/published/Supporting-Antimicrobial-Stewardship-in-Veterinary-Settings--Goals-for-Fiscal-Years-2019-2023.pdf.

**55** Dever, M.L. and Kahn, L.P. (2015). Decline in faecal worm egg couns in lambs suckling ewes treated with lipophilic anthelmintics; implications for hastening development of anthelmintic resistance. *Vet. Parasitol.* 209: 229–234.

**56** Tejedor-Junco, M.T., González-Martin, M., Bermeo-Garrido, E. et al. (2018). Doxycycline treatment for *Dirofilaria immitis* in dogs: impact on *Staphylococcus aureus* and *Enterococcus* antimicrobial resistance. *Vet. Res. Commun.* 42: 227–232.

**57** Pesavento, P.A., Hurley, K.F., Bannasch, M.J. et al. (2008). A clonal outbreak of acute fatal hemorrhagic pneumonia in intensively housed (shelter) dogs caused by *Streptococcus equi* subsp. *Zooepidemicus*. *Vet. Pathol.* 45: 51–53.

**58** Dillon, A.R., Tillson, D.M., Wooldridge, A. et al. (2014). Effects of intravenous and subcutaneous heartworm homogenate from doxycycline-treated and untreated donor dogs on bronchial reactivity and lung in cats. *Vet. Parasitol.* 206: 14–23.

**59** Liebenberg, J., Fourie, J., Lebon, W. et al. (2017). Assessment of the insecticidal activity of afoxolaner against *Aedes aegypti* in dogs treated with NexGard®. *Parasite* 24: 39.

**60** Miglianico, M., Eldering, M., Slater, H. et al. (2018). Repurposing isoxazoline veterinary drugs for control of vector-borne human diseases. *Proc. Natl. Acad. Sci. U.S.A.* 115 (29): e6920–e6926.

**61** Underwood, P.C. and Harwood, P.D. (1939). Survival and location of the microfilariae of *Dirofilaria immitis* in the dog. *J. Paraasitol.* 25 (1): 23–33.

**62** Bowman, D.D., Joahnson, R.C., Ulrich, M.E. et al. (1992). Effects of long-term administration of ivermectin and milbemycin oxime on circulating microfilariae and parasite antigenemia in dogs with patent heartworm infections. In: *Proceedings of the Heartworm Symposium '89* (ed. M.D. Soll), 151–158. Batavia, IL: American Heartworm Society.

**63** Bowman, D.D., Charles, S.D., Arther, R.G., and Settje, T. (2015). Laboratory evaluation of the efficacy of 10% imidacloprid + 2.5% moxidectin topical solution (Advantage® Multi, Advocate®) for the treatment of *Dirofilaria immitis* circulating microfilariae in dogs. *Parasitol. Res.* 114 (supplement 1): S165–S174.

**64** Harter, A. and von Samson-Himmelstjerna, G. (2002). Cyclooctadepsipeptides – a new class of anthelmintically active compounds. *Parasitol. Res.* 88: 481–488.

**65** Kaminsky, R., Ducray, P., Jung, M. et al. (2008). A new class of anthelmintics effective against drug-resistant nematodes. *Nature* 452: 06722.

**66** McCall, J.W., Dzimianski, M.T., McTier, T.L., Holmes, R.A., et al. (1990) Preliminary controlled experiments to prevent heartworm disease by seasonal IM injections of RM 340, in *Proceedings of the American Association of Veterinary Parasitologists 35th Annual Meeting. San Antonio, TX*; Abstract 68.

**67** Keister, D.M., Dzimianski, M.T., McTier, T.L. et al. (1992). Dose selection and confirmation of RM 340, a new filaricide for the treatment of dogs with immature and mature *Dirofilaria immitis*. In: *Proceedings of the Heartworm Symposium '92* (ed. M. Soll), 225–229. Batavia: American Heartworm Society.

**68** McCall, J.W., McTier, T.L., Dzimianski, M.T. et al. (1994). Clinical prophylactic activity of melarsomine dihydrochloride (RM 340) against *Dirofilaria immitis* in heartworm-naive beagles exposed to natural infection in three southeastern states. *Vet. Parasitol.* 55: 205–219.

**69** Godel, C., Kumar, S., Kousovoulos, G. et al. (2012). The genome of the heartworm, *Dirofilaria immitis*, reveals drug and vaccine targets. *FASEB J.* 26: 4650–4661.

**70** Hampshire, V.A. (2005). Evaluation of efficacy of heartworm preventive products at the FDA. *Vet. Parasitol.* 133: 191–195.

# 5

# Diagnosis and Assessment of Human Filarial Infections: Current Status and Challenges

*Charles D. Mackenzie*[1,*], *Ashley Souza*[1], *and Timothy G. Geary*[2,3]

[1] *Neglected Tropical Diseases Support Center, The Task Force for Global Health, 330 West Ponce de Leon Avenue, Decatur, GA 30030, USA*
[2] *McGill University, Institute of Parasitology, 21111 Lakeshore Road, Ste-Anne-de-Bellevue, H9X 3V9 QC, Canada*
[3] *Queen's University-Belfast, School of Biological Sciences, Microbes & Pathogen Biology, 19 Chlorine Gardens, Belfast BT9 5DL, Northern Ireland*

**Abstract**

The diagnosis of human filarial infections, despite important advances in recent years, remains in need of more practical and more informative improvements. Accurate diagnosis and assessment of these infections is vital for the medical management of individuals who become infected with a filarial parasite. However, it is also currently extremely important for the initiation, monitoring, and evaluation of the major elimination programs that are underway in endemic countries across the globe targeting the two most clinically significant human filarial diseases.

Identification and assessment of these infections have often been inhibited by clinically silent periods before pathognomonic presentations occur in an individual, thus placing emphasis on the need for increased specific and sensitive biomarkers as indicators of infection. In addition, valid and practical evaluation methods for monitoring large filariasis endemic populations are central to the road to success in global efforts to eliminate the transmission of onchocerciasis and eliminate lymphatic filariasis as a public health problem.

This chapter discusses aspects of diagnosis and assessment from a practical context, addresses both the needs and challenges that are faced in the development of functional diagnostic tools for filarial infections, and makes suggestions as to potential approaches for research in this area. This discussion is not intended to be a comprehensive review of all aspects of this wide and diverse subject; rather, it emphasizes the need to consider the biology of these parasites in developing new tests, the locations in which they are to be used, and sampling procedures that are acceptable and practical for the assessment of filariasis-endemic populations.

*Corresponding author.

*Human and Animal Filariases: Landscape, Challenges, and Control*, First Edition.
Edited by Ronald Kaminsky and Timothy G. Geary.

## 5.1 Introduction

Central to providing appropriate care for and prevention of human filariases is the ability to make the correct interpretation of a medical condition, i.e., the correct diagnosis and detection of the infection as the cause of disease. A misdiagnosis, or the use of and belief in poor detection systems, can lead to failure to succeed; possibly equally important is that failure to recognize that the information being used for diagnosis and assessment is in fact erroneous can have a major negative impact. These principles, and their inherent consequences, are especially relevant to the global effort in which national public health organizations are presently engaged to eliminate two major human filarial infections, onchocerciasis and lymphatic filariasis (LF), from all endemic areas of the world [1, 2].

The general purpose of this chapter is to discuss current diagnosis and assessment activities for the key filarial infections with regard to the examination of infected individuals as well as the all-important assessment of endemic populations. It is not our intention to extensively cover all areas of filarial diagnosis nor to provide an intensive historical account of the different approaches and techniques used or the science behind them, except where such information is relevant to the specific discussion.

The diagnosis and clinical assessment of filarial infections is challenged by the relatively complicated biology and pathogenesis of these infections. As parasites, filariae have the ability to invade their hosts and most often follow a slow course of development with relatively few sentinel clinical events that signal the presence of infection in most individuals. As with many nematodes that invade internal tissues of their hosts, the details of how filarial nematodes generally avoid immune and pathological responses remain a fascinating mystery [3].

There are major differences between diagnosing (identifying the presence of a parasite) and assessing (evaluation or estimation of the quantity or quality of infection) filarial infections in an individual compared to surveys of a large group of people (epidemiology). In terms of an individual, the *diagnosis* of the presence of a filarial infection is an important step that likely leads to therapy; *assessment*, on the other hand, is the continuing understanding of changes in the clinical condition over time related to the filarial infection and its treatment. There are indeed some similarities with the assessment of the status of an endemic population, but the latter requires a much more arithmetic approach – where statistics, probabilities, and the like are prominent [4]. However, the initial detection of the presence of infection or ongoing transmission that establishes endemicity in many cases uses the same procedures as diagnosing infection in an individual patient. In any discussion of diagnosis, the similarities and differences of these two situations with regard to diagnosis and assessment necessarily needs to be considered. In clinical diagnosis, depending on the supporting medical system – which clearly differs from country to country and from rural to urban areas of many countries – a range of diagnostic tools may be available for use as appropriate to the local situation. For epidemiological assessment, a single reliable test is needed, and the optimal diagnostic tool is a point-of-use test that is of low cost, easily transported (i.e., no cold chain necessary), readily available, and that provides accurate, specific, and reproducible results.

Scientific, biological, and practical issues currently challenge the interpretation of filarial infections, and it is important that these are understood and addressed if the lofty goals of elimination of LF as a public health problem [2], and elimination of transmission of onchocerciasis [1], are to be achieved. Although there are basic essential requirements for diagnostics, such as parasite species specificity, that apply to all levels and forms of testing needed, a major influencing fact is where these tests are to be employed; for field programs, they need to be simple, reliable, and robust. This requirement is important as these diseases commonly affect people living in rural areas, locations that often lack strong medical or laboratory support; this fact underscores the ongoing discussion regarding the importance of point-of-care (POC) testing (i.e., rapid diagnostic tests – RDTs) versus laboratory-based testing [5]. The correct diagnosis of the presence of filariasis is obviously important to an infected individual, especially if suffering from significant, and potentially permanent, clinical signs and symptoms. However, robust diagnostic procedures and techniques are also essential for the elimination of the important filarial infections in the ongoing global neglected tropical diseases (NTD) control programs. The definition of the presence of infection and the extent of ongoing transmission are the major indicators for chemotherapy-driven elimination programs for onchocerciasis and LF; the detection of infection in people (e.g., epidemiological levels) or in the vector (e.g., transmission levels) is central to success in these programs [1, 2].

## 5.2 Current Needs and Challenges Related to Different Locations and Situations

The types of diagnostic test or assessment procedures needed for filariases are very much driven by the status of the underlying science, and, as mentioned above, by the location where the test is to be used. Translation of new, improved serological approaches for diagnosis (commonly focused on the presence of a specific antigenic component of the target worm) to programmatic use in the field has been slow, especially for onchocerciasis and loiasis, although arguably less so for LF. The lack of robust POC serological tests has led to an emphasis on laboratory-based testing and a need to develop local laboratory capacity for testing for onchocerciasis [6]; LF diagnosis has benefitted from a robust POC lateral flow test strip that has been most useful and central to the success of that global program [7]. The combined difficulties of relatively poor-quality serological tests and the difficulty of using them in the field by program teams have hindered the mapping and assessment of both onchocerciasis and loiasis. These two diseases have special mapping needs: for onchocerciasis, it remains important to map hypoendemic areas that have never received chemotherapy in Mass Drug Administration (MDA) programs [8]. Loiasis, although not being targeted for elimination, is critically important in areas where it is co-endemic with the two infections targeted for elimination, onchocerciasis and LF [8–10]. MDA treatment of individuals carrying high levels of circulating *Loa*

*loa* microfilariae can induce serious adverse clinical events that can be fatal or lead to permanent central nervous system damage [11, 12]. Detecting these individuals and protecting them has been a major challenge to elimination programs in *L. loa* endemic areas.

Clinical signs and symptoms remain an important essential early component of diagnosis and assessment of filariasis, not only for infected individuals but also for endemic populations (Table 5.1). Signs and symptoms include the presence of highly pruritic papular eruptions and subdermal nodules (granulomas) for onchocerciasis, edematous limbs and local lymphadenopathy for LF, and passage of adult worms across the conjunctiva of the eye for loiasis. However, it should be noted that, although clinical signs often drive the need for definitive diagnosis in individual cases, these indicators of possible infection with a filarial worm have not been commonly used in programmatic epidemiological assessment of endemic areas. At least two exceptions to this situation were used early in the establishment of control programs. The first example is known as "onchocerciasis elimination mapping" (OEM), a procedure currently used in the global elimination program to determine the need for MDA in any area in an endemic country that has not previously received treatment [8]. The presence of onchocercal nodules and the presence of biting black-flies are decision-making observations that can activate further testing. The second example is the historical first recognition of the presence of a filarial infection in a country. This was often through recognition of the

**Table 5.1** Current approaches for diagnosis and assessment of human filariasis.

| Filarial infection | Clinical diagnosis in an individual | Community assessment | References |
|---|---|---|---|
| Onchocerciasis | • Travel history<br>• Nodule presence<br>• Onchodermatitis<br>• Ocular changes (Mf) | • Rapid Assessment nodules<br>• Antibody presence (Ov16, others<br>• Xenomonitoring (O150) | [13–16] |
| Lymphatic filariasis | • Early lymphadenopathy<br>• Fluid accumulation in the scrotum<br>• Ultrasound<br>• Exclusion of other diagnoses<br>• Serological testing | • Antibody surveys for Wb123, Bm13, etc.<br>• Xenomonitoring – not commonly used | [3, 17–19] |
| Loiasis | • Conjunctival adult worm<br>• Subcutaneous swelling<br>• Mf in blood smears | • Eye worm prevalence<br>• LoaScope surveys<br>• Post-ivermectin SAEs | [11, 20, 21] |
| Mansonellosis | • Blood sample for circulating Mf<br>• Molecular testing is available | • Not generally carried out but blood sampling has been used for field studies | [22, 23] |
| Dirofilariasis (zoonotic) | • Radiography | Not applicable | [24] |

pathognomonic clinical sign (e.g., nodules or severe skin disease) in an individual examined by a clinician, followed by an investigation of where the patient became infected, and the consequent definition of a new endemic zone, as happened in Guatemala, Ecuador, and Yemen [25–27].

Overshadowing the location for testing, or the format of the testing to be used, is the ever-present issue of cost, availability, and transport of the tests and necessary reagents. Filarial elimination programs almost exclusively target people who are financially poor, and, in addition, the clinical conditions associated with these infections are often not regarded by national health systems as being high priorities for local funding. Thus, there is a vital need to develop tests that are low cost, that do not involve costly production transport and delivery requirements, and have a long shelf life in tropical areas. The latter characteristic is essential because disease elimination programs in developing countries are often subjected to logistical delays for many reasons (e.g., heavy rains, civil disturbances, and inadequate transport equipment). Unused tests may accumulate in the field because of these delays and become unusable if they have a short shelf life. Experience in filarial control programs has shown this problem to be both common and a managerial challenge, with an unfortunate waste of both many POC tests and the anthelmintic drugs that provide the basis for control.

National elimination programs require that the diagnostic samples (e.g., dried blood samples [DBS] and vector insects) are tested in a timely manner and not stored for long periods of time. This is required both for quality assurance (QA) reasons and for optimal and timely use of the data the samples provide for programmatic decision-making. Important for maintaining high quality (optimal QA) of samples collected for laboratory analysis is following strict preservation and storage procedures, such as adequate desiccation and robust labeling. Timely processing of samples is essential for optimal program management and decision-making that often has very significant implications – for example, the costly implication of another annual round of MDA. Laboratory associated data drive the decisions and consequent advice provided by the supporting scientific committees to the programs and national health officials. Consequently, timely processing of samples that provide these data is essential. Although there have been valiant efforts to develop improved data management systems, for example, the ESPEN Portal [28], more efficient and usable data management systems need to be developed; these, in all likelihood, will utilize the powerful new mobile communication technology and aspects of artificial intelligence [29].

An important challenge, perhaps not yet of great focus for filarial control programs, but which will likely become increasingly important as countries approach elimination, is assessment of continuing suppression of transmission or disease levels after MDA treatment has stopped [30]. The optimal approach to this important "surveillance" step is under discussion at the global level. Countries that have sought to determine whether their often now disbanded national filarial control programs were correct in declaring success in elimination operations have taken different approaches depending on local circumstances [31]. However, more rigorous recommendations for post-elimination surveillance are needed for both targeted infections.

Lastly, while discussing challenges and needs for diagnosis and assessment of filariasis control programs, it is essential to address the issue of training and maintaining personnel. Filarial diseases and their accompanying global control programs are long-term efforts – onchocerciasis control efforts have been in place since 1974 – and consequently, there is a constant turnover of program staff, often with the loss of valuable experience and capability. It is important that training systems are in place so that newly appointed country personnel are given the appropriate skills to maintain these often long-drawn-out programs. National onchocerciasis elimination programs, for which basic entomological skills are very much needed, are now beginning to have trouble finding entomologists with the necessary expertise to fulfill the program's needs. Although much of the new testing technology involves modern approaches (e.g., molecular biology), there is still a need for the more basic skills of vector and parasite visual identification.

## 5.3 Filarial Infections: Current Approaches to Their Diagnosis

Filarial nematodes, as discussed elsewhere in this book [3], have adapted to generally live quietly within their hosts; however, consequential exceptions, as in most biological systems, do occur with severe clinical events in some of this group of infections. The primary human filariases are LF caused by *Wuchereria bancrofti*, *Brugia malayi*, and *Brugia timori*; onchocerciasis caused by *Onchocerca volvulus*; and loiasis caused by *L. loa*. Diagnosis of *Mansonella* sp. and human infections of *Dirofilaria immitis* (canine heart worm) typically occurs accidentally or as a result of clinical work up and will be mentioned only briefly in this chapter.

In humans, the filariases that cause the most obvious and well-described clinical changes are onchocerciasis (severe dermatitis and visual loss) and LF (lymphedema, dermal disfiguration, and hydrocele). Consequently, these two filariae are central to global efforts to reduce, control, and eliminate NTDs that cause suffering and disease in impoverished areas of the world [2]. Filariae as parasites are generally challenging to control, as they are typically robust and can live and produce larvae in their human hosts for relatively long periods of time.

Dermal onchocerciasis was arguably first described in West Africa during the British Naval efforts to block the Africa slavery trade in the 1870s, and the term "River Blindness" – the common name for this disease today – was first coined in what is now South Sudan in the late 1920s in relation to the severe eye disease found in individuals working and living along the Jur River in Bahr El Ghazal. The condition is still known today locally in Sudan as "Jur River Blindness." Egyptian hieroglyphs from 2000 BCE contain representations that are thought to depict lymphedema, the dramatic clinical manifestation of LF. Hydrocele development was first related to a filarial infection in the 1860s through the observation of microfilariae (mf) in hydrocele fluid isolated during surgery [32]. Elephantiasis was first associated with a filarial parasite in India in the 1870s, with adult worms first described by Bancroft in 1876 [33] and transmission by mosquitoes by Manson in 1877 [34].

Diagnosis of filarial parasites has historically relied on clinical criteria based on symptoms (hydrocele or swollen limbs in LF, and characteristic skin/eye pathology or the presence of subdermal nodules in onchocerciasis), usually followed by the detection of mf in blood (LF) or skin (onchocerciasis). Although definitive in trained hands, mf detection using microscopy requires invasive sampling, dedicated equipment, and specific technical ability, all of which can represent a significant operational challenge in many endemic settings. As is the case with other NTDs, research and development for more innovative diagnostic techniques and procedures for human filariasis is severely underfunded compared to those for pathogens that occur in wealthier countries. Recognizing this challenge, the World Health Organization (WHO) 2030 NTD Roadmap identified "diagnostics" as an area for priority action over the coming decade [35, 36].

### 5.3.1 Diagnosis of Onchocerciasis

Indication that an individual might be infected with *O. volvulus* most often comes from the appearance of a clinical sign or symptom, e.g., pruritic dermal changes, the appearance of a subcutaneous nodule, or less commonly the incidental finding of mf in the anterior chamber of the eye or associated with a snow-flake opacity in the cornea. The realization that onchocerciasis is endemic in a country often comes from incidental findings of clinicians who in the course of their general consultations encounter a patient who has signs and symptoms that could be due to *O. volvulus* infection. Through subsequent investigations (pathological analysis of a subcutaneous lump or of a skin biopsy), it is discovered to be onchocerciasis. A follow-up on the location where the patient may have been infected can lead to the discovery of a new endemic area – this happened in Guatemala, [25], Ecuador [26], and Yemen [27].

The presence of onchocercal nodules, as mentioned above, is definitive for the infection, and this indicator was, and still is, commonly used for diagnosis and assessment to identify geographic areas needing control of the disease at the population level. Onchocercal nodules are firm to hard swellings palpable in the dermis, usually 0.5–2 cm in diameter. They are most commonly found under the skin adjacent to the iliac crest, on the rib cage, or at the base of the spine. They can also be found on the bony prominences of the head, although this location is usually confined to children (Figure 5.1a). These nodules contain nests of adult *O. volvulus* (commonly four to six worms) lying in an immunologically active granuloma [3, 37]; the presence of the worm in a nodule confirms the diagnosis and distinguishes these from other subcutaneous lumps such as dermal cysts, cysticercal lesions, and the like. The presence of nodules has been used as a rapid approach for assessing the level of onchocerciasis in an endemic community - known as rapid epidemiologic assessment (REA), or rapid epidemiological mapping for onchocerciasis (REMO), in which estimation of the prevalence of onchocercal nodules in

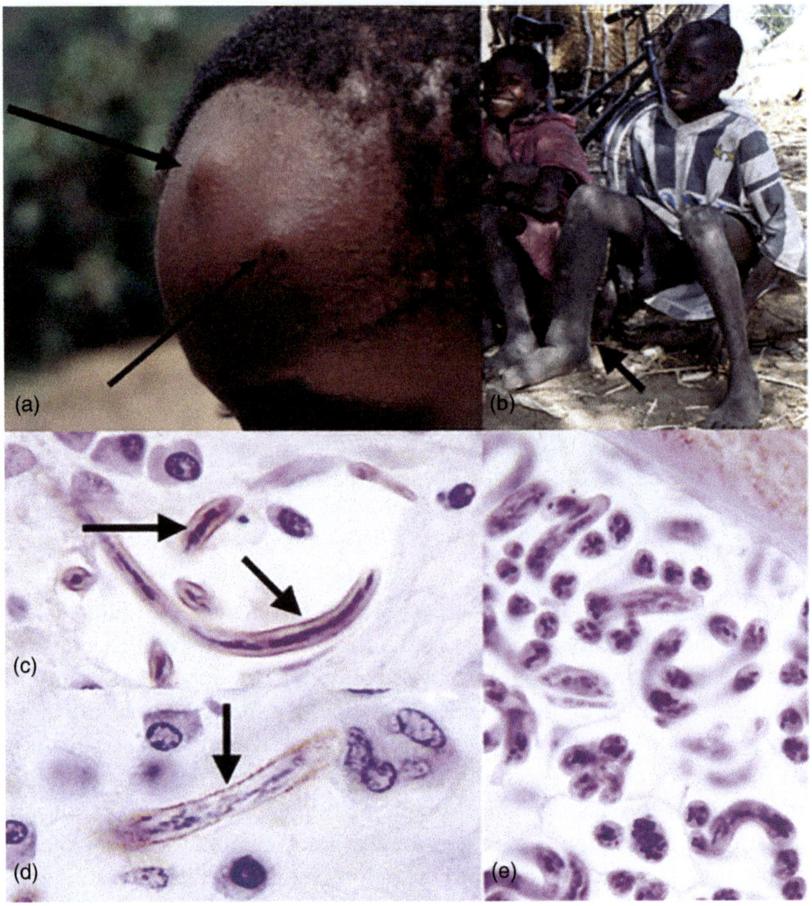


**Figure 5.1** Onchocerciasis. (a) Subcutaneous nodules on the head of a Cameroonian child. The presence of adult-containing granulomas is an indication of the presence of the infection in a geographic area. (b) Lymphedema of the leg in a Tanzanian child, the onset of which is an important diagnostic clinical indicator for LF, especially in children. (c and d) Immunocytochemical positivity for surface antigen (anti-Napsin A) on a newly emerged onchocercal microfilariae in a nodule. (e) This antigen is absent from the surface of intra-uterine microfilariae.

adult males of a community or district is made using simple palpation [38]. In the early days of the onchocerciasis control program, the aim was to intervene to reduce the prevalence of disease (as distinct from the current more stringent goal of eliminating transmission of infection) in any community with a nodule prevalence of 20% or above through MDA treatment with ivermectin [39]. The logic behind this metric is that the prevalence of palpable nodules in adult males is roughly equivalent to half the prevalence of positive skin snips in a population (20% nodules = 40–60% microfiladermia). This clinical tool was very useful for deciding to initiate an MDA program in an endemic area.

The standard method of identification of *O. volvulus* mf in the skin has been a skin snip, a small biopsy of the skin (dermis and upper dermis) approximately 4–5 mm diameter, commonly taken either using a corneoscleral punch (Holzer type) or, in the past, by elevating the skin with a pin and taking a skin shaving with a razor blade. The skin in the area of the iliac crest is commonly sampled, although other areas of the body have been used. This small skin sample is then incubated in normal saline for 4–24 hours (commonly in plastic micro-titer plates) and emerging mf counted using a dissecting microscope. This technique, although still used in some countries, has some important negative aspects that must be considered, aside from the obvious discomfort and pain experienced by the individual being examined. The procedure is invasive and can cause scarring at the skin snip site, but more importantly, especially in the era of HIV, the instruments used to perform the biopsy must be sterilized between samples. In addition, as some skin snips can be contaminated with blood, blood-circulating filariae may be caught in the sample, giving erroneous results regarding *O. volvulus*, although this is not common. Emphasizing that it is preferable to avoid using this invasive diagnostic technique when assessing endemic communities underscores the need to develop reliable serological assays in which the samples (finger-prick blood samples) can be taken more easily and safely.

It should also be emphasized that the skin snip biopsy technique does not detect mf that are not able to emerge from the skin; dead or significantly degenerating parasites will not emerge from the skin sample, and thus, a negative standard skin snip (i.e., no emerging mf) can give a false negative result if mf in the skin are dead or degenerating (e.g., patients with either reactive onchocercal dermatitis (Sowdah patients) or those following recent mf damaging chemotherapy) [3]. As an alternative, polymerase chain reaction (PCR)-based Nucleic Acid Amplification Techniques (NAAT) have been used to identify the presence of *O. volvulus* molecular fragments and intact parasites in skin biopsies [40]. However, this technique is not widely used in field programs. NAAT approaches are expensive and require specific expertise and facilities, and the concern remains that it is an invasive technique. If it is to be used, NAAT analyses of skin snips should perhaps be confined to essential research projects and the like.

Standard medical imaging techniques (ultrasound and X-ray) have been used in onchocerciasis, as have newer non-invasive techniques [41, 42]. However, it is important to highlight that the long-used visualization of mf in the anterior chamber and cornea using an ophthalmologic slit lamp is the quintessential non-invasive technique for actually seeing living nematodes *in situ* in humans. The sight of clumped, living mf wriggling in the anterior eye fluid of a heavily infected individual is undoubtedly horribly memorable to those who have experienced it. Fortunately, there are fewer locations in the world where this is still possible, and it is to be hoped that this phenomenon will not be present anywhere in the near future because of the success of MDA programs. Ultrasound procedures have been used to identify adult parasites in onchocercal nodules [41], but the granulomatous nature of these lesions tends to inhibit the ability to see parasite motion, and thus clear echogenic identification of the adult worm is often difficult. This contrasts with the situation with adult worms of LF, which are more easily seen with ultrasound as they lie in a fluid environment (the lymphatics) and are usually moving quite vigorously. Other techniques, such as optical coherence tomography (OCT), have been used [42] and are likely to provide useful information about the location of parasites and tissue changes in ocular and other tissues. Thermography, which has been used commonly to identify pathological lesions in veterinary medicine [43], has recently been suggested for the interpretation of clinical aspects of lymphedema in LF patients [44] and is under investigation to characterize onchocercal nodules. Thermographic findings with nodules are difficult to interpret: early results show considerable, presently undeciphered, variation in the heat maps between different nodules, especially those of different ages and pathological states (e.g., abscessed versus non-abscessed nodules).

With the skin snip technique facing significant criticism, serological approaches have become the primary approach for defining the presence of *O. volvulus* in an individual or a population. The only accepted serological test detects antibodies against *O. volvulus* [13]. These tests identify people who have been exposed to the parasite and not necessarily only those with an active infection; a test that could identify the latter would be more useful for programmatic management. For example, a test that identifies specific antigens that remain in the host for a short period of time once the parasite is eliminated or at least after the parasite's life cycle or viability has been interrupted, would be very valuable. The standard target used to determine antibody responses to *O. volvulus* has been Ov16, an antigen in the hypodermis, cuticle, and uterus of female worms; the IgG4 response to this antigen is regarded as a sensitive and specific indicator of prior exposure to onchocerciasis [45]. Much effort has been placed in developing both laboratory-based assays and POC tests, and a lateral flow assay has been developed for measuring Ov16 [14, 46]. Guidelines have also been developed for the use of this assay for the management of onchocerciasis programs, specifically the epidemiological assessment of infection levels in age groups [4]. For example, the confirmation of interruption of transmission in an area in children five to nine years of age requires testing for the presence of anti-Ov16 antibodies. In other situations, such as where the question is whether onchocerciasis exists in a new or long-treated community, adult residents – who are more likely to have been infected if the parasite is present – can be tested for Ov16 antibodies. A challenge that must be considered when using Ov16 as the only detection indicator is that not all people are likely able to mount antibody responses to the Ov16 antigen [47]; as high as 20% of a given population, because of the absence of the HLA type required for production of this specific antibody, may not mount effective antibody responses. A question that should be resolved is the duration for which Ov16 antibodies persist in the circulation after the individual is free of infection, sometimes predicted to be as long as 12 months. The answer to this question is essential when considering the use of Ov16 antibodies as an indicator of elimination of infection. It is likely that many factors influence the longevity of such an antibody response, e.g., the age of original infection, the extent of reinfection, etc. Thus, it is appropriate to be cautious when predicting how long Ov16 antibodies persist after infection is terminated.

More rigorous serological approaches that assess circulating antigens or other parasite components such as microRNAs [48] (see Section 5.4) have not yet been developed for onchocerciasis. The leading candidates for improving serological testing still involve assessing antibodies that recognize a wider range of onchocercal antigens defined from genetic mining, combining appropriate candidates into practical systems (e.g., lateral flow) that detect antibodies to two or three onchocercal antigens [15, 49]. The use of a combined antigen approach in the field could provide a more comprehensive picture of the endemicity of onchocerciasis. For example, incorporating additional antigens with Ov16 in rural Yemen provided a more extensive understanding of the endemicity of the infection [47].

Perhaps the most robust and crucial test now used to assess transmission of onchocerciasis in an endemic area focuses on demonstration of the presence, or more importantly the absence, of *O. volvulus* in the black-fly vector – the lack of parasites in the vector being evidence that transmission is not taking place. For many years, this assessment has been made using a NAAT (standard PCR) that detects an *Onchocerca*-specific DNA sequence in the heads of black flies collected in the field. Fly heads are tested separately from the body, as parasites present in the body of these *Simulium* vectors may not contribute to transmission, as they often do not migrate to the head for passage into a new host. This test is routinely carried out on batches of flies, i.e., is a pooled screening approach, and requires a trained technician and laboratory [16]. An essential component of this procedure is ensuring that all flies batched for PCR testing are the relevant species for transmission of onchocerciasis. Therefore, entomological experience in collection and identification of insects are essential. This expertise is becoming harder to source as trained entomologists are becoming scarce in endemic countries. Other challenges facing xenomonitoring include the supply chain of reagents needed for carrying out this laboratory-based test, and the difficulty in catching adequate numbers of flies for testing to enable programmatic decisions. In addition, there appears to be an overall reduction in the number of flies present in transmission areas in recent years, possibly because of global climate change; the use of new fly catching systems/traps and improved knowledge about breeding time periods and fly number peaks will likely be of assistance in resolving this challenge [50].

Updating the currently used PCR xenomonitoring test to the more efficient, reduced cost, qPCR technology has been recently achieved [51, 52]. Once this improved technology is fully field tested and essential training completed, it is expected that it will provide much-needed support for national programs at a time when the need for xenomonitoring is likely to increase significantly; more testing will be needed as endemic countries reach the point in their programs at which they need to assess the levels of parasite in vectors before stopping treatment or when they are undertaking essential surveillance activities.

### 5.3.2 Diagnosis of LF

Individuals are often first diagnosed as infected with LF parasites following the presentation of a typical clinical sign or symptom. For example, development of swelling of a limb or breast (Figure 5.1b), together with the enlargement of lymph nodes linked to the swelling, may lead to further investigation (e.g., examination of the blood and serological testing) that can definitively diagnose LF. As asymptomatic infections are common in this condition, the incidental detection of circulating mf during blood testing for other medical conditions can occur; understandably, this occurs most often in endemic countries or in patients who have visited an endemic country.

Ultrasound imaging of adult filarial worms lying in the lymphatics has become a standard indicator of lymphatic filarial infection; images of moving worms were termed the "filarial dance" sign [20, 53]. A common indicator to describe and assess improvements is the volume of an affected limb. Circumferences taken at different points on the limb were calculated by using a tape measure, and estimations of limb volume displacement have been used to describe the extent of swelling; however, the latter approach is generally regarded as rather cumbersome and is not often used. Lymphotech® is a digital system that provides numerical volume of limbs using a tablet computer program [54]; this methodology has been used to assess the volume of lymphedematous limbs in recent filarial chemotherapy studies [17].

The long-standing blood tests for LF are based on standardized thick blood smears that increase the chance of mf detection and provide more specific quantitation using Knott's concentration technique and a polycarbonate filter system [55]. These tests were often used in the field in the early days of global programs to eliminate LF, but the development of POC rapid tests, first the ICT (immunochromatographic test) system [7] and then the more recently used, and technically easier, Filariasis Test Strip (FTS) [18], replaced microscopy-based techniques. The ICT and FTS tests detect the presence of *W. bancrofti* antigen and produce much more robust data for control programs than has been experienced in onchocerciasis, for which the tests detect antibodies rather than antigens. One challenge with the original blood tests overcome by the new serological approaches is that, because of the diurnal periodicity of LF parasitemia, blood samples had to be drawn at night to detect mf, in general between 10 p.m. and 2 a.m., when the parasites are most likely to be present in the circulation. In contrast, the POC antigen detection tests can be used at any time of day.

Different antigens have been used for LF serology, developed initially from the different species that cause LF, produced by different techniques, and now used as recombinant antigens [56]. An ELISA test identifying IgG4 antibodies to the recombinant filarial antigen Bm14 has been explored, but this test suffers from cross-reactivity with other nematodes and thus has seen limited use in control programs, especially in countries or areas that are reaching very low prevalence [19, 57]. However, this test and another involving the antigen Bm33 provided data useful for programmatic monitoring [58]. The detection of antibodies against Wb123 has also been explored but is not yet seen as being a major tool for assisting elimination programs [59]. Xenomonitoring, which is vital to the assessment of onchocerciasis transmission, has not yet been a major requirement for the global elimination program for LF [21]. This is likely due, at least in part, to the wide diversity of mosquito species that transmit LF.

### 5.3.3 Diagnosis of Loiasis

Infection with *L. loa* was historically considered a minor condition, noted primarily for the development of "Calabar swellings" – subdermal swellings associated with degenerating adult worms. Adult worms migrate actively in subcutaneous tissues and when passing through the head can on occasion be seen passing across the ocular conjunctiva, unsurprisingly causing concern for the individual involved. This dramatic presentation led to this parasite being called "eye worm" [3]. Indeed, questioning a group of residents in a community as to whether they have experienced a worm moving across their conjunctivas can provide a reasonable estimation of the prevalence of this infection. This rapid assessment test is known as "RAPLOA." Typically, 80 people above the age of 15 are interviewed using clear photographs of *L. loa* worms in the eye to assist questioning. Early studies using the approach indicated that a prevalence of "eye worm" history of >40% identified communities that likely have residents with high circulating levels of *L. loa* mf.

High levels of circulating *L. loa* mf can predispose the carrier to very serious adverse clinical reactions after being treated with drugs that kill the mf, a critically important consideration for MDA programs that use ivermectin [11]. These reactions include severe central nervous system pathology resulting from vascular damage associated with parasite degeneration that blocks blood vessels and leads to irreversible damage to the surrounding tissues [12, 22]. Coma within 48 hours and death has occurred in many patients, and those that survive this initial phase commonly suffer from permanent neurological damage. Thus, it is essential not to treat individuals with high levels of circulating mf, and extensive studies have been carried out to develop a diagnostic procedure to identify those at risk [23]. In general, it is regarded that a circulating mf load of <30 000 mf/ml does not place the individual at risk of these potentially lethal adverse reactions, and a POC system known as "LoaScope" has been developed to assess parasite levels in communities [60]. This more informative approach has now replaced RAPLOA in most *L. loa* endemic areas, although RAPLOA is useful if there is a question in new areas concerning the possible presence of loiasis. Recently, a rapid test that detects the presence of *Loa* antibodies in an individual has been developed as a research tool [24] and is being considered for use in the field for filarial control and elimination programs.

It should be noted that loiasis is not a disease for which a major elimination program is in place; however, it is of significance where it is co-endemic with onchocerciasis and/or LF, both of which are under elimination programs that involve the distribution of microfilaricidal drugs, and thus residents of these areas are at risk of developing the loiasis-associated serious reactions.

### 5.3.4 Diagnosis of Mansonellosis

Three species of the filarial parasite genus *Mansonella* can infect humans [3] and are thought to cause mild clinical symptoms, with many infections being asymptomatic.

Diagnosis is usually established using regular blood smear examination or using Knott's concentration technique or membrane filtration, although one of the species (*Mansonella streptocerca*) is detected by a skin-snip technique similar to that formerly used for the diagnosis of onchocerciasis [61]. Molecular testing (NAAT) is generally used if there is a need to distinguish the parasites from other filariae [62], although this approach is not usually available in an endemic setting and is typically confined to research programs; however, a LAMP assay has been developed for field settings [63]. Serology is used but has limited value because of poor specificity (cross-reactivity with other filarial nematodes).

### 5.3.5 Diagnosis of Zoonotic Filariasis

Humans can become infected with the very common canine parasite *D. immitis*; although in almost all cases, the infection does not mature and remain as a third-stage larva or at most only develops to an early fourth-stage parasite. The most common presentation in humans is the presence of a sclerosing lesion in the lungs, which occurs when the parasite migrates to and becomes trapped in pulmonary tissues. The most common diagnosis is made when these lesions are detected on X-ray as a radiographically dense "coin" lesion [64]. As such, findings are often suspected to be early pulmonary neoplasia and are commonly surgically removed; subsequent pathological investigation then determines that these are due to the parasite and are not neoplastic. Other diagnostic approaches to investigate radiographic findings, such as filarial serology, are not commonly used.

## 5.4 New Approaches and Research Needed

Progress toward achieving the challenging elimination goal of onchocerciasis global control programs would be greatly enhanced by substantial improvement in our ability to assess parasite epidemiology in endemic populations through improved serological, or other liquid biopsy, tests. Arguably, onchocerciasis, although the first of the NTDs to be focused on for global control, is one of the hardest diseases for which to carry out the necessary steps to reduce its transmission; it has long been burdened by diagnostic difficulties in assessing endemic infections.

Without wishing any disrespect to scientists who have dedicated their research to this group of diseases, it is important to say that, overall, the research community's efforts, regarding both basic and applied aspects, have been poor with regard to filarial infections, certainly in comparison to efforts for other globally prevalent diseases and infections. Although there have been valiant efforts to include endemic country personnel in research on filarial infections, efforts that are very appropriate and should continue to be encouraged, more collaboration between countries with highly developed research communities and researchers in endemic countries is needed. The latter are especially important with regard to programmatic questions, for which their expertise and understanding is central to success. However, the technical and intellectual power present in the global scientific community has not been focused adequately on the needs of filarial infections and their control nor on most other parasitic diseases. There are, as to be expected, exceptions to this broad statement, but translation of research findings from concept to field implementation is frequently very slow or non-existent. There are different reasons for this overall situation: lack of flexible funding, a general lack of clarity of the needs, absence of good lateral thinking, and the like.

However, on a more positive note, the introduction of a few new approaches is slowly happening across different areas of filariasis control and elimination programs. Among these are the identification of new antigens for use in serological detection of onchocerciasis [15], the inclusion of mental health in the care for patients suffering from the worst clinical presentations of filariasis (filaria-induced dermatitis and onchocercal blindness) [65], and the use of newer imaging techniques to assess internal changes and presence of filarial parasites. However, arguably, the area that most needs more focused research and improvement is developing a field-usable rapid test for the diagnosis and assessment of onchocerciasis in endemic populations; improvement in this area would have a major impact on global control programs. In addition to innovative serological approaches for diagnosis, recent efforts have been made to discover non-protein biomarkers that could be exploited for this purpose. In particular, parasite-derived microRNAs (miRNAs) released into the circulation of infected individuals have been suggested as biomarkers of potential value for diagnosis of human filariases [66–68]. Stage-independent and stage-specific differences in the profile of secreted filarial miRNAs could provide a basis for identifying individuals harboring adult versus mf stages (or both) [69]. However, techniques for detecting and measuring miRNAs typically rely on PCR and so cannot be considered "field friendly." Efforts to develop parasite-derived miRNAs as biomarkers for onchocerciasis have not been rewarding [70, 71], and it is unlikely that much additional research will be devoted to their diagnostic utility.

Metabolites have also been investigated as potential biomarkers for diagnosis of filariases. Initial research identified a set of 14 potential biomarkers in plasma obtained from African onchocerciasis patients, although these were not consistently found in samples from Guatemalan patients [72]. Further work on plasma metabolomics identified increased concentrations of inosine and hypoxanthine (among others) as potential biomarkers of onchocerciasis [73, 74]. Obtaining plasma samples for metabolite analysis is still an invasive procedure, and efforts have been made for identifying onchocerciasis-associated metabolites in urine samples. A host-derived metabolite of the parasite neurotransmitter tyramine, *N*-acetyltyramine-*O*-glucuronide (NATOG), was first identified as a potential biomarker [75, 76], leading to the creation of a lateral immunoassay for detection of this molecule in urine [77]. Further work identified cinnamoylglycine as a potential urinary biomarker for *O. volvulus* infection [78]. However, additional studies have cast doubt on the general utility of either of these metabolites for diagnosis of onchocerciasis as part of control and elimination programs [74, 78–80], and more work is necessary to bring metabolomics into the diagnostic arena for human filariases, despite the inherent appeal of a urine-based platform.

It should also be mentioned that a common research objective in recent years has been the discovery of new chemotherapy agents, more specifically a macrofilaricide. This would undoubtedly be incredibly useful, as it might significantly shorten the time that chemotherapeutic treatments would need to be distributed to achieve elimination. In the meantime, taking lessons from veterinary medicine, investigating the possibility of using the more active microfilaricidal agent moxidectin [81] is likely to be another important scientific advance. This effective and safe anthelmintic has partially replaced ivermectin in the treatment of many animal parasitic infections.

One way to think about new diagnostic targets for filarial parasites is to approach this more from a biological point of view, in contrast to the more common genome screening and whole worm proteomics approaches (Figure 5.2). One of the most challenging areas of understanding the pathobiology of filariae is understanding the degeneration and death of these nematodes: how does one define that a worm is irreversibly degenerating? What are the specific indicators of this irreversibility? Do damaged worms have the ability to recover? How does one define an adult filarial worm as being "dead"? Answers to these questions are relevant not only for pathological assessment of tissues containing filarial parasites [3] but also relate to establishing practical indicators for assessing the status of an infection. For example, do irreversibly degenerating filariae release molecules that can be

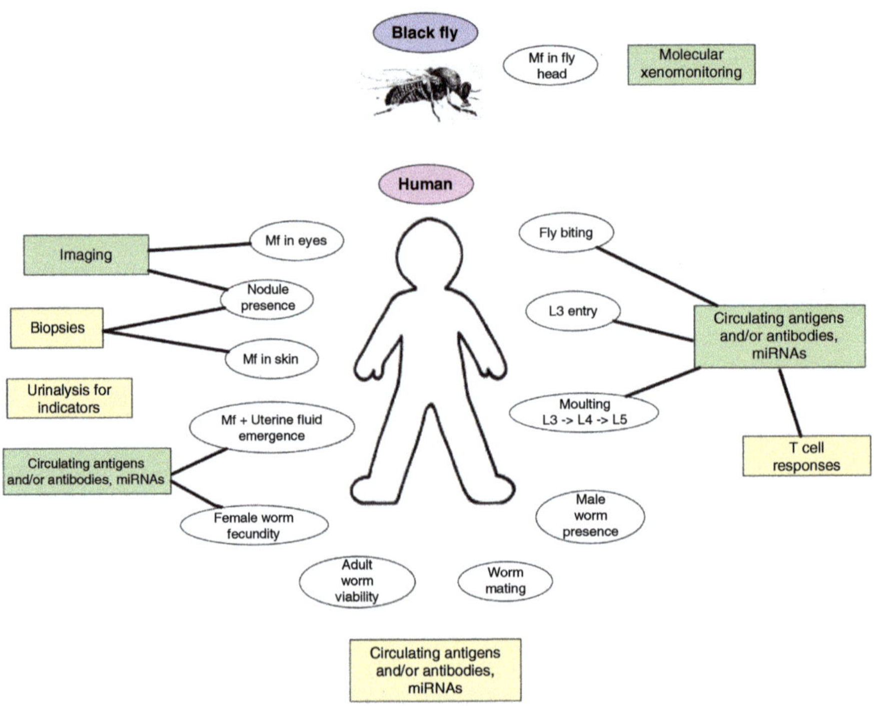


**Figure 5.2** Sites in the filarial life cycle where unique molecules might be present and serve as targets for new indicators for diagnosis and interpretation of the cycle status.

detected in blood, or in saliva, or even perhaps in conjunctival fluid in ocular onchocerciasis? Parallel with a need to improve assessment of the viability of parasites in our search for new assessment tools for filariasis is the development of more field-usable clinical assessment tools [49]. Assessing clinical improvements in the treatment of filarial diseases in endemic communities has not been a major goal, despite the fundamental aim of global control programs to reduce disease burdens. Disease assessment has always been largely the prerogative of trained clinicians; however, many individuals in public health systems that oversee filarial endemic areas – village health workers, for example – who, given clear instructions and guidelines, could quickly assess treated communities for changes in filarial disease prevalence. This is an advance that would in all likelihood greatly benefit these important health programs: demonstration of clear clinical improvement at a community level is always a strong advocacy message.

Filarial life cycles are complicated, and the parasite changes its form a number of times during its maturation and spread. Theoretically, as it changes from one form or stage to another, it is likely that different target components (proteins, antigens, etc.) are exposed on and in the parasite. Stage-specific antigenicity of the surface of nematodes has long been described [82], and characterizing newly exposed proteins or antigens is likely to reveal targets for new tests. An example of a newly presented antigen on *O. volvulus* mf is shown in Figure 5.1c–e; an epitope cross-reacting with Napsin A (Nap-A), a functional aspartic proteinase [83], is present on the mf surface as they leave the uterus and begin to move through the surrounding tissues away from the adult worm. Such newly revealed biochemical changes appearing as the worm matures could be important new indicator molecules that provide information on the status of infection. It is important to focus on indicators that reveal the presence of a reproducing worm, and arguably, given that the release of mf and associated uterine components induce significant host responses [84], the immunobiology of the uterus of filariae and its products are a fertile area for research. Being able to identify active production of new parasites through assessment of specific serological tests is of vital importance in assessing the status of an individual's infection. The continuing absence of reproduction of the parasite in an infected person, or indeed a community, would clearly indicate the success of chemotherapy, a status that is currently hard to assess.

The search for other indicators unique to different biological stages of the parasites is not a new concept. Using more modern scientific approaches, and keeping the specific biology of the worms in consideration, is likely to increase the probability of finding new ways to improve diagnosis and assessment. Filariae are complex organisms and the presence of unique molecules, which might be in relatively small relative quantities (such as a component of the anatomically thin surface of the worm), may be "masked" by more abundant, but less relevant, distracting molecules. Taking an approach that focuses on specific stages or specific organs within these complex worms may lead to new markers that relate to biological events that are practical and useful for assessment. Some possible biological targets are summarized in Figure 5.2.

Worm secretions are often used as antigens and undoubtedly are important, at least in the development of pathology and tissue changes associated with the presence of the parasite [84]. These products are important targets of investigation for diagnostic antigens; however, it is likely that there are practical differences between the products that a living healthy worm produces and those released from a degenerating parasite; this is indicated by the different tissue reactions seen surrounding living and degenerating filariae. It is therefore important to understand the differences in products, in antigens, and the biochemistry of a healthy living adult filarial worm compared with a degenerating or dying worm. As mentioned above, the definition of what marks the "death" of worm has never been clearly articulated [85].

Improved laboratory or POC testing needs to be linked to improved application of such tests in endemic, or potentially endemic, populations to ensure statistical validity; this includes both testing of humans (epidemiological testing) and testing of vectors (xenomonitoring). Robust community sampling protocols and careful recalculation of test cut-off points that define filarial control program success, i.e., breaking of transmission for onchocerciasis, and absence of a public health problem for LF, are urgently needed. The use of modern surveillance systems – spatial systems such as the Reveal technology and other similar technologies [86, 87] – will greatly assist, through improved micro-planning, both research and program implementation. It should be noted that supply chain issues are common challenges faced by national programs in their planning and implementation; interweaving up-to-date local geographic and environmental information (such as floods, population migration, etc.) with the logistics of distribution of drugs and assessment tools is essential for these programs. Geographic information, commonly provided by satellite imaging, is likely to be most useful for onchocerciasis and possibly for improved understanding of the distribution of loiasis; community prevalence of these two infections are closely related to vector presence. LF, with its diverse range of vector mosquitoes, may benefit less from geographic approaches, except for assisting micro-planning where it would be most useful.

Predictive modeling of global programs, an approach commonly used by international agencies that support global disease control and elimination programs for advocacy and planning efforts, relies on valid and relevant data. The current weaknesses in our testing protocols for filariasis directly affect predictions made by these models and indirectly affect the fiscal and moral global support for the programs. Improving reliability and accuracy of tests used to assess and manage filariasis control programs will assist in gaining wider support for these important health initiatives. The WHO and its various advisory committees, such as the Onchocerciasis Technical Sub-Committee (OTS) and the Diagnostics Technical Advisory Group (DTAG) [51, 88], have important roles to play in supporting, guiding, and approving new diagnostic and assessment tools. Country programs take advice from WHO seriously, and validation of success in national control and elimination programs is ultimately given by WHO. Notwithstanding this, new and innovative approaches to diagnosis are likely to come from the academic and research communities.

It is important to re-emphasize that research into NTDs, and especially filariases, is usually carried out by relatively small and dedicated disease-specific communities. It would be prudent to more actively seek assistance from the wider research community through partnerships between filarial experts and experts in other areas of science. Increased incentives are needed for the broader scientific community to focus on these often-forgotten diseases; this goal would be well served by more open discussion in the general scientific literature about the challenges and needs of filariasis research and control, as this discussion is typically confined to specialist groups, and rarely are these diseases and their scientific needs discussed widely.

## 5.5 General Comments

As one looks at the status of diagnosis and assessment of filarial infections (Table 5.2), it can be seen that, although certain areas of the three major infections have at least some adequate and practically useful testing systems, other aspects are still in great need of improvement. This is not to discount the great strides achieved in the overall reduction of the prevalence and incidence of LF and onchocerciasis across the world in the past 20 years or so [3]. In addition, there has been a most

**Table 5.2** Areas of research for improving diagnosis of human filarial infections.

| General area | Type test or focus area |
|---|---|
| Parasite biology | Develop stage/event-specific indicators |
| | Definition of the viability status of adult filarial worms. Indicators of degeneration and death. |
| | Understand better the biochemical plasticity of filarial worms – plasma metabolomics, etc. |
| Clinical interpretation | Improve non-invasive methodologies for locating and defining internal parasites |
| | Develop field-usable tools for messaging improvement in clinical disease |
| | Monitor and present clinical improvements as a result of the global filarial programs |
| Community assessment | Develop statistically sound, rapid community assessment tools |
| | Standardize the indicator and assessment tests and tools across the endemic areas |
| | Utilize newer technologies such as satellite mapping, artificial intelligence, and cloud-based data systems |
| Indicator systems | Focus of POC testing with supportive quality assurance |
| | Widen search of relevant indicators to unique, biologically relevant, molecules and small molecules (antigens, microRNAs, etc.) |

welcome and significant reduction in severity of two of the diseases these parasites induce [2]. However, as elimination programs achieve lower and lower levels of infection, the need for more robust tests becomes more important, and the need for effective sampling procedures used in these endemic communities becomes more vital to ultimate success.

Individual diagnosis of infection will in all likelihood continue to be influenced by the facilities available in the location in which the individual is found. For onchocerciasis, individual cases outside the known endemic countries (e.g., expatriate cases) are most likely to be recognized through dermatological presentations that can be easily diagnosed through history, clinical signs and symptoms together with laboratory testing. Previously unexposed individuals who become infected following visits to onchocerciasis-endemic areas often present with acute dermatological changes that respond well to anthelmintic treatment. A more difficult situation exists for individuals living in endemic areas who suffer from the severe effects of their infection, including intolerable pruritus, skin deformation, and purported epilepsy with onchocerciasis, as well as lymphedema, filarial dermatitis, and hydrocele with LF. The challenge for these patients is that the health systems in many endemic countries are not equipped to diagnose and provide care for these conditions. A bright light on this challenge, however, does exist through the patient care arm of the Global Programme to Eliminate Filariasis (GPELF), in which affected patients receive basic care for lymphedema and surgery for hydrocele conditions [89]. Although at present many individuals still need to be identified and cared for, the current WHO support specifically for LF-affected people is most welcome. As mentioned above, unfortunately, this is not the case for those still suffering from severe clinical onchocerciasis in most endemic countries; there are notable exceptions – e.g., Yemen's "sowdah" treatment program [90] and the anti-epilepsy drug treatment initiative in South Sudan [91].

The essential diagnosis-associated step that will herald ultimate success in the elimination of LF or onchocerciasis from an endemic area of a country is the judicious use of post-treatment and post-validation surveillance. This vital step requires the use of the most appropriate and reliable test, and that the testing is formally planned and carried out. Countries that have achieved validation of success have seen variable use of the tests and the manner in which post-treatment (post validation) surveillance is carried out. These issues have arisen in part due to the disbandment of the original MDA programs and their teams after validation success has been reached. In addition, there is significant cost and organizational component needed for these intermittent and often politically low-value assessments.

In the onchocerciasis world specifically (perhaps less so with LF), there is clearly a need to improve the quality and efficient collection of field-derived data, as discussed in this chapter. A range of changes could assist in achieving this improvement; however, there are differing opinions in the NTD community as to how to fulfill these needs. The ongoing discussion regarding the enhancement of laboratory support for programs [6] will certainly continue. However, a strong laboratory component for field filarial programs is most likely needed, even if the tests themselves become more and more POC tools rather than laboratory based. POC tests will always need to be validated and be monitored as part of a quality assurance system run and monitored by well-managed, active local laboratories.

## Abbreviations

| | |
|---|---|
| ELISA | enzyme-linked immunosorbent assay |
| ESPEN | expanded special project for elimination of neglected tropical diseases |
| FTS | filarial test strips |
| GPELF | global program for the elimination of lymphatic filariasis |
| HIV | human immunodeficiency virus |
| ICT | immunochromatographic test |
| LAMP | loop-mediated Isothermal Amplification |
| LF | lymphatic filariasis |
| MDA | mass drug administration |
| miRNAs | micro-ribonucleic acids |
| NAAT | nucleic acid amplification test |
| NTD | neglected tropical disease |
| OEM | onchocerciasis elimination mapping |
| OTS | onchocerciasis technical sub-committee |
| PCR | polymerase chain reaction |
| POC | point of care |
| POU | point of use |
| QA | quality assessment |
| qPCR | quantitative polymerase chain reaction |
| RAPLOA | rapid assessment procedure for loiasis |
| REA | rapid epidemiological assessment |
| REMO | rapid epidemiological mapping of onchocerciasis |
| RDT | rapid diagnostic test |
| WHO | World Health Organization |

## References

1 Mackenzie, C.D., Homeida, M.M., Hopkins, A. et al. (2012). Elimination of onchocerciasis from Africa: possible? *Trends Parasitol.* 28: 16–22. https://doi.org/10.1016/j.pt.2011.10.003.

2 World Health Organization (2021). Global programme to eliminate lymphatic filariasis: progress report, 2020. *Wkly Epidemiol. Rec.* 41 (96): 497–508. http://www.who.int/wer.

3 Mackenzie, C.D. (2022). Human filarial Infections: reflections on the current understanding of their importance, pathobiology, and management. In: *Human and Animal Filariases* (eds. R. Kaminsky and T.G. Geary), Chapter 3. Weinheim, Germany: Wiley-VCH.

4 World Health Organization. Onchocerciasis (2018). Guidelines for stopping mass drug administration and verifying elimination of human onchocerciasis. pp. 1–36 https://www.who.int/publications/i/item/9789241510011.

5 Vashist, S.K. (2017). Point-of-care diagnostics: recent advances and trends. *Biosensors* 7 (4): 62. https://doi.org/10.3390/bios7040062.

6 Shott, J., Ducker, C., Unnasch, T. et al. (2018) Establishing quality assured (QA) laboratory support for onchocerciasis elimination in Africa. *Int. Health*, 10 (suppl_1), i33–i39. https://doi.org/10.1093/inthealth/ihx059.

7 Weil, G.J., Lammie, P.J., and Weiss, N. (1997). The ICT filariasis test: a rapid-format antigen test for diagnosis of bancroftian filariasis. *Parasitol. Today* 13 (10): 401–404. https://doi.org/10.1016/s0169-4758(97)01130-7.

8 Hamill, L.C., Trotignon, G., Mackenzie, C.D. et al. (2022). Navigating the way to onchocerciasis elimination: the feasibility and affordability of onchocerciasis elimination mapping. *International Health* 14 (Supplement_1): i17–i23. https://doi.org/10.1093/inthealth/ihab083.

9 World Health Organization (2019). Report of the Third Meeting of. the WHO Onchocerciasis Technical Advisory Subgroup Geneva, 26–28 February 2019, pp 1–39. https://www.who.int/publications/i/item/9789240006638.

10 Kelly-Hope, L.A., Cano, J., Stanton, M.C. et al. (2014) Innovative tools for assessing risks for adverse events in areas of overlapping *Loa loa* and other filarial distributions: the application of micro-stratification mapping. *Parasites Vectors*, 7, 307 (2014). https://doi.org/10.1186/1756-3305-7-307.

11 Chippaux, J.P., Boussinesq, M., Gardon, J. et al. (1996) Severe adverse reaction risks during mass treatment with ivermectin in loiasis-endemic areas. *Parasitol. Today*, 12(11), 448–450. https://doi.org/10.1016/0169-4758(96)40006-0.

12 Wanji, S., Eyong, E.J., Tendongfor, N. et al (2017). Ivermectin treatment of *Loa loa* hyper-microfilaraemic baboons (*Papio anubis*): Assessment of microfilarial load reduction, haematological and biochemical parameters and histopathological changes following treatment. *PLoS Negl.Trop. Dis.*, 11, e0005576. https://doi.org/10.1371/journal.pntd.0005576.

13 Lobos, E., Weiss, N., Karam, M. et al. (1991). An immunogenic *Onchocerca volvulus* antigen: a specific and early marker of infection. *Science* 251: 1603–1605. pmid:2011741.

14 Golden, A., Stee, L C., Yokobe, L. et al. (2013) Extended result reading window in lateral flow tests detecting exposure to *Onchocerca volvulus*: a new technology to improve epidemiological surveillance tools. *PLoS One*, 8(7), e69231. https://doi.org/10.1371/journal.pone.0069231.

15 Bennuru, S., Oduro-Boateng, G., Osigwe, C. et al. (2020) Integrating multiple biomarkers to increase sensitivity for the detection of *Onchocerca volvulus* infection. *J. Infect. Dis.*, 221(11), 1805–1815. https://doi.org/10.1093/infdis/jiz307.

16 Pryce, J., Unnasch, T.R., Reimer, L.J. (2021) Evaluating the diagnostic test accuracy of molecular xenomonitoring methods for characterizing the community burden of onchocerciasis. *PLoS Negl.Trop. Dis.*, 15(10), e0009812. https://doi.org/10.1371/journal.pntd.0009812.

**17** Horton, J., Klarmann-Schulz, U., Stephens, M. et al. (2020) The design and development of a multicentric protocol to investigate the impact of adjunctive doxycycline on the management of peripheral lymphoedema caused by lymphatic filariasis and podoconiosis. *Parasites Vectors*, 13(1), 155. https://doi.org/10.1186/s13071-020-04024-2.

**18** Chesnais, C.B., Vlaminck, J., Kunyu-Shako, B. et al. (2016) Measurement of circulating filarial antigen levels in human blood with a point-of-care test strip and a portable spectrodensitometer. *Am. J. Trop. Med. Hyg.*, 94(6), 1324–1329. https://doi.org/10.4269/ajtmh.15-0916.

**19** Weil, G.J., Curtis, K.C., Fischer, P.U. et al. (2011) A multicenter evaluation of a new antibody test kit for lymphatic filariasis employing recombinant *Brugia malayi* antigen Bm-14. *Acta Trop.*, 120 (Suppl 1), S19–S22. https://doi.org/10.1016/j.actatropica.2010.04.010.

**20** Dreyer, G., Santos, A., Noroes, J. et al. (1998). Ultrasonographic detection of living adult *Wuchereria bancrofti* using a 3.5-MHz transducer. *Am. J. Trop. Med. Hyg.* 59 (3): 399–403. https://doi.org/10.4269/ajtmh.1998.59.399.

**21** Dorkenoo, M.A., de Souza, D.K., Apetogbo, Y. et al. (2018) Molecular xenomonitoring for post-validation surveillance of lymphatic filariasis in Togo: no evidence for active transmission. *Parasites Vectors*, 11, 52. https://doi.org/10.1186/s13071-017-2611-9.

**22** Kamgno, J., Boussinesq, M., Labrousse, F. et al. (2008). Encephalopathy after ivermectin treatment in a patient infected with *Loa loa* and *Plasmodium* spp. *Am. J. Trop. Med. Hyg.* 78: 546–551.

**23** Kamgno, J., Pion, S., Chesnais, C. et al. (2017) Test and treat strategy to combat onchocerciasis in *Loa loa* endemic regions. *N. Engl. J. Med.*, 377, 2044 –2052. https://doi.org/10.1056/NEJMoa1705026.

**24** Pedram, B., Pasquetto, V., Ji, Y. et al. (2017) A novel rapid test for detecting antibody responses to *Loa loa* infections. *PLoS Negl.Trop. Dis.*, 11(7), e0005741. https://doi.org/10.1371/journal.pntd.0005741.

**25** Robles, R. (1919). Onchocercose humaine au Guatemala produisant la cecité et "l'erysipé le du littoral" (Erisipela de la costa). *Bull. Soc. Pathol. Exot.* 12: 442–460.

**26** Azube, M.E., Carvajal, L., Zerega, F. (1981). First endemic focus of onchocerciasis discovered in Ecuador. https://iris.paho.org/bitstream/handle/10665.2/32419/13033.pdf?sequence=1&isAllowed=y.

**27** Fawdry, A.L. (1957). Onchocerciasis in South Arabia. *Trans. R. Soc. Trop. Med. Hyg.* 51: 253–256.

**28** World Health Organization – African Region (2021). WHO unveils new analytical tools on the ESPEN data portal. https://www.afro.who.int/news/who-unveils-new-analytical-tools-espen-data-portal.

**29** Serey, J., Quezada, L., Alfaro, M. et al. (2021) Artificial intelligence methodologies for data management. *Symmetry*, 13, 2040. https://doi.org/10.3390/sym13112040.

**30** Guevara, A., Salazar, E., Vicuña, Y. et al (2020) Use of Ov16-based serology for post-elimination surveillance of onchocerciasis in Ecuador. *Am. J. Trop. Med. Hyg.*, 103(4), 1569–1571. https://doi.org/10.4269/ajtmh.20-0082.

**31** Dorkenoo, M.A., Tchankoni, M.K., Yehadji, D. et al. (2021) Monitoring migrant groups as a post-validation surveillance approach to contain the potential reemergence of lymphatic filariasis in Togo. *Parasites Vectors*, 14, 134. https://doi.org/10.1186/s13071-021-04644-2.

**32** Demarquay, J.N. (1863). Sur une tumeur des bourses contenant un liquide laiteux (galactocèle de Vidal) et renferment des petits entres vermiformes que l'on peut considerée comme des helminthes hematoides a l'état d'embryon. *Gaz. Med.Paris* 18: 665–667.

**33** Cox, F.E. (2002). History of human parasitology. *Clin. Microbiol. Rev.* 15: 595–612. https://doi.org/10.1128/CMR.15.4.595-612.2002.

**34** Manson, P. (1878). On the development of Filaria sanguinis hominis and on the mosquito considered as a nurse. *J. Linn. Soc. Zool.* 14: 304–311.

**35** World Health Organization (2021). Ending the neglect to attain the Sustainable Development Goals: A road map for neglected tropical diseases 2021–2030. https://www.who.int/publications/i/item/9789240010352.

**36** Casulli A. (2021) New global targets for NTDs in the WHO roadmap 2021–2030. *PLoS Negl.Trop. Dis.*, 15(5), e0009373. https://doi.org/10.1371/journal.pntd.0009373.

**37** Mackenzie, C.D., Huntington, M.K., Wanji, S. et al. (2010). Association of adult *Onchocerca volvulus* with lymphatic vessels. *J. Parasitol.* 96: 219–221.

**38** Macé, J.M., Boussinesq, M., Ngoumou, P. et al. (1997) Country-wide rapid epidemiological mapping of onchocerciasis (REMO) in Cameroon. *Ann. Trop. Med. Parasitol.*, 91, 379–391. https://doi.org/10.1080/00034983.1997.11813153.

**39** Eyo, J., Onyishi, G., and Ugokwe, C. (2013). Rapid epidemiological assessment of onchocerciasis in a tropical semi-urban community, Enugu state. *Niger. Iran. J. Parasitol.* 8 (1): 145–151.

**40** Prince-Guerra, J.L., Cama, V.A., Wilson, N. (2018). Comparison of PCR methods for *Onchocerca volvulus* detection in skin snip biopsies from the Tshopo Province, Democratic Republic of the Congo. *Am. J. Trop. Med. Hyg.*, 98(5), 1427–1434. https://doi.org/10.4269/ajtmh.17-0809.

**41** Homeida, M., Mackenzie, C.D., Williams, J.F. et al. (1986). The detection of onchocercal nodules by ultrasound. *Trans. R. Soc. Trop. Med. Hyg.* 80: 570–571.

**42** Hong, A., Opoku, N., Kanza, E. et al. (2020). Novel use of optical coherence tomography in river blindness during treatment with ivermectin. *Invest. Ophthalmol. Visual Sci.* 61: 5370.

**43** Lahiri, B.B., Bagavathiappan, S., Jayakumar, T. et al. (2012) Medical applications of infrared thermography: a review. *Infrared Phys. Technol.*, 55(4), 221–235. https://doi.org/10.1016/j.infrared.2012.03.007.

**44** Krishnasastry, S. and Mackenzie, C.D. (2021) Alternative approaches to lymphoedema care in lymphatic filariasis. *PLoS Negl.Trop. Dis.*, 15, e0009293. https://doi.org/10.1371/journal.pntd.0009293.

**45** Coffeng, L.E., Stolk, W.A., Golden, A. et al. (2019) Predictive value of Ov16 antibody prevalence in different subpopulations for elimination of African onchocerciasis. *Am. J. Epidemiol.*, 188(9), 1723–1732. https://doi.org/10.1093/aje/kwz109.

**46** Steel, C., Golden, A., Stevens, E. et al. (2015) Rapid Point-of-Contact tool for mapping and integrated surveillance of *Wuchereria bancrofti* and *Onchocerca volvulus* infection. *Clin. Vaccine Immunol.*, 22, 896–901. https://doi.org/10.1128/CVI.00227-15.

**47** Mackenzie, C.D., Al-Kubati, A.S., Al-Qubati, Y. et al. (2018) A serological survey of human onchocerciasis in Yemen. *Am. J. Trop. Med. Hyg.*, 99, 1049 - 1052. https://doi.org/10.4269/ajtmh.18-0051.

**48** Behan-Braman, A., Weber, P.S., Tritten, L. et al. (2018) Further characterization of molecular markers in canine *Dirofilaria immitis* infection. *J. Parasitol.*, 104, 697–701. https://doi.org/10.1645/18-12.

**49** Mackenzie, C.D. (2020) A much-needed advance in the diagnosis of River Blindness. *J. Infect. Dis.*, 221, 1746–1748. https://doi.org/10.1093/infdis/jiz309.

**50** Loum, D., Cozart, D., Lakwo, T. et al. (2019) Optimization and evaluation of the Esperanza Window Trap to reduce biting rates of *Simulium damnosum* sensu lato in Northern Uganda. *PLoS Negl.Trop. Dis.*, 13(7), e0007558.https://doi.org/10.1371/journal.pntd.0007558.

**51** World Health Organization (2019). Report of the Third Meeting of the WHO Onchocerciasis Technical Advisory Subgroup Geneva, Switzerland, 26–28 February 2019. https://www.who.int/publications/i/item/9789240006638.

**52** Williams, S., Pilotte, N., Grant, J. et al. (2022). Real time PCT for xenomonitoring of onchocerciasis. *PLoS Negl.Trop. Dis.* in press.

**53** Mand, S., Marfo-Debrekyei, Y., Dittrich, M. et al. (2003). Animated documentation of the filaria dance sign (FDS) in bancroftian filariasis. *Filaria J.* 2 (1): 3. https://doi.org/10.1186/1475-2883-2-3.

**54** Zhou, C., Yahathugoda, C., De Silva, L. et al. (2019). Portable infrared imaging for longitudinal limb volume monitoring in patients with lymphatic filariasis. *PLoS Negl.Trop. Dis.* 13 (10): e0007762.

**55** Melrose, W., Turner, P.F., Pisters, P. et al. (2000). An improved Knott's concentration test for the detection of microfilariae. *Trans. R. Soc. Trop. Med. Hyg.* 94 (2): 176.

**56** Pastor, A.F., Silva, M.R., dos Santos, W.J.T. et al. (2021, 2021). Recombinant antigens used as diagnostic tools for lymphatic filariasis. *Parasites Vectors* 14 (1): 474. https://doi.org/10.1186/s13071-021-049. PMID: 34526120; PMCID: PMC8442287.

**57** Moss, D.M., Priest, J.W., Boyd, A. et al. (2011). Multiplex bead assay for serum samples from children in Haiti enrolled in a drug study for the treatment of lymphatic filariasis. *Am. J. Trop. Med. Hyg.* 85 (2): 229–237. https://doi.org/10.4269/ajtmh.2011.11-0029.

**58** Won, K.Y., Robinson, K., Hamlin, K.L. et al. (2018) Comparison of antigen and antibody responses in repeat lymphatic filariasis transmission assessment surveys in American Samoa. *PLoS Negl.Trop. Dis.*, 12(3), e0006347. https://doi.org/10.1371/.journal.pntd.0006347.

**59** Dorkenoo, A.M., Koba, A., Halatoko, W.A. et al. (2021). Assessment of the usefulness of anti-Wb123 antibody for post-elimination surveillance of lymphatic filariasis. *Parasites Vectors* 14: 23.

**60** Emukah, E., Rakers, L., Kahansim, B. et al. (2018). In Southern Nigeria, *Loa loa* blood microfilaria density is very low even in areas with high prevalence of loiasis: Results of a survey using the new LoaScope technology. *Am. J. Trop. Med. Hyg.* 99: 116–123.

**61** Rosenblatt, J.E., Reller, L.B., Weinstein, M.P. (2009) Laboratory diagnosis of infections due to blood and tissue parasites, *Clin. Inf. Dis.* 49, 1103–1108. https://doi.org/10.1086/605574.

**62** Medeiros, J.F., Almeida, T.A.P., Silva, L.B.T. et al. (2015). A field trial of a PCR-based *Mansonella ozzardi* diagnosis assay detects high-levels of submicroscopic *M. ozzardi* infections in both venous blood samples and FTA® card dried blood spots. *Parasites Vectors* 8: 280.

**63** Poole, C.B., Sinha, A., Ettwiller, L. et al. (2019) In silico identification of novel biomarkers and development of new rapid diagnostic tests for the filarial parasites *Mansonella perstans* and *Mansonella ozzardi*. *Sci. Rep.*, 9(1), 10275. https://doi.org/10.1038/s41598-019-46550-9.

**64** Rena, O., Leutner, M., Casadio, C. (2002) Human pulmonary dirofilariasis: uncommon cause of pulmonary coin-lesion. *Eur. J. Cardio-Thorac. Surg.*, 22, 157–159. https://doi.org/10.1016/S1010-7940(02)00221-X.

**65** Ton, T.G., Mackenzie, C.D., Molyneux D.H. (2015) The burden of mental health in lymphatic filariasis. *Infect. Dis. Poverty*, 4, 34. https://doi.org/10.1186/s40249-015-0068-7.

**66** Tritten, L., Burkman, E., Moorhead, A. et al. (2014). Detection of circulating parasite-derived microRNAs in filarial infections. *PLoS Negl. Trop. Dis.* 8: e0002971.

**67** Quintana, J.F., Makepeace, B.L., Babayan, S.A. et al. (2015). Extracellular *Onchocerca*-derived small RNAs in host nodules and blood. *Parasitol. Vectors* 8: 58. https://doi.org/10.1186/s13071-015-0656-1.

**68** Mu, Y., McManus, D.P., Gordon, C.A. et al. (2021). Parasitic helminth-derived microRNAs and extracellular vesicle cargos as biomarkers for helminthic infections. *Front. Cell. Infect. Microbiol.* 11: 708952. https://doi.org/10.3389/fcimb.2021.708952.

**69** Tritten, L., Clarke, D., Timmins, S. et al. (2016) *Dirofilaria immitis* exhibits sex-and stage- specific differences in excretory/secretory miRNA and protein profiles. *Vet. Parasitol.*, 232, 1–7. https://doi.org/10.1016/j.vetpar.2016.11.005.

**70** Lagatie, O., Batsa Debrah, L., Debrah, A. et al. (2017). Plasma-derived parasitic microRNAs have insufficient concentrations to be used as diagnostic biomarker for detection of *Onchocerca volvulus* infection or treatment monitoring using LNA-based RT-qPCR. *Parasitol. Res.* 116 (3): 1013–1022. https://doi.org/10.1007/s00436-017-5382-5.

**71** Macfarlane, C.L., Quek, S., Pionnier, N. et al. (2020). The insufficiency of circulating miRNA and DNA as diagnostic tools or as biomarkers of treatment efficacy for *Onchocerca volvulus*. *Sci. Rep.* 10 (1): 6672. https://doi.org/10.1038/s41598-020-63249-4.

**72** Denery, J.R., Nunes, A.A., Hixon, M.S. et al. (2010). Metabolomics-based discovery of diagnostic biomarkers for onchocerciasis. *PLoS Negl.Trop. Dis.* 4 (10): e834. https://doi.org/10.1371/journal.pntd.0000834.

**73** Bennuru, S., Lustigman, S., Abraham, D. et al. (2017). Metabolite profiling of infection-associated metabolic markers of onchocerciasis. *Mol. Biochem. Parasitol.* 215: 58–69. https://doi.org/10.1016/j.molbiopara.2017.01.008.

**74** Lagatie, O., Njumbe Ediage, E., Van Roosbroeck, D. et al. (2021). Multimodal biomarker discovery for active *Onchocerca volvulus* infection. *PLoS Negl.Trop. Dis.* 15 (11): e0009999. https://doi.org/10.1371/journal.pntd.0009999.

**75** Globisch, D., Moreno, A.Y., Hixon, M.S. et al. (2013). *Onchocerca volvulus* neurotransmitter tyramine is a biomarker for river blindness. *Proc. Natl. Acad. Sci. U.S.A.* 110: 4218–4223. https://doi.org/10.1073/pnas.1221969110.

**76** Globisch, D., Eubanks, L.M., Shirey, R.J. et al. (2017). Validation of onchocerciasis biomarker N-acetyltyramine-O-glucuronide (NATOG). *Bioorg. Med. Chem. Lett.* 27: 3436–3440. https://doi.org/10.1016/j.bmcl.2017.05.082.

**77** Shirey, R.J., Globisch, D., Eubanks, L.M. et al. (2018). Noninvasive urine biomarker lateral flow immunoassay for monitoring active Onchocerciasis. *ACS Infect. Dis.* 4: 1423–1431. https://doi.org/10.1021/acsinfecdis.8b00163.

**78** Wewer, V., Peisker, H., Gutbrod, K. et al. (2021). Urine metabolites for the identification of *Onchocerca volvulus* infections in patients from Cameroon. *Parasites Vectors* 14 (1): 397. https://doi.org/10.1186/s13071-021-04893-1.

**79** Lagatie, O., Njumbe Ediage, E., Batsa Debrah, L. et al. (2016). Evaluation of the diagnostic potential of urinary N-acetyltyramine-O,β-glucuronide (NATOG) as diagnostic biomarker for *Onchocerca volvulus* infection. *Parasites Vectors*, 23 9 (1): 302. https://doi.org/10.1186/s13071-016-1582-6.

**80** Hotterbeekx, A., Dusabimana, A., Mandro, M. et al. (2020). Urinary N-acetyltyramine-O,β-glucuronide in persons with Onchocerciasis-associated epilepsy. *Pathogens* 9: 191. https://doi.org/10.3390/pathogens9030191.

**81** Milton, P., Hamley, J., Walker, M. et al. (2020) Moxidectin: an oral treatment for human onchocerciasis. *Exp. Rev. Anti-infect. Ther.*, 18, 1–15. https://doi.org/10.1080/14787210.2020.1792772.

**82** Mackenzie, C.D., Preston, P.M., and Ogilvie, B.M. (1978). Immunological properties of the surface of parasitic nematodes. *Nature* 276: 826–828.

**83** Szecsi, P.B. (1992). The aspartic proteases. *Scand. J. Clin. Lab. Invest.* Suppl 210: 5–22.

**84** Weinkopff, T., Mackenzie, C., Eversole, R. et al. (2014). Molecular mechanisms of lymphangiectasia in lymphatic filariasis. *PLoS Negl.Trop. Dis.* 8 (7): e0002893. https://doi.org/10.137/journal.pntd.0002893.

**85** Mackenzie, C.D., Behan-Braman, A., Hauptman, J. et al. (2017) Assessing the viability and degeneration of the medically important filarial nematodes, in *Nematology – Concepts, Diagnosis and Control, Nematology*, (ed. M.M. Shah, M. Mahamood), InTech Open Publishing, Rijeka, Croatia, pp. 101–120, ISBN 978-953-51-3416-9, https://doi.org/10.5772/intechopen.69512.

**86** Mendes, A., Palmer, T., Berens, A. et al. (2021) Mapathons versus automated feature extraction: a comparative analysis for strengthening immunization microplanning. *Int. J. Health Geogr.*, 20, 27. https://doi.org/10.1186/s12942-021-00277-x.

**87** Cromwell, E.A., Osborne, J.C.P., Unnasch, T.R. et al. (2021). Predicting the environmental suitability for onchocerciasis in Africa as an aid to elimination planning. *PLoS Negl.Trop. Dis.* 15 (7): e0008824.

**88** Won, K.Y., Gass, K., Biamonte, M. et al. (2021) Diagnostics to support elimination of lymphatic filariasis-Development of two target product profiles. *PLoS Negl.Trop. Dis.*, 15(11), e0009968. https://doi.org/10.1371/journal.pntd.0009968.

**89** Mackenzie, C.D., Mante, S. (2021) Caring for patients in the global programme to eliminate lymphatic filariasis. *Int. Health*, 13(suppl 1), S48–S54. https://doi.org/10.1093/inthealth/ihaa080.

**90** Al-Kubati AS, Mackenzie, CD, Boakye D et al. (2018) Onchocerciasis in Yemen: moving forward towards an elimination programme. *Int. Health*, 10, i27–i32. https://doi.org/10.1093/inthealth/ihx066.

**91** Makuei, P.M., Sebit, M.R., and Dekker, M. (2021). Epilepsy in South Sudan: practical guidelines for better control. *South Sudan Med. J.* 14 (2): 38–42.

# 6

# Veterinary Diagnosis of Filarial Infection

*Christopher Evans[1], Nils Pilotte[2], Steven Williams[2], and Andrew Moorhead[1,*]*

[1] *University of Georgia, College of Veterinary Medicine, Department of Infectious Diseases, Athens, GA 30602, USA*
[2] *Smith College, Department of Biological Science, Northampton, MA 01063, USA*

**Abstract**

Filarial worms are a unique group of parasites with importance in both human and veterinary medicine. These parasites are typically long-lived and difficult to detect, often causing chronic disease states over a period of years and, for these reasons, effective diagnostic testing is crucial for their control. Adult filarial worms tend to occupy inaccessible anatomical sites within the host, but microfilariae disperse widely in the blood or skin to allow uptake and transmission by the hematophagous insects necessary to complete the life cycle, and the detection of this microscopic stage represents a fundamental form of diagnostic testing. Immunodiagnostic and DNA-based tests have since been developed for several filarial species, as well as methods for visualizing adult parasites *in situ*. All these techniques carry their own distinct strengths and weaknesses, so reliable diagnosis often requires a strategic combination of tests. Accurate diagnosis is important for potentially fatal infections like canine heartworm and is also essential for identifying emergent zoonoses, like *Onchocerca lupi*, and potential animal reservoirs, as with *Brugia malayi*. Accurate parasite detection and identification is useful not only in clinical settings but also greatly assists research efforts. This chapter will review the diagnostic methods available for some of the most common species of filarial nematodes in small animal veterinary medicine.

## 6.1 Introduction

Filarial nematodes (superfamily Filarioidea) comprise a group of parasites important to both human and veterinary medicine. They all rely on blood-feeding arthropods for transmission and generally live for long periods within the definitive host. Reproductive females release vermiform embryos (microfilariae) ovoviviparously, which migrate away from the adults to disperse in the bloodstream

*Corresponding author.

*Human and Animal Filariases: Landscape, Challenges, and Control*, First Edition.
Edited by Ronald Kaminsky and Timothy G. Geary.

or skin, depending on species, where they may be taken up by a suitable vector. Species that parasitize domestic and wild animals often also represent zoonotic threats, many of which are only recently being recognized. Infections with filarial worms are often asymptomatic and nonpathogenic, but overlooking their presence carries the risk that life-threatening conditions eventually develop. The canine heartworm (*Dirofilaria immitis*) is well known for causing cardiac failure in heavy and long-lived infections, standing as a compelling reason for regular testing. Meanwhile, the pathogenic potential of other species remains poorly understood, and it is here that accurate diagnosis can aid ongoing research.

The biology of filarial worms presents diagnostic challenges distinct from other nematodes. The techniques available today arose from advances in understanding of these parasites, including their life cycles, geographic ranges, and molecular characteristics. Although clinical practices may use certain tests to the virtual exclusion of others, an understanding of the technical strengths and limitations of those tests, as well as the range of existing alternatives, may offer insight into more effective diagnostic strategies. The purpose of this chapter is to review these methodologies and the species of filarial parasites for which they may be applied.

## 6.2 Diagnostic Methods

### 6.2.1 Microscopy-Based Methods

The simplest method for detecting a patent filarial infection, and one still frequently used in the clinic, is the direct blood smear. This technique requires only a drop of anticoagulated blood, which is placed on a glass slide, coverslipped, and examined by microscopy. The forms of the microfilariae are difficult to visualize directly in such a preparation, but their motility noticeably agitates the erythrocytes around them, and they can be visualized in this way. Direct smears may be useful when microfilaremia is high, but this method is susceptible to missing parasites at lower concentrations. This is especially problematic when microfilaremia is prone to periodic fluctuation.

Many species exhibit a phenomenon of circadian periodicity, in which microfilaremia in the peripheral blood rises and falls, sometimes dropping to undetectable levels. The periodicity of *D. immitis* varies with geographic location, with peak parasite circulation supposedly coinciding with the feeding habits of the prevailing vector species to optimize uptake and transmission [1–5]. Furthermore, an annual cycle tied to the passage of the seasons has been observed in *D. immitis*, with microfilaremia peaking in the summer months [6, 7]. *B. malayi*, which parasitizes humans, cats, and a number of other species, comprises distinct nocturnally periodic and subperiodic strains; interestingly, cats only seem to naturally acquire infections of the latter [8]. In fact, most of the parasites discussed in this chapter exhibit some form of periodicity. This has the potential to greatly complicate microfilaria testing, so concentration techniques were developed to detect parasites even at very low levels in the circulation.

The Knott test was first developed to detect human infections with *Wuchereria bancrofti*, the pronounced nocturnal periodicity of which had previously required that blood samples be drawn in the middle of the night. Briefly, in this procedure, 1 ml of anticoagulated venous blood is mixed with 10 ml of 2% formalin solution, which has the dual benefit of lysing erythrocytes to improve parasite visualization and preserving the sample for further examination. This mixture is centrifuged, the supernatant discarded, and the pellet stained with methylene blue or a similar dye. The stained blood sample can then be examined by microscopy, either coverslipped or after air drying. Knott's concentration technique (conventionally known as the "modified Knott test" in veterinary medicine) is advocated as the preferred method for detecting blood-dwelling microfilariae because it is simple, inexpensive, and standardized; the microfilariae observed from a modified Knott test can be measured and referenced against well-established diagnostic metrics (e.g. length and width) for species identification.

Nonetheless, the microfilariae of closely related species are often difficult to distinguish by morphology alone. In such cases, the localization of acid phosphatase activity can be used in diagnosis. This histochemical technique has been used to differentiate morphologically similar species for taxonomic purposes and to establish parameters for diagnostic reference [9, 10]. Over the past decade, it has seen renewed use in evaluating suspected *B. malayi* infections in dogs as a complement to morphological and molecular analysis [11–13].

It is possible to calculate the concentration of microfilariae in venous blood using methods such as the thick blood smear, in which known volumes (typically 20 μl) are stained and evaluated. Alternatively, the entire volume of the pellet derived from a Knott test can be measured and examined. While the usefulness of calculating an accurate microfilaria concentration is largely limited to research applications, there are a few situations where this proves helpful in the clinic. High microfilaremia is a risk factor for anaphylactic reactions against parasite antigens released during treatment, so this can inform therapeutic options. Another case is when employing the Microfilaria Suppression Test, which is a method for identifying likely cases of anthelmintic resistance in heartworm; in this test, microfilaria counts are taken prior to and one week following treatment with a microfilaricidal compound to determine drug efficacy [14, 15]. Beyond this, the observation of microfilariae in a patient is most useful to the clinician as a dichotomous diagnostic parameter and a means of species identification. Though it is tempting to make inferences about adult worm burden based on microfilaremia, the two do not always correlate well [16]. Symptomology and ultrasound visualization should instead be considered when estimating adult worm burden and, consequently, the likelihood of adverse reactions to treatment.

## 6.3 Immunodiagnostic Methods

In many settings, antigen testing has largely taken the place of routine microfilaria testing. These tests detect the presence of circulating adult worm antigen with a high level of sensitivity and specificity, independent of the presence or periodicity of microfilariae. A test based on the enzyme-linked immunosorbent assay (ELISA) was developed for the diagnosis and monitoring of human lymphatic filariasis 30 years ago, which has since been replaced by the immunographic card test and, ultimately, by the filariasis test strip still in use today [17, 18]. As serological diagnostic technology improved, a number of antigen tests were also developed for *D. immitis*, comprising various formats and all with high degrees of sensitivity and specificity. When this testing method was introduced, it offered a much-needed means of detecting amicrofilaremic infections, but over years of use, this role has expanded to that of a primary screening test [19].

Antibody detection tests also find use in both human and veterinary medicine. Antibody capture ELISA tests are used for lymphatic filariasis, which detect antibody against a filarial antigen (Bm14) indicative of both brugian and bancroftian infection [20, 21]. In addition, the *Brugia* Rapid test detects IgG4 antibodies against *Brugia* spp. in a robust, immunochromatographic format [22]. These tests, however, are subject to the limitation that antibodies remain present long after active infection and so cannot be used as a reliable indicator of current infection status. Because they are much better for assessing exposure, antibody tests are useful for monitoring endemicity, especially in children, who are not expected to be exposed in communities where the parasites have been successfully eradicated. In veterinary medicine, antibody tests are available for heartworm but are reserved for use in cats; owing to the typically low antigen levels in this species, cats are more prone to false-negative results via antigen testing, so antibody detection is favored for its sensitivity.

In all cases, timing is an important consideration in testing. Due to the relatively long developmental periods of filarial worms, the microfilariae or adult antigens on which testing methods rely may only reach detectable levels a matter of months after initial infection. In the case of heartworm, seven months of age is the earliest that testing is recommended for pups; any earlier and negative results provide little-to-no indication of absence of current infection. Similarly, when an animal starts preventive therapy, testing should follow six months later to ensure the absence of pre-existing infections that were mature enough to avoid clearance by the preventive. The interruption of prophylaxis allows opportunities for new infection, which follow-up testing helps detect. By ruling out the chance of a prepatent infection at the start of preventive therapy, the clinician is also able to detect potential lack of efficacy in the event that infection does later occur in the face of compliant drug treatment.

Of interest to the statistically minded reader, it may be noted that when performing tests with dichotomous outputs (like most of those mentioned thus far), the reliability of the result is affected by infection prevalence. The positive predictive value of a test (i.e., the likelihood that a patient with a positive test result is truly infected) varies with prevalence, approaching 100% as prevalence approaches 100%. Indeed, clinical data support this phenomenon in heartworm infection. In one study, an antigen test with established, high levels of sensitivity and specificity was assessed. In Louisiana, a state with high prevalence (499 per 10 000 tested), the positive predictive value of the test was 60%; that is, out of 100 positive-testing dogs, 60 were truly infected, while the remaining 40 were false positives. In Washington, a state with low prevalence (6 per 10 000 tested), the positive predictive value was a

**Table 6.1** Characteristics of microfilariae of veterinary significance.

| Species | Sheath | Length (μm) | Width (μm) | Head | Tail | Acid phosphatase | Location of microfilariae | Geographic location | References |
|---|---|---|---|---|---|---|---|---|---|
| *Dirofilaria immitis* | No | 295–325 | 5–7.5 | Tapered | Straight | Excretory pore, anal pore | Blood | Worldwide | [24, 25] |
| *Dirofilaria repens* | No | 368–380 | 5–9 | Blunt | Straight or hooked | Anal pore | Blood | Europe, Middle East | [24, 25] |
| *Dirofilaria striata* | No | 360–385 | 5–6 | Tapered | Curved | | Blood | North and South America | [25] |
| *Acanthocheilonema reconditum* | No | 230–288 | 4–6 | Blunt | Hooked or curved | Diffuse | Blood | Worldwide | [24–26] |
| *Acanthocheilonema dracunculoides* | No | 185–276 | 4.2–6 | Tapered | Straight, sharp | Excretory pore, innenkorper, anal pore | Blood | Africa, Europe, India | [24, 27, 28] |
| *Brugia pahangi* | Yes | 246–280 | 5–6 | Blunt, rounded | Tapered | Diffuse | Blood | India, southeast Asia | [29, 30] |
| *Brugia malayi* | Yes | 177–230 | 5–6 | | 2 Nuclei in tip | Excretory pore, anal pore (sometimes amphids, phasmids) | Blood | India, southeast Asia | [29] |
| *Brugia ceylonensis* | Yes | 220–275 | | | | | Blood | India, Sri Lanka | [31] |
| *Brugia patei* | Yes | similar to *B. malayi* | | | 2 Nuclei in tip | Cephalic vesicle, excretory pore, tail | Blood | Kenya (Pate Island) | [32] |
| *Brugia beaveri* | Yes | 285–325 | 4.5–6.5 | Blunt | | | Blood | USA (Louisiana) | [33] |
| *Onchocerca lupi* | No | 105–115 | 6 | Rounded | Pointed | | skin | Europe, Middle East, USA | [34] |
| *Cercopithifilaria grassi* | No | 610–644 | 12–25 | Bulbous | Slightly bent, conical | | Skin | Europe, Mediterranean basin | [35, 36] |
| *Cercopithifilaria bainae* | No | 170–196 | 6–7 | Slightly attenuated | Blunt | | Skin | North and South America | [35, 37] |
| *Cercopithifilaria* sp. II | No | 261–307 | 12–15 | Blunt | Attenuated | | Skin | Europe | [35, 36] |

mere 2% [23]. Confirmatory testing helps avoid this pitfall and is especially useful in regions with low rates of infection. When test results are unexpected given animal history, symptomology, and other diagnostic results, the test should be repeated; if results remain ambiguous, independent confirmation by a reference laboratory is recommended.

The specific needs of filarial diagnostics in the veterinary field differ from those in human medicine. The diagnosis of human filariases generally occurs on the community level as integral parts of elimination efforts utilizing mass drug administration. Veterinary practices, however, operate on a case-by-case basis, with diagnosis and treatment specific to each patient. As such, the diagnostic tests of choice may differ, but robust, accurate, and rapid methodologies are universally desirable, and each field may naturally borrow from the other where improvements are to be made. What follows is a review of diagnostic techniques used for some of the most common and potentially threatening filarial parasite infections in small animal veterinary medicine (Table 6.1).

## 6.4 Current Diagnostic Practice

### 6.4.1 *Dirofilaria immitis*

Undoubtedly, the most significant filarial worm in companion animal medicine, *D. immitis* (the causative agent of canine heartworm disease), is a mosquito-transmitted parasite primarily of the pulmonary arteries of the dog. In its most severe manifestations, the nematode's prolonged presence therein may ultimately result in cardiopulmonary failure and death and, for this reason, it is one of the most serious threats to canine health faced by clinicians. Cats are also susceptible to heartworm infection but suffer distinct pathologies and require alternate diagnostic methods. The distribution of *D. immitis* is cosmopolitan, encountered in temperate, tropical, and subtropical regions throughout the world, and although the parasite has been recovered from numerous species, wild canids (e.g., foxes, coyotes, and jackals) may represent meaningful reservoirs [38, 39]. A zoonotic potential exists wherever *D. immitis* is found, but the parasite does not develop to the adult stage in humans; while most infections are believed to be cleared without incident, larval migration to the pulmonary arterial tree may result in pulmonary dirofilariasis, giving rise to granuloma formation [40]. Zoonotic subcutaneous/ocular dirofilariasis is a rare outcome and more likely to be attributable to *Dirofilaria repens* in the Old World and other *Dirofilaria* spp. in the Americas [41].

Adult parasites reside in the pulmonary arteries and, in heavy infections, the right chambers of the heart and venae cavae. Adults have been shown to survive and reproduce for 7.5 years, releasing microfilariae into the bloodstream that may individually persist up to 2.5 years [42, 43]. The prepatent period is six months at the shortest, and more typically seven to nine months [44–46]. Mosquitoes of numerous genera serve as competent vectors, acquiring the infection by taking a blood meal from a microfilaremic host. Within the arthropod vector, microfilariae develop to infective third-stage larvae over the course of approximately 10–14 days, at which point they can be transmitted to a mammalian host during the mosquito's next blood meal [38].

The symptoms indicative of heartworm infection may not manifest for months or years depending on parasite burden, the animal's exercise habits, and individual physiological factors [47]. One common symptom with an earlier onset is chronic, persistent, nonproductive coughing, which may then be followed by dyspnea, weakness, and (more rarely) syncope. Higher-intensity and longer-lived infections are more likely to also present with anorexia, weight loss, ascites, edema, and acute pulmonary and cardiac signs [38, 41]. Treatment of mature *D. immitis* infection is not trivial and may be hazardous to the health of the animal; as such, prevention and early detection are universally encouraged.

A variety of tests are available for *D. immitis*, but because of their respective, non-overlapping strengths and weaknesses, multiple tests are often needed for a diagnosis. In the event of a positive antigen test result, for example, blood should always be examined for microfilariae; if this is also positive, the diagnosis is usually considered safely confirmed. However, when multiple tests are used, the opportunity for discrepancies naturally arises. In an animal that is antigen-positive and microfilaria-negative, the antigen test should be repeated with a kit of a different type (see Antigen Tests for the three available formats); this accounts for the possibility of a false-positive result. Conversely, an animal may test positive for microfilariae but negative for antigen in a number of scenarios (discussed in Microfilaria Tests), which may necessitate further testing. Ultimately, if results are unexpected or ambiguous even after test repetition, independent confirmation by a reference laboratory with more discriminating techniques is recommended.

#### 6.4.1.1 Microscopy-Based Tests

Though largely supplanted by antigen testing as a first measure for detecting *D. immitis* in dogs, routine microfilaria testing is recommended to better inform a diagnosis [48–50]. Furthermore, microfilaremia status is an important consideration in determining an animal's reservoir potential and the risk of reaction during microfilaricidal therapy. The modified Knott test [51, 52] is the preferred method for detecting microfilariae and performing morphometric evaluations. Briefly, one milliliter of venous blood is mixed with 10 ml 2% formalin and centrifuged. The resulting pellet is stained (e.g. methylene blue) and examined on a microscope slide. Being a centrifugation technique, this test has the advantage of being sensitive enough to demonstrate microfilariae present at low concentrations, helping to dampen the variables of parasite periodicity and blood collection time. Modified Knott tests performed with the standard 2% formalin fixation also benefit from the established reference measurements of blood-dwelling microfilariae, allowing morphometric species identification. This is especially valuable when a clinician must distinguish *D. immitis* from other co-endemic species that may confound a diagnosis and misdirect treatment, including *Acanthocheilonema reconditum*, *Acanthocheilonema dracunculoides*, *D. repens*, *Dirofilaria striata*, and *Cercopithifilaria grassi*. However, because the modified Knott test requires formalin for blood lysis and parasite fixation, some clinics may be deterred from regularly utilizing the technique; one study has validated acetic acid as a suitable alternative to formalin when morphometric evaluation is not required [53]. Histochemical staining for acid phosphatase activity can complement morphometric examination, using either an established protocol or a commercially available kit [10, 24, 54]. The microfilariae of *D. immitis* show two sites of acid phosphatase activity (near the anal and secretory pores), while those of *D. repens*, for example, show only one (near the anal pore); microfilariae of other species can be similarly distinguished.

Microfilariae may also be concentrated by filtration methods. While, at the time of this writing, no filtration test kit is currently marketed for *D. immitis*, procedures still exist that allow clinicians to make use of available components, including syringe filter holders and 5-μm disk filters [52, 55]. Whole blood must be lysed prior to filtration; 2% formalin is the recommended lysate, as it allows published morphometric standards to be referenced for species identification. Although exceptionally rare, cases are known in which parasite-contaminated lysate solution has resulted in false-positive results, so care must be taken with reagents if this method is selected.

A routine hematocrit test may also be used to detect microfilariae; parasites are concentrated in the buffy coat of a microhematocrit capillary tube and their movement can be visualized under low magnification [56]. This is less sensitive than the techniques described above and only slightly more sensitive than a non-concentrated, whole blood examination, the only real benefit being convenience. Parasites concentrated by this method can be further examined by smearing the buffy coat onto a microscope slide and staining.

The direct smear, in which a whole blood sample is placed directly on a microscope slide and coverslipped, has much poorer sensitivity than any of the concentration techniques and, thus, is more prone to false-negative results; direct smears potentially miss 19% of microfilaremic infections [57]. Nonetheless, this technique is the simplest of microfilaria tests and can quickly demonstrate parasites in animals with high microfilaremia. This can be an instructive demonstration for clients by allowing them to visualize live worms recovered from their animal. The direct smear can also help distinguish microfilariae of *D. immitis* and *A. reconditum*; the former exhibits non-progressive motility, while the latter may move intermittently with notable progressive motility, traversing the microscope field [52].

The key disadvantage of all microfilaria testing is an insensitivity to occult infections (i.e., cases in which mature parasites are present without circulating microfilariae), which represent an estimated 10–67% of heartworm infections [58]. Such infections may be prepatent, single-sex, or the result of drug or immune clearance [59]. Conversely, it is important to note that the presence of microfilariae does not necessarily indicate the presence of adult parasites. An animal with an incomplete history, for example, may have previously carried a mature infection that was either treated or cleared, but never received microfilaricidal treatment, in which case microfilaremia may persist. Alternatively, very young pups may test positive after acquiring microfilariae by the transplacental route [60]. And, of course, microfilariae of different species may be confused with *D. immitis* unless a more rigorous examination (e.g. morphological or molecular) is performed.

Due to the transient microfilaremia characteristic of infections in cats and ferrets, microfilaria tests are typically not useful for detecting heartworm in these species [61, 62]. Diagnosis must, instead, rely on serological testing.

#### 6.4.1.2 Antigen Tests

Several antigen tests are commercially available for *D. immitis*, all of which function by detecting a glycoprotein produced mainly in the reproductive tract of adult female parasites. In mature infections, this antigen can be found circulating throughout the bloodstream [63]. These tests are easily performed and require only a small quantity of serum, plasma, or anticoagulated whole blood, making them a popular choice for point-of-care screening. They have the key advantage of detecting occult (i.e. amicrofilaremic) infections, to which none of the microscopy tests described are sensitive. The numerous commercially available antigen test kits comprise three formats: the microtiter plate ELISA, the membrane-bound ELISA, and the lateral flow immunochromatographic assay.

The microtiter plate format for the ELISA is considered to be the most sensitive of the available antigen tests and yields colorimetric results that reflect the concentration of circulating antigen [64, 65]. Because this test relies on the interpretation of color intensity, only serum or plasma samples should be used; strongly hemolyzed or lipemic samples may contribute too much background color for a clear reading and should be avoided if at all possible. While a quantitative assessment of the assay's colorimetric results shows some correlation to adult female worm burden, there are several confounding factors (e.g., the age of adult females and transient spikes in antigenemia following worm death) that may influence these results [66–68]. As such, the use of ELISA color intensity alone to determine worm burden is discouraged and is best complemented with echocardiography and an assessment of symptomology.

The membrane-bound ELISA represents a simplification of the microtiter format. Results are either purely qualitative or may provide a "high/low" antigen scoring. A mixture of blood sample and antibody conjugate is all that needs to be added in these kits. The lateral flow immunochromatographic assay simplifies this process even further by incorporating the conjugate onto the capillary bed so that only the test sample needs to be added. In exchange for this convenience, however, lateral flow tests are reportedly less sensitive than either of the ELISA-based formats [65].

Antigen testing benefits from a high degree of specificity (at or near 100% for all available kits), and while sensitivity is also generally high for all test formats, it is subject to some variability based on the number of adult female worms present. In dogs with low worm burden (1–10 adults) sensitivity varies from 52% to 84%, while specificity remains high at 96% to 98% [65, 69]. As with microfilaria testing, timing is also important for informative results. Heartworm antigen can be most reliably detected at least eight months after initial infection; antigen detection may be inconsistent at five to seven months and is not usually possible in infections less than five months old [64]. It should also be noted that in dogs receiving macrocyclic lactone preventives heartworm antigen may not reach detectable levels until nine months after infection [64]. Detectable antigen levels usually precede microfilaremia, potentially allowing for earlier detection, but may also lag by a matter of weeks.

Because these tests rely on detectable levels of circulating adult female antigen, false-negative results may occur in prepatent infections, male-only infections, and infections with few adult females. A false-negative result may also occur when the target antigen is bound in host antigen/antibody immune complexes and is no longer available for detection [70, 71]. Earlier iterations of the heartworm tests available today included methods to encourage the dissociation of immune complexes (e.g. EDTA or heat treatment) as a routine step in sample preparation, but with improvements to sensitivity in the subsequent generations of these kits, such steps were removed from standard procedures out of an apparent obsolescence. More recently, however, heat treatment of samples has been demonstrated to increase the sensitivity of commercially available canine heartworm tests [72, 73] with similar findings in feline heartworm [74]. A recent study reported an increase in sensitivity from 90.7% to 98.4% in mature heartworm infections [75]. While the benefits of increasing sensitivity are clear, all antigen testing comes with the inherent risk of false positives; this same study reported a decrease in specificity from 97.8% to 96.1%. While rare, cross-reactivity in currently available heartworm tests has been demonstrated which has also been reported with *Angiostrongylus vasorum* [76, 77], *Spirocerca lupi* [78], *Acanthocheilonema ohendhali* [79], and *D. repens* [76]. When heartworm infection is suspected despite a negative antigen test result, heat treatment may help to ensure test accuracy, however, routine heat treatment is not currently recommended [48–50].

Unfortunately, antigen tests are not as successful at detecting *D. immitis* infection in cats as they are for dogs. Feline infections commonly involve few worms and/or immature worms and therefore are not as likely to produce detectable levels of antigen [80]. If an antigen test is still desired, the accuracy of diagnosis is greatly improved when complemented with antibody testing [81]. In ferrets, however, antigen testing remains a sensitive method of detection [61, 82]; antigen is detectable as early as four months after infection (one month earlier than in dogs and cats) likely due to greater concentration in a smaller blood volume [83]. Testing in both cats and ferrets can benefit from imaging techniques due to the complex nature of diagnosing heartworm infection in these species.

#### 6.4.1.3 Antibody Tests

Antibody testing has long been available for detecting heartworm infection in cats, a species in which antigen tests are more prone to false-negative results. This test detects circulating feline IgG antibodies specific to an antigen present in *D. immitis* adults and larvae of either sex, which may be detectable as early as two months after initial infection [84, 85]. The typically strong feline immune response enables the detection of even single-worm infections. It should be noted that antibody levels may persist after parasites have been cleared (either naturally or by preventives), so this technique is susceptible to false-positives [86]. As such, radiography and echocardiography should be considered as supplements to establishing a diagnosis.

Antibody tests were once marketed for use in dogs, comprising latex agglutination, ELISA, and indirect florescent antibody tests [19]. Due to relatively poorer accuracy, however, these tests have been replaced by more reliable antigen detection techniques.

#### 6.4.1.4 DNA-Based Tests

While not readily available to most clinicians, the sensitivity and specificity of polymerase chain reaction (PCR)-based techniques for *D. immitis* detection in dogs have been demonstrated in laboratory settings. PCR is capable of detecting microfilariae at very low concentrations and distinguishing closely related species [87–89]. Molecular identification of parasites is advantageous in cases where morphological characterization is difficult and a definitive diagnosis is desired. Such detection typically relies on DNA extraction from whole blood, facilitating recovery of genetic material from microfilariae present in the circulation. This is followed by PCR amplification of a DNA sequence specific to *D. immitis* if such a sequence is present within the sample. While it was historically believed that PCR-based detection of *D. immitis* required the presence of microfilariae, limited evidence has suggested that amplification of cell-free DNA may also be possible [90]. This supports the possibility that the target signal may be detectable even when samples are collected from a host harboring an occult infection, theoretically allowing for the detection of single worm infections, or sexually productive infections prior to the release of offspring into the circulation. If demonstrated to be reliable and able to be clinically standardized, such detection would represent a monumental advance for the field of heartworm diagnostics.

Current laboratory options for the DNA-based detection of *D. immitis* are limited to the growing body of assays published in the scientific literature [87–89, 91–93] and commercially available kits rated for general laboratory or research use only. As of this writing, clinically approved DNA-based diagnostic options were unavailable for the detection of *D. immitis*, leaving veterinary clinics and hospitals without an approved option for PCR-based confirmatory testing, and making such testing unrealistic in the vast majority of circumstances. This represents a considerable gap in the clinical capacity of the veterinary community. DNA-based options for *D. immitis* diagnosis at the point-of-treatment are even more limited, with a single diagnostic method described in the scientific literature [94]. Utilizing loop-mediated isothermal amplification (LAMP), this assay provides a proof-of-concept for techniques that bring DNA-based testing for *D. immitis* closer to the local clinic. However, such approaches will require significant standardization, and many hurdles remain before clinical use of any similar assay could be considered.

A unique strength of utilizing DNA-based detection approaches for *D. immitis* is the capacity to screen vector insects for the presence of pathogen. While some evidence has suggested that current PCR-based approaches for monitoring *Dirofilaria* may be ill-suited for the reliable testing of mosquito samples [95], assay-specific challenges can be easily overcome using advanced bioinformatic approaches to DNA target selection and assay design [96]. Studies have further demonstrated proof-of-concept for the detection of parasite in samples collected under field conditions from many geographic locations [97–100]. Therefore, while not useful in a direct clinical sense, vector surveillance provides researchers and veterinary health care officials with an indirect means of obtaining epidemiological data, amassing critical information pertaining to geographic ranges of infection, vector capacity of mosquito populations, and potential for disease spread. With ever growing importance being placed on the "One Health" approach to the integration of human and veterinary disease surveillance [101], technologies allowing for information gathering present unique options for the veterinary community and provide potential avenues for expanded cross-talk between human and animal health care professionals.

#### 6.4.1.5 Imaging

Visual assessment of animals suspected of heartworm infection can aid greatly in diagnosis. Cardiac ultrasound, significantly, can demonstrate current infections by visualizing adult parasites *in situ*. Live worms are highly echogenic and appear as short, parallel lines in the right cardiac chambers or the lumina of the connected vessels [102, 103]. It should be noted that the appearance and location of adult heartworms may be similar to that of the right ventricular chordae tendineae, so care must be taken to avoid misidentification.

Radiography may be useful as an adjunctive test by showing signs of heartworm disease, including enlargement of the right ventricle and pulmonary arteries and abnormal pulmonary patterns. If right-sided congestive heart failure is present, hepatomegaly and ascites may be observed [104]. Signs of disease, however, do not necessarily indicate current infection status [105].

The diagnosis of feline heartworm infection presents a special challenge. Several tests may be necessary in the event that no individual result is definitive and imaging techniques can contribute strongly to the accurate diagnosis of a suspected infection. Detecting the presence of worms in the pulmonary arteries by echocardiography or enlargement of the arterial vasculature, for example, would raise the suspicion of a heartworm etiology. Radiographic examination of heartworm-infected ferrets is more likely to reveal cardiomegaly and pleural effusion, while angiography may be used to visualize adult worms in the venae cavae [61, 106].

### 6.4.2 *Dirofilaria (Nochtiella) repens*

Unlike *D. immitis*, which is found worldwide, *D. repens* is endemic to Europe and the Middle East, with especially high prevalence in the Mediterranean Basin [39, 107]. As with all other known members of the genus (with the conspicuous exception of *D. immitis*), adult parasites localize in the subcutaneous tissues. This is usually non-pathogenic in dogs, but the zoonotic potential for *D. repens* is much higher than that of *D. immitis*. Human infection is usually characterized by subcutaneous nodules, but larva migrans-like symptoms may also occur and, notably, larvae may reach the eye, becoming visible in the conjunctiva. An increasing number of reports have described the presence of microfilariae in humans [108].

Adult parasites localize to the subcutaneous and intramuscular connective tissues of dogs and other carnivores (e.g. foxes, wolves, jackals, and weasels). Following a prepatent period of 164–238 days, microfilariae are released and enter the peripheral blood where they are taken up in a blood meal by the mosquito vector [109, 110]. From there, development to the infective stage requires approximately two weeks. Once in the mammalian host, larvae migrate through the subcutaneous and muscular connective tissues, developing to the adult stage. Adults usually live two to four years in this host but can survive as long as ten years [108].

Although less clinically significant than heartworm, the identification of *D. repens* in a dog may still have value apart from simply ruling out infection with the former; symptoms associated with microfilaria hypersensitivity can arise in untreated infections, including cutaneous erythema and ulcerative pruritic lesions [111]. While it has been suggested that the localization of these two *Dirofilaria* spp. in the canine host ensures little to no opportunity for antigen cross-reactivity [108], it has still be demonstrated for *D. immitis*-specific tests and, so, cannot be relied upon with total certainty [76]. While no serological kits exist for *D. repens*, diagnosis can be performed by a number of other means.

#### 6.4.2.1 Microscopy-Based Tests

Detection of microfilariae in whole blood is the most common and reliable means of diagnosis. Concentration methods (like the modified Knott test and filtration test described for *D. immitis*) are recommended. According to some studies, microfilaremia peaks in the late afternoon and evening, especially after meals, so sensitivity may be increased by timing tests with this circadian periodicity [112, 113]. The stained microfilariae of *D. repens* can be differentiated morphologically from other species based on, for example, head shape, tail shape, and overall length. Treatment of microfilariae with 2% formalin sometimes produces a distinctive artifact in tail morphology in this and other species, becoming curved in a "button hook" or "umbrella" shape that can aid identification; note that *A. reconditum* shares head and tail features with this species. Staining for localized acid phosphatase activity may also be diagnostic, especially in cases of ambiguous parasite morphology.

#### 6.4.2.2 Recovery of Adults

The surgical removal of adult worms in subcutaneous nodules can be diagnostic. These nematodes are long, tapered at both ends, and whitish in color; adult female *D. repens* measure 100–170 mm in length and 4.6–6.5 mm in width, whereas males measure 48–70 mm in length and 3.7–4.5 mm in width. The parasite cuticle also bears distinctive longitudinal ridges absent in *D. immitis* [107]. Most nodules contain only a single intact worm, and in cases where dermatitis is present (either locally or generally), it may not be possible to recover any adults [114].

#### 6.4.2.3 Cytology

Samples from suspected parasite-bearing nodules can be taken by fine-needle aspiration and examined by cytological methods. Infected nodules typically present a mixed inflammatory infiltrate with or without eosinophilia. Aspirate samples may also include uterus fragments, microfilariae, or multinucleated morulae of the pre-microfilarial stage [107, 114]. Worm sections, if present, can also be visualized in histological preparations.

#### 6.4.2.4 DNA-Based Tests

If needed, PCR-based testing for adult worms, microfilariae, and vector-borne larval stages can be performed to confirm species identity, and both multiplexed [89, 91, 92, 115] and singleplex [116] PCR-based assays capable of specifically detecting *D. repens* have been described in the scientific literature. Both intact and partial specimens, as well as microfilariae in whole blood or filtrate, can be used for testing. However, even when a PCR result is positive, it is recommended that cytological findings in nodular lesions be used to support a diagnosis [117]. This speaks to the current lack of clinical approval for all such assays, pointing to a gap in the clinical testing toolkit.

Similar to the research environment surrounding *D. immitis* diagnostic development, at least one assay has been described with the capacity to facilitate DNA-based detection of *D. repens* at the point of collection [118]. However, development to date has been limited to the academic setting, and while novel and worthy of pursuit, point-of-collection-based testing for *D. repens* has garnered limited traction in the research community. Again, similar to the work done by the *D. immitis* community, proof-of-concept work for the PCR-based monitoring of mosquitoes for the presence of *D. repens* has been performed in a variety of settings [97, 99, 119, 120], but the feasibility of expanding and standardizing this work has been questioned [95].

#### 6.4.2.5 Imaging

Like *D. immitis*, the cuticles of adult worms are highly echogenic and can be distinguished by ultrasound of nodules as a less invasive alternative to the tests described above. The worms appear as double linear parallel structures, typical of mature filarial nematodes [114].

### 6.4.3 *Acanthocheilonema (Dipetalonema) reconditum*

Though nonpathogenic and of minimal medical concern, *A. reconditum* is a common parasite of dogs and its microfilariae must be distinguished from other, less benign parasites. As its specific epithet suggests, the biology of this nematode remains poorly studied and many significant details lacking to this day. Despite this, the parasite has a global distribution and in some regions (including the Mediterranean Basin, South America, southern Africa, and Oceania) may be the most prevalent filarioid infecting dogs [121]. The parasite is found in red foxes, but their role as a reservoir remains unknown [122]. In spite of its prevalence, only one report exists to date of *A. reconditum* parasitizing humans [123] and, as such, it is not considered a zoonotic risk.

Adult worms are found in subcutaneous connective tissue, localizing mainly to the limbs and dorsal regions [124]. After a prepatent period of 67–101 days, females release microfilariae directly into the bloodstream [125, 126]. Microfilariae are taken up in a blood meal by their arthropod vector, either fleas (*Ctenocephalides*, *Pulex*, and *Echidnophaga* spp.) or lice (*Heterodoxus* and *Linognathus* spp.) [121, 124]; the purported role of ixodid ticks as competent vectors has been rejected. Development of larvae to the infective stage in *Ctenocephalides felis* occurs over the course of 7–15 days, depending on environmental temperature [126, 127].

The tick-transmitted filarial nematode *A. dracunculoides* may present similar diagnostic challenges as *A. reconditum*. Adult worms inhabit the abdominal cavity of dogs and release microfilariae into the peripheral circulation, which may be confused with those of *D. immitis* [128]. This species is encountered only in the Old World.

#### 6.4.3.1 Microscopy-Based Tests

Detection of *A. reconditum* microfilariae is as for *D. immitis*, though infections with this parasite are usually characterized by much lower concentrations than the latter. Mixed infections do occur, and a heavy microfilaremia should be taken as indicative of this. The modified Knott test or filter test may be used, preferably with 2% formalin as morphological analysis is a key means of species identification. The head of the microfilaria is blunt compared to *D. immitis*, which is slightly tapered, and the tail exhibits a distinctive curve ("button hook" or "umbrella" shape); note that both of these features are also present in *D. repens* microfilariae. The most recent literature suggests that the microfilariae of *A. reconditum* exhibit little if any periodicity, which seems to agree with the lack of a pronounced circadian rhythm in the feeding habits of the flea vector [121, 129]. Blood for testing can thus be drawn any time of day.

Histochemical staining for acid phosphatase activity using a commercially available kit shows diffuse patterning for *A. reconditum*. This can supplement other species identification methods because *D. immitis* exhibits two foci of activity, while *A. dracunculoides* exhibits three [24].

The direct smear also has utility in species identification. As mentioned in Microscopy-based Tests for *D. immitis*, that parasite exhibits non-progressive motility, while *A. reconditum* moves with notable progressive motility, traversing the microscope field [52]. By their nature, direct smears are less sensitive than concentration techniques, especially in samples with <10 microfilariae per milliliter [57]. This is particularly relevant in infections with parasites, such as *A. reconditum* and *A. dracunculoides*, with characteristically low microfilaria concentrations [121, 130, 131]. As such, use of the direct smear alone for the detection of these parasites is not recommended.

#### 6.4.3.2 Serological and Molecular Tests

Morphological differentiation of *A. reconditum* and *D. immitis* can be challenging and may not be possible for clinicians without specific training. As such, if species confirmation is needed, PCR-based testing may be the most useful option [89, 132]. Due to overlapping microfilaria size ranges with *A. dracunculoides*, PCR may also be required to distinguish these species [24]. However, as is the case with *Dirofilaria* spp., clinically approved DNA-based tests do not exist, despite the availability of testing options within the published literature [88, 89, 133, 134].

The development of a clinically approved PCR-based test for the detection of *A. reconditum*, or of a multiplexed test enabling the simultaneous detection of both *D. immitis* and *A. reconditum*, would be of great benefit to the veterinary community, as diagnosis based on morphology alone may result in misdiagnosis, particularly when infections are light or the condition of the microfilariae within a sample is suboptimal [135]. While of reduced concern using current methods, historical reports of possible cross-reactivity of *A. reconditum* when testing for *D. immitis* using serological methods also exist [136].

### 6.4.4 *Brugia pahangi*

A parasite naturally infecting cats, *Brugia pahangi*, is endemic to southeast Asia and India with a reported prevalence of 11% to 25% in feline populations [11, 137–139]. Adult worms find their predilection site in the lymphatics, causing inflammation and fibrosis therein [140]. Dogs are also permissive to infection; while largely asymptomatic, the presence of worms may cause lymphadenopathy and lymphedema in canine hosts [141]. Considering the similarity of microfilariae from this species and the human-parasitic *B. malayi*, determining the extent of zoonotic potential is difficult, however, molecular tests have identified *B. pahangi* in cases of human filariasis [142]. Numerous wildlife hosts have been reported (e.g. primates, wild felids, and civets) that may serve as sylvatic reservoirs [143].

Both larval and adult stages of *B. pahangi* are found occupying the lymphatics, though they have also been recovered from subcutaneous tissues. In dogs, adults are most commonly recovered from the mandibular, retropharyngeal, and axillary lymphatics [144]. Adult males are 17.4–20 mm in length, whereas females reach 38–63 mm [140]. The prepatent period in cats is 69–96 days [145]. Microfilariae reach the peripheral circulation where they are taken up by the mosquito vector in a blood meal (primarily *Mansonia*, *Anopheles*, and *Armigeres* spp.) [146]. While *B. pahangi* is not present in the United States, mosquitoes including *Anopheles* and *Psorophora* spp. endemic to Louisiana have been identified as potential vectors [30].

Other, less common *Brugia* spp. have been identified in dogs and cats, which may be mistaken for *B. pahangi* on blood smears. *Brugia ceylonensis* is a parasite found in dogs that has some zoonotic potential; it is endemic to Sri Lanka with one survey putting prevalence in the canine population at 7% [31]. Distinguishing diagnostic features (e.g. staining features and gene accessions for PCR) is currently lacking, complicating definitive identification of this species [11]. *Brugia beaveri* is found in the southeastern United States; it has been identified in cats, with raccoons serving as the most significant natural host. Species identification relies on morphological characteristics [33, 147]. *Brugia patei* is found in east Africa, where it is known to infect dogs, cats, and genets. Though very little information exists on this species, microfilariae are described as morphologically similar to *B. malayi* and *B. pahangi* [32]. *B. malayi* is discussed separately in the following section.

#### 6.4.4.1 Microscopy-Based Tests

Detection of *B. pahangi* microfilariae in the blood is achieved by the same methods as for *D. immitis*. A concentration technique, like the modified Knott test, is preferred as it allows morphological examination. Microfilariae can be found in the peripheral blood at any given time, but some studies report a subperiodic nature in the infection, peaking either during the day or night [148, 149].

The microfilariae of *Brugia* spp., unlike any of the other filarial parasites discussed here, possess a distinctive sheath derived from embryonic eggshell [150]. While the sheath is lost by microfilariae of other species in the uterus, *Brugia* spp. retain it until ingested by the mosquito vector and undergoes penetration of the midgut. With staining, the structure of the sheath becomes very apparent and, combined with morphometric findings, is useful for diagnosis. Staining for acid phosphatase activity can also aid in identification; rather than foci of activity, *B. pahangi* is characterized by diffuse staining throughout the length of the worm [151].

#### 6.4.4.2 Serological and Molecular Tests

ELISA and immunoprecipitation have been used experimentally to detect circulating *B. pahangi* antigen in cats, but no commercially available test exists [152, 153]. PCR-based approaches have also been described in the scientific literature that may allow for species-specific screening and diagnosis. However, many of the assays proposed to be capable of differentiating *B. pahangi* from other species of *Brugia* require the use of cumbersome techniques, such as PCR coupled with restriction fragment length polymorphism (RFLP) analysis [154, 155] or the cloning and sequencing of PCR products in an attempt to differentiate closely related *Brugia* spp. [12]. While such approaches have been employed in operational research settings [11, 13, 156, 157], PCR–RFLP is prone to sample contamination-related error and cloning/sequencing requires a significant level of technical expertise and infrastructure. Database-derived sequences available for comparative analysis of results are also imperfect, with questionable annotation in some instances. Thus, such assays are extremely difficult to standardize and are poorly suited for use at the level of the local clinic. While the development of assays utilizing increasingly automated techniques such as high-resolution melting real-time PCR (HRM real-time PCR) [158–160] has helped to overcome some of the technical challenges associated with PCR–RFLP and sequencing-based approaches, these techniques are costly and ill-suited to use outside of a reference or research laboratory. Furthermore, standardization and clinical approval are currently lacking.

### 6.4.5 *Brugia malayi*

This is a parasite of notable concern in human medicine, occurring in India and southeast Asia where it is one of the principal causative agents of lymphatic filariasis. But, cats also serve as competent hosts for *B. malayi*, with prevalence reaching as high as 20% in endemic feline populations [139, 161]. Primates, wild felids, civets, and pangolins have been found to carry infections and may serve as reservoirs [162].

As with *B. pahangi*, larvae and adults localize in the lymphatics [163]; adult males are 13–23 mm long, whereas females are 43–55 mm long [164]. Here, adult worms may cause pathology by occluding the lymphatic vessels. In experimentally induced infections with *B. malayi*, cats have been observed to develop gross dependent limb edema [165]. The prepatent period in feline hosts ranges from 70 to 147 days [166, 167]. Mosquitoes draw microfilariae from the peripheral blood, which, after approximately 10 days, develop to infective third-stage larvae.

It is important to distinguish *B. malayi* from other species with less zoonotic potential. While a number of cases of *B. malayi* infection have been reported in dogs [11–13, 157, 168], the limitations of contemporary testing methods leaves some ambiguity. Differentiation from species more likely to infect canids, like *B. pahangi* and *B. ceylonensis*, is required to confidently reach such a diagnosis.

#### 6.4.5.1 Microscopy-Based Tests

Microfilariae can be detected using the same methods as for *B. pahangi*, the modified Knott test being recommended. The microfilaremia in feline *B. malayi* infection is typically lower than in *B. pahangi* infection [169]. Though a nocturnally periodic strain of *B. malayi* exists in humans, cats are only known to be naturally infected with the subperiodic strain of this parasite. Furthermore, experimentally induced infections of cats with the periodic strain produce subperiodic microfilaremias in this host [8, 143]. Therefore, feline blood samples drawn at any time of day can be expected to yield reasonable results.

The microfilariae of *B. malayi* are morphologically similar to other *Brugia* spp. Staining can be used to reveal the innenkorper (central viscus), which is shorter in *B. malayi* than *B. pahangi* [170]. Acid phosphatase activity can also distinguish these two species; two foci of staining (excretory and anal pores) are typical in *B. malayi* microfilariae, where *B. pahangi* is characterized by diffuse staining throughout the length of the body [151].

#### 6.4.5.2 Serological and Molecular Tests

An indirect immunofluorescence test has been described for feline infection with *B. malayi* [171]. This test was reported to be more sensitive to microfilaremic infections than amicrofilaremic infections.

Gold standard DNA-based detection of *Brugia* infections in humans employs assays targeting the *Hha*I repeat sequence [172, 173]. While these assays demonstrate exceptional sensitivity, they are specific only at the genus level. Because *B. pahangi* is generally believed to pose a limited zoonotic risk to humans, the inability to differentiate various species of *Brugia* is not generally of concern when testing human samples. However, when working with animal populations suspected of serving as reservoir hosts, the capacity for species-level differentiation becomes critical.

While felids constitute a well-documented reservoir population for *B. malayi*, the possible role played by canids as a reservoir host remains an open question. Numerous studies have proposed to demonstrate the existence of *B. malayi* infection in wild canids [11–13, 157, 168]. However, conclusions have been based on the imperfect methods available to the researchers, and therefore, significant ambiguity remains. As such, both the human and veterinary communities would benefit greatly from the development of an assay that is capable of reliably differentiating *B. malayi* from *B. pahangi*. Standardization of one or more of the HRM real-time PCR assays described in the scientific literature [158–160] could prove capable of filling this void, but such assays have not yet been used to incriminate canids as a reservoir host for *B. malayi*. A conventional reverse transcriptase PCR reaction has also been described with the capacity to differentiate *B. malayi* from *B. pahangi* through the exploitation of a single nucleotide polymorphism in the TC8100 gene transcript [174]. At present, however, this assay remains largely untested under field conditions.

Similar to the PCR-based methods described above, point-of-collection-based assays for the detection of *Brugia* are limited to genus-level discrimination. Assays utilizing LAMP [175], helicase-dependent amplification [176], and mini-PCR [177] have been described.

### 6.4.6 *Onchocerca lupi*

The medical significance of this parasite has only recently been appreciated. *Onchocerca lupi* infection has been described in dogs and cats and is now also deemed an emerging zoonosis [178, 179]. It has been reported in the southwestern United States, Europe, and the Middle East, as well as in animals with a history of travel to those regions [180–183]. Infected dogs are usually asymptomatic but can present with ocular signs, including subconjunctival and episcleral nodules. Diagnosis of this infection may be accomplished by detection of nodules or parasites on ophthalmic evaluation. If adult worms develop in the retrobulbar space, however, no obvious signs of infection may be present [184]. In zoonotic infections, humans may develop spinal, orbital, and subdermal nodules. Less invasive, subconjunctival manifestations have only been reported in Europe [179]. This parasite was first identified in a steppe wolf, so wild canids are suspected as potential reservoirs [185].

Adults of this species are typically found in nodules associate with ocular connective tissue. Microfilariae, in contrast to the other filarial species described, are found in the skin of their host where they are available to hematophagous vectors. The prepatent period remains unknown. Microfilariae of *O. lupi* localize preferentially to the ears, nose, and intrascapular, periocular, and umbilical regions [186]. As of this writing, only the black fly *Simulium tribulatum* has been described as a putative vector, though biting midges and other *Simulium* spp. have been considered [187].

### 6.4.7 Microscopy-Based Tests

Due to their subcutaneous localization, skin biopsies are necessary to collect microfilariae. This is a more invasive and labor-intensive technique than the routine blood sampling required for other microfilaria tests. Briefly, a skin biopsy is collected either by skin snip (i.e. using a scalpel blade) or a biopsy punch (8 mm is preferred). The sample is then soaked in saline solution at 37 °C for 6–12 hours, after which the sediment is examined by microscopy for the presence of microfilariae that have emerged. This method is also useful for monitoring microfilaridermia during treatment. As mentioned prior, microfilariae may be recovered at greater concentrations from some anatomical regions, including the ears, nose, and intrascapular, periocular, and umbilical regions. Concentration may also depend on periodicity, with one study reporting greatest mean microfilarial recovery in the afternoon [186]. As with all filarial infections, microfilaria tests are subject to false-negative results when infections are in the prepatent phase.

The microfilariae of *O. lupi* and *Cercopithifilaria* spp. are both recoverable by skin biopsies and may need to be differentiated [188]. The latter comprise three species of veterinary concern encountered in Europe (*Cercopithifilaria grassii*, *Cercopithifilaria bainae*, and another incompletely described species) and *C. bainae* in Brazil and the United States, all vectored by the brown dog tick (*Rhipicephalus sanguineus*) [36, 189–191]. These species can be morphometrically distinguished, *O. lupi* microfilariae being consistently shorter in length.

#### 6.4.7.1 Recovery and Imaging of Adults

In the event that nodules are removed from the site of infection, adults of *O. lupi* can be identified by histopathological examination. Characteristic cuticular patterns can be used for identification with staining (e.g. hematoxylin and eosin) and light microscopy; the *O. lupi* cuticle exhibits a pair of inner transverse striae between each pair of outer cuticular ridges [192].

Ultrasound can be used as a minimally invasive means of visualizing adult worms, which appear as hyperechoic round lesions in the periorbital tissue. This carries the advantage of revealing parasites in the retrobulbar space, which may be missed by ophthalmic examination. Neither MRI nor CT scanning are found to be as useful as ultrasound [193].

#### 6.4.7.2 Molecular Tests

Both conventional and real-time PCR techniques are available for the molecular identification of *O. lupi* [194–196]. As these assays target DNA elements of the *O. lupi* genome, they allow for the detection of both adult and larval parasite stages. Currently, serological methods for the detection of *O. lupi* are lacking. Therefore, while not yet standardized or approved for clinical use, DNA-based options are an attractive strategy and PCR assays have been used to identify the presence of infection in multiple geographic locations [187, 192, 197]. These methods have also proven useful for the detection of the pathogen within vector insects [187, 196]. As efforts to identify and incriminate competent species of vectors remain ongoing, the high-throughput nature and specificity of detection makes such DNA-based assays increasingly attractive as vector surveillance tools. Only through the successful completion of properly conducted incrimination studies can potential ranges of infection be determined and the risk of disease spread be assessed.

## 6.5 Conclusions

Over a century of research, the diagnostic methods developed for filarial nematodes have been varied and numerous, taking advantage of their particular life cycles and biological characteristics. For most species of veterinary importance, multiple tests exist that provide complementary means of detection and identification (i.e., the pairing of microscopy-based techniques with molecular techniques), allowing for a well-informed diagnosis. As technologies develop, however, new options may become more appealing. DNA-based tests are desirable for their sensitivity and specificity, overcoming some of the persistent issues in immunodiagnostic methods. The benefits of developing and validating such tests are clear, and current work may allow DNA-based identification of more species, and in formats more readily available to clinicians. This may be especially beneficial for work with less well-studied filarial species, for which validated testing is particularly lacking. With the advent of rapid, molecular techniques, however, the importance of microfilaria detection might easily be discounted. Nevertheless, direct observation of microfilariae provides important information on the present status of infection, likelihood of adverse reactions, and reservoir potential; as such, it should regularly accompany other approaches for the most reliable, informative diagnostic picture.

The characteristics most desirable in a diagnostic test may differ between human and veterinary medicine, the former focusing on community-level treatment with the ultimate goal of complete elimination, and the latter operating on a patient-by-patient basis. Similarly, some tests require sensitive reagents and/or equipment and may be suitable for a clinical setting but not fieldwork. Despite such differences, techniques for the detection and treatment of filarial parasites developed in the veterinary arena have seen use in human medicine, and vice versa. Rapid and reliable methodologies are universally sought after, and advancements made in the diagnosis of one filarial species, or in one particular host, may beneficially translate to others.

## References


1 Rhee, J.K., Yang, S.S., and Kim, H.C. (1998). Periodicity exhibited by *Dirofilaria immitis* microfilariae identified in dogs of Korea. *Korean J. Parasitol.* The Korean Society for Parasitology 36: 235–239. Available from: http://www.ncbi.nlm.nih.gov/pmc/articles/PMC2732962/.

2 Church, E.M., Georgi, J.R., and Robson, D.S. (1976). Analysis of the microfilarial periodicity of *Dirofilaria immitis*. *Cornell Vet. U. S.* 66: 333–346.

3 Angus, B.M. (1981). Periodicity exhibited by microfilariae of *Dirofilaria immitis* in south east Queensland. *Aust. Vet. J.* Blackwell Publishing Ltd 57: 101–102. http://dx.doi.org/10.1111/j.1751-0813.1981.tb00464.x.

4 Ranjbar-Bahadori, S., Veshgini, A., Shirani, D. et al. (2011). Epidemiological aspects of canine dirofilariasis in the north of Iran. *Iran. J. Parasitol.* Tehran University of Medical Sciences 6: 73–80. https://pubmed.ncbi.nlm.nih.gov/22347277.

5 Matola, Y.G. (1991). Periodicity of *Dirofilaria immitis* microfilariae in a dog from Muheza district, Tanzania. *J. Helminthol. Engl.* 65: 76–78.

6 Kume, S. (1974, 16–17 March, 1974, Auburn, Alabama, USA. (Compiled Amer. Hear. Soc.). Bonner Springs, Kansas 66012: VM Publishing, Inc.; 1975). Experimental observations on seasonal periodicity of microfilariae. In: *Proc. Hear. Symp.*, 26–31.

7 Sawyer, T.K. (1974, 16–17 March, 1974, Auburn, Alabama, USA. (Compiled Amer. Hear. Soc.). Bonner Springs, Kansas 66012). Seasonal fluctuations of microfilariae in two dogs naturally infected with *Dirofilaria immitis*. In: *Proc. Hear. Symp.*, 23–25. VM Publishing, Inc.; 1975.

8 Laing, A.B.G. (1961). Influence of the animal host on the microfilarial periodicity of *Brugia malayi*. *Trans. R. Soc. Trop. Med. Hyg.* 55: 558.

9 Chalifoux, L.V., Hunt, R.D., Garcia, F.G. et al. (1973). Filariasis in New World monkeys: histochemical differentiation of circulating microfilariae. *Lab Anim. Sci.* United States 23: 211–220.

10 Chalifoux, L. and Hunt, R.D. (1971). Histochemical differentiation of *Dirofilaria immitis* and *Dipetalonema reconditum*. *J. Am. Vet. Med. Assoc.* 158: 601–605.

11 Ravindran, R., Varghese, S., Nair, S.N. et al. (2014). Canine Filarial Infections in a Human *Brugia malayi* Endemic Area of India. Braga FR, editor. *Biomed. Res. Int.* Hindawi Publishing Corporation 2014: 630160. Available from: https://doi.org/10.1155/2014/630160.

12 Ambily, V.R., Pillai, U.N., Arun, R. et al. (2011). Detection of human filarial parasite *Brugia malayi* in dogs by histochemical staining and molecular techniques. *Vet. Parasitol.* Netherlands 181: 210–214.

13 Chirayath, D., Alex, P.C., George, S. et al. (2017). Identification of *Brugia malayi* in dogs in Kerala. *India. Trop. Biomed.* 34: 804–814.

14 Moorhead, A.R., Evans, C.C., and Kaplan, R.M. (2017). A diagnostic algorithm for evaluating cases of potential macrocyclic lactone-resistant heartworm. *Parasites Vectors.* 10.

15 Geary, T.G., Bourguinat, C., and Prichard, R.K. (2011). Evidence for macrocyclic lactone anthelmintic resistance in *Dirofilaria immitis*. *Top Companion Anim. Med.* United States 26: 186–192.

16 Pennington, N.E. and Phelps, C.A. (1969). Canine filariasis on Okinawa, Ryukyu Islands. *J. Med. Entomol.* England 6: 59–67.

17 More, S.J. and Copeman, D.B. (1990). A highly specific and sensitive monoclonal antibody-based ELISA for the detection of circulating antigen in bancroftian filariasis. *Trop. Med. Parasitol.* 41: 403–406. Available from: http://europepmc.org/abstract/MED/2075384.

18 WHO (2015). Global programme to eliminate lymphatic filariasis: progress report, 2014. *Wkly Epidemiol. Rec. Relev épidémiologique Hebd* 90: 489–504.

19 Atwell, R.B. (2018). Clinical signs and diagnosis of canine dirofilariasis. In: *Dirofilariasis* (eds. B. PFL and R.B. Atwell), 61–81. Boca Raton, FL: CRC Press, Inc.

20 Weil, G.J. and Ramzy, R.M.R. (2007). Diagnostic tools for filariasis elimination programs. *Trends Parasitol.* England 23: 78–82.

21 Weil, G.J., Curtis, K.C., Fischer, P.U. et al. (2011). A multicenter evaluation of a new antibody test kit for lymphatic filariasis employing recombinant *Brugia malayi* antigen Bm-14. *Acta Trop.* Elsevier 120: S19–S22.

22 Rahmah, N., Shenoy, R.K., Nutman, T.B. et al. (2003). Multicentre laboratory evaluation of *Brugia* Rapid dipstick test for detection of brugian filariasis. *Trop. Med. Int. Heal.* Wiley Online Library 8: 895–900.

**23** Shearer, P. (2011). BARK report: literature review-heartworm disease. *BARK Rep. Lit. Rev. Dis.* Banfield Pet Hospital Available from: https://www.banfield.com/getmedia/e456eec3-77f2-46f1-b302-87a17a8fba0a/5906a327-4e82-49f7-9432-e7da7fc568a9-pdf0 (accessed 29 May 2020).

**24** Magnis, J., Lorentz, S., Guardone, L. et al. (2013). Morphometric analyses of canine blood microfilariae isolated by the Knott's test enables *Dirofilaria immitis* and *D. repens* species-specific and *Acanthocheilonema* (syn. *Dipetalonema*) genus-specific diagnosis. *Parasites Vectors* 6: 48. http://dx.doi.org/10.1186/1756-3305-6-48.

**25** Redington, B.C., Jackson, R.F., Seymour, W.G., and Otto, G.F. (1978). The various microfilariae found in dogs in the United States. In: *Proc Hear Symp KS* (ed. G.F. Otto), 14–21. Atlanta, GA: VM Publ Co, Bonner Springs.

**26** Pennington, N.E. (1971). *Arthropod vectors, cyclodevelopment and prepatent period of Dipetalonema reconditum (Grassi) and the incidence of canine filariasis and ectoparasites in North-central Oklahoma*. Oklahoma State University.

**27** Ortega-Mora, L.M., Gomez-Bautista, M., and Rojo-Vazquez, F.A. (1989). The acid phosphatase activity and the morphological characteristic of *Dipetalonema dracunculoides* (Cobbold, 1870) microfilariae. *Vet. Parasitol.* 33 http://dx.doi.org/10.1016/0304-4017(89)90066-6.

**28** Schwan, E.V. and Schroter, F.G. (2006). First record of *Acanthocheilonema dracunculoides* from domestic dogs in Namibia. *J. S. Afr. Vet. Assoc.* AOSIS 77: 220–221.

**29** Yen, P.K.F. and Mak, J.W. (1978). Histochemical differentiation of *Brugia*, *Wuchereria*, *Dirofilaria* and *Breinlia* microfilariae. *Ann. Trop. Med. Parasitol.* Taylor & Francis 72: 157–162.

**30** Schacher, J.F. (1962). Morphology of the microfilaria of *Brugia pahangi* and of the larval stages in the mosquito. *J. Parasitol.*United States 48: 679–692.

**31** Rajapakshe, R.P.A.S., Perera, W.S.R., Ihalamulla, R.L. et al. (2005). Study of dirofilariasis in a selected area in the Western Province. *Ceylon Med. J.* Sri Lanka 50: 58–61.

**32** Buckley, J.J.C., Nelson, G.S., and Heisch, R.B. (1958). On *Wuchereria patei* n. sp. from the Lymphatics of Cats, Dogs and Genet Gats on Pate Island, Kenya. *J. Helminthol.* Cambridge University Press 32: 73–80.

**33** Ash, L.R. and Little, M.D. (1964). *Brugia beaveri* sp. n.(Nematoda: Filarioidea) from the raccoon (*Procyon lotor*) in Louisiana. *J. Parasitol.* JSTOR: 119–123.

**34** Mutafchiev, Y., Dantas-Torres, F., Giannelli, A. et al. (2013). Redescription of *Onchocerca lupi* (Spirurida: Onchocercidae) with histopathological observations. *Parasites Vectors* 6: 309. https://doi.org/10.1186/1756-3305-6-309.

**35** Cortes, H.C.E., Cardoso, L., Giannelli, A. et al. (2014). Diversity of *Cercopithifilaria* species in dogs from Portugal. *Parasites Vectors*. BioMed Central 7: 261.

**36** Otranto, D., Brianti, E., Dantas-Torres, F. et al. (2013). Species diversity of dermal microfilariae of the genus *Cercopithifilaria* infesting dogs in the Mediterranean region. *Parasitology*. England 140: 99–108.

**37** Almeida, G.L.G. and Vicente, J.J. (1984). *Cercopithifilaria bainae* sp. n. parasita de *Canis familiaris* (L.)(Nematoda, Filarioidea). *Atas Soc. Biol. Rio. Janeiro* 24: 18.

**38** McCall, J.W., Genchi, C., Kramer, L.H. et al. (2008). Chapter 4: Heartworm Disease in Animals and Humans. *Adv. Parasitol.* Academic Press 66: 193–285. Available from: https://www.sciencedirect.com/science/article/abs/pii/S0065308X08002042.

**39** Genchi, C. and Kramer, L.H. (2020). The prevalence of *Dirofilaria immitis* and *D. repens* in the Old World. *Vet. Parasitol.* Elsevier 280: 108995. Available from: https://www.sciencedirect.com/science/article/abs/pii/S0304401719302766.

**40** Orihel, T.C. and Eberhard, M.L. (1998). Zoonotic filariasis. *Clin. Microbiol. Rev.* United States 11: 366–381.

**41** Simon, F., Siles-Lucas, M., Morchon, R. et al. (2012). Human and animal dirofilariasis: the emergence of a zoonotic mosaic. *Clin. Microbiol. Rev.* United States 25: 507–544.

**42** Newton, W.L. (1968). Longevity of an experimental infection with *Dirofilaria immitis* in a dog. *J. Parasitol.* 54: 187–188.

**43** Underwood, P.G. and Harwood, P.D. (1939). Survival and location of the microfilariae of *Dirofilaria immitis* in the dog. *J. Parasitol.* 25: 23–33.

**44** Orihel, T.C. (1961). Morphology of the larval stages of *Dirofilaria immitis* in the dog. *J. Parasitol.* United States 47: 251–262.

**45** Kotani, T. and Powers, K.G. (1982). Developmental stages of *Dirofilaria immitis* in the dog. *Am. J. Vet. Res.* United States 43: 2199–2206.

**46** McCall, J.W., Genchi, C., Kramer, L. et al. (2008). Heartworm and *Wolbachia*: therapeutic implications. *Vet. Parasitol.* 158: 204–214. Available from: http://www.sciencedirect.com/science/article/pii/S0304401708004500.

**47** Dillon, R., Brawner, W.R., and Hanrahan, L. (1995). Influence of number of parasites and exercise on the severity of heartworm disease in dogs. In: *Proc Hear Symp '95* (eds. M.D. Soll and D.H. Knight), 113. Batavia, IL: American Heartworm Society.

**48** American Heartworm Society (2020). Current Canine Guidelines for the Prevention, Diagnosis, and Management of Heartworm (*Dirofilaria immitis*) Infection in Dogs. Wilmington, DE.

**49** Companion Animal Parasite Council. *CAPC guidelines on heartworm for dogs.* 2016. Available from: https://capcvet.org/guidelines/heartworm/ (accessed 21 March 2020).

**50** European Society of Dirofilariasis and Angiostrongylosis. *Guidelines for Clinical Management of Canine Heartworm Disease.* 2017.

**51** Knott, J. (1939). A method for making microfilarial surveys on day blood. *Trans. R. Soc. Trop. Med. Hyg.* No longer published by Elsevier 33: 191–196. Available from: https://www.sciencedirect.com/science/article/pii/S003592033990101X.

**52** Zajac, A.M., Conboy, G.A., Greiner, E.C. et al. (2006). *Veterinary Clinical Parasitology*, 8e (eds. A.M. Zajac and G.A. Conboy). Wiley-Blackwell: Danvers, MA.

**53** Evans, C.C., Bradner, J.L., Savadelis, M.D. et al. (2019). Acetic acid as an alternative reagent in the modified Knott test. *Vet. Parasitol.* Elsevier 276: 108975. Available from: https://www.sciencedirect.com/science/article/abs/pii/S0304401719302560.

**54** Peribanez, M.A., Lucientes, J., Arce, S. et al. (2001). Histochemical differentiation of Dirofilaria immitis, *Dirofilaria repens* and *Acanthocheilonema dracunculoides* microfilariae by staining with a commercial kit, Leucognost-SP®. *Vet. Parasitol.*: 102. Available from: http://dx.doi.org/10.1016/S0304-4017(01)00516-7.

**55** Tilley, L.P. and Wilkins, R.J. (1974). The Difil Test kit for detection of canine heartworm microfilariae. *Vet. Med. Small Anim. Clin.* United States 69: 288–294.

**56** Collins, J.D. (1971). The detection of microfilariae using the capillary haematocrit tube method. *Trop. Anim. Health Prod.* 3: 23–25. Available from: https://doi.org/10.1007/BF02356680.

**57** Courtney, C.H. and Zeng, Q.Y. (2001). Relationship between microfilaria count and sensitivity of the direct smear for diagnosis of canine dirofilariosis. *Vet. Parasitol.*: 94. Available from: http://dx.doi.org/10.1016/S0304-4017(00)00377-0.

**58** Calvert, C.A. (1987). Confirming a diagnosis of heartworm infection in dogs. *Vet. Med.* 82: 232–237.

**59** Rawlings, C.A., Dawe, D.L., McCall, J.W. et al. (1982). Four types of occult *Dirofilaria immitis* infection in dogs. *J. Am. Vet. Med. Assoc.* 180: 1323–1326. Available from: http://europepmc.org/abstract/MED/7096174.

**60** Todd, K.S.J. and Howland, T.P. (1983). Transplacental transmission of *Dirofilaria immitis* microfilariae in the dog. *J. Parasitol.* United States 69: 371.

**61** Kemmerer, D.W. (1998). Heartworm disease in the domestic ferret. In: *Recent Adv Hear Dis Symp* (eds. R.L. Seward and D.H. Knight), 87–89. Batavia, IL: American Heartworm Society.

**62** Donahoe, J.M. (1975). Experimental infection of cats with *Dirofilaria immitis. J. Parasitol.* United States 61: 599–605.

**63** Courtney, C.H. and Cornell, J.A. (1990). Evaluation of heartworm immunodiagnostic tests. *J. Am. Vet. Med. Assoc.* United States 197: 724–729.

**64** McCall, J.W., Supakorndej, N., Donoghue, A.R. et al. (2001). Evaluation of the performance of canine heartworm antigen test kits licensed for use by veterinarians and canine heartworm antigen tests conducted by diagnostic laboratories. In: *Recent Adv Hear Dis Symp* (eds. R.L. Seward and D.H. Knight), 97–104. Batavia, IL: American Heartworm Society.

**65** Courtney, C.H. and Zeng, Q.Y. (2001). Comparison of heartworm antigen test kit performance in dogs having low heartworm burdens. *Vet. Parasitol.* Elsevier 96: 317–322. Available from: https://www.sciencedirect.com/science/article/abs/pii/S0304401701003740.

**66** Grieve, R.B. and Knight, D.H. (1985). Anti-*Dirofilaria immitis* antibody levels before and after anthelmintic treatment of experimentally infected dogs. *J. Parasitol.* United States 71: 56–61.

**67** Wang, L.C. (1998). Comparison of a whole-blood agglutination test and an ELISA for the detection of the antigens of *Dirofilaria immitis* in dogs. *Ann. Trop. Med. Parasitol.* England 92: 73–77.

**68** Courtney, C.H. and Zeng, Q.Y. (1987). Predicting heartworm burdens with a heartworm antigen test kit. *J. Am. Anim. Hosp. Assoc.* 23: 387–390.

**69** Atkins, C.E. (2003). Comparison of results of three commercial heartworm antigen test kits in dogs with low heartworm burdens. *J. Am. Vet. Med. Assoc.* United States 222: 1221–1223.

**70** Tonelli, Q.J. and Quentin, A.B. (1989). Factors affecting the accuracy of enzyme immunoassays for *Dirofilaria immitis* adult antigen. In: *Proc Hear Symp '89* (ed. M.D. Soll), 161–165. Batavia, IL: American Heartworm Society.

**71** Drake, J., Gruntmeir, J., Merritt, H. et al. (2015). False negative antigen tests in dogs infected with heartworm and placed on macrocyclic lactone preventives. *Parasites Vectors*. England 8: 68.

**72** Little, S.E., Munzing, C., Heise, S.R. et al. (2014). Pre-treatment with heat facilitates detection of antigen of *Dirofilaria immitis* in canine samples. *Vet. Parasitol.* Netherlands 203: 250–252.

**73** Velasquez, L., Blagburn, B.L., Duncan-Decoq, R. et al. (2014). Increased prevalence of *Dirofilaria immitis* antigen in canine samples after heat treatment. *Vet. Parasitol.* Netherlands 206: 67–70.

**74** Little, S.E., Raymond, M.R., Thomas, J.E. et al. (2014). Heat treatment prior to testing allows detection of antigen of *Dirofilaria immitis* in feline serum. *Parasites Vectors*. England 7: 1.

**75** Gruntmeir, J.M., Long, M.T., Blagburn, B.L., and Walden, H.S. (2020). Canine heartworm and heat treatment: an evaluation using a well based enzyme-linked immunosorbent assay (ELISA) and canine sera with confirmed heartworm infection status. *Vet. Parasitol.* 283: 109169. Available from: http://www.sciencedirect.com/science/article/pii/S0304401720301497.

**76** Venco, L., Manzocchi, S., Genchi, M., and Kramer, L.H. (2017). Heat treatment and false-positive heartworm antigen testing in ex vivo parasites and dogs naturally infected by *Dirofilaria repens* and *Angiostrongylus vasorum*. *Parasites Vectors* 10: 476. https://doi.org/10.1186/s13071-017-2444-6.

**77** Schnyder, M. and Deplazes, P. (2012). Cross-reactions of sera from dogs infected with *Angiostrongylus vasorum* in commercially available *Dirofilaria immitis* test kits. *Parasites Vectors*. England 5: 258.

**78** Aroch, I., Rojas, A., Slon, P. et al. (2015). Serological cross-reactivity of three commercial in-house immunoassays for detection of *Dirofilaria immitis* antigens with *Spirocerca lupi* in dogs with benign esophageal spirocercosis. *Vet. Parasitol.* Netherlands 211: 303–305.

**79** Krucik, D.D.R., Van Bonn, W., and Johnson, S.P. (2016). Association between positive canine heartworm (*Dirofilaria immitis* ) antigen results and presence of *Acanthocheilonema odenhali* microfilaria in California sea lions (*Zalophus californius*). *J. Zoo. Wildl. Med.* United States 47: 25–28.

**80** Berdoulay, P., Levy, J.K., Snyder, P.S. et al. (2004). Comparison of serological tests for the detection of natural heartworm infection in cats. *J. Am. Anim. Hosp. Assoc.* United States 40: 376–384.

**81** Snyder, P.S., Levy, J.K., Salute, M.E. et al. (2000). Performance of serologic tests used to detect heartworm infection in cats. *J Am Vet Med Assoc.* United States 216: 693–700.

**82** Supakorndej, P. (1992). Biology, diagnosis, and prevention of heartworm infection in ferrets. In: *Proc Hear Symp '92* (ed. M.D. Soll), 59–69. Batavia, IL: American Heartworm Society.

**83** Supakorndej, P., McCall, J.W., and Jun, J.J. (1994). Early migration and development of *Dirofilaria immitis* in the ferret, *Mustela putorius furo*. *J. Parasitol.* [The American Society of Parasitologists, Allen Press] 80: 237–244. Available from: http://www.jstor.org/stable/3283753 (accessed 25 March 2020).

**84** Prieto, G., Simón, F., Genchi, C. et al. (1999). Utility of adult antigens of *Dirofilaria immitis* for the early detection of dirofilariosis and for the evaluation of chemoprophylactic treatment in experimentally infected cats. *Vet. Parasitol.* Elsevier 86: 5–13.

**85** Prieto, G., Ceciliani, F., Venco, L. et al. (2002). Feline dirofilariosis: antibody response to antigenic fractions containing specific 20 to 30 kDa polypeptides from the adult *Dirofilaria immitis* somatic antigen. *Vet. Parasitol.* Elsevier 103: 341–353. Available from: https://www.sciencedirect.com/science/article/abs/pii/S0304401701005404.

**86** McCall, J.W., Suprakorndej, N., McCall, S.D., and Mansour, A.E. (2001). Evaluation of feline heartworm antibody test kits and diagnostic laboratory tests. In: *Recent Adv Hear Dis Symp* (eds. R.L. Seward and D.H. Knight), 125–133. Batavia, IL: American Heartworm Society.

**87** Gioia, G., Lecova, L., Genchi, M. et al. (2010). Highly sensitive multiplex PCR for simultaneous detection and discrimination of *Dirofilaria immitis* and *Dirofilaria repens* in canine peripheral blood. *Vet. Parasitol.* Netherlands 172: 160–163.

**88** Rishniw, M., Barr, S.C., Simpson, K.W. et al. (2006). Discrimination between six species of canine microfilariae by a single polymerase chain reaction. *Vet. Parasitol.* 135 Available from: http://dx.doi.org/10.1016/j.vetpar.2005.10.013.

**89** Latrofa, M.S., Weigl, S., Dantas-Torres, F. et al. (2012). A multiplex PCR for the simultaneous detection of species of filarioids infesting dogs. *Acta Trop* 122 Available from: http://dx.doi.org/10.1016/j.actatropica.2012.01.006.

**90** Oi, M., Sato, Y., Nakagaki, K., and Nogami, S. (2015). Detection of *Dirofilaria immitis* DNA in host serum by nested PCR. *Parasitol Res.* Germany 114: 3645–3648.

**91** Tahir, D., Bittar, F., Barré-Cardi, H. et al. (2017). Molecular survey of *Dirofilaria immitis* and *Dirofilaria repens* by new real-time TaqMan® PCR assay in dogs and mosquitoes (Diptera: Culicidae) in Corsica (France). *Vet Parasitol.* Elsevier 235: 1–7.

**92** Albonico, F., Loiacono, M., Gioia, G. et al. (2014). Rapid differentiation of *Dirofilaria immitis* and *Dirofilaria repens* in canine peripheral blood by real-time PCR coupled to high resolution melting analysis. *Vet. Parasitol.* Elsevier 200: 128–132.

**93** In Young, O.H., Kyung Tae, K.I.M., and Ho, J.S. (2017). Molecular detection of *Dirofilaria immitis* specific gene from infected dog blood sample using polymerase chain reaction. *Iran. J. Parasitol.* Tehran University of Medical Sciences 12: 433.

**94** Aonuma, H., Yoshimura, A., Perera, N. et al. (2009). Loop-mediated isothermal amplification applied to filarial parasites detection in the mosquito vectors: *Dirofilaria immitis* as a study model. *Parasites Vectors.* BioMed Central 2: 1–7.

**95** Masny, A., Sałamatin, R., Rozej-Bielicka, W., and Golab, E. (2016). Is molecular xenomonitoring of mosquitoes for *Dirofilaria repens* suitable for dirofilariosis surveillance in endemic regions? *Parasitol. Res.* Springer 115: 511–525.

**96** Grant, J.R., Pilotte, N., and Williams, S.A. (2019). A case for using genomics and a bioinformatics pipeline to develop sensitive and species-specific PCR-based diagnostics for soil-transmitted helminths. *Front. Genet. Frontiers*; Springer 10: 883.

**97** Tomazatos, A., Cadar, D., Török, E. et al. (2018). Circulation of *Dirofilaria immitis* and *Dirofilaria repens* in the Danube Delta Biosphere Reserve, Romania. *Parasites Vectors.* BioMed Central 11: 392.

**98** Torres-Chable, O.M., Baak-Baak, C.M., Cigarroa-Toledo, N. et al. (2018). Molecular detection of *Dirofilaria immitis* in dogs and mosquitoes in Tabasco, Mexico. *J Vector Borne Dis.* Medknow Publications 55: 151.

**99** Şuleşco, T., Volkova, T., Yashkova, S. et al. (2016). Detection of *Dirofilaria repens* and *Dirofilaria immitis* DNA in mosquitoes from Belarus. *Parasitol. Res.*Springer 115: 3535–3541.

**100** Mckay, T., Bianco, T., Rhodes, L., and Barnett, S. (2013). Prevalence of *Dirofilaria immitis* (Nematoda: Filarioidea) in mosquitoes from northeast Arkansas, the United States. *J. Med. Entomol.* Oxford University Press Oxford, UK 50: 871–878.

**101** Tahir, D., Davoust, B., and Parola, P. (2019). Vector-borne nematode diseases in pets and humans in the Mediterranean Basin: an update. *Vet World.* Veterinary World 12: 1630.

**102** Badertscher, R.R. 2nd, Losonsky, J.M., Paul, A.J., and Kneller, S.K. (1988). Two-dimensional echocardiography for diagnosis of dirofilariasis in nine dogs. *J. Am. Vet. Med. Assoc.* United States 193: 843–846.

**103** Moise, N.S. (1988). Echocardigraphy. In: *Canine and Feline Cardiography* (ed. P.R. Fox), 113–156. New York, NY: Churchill Livinstone, Inc.

**104** Thrall, D.E. (2018). Canine and Feline Lung. In: *Textbook of Veterinary Diagnostic Radiology*, 7e (ed. D.E. Thrall), 710–734. Philadelphia, PA: Elsevier.

**105** Venco, L., Genchi, C., Vigevani Colson, P., and Kramer, L. (2003). Relative utility of echocardiography, radiography, serologic testing and microfilariae counts to predict adult worm burden in dogs naturally infected with heartworms. In: *Recent Advances in Heartworm Disease: Symposium* (eds. R.L. Seward and D.H. Knight), 111–124. Batavia, IL: American Heartworm Society.

**106** Supakorndej, P., Lewis, R.E., McCall, J.W. et al. (1995). Radiographic and angiographic evaluations of ferrets experimentally infected with *Dirofilaria immitis. Vet. Radiol. Ultrasound.* Wiley 36: 23–29. Available from: https://doi.org/10.1111/j.1740-8261.1995.tb00208.x.

**107** Capelli, G., Genchi, C., Baneth, G. et al. (2018). Recent advances on *Dirofilaria repens* in dogs and humans in Europe. *Parasites Vectors.* BioMed Central 11: 663.

**108** Genchi, C. and Kramer, L. (2017). Subcutaneous dirofilariosis (*Dirofilaria repens*): an infection spreading throughout the old world. *Parasites Vectors.* BioMed Central 10: –517.

**109** Petry, G., Genchi, M., Schmidt, H. et al. (2015). Evaluation of the adulticidal efficacy of imidacloprid 10%/moxidectin 2.5%(w/v) spot-on (Advocate®, Advantage® Multi) against *Dirofilaria repens* in experimentally infected dogs. *Parasitol. Res.* Springer 114: 131–144.

**110** Webber, W.A.F. and Hawking, F. (1955). Experimental maintenance of *Dirofilaria repens* and *D. immitis* in dogs. *Exp. Parasitol.* 4: 143–164. Available from: http://linkinghub.elsevier.com/retrieve/pii/0014489455900072.

**111** Otranto, D., Dantas-Torres, F., Brianti, E. et al. (2013). Vector-borne helminths of dogs and humans in Europe. *Parasites Vectors.* BioMed Central 6: 16.

**112** Ionica, A.M., Matei, I.A., D'Amico, G. et al. (2017). *Dirofilaria immitis* and *D. repens* show circadian co-periodicity in naturally co-infected dogs. *Parasites Vectors.* England 10: 116.

**113** Di Cesare, A., Otranto, D., Di Giulio, E. et al. (2013). Microfilarial periodicity of *Dirofilaria repens* in naturally infested dogs. *Parasitol Res.* 112: 4273–4279.

**114** Giori, L., Garbagnoli, V., Venco, L. et al. (2010). What is your diagnosis? Fine-needle aspirate from subcutaneous mass in a dog. *Vet. Clin. Pathol.* Wiley 39: 255–256.

**115** Laidoudi, Y., Davoust, B., Varloud, M. et al. (2020). Development of a multiplex qPCR-based approach for the diagnosis of *Dirofilaria immitis, D. repens* and *Acanthocheilonema reconditum. Parasites Vectors.* BioMed Central 13: 1–15.

**116** Vakalis, N., Spanakos, G., Patsoula, E., and Vamvakopoulos, N.C. (1999). Improved detection of *Dirofilaria repens* DNA by direct polymerase chain reaction. *Parasitol. Int.* Elsevier 48: 145–150.

**117** Manzocchi, S., Venco, L., Piseddu, E. et al. (2017). Positive PCR alone should not be considered sufficient to establish *Dirofilaria repens* as the cause of subcutaneous nodular lesions in the absence of a clear cytologic picture. *Vet. Clin. Pathol.* United States: 389–390.

**118** Raele, D.A., Pugliese, N., Galante, D. et al. (2016). Development and application of a loop-mediated isothermal amplification (LAMP) approach for the rapid detection of *Dirofilaria repens* from biological samples. *PLoS Negl. Trop. Dis.* Public Library of Science San Francisco, CA USA 10: e0004789.

**119** Latrofa, M.S., Montarsi, F., Ciocchetta, S. et al. (2012). Molecular xenomonitoring of *Dirofilaria immitis* and *Dirofilaria repens* in mosquitoes from north-eastern Italy by real-time PCR coupled with melting curve analysis. *Parasites Vectors.* Springer 5: 76.

**120** Dyab, A.K., Galal, L.A., Mahmoud, A.E.-S., and Mokhtar, Y. (2015). Xenomonitoring of different filarial nematodes using single and multiplex PCR in mosquitoes from Assiut Governorate, Egypt. *Korean J. Parasitol.* Korean Society for Parasitology 53: –77.

**121** Brianti, E., Gaglio, G., Napoli, E. et al. (2012). New insights into the ecology and biology of *Acanthocheilonema reconditum* (Grassi, 1889) causing canine subcutaneous filariosis. *Parasitology*: 139. Available from: http://dx.doi.org/10.1017/S0031182011002198.

**122** Ionică, A.M., Matei, I.A., D'Amico, G. et al. (2017). Filarioid infections in wild carnivores: a multispecies survey in Romania. *Parasites Vectors* 10: 332. Available from: https://doi.org/10.1186/s13071-017-2269-3.

**123** Huynh, T., Thean, J., and Maini, R. (2001). *Dipetalonema reconditum* in the human eye. *Br. J. Ophthalmol.* 85 Available from: http://dx.doi.org/10.1136/bjo.85.11.1384i.

**124** Nelson, G.S. (1962). *Dipetalonema reconditum* (Grassi, 1889) from the dog with a note on its development in the flea, *Ctenocephalides felis* and the louse, *Heterodoxus spiniger*. *J. Helminthol.* Cambridge University Press 36: 297–308.

**125** Lindemann, B.A. and McCall, J.W. (1984). Experimental *Dipetalonema reconditum* infections in dogs. *J. Parasitol.* 70: 167.

**126** Farnell, D.R. and Faulkner, D.R. (1978). Prepatent period of *Dipetalonema reconditum* in experimentally-infected dogs. *J Parasitol.* JSTOR 64: 565–567.

**127** Napoli, E., Brianti, E., Falsone, L. et al. (2014). Development of *Acanthocheilonema reconditum* (Spirurida, Onchocercidae) in the cat flea *Ctenocephalides felis* (Siphonaptera, Pulicidae). *Parasitology*. England 141: 1718–1725.

**128** Muñoz, C., Gonzálvez, M., Rojas, A. et al. (2020). Massive microfilaremia in a dog subclinically infected with *Acanthocheilonema dracunculoides*. *Parasitol. Int.* 76: 102070. Available from: http://www.sciencedirect.com/science/article/pii/S1383576920300209.

**129** Koehler, P.G., Leppla, N.C., and Patterson, R.S. (1989). Circadian rhythm of cat flea (Siphonaptera: Pulicidae) locomotion unaffected by ultrasound. *J. Econ. Entomol.* England 82: 516–518.

**130** Cringoli, G., Rinaldi, L., Veneziano, V., and Capelli, G. (2001). A prevalence survey and risk analysis of filariosis in dogs from the Mt. Vesuvius area of southern Italy. *Vet. Parasitol.* Elsevier 102: 243–252.

**131** Giannetto, S., Poglayen, G., Gaglio, G. et al. (2003). *Dipetalonema dracunculoides* (Nematoda: Onchocercidae): first report in dog in Italy. *Parasite* 10: 188.

**132** Engelmann, A.M., Schafer, A.S., Lhamas, C.L. et al. (2019). Morphological and molecular identification of *Acanthocheilonema reconditum* in a canine. *Comp. Clin. Path.* 28: 271–274. Available from: https://doi.org/10.1007/s00580-018-2859-2.

**133** Mar, P.-H., Yang, I.-C., Chang, G.-N., and Fei, A.C.-Y. (2002). Specific polymerase chain reaction for differential diagnosis of *Dirofilaria immitis* and *Dipetalonema reconditum* using primers derived from internal transcribed spacer region 2 (ITS2). *Vet. Parasitol.* Elsevier 106: 243–252.

**134** Wongkamchai, S., Nochote, H., Foongladda, S. et al. (2014). A high resolution melting real time PCR for mapping of filaria infection in domestic cats living in brugian filariosis-endemic areas. *Vet. Parasitol.* Elsevier 201: 120–127.

**135** Little, S., Saleh, M., Wohltjen, M., and Nagamori, Y. (2018). Prime detection of *Dirofilaria immitis*: understanding the influence of blocked antigen on heartworm test performance. *Parasites Vectors*. Springer 11: 186.

**136** Gillis, J.M., Smith, R.D., and Todd, K.S. Jr., (1984). Diagnostic criteria for an enzyme-linked immunosorbent assay for occult heartworm disease: standardization of the test system in naturally exposed dogs. *Am. J. Vet. Res.* 45: 2289–2292.

**137** Mak, J.W., Yen, P.K., Lim, K.C., and Ramiah, N. (1980). Zoonotic implications of cats and dogs in filarial transmission in Peninsular Malaysia. *Trop. Geogr. Med.* Netherlands 32: 259–264.

**138** Chungpivat, S. and Sucharit, S. (1993). Microfilariae in cats in Bangkok. *Thai. J. Vet. Med.* 23: 75–87.

**139** Palmieri, J.R., Masbar, S., Marwoto, H.A. et al. (1985). The domestic cat as a host for *Brugian filariasis* in South Kalimantan (Borneo). *Indonesia J. Helminthol.* Cambridge University Press 59: 277–281.

**140** Schacher, J.F. (1962). Developmental stages of *Brugia pahangi* in the final host. *J. Parasitol.* United States 48: 693–706.

**141** Snowden, K.F. and Hammerberg, B. (1989). The lymphatic pathology of chronic *Brugia pahangi* infection in the dog. *Trans. R. Soc. Trop. Med. Hyg.* England 83: 670–678.

**142** Tan, L.H., Fong, M.Y., Mahmud, R. et al. (2011). Zoonotic *Brugia pahangi* filariasis in a suburbia of Kuala Lumpur City, Malaysia. *Parasitol. Int.* 60: 111–113. Available from: http://www.sciencedirect.com/science/article/pii/S1383576910001558.

**143** Denham, D.A. and McGreevy, P.B. (1977). *Brugian filariasis*: epidemiological and experimental studies. *Adv. Parasitol.* England 15: 243–309.

**144** Megat Abd Rani, P.A., Irwin, P.J., Gatne, M. et al. (2010). Canine vector-borne diseases in India: a review of the literature and identification of existing knowledge gaps. *Parasites Vectors* 3: 28. Available from: https://doi.org/10.1186/1756-3305-3-28.

**145** Denham, D.A. (1974). Studies with *Brugia pahangi*. 6. The susceptibility of male and female cats to infection. *J. Parasitol.* United States 60: 642.

**146** Edeson, J.F.B., Wharton, R.H., and Laing, A.B.G. (1960). A preliminary account of the transmission, maintenance and laboratory vectors of *Brugia pahangi*. *Trans. R. Soc. Trop. Med. Hyg.* 54: 439–449. Available from: http://www.sciencedirect.com/science/article/pii/0035920360900894.

**147** Harbut, C.L. and Orihel, T.C. (1995). *Brugia beaveri*: microscopic morphology in host tissues and observations on its life history. *J. Parasitol.* JSTOR: 239–243.

**148** Chungpivat, S. and Sucharit, S. (1990). Microfilarial periodicity of *Brugia pahangi* in naturally infected cats in Bangkok. *Thai. J. Vet. Med.* 20: 239–245.

**149** Sucharit, S. (1973). *Brugia pahangi* in small laboratory animals: the micro-filarial periodicity. *Southeast Asian J. Trop. Med. Public Health* 4: 492–497.

**150** Rogers, R., Ellis, D.S., and Denham, D.A. (1976). Studies with *Brugia pahangi*. 14. Intrauterine development of the microfilaria and a comparison with other filarial species. *J. Helminthol.* England 50: 251–257.

**151** Redington, B.C., Montgomery, C.A., Jervis, H.R., and Hockmeyer, W.T. (1975). Histochemical differentiation of the microfilariae of *Brugia pahangi* and sub-periodic *Brugia malayi*. *Ann. Trop. Med. Parasitol.* Taylor & Francis 69: 489–492.

**152** Au, A.C., Denham, D.A., Steward, M.W. et al. (1981). Detection of circulating antigens and immune complexes in feline and human lymphatic filariasis. *Southeast Asian J Trop Med Public Health* 12: 492–498. Available from: http://europepmc.org/abstract/MED/7046073 (accessed 20 April 2020).

**153** Kumar, H., Baldwin, C., Birch, D.W. et al. (1991). Circulating filarial antigen in cats infected with *Brugia pahangi* is indicative of the presence of adult worms. *Parasite Immunol.* Wiley Online Library 13: 405–412.

**154** Nuchprayoon, S., Sangprakarn, S., Junpee, A. et al. (2003). Differentiation of *Brugia malayi* and *Brugia pahangi* by PCR-RFLP of ITS1 and ITS2. *Southeast Asian J. Trop. Med. Public Health*. Seameo Regional Tropical Medicine & Public Health 34: 67–73.

**155** Nuchprayoon, S., Junpee, A., Poovorawan, Y., and Scott, A.L. (2005). Detection and differentiation of filarial parasites by universal primers and polymerase chain reaction–restriction fragment length polymorphism analysis. *Am. J. Trop. Med. Hyg.* ASTMH 73: 895–900.

**156** Nuchprayoon, S., Junpee, A., Nithiuthai, S. et al. (2006). Detection of filarial parasites in domestic cats by PCR-RFLP of ITS1. *Vet. Parasitol.* Elsevier 140: 366–372.

**157** Satjawongvanit, H., Phumee, A., Tiawsirisup, S. et al. (2019). Molecular analysis of canine filaria and its *Wolbachia* endosymbionts in domestic dogs collected from two animal university hospitals in Bangkok Metropolitan Region, Thailand. *Pathogens (Basel, Switzerland)*. Switzerland: 8.

**158** Areekit, S., Kanjanavas, P., Pakpitchareon, A. et al. (2009). High resolution melting real-time PCR for rapid discrimination between *Brugia malayi* and *Brugia pahangi*. *J. Med. Assoc.* Thailand= Chotmaihet thangphaet 92: S24–S28.

**159** Wongkamchai, S., Monkong, N., Mahannol, P. et al. (2013). Rapid detection and identification of *Brugia malayi*, *B. pahangi*, and *Dirofilaria immitis* by high-resolution melting assay. *Vector-Borne Zoonotic Dis.* Mary Ann Liebert, Inc. 140 Huguenot Street, 3rd Floor New Rochelle, NY 10801 USA 13: 31–36.

**160** Thanchomnang, T., Intapan, P.M., Chungpivat, S. et al. (2010). Differential detection of *Brugia malayi* and *Brugia pahangi* by real-time fluorescence resonance energy transfer PCR and its evaluation for diagnosis of *B. pahangi*-infected dogs. *Parasitol. Res.* Germany 106: 621–625.

**161** Dondero, T.J. Jr., and Menon, V.V.V. (1972). Clinical epidemiology of filariasis due to *Brugia malayi* on a rubber estate in West Malaysia. *Southeast Asian J. Trop. Med. Public Health* 3: 355–365.

**162** Laing, A.B.G., Edeson, J.F.B., and Wharton, R.H. (1960). Studies on filariasis in Malaya: the vertebrate hosts of *Brugia malayi* and *B. pahangi*. *Ann. Trop. Med. Parasitol.* Taylor & Francis 54: 92–99.

**163** Ahmed, S.S. (1966). Location of developing and adult worms of *Brugia* sp. in naturally and experimentally infected animals. *J. Trop. Med. Hyg.* 69: 291–293.

**164** Buckley, J.J.C. and Edeson, J.F.B. (1956). On the adult morphology of *Wuchereria* sp.(*malayi*?) from a monkey (*Macaca irus*) and from cats in Malaya, and on *Wuchereria pahangi* n. sp. from a dog and a cat. *J. Helminthol.* Cambridge University Press 30: 1–20.

**165** Folse, D.S. and Ewert, A. (1988). Edema resulting from experimental filariasis. *Lymphology* 21: 244–247.

**166** Ewert, A., Folse, D., Hillman, G., and Wang, Y.-X. (1983). Effect of diethylcarbamazine citrate on *Brugia malayi* infections in cats following daily, weekly, or monthly administration. *Am. J. Trop. Med. Hyg.* ASTMH 32: 385–391.

**167** Edeson, J.F.B. and Wharton, E.H. (1957). The transmission of *Wuchereria malayi* from man to the domestic cat. *Trans. R Soc. Trop. Med. Hyg.* 51: 366–370.

**168** Mallawarachchi, C.H., Chandrasena, N.T.G.A., Wickramasinghe, S. et al. (2018). A preliminary survey of filarial parasites in dogs and cats in Sri Lanka. *PLoS One*. Public Library of Science 13.

**169** Burren, C.H. (1972). The behaviour of *Brugia malayi* microfilariae in experimentally infected domestic cats. *Ann. Trop. Med. Parasitol.* Taylor & Francis 66: 235–242.

**170** Sivanandam, S. and Fredericks, H.J. (1966). The "Innenkorper" in differentiation between the microfilanriae of *Brugia pahangi* and *B. malayi* (sub-periodic form). *Med. J. Malaya* 20: 337.

**171** Prasomsitti, P., Mak, J.W., Sucharit, P., and Liew, L.M. (1983). Detection of antibodies in cats infected with filarial parasites by the indirect immunofluorescence technique. *Southeast Asian J. Trop. Med. Public Health.* 14: 353–356.

**172** Lizotte, M.R., Supali, T., Partono, F., and Williams, S.A. (1994). A polymerase chain reaction assay for the detection of *Brugia malayi* in blood. *Am. J. Trop. Med. Hyg.* ASTMH 51: 314–321.

**173** Rao, R.U., Weil, G.J., Fischer, K. et al. (2006). Detection of *Brugia* parasite DNA in human blood by real-time PCR. *J. Clin. Microbiol. Am. Soc. Microbiol.* 44: 3887–3893.

**174** Laney, S.J., Buttaro, C.J., Visconti, S. et al. (2008). A reverse transcriptase-PCR assay for detecting filarial infective larvae in mosquitoes. *PLoS Negl. Trop. Dis.* Public Library of Science 2: e251.

**175** Poole, C.B., Tanner, N.A., Zhang, Y. et al. (2012). Diagnosis of brugian filariasis by loop-mediated isothermal amplification. *PLoS Negl. Trop. Dis.* Public Library of Science 6: e1948. http://dx.doi.org/10.1371%2Fjournal.pntd.0001948.

**176** Vincent, M., Xu, Y., and Kong, H. (2004). Helicase-dependent isothermal DNA amplification. *EMBO Rep.* Wiley, Ltd Chichester, UK 5: 795–800.

**177** Zaky, W.I., Tomaino, F.R., Pilotte, N. et al. (2018). Backpack PCR: a point-of-collection diagnostic platform for the rapid detection of *Brugia* parasites in mosquitoes. *PLoS Negl. Trop. Dis.* Public Library of Science 12: e0006962. https://doi.org/10.1371/journal.pntd.0006962.

**178** Otranto, D., Dantas-Torres, F., Cebeci, Z. et al. (2012). Human ocular filariasis: further evidence on the zoonotic role of *Onchocerca lupi*. *Parasites Vectors.* England 5: 84.

**179** Cantey, P.T., Weeks, J., Edwards, M. et al. (2016). The emergence of zoonotic *Onchocerca lupi* infection in the United States–a case-series. *Clin. Infect. Dis.* Oxford University Press 62: 778–783.

**180** Labelle, A.L., Daniels, J.B., Dix, M., and Labelle, P. (2011). *Onchocerca lupi* causing ocular disease in two cats. *Vet. Ophthalmol.* England 14 (Suppl 1): 105–110.

**181** Labelle, A.L., Maddox, C.W., Daniels, J.B. et al. (2013). Canine ocular onchocercosis in the United States is associated with *Onchocerca lupi*. *Vet. Parasitol.* Elsevier 193: 297–301.

**182** Hermosilla, C., Hetzel, U., Bausch, M. et al. (2005). First autochthonous case of canine ocular onchocercosis in Germany. *Vet Rec Ed.* London: The Association 156: 450–451.

**183** Mowlavi, G., Farzbod, F., Kheirkhah, A. et al. (2014). Human ocular onchocerciasis caused by *Onchocerca lupi* (Spirurida, Onchocercidae) in Iran. *J. Helminthol.* Cambridge University Press 88: 250–255.

**184** Otranto, D., Giannelli, A., Scotty Trumble, N. et al. (2015). Clinical case presentation and a review of the literature of canine onchocercosis by *Onchocerca lupi* in the United States. *Parasites Vectors* 8: 89. Available from: https://doi.org/10.1186/s13071-015-0699-3.

**185** Rodonaja, T.E. (1967). A new species of Nematode, *Onchocerca lupi* n. sp., from *Canis lupus cubanensis. Soobshchenyia Akad Nauk Gruz SSR.* 45: 715–719.

**186** Otranto, D., Dantas-Torres, F., Giannelli, A. et al. (2013). Cutaneous distribution and circadian rhythm of *Onchocerca lupi* microfilariae in dogs. *PLoS Negl. Trop. Dis.* Public Library of Science 7: e2585–e2585. Available from: https://pubmed.ncbi.nlm.nih.gov/24349594.

**187** Hassan, H.K., Bolcen, S., Kubofcik, J. et al. (2015). Isolation of *Onchocerca lupi* in dogs and black flies, California, USA. *Emerg. Infect. Dis.* Centers for Disease Control and Prevention 21: 789.

**188** Otranto, D., Brianti, E., Abramo, F. et al. (2012). Cutaneous distribution and localization of *Cercopithifilaria* sp. microfilariae in dogs. *Vet. Parasitol.* Netherlands 190: 143–150.

**189** Boyd, M., Santoro, D., Craft, W.F. et al. (2019). Dermatitis caused by autochthonous *Cercopithifilaria bainae* from a dog in Florida, USA: clinical, histological and parasitological diagnosis and treatment. *Vet. Dermatol.* Wiley Online Library 30: 68–e20.

**190** Lineberry, M.W., Sundstrom, K.D., Little, S.E. et al. (2020). Detection of *Cercopithifilaria bainae* infection in shelter dogs and ticks in Oklahoma, USA. *Parasites Vectors* 13. Article number 216,: 1–6.

**191** Otranto, D., Varcasia, A., Solinas, C. et al. (2013). Redescription of *Cercopithifilaria bainae* Almeida & Vicente, 1984 (Spirurida, Onchocercidae) from a dog in Sardinia, Italy. *Parasites Vectors.* Springer 6: 132.

**192** Verocai, G.G., Conboy, G., Lejeune, M. et al. (2016). *Onchocerca lupi* nematodes in dogs exported from the United States into Canada. *Emerg. Infect. Dis.* Centers for Disease Control and Prevention 22: 1477–1479. Available from: https://pubmed.ncbi.nlm.nih.gov/27434170.

**193** Franchini, D., Giannelli, A., Di Paola, G. et al. (2014). Image diagnosis of zoonotic onchocercosis by *Onchocerca lupi*. *Vet. Parasitol.* Elsevier 203: 91–95.

**194** Otranto, D., Sakru, N., Testini, G. et al. (2011). First evidence of human zoonotic infection by *Onchocerca lupi* (Spirurida, Onchocercidae). *Am J Trop Med Hyg.* ASTMH 84: 55–58.

**195** Sréter-Lancz, Z., Széll, Z., and Sréter, T. (2007). Molecular genetic comparison of *Onchocerca* sp. infecting dogs in Europe with other spirurid nematodes including *Onchocerca lienalis*. *Vet. Parasitol.* Elsevier 148: 365–370.

**196** Latrofa, M.S., Annoscia, G., Colella, V. et al. (2018). A real-time PCR tool for the surveillance of zoonotic *Onchocerca lupi* in dogs, cats and potential vectors. *PLoS Negl. Trop. Dis.* Public Library of Science 12: e0006402.

**197** Miró, G., Montoya, A., Checa, R. et al. (2016). First detection of *Onchocerca lupi* infection in dogs in southern Spain. *Parasites Vectors* BioMed Central 9: 1–3.

# 7

# Antifilarial Chemotherapy: Current Options for Humans

*Sabine Specht*[1,*], *Joseph Kamgno*[2,3], *and Timothy G. Geary*[4,5]

[1] *Drugs for Neglected Diseases initiative, 15 Chemin Camille-Vidart, 1202 Geneva, Switzerland*
[2] *Epidemiology and Biostatistics, Centre for Research on Filariasis and other Tropical Diseases (CRFilMT), Street 1.411, Fouda Quarter Yaounde, Cameroon*
[3] *University of Yaoundé I, Department of Public Health, Faculty of Medicine and Biomedical Sciences, P.O. Box 1364, Yaoundé, Cameroon*
[4] *Queen's University – Belfast, School of Biological Sciences, 19 Chlorine Gardens, Belfast BT9 5DL, UK*
[5] *McGill University, Institute of Parasitology, 21111 Lakeshore Road, Ste-Anne-de-Bellevue, H9X 3V9 QC, Canada*

**Abstract**

Treatment options for the major human filariases are predicated on an understanding of the diseases themselves and the history of control efforts. In this chapter, we provide an integrated summary of onchocerciasis, lymphatic filariasis, and loaisis based on consideration of pathology and diagnosis as a basis for therapy. The origins of antifilarial drugs and their incorporation into treatment and control programs are reviewed. We provide a summary of currently recommended chemotherapeutic regimens for Mass Drug Administration as well as individual treatment in non-endemic regions and comment on remaining challenges in this area.

## 7.1 Introduction

Humans can be parasitized by several species of filarial nematodes. Of them, three have attracted most chemotherapeutic attention: parasites that cause lymphatic filariasis (LF), including most prominently *Wuchereria bancrofti* and to a lesser extent *Brugia timori* and *Brugia malayi*; the parasite that causes onchocerciasis (river blindness, *Onchocerca volvulus*); and *Loa loa*, also called the African eye worm. Infections with *Mansonella* spp. (*Mansonella ozzardi, M. streptocerca, M. perstans*) are mostly asymptomatic and are rather poorly studied in humans. Of these diseases, onchocerciasis and LF are classified as neglected tropical diseases (NTDs) and are found almost exclusively in tropical areas, primarily in resource-limited regions.

*Corresponding author.

*Human and Animal Filariases: Landscape, Challenges, and Control*, First Edition.
Edited by Ronald Kaminsky and Timothy G. Geary.

Filariae are two-host parasites; humans are the definitive hosts in which sexual reproduction occurs, and an arthropod intermediate host, in which parasite development occurs, is required to transmit the parasite from one person to another [1]. Parasites that cause LF are transmitted by multiple species of mosquito in two genera (*Anopheles* and *Culex*), while onchocerciasis and loiasis are transmitted by flies in the genera *Simulium* and *Chrysops*, respectively. These arthropod vectors acquire L1 larval stages (microfilariae; mf) during a blood meal from the human host; L1 larvae subsequently molt twice to the L3 (infective) stage in the vector, from which they are introduced into a new host during a blood meal and subsequently develop to adult stages. L3 larvae molt inside the human host to L4 and then L5 stages before maturing to adults over a period of months.

Humans appear to be the exclusive hosts in natural settings for *O. volvulus*, *W. bancrofti*, *B. timori*, and *L. loa*, whereas *B. malayi* can infect other mammals (most notably felines). Accidental infections may also occur with other filariae, for example, with the canine heartworm *Dirofilaria immitis*, but these infections almost never produce fertile adult parasites.

As noted, filarial infections are almost entirely found in people living in areas with limited healthcare resources, where they compete with more urgent needs for diagnostic and therapeutic attention. As a result, and due to the fact that filarial diseases are transmissible and most of the time highly prevalent in a locale, antifilarial chemotherapy has been primarily directed at community-level control rather than treatment of individuals following diagnosis. Current treatment of river blindness and LF relies on preventive chemotherapy (PC), in which medicines are administered annually or biannually. Ivermectin (IVM) is given alone for onchocerciasis and in combination with albendazole (ALB) for LF in onchocerciasis regions, with diethylcarbamazine (DEC) given in combination with IVM and ALB where onchocerciasis is absent. The strategy in place in Africa for onchocerciasis is Community Directed Treatment with Ivermectin (CDTI) developed by the former African Program for Onchocerciasis Control (APOC). The objective for control is reduction of larval stages (mf) of the parasites in humans and consequently reduction of transmission among populations at risk of infection or illness. In onchocerciasis, in which pathology is primarily caused by mf, this strategy also markedly reduces blindness and skin disease. Morbidity management is also a component of control for LF (https://www.who.int/neglected_diseases/resources/9789240696471/en/), leading to programs in which entire communities in endemic areas are treated without individual diagnosis.

These efforts are coordinated by several stakeholders, including endemic countries, funders, donor countries, pharmaceutical companies, and the WHO (ESPEN|Expanded Special Project for Elimination of Neglected Tropical Diseases). Mass drug administration (MDA) campaigns of chemotherapy in humans have demonstrated the potential to lead to the elimination of these parasites as public health problems and perhaps even their elimination.

The onchocerciasis control program was designed to accomplish two aims. The first was to eliminate transmission of the parasite that causes onchocerciasis by markedly reducing the population of mf in treated populations, and secondly to reduce overall pathology in patients; targeting mf abundance had the important aim of preventing ocular and dermal pathology, with special emphasis on the prevention of blindness. Both aims were achieved by the prolonged suppression of mf in the blood or skin of infected individuals through yearly or twice-yearly oral doses of IVM.

The LF program was also built around two main objectives: the reduction of mf in treated subject to reduce transmission with one or twice a year mass treatment with DEC or IVM combined with ALB (now all three in combination; see below). The second objective was morbidity management of LF-related lesions (elephantiasis and hydrocele).

Both programs have had marked benefits on the prevalence of pathological manifestations of these infections and have led to the elimination of LF and onchocerciasis as public health problems in many areas. Indeed, MDA programs have almost eradicated *O. volvulus* in the Americas [2]. Despite these efforts, elimination has not yet been achieved and challenges for controlling these diseases remain. A major limitation to success is that adult filaroid parasites can live for almost 15 years in the human host. Macrofilaricidal drugs can speed up the time to elimination. In this regard, doxycycline kills the *Wolbachia* endosymbiotic bacteria of *O. volvulus* and LF worms, leading to sterilization and death. However, a constraint with this drug is the long duration of treatment needed for efficacy (four to six weeks). This constraint means that doxycycline cannot easily be used for MDA. The combination of IVM + DEC + ALB appears to be macrofilaricidal in LF (see below); if consistently confirmed in LF areas, this strategy would clearly accelerate progress toward elimination.

In April 2020, the WHO published a new road map for NTDs for the period 2021–2030 [3], which aims to guide countries to achieve Sustainable Development Goal 3.3: to ensure healthy lives and promote well-being for all at all ages and to end endemics of NTDs by 2030. It proposes important shifts for NTD programs, which will be fundamental to sustain progress for NTD control, including filarial diseases. The need for a more holistic approach to address cross-cutting development issues and strong country ownership, including domestic funding, has been recognized, and programs and activities to fight NTDs should be further combined to achieve maximum return on investment. This has become even more obvious during the current SARS-CoV-2 pandemic, in which tremendous financial and strategic investments are being made that could potentially be useful for NTDs as well.

LF has been eliminated as a public health problem in 17 countries as of 2019. The new roadmap targets 58 countries (81% of the total). Three critical actions are identified:

(1) Start MDA in all endemic districts and strengthen MDA in all settings. Implement improved interventions where appropriate (e.g. three-drug treatment in settings that qualify; strategies for hotspots).
(2) Improve capacity for morbidity management and disability prevention; prioritize in primary health care and as part of universal health coverage.
(3) Improve diagnostics, strengthen criteria for stopping MDA, establish post-MDA and post-validation surveillance standards; update guidelines with new tools and strategies as appropriate.

Onchocerciasis has been eliminated from four countries in the Region of the Americas and is now targeted for elimination (interruption of transmission) in

12 countries (31% of countries in which the parasite is transmitted). Three critical actions are identified:

(1) Start MDA in all endemic areas after mapping, improve delivery of current MDA programs and implement alternative strategies where appropriate.
(2) Develop improved diagnostics to facilitate mapping and decisions to eliminate transmission; develop improved diagnostic strategy for *L. loa*; increase program capacity to perform entomological and laboratory diagnostics.
(3) Develop a macrofilaricidal drug and diagnostic tool or other elimination strategies to accelerate interruption of transmission; design a case management strategy; develop and implement elimination strategies for *L. loa* co-endemic areas where onchocerciasis is hypoendemic.

A future challenge will be improved mapping of cases with a sensitive and specific diagnostic tool to allow a more targeted approach for treatment, especially for onchocerciasis, for which no sensitive, noninvasive diagnostic tools are available. This is a fundamental requirement to reduce costs to country programs, pharmaceutical partners, and international donors.

Loiasis, however, is not yet incorporated into the roadmap, despite its clinical manifestations, such as localized angioedema (Calabar swelling), itching, muscle pain, neurological, cardiac and renal complications (see below). The same status applies to mansonellosis, which, in contrast to other filarial diseases, generally appears to be asymptomatic, particularly in individuals living in endemic regions.

In this chapter, we discuss disease symptoms, drugs available for use in humans, and current treatment approaches.

## 7.2 Clinical Presentation

### 7.2.1 Lymphatic Filariasis

Adult LF parasites live in nests in lymphatic vessels, often in the groin region, where they disrupt the function of the lymphatic system, sometimes resulting in lymphangitis and progressively massive swelling of downstream limbs or tissues. Adult parasites are the primary source of pathology. Microfilariae released from fertilized females enter the bloodstream, where they are available to mosquito vectors, but are not associated with significant harm to the infected individual.

LF varies from subclinical infection to chronic manifestations, resulting in lymphatic damage and dysfunction leading to recurrent swelling and disfigurement of the limbs (elephantiasis), genitalia (hydrocele) in men, and sometimes breasts in women. The three species cause similar signs and symptoms, although urogenital disease and chyluria do not occur in *Brugia* infections.

Most individuals with patent infections documented by the presence of microfilaremia do not have clinically overt manifestations of a lymphatic pathologic process. Nevertheless, imaging studies indicate that asymptomatic infected adults and children may have compromised lymphatic function [4–7]. Overt sequalae first become apparent during adolescence and early adulthood, often as acute adenolymphangitis (AFL, acute filarial lymphangitis) with fever and swelling of the leg, arm, or the male genitalia. These episodes may last up to a week in persons who have been previously asymptomatic or longer in persons who have experienced repeated attacks. Pathology that progresses toward the lymph node from a peripheral site is believed to be due to secondary bacterial infections and not inflammation elicited by filarial worms [8]. Repeated episodes of AFL and reversible episodes of lymphedema often precede the development of chronic lymphatic pathology that includes lymphedema of the legs, arms, breasts, and chronic disfigurement of the male genitalia.

Genital manifestations are more common in *W. bancrofti* infection than in *Brugia* spp. infection. Genital involvement includes hydrocele and other forms of genital disease, such as funiculitis, epididymitis, orchitis, and lymphedema of the scrotum [9] in men, and vulvar lymphedema in women [10]. The pathogenesis of these manifestations of lymphatic filariasis is poorly understood [7, 11], but is related to lymphatic damage caused by living adult *W. bancrofti*. Chyluria, resulting from rupture of dilated retroperitoneal lymphatic structures into the renal pelvis, is a rare but serious manifestation of LF. Albeit itself painless, weight loss and malnutrition can occur because of the loss of chyle, which contains dietary lipids, proteins, and vitamins. The frequency of chyluria in filariasis is not known, but it is lower than appearance of lymphedema or hydrocele.

Lymphedema occurs in the legs, scrotum, penis, breast, and arms. It is more common in the lower extremities with asymmetric involvement. In bancroftian filariasis, lymphedema can involve the entire limb, whereas in brugian filariasis, the swelling is restricted to the distal extremities below the knee or elbow. The most severe cases are referred to as elephantiasis [12].

Tropical pulmonary eosinophilia (TPE), another manifestation, is associated with *W. bancrofti* and *B. malayi* infection. Patients are typically middle-aged men who present with nocturnal asthma, cough, fever, and weight loss. The pathogenesis of TPE results from an exaggerated allergic hypersensitivity response to mf as they migrate through pulmonary blood vessels. It is most commonly observed in South and Southeast Asia and endemic areas of Brazil and Guyana.

### 7.2.2 Onchocerciasis

Adult *O. volvulus* reside primarily in subcutaneous and deep tissue nodules and produce first stage larvae (mf) [7]. Symptoms develop over many years into chronic skin and ocular manifestation. The immunological response to the death of mf is responsible for the disease symptoms, ranging from skin itching and impaired vision to disfiguring skin lesions and blindness based on the location of mf.

Adult females, alone or in groups, reside in these nodules, which are visited by males. Shed mf migrate to the skin, from which they may be acquired by a black fly vector. Unlike LF parasites, adult *O. volvulus* cause little overt pathology. Instead, immune-mediated killing of mf causes skin and eye damage, defining the pathology of the infection. In the skin, these include acute and chronic papular dermatitis, lichenified dermatitis, atrophy, and depigmentation [7]. The earliest and most troublesome symptom of onchocercal skin involvement is itching [13]. Skin changes have been classified as acute papular onchodermatitis (APOD), chronic papular onchodermatitis (CPOD), localized onchodermatitis, atrophy, and pigmentation changes [14]. More than one type of skin disease can coexist. Atrophy of the skin is relatively common in areas of high endemicity, showing many characteristics of aging, such as loss of elasticity, and the skin appears excessively wrinkled. Hair may be lost and sweating reduced. Depigmentation, "leopard skin," is associated with patches of complete pigment loss. A rare manifestation is the "hanging groin" or "adenolymphocele" that is observed only in heavy and long-standing infections. The infection may also cause inguinal lymph node fibrosis and atrophy of the overlying skin that leads to the appearance of hanging groin [15].

Ocular disease in onchocerciasis is characterized by pathology in both anterior and posterior segments of the eye [7]. Early stages typically begin with the appearance of anterior punctate lesions consisting of opacities of the superficial corneal stroma. Up to 100 or more opacities measuring 0.5 mm in diameter may be observed [15]. These lesions result from local inflammation around dead corneal mf and heal without scarring. Anterior punctate keratitis can be asymptomatic or accompanied by conjunctival injection, chemosis, limbitis, and epiphora. Long-standing and heavy infections can result in massive mf invasion of the cornea and other anterior regions, leading to later stages of disease, including sclerosing keratitis, a progressive irreversible scarring of the cornea [16]. Sclerosing keratitis and accompanying anterior pathologies (e.g., anterior uveitis, iris atrophy, anterior and posterior synechiae, secondary cataract, and glaucoma) usually develop after the second decade, resulting in blindness. Progressive pathologic changes predominantly in the posterior segment of the eye also account for much of onchocercal-induced blindness. These include retinal pigmentary disturbances, chorioretinitis, chorioretinal atrophy, subretinal fibrosis, and, finally, optic nerve atrophy [17, 18]. Onchocercal chorioretinitis appears to be a chronic, indolent, low-grade, progressive inflammation rather than an active, fulminant disease. In both anterior and posterior segments, ocular pathology is caused by an immunological response to the invasion and local death of mf [19]. The presence of mf in the anterior chamber is associated with increased risk of serious eye lesions. The presence of many mf in the anterior chamber seems to be a precursor to the development of posterior segment lesions [20]. Finally, higher skin mf levels, particularly >10 mf/mg, are associated with optic nerve disease [19]. Thus, individuals with greater-intensity infections, i.e., high densities of mf, are more likely to develop severe ocular disease.

Other associated conditions have been described, including a form of dwarfism and epilepsy. A related epidemic form of epilepsy, termed nodding syndrome, has recently gained attention and may be related to *O. volvulus*–induced antibodies that cross-react with proteins expressed on the surface of neurons [7].

### 7.2.3 Loiasis

Adult worms migrate through subcutaneous tissues and sometimes beneath the conjunctiva (hence the popular name *eye worm*). The clinical spectrum of loiasis is broad, ranging from asymptomatic infection to life-threatening complications. Subconjunctival migration of adult worms is usually accompanied by transient swelling of the lid and intense conjunctivitis. Localized angioedema (Calabar swelling) is a typical sign. Although most episodes resolve spontaneously and completely, rare cases of retinal artery occlusion and macular retinopathy due to aberrant migration of the adult parasite have been reported [21]. Atypical symptoms, such as cardiac, respiratory, renal, neurological, gastrointestinal, and other pathologies, including early mortality in heavily infected individuals, have been underestimated, as recent studies show [22, 23]. Except for eye worm and renal abnormalities, clinical signs and symptoms are more common in visitors to *Loa*-endemic areas than in indigenous people and reflect a heightened immune response to the parasite. Conversely, mf are detectable in peripheral blood of most endemic individuals with loiasis but are rare in infected visitors [24]. Familial clustering of microfilarial density in an endemic area in Cameroon suggests that host genetics play a role [25].

Although serious complications of loiasis are rare, a recent study demonstrated increased risk of death in infected individuals with high levels of microfilaremia [23]. In persons with high-level microfilaremia, the most serious complication of *L. loa* infection is meningoencephalitis, triggered during treatment with DEC or IVM, but not ABZ, possibly due to the immune-mediated, rapid destruction of large numbers of mf and the mechanical effects of mf paralyzed or killed by treatment that create emboli of high numbers of mf in the capillary circulation [26, 27].

### 7.2.4 Mansonellosis

In contrast to other filarial diseases, the clinical presentation of *Mansonella* spp. infections generally appears milder, and thus this parasite is not included in general strategies to eliminate filariases. Most infections with *M. perstans* are believed to be asymptomatic. When symptoms occur, they appear related to migration of adult worms and include transient subcutaneous swellings (similar to those caused by *L. loa*), pericarditis and pleuritis, and ocular symptoms (e.g., impaired visual acuity) if mf enter the eye. Nonspecific symptoms, including fever, fatigue, pruritus, arthralgias, and abdominal pain, may occur. Headache and neuropsychiatric symptoms have also been reported. Signs may include lymphadenopathy and eosinophilia. Mf of *M. ozzardi* circulate in the skin and in the blood, whereas *M. streptocerca* has dermal mf; hence if symptoms develop, patients present pruritus, rash or unexplained eosinophilia, edema, lymphadenopathy, articular pains, fever, headache, vertigo, and pulmonary symptoms in the case of *M. ozzardi* [28].

## 7.3 Diagnosis

Filariases are diagnosed by obtaining an appropriate exposure history in areas where the disease is endemic in combination with pathological signs and symptoms. This has to be certified via a definitive diagnosis, traditionally made by microscopic detection of the parasite, with some assays available for the detection of parasite antigens or DNA, or immunoglobulin responses to the parasite [29]. Diagnostics support individual-level treatment choices, inform population-level decisions on changing treatment frequency or stopping mass treatment, enable disease surveillance, and provide confidence in validating or verifying elimination or certifying eradication. Although classical clinical and microscopic techniques are often adequate for mapping disease distribution and for monitoring the progress of most NTD interventions, the need for improved diagnostics comes into much sharper focus as infection prevalence declines and elimination or eradication becomes a possibility.

### 7.3.1 Microscopic Detection

The standard detection of blood-dwelling mf (W. *bancrofti*, *Brugia* spp., *L. loa*) is made in a thick blood film stained with Giemsa or hematoxylin, which allows filarial speciation. Blood should be drawn according to the periodicity of the suspected species (with adjustment for recent [<2 weeks] travel from a different time zone) to coincide with peak microfilarial levels. Although mf can be detected in thick blood smears stained with Giemsa or Wright's stain, concentration techniques, including Knott's concentration, saponin lysis, and nucleopore filtration of anticoagulated blood, are useful in patients with low numbers of circulating mf. Skin dwelling mf (*O. volvulus*, *M. streptocerca*) are diagnosed by skin snips using a corneoscleral punch. The skin snip is incubated in saline and examined using an optical microscope to detect emergent mf. Ocular mf can be detected by slit lamp examination.

Microfilariae can be distinguished from each other by differences in periodicity, size, presence or absence of a sheath, and positioning of nuclei extending to the tip of the tail [30]. Rapid point-of-care quantification of *L. loa* mf in blood has recently been demonstrated using a cell phone–based video-microscope and can be used to identify individuals with high microfilarial levels at risk for severe adverse effects of microfilaricidal therapy [31].

While microscopic identification of parasites is highly specific, sensitivity declines when the prevalence and intensity of the infection are very low, and co-infection can complicate detection. For example, in loaisis-endemic areas, *L. loa* mf appear in skin snips in heavily infected patients [32].

### 7.3.2 Serological Detection

Detection of circulating filarial antigen is the most widely used method for diagnosis of *W. bancrofti*. Plasma or serum can be tested using a commercially available rapid-format immunochromatographic test strip [33]. This test is semiquantitative, more sensitive than microscopic detection of mf and can be performed at any time of the day. This test is specific for *W. bancrofti*, except among individuals co-infected with a heavy burden of *L. loa* [34]. In contrast to diagnosing an active infection, exposure to filarial parasites is measured in most of the other available rapid diagnostic tests (RDTs) specific for human IgG4, as it is the case for the *L. loa* RDT [35], brugian filariasis [36], and onchocerciasis [37]. It has been widely recognized that more sensitive and specific tools are needed to support elimination programs.

A low-molecular-weight antigenic protein termed Ov16 has been used to develop an enzyme-linked immunosorbent assay (ELISA) and RDT for diagnosis of onchocerciasis [37]. Nevertheless, due to limitations of these tools, including that they detect past infections, more sensitive and specific tests are required to ensure disease elimination, optimally with the ability to identify active infections (i.e., detection of parasite antigens instead of antibodies that recognize them). The main diagnostic tool for loiasis is the blood smear or preferably calibrated blood smear with blood collected during the day (10 a.m.–2 p.m.) and stained with Giemsa. The Loa Antibody Rapid Test is also available with high sensitivity and medium specificity [35].

Antigen- or antibody-based assays are the best methods for diagnosis of travelers; they are simple to perform and sensitive, and most laboratories outside of filarial-endemic areas do not have experience detecting mf. Immunodiagnostic tests are useful for confirming the diagnosis of filariasis in travelers from endemic areas who have characteristic clinical symptoms or unexplained eosinophilia without detectable blood mf.

## 7.4 History of Anthelmintics Currently Used for MDA in Human Filariases

### 7.4.1 Diethylcarbamazine

Drug discovery and development are almost entirely driven by commercial interest. Particularly in low resource settings, this creates a tremendous imbalance in the availability of drugs. During and following World War II, faced by the challenges of tropical parasitic diseases encountered by the military, government, academic, and industrial laboratories discovered and developed a range of antiparasitic drugs for human use.

For filariasis, antimonials and arsenicals had been used without satisfactory results. The first effective drug on the scene was the piperazine derivative, 1-diethyl-carbamyl-4-methyl-piperazine hydrochloride (DEC; Hetrazan), introduced in 1947. DEC showed remarkable efficacy in experimental filarial infections, i.e., *Litomosoides sigmodontis* (formerly: *Litomosoides carinii*) and *Dirofilaria immitis*, as well as in humans infected with *W. bancrofti* (reviewed in [38]). Treatment markedly reduced circulating mf. In 1950, a citrate salt (Supatonin) was synthesized [39], and its microfilaricidal effects were confirmed (reviewed in [40]). It was, however, quickly discovered that DEC administration to onchocerciasis patients leads inevitably to adverse reactions that could be life-threatening in heavily infected patients (reviewed in [7, 41]). This reaction was described in 1948 by Luis Mazzotti in patients after treatment with DEC [42], including dermal, lymphatic, musculoskeletal, and cardiovascular symptoms. A similar reaction was observed in animal models [43]. Pretreatment with corticosteroids reduced clinical symptoms during DEC treatment, but at the same time reduced efficacy [44, 45], pointing to the involvement of the immune system in rapid drug-induced clearance of mf.

Studies in *Brugia*-infected individuals supported the positive effects of DEC [46]. Many important studies were carried out in the South Pacific islands, where more than 10 000 American soldiers suffered clinical filariasis due to nocturnal periodic *W. bancrofti* during World War II. These studies addressed optimization of the dose and treatment schedule of DEC; based on the results, control programs with various regimens were implemented in different endemic countries. By the mid-1970s, the standard DEC regimen for *W. bancrofti* infection in mass treatment had been established: a total dose of 72 mg/kg given in 12 divided doses, once weekly or monthly at 6 mg/kg each [47]. However, difficulties in multidose treatment were encountered in many LF-endemic communities, especially in the WHO/Samoa Filariasis Research project, in which a year-long multidose MDA regimen was conducted twice in 1956/1966 and 1971 utilizing limited health resources, making the government reluctant to repeat this procedure. This was the time when an annual single 6 mg/kg dose of DEC was reported successful in French Polynesia [48]. This was repeated in Samoa when 6 mg/kg was delivered to the entire population, except infants under the age of 1, pregnant women, and sick and elderly people [49]. A similar program in Fiji confirmed the effectiveness of annual single-dose MDA. Between 1979 and 1981, doses of 4–8 mg/kg DEC were compared, and based on higher incidence of adverse events at 8 mg/kg [48], the Western Pacific Regional Office of the WHO implemented a national MDA program using 6 mg/kg DEC.

Despite showing that a single-dose regimen is effective, the multidose treatment schedule remained in use for many years. Kimura concluded that researchers at that time focused more on the cure of infection in the early stage of dosing trials and thus were in favor of multi-dosing [50]. In contrast, a small number of researchers, especially those working in less developed settings, paid more attention to the feasibility of a treatment scheme. It became obvious that an annual scheme was much easier and more practical than the multidose scheme. Therefore, a regimen that was more operationally feasible was selected instead of one that was potentially more effective [50]. In 1984, the WHO Expert Committee of Filariasis recognized the effectiveness of annual treatments of 6 mg/kg DEC for LF control [51].

The extraordinary pharmaceutical stability of DEC under lab conditions, including autoclaving, made its use as a salt additive a treatment option. In 1967, at a time when LF was being addressed primarily as a clinical problem by case identification and standard treatment was a 12-day regimen, initial studies were done using DEC-fortified salt. Efficacy of DEC-fortified salt is well-established, and China and India demonstrated success with DEC salt intervention at the programmatic level (reviewed in [52]). Advantages of this strategy include that it causes few or no adverse events compared with DEC tablets. Whether DEC-fortified salt could be used for onchocerciasis without inducing side effects has not been fully elucidated. DEC salt, however, despite its success, has never been adopted by MDA programs. As for all microfilaricidal drugs, it may be contraindicated in areas co-endemic for *L. loa*.

### 7.4.2 Ivermectin

*Streptomyces avermitilis*, a microbe that produces anthelmintic avermectins, was isolated from a soil sample obtained on a Japanese golf course in a project led by Satoshi Omura at the Kitasato Institute, and the commercial product IVM was then generated and developed by a team led by William Campbell at Merck, both of whom were acknowledged with the Nobel Prize in Medicine in 2015 for the impact of their discovery on global health. In the mid-1980s, IVM became the first-choice drug for onchocerciasis due to its safety profile in microfilaremic patients. Several studies led to its registration, comparing its efficacy against DEC and identifying the dose necessary [53–57]. The results suggested that there is no advantage in administering doses >150 μg/kg and that the higher dose of 200 μg/kg may be associated with an increased incidence of adverse effects. Additional data and a longer follow-up period would be needed to clearly assess the performance of this dose compared with 100 μg/kg. Thus, 150 μg/kg was chosen as the dose of IVM for use in onchocerciasis control programs. A higher dose (800 μg/kg) was tested for macrofilaricidal efficacy; IVM at 150 μg/kg or 800 μg/kg given every three months had some macrofilaricidal effects, but of insufficient magnitude to warrant incorporation into MDA programs [58].

IVM was introduced in 1988 for LF based on efficacy against *W. bancrofti* infection [59]. Since efficacy of a single dose ranging from 25 to 200 μg/kg was remarkable, and often better than obtained with DEC, it was suggested as a replacement for DEC [60]. Further studies identified that the ability of IVM to reduce microfilaremia was higher than DEC during the first year, but slightly less so during the second year. Higher efficacy of DEC in LF was also reported in Brazil [61]. This study investigated the filarial dance sign associated with the presence of adult parasites in a lymph "nest," and no change was observed after IVM; it was concluded that IVM was not macrofilaricidal in LF. Nevertheless, in all studies performed with IVM, it appeared to be a potent microfilaricide with a good safety profile, with the exception of patients harboring very high *L. loa* microfilarial load (>30 000 mf/ml), in whom cases of Severe Adverse Events (SAEs) were described, including fatalities [26 and see below].

### 7.4.3 Albendazole

Benzimidazoles were originally developed as plant fungicides and later as veterinary anthelmintics, with the first developed and licensed for human use being thiabendazole in 1962. Since then, four other benzimidazoles (mebendazole, ABZ, flubendazole, and triclabendazole) have been licensed for human use in various parts of the world; ABZ is the only one approved for use in filariases. An initial trial in LF showed that ABZ (600 mg) combined with IVM (400 μg/kg) was initially more effective than the combination of ABZ (600 mg) plus DEC (6 mg/kg), but with no difference in mf reduction after 15 months. The extent of the therapeutic effect of the addition of ABZ to either DEC or IVM on filarial parasites remains unclear, as both beneficial effects [62–64] and no additive effects have been reported [65–68]. A recent Cochrane analysis revealed no additional benefit of ABZ in combination with DEC three years after treatment [69], although a recent study in DRC showed positive effects of semi-annual ABZ over several years against *W. bancrofti* [70]. Since ABZ is highly effective against gastrointestinal helminths, its routine inclusion in LF control programs is recognized for providing additional benefits of broader health significance, including increased compliance and potential help against the development of resistance.

Thus, throughout the world, except in loiasis or onchocerciasis zones of Africa, a combination of either IVM/ABZ (Africa) or DEC/ABZ was used for MDA for LF in areas where onchocerciasis is not prevalent (https://apps.who.int/iris/handle/10665/259381) in the following regimens:

Annual single-dose treatment (for four to six years) with IVM (200 μg/kg) or DEC (6 mg/kg) coadministered with ABZ (400 mg). Also applicable is a one-drug regimen with DEC (6 mg/kg) alone or DEC-fortified salt (0.2–0.4% w/w) substituted for regular table/cooking salt for 6–12 months. As noted, ALB (400 mg) alone administered twice a year appears to have some efficacy against LF and can be recommended for the treatment of this disease in loiasis-endemic areas to avoid post treatment SAEs due to IVM [70] if confirmed in additional studies. Very importantly, the discovery that a single treatment with the combination of ABZ, DEC, and IVM administered simultaneously may have high macrofilaricidal efficacy in LF [71] has enormous implications for the time needed to achieve elimination of these parasites and has been adopted in the guidelines for LF elimination. Confirmation of efficacy in additional studies is awaited. Whether this regimen can be safely used in onchocerciasis endemic areas is currently under investigation (clinicaltrials.gov/ NCT04188301).

## 7.5 Chemotherapy of Human Filariases

Current drug treatment options for filariases are summarized in Table 7.1.

### 7.5.1 Control of Lymphatic Filariasis

Prior to the initiation of organized control programs, elimination efforts were performed locally. The first countries to address LF were India and China [72], recognizing the extent of the health and economic costs of this disease on national well-being as a justification for prioritizing its elimination. China's basic strategy starting in the 1950s was to test at-risk populations for mf in the blood (particularly challenging because of the parasite's nocturnal periodicity), treat mf-positive individuals with DEC for multiple weeks, and then ensure that the entire populations of its 15 endemic provinces (330 million people) either received single treatments of DEC or utilized DEC-fortified table salt routinely. These principal approaches were embedded in a highly organized and regimented framework of rigorous monitoring, evaluation, data management, logistics, and social science. In the 1980s, countries including Japan, Korea, the Philippines, the Pacific Island countries, Malaysia, Thailand, Indonesia, Egypt, and Brazil followed and initiated LF control programs based on DEC administration [73]. DEC was effective everywhere, but the optimal regimen for reducing mf or killing adult worms remained controversial until studies in the Pacific, Brazil, and elsewhere showed that a single dose of DEC (6 mg/kg) was essentially equivalent to the same dose repeated daily for two or three weeks, with both regimens leading consistently to partial clearance of microfilaremia and partial killing of adult worms [73–75].

In 1997, the World Health Assembly made a resolution to eliminate LF as a public health problem. Consequently, the Global Programme to Eliminate Lymphatic Filariasis (GPELF) was organized and started its worldwide activities in 2000, with the target of elimination by 2020. The Global Alliance to Eliminate Lymphatic Filariasis (GAELF) was formed the same year to support this unprecedented global program.

**Table 7.1** Treatment options and clinical pharmacology of antifilarial drugs.

| Drug | *O. volvulus* | *Loa loa* | Lymphatic filariasis | Contraindications |
|---|---|---|---|---|
| DEC | Adult: –<br>Mf: ++ | Adults: +<br>Mf: ++ | Adults: +<br>Mf: ++ | Onchocerciasis, loaisis[a)] |
| ABZ | Adults: –<br>Mf: - | Adults: (+)<br>Mf: - | Adults: (+)<br>Mf: – | |
| IVM[b)] | Adults: (+)<br>Mf: ++ | Adults: –<br>Mf: ++ | Adults: –<br>Mf: ++ | Loiasis |
| DOX[c)] | Adults: ++<br>Mf: - | No *Wb* | Adults: ++<br>Mf: – | |
| MOX | Adults: –<br>Mf: ++ | Not yet tested | Not yet tested | Loiasis |

**Drugs and drug regimens**
Lymphatic filariasis

| | |
|---|---|
| MDA | [103]<br>IVM 200 μg/kg + ABZ 400 mg or DEC 6 mg/kg + ABZ 400 mg<br>Recently recommended: DEC 6 mg/kg + ABZ 400 mg + IVM 200 μg/kg |
| US | [104]<br>*Adult*:<br>Day 1: DEC 1 × 50 mg p.o. after meals<br>Day 2: DEC 3 × 50 mg p.o.<br>Day 3: DEC 3 × 100 mg p.o.<br>Day 4–14: DEC 3 × 2 mg/kg/d p.o.<br>*Pediatric*:<br>Day 1: DEC 1 × 1 mg/kg p.o. after meals<br>Day 2: DEC 3 × 1 mg/kg p.o.<br>Day 3: DEC 3 × 1–2 mg/kg p.o.<br>Day 4–14: DEC 3 × 2 mg/kg/d p.o.<br>[12, 105]<br>DEC 6 mg/kg for 1 or 12 d<br>Alternative: DOX 200 mg/d for 4–6 wk<br>[106]<br>DEC 3 × 2 mg/kg/d for 12 d<br>Alternative: DEC 6 mg/kg single dose<br>Alternative: DOX 100 mg orally twice a day for 4–6 wk (could be combined with DEC)<br>[107]<br>DOX 2 × 100 mg/d p.o. for 4–6 wk, followed by ABZ 1 × 400 mg p.o. and DEC 1 × 6 mg/kg p.o. |
| Europe | [108]<br>DOX 200 mg/d p.o. for 4 wk, preferred over DEC<br>Alternative: DEC 6 mg/kg for 12 d<br>*Brugia*: total dose: 36 mg/kg for 7 d<br>*Wuchereria*: total dose: 72 mg/kg for 14 d<br>In case of high mf levels, slowly increasing dose:<br>Day 1: DEC 1 × 50 mg p.o.<br>Day 2: DEC 3 × 50 mg p.o.<br>Day 3: DEC 3 × 100 mg p.o. |

**Table 7.1** (Continued)

| | |
|---|---|
| | Day 4-[d]: DEC 3 × 150 mg/kg/d p.o.<br>For the treatment of allergic reactions:<br>3 × 1–2 mg betamethasone, starting 2 d prior to DEC, stepwise dose reduction after 7 d<br>IVM, if DOX or DEC is contraindicated |
| Tropical pulmonary eosinophilia | |
| US | [103]<br>DEC 2 × 3 mg/kg/d p.o. for 14 d |
| Europe | [106–108]<br>DEC 3 × 2 mg/kg/d p.o. 12 or 14 to 21 d |
| Onchocerciasis | |
| MDA | IVM 150 µg/kg |
| US | [109]<br>To kill mf: IVM 150 µg/kg every 6 mo<br>To kill adult worms: DOX 200 mg/d for 6 wk<br>[15]<br>DOX 200 mg/d for 6 wk + IVM 150 µg/kg |
| Europe | [110]<br>IVM 150 µg/kg[e] every 6 mo<br>Alternative: DOX 100 or 200 mg/d for 6 wk<br>[107]<br>DOX 200 mg/d for 6 wk + IVM 200 µg/kg p.o. single dose after 4 mo<br>[108]<br>DOX 200 mg/d for 6 wk + 2× IVM 100–200 µg/kg single dose |
| Loiasis | |
| MDA | Not treated |
| US | [21, 111]<br>Symptomatic loiasis with mf/ml <8000:<br>DEC 8–10 mg/kg/d p.o. divided three times daily for 21 d<br>Symptomatic loiasis with mf/mL <8000 and failed two rounds DEC or symptomatic loiasis with mf/ml ≥8000 to reduce level to <8000 prior to treatment with DEC: ABZ 2 × 200 mg/d for 21 d followed by DEC<br>Symptomatic loiasis with mf/ml ≥8000 alternative: Apheresis followed by DEC |
| US | [112]<br>Loiasis with mf/ml <8000:<br>Day 1: DEC 1 × 50 mg p.o.<br>Day 2: DEC 3 × 50 mg p.o.<br>Day 4–21: DEC 3 × 2.7–3.3 mg/kg/d<br>Loiasis with mf/ml >8000: ABZ 2 × 200 mg/d p.o. for 21 d to reduce microfilaremia <1000 mf/ml before initiating DEC<br>Alternative: Apheresis followed by DEC |

**Table 7.1** (Continued)

| | |
|---|---|
| | [107]<br>Loiasis with mf/ml <1000:<br>Day 1: DEC 1 × 25 mg p.o.<br>Day 2: DEC 1 × 50 mg p.o.<br>Day 3: DEC 3 × 50 mg p.o.<br>Day 4: DEC 3 × 100 mg p.o.<br>Day 5–21: DEC 3 × 150 mg/d<br>Loiasis with mf/ml >1000: ABZ 2 × 200 mg/d p.o. for 21 d to reduce microfilaremia <1000 mf/ml before initiating DEC |
| Europe | [108]<br>Loiasis with mf/mL <1,000:<br>Day 1: DEC 1 × 1 mg/kg p.o.<br>Day 2: DEC 1 × 3 mg/kg mg p.o.<br>Day 3: DEC 6 mg/kg p.o.<br>Day 4–21: DEC 9 mg/kg/d<br>Loiasis with mf/mL >8000: ABZ 2 × 200 mg/d p.o. for 21 d to reduce microfilaremia <1000 mf/ml before initiating DEC<br>Alternative: Apheresis followed by DEC<br><br>Loiasis with mf/ml <8000:<br>IVM 150–200 μg/kg p.o. single dose to reduce microfilaremia < 1000 mf/ml |
| *Mansonella perstans* | |
| MDA | Not treated |
| US | [28]<br>DEC 2 × 200 mg + MBZ 100 mg daily p.o. for 21 d<br>In areas where *M. perstans* harbor *Wolbachia*: DOX 200 mg/d p.o. for 6 wk |
| Europe | [107]<br>MBZ 100–200 mg daily p.o. for 21 d +<br>Day 1: DEC 1 × 25 mg p.o.<br>Day 2: DEC 1 × 50 mg p.o.<br>Day 3: DEC 3 × 50 mg p.o.<br>Day 4: DEC 3 × 100 mg p.o.<br>Day 5–21: DEC 3 × 150 mg/d<br>In areas where *M. perstans* harbor *Wolbachia*: DOX 2 × 100 mg/d p.o. for 6 wk<br><br>[108]<br>Without symptoms: no treatment required.<br>With symptoms: DEC (total dose 75 mg/kg)<br>Alternative: mebendazole 200–500 mg/d for 30 d |
| *Mansonella streptocerca* | |
| US | [28]<br>DEC 6 mg/kg/d for 14–21 d and/or IVM 1 × 150 μg/kg to reduce mf load, if symptomatic |

**Table 7.1** (Continued)

| | |
|---|---|
| Europe | [107]<br>Day 1: DEC 1 × 25 mg p.o.<br>Day 2: DEC 1 × 50 mg p.o.<br>Day 3: DEC 3 × 50 mg p.o.<br>Day 4: DEC 3 × 100 mg p.o.<br>Day 5–21: DEC 3 × 150 mg/d<br>+ IVM 1 × 200 μg/kg to reduce the mf load, if symptomatic |
| *Mansonella ozzardi*[f)] | |
| US | [28]<br>IVM 1 × 150 μg/kg |
| Europe | [107]<br>DOX 2 × 100 mg/d p.o. for 6 wk + IVM 1 × 200 μg/kg to reduce the mf load, if symptomatic<br>[108]<br>DEC is effective against mf and adult worms |
| **Clinical pharmacology** | |
| Diethylcarbamazine | |
| Trade names | Hetrazan, Carbilazine, Caricide, Cypip, Ethrodyl. Notezine, Spatonin, Banocide Forte, and Eofil |
| MDA dose | 6 mg/kg |
| $C_{max}$ | 80–200 ng/ml |
| $T_{max}$ | 1–2 h |
| $T_{1/2}$ | 8 h |
| Food effect | No food effect observed, recommended to take after meal |
| DDI[g)] | None significant reported |
| Albendazole | |
| Trade names | Zentel, Escazole, Albenza, Valbazel, and many others |
| MDA dose | 400 mg |
| $C_{max}$ | 1310 ng/ml (range 460–1580 ng/ml) |
| $T_{max}$ | 2–5 h |
| $T_{1/2}$ | 8–12 h |
| Food effect | Dose proportional increase up to fivefold should be taken with food |
| DDI | Dexamethasone, praziquantel, cimetidine, theophylline, grapefruit juice, carbamazepine, phenytoin, phenobarbital, clozapine, and other drugs metabolized through CYP450 family (2D6, 1A2) |

**Table 7.1** (Continued)

| | |
|---|---|
| Ivermectin | |
| Trade names | Stromectol, Mectizan, Scabioral, Revectina |
| MDA dose | 150–200 μg/kg |
| $C_{max}$ | 46 ng/ml (range 16.4–101.1 ng/ml) |
| $T_{max}$ | 4 h |
| $T_{1/2}$ | 18 h |
| Food effect | Absorption improved with high-fat meal up to 2.5-fold should be taken without food |
| DDI | Barbiturates (such as phenobarbital, butalbital), benzodiazepines (such as clonazepam, lorazepam), sodium oxybate (GHB), valproic acid, and other drugs metabolized through CYP450 family (3A4) and P-gp |

a) DEC used with caution in loiasis, as severe adverse effects have been reported.
b) IVM not approved for use in children weighing less than 15 kg or under the age of 5 years.
c) Doxycycline is contraindicated in children under the age of 9 years and in pregnant women.
d) Duration of DEC treatment depends on parasite species, up to 14 days.
e) IVM may not be registered for human use for this indication in every country.
f) Efficacy of DOX has not been studied in this infection.
g) Drug–drug interactions.

The alliance includes health ministries of endemic countries, UN agencies with the WHO as secretariat, the private sector, NGOs, academia, and government bodies. It is essential to recognize that these programs would not be feasible without the essential drugs being supplied free of charge, namely GSK (ABZ), Merck & Co (IVM), and Eisai (DEC).

In MDA, all people living in an endemic area are treated, with or without diagnosis of an ongoing infection. The basic strategy is to conduct annual single-dose MDA campaigns for four to six years. In 2000–2007, at least 570 million individuals were treated in 48 of 83 endemic countries. The drugs used for MDA are the combination of DEC (6 mg/kg) and ABZ (400 mg) in onchocerciasis-free areas, and IVM (200 μg/kg) and ABZ (400 mg) in onchocerciasis-endemic areas of Africa. The reason for the use of two separate regimens is that DEC can cause severe adverse reactions if administered to *Onchocerca*-infected individuals. As noted above, a triple drug combination (IVM + DEC + ALB; IDA) has recently been recommended by WHO for LF treatment in non-onchocerciasis areas [71] and is being trialed in patients with low levels of *O. volvulus* microfilaridermia to determine if it is safe and macrofilaricidal in onchocerciasis.

It must still be recognized that many infected people are asymptomatic despite large numbers of blood-borne mf and that administration of microfilaricidal agents, including DEC and IVM, can cause sometimes fatal SAEs. This risk in populations co-infected with multiple filariae, especially *L. loa*, continues to pose challenges for global elimination programs for LF and onchocerciasis.

The GPELF has achieved impressive results in terms of parasitological cure/improvement, clinical benefits, social and economic impacts, etc. Of note is the program's success in mobilizing hundreds of millions of local people, who not only took drugs but also actively supported MDA programs as drug distributors and volunteers. Beyond filariasis, the role volunteers can play in supplementing rural health services is now a topic of discussion and a source of hope for new and more sustainable systems.

### 7.5.2 Control of Onchocerciasis

Historical chemotherapeutic treatments for onchocerciasis were of limited efficacy and safety, including suramin and DEC. Neither intervention was suitable for MDA, and both drugs have been removed from use for this indication. The situation changed with the introduction of IVM. Based on its activity against filarial parasites of veterinary significance, especially *Dirofilaria immitis* [76] and in multiple experimental animal models, IVM was tested in human onchocerciasis patients and shown to be effective as a prolonged-action microfilaricide after a single oral dose with few side effects [77, 78]. Adult worms examined after nodulectomy were found to have prolonged suppression of mf production. Realization that a single yearly treatment with IVM could markedly limit transmission and pathology led to the decision by Merck and Co. to donate the drug in the Mectizan Donation Program (MDP) for MDA campaigns to control onchocerciasis.

Adoption of IVM for MDA was swift and has risen to the level of 300 million doses per year (https://mectizan.org/what/overview/). Although the dose remains the same (150 μg/kg), some programs, particularly in the Americas, adopted a biannual rather than annual administration scheme to reduce time to elimination by keeping mf levels very low [79]. Notable achievements toward control and elimination of onchocerciasis have been made in many countries through MDA campaigns [80], though reaching the current goals for global elimination is thought to be unlikely to be met through IVM distribution alone [81].

Current guidelines recommend annual (or biannual) treatment of every individual in targeted communities with an oral tablet of IVM (Mectizan) based on height. The current height-based IVM dosing has a range of 3–12 mg for four different height groups (90–119 cm, 120–140 cm, 141–158 cm, and >158 cm). Children below 15 kg and visibly pregnant women are excluded from MDA [82].

Recognition that SAEs, including coma and death, can occur in individuals harboring high levels of *L. loa* microfilaremia who were treated with IVM [26, 83] forestalled extension of MDA programs to loaisis regions. Efforts to perform MDA in these areas rely on individual diagnosis of *L. loa* microfilaremia using microscopic assays, but since these areas were previously excluded from MDA, they contribute to incomplete coverage of current MDA programs. A new strategy termed "Test and Not Treat" (TaNT), based on the use of the LoaScope to estimate *L. loa* microfilaraemia before treatment to exclude heavily infected patients, can allow safe implementation of MDA in loiasis endemic areas [84].

### 7.5.3 Recent Developments

Two other medicines are relevant for treatment of filariases: doxycycline (DOX) and moxidectin. *Wolbachia* symbionts are essential for most filarial parasites [85]. DOX was approved by the US FDA in 1967 for the treatment of bacterial infections of humans and clinicians have a large body of experience in its use. The discovery that the parasites that cause onchocerciasis and LF harbor *Wolbachia* endosymbionts led to evaluation of approved antibiotics for utility in these infections. Long-term sterilization and macrofilaricidal efficacy were shown in animal models, which was translated into clinical trials in human onchocerciasis and LF. A meta-analysis of the available data from clinical trials on onchocerciasis [86] has shown that the three tested regimens, DOX (200 mg/day) for four weeks, 200 mg/day for six weeks [87], and 100 mg/day for five weeks [88], are broadly equivalent, in particular with regard to sterilizing effects on adult worms; therefore, DOX treatment with 200 mg/day for four weeks can be considered as standard anti-*Wolbachia* treatment. With its lack of a direct microfilaricidal effect, DOX causes fewer parasite-related side effects than DEC or IVM. Shorter courses of DOX are not sufficient for adult worm killing, but do lead to long-term inhibition of the production of new mf. Similar results were found for LF infections with a treatment duration of four weeks [89]. Anti-*Wolbachia* therapy has been shown in a pilot trial to improve lymphatic pathology and decrease severity of lymphoedema and hydrocele in LF patients, which is currently being further investigated in a large multi-center trial [90]. Community-level compliance with a six-week, 100 mg per day regiment has been demonstrated [91] and suggests the feasibility of treatment implementation even if some logistical difficulties remain for the follow-up of such a long regimen at the community level.

Since DOX lacks acute microfilaricidal activity and, in addition, *L. loa* do not harbor *Wolbachia*, DOX does not pose a threat of SAEs in high-risk *Loa* patients. Despite its long treatment duration, it is currently being tested as an alternative treatment for *L. loa*-coinfected patients with high *Loa* mf loads in clinical trials in Cameroon (Kamgno et al., personal observation).

Antibiotics are not as commonly used in many onchocerciasis-endemic regions as in wealthy countries, partially allaying fears of selecting antibiotic-resistant strains of bacterial pathogens during intensive, prolonged use of DOX for onchocerciasis. The necessity for prolonged daily dosing has limited its incorporation in onchocerciasis control campaigns and has led to research to identify antibiotics that are equally effective in shorter (i.e., <2 weeks) regimens [92, 93]. Should resistance to IVM become a therapeutic problem [94], implementation of a DOX regimen could prevent the spread of resistant parasites; this has not yet become necessary. As the treatment regimen is rather long and the drug is contraindicated in children <9 years of age and in pregnant or breastfeeding women, DOX is not suitable for MDA programs, but should be considered for individual treatment or for use in difficult to treat areas.

Moxidectin, like IVM, is a macrocyclic lactone endectocide. It is significantly more lipophilic than IVM and has a considerably longer serum half-life in humans [95, 96]. It also is more potent than IVM against many species of filarial parasites. Clinical trials in onchocerciasis patients showed that a single oral dose of moxidectin safely provided longer suppression of microfilaridermia than IVM [97], and the drug received FDA approval at a dose of 8 mg for this indication in 2018. It remains to be seen how moxidectin will be employed in MDA programs for onchocerciasis control.

Modeling studies suggest that its superior efficacy could accelerate progress to elimination [98]. Currently, a large post-registration trial is being set up to investigate the safety of moxidectin for use in MDA programs (clinicalicaltrials.gov/NCT04311671).

Safety of moxidectin in special populations is currently being investigated, e.g. children (clinicalicaltrials.gov/NCT03962062) and those residing in *Loa*-endemic regions (clinicalicaltrials.gov/NCT04049851). It remains to be elucidated how a drug donation program, if organized, will intersect with the existing Mectizan system for distribution of IVM.

### 7.5.4 Treatment of Travelers, Expatriates, Immigrants, and Refugees

With the rise in global migration and travel to remote destinations, filariases can occur in non-endemic areas in immigrants, refugees, and travelers. Imported cases seem to be decreasing, which could be due to the success of the global control programs [99, 100]. The risk of infection and development of chronic manifestations is very low for the traveler, as the parasite is rather poorly transmitted to humans by infected vectors [100], typically requiring prolonged exposure. Unlike the situation in endemic countries, readily available treatment in the Western world allows definitive diagnosis and treatment to cure the patient, without the risk of continuous exposure and reinfection. Clinicians should be familiar with imported NTDs, including LF, onchocerciasis, and loiasis. Differential diagnosis, travel history assessment, and a definitive diagnostic test should be done. Treatment guidelines do not officially exist; however, recommendations in Western countries are based on the evidence described in this and other chapters of this book. They may vary from country to country. The following represent our understanding of the consensus recommendations.

For LF, DEC (6 mg/kg) is the drug of choice. Since DEC is no longer registered, it is not freely available in the United States or Europe, but can be obtained from the US-Centers for Disease Control or international pharmacies. Common side effects are dizziness, nausea, fever, headache, or pain in muscles and joints and are common especially in those patients with evident parasitemia. As for MDA, co-infection with onchocerciasis warrants an alternative strategy and IVM (200 μg/kg) is recommended. Although official recommendations are awaited, the triple combination of DEC (6 mg/kg), IVM (200 μg/kg), and ALB (400 mg) can provide sustained reduction in mf levels (up to three years). This regimen was adopted by the WHO for MDA outside sub-Saharan Africa, and it could be reasonable to use this regimen at six-month intervals until patients are amicrofilaremic and circulating antigen negative [12]. Due to its macrofilaricidal potential and the added benefit of halting or reducing lymphedema, a four-week course of DOX (200 mg/day) could also be considered in symptomatic patients.

For TPE, treatment with DEC leads to symptomatic improvement and reduces eosinophilia and immunoglobulin E levels. Retreatment may be necessary in some cases. If the infection is not treated, progressive interstitial fibrosis and restrictive lung disease may develop [101].

Patients with onchocerciasis should be treated to prevent chronic disease development. IVM (150 μg/kg) is recommended to remove mf, but some countries suggest the combination of IVM and DOX (four to six weeks, 200 mg) to remove mf while slowly killing adult worms. Surgical removal of onchocerciasis nodules can be indicated when high numbers of mf are being produced or the nodule is located close to the eye and is often done whenever palpable nodules are present. Contraindications for IVM treatment are conditions associated with an impaired blood–brain barrier because penetration of the drug into the central nervous system can cause lethargy, ataxia, tremors, and death [15]. The drug is not approved for use in children <5 years of age or weighing less than 15 kg, pregnant women, and mothers nursing infants during the first week of life. Existing limited data, however, suggest that oral IVM in children weighing less than 15 kg has an acceptable safety profile. Data from well-designed clinical trials are needed to provide further assurance. IVM has no significant drug–drug interactions. The presence of heavy *L. loa* infections (mf densities >20 000 mf/ml) is a contraindication to the use of IVM and loiasis should be ruled out before IVM treatment; if present, DOX is the treatment of choice.

For loiasis, DEC treatment (8–10 mg/kg) for 12 days is often recommended, when co-infection with onchocerciasis is excluded and *L. loa* microfilaremia is low, given the risk of drug-induced SAEs [27]. Some countries suggest a gradual dose increase of DEC over several days, which can be followed by a single 400 mg dose of ABZ, depending on the intensity of the infection. Mild side effects, including Calabar swellings, urticaria, arthralgias, fever, and tenderness, are common during the first few days of DEC treatment and patients generally respond to antihistamines or a short course of corticosteroids. In cases with higher mf loads, although advocated in the past, neither a gradual increase in DEC dose nor corticosteroid pretreatment is completely effective in preventing complications [21]. Cytapheresis can be used to reduce the mf load before initiation of DEC [102]. Alternative recommendations are the use of ALB (200 mg BID for three weeks), which acts more slowly than DEC. Shorter regimens with higher doses seem to be less effective. Subconjunctival migrating adult worms can be removed under local anesthesia. DOX is ineffective in loiasis due to the absence of *Wolbachia*. Loiasis has significant clinical and community impacts, but chemotherapeutic options for it remain more than suboptimal. Drug discovery is unquestionably needed for this disease.

## 7.6 Conclusions

In this chapter, we integrate what is known about the major human filarial diseases with a discussion of current treatment paradigms using antifilarial drugs. In a global effort to eliminate filarial diseases, successful MDA programs were established and have reached elimination of filariasis in some, but not all areas and countries. Whereas drug discovery was formerly driven by needs and feasibility and often followed trial-and-error approaches, standard drug discovery approaches guided by a formal Target Product Profile (TPP) were not performed. Even now, de novo drug discovery programs for human filariases are too rare due to funding limitations in this area. Despite these shortcomings, new drugs or drug regimens await proof of safety and efficacy. New drug regimens with superior safety and macrofilaricidal efficacy would accelerate elimination, as more than one tool should be in place to be able to react to possible drug resistance and to satisfy the need for therapeutic alternatives at the end stage of control programs.

## References


**1** Mäser, P. (2022). Filariae as organisms. In: *Human and Animal Filariases* (ed. R. Kaminsky and T.G. Geary), Chapter 2. Weinheim, Germany: Wiley-VCH.

**2** Sauerbrey, M., Rakers, L.J., and Richards, F.O. (2018). Progress toward elimination of onchocerciasis in the Americas. *Int. Health* 10: i71–i78.

**3** World Health Organization (2020). *Ending the neglect to attain the Sustainable Development Goals: a road map for neglected tropical diseases 2021-2030*. Geneva: World Health Organization.

**4** Figueredo-Silva, J. and Dreyer, G. (2005). Bancroftian filariasis in children and adolescents: clinical-pathological observations in 22 cases from an endemic area. *Ann. Trop. Med. Parasitol.* 99: 759–769.

**5** Fox, L.M., Furness, B.W., Haser, J.K. et al. (2005). Ultrasonographic examination of Haitian children with lymphatic filariasis: a longitudinal assessment in the context of antifilarial drug treatment. *Am. J. Trop. Med. Hyg.* 72: 642–648.

**6** Shenoy, R.K., Suma, T.K., Kumaraswami, V. et al. (2008). Lymphoscintigraphic evidence of lymph vessel dilation in the limbs of children with *Brugia malayi* infection. *J. Commun. Dis.* 40: 91–100.

**7** Mackenzie, C.D. (2022). Human filarial infections: reflections on the current understanding of their importance, pathobiology and management. In: *Human and Animal Filariases* (ed. R. Kaminsky and T.G. Geary), Chapter 3. Weinheim, Germany: Wiley-VCH.

**8** Dreyer, G., Medeiros, Z., Netto, M.J. et al. (1999). Acute attacks in the extremities of persons living in an area endemic for bancroftian filariasis: differentiation of two syndromes. *Trans. R. Soc. Trop. Med. Hyg.* 93: 413–417.

**9** Aguiar-Santos, A.M., Leal-Cruz, M., Netto, M.J. et al. (2009). Lymph scrotum: an unusual urological presentation of lymphatic filariasis. A case series study. *Rev. Inst. Med. Trop. Sao Paulo* 51: 179–183.

**10** Adesiyun, A.G. and Samaila, M.O. (2008). Huge filarial elephantiasis vulvae in a Nigerian woman with subfertility. *Arch. Gynecol. Obstet.* 278: 597–600.

**11** Noroes, J., Addiss, D., Cedenho, A. et al. (2003). Pathogenesis of filarial hydrocele: risk associated with intrascrotal nodules caused by death of adult *Wuchereria bancrofti*. *Trans. R. Soc. Trop. Med. Hyg.* 97: 561–566.

**12** King, C.L. (2020). Lymphatic filariasis. In: *Hunter's Tropical Medicine and Emerging Infectious Diseases, 10*e (ed. E.T. Ryan, D.R. Hill, T. Solomon, et al.), 851–858. London: Elsevier.

**13** Murdoch, M.E. (2010). Onchodermatitis. *Curr. Opin. Infect. Dis.* 23: 124–131.

**14** Murdoch, M.E., Hay, R.J., Mackenzie, C.D. et al. (1993). A clinical classification and grading system of the cutaneous changes in onchocerciasis. *Br. J. Dermatol.* 129: 260–269.

**15** Nutman, T.B. (2020). Onchocerciasis. In: *Hunter's Tropical Medicine and Emerging Infectious Diseases, 10*e (ed. E.T. Ryan, D.R. Hill, T. Solomon, et al.), 864–871. London: Elsevier.

**16** Budden, F.H. (1976). The natural history of ocular onchocerciasis over a period of 14-15 years and the effect on this of a single course of suramin therapy. *Trans. R. Soc. Trop. Med. Hyg.* 70: 484–491.

**17** Bird, A.C., Anderson, J., and Fuglsang, H. (1976). Morphology of posterior segment lesions of the eye in patients with onchocerciasis. *Br. J. Ophthalmol.* 60: 2–20.

**18** Semba, R.D., Murphy, R.P., Newland, H.S. et al. (1990). Longitudinal study of lesions of the posterior segment in onchocerciasis. *Ophthalmology* 97: 1334–1341.

**19** Thylefors, B. (1978). Ocular onchocerciasis. *Bull. WHO* 56: 63–73.

**20** Budden, F.H. (1957). Natural history of onchocerciasis. *Brit. J. Ophthal.* 41: 214–227.

**21** Kamgno, J. and Klion, A.D. (2020). Loiasis. In: *Hunter's Tropical Medicine and Emerging Infectious Diseases, 10*e (ed. E.T. Ryan, D.R. Hill, T. Solomon, et al.), 859–863. London: Elsevier.

**22** Buell, K.G., Whittaker, C., Chesnais, C.B. et al. (2019). Atypical clinical manifestations of loiasis and their relevance for endemic populations. *Open Forum. Infect. Dis.* 6: ofz417.

**23** Chesnais, C.B., Takougang, I., Paguele, M. et al. (2017). Excess mortality associated with loiasis: a retrospective population-based cohort study. *Lancet Infect. Dis.* 17: 108–116.

**24** Herrick, J.A., Metenou, S., Makiya, M.A. et al. (2015). Eosinophil-associated processes underlie differences in clinical presentation of loiasis between temporary residents and those indigenous to *Loa*-endemic areas. *Clin. Infect. Dis.* 60: 55–63.

**25** Eyebe, S., Sabbagh, A., Pion, S.D. et al. (2018). Familial aggregation and heritability of *Loa loa* microfilaremia. *Clin. Infect. Dis.* 66: 751–757.

**26** Boussinesq, M., Gardon, J., Gardon-Wendel, N., and Chippaux, J.P. (2003). Clinical picture, epidemiology and outcome of *Loa*-associated serious adverse events related to mass ivermectin treatment of onchocerciasis in Cameroon. *Filaria J.* 2 (Suppl 1): S4.

**27** Carme, B., Boulesteix, J., Boutes, H., and Puruehnce, M.F. (1991). Five cases of encephalitis during treatment of loiasis with diethylcarbamazine. *Am. J. Trop. Med. Hyg.* 44: 684–690.

**28** Coulibaly, Y.I. and Klion, A.D. (2020). Miscellaneous filariae. In: *Hunter's Tropical Medicine and Emerging Infectious Diseases, 10*e (ed. E.T. Ryan, D.R. Hill, T. Solomon, et al.), 872–877. London: Elsevier.

**29** Mackenzie, C.D., Souza, A., and Geary, T.G. (2022). Diagnosis and assessment of human filarial infections. In: *Human and Animal Filariases* (ed. R. Kaminsky and T.G. Geary), Chapter 5. Weinheim, Germany: Wiley-VCH.

**30** Eberhard, M.L. and Lammie, P.J. (1991). Laboratory diagnosis of filariasis. *Clin. Lab. Med.* 11: 977–1010.

**31** D'Ambrosio, M.V., Bakalar, M., Bennuru, S. et al. (2015). Point-of-care quantification of blood-borne filarial parasites with a mobile phone microscope. *Sci. Transl. Med.* 7: 286re4.

**32** Nana-Djeunga, H.C., Fossuo-Thotchum, F., Pion, S.D. et al. (2019). *Loa loa* microfilariae in skin snips: Consequences for onchocerciasis monitoring and evaluation in *L. loa*-endemic areas. *Clin. Infect. Dis.* 69: 1628–1630.

**33** Chesnais, C.B., Vlaminck, J., Kunyu-Shako, B. et al. (2016). Measurement of circulating filarial antigen levels in human blood with a point-of-care test strip and a portable spectrodensitometer. *Am. J. Trop. Med. Hyg.* 94: 1324–1329.

**34** Pion, S.D., Montavon, C., Chesnais, C.B. et al. (2016). Positivity of antigen tests used for diagnosis of lymphatic filariasis in individuals without *Wuchereria bancrofti* infection but with high *Loa loa* microfilaremia. *Am. J. Trop. Med. Hyg.* 95: 1417–1423.

**35** Gobbi, F., Buonfrate, D., Boussinesq, M. et al. (2020). Performance of two serodiagnostic tests for loiasis in a non-endemic area. *PLoS Negl.Trop. Dis.* 14: e0008187.

**36** Rahmah, N., Shenoy, R.K., Nutman, T.B. et al. (2003). Multicentre laboratory evaluation of *Brugia* rapid dipstick test for detection of brugian filariasis. *Trop. Med. In.t Health* 8: 895–900.

**37** Unnasch, T.R., Golden, A., Cama, V., and Cantey, P.T. (2018). Diagnostics for onchocerciasis in the era of elimination. *Int. Health* 10 (Suppl_1): i20–i26.

**38** Hewitt, R.I. and White, D.E. (1948). Parasitology of piperazines in the treatment of filariasis. *Ann. N.Y. Acad. Sci.* 50: 128–140.

**39** Hawking, F. (1979). Diethylcarbamazine and new compounds for the treatment of filariasis. *Adv. Pharmacol. Chemother.* 16: 129–194.

**40** Otsuji, Y. (2011). History, epidemiology and control of filariasis. *Trop. Med. Health* 39: 3–13.

**41** Awadzi, K. and Gilles, H.M. (1992). Diethylcarbamazine in the treatment of patients with onchocerciasis. *Br. J. Clin. Pharmacol.* 34: 281–288.

**42** Mazzotti, L. (1948). Observations on onchocerciasis in Mexico. *Medicina (Mex)* 28: 217–224.

**43** Bain, O., Vuong, P.N., Petit, G. et al. (1987). Cutaneous changes induced by a dose of DEC in rodents with skin microfilaria: relevance of these phenomena for understanding the Mazzotti reaction and the pathogenesis of human onchocerciasis. *C. R. Acad. Sci. III* 304: 133–138.

**44** Awadzi, K., Orme, M.L., Breckenridge, A.M., and Gilles, H.M. (1982). The chemotherapy of onchocerciasis VII. The effect of prednisone on the Mazzotti reaction. *Ann. Trop. Med. Parasitol.* 76: 331–338.

**45** Stingl, P., Pierce, P.F., Connor, D.H. et al. (1988). Does dexamethasone suppress the Mazzotti reaction in patients with onchocerciasis? *Acta Trop.* 45: 77–85.

**46** Wilson, T. (1950). Hetrazan in the treatment of filariasis due to *Wuchereria malayi. Trans. R. Soc. Trop. Med. Hyg.* 44: 49–66.
**47** WHO Expert Committee on Filariasis & World Health Organization (1974). *WHO Expert Committee on Filariasis: third report [of a meeting held in Athens from 8 to 13 October 1973]*. Geneva: World Health Organization.
**48** Laigret, J., Fagneaux, G., and Tuira, E. (1978). Progress in the use of diethylcarbamazine in the drug therapy of lymphatic filariasis caused by *Wuchereria bancrofti* var. pacifica: administration in widely spaced doses. *Bull. WHO* 56: 985–990.
**49** Kimura, E., Penaia, L., and Spears, G.F. (1985). The efficacy of annual single-dose treatment with diethylcarbamazine citrate against diurnally subperiodic bancroftian filariasis in Samoa. *Bull. WHO* 63: 1097–1106.
**50** Kimura, E. (2011). The Global Programme to Eliminate Lymphatic Filariasis: history and achievements with special reference to annual single-dose treatment with diethylcarbamazine in Samoa and Fiji. *Trop. Med. Health* 39: 17–30.
**51** World Health Organization (1984). Lymphatic filariasis. Fourth report of the WHO Expert Committee on Filariasis. *World Health Organ. Tech. Rep. Ser.* 702: 3–112.
**52** Lammie, P., Milner, T., and Houston, R. (2007). Unfulfilled potential: using diethylcarbamazine-fortified salt to eliminate lymphatic filariasis. *Bull. WHO* 85: 545–549.
**53** Greene, B.M., Taylor, H.R., Cupp, E.W. et al. (1985). Comparison of ivermectin and diethylcarbamazine in the treatment of onchocerciasis. *N. Engl. J. Med.* 313: 133–138.
**54** Lariviere, M., Vingtain, P., Aziz, M. et al. (1985). Double-blind study of ivermectin and diethylcarbamazine in African onchocerciasis patients with ocular involvement. *Lancet* 2 (8448): 174–177.
**55** Diallo, S., Aziz, M.A., Lariviere, M. et al. (1986). A double-blind comparison of the efficacy and safety of ivermectin and diethylcarbamazine in a placebo controlled study of Senegalese patients with onchocerciasis. *Trans. R. Soc. Trop. Med. Hyg.* 80: 927–934.
**56** Awadzi, K., Dadzie, K.Y., Schulz-Key, H. et al. (2016). The chemotherapy of onchocerciasis XI A double-blind comparative study of ivermectin, diethylcarbamazine and placebo in human onchocerciasis in Northern Ghana. *Ann. Trop. Med. Parasitol.* 80: 433–442.
**57** White, A.T., Newland, H.S., Taylor, H.R. et al. (1987). Controlled trial and dose-finding study of ivermectin for treatment of onchocerciasis. *J. Infect. Dis.* 156: 463–470.
**58** Gardon, J., Boussinesq, M., Kamgno, J. et al. (2002). Effects of standard and high doses of ivermectin on adult worms of *Onchocerca volvulus*: a randomised controlled trial. *Lancet* 360: 203–210.
**59** Kumaraswami, V., Ottesen, E.A., Vijayasekaran, V. et al. (1988). Ivermectin for the treatment of *Wuchereria bancrofti* filariasis. Efficacy and adverse reactions. *JAMA* 259: 3150–3153.

**60** Ottesen, E.A., Vijayasekaran, V., Kumaraswami, V. et al. (1990). A controlled trial of ivermectin and diethylcarbamazine in lymphatic filariasis. *N. Engl. J. Med.* 322: 1113–1117.

**61** Dreyer, G., Coutinho, A., Miranda, D. et al. (1995). Treatment of bancroftian filariasis in Recife, Brazil: a two-year comparative study of the efficacy of single treatments with ivermectin or diethylcarbamazine. *Trans. R. Soc. Trop. Med. Hyg.* 89: 98–102.

**62** El Setouhy, M., Ramzy, R.M., Ahmed, E.S. et al. (2004). A randomized clinical trial comparing single- and multi-dose combination therapy with diethylcarbamazine and albendazole for treatment of bancroftian filariasis. *Am. J. Trop. Med. Hyg.* 70: 191–196.

**63** Gyapong, J.O., Kumaraswami, V., Biswas, G., and Ottesen, E.A. (2005). Treatment strategies underpinning the global programme to eliminate lymphatic filariasis. *Expert Opin. Pharmacother.* 6: 179–200.

**64** Sunish, I.P., Rajendran, R., Mani, T.R. et al. (2006). Impact of single dose of diethylcarbamazine and other antifilarial drug combinations on bancroftian filarial infection variables: assessment after 2 years. *Parasitol. Int.* 55: 233–236.

**65** Dunyo, S.K., Nkrumah, F.K., and Simonsen, P.E. (2000). Single-dose treatment of *Wuchereria bancrofti* infections with ivermectin and albendazole alone or in combination: evaluation of the potential for control at 12 months after treatment. *Trans. R. Soc. Trop. Med. Hyg.* 94: 437–443.

**66** Dunyo, S.K., Nkrumah, F.K., and Simonsen, P.E. (2000). A randomized double-blind placebo-controlled field trial of ivermectin and albendazole alone and in combination for the treatment of lymphatic filariasis in Ghana. *Trans. R. Soc. Trop. Med. Hyg.* 94: 205–211.

**67** Pani, S., Subramanyam, R.G., Das, L. et al. (2002). Tolerability and efficacy of single dose albendazole, diethylcarbamazine citrate (DEC) or co-administration of albendazole with DEC in the clearance of *Wuchereria bancrofti* in asymptomatic microfilaraemic volunteers in Pondicherry, South India: a hospital-based study. *Filaria J.* 1: 1.

**68** Rizzo, J.A., Belo, C., Lins, R., and Dreyer, G. (2007). Children and adolescents infected with *Wuchereria bancrofti* in Greater Recife, Brazil: a randomized, year-long clinical trial of single treatments with diethylcarbamazine or diethylcarbamazine-albendazole. *Ann. Trop. Med. Parasitol.* 101: 423–433.

**69** Macfarlane, C.L., Budhathoki, S.S., Johnson, S. et al. (2019, 2019). Albendazole alone or in combination with microfilaricidal drugs for lymphatic filariasis. *Cochrane Database Syst. Rev.* (1): (Art. No.: CD003753).

**70** Pion, S.D.S., Chesnais, C.B., Awaca-Uvon, N.P. et al. (2020). The impact of four years of semiannual treatments with albendazole alone on lymphatic filariasis and soil-transmitted helminth infections: a community-based study in the Democratic Republic of the Congo. *PLoS Negl.Trop. Dis.* 14: 0008322.

**71** Weil, G.J., Jacobson, J.A., and King, J.D. (2021). A triple-drug treatment regimen to accelerate elimination of lymphatic filariasis: From conception to delivery. *Int. Health* 13 (Suppl 1): S60–S64.

**72** Hawking, F. (1962). A review of progress in the chemotherapy and control of filariasis since 1955. *Bull. WHO* 27: 555–568.
**73** Ottesen, E.A. (1985). Efficacy of diethylcarbamazine in eradicating infection with lymphatic-dwelling filariae in humans. *Rev. Infect. Dis.* 7: 341–356.
**74** Sasa, M. (1963). Pilot experiments in the control of bancroftian filariasis in Japan and Ryukyu. *Bull. WHO* 28: 437–454.
**75** Noroes, J., Dreyer, G., Santos, A. et al. (1997). Assessment of the efficacy of diethylcarbamazine on adult *Wuchereria bancrofti in vivo. Trans. R. Soc. Trop. Med. Hyg.* 91: 78–81.
**76** Campbell, W.C. (1983). Progress and prospects in the chemotherapy of nematode infections of man and other animals. *J. Nematol.* 15: 608–615.
**77** Aziz, M.A., Diallo, S., Lariviere, M. et al. (1982). Ivermectin in onchocerciasis. *Lancet* 2 (8313): 1456–1457.
**78** Aziz, M.A., Diallo, S., Diop, I.M. et al. (1982). Efficacy and tolerance of ivermectin in human onchocerciasis. *Lancet* 2 (8291): 171–173.
**79** Boussinesq, M., Fobi, G., and Kuesel, A.C. (2018). Alternative treatment strategies to accelerate the elimination of onchocerciasis. *Int. Health* 10: i40–i48.
**80** Lakwo, T., Oguttu, D., Ukety, T. et al. (2020). Onchocerciasis elimination: Progress and challenges. *Res. Rep. Trop. Med.* 11: 81–95.
**81** Gebrezgabiher, G., Mekonnen, Z., Yewhalaw, D., and Hailu, A. (2019). Reaching the last mile: main challenges relating to and recommendations to accelerate onchocerciasis elimination in Africa. *Infect. Dis. Pov.* 8: 60.
**82** Crompton, D.W.T, World Health Organization (2006). *Preventive chemotherapy in human helminthiasis: coordinated use of anthelminthic drugs in control interventions: a manual for health professionals and programme managers*. Geneva: World Health Organization.
**83** Chesnais, C.B., Pion, S.D., Boulle, C. et al. (2020). Individual risk of post-ivermectin serious adverse events in subjects infected with *Loa loa*. *EClinMed* 28: 100582.
**84** Kamgno, J., Pion, S.D., Chesnais, C.B. et al. (2017). A Test-and-Not-Treat strategy for onchocerciasis in *Loa loa*-endemic areas. *N. Engl. J. Med.* 377: 2044–2052.
**85** Hübner, M.P, Pfarr, K., Hoerauf, A. (2022) *Wolbachia* endosymbionts as treatment targets for filarial diseases, in *Human and Animal Filariases*, (eds. R. Kaminsky and T.G. Geary), Chapter 24. Weinheim, Germany: Wiley-VCH.
**86** Walker, M., Specht, S., Churcher, T.S. et al. (2015). Therapeutic efficacy and macrofilaricidal activity of doxycycline for the treatment of river blindness. *Clin. Infect. Dis.* 60: 1199–2007.
**87** Hoerauf, A., Specht, S., Buttner, M. et al. (2008). *Wolbachia* endobacteria depletion by doxycycline as antifilarial therapy has macrofilaricidal activity in onchocerciasis: a randomized placebo-controlled study. *Med. Microbiol. Immunol.* 197: 295–311.
**88** Hoerauf, A., Specht, S., Marfo-Debrekyei, Y. et al. (2009). Efficacy of 5-week doxycycline treatment on adult *Onchocerca volvulus*. *Parasitol. Res.* 104: 437–447.

**89** Debrah, A.Y., Mand, S., Marfo-Debrekyei, Y. et al. (2007). Macrofilaricidal effect of 4 weeks of treatment with doxycycline on *Wuchereria bancrofti*. *Trop. Med. In.t Health* 12: 1433–1441.

**90** Horton, J., Klarmann-Schulz, U., Stephens, M. et al. (2020). The design and development of a multicentric protocol to investigate the impact of adjunctive doxycycline on the management of peripheral lymphoedema caused by lymphatic filariasis and podoconiosis. *Parasites Vectors* 13: 155.

**91** Wanji, S., Tendongfor, N., Nji, T. et al. (2009). Community-directed delivery of doxycycline for the treatment of onchocerciasis in areas of co-endemicity with loiasis in Cameroon. *Parasites Vectors* 2: 39.

**92** Hawryluk, N. (2022). The antifilarial drug pipeline. In: *Human and Animal Filariases* (ed. R. Kaminsky and T.G. Geary), Chapter 18. Weinheim, Germany: Wiley-VCH.

**93** Geary, T.G., Long, A., and Tritten, L. (2022). The antifilarial drug pipeline, in *Advances in Control of Heartworm and Human Filariases*, (eds. R. Kaminsky and T.G. Geary), Chapter 10. Weinheim, Germany: Wiley-VCH.

**94** Prichard, R.K. (2022). Drug resistance in filariae. In: *Advances in Control of Heartworm and Human Filariases* (ed. R. Kaminsky and T.G. Geary), Chapter 11. Weinheim, Germany: Wiley-VCH.

**95** Cotreau, M.M., Warren, S., Ryan, J.L. et al. (2003). The antiparasitic moxidectin: safety, tolerability, and pharmacokinetics in humans. *J. Clin. Pharmacol.* 43: 1108–1115.

**96** Prichard, R.K. and Geary, T.G. (2019). Perspectives on the utility of moxidectin for the control of parasitic nematodes in the face of developing anthelmintic resistance. *Int. J. Parasitol. Drugs Drug Resist.* 10: 69–83.

**97** Opoku, N.O., Bakajika, D.K., Kanza, E.M. et al. (2018). Single dose moxidectin versus ivermectin for *Onchocerca volvulus* infection in Ghana, Liberia, and the Democratic Republic of the Congo: a randomised, controlled, double-blind phase 3 trial. *Lancet* 392 (10154): 1207–1216.

**98** Turner, H.C., Walker, M., Attah, S.K. et al. (2015). The potential impact of moxidectin on onchocerciasis elimination in Africa: an economic evaluation based on the Phase II clinical trial data. *Parasites Vectors* 8: 167.

**99** Showler, A.J. and Nutman, T.B. (2018). Imported onchocerciasis in migrants and travelers. *Curr. Opin. Infect. Dis.* 31: 393–398.

**100** Lipner, E.M., Law, M.A., Barnett, E. et al. (2007). Filariasis in travelers presenting to the GeoSentinel Surveillance Network. *PLoS Negl.Trop. Dis.* 1: e88.

**101** Babu, S. and Nutman, T.B. (2012). Immunopathogenesis of lymphatic filarial disease. *Semin Immunopathol.* 34: 847–861.

**102** Chandenier, J., Pillier-Loriette, C., Datry, A. et al. (1987). Value of cytapheresis in the treatment of loaiasis with high blood microfilaria levels. Results in 7 cases. *Bull. Soc. Pathol. Exot. Filiales* 80: 624–633.

**103** King, J.D. (2017). *Lymphatic Filariasis*. WHO/HTM/NTD/PCT/2017.07: ISBN: 978 92 4 15016 1.

**104** Cunha, J.P. (2010). *Diethylcarbamazine*, https://www.rxlist.com/consumer_hetrazan_diethylcarbamazine/drugs-condition.html (accessed 14 March 2021).

**105** US Centers for Disease Control. (2018). Treatment of Lymphatic Filariasis. https://www.cdc.gov/parasites/lymphaticfilariasis/treatment.html (accessed 14 March 2021).

**106** Pearson, R.D. (2017). Lymphatische Filariose (*W. bancrofti, B. malayi*). https://www.msdmanuals.com/de/profi/infektionskrankheiten/nematoden-rundw%C3%BCrmer/lymphatische-filariose-w-bancrofti-b-malayi#:~:text=Reaktionen%20hervorrufen%20kann-,Behandlung%20der%20akuten%20lymphatischen%20Filariose,das%2012%2DTage%2DRegime (accessed 14 March 2021).

**107** Neumayr, A. (2018). *Antiparasitic Treatment Recommendations – A Practical Guide to Clinical Parasitology*, 2e. Hamburg: Tredition GmbH.

**108** Hörauf, A. and Burchard, G.-D. (2010). Filariosen und Drakunkulose. In: *Tropenmedizin in Klinik und Praxis* (ed. T. Löscher and G.-D. Burchard), 740–169. Stuttgart: Georg Thieme Verlag.

**109** US Centers for Disease Control. (2021). *Onchocerciasis*. https://www.cdc.gov/parasites/onchocerciasis/health_professionals/index.html#tx (accessed 14 March 2021).

**110** Pearson, R.D. (2017). Onchocerciasis (River Blindness). https://www.msdmanuals.com/de/profi/infektionskrankheiten/nematoden-rundw%C3%BCrmer/onchozerkose-flussblindheit (accessed 14 March 2021).

**111** US Centers for Disease Control. (2021). Treatment of Onchocerciasis. https://www.cdc.gov/parasites/loiasis/health_professionals/index.html#tx (accessed 14 March 2021).

**112** Pearson, R.D. (2017). Loiasis, https://www.msdmanuals.com/de/profi/infektionskrankheiten/nematoden-rundw%C3%BCrmer/loiasis (accessed 14 March 2021).

# 8

# Antifilarial Chemotherapy: Current Options in Veterinary Medicine

*Jennifer Ketzis[1] and Christian Epe[2,*]*

*[1]Ross University School of Veterinary Medicine, Biomedical Sciences, PO Box 334, Basseterre, St. Kitts, West Indies*
*[2]Boehringer Ingelheim Animal Health, Binger Str. 173, 55216, Ingelheim am Rhein, Germany*

**Abstract**

Chemotherapy options for the prevention of adult *Dirofilaria immitis* infections and other filarial species in dogs and cats focus primarily on the use of macrocyclic lactones. While macrocyclic lactones are considered safe when used as per their registered product labels, veterinarians should be aware of their basic mechanisms of action and metabolism to understand potential differences in efficacy and safety in the individual animal. With the advent of *D. immitis* resistance to macrocyclic lactones and the lack of preventatives or treatments for many other filarial species, supportive measures to decrease exposure to the intermediate host are growing in importance. Chemotherapeutic treatment of adult *D. immitis* infections is only available for dogs and uses melarsomine as the basis. Treatment guidelines differ with regard to the use of antibiotics and treatment regimen but result in similar efficacy and provide veterinarians with options that can be adapted based on the client and patient.

## 8.1 Introduction

This chapter on current veterinary options for antifilarial chemotherapy and metaphylaxis focuses on the most prominent filarial species in dogs and cats, the heartworm *Dirofilaria immitis*. Its geographic distribution has changed importantly over the years, and it is now ubiquitous in the USA, present in southern regions of Canada, the Caribbean, Central and South America, Europe (primarily around the Mediterranean Basin), and focal areas in Africa, Japan, and Asia. Chemotherapy for *Dirofilaria repens*, a species closely related to *D. immitis* and for which there is some overlap in distribution, is briefly addressed.

Other filarial species of canids include, but are not limited to, *Acanthocheilonema* spp., *Brugia malayi* and *Brugia pahangi, Cercopithifilaria* spp., *Dipetalonema* spp., and *Onchocerca lupi*, that latter that also can infect cats. Compared to *D. immitis* and

*Corresponding author.

*Human and Animal Filariases: Landscape, Challenges, and Control*, First Edition.
Edited by Ronald Kaminsky and Timothy G. Geary.

*D. repens*, these are of lesser importance for the veterinary field, as they either are confined to circumscribed endemic areas or are of lower prevalence and only rarely found in veterinary practice. Expanding from dogs and cats, many other filarial species are of veterinary importance, including those that infect ungulates, such as species in the genera *Setaria, Elaeophora*, and *Onchocerca*, and equids, including *Parafilaria multipapillosa*, which causes summer bleeding, and bovids, including species in the genera *Stephanofilaria* and *Parafilaria*, among others. As this book focuses predominately on the major human and canine filariae, of which *D. immitis* and *D. repens* are the most significant for canids, other filarial parasites of dogs will be mentioned only briefly and those of other nonhuman hosts will not be discussed.

## 8.2 Chemotherapy of *Dirofilaria immitis*

Chemotherapeutic options for the prevention and treatment of adult *D. immitis* infections are limited and depend on the stage of the parasite. The available chemotherapeutics can be classified in three general categories. Preventatives are those with efficacy directed at third and/or fourth stage larvae (L3, L4), primarily macrocyclic lactones (MLs) but also diethylcarbamazine (DEC). The drugs prevent the development of larvae to adult stages. They also block production of microfilariae (mf, L1) for several months following a single dose, but this action provides little clinical benefit. Macrofilaricides or adulticides have efficacy against immature adult ("fifth stage larvae"; L5) and adult *D. immitis*, of which there is one, melarsomine, which is used to eliminate adult *D. immitis* and consequently limit pathology caused by these parasites. Microfilaricides have efficacy against mf and are used to prevent intermediate hosts from becoming infected, diminishing the local distribution or prevalence of *D. immitis*. Only MLs are used for this purpose, and only one (moxidectin, MOX) is registered for this indication. In addition, antibiotics that target *Wolbachia*, symbiotic bacteria of filarioids, are used with the goal of inhibiting adult female worm reproduction and production of mf, decreasing pathology when eliminating adult *D. immitis* infections with melarsomine, and decreasing the development in mosquitoes of ingested mf acquired from treated definitive hosts to infectious L3. Repellents and ectoparasiticides that decrease the transfer of L3 to definitive hosts and mf to intermediate hosts also can be used in a comprehensive control program for *D. immitis*.

Guidelines for the prevention of adult *D. immitis* infections in dogs and cats and for treatment of adult infections in dogs are provided by the American Heartworm Society (AHS) [1], the Companion Animal Parasite Council (CAPC) [2], both based in the USA, the European Scientific Counsel for Companion Animal Parasites (ESCCAP) [3], and the Tropical Council for Companion Animal Parasites (TroCCAP) [4]. A core component of all guidelines is the administration of only registered (i.e. registered by the Food and Drug Administration (FDA) for USA, European Medical Agency (EMA) for the European Union, and respective agencies for other geographic regions with use for *D. immitis* in cats and dogs specified on the product label) ML-based products for prevention of the development of adult heartworm infections and to enhance pet-owner compliance with correct administration methods and frequency, the latter being particularly important in light of the documented existence of resistant *D. immitis* subpopulations in certain geographic regions (see Chapter 11 in this book). ESCCAP recommends year-round application in hyperendemic regions and seasonal use in areas in which low temperatures eliminate intermediate hosts, starting ML prevention before the beginning of the mosquito season in spring and continuing until the end of the transmission period in late autumn. AHS and CAPC embrace a year-round prevention program with MLs to maximize efficacy in the face of resistance and due to the pockets of mosquitoes that can survive and potentially transmit *D. immitis* even outside of the mosquito season in endemic regions of the USA [1, 2]. TroCCAP also recommends year-round use with a focus on using broad-spectrum ML-based products, due to the general parasite pressures in tropical regions. It should be noted, however, that while year-round use is recommended in endemic areas, this can be in contrast to the label for many products, which state that the drug is to be used during the transmission season.

In addition to the use of MLs, the guidelines [1–3] recommend reduction of exposure to mosquitoes. This includes the application of registered (e.g. Environmental Protection Agency [EPA] for USA) mosquito repellents/ectoparasiticides to dogs, reducing outdoor exposure during key mosquito feeding periods by containing dogs and cats indoors during these times and the use of standard environmental control practices for mosquitoes and their breeding environments (e.g. decreasing standing water in kennels and other areas). The ESCCAP [3] guidelines also refer to an integrated parasite control concept that considers the best use of combination products.

Guidelines for holiday travel with animals from nonendemic to endemic areas (e.g. from the United Kingdom to Italy or Canada to the southern USA) recommend that, if travel is for less than one month, administration of one dose of a preventative after return from the journey to the place of residence is sufficient [5]. These guidelines, however, do not consider that some product labels state that three or more applications are needed for the expected efficacy, nor do they consider the situation in the region visited by the dog. The region visited, the season, and the health of the dog should be considered in the decision to use one or more treatments. As well, knowledge of mosquito species in the home region and the risk of an infected dog becoming a reservoir for mosquito infection should be considered in the assessment of the treatment regimen.

All available *D. immitis* treatment and prevention guidelines refer to appropriate diagnostics, which are covered elsewhere in this book (Chapter 6). Each prevention and treatment program should include diagnostics, both prior to implementation and during the course of the program to monitor program success or gaps. Testing at least once a year using antigen and mf tests is recommended in all the guidelines. In highly endemic areas, testing twice per year is encouraged and testing before changing preventative products is recommended. While not specified in the guidelines, dogs relocated (with owners or via rehoming programs) should be carefully assessed for *D. immitis* infection to prevent spreading the parasite to new areas [6].

In the case of established immature adult and adult *D. immitis* infections, AHS and CAPC have developed guidelines for treatment, with TroCCAP, ESCCAP, and other regional and country specific guidelines typically referencing the AHS guidelines. These guideline recommendations are discussed in relation to melarsomine below. All the recommendations in the guidelines require specific adaptation by the veterinarian to the individual case and the epidemiological and biological situation of the respective animal, its environs and travel history.

### 8.2.1 Chemotherapeutics for L3/L4: Macrocyclic Lactones (MLs)

The first and foremost step in any program for *D. immitis* control as recommended by AHS, CAPC, ESCCAP, and TroCCAP is metaphylaxis, a strategy that does not block infection per se but instead prevents development of larval stages to adults in the definitive host; this is now almost entirely achieved through the use of MLs. Use of highly potent agents eliminates L3 and L4 stages as the primary therapeutic action preventing the development of adult worms, i.e. "prevention of heartworm disease." Typically, these treatments are referred to as "preventatives" in the veterinary clinical setting. Most MLs are administered on a monthly basis orally or topically; in addition, a 12-week topically applied formulation is available for cats and slow release injectable 6- and 12-month formulations are available for dogs. The product labels for the ML preventatives administered monthly have typically recommended a final administration one month after the end of the mosquito season/*D. immitis* transmission period (if not used year-round). However, the occurrence of *D. immitis* biotypes less sensitive to MLs may be reflected in more recent product approvals in the USA, with labels of these products stating that administration should be maintained for three to five months post-exposure. Use of increased doses as a means to cope with resistance is an ongoing area of research; however, for many MLs, the dose used is not the minimum efficacious dose for *D. immitis* and is instead based on other parasites such as hookworms and fleas [7, 8]. Regardless of concerns with resistance, MLs still provide a high level of protection to infection; hence, their use is still the first line of defense against the development of adult *D. immitis* infections in dogs and cats.

Veterinarians have a wide and ever-expanding choice of ML formulations and combinations with other actives to suit their geographical location in terms of common parasites and client preferences with regard to route and frequency of administration. A brief summary of the MLs available is presented in Table 8.1, including some unique features for each that could be of interest to veterinarians in selecting a product for a particular patient. In addition, aspects of the mechanisms of action, toxicity, and pharmacokinetics of MLs are useful in understanding variations in efficacy and safety in an individual dog or cat.

MLs act as agonists of glutamate-gated chloride channels. The number and location of these channels differ between nematode species and life stages and determine the type of physiological reaction exhibited by the parasite to ML exposure. In the case of *D. immitis* L3 and L4, these channels are around the excretory vesicles. The MLs inhibit secretion of parasite-derived molecules that enable the larvae to evade the host's immune system [14–16]; see also Chapter 10 in this book**. Hence, technically, the MLs do not kill L3 and L4; instead, they "unmask"

**Table 8.1** Overview of macrocyclic lactones used in veterinary medicine for *Dirofilaria immitis*: Points to consider in product selection in a veterinary clinic.

**Ivermectin** (IVM)

Dogs (oral, 6 μg/kg BW monthly for dogs over 6 wk of age; topical, 80 μg/kg BW)

- Lower doses of IVM, such as 3–3.3 μg/kg BW orally, have achieved 100% efficacy.
- Selection of 6 μg/kg BW orally was based on 100% efficacy in three experimental studies with administration either 45 or 60 d post-inoculation, thus providing efficacy if a dose was inadvertently delayed or missed [9].
- Higher doses and combinations with other anthelmintics expand the spectrum of activity of products.

Cats (oral, 24 μg/kg BW monthly for cats over 6 wk of age)

- Dose is based on *Ancylostoma tubaeforme* and not *D. immitis*.

**Selamectin** (SLM)

Dogs and cats (oral; 6 mg/kg BW monthly for dogs >6 wk of age and cats >6 or 8 wk of age)

- 3 mg/kg BW effective for *D. immitis* 45 d post inoculation with L3 and 6 mg/kg BW effective 60 d post inoculation with L3; higher doses and longer time periods not tested with either dose during registration of the originally registered product.
- Dose selected based on flea efficacy.
- Has one of the broader safety margins of the MLs.
- Efficacy against some other nematodes and fleas and combination with sarolaner expands ectoparasite activity of products.

**Milbemycin oxime (MBO)**

Dogs (oral; 0.5–0.75 mg/kg BW monthly in dogs ≥4 wk of age and ≥ 0.9 kg BW or ≥ 2 wk of age and ≥ 0.5 kg BW)

- Dose selected based on *Ancylostoma caninum*.
- Lower doses at 5 monthly administrations 100% effective against L3/L4 of *D. immitis*.
- Product dose 95% effective at 60- and 90-d post inoculation.
- At the dose used in *D. immitis* preventatives, MBO is efficacious against several other nematodes and has been combined with praziquantel for cestode control and with a variety of ectoparasiticides for flea control (afoxolaner, lufeneron).

Cats (oral; 2.0 mg/kg BW for cats ≥6 wk of age and ≥0.68 or ≥0.5 kg BW; minimum BW depends on the country of registration)

- Dose selected based on hookworms (*A. tubaeforme*) and roundworms (*T. cati*); 100% efficacy versus *D. immitis* at 0.5–0.9 mg/kg [10].

**Moxidectin** (MOX)

Dogs (topical: 2.5 mg/kg BW monthly for dogs ≥7 wk of age; oral: 0.24 mg/kg BW monthly ≥8 wk of age and ≥ 1.25 kg/BW; injectable: slow-release 6- (0.17 mg/kg BW) and 12-mo (0.23 mg/kg BW) formulations for dogs ≥6 mo of age)

Cats (topical: 1 mg/kg BW monthly for cats ≥9 wk of age; topical: 2 mg/kg BW every 12 wk for cats ≥9 wk of age and ≥1.2 kg BW)

- The most lipophilic and broadest in spectrum of the MLs.
- The mean elimination time and distribution volume is high, resulting in the ability to reach a steady state providing persistent activity [11–13].
- Monthly topically applied formulations have activity against several nematode species.
- Available in combinations with imidacloprid for fleas for cats and dogs; fluralaner for fleas and ticks for cats; and sarolaner and pyrantel embonate for fleas and additional nematodes for dogs.
- Some MOX-containing products also are registered for ferrets.

**Eprinomectin**

Cats (topical, 0.5 mg/kg BW at and above 7 wk of age and > 0.6 kg BW)

- Relatively new as an active for the treatment of L3/L4 *D. immitis*.
- Eprinomectin provides treatment and control for some gastrointestinal nematodes.
- Combinations available with fipronil and (*S*)-methoprene for ectoparasite control and praziquantel for cestodes.

Unless otherwise referenced, data are from the USA Food and Drug Administration New Animal Drug Application freedom of information summary for the first registered product containing each ML or from the European Product Assessment Report for the relevant ML containing products. Different doses and minimum age or BW reflect different country registrations.

the larvae to the host's immune system resulting in larval death due to an effective host immune response. The ML-*D. immitis*–host immune reaction interaction is complex, and several aspects require further elucidation. Under natural infection conditions, only some transmitted L3 develop into adult nematodes, resulting in a dog–heartworm balance in endemic regions [17, 18], suggesting that immunity is not only important after ML administration. This "balance" does not indicate a lack of pathology, but that adult heartworm numbers are limited. A study with cats indicated that retrovirus infections (FIV, FeLV) were risk factors for *D. immitis*, again demonstrating the importance of the host immune system [19]. From a clinical perspective, an individual dog's or cat's immune status, particularly in the case of substantial immune dysfunction, may need to be considered when assessing efficacy expectations with ML use in the context of heartworm prevention.

As a group, MLs are lipophilic and have a large volume of distribution, although there are differences between the individual MLs. Time to maximum concentration and mean residence time are believed to be influenced by the lipophilicity and efflux potential via ABC transporters, with body fat composition also potentially influencing $C_{max}$, $T_{max}$, and $T_{1/2}$ [20, 21]. In dogs and cats with poor body condition, $C_{max}$ would still be reached; however, potentially $T_{max}$ and $T_{1/2}$ could be shorter with less adipose tissue serving as a reservoir. In obese animals, $T_{max}$ and $T_{1/2}$ could be extended while $C_{max}$ could be lowered [20, 22]. Given that most ML preventatives are dosed above the minimum level required for efficacy against *D. immitis* and have wide safety margins, it is very unlikely that these changes in concentration would affect prevention efficacy or safety in malnourished or obese dogs and cats [9, 23]. However, there are a few case reports and single adverse events in studies suggesting that, even with the wide safety margins, statements on product labels regarding animal body condition should not be taken lightly [8, 24].

All ML-containing products registered for use in dogs and cats are considered safe when used as per the label, even in dogs deficient in P-glycoprotein drug extrusion pumps due to an ABCB1 (MDR1) loss-of-function gene mutation. Issues and concerns may arise when these dogs are not of normal body condition. Less is known about ABCB1 gene mutations in cats, and cat colonies with these mutations are not available for assessing safety of ML-containing products; therefore, an ABCB1 mutation cannot be ruled out if a neurological adverse event occurs after ML application to a cat [25].

A more in-depth understanding of these aspects of MLs are provided in other chapters within this book as well as several review articles [15–17, 20–22].

### 8.2.2 Chemotherapeutics for L3: Diethylcarbamazine (DEC)

While MLs are the primary chemotherapeutics for prevention of adult *D. immitis* development in dogs, DEC, the first commercialized compound for "prevention," should not be forgotten. Although never used in cats, registered products are still used and available for dogs in some countries (e.g., Dimmitrol, Mavlab Pty Ltd., Australia), and veterinarians can prescribe human DEC formulations. Reasons for use vary and include concerns about ML toxicity in dogs with ABCB1 mutations, cost, and pet owner's personal preferences; hence, its efficacy and safety require at least some attention even if not typically considered a current chemotherapeutic. For a historical perspective on DEC, see Rawlings [26].

DEC, administered orally at a rate of 6.6 mg/kg BW, has activity only against the L3 of *D. immitis* and continuous exposure is thus required for efficacy. Per the product label, DEC must be administered daily starting at or before exposure begins (mosquito emergence) and continuing for several weeks after exposure ends. Omitting even one or two doses impacts protection of the dog against *D. immitis* even when administered at higher doses [26–29]. In addition to the need to ensure that the pet owner administers DEC at beginning or prior to the heartworm transmission season and continues for several weeks afterward, without missing any doses, prescribing veterinarians must be aware of safety issues with DEC. The primary concern with DEC administration is in dogs that have mf. Anaphylaxis in mf-positive dogs can be fatal [30, 31]. Any dog treated with DEC must be confirmed negative for mf and adult worm infections prior to administration. Additional testing in season could be warranted if the season is longer than six months, given that early detection of prevention failure is critical to avoid shock from DEC use when mf are present.

### 8.2.3 Chemotherapeutics and Treatment Protocols for Immature (L5) and Mature *D. immitis*: Dogs

Several approaches to the treatment of patent *D. immitis* infections in dogs rely on the use of melarsomine. "Slow kill" methods that rely on MLs are no longer recommended in any of the established guidelines, which all endorse the use of melarsomine injections, incorporate treatment for mf and support the administration of a metaphylaxis regimen as soon as feasible to prevent more L3/L4 from developing to adult worms. The scientific community differs slightly in their interpretation of data on the optimal timing of melarsomine injections and the timing or length of use of doxycycline or other antibiotics to eliminate *Wolbachia* prior to adulticide treatment, resulting in some differences in the guidelines. However, all guidelines use the registered dosing regimens of melarsomine injections as the basis with the same expected efficacy outcome [32–34]. Several adjunct treatments are used to stabilize dogs prior to and during treatment (e.g. prednisone), which are patient- and case-specific. These treatments have no efficacy against *D. immitis* and hence are not discussed here.

The treatment of patent infections in cats differs from that in dogs. It does not include the use of melarsomine due to safety issues and does not focus on mf treatment due to the transient nature of mf in cats. Most treatment recommendations are instead focused on managing clinical signs and symptoms.

In some situations, although beyond the scope of this chapter, surgical removal of adult *D. immitis* from cats and dogs is recommended, particularly for animals exhibiting the caval syndrome, a life-threatening illness caused by a mass of adult *D. immitis* located aberrantly in the right atrium, the ventricle, and often the vena cava. The worm mass, which interferes with closure of the tricuspid valve and impedes normal flow of blood through the right heart, leading to cardiovascular collapse, must be promptly surgically extracted to re-establish normal cardiac function [35].

#### 8.2.3.1 Non-Melarsomine Treatment: "Slow Kill"

"Slow kill" methods that rely on the MLs IVM or MOX in combination with doxycycline are largely discouraged and are no longer advocated in any treatment guidelines due to the potential to add to resistance development. However, in situations in which melarsomine is unavailable or in which the pet owner declines melarsomine treatment, "slow kill" (with or without doxycycline) may be the only alternative [36]. In these methods, the dog is treated monthly with the preventative dose of IVM or MOX and typically administered doxycycline daily for 30 or more days. This approach can require extensive time before adult *D. immitis* are eliminated and the pathological changes due to heartworm disease might not regress during this period [37–42].

Given the cost of doxycycline, this method might not, in the long term, be more economical than melarsomine treatment. In addition, the continuing presence of adult heartworms in the pulmonary arteries is not beneficial to the health of the dog. Also, based on a small study, many owners are not compliant with the "slow kill" treatment protocol, likely decreasing the efficacy and increasing the health risks [43]. Lastly, there are concerns about overuse of antibiotics in companion animals, a special concern with doxycycline given its importance in human medicine.

#### 8.2.3.2 Doxycycline, Minocycline, and the Endosymbiont *Wolbachia*

*Wolbachia* are obligate, intracellular, gram-negative, endo-symbiotic bacteria, which have been implicated as crucial in the pathogenesis of filarial diseases. All stages of *D. immitis* have *Wolbachia* and reduction of the *Wolbachia* population via antibiotic treatment has been linked to a decrease in production of mf and the ability of mf to develop into L3 in the mosquito intermediate host [44–46]. In addition, studies by Kramer et al. [32, 47], among others, have indicated that killing *Wolbachia* prior to adulticide treatment with melarsomine can decrease the pathological signs related to the death of adult worms, although the mechanisms of the pathology are still to be determined [48]. Doxycycline has been identified as a treatment for *Wolbachia* prior to administration of melarsomine; however, the effect on treatment outcome and pathology has primarily been demonstrated when used in combination with an ML and not when doxycycline has been used on its own [32]. Minocycline also has been suggested as a means of killing *Wolbachia*. Studies with mouse models suggest minocycline could be more potent than doxycycline, although studies with dogs did not support this conclusion [49–51].

The AHS guidelines recommend daily administration of doxycycline (10 mg/kg BW twice daily) for four weeks followed by a 30-day waiting period before the administration of melarsomine to support the adulticide therapy. While limited data are available on minocycline compared to doxycycline, it is considered an alternative when doxycycline is not available, with the dosage being the same as for doxycycline [1]. Both the ESCCAP and TroCCAP guidelines reference the AHS guidelines for treatment protocols. The CAPC guidelines differ slightly in that they state that data indicate that doxycycline improves adulticide efficacy and treatment outcome and outline different approaches, but do not have the same prescriptive approach as AHS. The limited literature available reports improvement in outcome of dogs administered doxycycline prior to melarsomine administration, with less frequent respiratory complications [33, 34, 44, 52]. However, there has been no systematic evaluation of the minimal effective dose of doxycycline nor the optimum length of treatment prior to melarsomine administration. One study indicates that lower doses might be as effective as the currently recommended dose, and in cases in which dogs cannot tolerate 10 mg/kg, 5 mg/kg are suggested as an alternative [53]. While AHS recommends administration for four weeks, shorter periods of one to three weeks and administration during the melarsomine treatment (e.g. starting before a melarsomine injection and continuing for a few days after) have not been assessed to determine their efficacy and impact on pathology in the dog undergoing *D. immitis* adulticide treatment. There are concerns that the long-term administration of doxycycline, as well as other antibiotics, could contribute to selection for resistance of other bacterial pathogens and, while not raised as an issue specifically for doxycycline, there is generally increased concern about the impact of long-term antibiotic use on the gut microbiome [54, 55]. These concerns argue for the need to assess shorter treatment periods and timing of administration in relation to melarsomine treatment.

#### 8.2.3.3 Melarsomine

The only approved and available adulticide for dogs with patent *D. immitis* infections is melarsomine dihydrochloride, a melaminyl thioarsenite (14.95% arsenic), which is registered for use against four months old (immature adult) and older (adult) *D. immitis*. It is administered intramuscularly in the lumbar region (lumbar 3–5) at a rate of 2.5 mg/kg BW with the registered products having different dosing regimens for dogs with class 1 and 2 versus class 3 heartworm disease. It is not recommended or registered for the treatment of class 4 heartworm disease. The classes of heartworm disease referenced on the label for registered melarsomine products refer to stage 1 mild, stage 2 moderate, stage 3 severe, and stage 4 caval syndrome (for classification description see Chapter 3 in this book).

Before treatment commences, the dog must be stabilized in regard to the heart condition. For class 1 and 2 heartworm diseases, the regimen as per the label of registered melarsomine products is two injections 24 hours apart. In the case of class 3 heartworm disease, one injection is administered followed one month later with two injections 24 hour apart. This three-injection approach is believed to decrease the severity and complications of thromboembolic events and the occurrence and severity of pulmonary hypertension and allows time for recovery between the initial single injection, which results in death of a proportion of the adult *D. immitis* (primarily male worms) and the later set of two injections. Both the two and three injection treatment regimens developed were demonstrated to result in >95% reduction of immature adult and adult *D. immitis* in >90% of the dogs, although the three-injection regimen is believed to obtain higher efficacy [56, 57]. A potential misconception is that melarsomine treatment provides 100% efficacy, which is not the case. The efficacy threshold for adulticide *D. immitis* treatments is in accordance with the registration requirements for other anthelmintics as set by VICH GL 7 [58].

Shorter injection intervals of 3 and 6 hours provide higher efficacy than a 24 hours interval. However, these intervals provide less time for veterinarians to address adverse events and the impact from large numbers of *D. immitis* dying in a short time frame [57, 59]. It is possible that slightly higher doses could increase efficacy. However, the safety margin is already narrow with the recommended dose; doses between 7.5 and 10 mg/kg result in toxicity [57, 60]. British anti-lewisite (BAL; Dimercaprol) at 3 mg/kg can be used in cases of adverse reactions, including overdose [57, 61]. Occasionally, neurological signs have occurred posttreatment with the dog recovering without interventions; these have been attributed to the injection site location and not the *D. immitis* or melarsomine dose administered [62].

All current guidelines [1–4] support the three-dose regimen for classes 1, 2, and 3 heartworm disease due to a potentially better clinical outcome and higher efficacy [63, 64]. The AHS [1] (referenced by ESCCAP [3] and TroCCAP [4]) recommend placing the dog on an ML preventative, 30 days of doxycycline administration followed by a 30-day waiting period to provide time for *Wolbachia* metabolites and antigens to be cleared and the adult *D. immitis* to weaken and then starting the three-dose melarsomine regimen. From initiation of treatment with ML and doxycycline to time of final melarsomine injections is 90 days; the concept is that older L4 *D. immitis* present at the time of treatment initiation that were not killed by the ML preventative will be susceptible to melarsomine, addressing any treatment gap between the stages impacted by MLs and those impacted by melarsomine. The potential benefits of this approach are as follows:

- higher efficacy and less likely to require further melarsomine treatments; and
- less pathology because of *Wolbachia* elimination.

While this regimen is reported to work, there are limited systematic and statistically meaningful comparative studies to other approaches. A concern is that much of the pathology and risk of complications of thromboembolic events have been attributed to total worm mass, typically reflected in total number of worms. While the use of doxycycline could decrease growth of the adult worms present at diagnosis, the >60-day waiting period from diagnosis to first melarsomine injection, allowing more larvae to develop to immature adults and adults, could increase total mass. In addition, this treatment protocol is lengthy and owner compliance, especially regarding exercise restriction, might be reduced given the length of the full treatment program. Lastly, the inclusion of doxycycline might be cost prohibitive for some owners.

There has been some recent interest in exploring options for shortening the AHS recommended treatment approach. One option explored in a 76-dog study was the omission of the 30-day waiting period between the end of doxycycline treatment and the first melarsomine injection [65]. In that study, all dogs became antigen- and microfilariae-negative with no severe or moderate side effects, suggesting that there is room to explore modifications to the AHS guidelines.

A concern of having a shorter time from treatment initiation to final treatment, omitting either the 30-day waiting period between the end of doxycycline treatment and the first melarsomine injection or omitting doxycycline altogether, may raise concerns with regard to the susceptibility gap (i.e. the *D. immitis* stages susceptible to neither MLs nor melarsomine) [66]. Most MLs, however, when used continuously have high efficacy (>96%) against two- and three-month-old larvae and melarsomine has high efficacy against two-month-old larvae [8, 9, 23, 65, 67], suggesting that concerns of a susceptibility gap are not well supported.

In July 2020, CAPC [2] updated its guidelines to recognize that data indicate that doxycycline could improve the treatment outcome. Prior to this update, CAPC prioritized immediate treatment using the three-dose melarsomine regimen to limit worm mass and the time the dog harbors adult worms over delays encountered with use of an antibiotic for *Wolbachia,* citing that while there were potential benefits to including an antibiotic in the treatment protocol, further data were needed regarding dose, length of treatment, and impact on patient outcome. The potential benefits of this approach of immediate treatment are as follows:

- less time consuming;
- may have higher owner compliance due to shorter length;
- less expensive with no or a short dose of doxycycline or another antibiotic; and
- limits time mf are exposed to a ML.

As of July 2020, CAPC still emphasizes the need to begin "adulticide as soon as is medically practical." However, CAPC now also states that while research continues, data indicate that doxycycline does improve the outcome of treatment. CAPC also recognizes the differing opinions regarding the treatment gap and, while advocating no prescriptive treatment program such as that of AHS, does briefly describe the AHS protocol as well as the shortened protocol that omits the waiting period between doxycycline administration and the first melarsomine injection. In essence, the updated CAPC guideline supports antibiotic use with no prescribed pattern while still emphasizing adulticide treatment as soon as feasible, leaving the actual approach more flexible and adaptable to the individual patient.

Regardless of the approach used, melarsomine efficacy is not always complete. The label for melarsomine products suggests antigen testing four months after the final injection and administering two additional injections 24 hour apart if needed, dependent on the condition of the dog. However, while antigen testing typically is negative four months after treatment, it can be positive for longer. Therefore, the CAPC guidelines recommend retesting six to seven months later, while the AHS guideline recommends retesting nine months later.

### 8.2.4 Chemotherapeutics and Treatment Protocols for Immature (L5) and Mature *D. immitis*: Cats

The clinical presentation, diagnosis, and treatment of *D. immitis* in cats differs from dogs for several reasons. Cats typically do not have many adult *D. immitis*, impeding diagnosis with antigen testing. Instead, cats with immature adult *D. immitis* can present with heartworm-associate respiratory disease (HARD) [68]. While melarsomine is effective against this stage of *D. immitis*, it is not registered or safe for use in cats, and no treatment is approved for immature and adult *D. immitis* in cats. Thiacetarsamide, which is no longer readily available, was used in cats previously, albeit with high risk of toxicity and uncertain efficacy. Thiacetarsamide was also previously used in dogs, but was replaced with melarsomine due to its higher efficacy and better safety margin [69, 70].

All the guidelines present similar approaches for handling infections in cats, with a focus on minimizing clinical signs and pathology from immature adult and adult worms using corticosteroids and bronchodilators as well as oxygen therapy as per the individual cat's needs. Treatment for mf is not recommended, primarily because mf do not typically persist in cats. Positive cats should be maintained on a metaphylaxis regimen to prevent more *D. immitis* from developing. While this can result in a "slow kill," the risk in regard to selection for and spread of resistance is lower since mf are rarely produced. The AHS guideline also mentions the potential use of doxycycline, but no studies are available to document efficacy.

### 8.2.5 Other Chemotherapy: Treating Microfilaria (L1)

A complete treatment program for a dog with reproducing adult *D. immitis* must include treatment for mf [1–4]. Microfilariae can live for more than two years [71, 72], which could lead to incongruent results after elimination of adult *D. immitis* during annual diagnostic tests prior to ML metaphylaxis prescriptions. Also, dogs with circulating mf can be a reservoir of infection for mosquitoes and hence are a risk factor for other dogs/cats in the area. Lastly, presence of mf while using a ML metaphylaxis could select for resistance. Only one ML product, containing MOX, is registered (in the USA) for use against circulating mf. Other MLs, when used at the metaphylaxis dose, can have some impact on circulating mf. However, typically a higher dose is needed for effective treatment of circulating mf. Doxycycline also has been used to impact mf via its activity on their *Wolbachia*, which can result in inhibited development to L3 in the mosquito intermediate host. However, the effect of doxycycline on development to L3 is not 100% [44, 45].

MOX in the Advantage Multi®/Advocate® formulation is currently the only product registered for mf treatment, and both CAPC and AHS recommend that if mf are still present after adulticide treatment, MOX, given its registration status, should be used. Prior to adulticide treatment, the guidelines are less prescriptive. The AHS guideline relies on the use of doxycycline in combination with an ML preventative to address mf prior to adulticide treatment, while the CAPC guideline updated in 2020 changed its recommendation from MOX to no specific ML prior to adulticide treatment.

### 8.2.6 Other Measures: Decreasing Exposure

Given the development of *D. immitis* populations resistant to ML preventatives, other means of decreasing exposure of dogs and cats to infection should be considered. All the guidelines encourage reducing mosquito exposure, with AHS and ESCCAP outlining specific measures, such as keeping dogs and cats inside or within mosquito-proof screened enclosures during times of the day when mosquitoes are most abundant. However, it must be noted that in one study, >25% of *D. immitis* positive cats were characterized as indoor only and that mosquitoes can be found inside homes [73]. From a chemotherapeutic perspective, using topical products (spot-ons or collars) or oral repellents and insecticides should be encouraged. This requires a shift in reliance solely on ML preventatives and has the added benefit of control of ectoparasites, potentially limiting transmission of other vector-borne diseases. Studies by McCall et al. [74, 75] demonstrated that a combination of ectoparasiticides and ML preventatives lowers the development of adult *D. immitis*-resistant isolates. Ectoparasiticides tested in this regard or having a label claim for repelling mosquitoes include, but are not limited to, topically applied dinotefuran 4.95% w/w, pyriproxyfen 0.44% w/w, and permethrin 36.08% w/w (Vectra® 3D) [74]; topically applied imidacloprid 8.8% w/w, permethrin 44% w/w, and pyriproxyfen 0.44% w/w (K9 Advantix® II); and collars with deltamethrin (e.g. Scalibor®).

### 8.2.7 Role of ML Lack of Expected Efficacy (LOE) and Resistance

The expectation for ML-based preventative products has been 100% efficacy against *D. immitis* L3 and L4 of 30 days of age, demonstrated by the lack of development of adult worms in dogs and cats after inoculation of L3. This expectation did not take in to account the genetic variation or lack thereof in isolates used for efficacy testing and the representativeness of these isolates [76], and the claim for 100% efficacy might not have actually ever been reached [77]. In the early 2000s, an increase in LOE reports for ML preventatives in the Mississippi Delta region of the USA brought attention to the potential occurrence of *D. immitis* resistant to MLs [78–80] and the presence of biotypes that are resistant to all MLs is now acknowledged [81–84].

The general belief is that resistant or less-sensitive biotypes are currently primarily geographically limited to the Mississippi Delta region, with most clinical cases outside of this region being traced back to that region (i.e. via inter-state adoptions or movement of pet owners). However, the true limitation of resistant biotypes to this region is not known, with some isolates originating outside of the Mississippi Delta region. LOE cases have also been reported from Brazil [84, 85], and while there has been limited investigation into resistant biotypes in Europe, there are concerns of potential occurrence [80, 84, 86].

Regardless of the current distribution of resistant biotypes and their cause (a reflection of the natural variation in ML susceptibility or directly arising due to selection of resistance alleles resulting from intensive ML use), it is clear that *D. immitis* is not exempt from anthelmintic resistance to MLs, which should be used wisely. Veterinarians should be aware of differences between countries regarding monthly use patterns on the label of ML registered preventatives, changes to labels, and the requirement for three to six months of use (versus one month) for 100% efficacy for some products [87, 88].

There are proposals and research regarding changes in dosing of MLs to slow down the development of resistance [89]. The use of MOX, specifically the injectable 6- and

12-month formulations developed for long-term control to increase owner compliance and the so-called "forward killing" effect, has been discussed as a means to cope with *D. immitis* resistance [89–91]. It is important to note, though, that all MLs at the current doses provide efficacy beyond 30 days and few of the registered doses are actually at the minimal level needed for efficacy (see Table 8.1). Therefore, increasing doses or using formulations that could improve owner compliance will, in general, only delay or slow the development of resistance. As all preventatives belong to the ML class, the risk of class-wide cross-resistance must be assumed as likely, as shown for drug classes with gastrointestinal nematodes in livestock [92, 93].

Determining if there is L3/L4 resistance to MLs or if failure is due to compliance or incorrect administration in a particular case is challenging and, due to lack of tools for direct and specific diagnostic, not easy or even possible in a clinical setting. Careful record keeping of testing for adult *D. immitis* infections and frequency of preventative prescriptions are needed. In addition, reviewing the administration method with the client for correct application of topically administered products (location and skin contact) and ensuring instructions are followed for oral products (e.g. regarding food and encouraging dogs to chew some of the chewable formulations) is needed to achieve the expected efficacy. Prior to any change in preventative product, the dog or cat should be confirmed negative for an adult *D. immitis* infection and a follow-up test performed six months after changing products. While not definitive, a positive test at six months or less after changing product would suggest a LOE with the prior product used. Molecular tests have been described [84, 94], which offer additional tools to elucidate resistance based on single nucleotide polymorphism (SNP) profiles. However, these tests are not routinely available to the veterinary practitioner and are currently more useful for epidemiological studies to understand and confirm the presence and extent of resistance in geographic areas.

ML preventatives in almost all cases still achieve 100% efficacy when used as per the product label. The important take-home message is that true efficacy is a range and based on this range, one can expect some adult *D. immitis*, albeit very low numbers, to develop in some dogs even under the best preventative programs. Following label instructions, ensuring correct administration of products, and considering the health status of the dog or cat could all assist in ensuring the lowest number of these infections.

## 8.3 Chemotherapy of *Dirofilaria repens*

Treatment of *D. repens* in dogs has grown in importance due to its zoonotic potential, the increasing prevalence and distribution of this parasite in Europe and the potential for *D. repens* mf to be confused with those of *D. immitis*. The primary goals of *D. repens* treatments are to prevent mf transfer to intermediate hosts, prevent the introduction of *D. repens* to nonendemic regions, and ultimately to prevent zoonotic infections [3, 95–97]. Only MOX in a topical formulation is registered with claims against *D. repens* with the dose being the same as for *D. immitis* prevention [98]. As a preventative, application is started one month prior to the transmission season and continued for one month after the transmission season ends. As an adulticide, six months of continuous use is required and four months of continuous use results in suppression or reduction of circulating mf [99–103]. The injectable formulation of MOX also has been assessed for *D. repens* control, with efficacy similar to that of the topical formulation, but it is not registered for this use [104, 105].

Other approaches using ML regimens similar to those used for *D. immitis* have been assessed for prevention of adult *D. repens* and treatment of mf. However, most of these have been used in limited studies or are based on case reports, with insufficient data to support their use. Efficacy achieved is <100% and, for some, below the minimum 90% expected for anthelmintics [106–109]. Melarsomine also has been investigated, using a two-injection protocol at 2.5 mg/kg. In one case report, it was combined with doramectin (0.4 mg/kg subcutaneously once five days after melarsomine) and resulted in a complete cure with no *D. repens* at euthanasia (performed for an unrelated reason) [110]. In a study with natural infections, melarsomine followed with MOX (three applications) eliminated mf from 35 of 36 dogs [96]. A combination of IVM (6 μg/kg every 15 days for 6 months) with doxycycline (10 mg/kg daily for 30 days) also has been used to eliminate adult *D. repens* and mf. These approaches raise the same concerns as with the use of "slow kill" for *D. immitis* [111].

## 8.4 Chemotherapy of Other Filarioidea of Dogs and Cats

Other filarioids that receive some attention in dogs are *Acanthocheilonema reconditum* (formerly *Dipetalonema reconditum*) and *Acanthocheilonema dracunculoides* due to the potential to confuse their mf with those of *D. immitis* or *D. repens*; *O. lupi* (also in cats) due to the ocular lesions and zoonotic potential; *Cercopithifilaria* spp. for which the clinical significance is not well understood and are not further mentioned here; and *B. malayi* and *B. pahangi*, which cause lymphatic filariasis in humans and for which dogs are suspected to serve as a reservoir host.

*A. reconditum* and *A. dracunculoides* lack clinical significance and no chemotherapy is needed, although distinguishing their mf from *D. immitis* mf and *D. repens* mf is important in diagnostics and due to the potential for adverse reactions with mass *Acanthocheilonema* spp. mf death after ML use [108, 112]. Use of insecticides for vector control (fleas and lice) is recommended to prevent *A. reconditum* infections [2].

For *O. lupi*, Otranto et al. [113] provide data on several cases. Treatments used included IVM at 400–600 μg/kg daily, every other day or monthly and MOX at 220 μg/kg subcutaneously once monthly for four or five months, with or without daily doxycycline administration. However, the most effective treatment reported to date is the manual removal of *O. lupi* from the eyes [113–115]. *Brugia malayi* and *B. pahangi*, restricted to Southeast Asia and India, rarely result in clinical signs in infected dogs. Efficacious treatment has been reported for MOX, SLM, doramectin, and IVM, and vector control to minimize exposure is recommended [4, 116].

## 8.5 Conclusions

Current chemotherapy for *D. immitis* continues to rely on MLs to prevent adult worm development and to eliminate circulating mf and on melarsomine to kill immature adult and adult worms. For filarioids other than *D. immitis*, there are either limited data on prevention or treatment or, in the case of *D. repens*, only one registered product, with most other options providing subpar efficacy.

Changes in prevention protocols regarding doses and number of monthly administrations required to maximize efficacy against *D. immitis* are ongoing, and careful attention should be paid to the specific use pattern on product labels. Monitoring of resistance development and tracking to determine if it is of local origin or imported are critical, and veterinarians play a key role in this process. Identifying cases of ML resistance and eliminating the mf and adult worm infections are critical to maintain the efficacy of the MLs and to slow distribution or prevalence increases of less susceptible biotypes.

Efficacy of MLs for the prevention of adult *D. immitis* infections is still very high, even with the presence of less susceptible biotypes. However, the future reality is that eventually 100% efficacy might not be consistently achievable, depending on the changing prevalence of less susceptible biotypes and their distribution. The number of dogs and cats with adult *D. immitis* infections could, therefore, increase over time. Consideration of additional control measures, such as repellents to decrease exposure to intermediate hosts is a paradigm shift, but one that is becoming reality.

Several protocols are available for the treatment of adult *D. immitis* infections. Further research is needed to determine the most effective use of antibiotics in adulticide protocols, considering both the impact on patient outcome and good stewardship of antibiotics. Given that all clients might not be able to afford a lengthy regimen of antibiotics prior to use of melarsomine and the fact that doxycycline is not always available, veterinarians have options to use shorter treatment regimens and contribute to the body of evidence-based medicine.

## References

1 American Heartworm Society (2020). Current canine guidelines for the prevention, diagnosis, and management of heartworm (*Dirofilaria immitis*) infection in dogs. https://d3ft8sckhnqim2.cloudfront.net/images/pdf/2020_AHS_Canine_Guidelines.pdf?1580934824 (accessed 16 April 2020).

2 Companion Animal Parasite Council (2016/2020). Heartworm for Dog. https://capcvet.org/guidelines/heartworm (accessed 16 April 2020/11 August 2020).

3 European Scientific Counsel Companion Animal Parasites (2019). Control of vector-borne diseases in dogs and cats. https://www.esccap.org/uploads/docs/znkh6j1d_0775_ESCCAP_Guideline_GL5_v8_1p.pdf (accessed 9 April 2020).

4 Tropical Council for Companion Animal Parasites (2019). TroCCAP Guidelines for the diagnosis, treatment and control of canine endoparasites in the tropics, 2nd Ed March 2019. https://www.troccap.com/2017press/wp-content/uploads/2019/05/TroCCAP_Canine_Endo_Guidelines_English_Ver2.pdf (accessed 9 April 2020).

5 EV/ESDA. European Society of Dirofilariosis and Angiostrongylosis (ESDA) (2017). Guidelines for clinical management of canine heartworm disease. https://www.esda.vet/wp-content/uploads/2017/11/guidelinesfor-clinical-management-of-canineheartworm-disease.pdf (accessed 9 April 2020).

6 Drake, J. and Wiseman, S. (2018). Increasing incidence of *Dirofilaria immitis* in dogs in USA with focus on the southeast region 2013–2016. *Parasite Vectors* 11 (1): 1–7.

7 NADA 141-078 (1996). Freedom of Information Summary. Supplemental New Animal Drug Application Heartgard® (ivermectin) for cats. https://animaldrugsatfda.fda.gov/adafda/app/search/public/document/downloadFoi/613 (accessed 9 April 2020).

8 NADA 141-152 (2000). Freedom of Information Summary. Supplemental New Animal Drug Application Revolution® (selamectin). https://animaldrugsatfda.fda.gov/adafda/views/#/home/previewsearch/141-152; https://animaldrugsatfda.fda.gov/adafda/app/search/public/document/downloadFoi/671 (accessed 9 April 2020).

9 NADA 138-412 (1987). Freedom of information summary. New Animal Drug Application Heartgard-30® (ivermectin). https://animaldrugsatfda.fda.gov/adafda/app/search/public/document/downloadFoi/420 (accessed 9 April 2020).

10 Stewart, V.A., Hepler, D.I., and Grieve, R.B. (1992). Efficacy of milbemycin oxime in chemoprophylaxis of dirofilariasis in cats. *Am. J. Vet. Res.* 53: 2274–2277.

11 EMA/359888/2019 (2019). Veterinary Medicines Division Committee for Medicinal Products for Veterinary Use. CVMP assessment report for a type II variation for Advocate (EMEA/V/C/000076/II/0041/G) International non-proprietary name: imidacloprid/moxidectin. https://www.ema.europa.eu/en/documents/variation-report/advocate-v-c-76-ii-0041-g-epar-assessment-report-variation_en.pdf (accessed 16 April 2020).

12 Bowman, D.D., Grazette, A.R., Basel, C. et al. (2016). Protection of dogs against canine heartworm infection 28 days after four monthly treatments with Advantage Multi® for dogs. *Parasite Vectors* 9: 12.

13 Little, S.E., Hostetler, J.A., Thomas, J.E., and Bailey, K.L. (2015). Moxidectin steady state prior to inoculation protects cats from subsequent, repeated infection with *Dirofilaria immitis*. *Parasite Vectors* 8: 107.

14 Geary, T.G. and Moreno, Y. (2012). Macrocyclic lactone anthelmintics: spectrum of activity and mechanism of action. *Current Pharm. Biotechnol.* 13: 866–872.

15 Carithers, D.S. (2017). Examining the role of macrolides and host immunity in combatting filarial parasites. *Parasite Vectors* 10: 1–13.

16 Berrafato, T., Coates, R., Reaves, B.J. et al. (2019). Macrocyclic lactone anthelmintic-induced leukocyte binding to *Dirofilaria immitis* microfilariae:

influence of the drug resistance status of the parasite. *Int. J. Parasitol.: Drug. Drug Resist.* 10: 45–50.

**17** Capelli, G., Poglayen, G., Bertotti, F. et al. (1996). The host-parasite relationship in canine heartworm infection in a hyperendemic area of Italy. *Vet. Res. Commun.* 20: 320–330.

**18** Simón, F., Genchi, C., Prierto, G., and Allende, E. (2001). Immunity in the vertebrate hosts. In: *Heartworm Infections in Humans and Animals* (eds. F. Simon and C. Genchi), 83–101. Salamanca: Ediciones Universidad de Salamanca.

**19** Levy, J.K., Burling, A.N., Crandall, M.M. et al. (2017). Seroprevalence of heartworm infection, risk factors for seropositivity, and frequency of prescribing heartworm preventives for cats in the United States and Canada. *JAVMA* 250: 873–880.

**20** McKellar, Q.A. and Gokbulut, C. (2012). Pharmacokinetic features of the antiparasitic macrocyclic lactones. *Current Pharm. Biotechnol.* 13: 888–911.

**21** Prichard, R., Ménez, C., and Lespine, A. (2012). Moxidectin and the avermectins: consanguinity but not identity. *Int. J. Parasitol.: Drugs Drug Resist.* 2: 134–153.

**22** Merola, V.M. and Eubig, P.A. (2018). Toxicology of avermectins and milbemycins (macrocyclic lactones) and the role of p-glycoprotein in dogs and cats. *Vet. Clin. N. Am.: Small Anim. Pract.* 48: 991–1012.

**23** NADA 141-251 (2006). Freedom of information Summary. New Animal Drug Application. Advantage Multi® for Dogs (moxidectin). https://animaldrugsatfda.fda.gov/adafda/app/search/public/document/downloadFoi/803 (accessed 9 April 2020).

**24** See, A.M., McGill, S.E., Raisis, A.L., and Swindells, K.L. (2009). Toxicity in three dogs from accidental oral administration of a topical endectocide containing moxidectin and imidacloprid. *Aust. Vet. J.* 87: 334–337.

**25** Mealey, K.L. and Burke, N.S. (2015). Identification of a nonsense mutation in feline ABCB1. *J. Vet. Pharmacol. Ther.* 38: 429–433.

**26** Rawlings, C.A. (1986). Prevention of heartworm infection. In: *Heartworm Disease in Dogs and Cats* (ed. C.A. Rawlings), 255–280. Philadelphia, PA: WB Saunders Co.

**27** Kume, S., Ohishi, I., and Kobayashi, S. (1962). Prophylactic therapy against the developing stages of *Dirofilaria immitis*. *Am. J. Vet. Res.* 23: 1257–1260.

**28** Kume, S., Ohishi, I., and Kobayashi, S. (1967). Prophylactic therapy against the developing stages of *Dirofilaria immitis* supplemental studies. *Am. J. Vet. Res.* 28: 975–978.

**29** Fowler, J.L., Warne, R.J., Furusho, Y., and Sugiyama, H. (1970). Testing fenthion, dichlorvos, and diethylcarbamazine for prophylactic effects against the developing stages of *Dirofilaria immitis*. *Am. J. Vet. Res.* 31: 903–906.

**30** Boreham, P.F., Atwell, R.B., and Euclid, J.M. (1985). Studies on the mechanism of the DEC reaction in dogs infected with *Dirofilaria immitis*. *Int. J. Parasitol.* 15: 543–549.

**31** Hamilton, R.G., Wagner, E., April, M. et al. (1986). *Dirofilaria immitis*: diethylcarbamazine-induced anaphylactoid reactions in infected dogs. *Exp. Parasitol.* 61: 405–420.

**32** Kramer, L., Grandi, G., Leoni, M. et al. (2008). *Wolbachia* and its influence on the pathology and immunology of *Dirofilaria immitis* infection. *Vet. Parasitol.* 158: 191–195.

**33** Carretón, E., Morchón, R., Simón, F., and Juste, M.C. (2014). Evaluation of cardiopulmonary biomarkers during classic adulticide treatment versus the American Heartworm Society recommended treatment protocol in dogs infected by *Dirofilaria immitis*. *Vet. Parasitol.* 206: 55–59.

**34** Nelson, C.T., Myrick, E.S., and Nelson, T.A. (2017). Clinical benefits of incorporating doxycycline into a canine heartworm treatment protocol. *Parasite Vectors* 10 (S2): 181–184.

**35** Jones, S.L. (2016). Canine caval syndrome series – Part 3: management of caval syndrome. *Today's Vet. Pract.,* http://tvpjournal.com May/June 2016.

**36** Ku, T.N. (2017). Investigating management choices for canine heartworm disease in northern Mississippi. *Parasite Vectors* 10 (S2): 193–199.

**37** Grandi, G., Quintavalla, C., Mavropoulou, A. et al. (2010). A combination of doxycycline and ivermectin is adulticidal in dogs with naturally acquired heartworm disease (*Dirofilaria immitis*). *Vet. Parasitol.* 169: 347–351.

**38** Choi, M., Yoon, W.K., Suh, S.I., and Hyun, C. (2017). Assessment of clinical outcome in dogs naturally infected with *Dirofilaria immitis* after AHS protocol vs slow kill method. *J. Vet. Clin.* 34: 1–6.

**39** Kramer, L., Grandi, G., Passeri, B. et al. (2011). Evaluation of lung pathology in *Dirofilaria immitis* experimentally infected dogs treated with doxycycline or a combination of doxycycline and ivermectin before administration of melarsomine dihydrochloride. *Vet. Parasitol.* 176: 357–360.

**40** Yoon, W.K., Kim, Y.W., Suh, S.I., and Hyun, C. (2017). Evaluation of cardiopulmonary and inflammatory markers in dogs with heartworm infection treated using the slow kill method. *Vet. Parasitol.* 244: 35–38.

**41** Genchi, M., Vismarra, A., Lucchetti, C. et al. (2019). Efficacy of imidacloprid 10%/moxidectin 2.5% spot on (Advocate®, Advantage Multi®) and doxycycline for the treatment of natural *Dirofilaria immitis* infections in dogs. *Vet. Parasitol.* 273: 11–16.

**42** Ames, M.K., VanVranken, P., Evans, C., and Atkins, C.E. (2020). Non-arsenical heartworm adulticide therapy using topical moxidectin-imidacloprid and doxycycline: a prospective case series. *Vet. Parasitol.* 282: 109099.

**43** Drake, J., Gruntmeir, J., Merritt, H. et al. (2015). False negative antigen tests in dogs infected with heartworm and placed on macrocyclic lactone preventives. *Parasite Vectors* 8: 68.

**44** McCall, J.W., Genchi, C., Kramer, L. et al. (2008). Heartworm and *Wolbachia*: therapeutic implications. *Vet. Parasitol.* 158: 204–214.

**45** McCall, J.W., Kramer, L., Genchi, C. et al. (2014). Effects of doxycycline on heartworm embryogenesis, transmission, circulating microfilaria, and adult worms in microfilaremic dogs. *Vet. Parasitol.* 206: 5–13.

**46** Kramer, L., Crosara, S., Gnudi, G. et al. (2018). *Wolbachia*, doxycycline and macrocyclic lactones: new prospects in the treatment of canine heartworm disease. *Vet. Parasitol.* 254: 95–97.

**47** Kramer, L.H., Tamarozzi, F., Morchón, R. et al. (2005). Immune response to and tissue localization of the *Wolbachia* surface protein (WSP) in dogs with natural heartworm (*Dirofilaria immitis*) infection. *Vet. Immunol. Immunopathol.* 106: 303–308.

**48** Diosdado, A., Gómez, P.J., Morchón, R. et al. (2017). Interaction between *Wolbachia* and the fibrinolytic system as a possible pathological mechanism in cardiopulmonary dirofilariosis. *Vet. Parasitol.* 247: 64–69.

**49** Sharma, R., Jayoussi, G.A.I., Tyrer, H.E. et al. (2016). Minocycline as a re-purposed anti-*Wolbachia* macrofilaricide: superiority compared with doxycycline regimens in a murine infection model of human lymphatic filariasis. *Sci. Rep.* 6: 23458.

**50** Papich, M.G. (2017). Considerations for using minocycline vs doxycycline for treatment of canine heartworm disease. *Parasite Vectors* 10 (S2): 185–191.

**51** Savadelis, M.D., Day, K.M., Bradner, J.L. et al. (2018). Efficacy and side effects of doxycycline versus minocycline in the three-dose melarsomine canine adulticide heartworm treatment protocol. *Parasite Vectors* 11: 671.

**52** Serrano-Parreño, B., Carretón, E., Caro-Vadillo, A. et al. (2017). Pulmonary hypertension in dogs with heartworm before and after the adulticide protocol recommended by the American Heartworm Society. *Vet. Parasitol.* 236: 34–37.

**53** Carretón, E., Morchón, R., Falcón-Cordón, Y. et al. (2020). Evaluation of different dosages of doxycycline during the adulticide treatment of heartworm (*Dirofilaria immitis*) in dogs. *Vet. Parasitol.* 283: 109–141.

**54** Tejedor-Junco, M.T., González-Martín, M., Bermeo-Garrido, E. et al. (2018). Doxycycline treatment for *Dirofilaria immitis* in dogs: impact on *Staphylococcus aureus* and *Enterococcus* antimicrobial resistance. *Vet. Res. Commun.* 42: 227–232.

**55** Pilla, R. and Suchodolski, J.S. (2020). The role of the canine gut microbiome and metabolome in health and gastrointestinal disease. *Front. Vet. Sci.* 6: 498.

**56** NADA 141-042 (1995). Freedom of information summary. New Animal Drug Application Immiticide® (melarsomine dihydrochloride).

**57** Vezzoni, A., Genchi, C., and Raynaud, J.-P. (1992). Adulticide efficacy of RM 340 in dogs with mild and severe natural infections. In: *Proceedings of the Heartworm Symposium*, March 27–29, Austin, TX, 231–240. Batavia, IL: American Heartworm Society.

**58** VICH GL 7 (2000). Efficacy of Anthelmintics: general requirements. International Cooperation on Harmonisation of Technical Requirements for Registration of Veterinary Medicinal Products. https://vichsec.org/en/guidelines/pharmaceuticals/pharma-efficacy/anthelmintics.html (accessed 18 April 2020).

**59** Keister, D.M., Dzimianski, M.T., McTier, T.L. et al. (1992). Dose selection and confirmation of RM340, a new filaricide for the treatment of dogs with immature and mature *Dirofilaria immitis*. In: *Proceedings of the Heartworm Symposium*, 27–29 March, Austin, TX, 225–229. Batavia, IL: American Heartworm Society.

**60** Toutain, P.L. and Raynaud, J.P. (1988). Pharmacokinetics and clinical pharmacology in dogs of a new heartworm adulticide (RM 340). In: *Proceedings of the 33rd Annual Meeting of the American Association of Veterinary Parasitologists*, 17–18 July, Portland, OR, 46. American Association of Veteirnay Parasitologists.

**61** Atwell, R.B., Sheridan, A.D., Buoro, I.B.J. et al. (1989). Effective reversal of induced arsenic toxicity using BAL therapy. In: *Proceedings of the Heartworm Symposium*, 17–19 March, Charleston, SC. Washington, DC: American Heartworm Society.

**62** Hettlich, B.F., Ryan, K., Bergman, R.L. et al. (2003). Neurologic complications after melarsomine dihydrochloride treatment for *Dirofilaria immitis* in three dogs. *JAVMA* 223: 1456–1461.

**63** Case, J.L., Tanner, P.A., Keister, D.M., and Meo, N.J. (1995). A clinical field trial of melarsomine dihydrochloride (RM340) in dogs with severe (class 3) heartworm disease. In: *Proceedings of the Heartworm Symposium*, 31 March-2 April, Auburn, AL, 243–250. Batavia, IL: American Heartworm Society.

**64** Carretón, E., Morchón, R., González-Miguel, J. et al. (2013). Utility of cardiac biomarkers during adulticide treatment of heartworm disease (*Dirofilaria immitis*) in dogs. *Vet. Parasitol.* 197: 244–250.

**65** Carretón, E., Falcón-Cordón, Y., Falcón-Cordón, S. et al. (2019). Variation of the adulticide protocol for the treatment of canine heartworm infection: can it be shorter? *Vet. Parasitol.* 271: 54–56.

**66** American Heartworm Society (2014). Current canine guidelines for the prevention, diagnosis, and management of heartworm (*Dirofilaria immitis*) infection in dogs. https://www.heartwormsociety.org/images/pdf/2014-AHS-Canine-Guidelines.pdf (accessed 16 April 2020).

**67** Bowman, D.D. and Drake, J. (2017). Examination of the "susceptibility gap" in the treatment of canine heartworm infection. *Parasite Vectors* 10 (S2): 65–73.

**68** Dillon, A.R., Blagburn, B.L., Tillson, M. et al. (2017). Heartworm-associated respiratory disease (HARD) induced by immature adult *Dirofilaria immitis* in cats. *Parasite Vectors* 10 (S2): 514.

**69** Raynaud, J.P. (1992). Thiacetarsamide (adulticide) versus melarsomine (RM 340) developed as macrofilaricide (adulticide and larvicide) to cure canine heartworm infection in dogs. *Ann. Rech. Vet.* 23: 1–25.

**70** Sundlof, S.F., Courtney, C.H., and King, R.R. (1989). Laboratory methods for evaluating the effectiveness of thiacetarsamide in treating canine heartworm disease. In: *Proceedings of the Heartworm Symposium*, March 17–19, Charleston, SC. Washington, DC: American Heartworm Society.

**71** Underwood, P.C. and Harwood, P.D. (1939). Survival and location of the microfilariae of *Dirofilaria immitis* in the dog. *J. Parasitol.* 25 (1): 23–33.

**72** Bourguinat, C., Keller, K., Bhan, A. et al. (2011). Macrocyclic lactone resistance in *Dirofilaria immitis*. *Vet. Parasitol.* 181: 388–392.

**73** Atkins, C.E., DeFrancesco, T.C., Coats, J.R. et al. (2000). Heartworm infection in cats: 50 cases (1985–1997). *JAVMA* 217: 355–358.

**74** McCall, J.W., Hodgkins, E., Varloud, M. et al. (2017). Blocking the transmission of heartworm (*Dirofilaria immitis*) to mosquitoes (*Aedes aegypti*) by weekly exposure for one month to microfilaremic dogs treated once topically with dinotefuran-permethrin-pyriproxyfen. *Parasite Vectors* 10 (S2): 87–95.

**75** McCall, J.W., Varloud, M., Hodgkins, E. et al. (2017). Shifting the paradigm in *Dirofilaria immitis* prevention: blocking transmission from mosquitoes to dogs using repellents/insecticides and macrocyclic lactone prevention as part of a multimodal approach. *Parasite Vectors* 10: 525.

**76** Vidyashankar, A.N., Jimenez Castro, P.D., and Kaplan, R.M. (2017). A statistical approach for evaluating the effectiveness of heartworm preventive drugs: what does 100% efficacy really mean? *Parasite Vectors* 10 (S2): 97–108.

**77** Bowman, D.D. (2012). Heartworm, macrocyclic lactones, and the specter of resistance to prevention in the United States. *Parasite Vectors* 5: 138.

**78** Hampshire, V.A. (2005). Evaluation of efficacy of heartworm preventive products at the FDA. *Vet. Parasitol.* 133: 191–195.

**79** Atkins, C.E., Murray, M.J., Olavessen, L.J. et al. (2014). Heartworm lack of effectiveness claims in the Mississippi delta - computerized analysis of owner compliance 2004–2011. *Vet. Parasitol.* 206: 106–113.

**80** Bourguinat, C., Lee, A.C., Lizundia, R. et al. (2015). Macrocyclic lactone resistance in *Dirofilaria immitis*. Failure of heartworm preventives and investigation of genetic markers for resistance. *Vet. Parasitol* 210: 167–178.

**81** Blagburn, B.L., Dillon, A.R., Arther, R.G. et al. (2011). Comparative efficacy of four commercially available heartworm preventive products against the MP3 laboratory strain of *Dirofilaria immitis*. *Vet. Parasitol.* 176: 189–194.

**82** Blagburn, B.L., Arther, R.G., Dillon, A.R. et al. (2016). Efficacy of four commercially available heartworm preventive products against the JYD-34 laboratory strain of *Dirofilaria immitis*. *Parasite Vectors* 9: 191.

**83** Ballesteros, C., Pulaski, C.N., Bourguinat, C. et al. (2018). Clinical validation of molecular markers of macrocyclic lactone resistance in *Dirofilaria immitis*. *Int. J. Parasitol.: Drug. Drug Resist.* 8 (3): 596–606.

**84** Bourguinat, C., Keller, K., Xia, J. et al. (2017). Genetic profiles of ten *Dirofilaria immitis* isolates susceptible or resistant to macrocyclic lactone heartworm preventives. *Parasite Vectors* 10: 504.

**85** Labarthe, N.V., Willi, L.M.V., Paiva, J.P. et al. (2015). Chemoprophylaxis of *Dirofilaria immitis* (Leidy 1856) infection at a high challenge environment. *Parasite Vectors* 8: 523.

**86** Mendoza-Roldan, J., Benelli, G., Panarese, R. et al. (2020). *Leishmania infantum* and *Dirofilaria immitis* infections in Italy, 2009–2019: changing distribution patterns. *Parasite Vectors* 13 (1): 1–8.

**87** NADA 141-321 (2011). Freedom of information Summary. New Animal Drug Application. Trifexis® (milbemycin oxime). https://animaldrugsatfda.fda.gov/adafda/app/search/public/document/downloadFoi/878 (accessed 9 April 2020).

**88** NADA 141-338 (2012). Freedom of information Summary. New Animal Drug Application. Interceptor® Spectrum (milbemycin oxime). https://animaldrugsatfda.fda.gov/adafda/app/search/public/document/downloadFoi/895 (accessed 9 April 2020).

**89** Prichard, R.K. and Geary, T.G. (2019). Perspectives on the utility of moxidectin for the control of parasitic nematodes in the face of developing anthelmintic resistance. *Int. J. Parasitol.: Drugs Drug Resist.* 10: 69–83.

**90** Bowman, D.D., Ohmes, C.M., Hostetler, J.A. et al. (2017). Efficacy of 10% imidacloprid + 2.5% moxidectin topical solution (Advantage Multi® for dogs) for the prevention of heartworm disease and infection all month long. *Parasite Vectors* 10 (S2): 59–64.

**91** Bowman, D.D., McTier, T.L., Adams, E.L. et al. (2017). Evaluation of the efficacy of ProHeart® 6 (moxidectin) against a resistant isolate of *Dirofilaria immitis* (JYD-34) in dogs. *Parasite Vectors* 10 (S2): 53–57.

**92** Sutherland, I.A. and Leathwick, D.M. (2011). Anthelmintic resistance in nematode parasites of cattle: a global issue? *Trends Parasitol.* 27: 176–181.

**93** Jabbar, A., Iqbala, Z., Kerboeuf, D. et al. (2006). Anthelmintic resistance: the state of play revisited. *Life Sci.* 79: 2413–2431.

**94** Prichard, R.K., Bourguinat, C., and Geary, T.G. (2014). Markers to predict macrocyclic lactone drug resistance in *Dirofilaria immitis*, the causative agent of heartworm disease. US Patent US20180187264A1.

**95** Capelli, G., Genchi, C., Baneth, G. et al. (2018). Recent advances on *Dirofilaria repens* in dogs and humans in Europe. *Parasite Vectors* 11 (1): 1–21.

**96** Pingen, C.H., Lorentz, S., Magnis, J. et al. (2009). Successful treatment of *Dirofilaria repens* infections in dogs with melarsomine (Immiticide®, Merial) against adults and a combination of moxidectin 2.5%/imidacloprid 10% (Advocate®, Bayer) against microfilaria. In: *Poster, 22nd Conference of the World Association for the Advancement of Veterinary Parasitology*, 9–13 August 2009, Calgary. CDN.

**97** Sonnberger, K., Duscher, G.G., Fuehrer, H.P., and Leschnik, M. (2020). Current trends in canine dirofilariosis in Austria-do we face a pre-endemic status? *Parasitol. Res.* 119: 1001–1009.

**98** EMEA/V/C/000076 - IB/0042.202 (2021). Veterinary Medicines Division Committee for Medicinal Products for Veterinary Use. Advocate EPAR: Product Information Annex 1 Summary of product characteristics. Summary International non-proprietary name: imidacloprid/moxidectin. https://www.ema.europa.eu/en/documents/product-information/advocate-epar-product-information_en.pdf (accessed 9 April 2021).

**99** Fok, E., Jacsó, O., Szebeni, Z. et al. (2010). Elimination of *Dirofilaria* (syn. *Nochtiella*) *repens* microfilariae in dogs with monthly treatments of moxidectin 2.5%/imidacloprid 10% (Advocate, Bayer) spot-on. *Parasitol. Res.* 106: 1141–1149.

**100** Traversa, D., Aste, G., Di Cesare, A. et al. (2011). Efficacy of a single administration of a spot-on solution containing imidacloprid 10%/moxidectin 2.5% in eliminating *Dirofilaria repens* microfilariae in naturally infected dogs. *Vet. Parasitol.* 179: 107–112.

**101** Hellmann, K., Heine, J., Braun, G. et al. (2011). Evaluation of the therapeutic and preventive efficacy of 2.5% moxidectin/10% imidacloprid (Advocate®, Bayer Animal Health) in dogs naturally infected or at risk of natural infection by *Dirofilaria repens*. *Parasitol. Res.* 109: S77–S86.

**102** Genchi, C., Genchi, M., Petry, G. et al. (2013). Evaluation of the efficacy of imidacloprid 10% / moxidectin 2.5% (Advocate®, Advantage® Multi, Bayer) for the prevention of *Dirofilaria repens* infection in dogs. *Parasitol. Res.* 112 (S1): 81–89.

**103** Petry, G., Genchi, M., Schmidt, H. et al. (2015). Evaluation of the adulticidal efficacy of imidacloprid 10%/moxidectin 2.5% (w/v) spot on (Advocate®, Advantage® Multi) against *Dirofilaria repens* in experimentally infected dogs. *Parasitol. Res.* 114: 131–144.

**104** Rossi, L., Ferroglio, E., and Agostini, A. (2004). Use of an injectable, sustained-release formulation of moxidectin to prevent canine subcutaneous dirofilariosis. *Vet. Rec.* 154: 26–27.

**105** Genchi, M., Pengo, G., and Genchi, C. (2010). Efficacy of moxidectin microsphere sustained release formulation for the prevention of subcutaneous filarial (*Dirofilaria repens*) infection in dogs. *Vet. Parasitol.* 170: 167–169.

**106** Cancrini, G., Tassi, P., and Coluzzi, M. (1989). Ivermectin against larval stages of *Dirofilaria repens* in dogs. *Parassitologia* 31: 177–182. [in Italian].

**107** Jacsó, O., Fok, E., Kiss, G. et al. (2010). Preliminary findings on the efficacy of selamectin in the treatment of dogs naturally infected with *Dirofilaria repens*. *Acta Vet. Hung.* 58: 405–412.

**108** Pantchev, N., Etzold, M., Daugschies, A., and Dyachenko, V. (2011). Diagnosis of imported canine filarial infections in Germany 2008–2010. *Parasitol. Res.* 109 (S1): 61–76.

**109** Di Cesare, A., Braun, G., Di Giulio, E. et al. (2014). Field clinical study evaluating the efficacy and safety of an oral formulation containing milbemycin oxime/praziquantel (Milbemax®, Novartis Animal Health) in the chemoprevention of the zoonotic canine infection by *Dirofilaria repens*. *Parasite Vectors* 7: 347.

**110** Baneth, G., Volansky, Z., Anug, Y. et al. (2002). *Dirofilaria repens* infection in a dog: diagnosis and treatment with melarsomine and doramectin. *Vet. Parasitol.* 105: 173–178.

**111** Giannelli, A., Ramos, R.A.N., Traversa, D. et al. (2013). Treatment of *Dirofilaria repens* microfilaraemia with a combination of doxycycline hyclate and ivermectin. *Vet. Parasitol.* 197: 702–704.

**112** Muñoz, C., Gonzálvez, M., Rojas, A. et al. (2020). Massive microfilaremia in a dog subclinically infected with *Acanthocheilonema dracunculoides*. *Parasitol. Int.* 76: 102070.

**113** Otranto, D., Giannelli, A., Trumble, N.S. et al. (2015). Clinical case presentation and a review of the literature of canine onchocercosis by *Onchocerca lupi* in the United States. *Parasite Vectors* 8: 89.

**114** Komnenou, A., Eberhard, M.L., Kaldrymidou, E. et al. (2002). Subconjunctival filariasis due to *Onchocerca* sp. in dogs: report of 23 cases in Greece. *Vet. Ophthalmol.* 5: 119–126.

**115** Labelle, A.L., Daniels, J.B., Dix, M., and Labelle, P. (2011). *Onchocerca lupi* causing ocular disease in two cats. *Vet. Ophthalmol.* 14: 105–110.

**116** Ravindran, R., Varghese, S., Nair, S.N. et al. (2014). Canine filarial infections in a human *Brugia malayi* endemic area of India. *Biomed. Res. Int.* 2014: 630160.

# 9

# Heartworm Disease – Intervention and Industry Perspective[1]

*Sandra Noack[1], John Harrington[2], Douglas S. Carithers[3], Ronald Kaminsky[4], and Paul M. Selzer[1,*]*

[1] *Boehringer Ingelheim Animal Health, Binger Str. 173, Ingelheim am Rhein 55216, Germany*
[2] *Boehringer Ingelheim Animal Health, 1730 Olympic Drive, Athens GA 30601, USA*
[3] *Boehringer Ingelheim Animal Health, 3239 Satellite Blvd, Duluth GA 30096, USA*
[4] *paraC Consulting, Altenstein 13, Häg-Ehrsberg 79685, Germany*

**Abstract**

*Dirofilaria immitis*, also known as heartworm, is a major parasitic threat for dogs and cats around the world. Because of its impact on the health and welfare of companion animals, heartworm disease is of huge veterinary and economic importance especially in North America, Europe, Asia, and Australia. Within the animal health market, many different heartworm preventive products are available, all of which contain active components of the same drug class, the macrocyclic lactones. In addition to compliance issues, such as under-dosing or irregular treatment intervals, the occurrence of drug-resistant heartworms within populations in the Mississippi River (United States) delta areas adds to the failure of preventive treatments. Here, we show possibilities for interventions within the parasite life cycle and provide an overview on the prevalence of *Dirofilaria* spp., on current disease control measures, and available drugs and products.

## 9.1 Introduction

Companion animals, specifically dogs and cats, are hosts to a variety of external and internal parasites [1]. One of the most important endoparasites in companion animal health, from both a pathologic and an economic perspective, is *Dirofilaria immitis*, the filarial nematode parasite that causes heartworm disease. In dogs, the disease is caused by young adult and adult parasites provoking pathology in the pulmonary arteries. Canines act as the definitive host, so sexual reproduction occurs in the pulmonary arteries, and microfilariae are released into the circulatory system [1–3]. *D. immitis* infections also occur in cats, and the disease is usually more severe in

*Corresponding author.
[1]This article is a condensed part of a pre-published comprehensive review in IJDDR 16 (2021) 65-89.

*Human and Animal Filariases: Landscape, Challenges, and Control*, First Edition.
Edited by Ronald Kaminsky and Timothy G. Geary.

this atypical host. In cats, severe disease, or even death, can be caused by just a few developing immature or adult filariae. The adult worms are often shorter than those found in dogs and rarely produce microfilariae [4]. The life cycle of *D. immitis* is similar to that of the pathogenic filarial parasites of humans: *Onchocerca volvulus*, *Brugia malayi*, and *Wucheria bancrofti*, in that they are all transmitted by arthropod vectors [5]. *D. immitis* is distributed across the globe, being endemic in countries on six continents [6]. From an animal health perspective, dogs, and to a lesser extend cats, are the most important animals among the numerous mammalian hosts that are infected by *D. immitis* [1, 7, 8]. Thus, this chapter will focus primarily on heartworm disease in dogs and include relevant background on cats where appropriate.

Treatment of an established *D. immitis* infection in dogs requires a prolonged regimen of medication, exercise restriction, and sometimes even surgery (detailed in Chapter 8) [3, 9, 10]. Therefore, current practice is to control heartworm disease through prevention, which utilizes a single class of drugs, the macrocyclic lactones (MLs), including ivermectin, abamectin, eprinomectin, milbemycin oxime, selamectin and moxidectin [11]. However, prevention of heartworm disease in dogs may be compromised by ML-resistant *D. immitis* [12, 13].

In this chapter, we outline the industry's perspective on heartworm control, the potential chemical points of intervention and currently available treatment options. In addition to the general distribution of *D. immitis*, reasons for both the increasing and decreasing prevalence in certain geographic areas will be discussed.

As a note for the reader and in accordance with the World Association for the Advancement of Veterinary Parasitology (WAAVP, https://www.waavp.org), we used the term "dirofilariosis" [14]. However, in many publications, the disease is also referred to as "dirofilariasis" or occasionally also as "dirofilarosis," which is important to consider when reviewing the literature [15, 16].

## 9.2 Heartworm Biology

Like the majority of filariae, *D. immitis* has no free-living stages and exhibits a complex lifecycle involving multiple developmental stages in both the definitive mammalian host and the mosquito vector (Figure 9.1). The adult worms (macrofilariae) live as obligate endoparasites mainly in the lobar arteries and the main pulmonary artery of the canine hosts. In dogs with high worm burden, adult *D. immitis* can also be found in the right ventricle. Females measure up to 30 cm and males around 18 cm. Females are ovoviviparous; they release sheathless microfilariae in the blood, which are 250–300 μm in size. Various mosquito species can acquire the *D. immitis* microfilarial stage in a blood meal from an infected host. Once ingested by a mosquito, the microfilariae migrate within hours from the midgut to the Malpighian tubules where they morphologically change into various "sausage" forms representing the first-stage larvae [18, 19]. However, available reports for *D. immitis* described these changes using *in vitro* cultivated microfilariae, which were unable to develop in those systems beyond the L1 stage [19]. Nevertheless, the principle developmental stages of *D. immitis* in mosquitoes are confirmed by observations on the related species *B. malayi* [20]. Filariae harvested from infected mosquitoes at various times

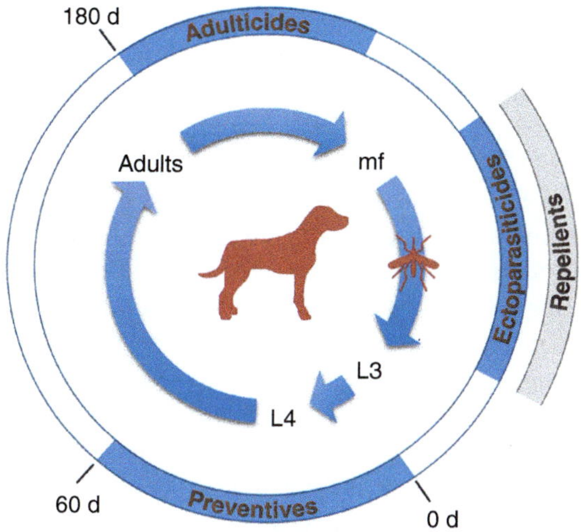


**Figure 9.1** ***Dirofilaria immitis*** **life cycle and chemical intervention periods**. The inner circle represents the life cycle of *D. immitis* within the mammalian host (dog) and the vector (mosquito). The length of the arrows approximately indicates the development time of each stage. During a blood meal from an infected dog, microfilariae (mf) circulating in the blood are ingested by a mosquito. In the vector, they develop into the larval (L) stages from L1 to infective L3, which can be transmitted during another blood meal to a mammalian host. Within the host, they develop rapidly into L4 and finally to adult female or male parasite that reside and mate in the heart (vascular and cardiopulmonary system, right heart chamber). The outer circle shows prevention and treatment options depending on the stage of development of the parasite. MLs are used as preventive treatment up to 60 days (d) post-infection (0 d) against *D. immitis* L3 and L4 larval stages. After that period, efficacy of MLs is no longer 100% [17]. Melarsomine, the only registered heartworm adulticide, is efficacious against adult *D. immitis*, which can be diagnosed around 180 days post-infection. Melarsomine, administered in two doses 24 hours apart, also has substantial efficacy against juveniles, as shown in controlled studies summarized by Bowman and Drake [17]. However, melarsomine is not recommended to be used as a preventive or prior to definitive diagnosis of heartworm infection, but only for treatment of adult heartworms. Ectoparasiticides or repellents can be used to prevent mosquitos from feeding on dogs and cats, reducing the potential for infection. Source: Modified from CAPC, Heartworm.

can be differentiated into microfilariae, which migrate and develop within hours into intracellular L1 stages. These shorter and non-feeding L1 stages undergo a first molt, resulting in L2 stages that remain intracellular. They molt for a second time to become infective *B. malayi* L3 stages [20], which can be transmitted to another mammalian host. Development of *D. immitis* and migration in the canine host, as well as the host immune response, are largely uncharacterized until adults appear in the pulmonary arteries approximately six months after infection [3, 6, 21, 22].

Cats are less suitable hosts for *D. immitis* than dogs [2], and most worms in cats do not develop to the adult stage. Those cats with adult *D. immitis* usually harbor only one to three worms. These worms are also smaller than those found in dogs and rarely produce microfilariae [4, 6] which complicates the diagnosis [23], and therefore, cats are infected but remain often undiagnosed (American Heartworm Society Heartworm Basics https://www.heartwormsociety.org/pet-owner-resources/heartworm-basics accessed November 25th, 2020).

## 9.3 Prevalence

*Dirofilaria immitis* infections in dogs and cats have been identified throughout the world in tropical and temperate regions [6, 24], while the occurrence of *D. repens* is restricted to the Old World (Figure 9.2). A few anecdotal reports suggest the presence of the related species (not found in the heart) *Dirofilaria repens* in Mexico and Chile as well [25, 26]. Two opposing phenomena influence the prevalence and spread of dirofilariosis. The prevalence of *Dirofilaria* infections appears to be increasing worldwide mainly because of climate changes and the accompanying spread of competent mosquito species such as *Aedes albopictus* and *Ae. koreicus* [27]. In contrast, heartworm prevalence is decreasing in some regions like Japan [28] and Northern Italy, likely because of a higher awareness and intensified control of the disease [24, 29]. The latter observation may indicate that the distribution and thus the risk of heartworm infection could be reduced by higher awareness of veterinary practitioners and a concomitant increase in preventive treatment of dogs.

However, decreasing heartworm prevalence, particularly in the Mediterranean, may also be because of overall reduction of mosquito populations. The general correlation of mosquito abundance and risk of disease transmission is well established for malaria [30], and it was demonstrated that, depending on local conditions for mosquito populations, there is less or more risk of malaria transmission [31]. Efforts to eliminate malaria in Europe and the Mediterranean date back as far as the late nineteenth century, with widespread achievement of elimination in the twentieth century [32]. Today, an additional important effect is the ongoing reduction of insect populations because of industrialization, urbanization, and application of insecticides [33]. Thus, heartworm prevalence may continue to decline in some endemic areas because of reduction of the vector population in

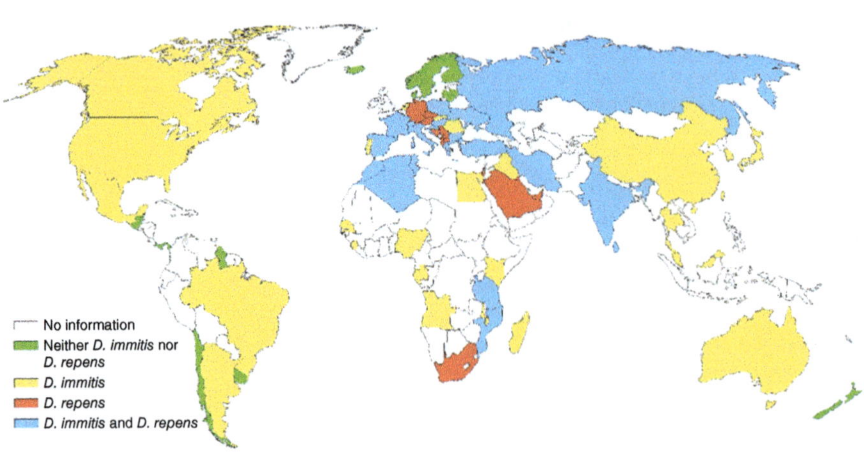


**Figure 9.2 Presence of *D. immitis* and *D. repens* infections throughout the world.** Analyzing the number of dogs at risk for *Dirofilaria* infection, in Asia approximately 148 million dogs are at risk, in Latin America as well as in Europe approximately 98 million dogs each, in North America approximately 80 million, in Africa approximately 50 million, and in Oceania approximately 6 million; Boehringer Ingelheim internal analysis. Source: Based on [6, 24–26].

these regions. In contrast, climate changes will lead to a northward spread of the occurrences of mosquito vectors, leading to a higher disease risk in presently non-endemic areas [34–36]. This forecasts the spread of *Anopheles* spp. and, as a result, malaria could also mimic the prevalence of dirofilariosis, assuming that the mosquito migration includes species that are competent intermediate hosts for *Dirofilaria* spp.

The prevalence of *D. immitis* in cats is estimated to be 5–20 fold lower than in dogs [37, 38]. In a heartworm endemic area in northern Italy, a prevalence of 29% in dogs and 4.7% in cats was shown [39]. A similar ratio of prevalence rates was detected in central Italy with 5.6% in dogs and 1.6% in cats [40].

In dogs, *D. immitis* is prevalent in all the Americas with a few exceptions such as Chile, where no heartworm cases were found in surveys [6, 41–43]. In the United States, the mean prevalence rates are generally between 1% and 12% [44–46] but can be locally quite high. Florida, the most southeastern state in the United States, exhibits a 28% prevalence rate [47], and rates as high as 48% were observed in Gulf Coast regions [2, 48]. These reports have been confirmed by more recent surveys of the American Heartworm Society (https://www.heartwormsociety.org/in-the-news/558-ahs-announces-findings-of-2019-heartworm-incidence-survey, accessed November 25th, 2020), which revealed that the top five states in heartworm incidence for 2019 were in the Southeast (Mississippi, Louisiana, South Carolina, Arkansas, and Alabama) (Figure 9.3), similar to the report of Little [46], which showed highest incidence in Mississippi, Louisiana, Arkansas, Alabama, and Texas. The general distribution has not changed compared to 2016. No state is free of *D. immitis* infections [46]. The prevalence rates are much lower in

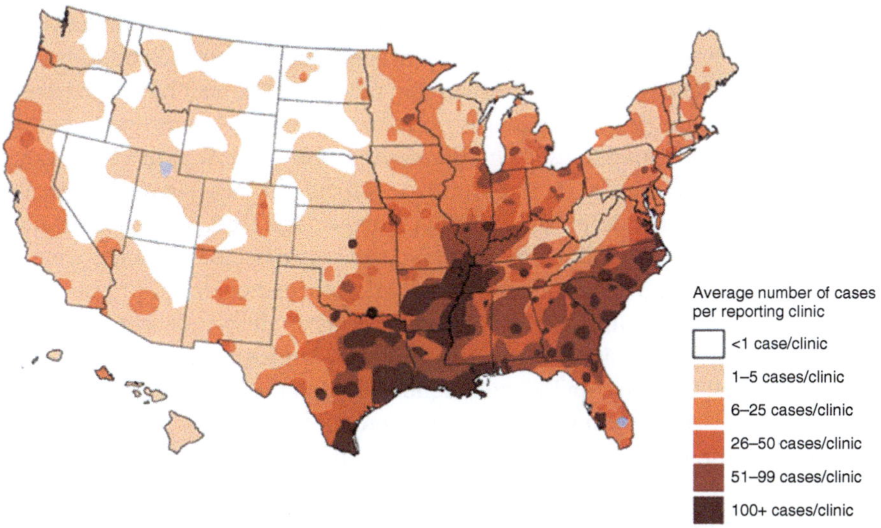


**Figure 9.3 2019 Heartworm Incidence Survey, American Heartworm Society.** Heartworm incidence as shown in this map is based on the average number of cases per reporting clinic in 2019. Some remote regions of the United States lack veterinary clinics; therefore, we have no reported cases from these areas. Source: Used with permission from the American Heartworm Society https://www.heartwormsociety.org/.

regions with colder temperatures. In Canada, for example, a study conducted in 1993 determined the prevalence as 0.24% [49]. A more recent study revealed a prevalence of 3.9% in shelter dogs in Ontario, Canada [50]. Climate change may also drive an increase in heartworm prevalence in previously preclusive cold regions [51].

No detailed European surveys on the distribution of *D. immitis* have been reported, but surveys conducted at the national or regional level are available. A few surveys have also focused on *D. repens* in Europe because of the zoonotic potential in humans, even though the infection in dogs is often asymptomatic [52, 53]. Furthermore, a substantial decrease in *D. immitis* infections has been observed in some endemic areas such as Northern Italy and the Canary Islands (Spain) [24]. In contrast to the reduction of prevalence in those areas, increased transmission in Central and Northern Europe has been observed and may be attributed to climate change [6, 54, 55]. Even in areas as far north as Finland, Estonia, and Siberia, autochthonous cases have been reported [24, 56, 57]. An additional factor contributing to the spread of dirofilariosis is the movement of positive dogs from endemic countries to formerly heartworm-free countries such as Germany [58]. In addition, increasing occurrence and climate-change-driven spread of reservoir hosts in wildlife, e.g., the golden jackal (*Canis aureus*), seem to play a significant role, too [55]. A special case appears to be Austria, where only recently the introduction of *D. repens* has been confirmed in mammals and in the mosquito vectors *Anopheles algeriensis* and *A. maculipennis* [59]. Most cases of dirofilariosis were imported cases, but climate analysis indicates that *D. immitis* has the capacity to establish itself in the lowland regions of Austria, given that the host and a number of competent culicid vectors are present [59].

In Australia, *D. immitis* was reported in all states with historical prevalence up to 100% in the Northern Territory [60]. Today, prevalence is considered to be low throughout Australia [61]. Dirofilariosis is also present in the near and far East and in Asia. In China, the prevalence ranges from 2% to 15% [62]. Interestingly, a novel *Dirofilaria* species, *Candidatus D. hongkongensis*, was identified in Hong Kong [63, 64], which also occurs in India [65]. In Japan, the heartworm prevalence decreased within a decade by about half in shelter dogs, from 46% in 1999–2001 to 23% in 2009–2011 [28]. Notably, no *D. immitis* was detected in Israel, while it was observed in other Middle East countries [24]. Information on the prevalence of dirofilariosis in Africa is limited [2]. However, reports on the presence of *D. immitis* and *D. repens* are increasing in recent years. Dirofilariosis has been observed in Tunisia, Algeria, Tanzania, and Mozambique, although the more dominant filarial species appears to be *Acanthocheilonema dracunculoides* [24].

There are more than 470 million dogs and 370 million cats worldwide (https://www.statista.com/statistics/1044386/dog-and-cat-pet-population-worldwide/ accessed October 10th, 2020), and the populations continue growing. More than 200 million dogs live in North America, Europe, Australia, and Japan, where prophylaxis and treatment probability are expected to be high (Boehringer Ingelheim internal analysis), making heartworm prevention a highly attractive market segment.

## 9.4 Disease Control

Heartworm control relies on antifilarial chemotherapy mainly through prevention, which utilizes a single class of drugs, the MLs, namely, ivermectin, milbemycin oxime, moxidectin, abamectin, eprinomectin, and selamectin [11]. Curative treatment of adult *D. immitis* depends largely on one drug, melarsomine dihydrochloride, while doxycycline as an anti-*Wolbachia* drug is under investigation.

### 9.4.1 Macrocyclic Lactones

Starting with the introduction of ivermectin in the canine market in 1987 (Heartgard®), heartworm prevention is now achieved almost solely through regular administration of active pharmaceutical ingredients (APIs) from the same chemical class, the MLs. The first MLs active against parasites – the avermectins, fermentation products of *Streptomyces avermitilis* – were discovered in 1975 from a soil sample collected in Japan [66, 67]. Subsequently, in 2015, half of the Nobel Prize in Physiology or Medicine was awarded jointly to Campbell and Ōmura for their outstanding contribution to the discovery of the avermectins [68].

The MLs can be divided into two groups, the avermectins (abamectin, ivermectin, eprinomectin, and selamectin) and the milbemycins (milbemycin oxime and moxidectin) [69, 70]. All contain a common 16-member ML ring. The main structural difference between both classes resides at C13 of the macrocyclic ring: avermectins contain sugar residues, whereas milbemycins are protonated (Figure 9.4). For further details on classification of the different MLs, for example, the differences between avermectin 1- and 2-subsets or A and B series, please refer to Shoop et al. [69] and Prichard and Geary [71].

#### 9.4.1.1 Ivermectin

The soil-dwelling bacterium *Streptomyces avermitilis* was discovered by Satoshi Ōmura in a soil sample from Kawana on the southeast coast of Honshu, Japan, in 1973. As anecdotes are told, Ōmura always carried plastic bags with him to collect specimen and found his most promising sample in the woods next to a golf course [68]. Extracts from cultures from this strain were sent to Merck laboratories to be tested in anthelmintic screens in 1974, where it showed promising activity against nematodes and many ectoparasites [66, 67, 72, 73]. William C. Campbell had the active components purified and identified the avermectins in 1975. The more effective chemical derivative ivermectin was subsequently commercialized, entering the Animal Health market in 1981 initially for livestock [68, 74, 75]. The drug's potential in human health to fight onchocerciasis was confirmed a few years later, and it was registered in 1987 and immediately provided free of charge for control of river blindness (branded as Mectizan®) [76, 77] (http://www.mectizan.org/resources/2014-annual-highlights accessed November 27th, 2020). The World Health Organization lists ivermectin among the essential medicines [78], and mass drug administration campaigns in Africa rely on its efficacy to control human filarial parasites [79].

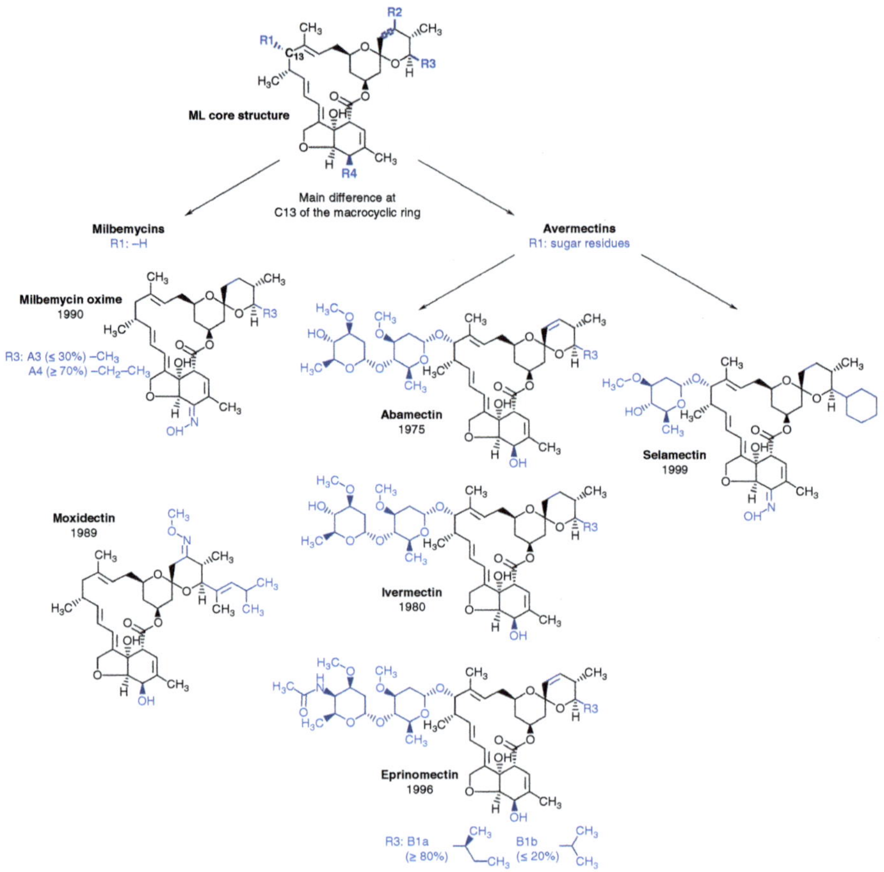


**Figure 9.4 Structure of MLs marketed against heartworm infections.** The macrocyclic core ring structure (top) indicates regions where MLs differ from each other. C13 is marked and highlighted in bold, where the main difference between the two major classes of MLs resides. Residues specific for each individual ML are visualized in blue. Source: Modified from Prichard and Geary [71].

Ivermectin is a chemically modified, dihydro derivative of naturally produced avermectin B1 composed of >80% 22,23-dihydro-avermectin B1a and <20% 22,23-dihydro-avermectin B1b at the initial launch (Figure 9.6) [80, 81], whereas ratio is now 90% to 10% (https://online.uspnf.com DocID: GUID-2506EE29-023C-4689-BE0C-392C296F1803_4_en-US). It shows activity against a broad spectrum of parasitic nematodes after both oral and parenteral administration, but not against cestodes or trematodes. In addition, it has activity against arthropods such as fleas, lice, mites, and some tick species [82, 83]. Although it is effective against microfilariae, L3 and L4 stages, it is not lethal for adult filariae but does reduce fertility [83]. Ivermectin is marketed as oral, topical, and injectable formulations, including long-acting injectables and boluses against endo- and ectoparasites of animals [84, 85].

#### 9.4.1.2 Eprinomectin

Eprinomectin or 4″-epi-acetylamino-4″-deoxyavermectin B1 was developed exclusively for veterinary medicine as the first topical endectocide for all cattle, including lactating animals [86, 87]. It is a semisynthetic derivative of avermectin B1 or abamectin, consisting of two homologs, B1a (not less than 90%) and B1b (not more than 10%), which differ by a methylene group. Eprinomectin first entered the market in a topical formulation against internal and external parasites of cattle including lactating cows [87–92]. As it showed good bioavailability and systemic activity in cats following topical application [93], it was included in a topical endectoparasiticide combination product together with fipronil, (*S*)-methoprene, and praziquantel for cats (Broadline®) [94, 95].

#### 9.4.1.3 Abamectin

Abamectin remains the only ML used in both animal health and crop protection [96]. It consists of avermectin B1a (>90%) and avermectin B1b (<10%). Abamectin is approved in Australia for preventive use against heartworm in dogs, however only in endoparasiticide combination products that include oxibendazole and praziquantel (https://portal.apvma.gov.au/pubcris accessed November 27th, 2020).

#### 9.4.1.4 Selamectin

Selamectin is a semisynthetic monosaccharide oxime derivative of doramectin (25-cyclohexyl-25-de(1-methylpropyl)-5-deoxy-22,23-dihydro-5-(hydroxyimino)-avermectin B1 monosaccharide). Doramectin is the most potent nematicide in a series of new avermectins prepared by mutational biosynthesis, having a cyclohexyl group in the C25 position of the avermectin ring [97, 98]. Selamectin was selected for its efficacy against *D. immitis*, gastrointestinal nematodes, fleas, and ticks in 1999 [71, 99, 100]. It is available as topical formulation for dogs and cats, while doramectin is marketed for ruminants and swine only.

#### 9.4.1.5 Milbemycin Oxime

The milbemycins were initially isolated in 1967 as fermentation products from *Streptomyces hygroscopicus*, and subsequently from *Streptomyces cyaneogriseus* in 1983, that displayed very high acaricidal activity [71, 85]. Structure elucidation, in 1972 by Sankyo scientists revealed the 16-membered ML structure of the active compound family of milbemycins, from which the anthelmintic milbemycin oxime (6R, 25R)-5-demethoxy-28-deoxy-6, 28-epoxy-5-hydroxyimino-25-ethyl/methyl milbemycin) was derived [85, 101, 102]. Milbemycin oxime is available in an oral formulation, consisting of a mixture of 70–80% milbemycin A4 oxime and 30–20% milbemycin A3 oxime, for heartworm prevention in dogs and cats. In addition, milbemycin oxime shows efficacy against immature and adult stages of other parasitic roundworms, hookworms, whipworms, lungworms, and mites [85, 103, 104].

#### 9.4.1.6 Moxidectin

Exploration of fermentation products from *Streptomyces cyaneogriseus* in 1983 revealed not only a new source of milbemycin but also the new ML nemadectin (F-29249α) [105, 106]. Addition of a methoxime moiety at C-23 and a substituted olefinic side chain at the 25-position to nemadectin yields moxidectin [107]. Heartworm prophylaxis products administer moxidectin orally, topically, or as an injectable. Moxidectin has been approved for human use against river blindness (https://www.fda.gov/drugs/drug-approvals-and-databases/drug-trials-snapshots-moxidectin accessed November 25th, 2020).

Compared to the avermectins, moxidectin inhibits Pgp-mediated rhodamine123 transport with 10 times lower potency [108]. Moxidectin is very lipophilic and has a long half-life, which makes it particularly suitable for long-acting injectable formulations, e.g., ProHeart-6® and ProHeart-12® preventives in canines, and Cydectin® LA for prevention and treatment of gastrointestinal nematodes in cattle [71, 85]. Topical moxidectin products obtained FDA approval for elimination of microfilariae in heartworm-positive dogs, diminishing adverse reactions, which can occur because of high microfilarial counts in infected dogs [109]. A combination product containing moxidectin, sarolaner, and pyrantel to obtain protection against endo- and ectoparasites has been marketed recently (Simparica® Trio) [110, 111].

## 9.4.2 Non-macrocyclic Lactone Treatments

### 9.4.2.1 Diethylcarbamazine Citrate

Diethylcarbamazine citrate (DEC) (Figure 9.5) is the oldest heartworm preventive and was discovered in 1947 as a derivative of piperazine. It shows both microfilaricidal and adulticidal activity, presumably by increasing filarial susceptibility to innate immune attack [112, 113]. It was first used to control human filariosis [114]

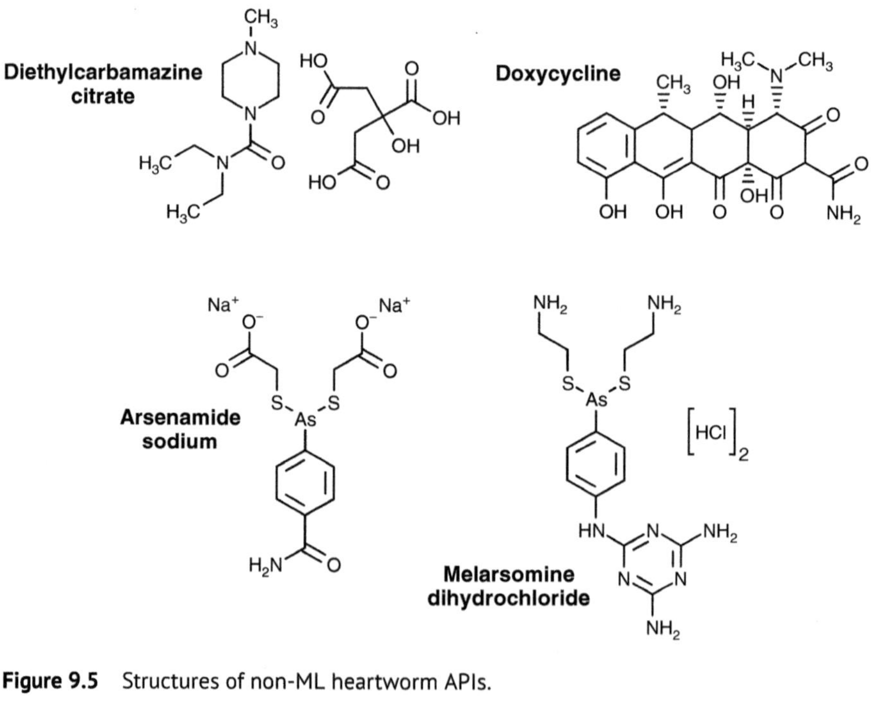


**Figure 9.5** Structures of non-ML heartworm APIs.

and made it to the animal health market in products for heartworm prophylaxis in the 1960s [115, 116]. In contrast to other preventives, it has to be given daily (https://apvma.gov.au accessed December 2nd, 2020).

As DEC is on the World Health Organization's List of Essential Medicines [78] for treatment of filariosis including lymphatic filariosis, tropical pulmonary eosinophilia, and loiasis [117], its use in animal health has been limited, with only a few products still marketed.

#### 9.4.2.2 Adulticide Treatment Using Arsenamide Sodium and Melarsomine Dihydrochloride

The adulticide arsenamide (thiacetarsamide) sodium (Caparsolate®) (Figure 9.5) was used for treatment of adult *D. immitis* beginning in the 1940s. The treatment needed to be administered intravenously, and dogs had to be hospitalized during initial treatment to handle possible hepatotoxic and nephrotoxic side effects [118]. In the 1990s, melarsomine dihydrochloride (Immiticide®) (Figure 9.5) supplanted thiacetarsamide as an adulticide, as it provided easier administration as well as increased safety [118–121]. Melarsomine dihydrochloride is now the first-line adulticide treatment for heartworm infections. Even though this therapy reduces the need for hospitalization of dogs, strict exercise restriction is required to limit thromboembolic effects [17, 118, 122, 123].

To achieve complete elimination of adult heartworm infections, both the American Heartworm Society and the European Society of Dirofilariosis and Angiostrongylosis propose protocols with two to three month pre-treatment with an ML combined with an antibiotic against *Wolbachia* (such as doxycycline, see below) before the administration of three doses of Immiticide® [123, 124]. As some studies provided evidence that melarsomine dihydrochloride may be effective against worms two to four months of age, the adulticidal treatment protocol might be improved further [10, 122, 123, 125, 126].

Adulticide therapy using melarsomine is not considered safe for cats, as worm death in cats is associated with a high risk for pulmonary thromboembolism and anaphylactic reactions [127]. Surgery known as worm embolectomy is an alternative to relying on self-cure, which can occur within 18–48 months, while carefully monitoring disease progression [124, 127]. Surgical extraction of adult heartworms in dogs remains the only solution for heavily infected dogs manifesting clinical signs of caval syndrome, as dying or dead adult heartworms obstruct blood flow and degrading worms cause inflammatory reactions [128–130].

#### 9.4.2.3 Doxycycline for Supportive Treatment

Doxycycline-mediated clearance of *Wolbachia* in *Onchocerca* or *Dirofilaria* infections demonstrated that *Wolbachia* are required for filarial larval development, embryogenesis, and long-term viability [131]. Treatment with doxycycline (Figure 9.5) killed third- and fourth-stage heartworm larvae in experimentally infected dogs [132]. However, a major drawback of doxycycline is the long treatment duration needed to eliminate the required 90% of *Wolbachia* for a sustainable effect [131]. In addition, long-term application of doxycycline in dogs is often associated with low tolerability and severe gastrointestinal side effects [133]. Nevertheless, for adulticidal treatment, a combination of doxycycline with monthly doses of ivermectin showed superior microfilaricidal and adulticidal efficacy compared to the drugs given alone [134, 135]. Moreover, doxycycline seems to enable a shorter treatment regimen to eliminate *Wolbachia* before the first adulticidal dose of melarsomine dihydrochloride [10]. Reducing the burden of *Wolbachia* in *D. immitis* before adulticide treatment proved to be more efficacious with fewer inflammatory reactions and lower risk of fatal pulmonary thromboembolisms [134, 136, 137]. The combination of doxycycline, ivermectin, and melarsomine significantly reduced the severity of arterial lesions and thrombi [136]. The American Heartworm Society [123] recommends a therapy including ivermectin or moxidectin, doxycycline, and melarsomine.

### 9.4.3 *Mdr1* Mutations in Collies and Related Breads

Although in general MLs are considered to be safe for most mammals, some dog breeds, including collies and shepherds, are prone to moderate to severe neurological effects. The genetic reason behind this susceptibility is a 4 bp deletion mutation leading to a frame shift in the multidrug resistant (*mdr1*) transporter gene (nt230 (del4) *mdr1* mutation; Figure 9.6) [138, 139]. The *mdr1* (del4) mutation in these dogs can be tracked back to a common ancestor in Great Britain around 1873, before formal breeds were registered and genetically isolated [140, 141].

The P-glycoprotein MDR1 belongs to a family of membrane-bound ATP-binding cassette transporters (ABC transporters) [142] and acts as a drug efflux pump across the blood–brain barrier. MDR1 plays an important role in the elimination of many drugs from the mammalian central nervous system, including humans (Figure 9.7) [143]. The channel was first isolated and characterized from Chinese hamster ovary cells that had developed resistance to cancer chemotherapy drugs by overexpression of MDR1 [144]. Although mice with a deficient *mdr1* gene showed no obvious phenotype in general, all mice of a colony infested with mites showed enhanced drug sensitivity and subsequently died after treatment with ivermectin [145]. Not only MLs but also many structurally unrelated drugs, toxins, and xenobiotics

```
Coding sequences
mdr1 wt     199 CTCATGATGCTGGTTTTTGGAAACATGACAGATAGCTTTGCAAATGCAGGAATTTCAAGAAACAAAACTTTTCCAGTTATAATT...
                |||||||||||||||||||||||||||||||    ||||||||||||||||||||||||||||||||||||||||||||||||||
mdr1 del4   205 CTCATGATGCTGGTTTTTGGAAACATGACA----GCTTTGCAAATGCAGGAATTTCAAGAAACAAAACTTTTCCAGTTATAA

Protein transcripts
MDR1 wt     67 L  M  M  L  V  F  G  N  M  T  D  S  F  A  N  A  G  I  S  R  N  K  T  F  P  V  I  I...
               |  |  |  |  |  |  |  |  |  |
MDR1 del4   67 L  M  M  L  V  F  G  N  M  T     A  L  Q  M  Q  E  F  Q  E  T  K  L  F  Q  L  Stop
```

**Figure 9.6 Local alignment of *mdr*1 wt and mutant *mdr1* del4**. *Mdr1* wt from *Canis lupus familiaris* (Genbank accession number DQ068953.1) and *mdr1* nt230 (del4) (Genbank accession number AJ419568.1) coding sequences as well as their transcripts were aligned. The nt230(del4) *mdr1* deletion leads to a frame shift at amino acid position 75, resulting in a truncated, non-functional MDR1 protein because of the premature stop codon (TAA marked in red) after amino acid position 91. Source: Based on Campbell [80] and Campbell et al. [81].

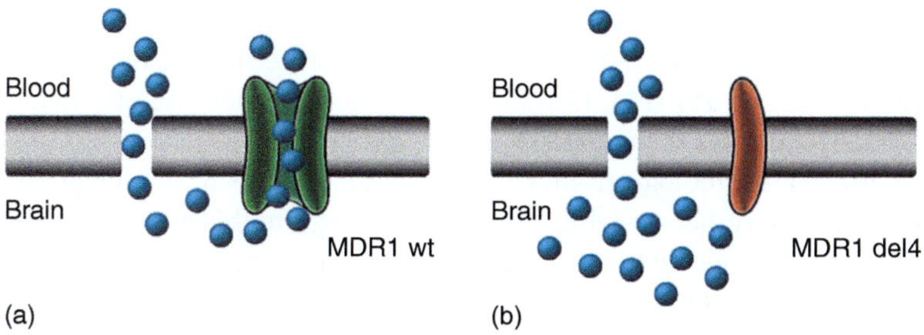


**Figure 9.7 Consequence of mdr1 mutation for drug exposure in the central nervous system.** (a) Functional MDR1 receptor actively lowers the concentration of APIs within the central nervous system, while (b) *mdr1*- mutation nt230 (del4) leading to a non-functional, truncated MDR1 channel results in accumulation of APIs within brain cells, thus increasing susceptibility to neurotoxic side effects.

can be substrates for MDR1, with distinct affinities and binding modes for different classes of substrates [146, 147].

Sensitivity increases for heterozygous *mdr1* (+/−) but is especially evident for homozygous *mdr1* (−/−) dogs that lack expression of functional MDR1 [143, 148]. Although all marketed products provide ML doses for heartworm prevention that are well tolerated by *mdr1* (del4)-deficient dogs, it is advised to genetically test collies, shepherds, and related breeds for *mdr1* (del4) mutations before treatment with MLs [149, 150]. The prevalence of at least one *mdr1* (del4) allele, either as *mdr1* (+/−) or *mdr1* (−/−), can be as high as 75% for collies (Table 9.1) but is almost nonexistent for other breeds, including breeds that share some history with the affected dog

**Table 9.1** Allelic frequency of *mdr1* (del4) mutation in dog breeds worldwide.

| Dog breed | Range of *mdr1* (del4) allelic frequency (%) |
|---|---|
| Collie | 48–75 |
| Longhaired Whippet | 24–45 |
| Shetland Sheepdog | 7–36 |
| (Miniature) Australian Shepherd | 16–54 |
| Australian Shepherd | 17–46 |
| White Swiss Shepherd | 7–16 |
| Old English Sheepdog | 0–11 |
| English Shepherd | 7 |
| German Shepherd | 0–6 |
| Border Collie | 0–4 |

Source: Data retrieved from Refs [141, 149, 151–155]. As data availability and sample size differ widely, frequencies have not been listed for all breeds.

breads (e.g. Bearded Collie, Anatolian Shepherd Dog, Greyhound, and Belgian Tervuren). It is estimated that about 1–2% of all dogs in the northern hemisphere carry such a mutation [139–141, 151–153, 156–158]. Across geographies, the prevalence of MDR1 deficiency in collies is comparable [149].

Today, with further technological advancement and the rise of an integrated health management, it is rather straightforward to assay dogs for this mutation using genetic tests on cells obtained by cheek swabbing or a blood sample [150, 159, 160]. It is not only reasonable to know the risk of one's own dog before treatment with MLs, but this information is also used by many dog breeders, selecting for *mdr wt* dogs to increase the value of the pups and thus hopefully outbreeding the *mdr1* (del4) mutation in the future.

In addition to the nt230 (del4) *mdr1* mutation, more than 30 single nucleotide polymorphisms have been identified in the canine MDR1 gene, which might affect the transport function or expression level [149]. One can speculate that these polymorphisms might be the reason for increased drug sensitivity observed in some dogs that lack the deletion.

### 9.4.4 Marketed Products

Most heartworm preventives today are available as oral formulations, while only a few topicals and two injectable formulations are marketed. To illustrate distinctions in differently regulated markets, we focus on marketed products in the United States, Europe, Japan, and Australia. In these geographies, heartworm-active APIs take different market shares – either based on local registrations, differences in marketing, or customer preferences (Figure 9.8).

#### 9.4.4.1 United States

International Cooperation on Harmonisation of Technical Requirements for Registration of Veterinary Medicinal Products (VICH) guidelines for the registration of anthelmintic products have been largely adopted by the US Food & Drug Administration (FDA) in its Guidance for Industry (GFI) system, e.g. GFI #90 (VICH GL7) Efficacy of Anthelmintics: General Requirements, GFI #111 (VICH GL19) Efficacy of Anthelmintics: Specific Recommendations for Canine, and GFI #113 (VICH GL20) Efficacy of Anthelmintics: Specific Recommendations for Feline (https://www.fda.gov/animal-veterinary/guidance-regulations/guidance-industry accessed November 30th, 2020). At present, two laboratory dose confirmation studies and one multisite field safety and effectiveness study must be conducted to demonstrate heartworm preventive efficacy, following the principles of Good Clinical Practice (GCP) as described in GFI #85 (VICH GL9) "Good Clinical Practice." The FDA has historically required 100% efficacy in these studies for registration. The Center for Veterinary Medicine (CVM) is currently evaluating alternative approaches for the design of studies conducted to show effectiveness [161].

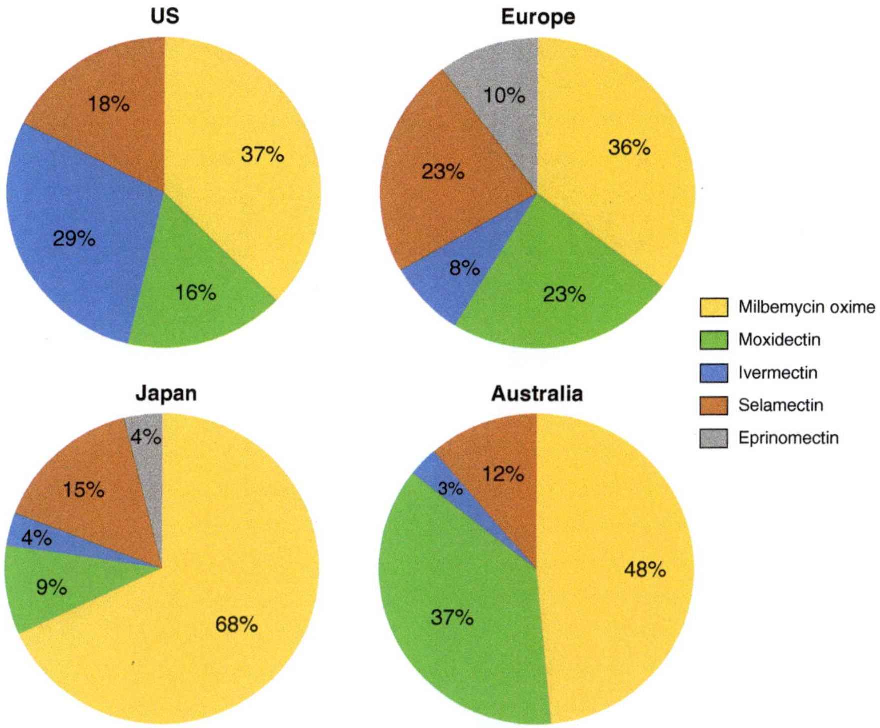


**Figure 9.8 Market shares of products based on specific APIs differ across geographies.** As not all approved products are marketed anymore or in all geographies or might be marketed via channels not captured for this analysis, the API market analysis might be slightly skewed. One example is products containing DEC, a rather old API that has been replaced in more modern products. Source: Data based on BI internal market analysis, AH market 2019.

Although there are several approved heartworm preventives for dogs and cats, only one has been approved for ferrets to date (Advantage Multi® for Cats). Table 9.2 summarizes the products approved for use in the United States.

#### 9.4.4.2 European Union

The EU Regulation 2019/6 currently governs the centralized marketing authorization procedure for both human and veterinary medicines (amending EU Regulation 726/2004 relating to authorization and supervision of veterinary medicines) (https://eur-lex.europa.eu/eli/reg/2019/6/oj accessed November 30th, 2020), while national registrations can be requested from the respective national competent authorities. In addition, the respective VICH guidelines have to be followed: VICH GL7 Efficacy of Anthelmintics: General Requirements, VICH GL19 Efficacy of Anthelmintics: Specific Recommendations for Canine, and VICH GL20 Efficacy of Anthelmintics: Specific Recommendations for Feline (https://vichsec.org/en/

**Table 9.2** APIs and products approved in the United States for prevention of heartworm disease or treatment of heartworm infections.

| Heartworm-active API | Species | Route of application | Trade names | Combination product with | Company |
|---|---|---|---|---|---|
| Diethylcarbamazine citrate | Dog, cat | Oral | Carbam<br>Diro-Form<br>Dirocide/Filaribits/Pet-Dec<br>Filban<br>Nemacide | | Bimeda AH<br>Lloyd<br>Zoetis<br>Intervet<br>Cronus Pharma |
| | Dog | Oral | Filaribits Plus | Oxibendazole | Zoetis |
| Eprinomectin | Cat | Topical | Centragard | Praziquantel | Boehringer Ingelheim AH |
| Ivermectin | Dog, cat | Oral | Heartgard<br>generics thereof: Iverhart<br>Ivermectin | | Boehringer Ingelheim AH<br>Virbac AH<br>Cronus Pharma |
| | Dog | Topical | Advantage DUO | Imidacloprid | Elanco |
| | Dog | Oral | Heartgard Plus<br>generics thereof: Iverhart Plus<br>Tri-Heart Plus | Pyrantel Pamoate | Boehringer Ingelheim AH<br>Virbac AH<br>Heska |
| | Dog | Oral | Panacur Plus | Praziquantel,<br>Fenbendazole, | Intervet |
| | Dog | Oral | Iverhart Max | Praziquantel,<br>Pyrantel pamoate | Virbac AH |
| Milbemycin oxime | Dog, cat | Oral | Interceptor<br>generic thereof:<br>MilbeGuard | | Elanco<br>Ceva Sante Animale |
| | Dog | Oral | Sentinel | Lufenuron | Intervet |
| | Dog | Oral | Sentinel Spectrum | Lufenuron,<br>Praziquantel | Intervet |
| | Dog | Oral | Interceptor Plus | Praziquantel | Elanco |
| | Dog | Oral | Trifexis | Spinosad | Elanco |

| | | | | | |
|---|---|---|---|---|---|
| Moxidectin | Dog | Oral s.c. Topical | ProHeart<br>Proheart 6/12<br>Coraxis | | Zoetis<br><br>Elanco |
| | Dog, cat, ferret | Topical | Advantage Multi<br>generic thereof: Imoxi | Imidacloprid | Elanco<br>Vetoquinol |
| | Cat | Topical | Bravecto Plus | Fluralaner | Intervet |
| | Dog | Oral | Simparica Trio | Sarolaner, Pyrantel pamoate | Zoetis |
| Selamectin | Dog, cat | Topical | Revolution<br>generic thereof: Revolt, Selarid, Senergy | | Zoetis<br>Aurora Pharmaceutical, Norbrook Laboratories, Chanelle Pharmaceuticals |
| | Cat | Topical | Revolution Plus | Sarolaner | Zoetis |
| Arsenamide sodium | Dog | i.v. | Caparsolate Sodium<br>not marketed anymore | | Boehringer Ingelheim AH |
| Melarsomine dihydrochloride | Dog | i.m. | Immiticide<br>generic thereof: Diroban | | Boehringer Ingelheim AH<br>Anzac AH |

Preventives dominate the market; only three products are approved for adulticidal treatment (see the last two rows of the table).
Source: Data retrieved from: U.S. Food & Drug Administration https://animaldrugsatfda.fda.gov/adafda/views/#/search, accessed November 30th 2020; AH – Animal Health.

guidelines/pharmaceuticals/pharma-efficacy/anthelmintics accessed November 30th, 2020). In general, most new, innovative medicines are submitted to the European Medicines Agency (EMA) for centralized authorization, while most generics and over-the-counter medicines use national marketing authorization. Expansion of marketing authorization to other EU member states can be obtained by either a mutual recognition procedure or a centralized procedure. Nevertheless, data requirements and standards for authorization of medicines in the EU are the same, irrespective of the authorization route. Combination products dominate the European market for heartworm prevention (Table 9.3).

#### 9.4.4.3 Japan

For Japan, the Ministry of Agriculture, Forestry and Fisheries (MAFF) holds jurisdiction over affairs concerning veterinary medicinal products. Regulations include the Law and the Enforcement Ordinance of the Law for Ensuring the Quality, Efficacy, and Safety of Drugs and Medical Devices, (Enforcement Ordinance No. 11, 1961), and the Control Regulations of Veterinary Medical Products (Control Regulations, Ministerial Ordinance No. 107, 2004). Besides local guidelines established for registration studies by MAFF, further globally harmonized VICH guidelines on quality, safety, and efficacy have to be considered for registration (http://www.maff.go.jp/nval/ accessed November 30th, 2020). Preventive treatment using oral products containing only one API dominate the market, including many generics (Table 9.4). Adulticidal products based on melarsomine dihydrochloride (Immiticide and generics thereof) have been authorized but are not available anymore in Japan, as their production and sales have been discontinued.

#### 9.4.4.4 Australia

All agricultural and veterinary chemical products sold in Australia have to be registered by the Australian Pesticides and Veterinary Medicines Authority (APVMA). The "Efficacy and target animal safety general guideline (Part 8)" in conjunction with the adopted VICH guidelines as well as the World Association for the Advancement of Veterinary Parasitology (WAAVP) guidelines have to be followed. More than 95% efficacy against *D. immitis* is required for registration.

Although preventive products containing abamectin and DEC as well as melarsomine are available only for dogs, all other products are also approved for use in cats. Some moxidectin-based products are also approved for use in ferrets, while some selamectin-based products can also be used in rabbits. For adulticidal treatment, only melarsomine dihydrochloride as Immiticide is registered (Table 9.5).

**Table 9.3** APIs and products approved in Europe for prevention of heartworm disease or treatment of heartworm infections.

| Heartworm-active API | Species | Route of application | Trade names | Combination product with | Company |
|---|---|---|---|---|---|
| Eprinomectin | Cat | Topical | NexGard Combo | Esafoxolaner, Praziquantel | Boehringer Ingelheim AH |
| | Cat | Topical | Broadline | Fipronil, Praziquantel, (*S*)-Methoprene | Boehringer Ingelheim AH |
| Ivermectin | Dog | s.c. | Guardian inj. | | Elanco |
| | Dog | Oral | Heartgard/Cardotek plus | Pyrantel | Boehringer Ingelheim AH |
| Milbemycin oxime | Dog | Oral | Program plus | Lufenuron | Elanco |
| | Dog, cat | Oral | Milbemax<br>generics thereof:<br>Milbactor, Milprazon, Milquantel<br>Milpro | Praziquantel | Elanco<br>Krka<br>Virbac |
| | Dog | Oral | Nexgard Spectra | Afoxolaner | Boehringer Ingelheim AH |
| | Dog | Oral | Trifexis | Spinosad | Elanco |
| Moxidectin | Dog | s.c. | Afilaria | | Support Pharma, Fatro |
| | Dog, cat, ferret | Topical | Advocate/Prinovox | Imidacloprid | Bayer AH |
| | Cat | Topical | Bravecto Plus | Fluralaner | Intervet |
| | Dog | Oral | Simparica Trio | Sarolaner, Pyrantel embonate | Zoetis |
| Selamectin | Dog, cat | Topical | Stronghold<br>generic thereof:<br>Chanhold, Evicto | | Zoetis<br>Chanelle Pharmaceuticals, Virbac AH |
| | Cat | Topical | Stronghold Plus<br>Felisecto Plus | Sarolaner | Zoetis |
| Melarsomine dihydrochloride | Dog | i.m. | Immiticide | | Boehringer Ingelheim AH |

Preventives dominate the market; only one product is approved for adulticidal treatment (see the last row of the table).
Source: Data retrieved from: EMA Europa Veterinary https://www.ema.europa.eu/en/medicines/field_ema_web_categories%253Aname_field/Veterinary, HMA Heads of Medicines Agencies https://mri.cts-mrp.eu/veterinary/ and CIMAVET https://cimavet.aemps.es/cimavet/publico/home.html accessed November 30th 2020; AH – Animal Health.

**Table 9.4** APIs and products approved in Japan for prevention of heartworm disease or treatment of heartworm infections.

| Heartworm-active API | Species | Route of application | Trade name | Combination product with | Company |
|---|---|---|---|---|---|
| Eprinomectin | Cat | Topical | Broadline | Fipronil, Praziquantel, (*S*)-Methoprene | Boehringer Ingelheim AH |
| Ivermectin | Dog | Oral | Cardomec<br>generics thereof: Azavasuca<br>Heartmectin<br>Panamectin | | Boehringer Ingelheim AH<br>Nissin Pharmaceutical<br>ASKA AH<br>Meiji Seika Pharma |
| | Dog | Oral | Cardomec P<br>generics thereof: Iverguard P<br>Ivermec DSP, PI<br>Panamectin | Pyrantel pamoate | Boehringer Ingelheim AH<br>Kyoritsu Seiyaku<br>Fujita Pharmaceutical<br>Meiji Seika Pharma |
| Milbemycin oxime | Dog | Oral | Milbemycin A<br>generics thereof: Milbeguard<br>Milbejelly<br>Milbemycin | | Elanco<br>SAMPO Pharm<br>Meiji Seika Pharma<br>Fujita Pharmaceutical |
| | Dog | Oral | Interceptor S | Praziquantel | Elanco |
| | Dog, cat | Oral | Milbemax | Praziquantel | Elanco |
| | Dog | Oral | Nexgard Spectra | Afoxolaner | Boehringer Ingelheim AH |
| | Dog | Oral | Panoramis | Spinosad | Elanco |
| | Dog | Oral | Systec | Lufenuron | Elanco |
| Moxidectin | Dog | Oral | Moxidec<br>generics thereof: Moxiguard<br>Moxiheart | | Zoetis<br>SAMPO Pharm<br>Fujita Pharmaceutical |
| | Dog | s.c. | Proheart-12 | | Zoetis |
| | Dog, cat | Topical | Advocate | Imidacloprid | Bayer Yakuhin |
| Selamectin | Dog, cat | Topical | Revolution | | Zoetis |
| | Cat | Topical | Revolution plus | Sarolaner | Zoetis |
| Melarsomine dihydrochloride | Dog | i.m. | Immiticide | Melarsomine dihydrochloride | Boehringer Ingelheim AH<br>Kyoritsu Pharmaceutical |

Preventives dominate the market; only one product is approved for adulticidal treatment (see the last row of the table).
Source: Data retrieved from: MAFF National Veterinary Assay Laboratory http://www.maff.go.jp/nval/ accessed November 30th, 2020; AH – Animal Health.

**Table 9.5** APIs and products approved in Australia for prevention of heartworm disease or treatment of heartworm infections.

| Heartworm active API | Species | Route of application | Trade names | Combination product with | Company |
|---|---|---|---|---|---|
| Abamectin | Dog | Oral | Virbac Canimax | Praziquantel | Virbac |
| | | | Purina Total Care Heartwormer and Allwormer | Oxibendazole | Nestle Purina Petcare |
| Diethylcarbamazine citrate | Dog | Oral | Aristopet/Vitapet Heartworm Dimmitrol | | Aristopet |
| | | | | | Mavlab |
| Ivermectin | Dog | Oral | Exelpet EZY - Heartworm | | Exelpet Products/Mars |
| | | | Heartgard 30 | | Boehringer Ingelheim AH |
| | | | Heartworm Soluble | | Arkolette |
| | | | I Love My Pet Heartworm Chewables | | My Pet Products Australia |
| | | | Nuheart | | |
| | | | | | Bocko/Flexsky in partnership |
| | | | Saint Bernard Petcare | | Australian Pharmavet |
| | | | Soluble Heartworm Valuheart | | Contract Manufacturing |
| | | | Vetafarm Heart Gold | | |
| | | | | | Jurox & Zoo Pets |
| | | | | | Vetafarm |
| | | | Exelpet EZY - Heartwormer + Intestinal All-Wormer | Oxantel embonate | Exelpet Products/Mars |
| | | | Guardian Complete | Pyrantel embonate | Intervet |
| | | | Popantel Allwormer Plus Heartworm | Praziquantel | Jurox |
| | | | Heartgard 30 Plus | Pyrantel embonate | Boehringer Ingelheim AH |
| | | | Startgard Plus for Puppies (Heartgard 30 Plus + Frontline Plus combi pack) | | |
| | Cat | Oral | Heartgard 30 FX | | Boehringer Ingelheim AH |
| | | | Startgard Plus for Kittens (Heartgard 30 FX + Frontline plus combi pack) | | |
| Milbemycin oxime | Dog | Oral | NexGard Spectra | Afoxolaner | Boehringer Ingelheim AH |
| | | | Interceptor Spectrum, | Praziquantel | Elanco |
| | | | Purina Total Care Heartwormer & Allwormer | | |

*(Continued)*

**Table 9.5** (Continued)

| Heartworm active API | Species | Route of application | Trade names | Combination product with | Company |
|---|---|---|---|---|---|
| | | | Purina Total Care Heartwormer, Allwormer & Flea Control, | Praziquantel, Lufenuron, | Elanco |
| | | | Sentinel Spectrum | | |
| | | | Panoramis, Trifexis | Spinosad | Elanco |
| | Dog, cat | Oral | Milbemax<br>Milpro | Praziquantel | Elanco<br>Virbac |
| Moxidectin | Dog | Oral | Proheart | | Zoetis |
| | Dog | s.c. | Proheart SR-12 Injection | | Zoetis |
| | Dog | Topical | Vets Choice for Fleas, Heartworm and Worms | Imidacloprid | Elanco |
| | Dog, cat, ferret | Topical | Advantage Advocate Exelpet Vet Series Flea, Intestinal & Heartworm | Imidacloprid | Elanco |
| | | | Exi-Flea Plus<br>Moxiclear | | Abbey Laboratories<br>Norbrook Laboratories |
| | Dog, cat | Topical | Aristopet AH fleas, heartworm and worms | Imidacloprid | Aristopet |
| | | | Neovet<br>Wagg & Purr Fleas, Heartworm & Worms | | Shanghai Neway AH<br>Avet Health |
| | Cat | Topical | Bravecto Plus | Fluralaner | Intervet |
| | Dog | Oral | Simparica Trio | Sarolaner, Pyrantel embonate | Zoetis |
| Selamectin | Dog, cat, rabbit | Topical | Evicto Neovela | | Virbac Shanghai Neway AH |
| | | | Purevet/Revolution<br>Selapro<br>Wagg & Purr Fleas & Heartworm | | Zoetis<br>Norbrook Laboratories<br>Avet Health |
| | Cat | Topical | Revolution Plus | Sarolaner | Zoetis |
| Melarsomine dihydrochloride | | | Immiticide Canine Heartworm Treatment | | Boehringer Ingelheim AH |

Preventives dominate the market; only one product is approved for adulticidal treatment (see the last row of the table).
Source: Data retrieved from: Australian Pesticides and Veterinary Medicines Authority https://apvma.gov.au accessed November 30th, 2020; AH – Animal Health.

## 9.5 Conclusion

Heartworm disease is a serious threat for dogs and cats in many parts of the world. Because it is unlikely that heartworm prevention could be achieved solely by improved vector control, preventive treatment based on actives of the ML class remains the mainstay of control. However, ML-resistant populations have been reported from the Mississippi River delta areas in the United States. Fortunately, for the majority of regions, available drugs are still of high value. With confirmed resistance to all MLs, it is evident that new control methods are urgently needed. Continuous progression in technology and science enables new innovative approaches to search for novel drugs affecting filarial-specific targets. For example, focused assays have been developed to discover anti-*Wolbachia* compounds with an indirect but finally lethal effect on *D. immitis*. Apart from the search for novel APIs, the time and technology also seem ripe to discover highly effective vaccines. Antifilarial vaccines would have the potential to change heartworm control substantially. Such a vaccine may become part of a more holistic heartworm control program when combined with preventive drugs such as MLs or yet to be discovered actives. Finally, it is important to emphasize compliance with veterinarians and pet owners to provide sufficient protection of the animals and to delay development and spread of drug-resistant *D. immitis* populations. Therefore, ideal novel products should be sustainable and highly effective and should possess a convenient route of administration to encourage owner compliance with required dosing regimens. Covering a broader range of endo- and ectoparasites within one product is a desired add-on.

## Acknowledgment

We are very grateful to Hiroki Maeda, Charles Q. Meng, and Cara A. Noack for their technical support in preparing tables and figures. We thank Frédéric Beugnet and Steffen Rehbein for helpful and constructive discussions.

## References

**1** Selzer, P.M. and Epe, C. (2021). Antiparasitics in animal health: quo vadis? *Trends Parasitol.* 37 (1): 77–89.

**2** McCall, J.W., Genchi, C., Kramer et al. (2008). Heartworm disease in animals and humans. *Adv. Parasitol.* 66: 193–285.

**3** Bowman, D.D. and Atkins, C.E. (2009). Heartworm biology, treatment, and control. *Vet. Clin. North Am. Small Anim. Pract.* 39 (6): 1127–1158. vii.

**4** Venco, L., Marchesotti, F., and Manzocchi, S. (2015). Feline heartworm disease: A'Rubik's-cube-like' diagnostic and therapeutic challenge. *J. Vet. Cardiol.* 17 (Suppl 1): S190–S201.

**5** Tahir, D., Davoust, B., and Parola, P. (2019). Vector-borne nematode diseases in pets and humans in the Mediterranean Basin: An update. *Vet. World* 12 (10): 1630–1643.

**6** Simón, F., Siles-Lucas, M., Morchón, R. et al. (2012). Human and animal dirofilariasis: the emergence of a zoonotic mosaic. *Clin. Microbiol. Rev.* 25 (3): 507–544.

**7** Simón, F., Morchón, R., González-Miguel, J. et al. (2009). What is new about animal and human dirofilariosis? *Trends Parasitol.* 25 (9): 404–409.

**8** Moroni, B., Rossi, L., Meneguz, P.G. et al. (2020). *Dirofilaria immitis* in wolves recolonizing northern Italy: are wolves competent hosts? *Parasite Vectors* 13 (1): 482.

**9** Alho, A.M., Fiarresga, A., Landum et al. (2016). A homemade snare: an alternative method for mechanical removal of *Dirofilaria immitis* in dogs. *Vet. Med. Int.* https://doi.org/10.1155/2016/5780408.

**10** Carretón, E., Falcón-Cordón, Y., Falcón-Cordón, S. et al. (2019). Variation of the adulticide protocol for the treatment of canine heartworm infection: can it be shorter? *Vet. Parasitol.* 271: 54–56.

**11** Wolstenholme, A.J., Maclean, M.J., Coates, R. et al. (2016). How do the macrocyclic lactones kill filarial nematode larvae? *Invert. Neurosci.* 16 (3): 7.

**12** Hampshire, V.A. (2005). Evaluation of efficacy of heartworm preventive products at the FDA. *Vet. Parasitol.* 133 (2): 191–195.

**13** Bourguinat, C., Lee, A.C., Lizundia, R. et al. (2015). Macrocyclic lactone resistance in *Dirofilaria immitis*: failure of heartworm preventives and investigation of genetic markers for resistance. *Vet. Parasitol.* 210 (3-4): 167–178.

**14** Kassai, T., Campillo, M.C.d., Euzeby, J. et al. (1988). Standardized nomenclature of animal parasitic diseases (SNOAPAD). *Vet. Parasitol.* 29 (4): 299–326.

**15** Ashford, R.W. (2001). Current usage of nomenclature for parasitic diseases, with special reference to those involving arthropods. *Med. Vet. Entomol.* 15 (2): 121–125.

**16** Kassai, T. (2006). The impact on database searching arising from inconsistency in the nomenclature of parasitic diseases. *Vet. Parasitol.* 138 (3-4): 358–361.

**17** Rawlings, C.A., Tonelli, Q., Lewis, R.E., and Duncan, J.R. (1993). Semiquantitative test for *Dirofilaria immitis* as a predictor of thromboembolic complications associated with heartworm treatment in dogs. *Am. J. Vet. Res.* 54 (6): 914–919.

**18** Sawyer, T.K. and Weinstein, P.P. (1963). The in vitro development of microfilariae of the dog heartworm *Dirofilaria immitis* to the "sausage-form". *J. Parasitol.* 49: 218–224.

**19** Shang Kuan, T.C. and Prichard, R.K. (2020). Developmental regulation of *Dirofilaria immitis* microfilariae and evaluation of ecdysone signaling pathway transcript level using droplet digital PCR. *Parasite Vectors* 13 (1): 614.

**20** Erickson, S.M., Xi, Z., Mayhew, G.F. et al. (2009). Mosquito infection responses to developing filarial worms. *PLoS Negl.Trop. Dis.* 3 (10): e529.

**21** Deplazes, P., Eckert, J., Mathis, A. et al. (2016). *Parasitology in Veterinary Medicine*. The Netherlands https://doi.org/10.3920/978-90-8686-274-0: Wageningen Academic Publishers.

**22** Mehlhorn, H. (2016). *Encyclopedia of Parasitology*, 4e. Springer Berlin Heidelberg.

**23** Bowman, D.D., Hendrix, C.M., Lindsay, D.S., and Barr, S.C. (2002). *Feline Clinical Parasitology*, 1e, 469. Wiley-Blackwell.

**24** Genchi, C. and Kramer, L.H. (2019). The prevalence of *Dirofilaria immitis* and *D. repens* in the old World. *Vet. Parasitol.* 280: https://doi.org/10.1016/j.vetpar.2019.108995.

**25** López, J., Valiente-Echeverría, F., Carrasco, M. et al. (2012). Morphological and molecular identification of canine filariae in a semi-rural district of the Metropolitan Region in Chile. *Rev. Chil. Infectol.* 29 (3): 248–289.

**26** Ramos-Lopez, S., León-Galván, M.F., Salas-Alatorre, M. et al. (2016). First molecular identification of *Dirofilaria repens* in a dog blood sample from Guanajuato, Mexico. *Vector Borne Zoonotic Dis.* 16 (11): 734–736.

**27** Montarsi, F., Ciocchetta, S., Devine, G. et al. (2015). Development of *Dirofilaria immitis* within the mosquito Aedes (Finlaya) koreicus, a new invasive species for Europe. *Parasite Vectors* 8: 177.

**28** Oi, M., Yoshikawa, S., Ichikawa, Y. et al. (2014). Prevalence of *Dirofilaria immitis* among shelter dogs in Tokyo, Japan, after a decade: comparison of 1999-2001 and 2009-2011. *Parasite* 21: 10.

**29** Mendoza-Roldan, J., Benelli, G., Panarese, R. et al. (2020). *Leishmania infantum* and *Dirofilaria immitis* infections in Italy, 2009-2019: changing distribution patterns. *Parasite Vectors* 13 (1): 193.

**30** Kitron, U. and Spielman, A. (1989). Suppression of transmission of malaria through source reduction: antianopheline measures applied in Israel, the United States, and Italy. *Rev. Infect. Dis.* 11 (3): 391–406.

**31** Parham, P.E. and Michael, E. (2010). Modelling climate change and malaria transmission. *Adv. Exp. Med. Biol.* 673: 184–199.

**32** Feachem, R.G., Phillips, A.A., Hwang, J. et al. (2010). Shrinking the malaria map: progress and prospects. *Lancet* 376 (9752): 1566–1578.

**33** Goulson, D., Nicholls, E., Botías, C., and Rotheray, E.L. (2015). Bee declines driven by combined stress from parasites, pesticides, and lack of flowers. *Science* 347 (6229): 1255957.

**34** Medlock, J.M., Hansford, K.M., Schaffner, F. et al. (2012). A review of the invasive mosquitoes in Europe: ecology, public health risks, and control options. *Vector Borne Zoonotic Dis.* 12 (6): 435–447.

**35** Cella, W., Baia-da-Silva, D.C., Melo, G.C. et al. (2019). Do climate changes alter the distribution and transmission of malaria? Evidence assessment and recommendations for future studies. *Rev. Soc. Bras. Med. Trop.* 52: e20190308.

**36** Hertig, E. (2019). Distribution of *Anopheles* vectors and potential malaria transmission stability in Europe and the Mediterranean area under future climate change. *Parasite Vectors* 12 (1): 18.

**37** Montoya-Alonso, J.A., Morchón, R., Falcón-Cordón, Y. et al. (2017). Prevalence of heartworm in dogs and cats of Madrid, Spain. *Parasite Vectors* 10 (1): 354.

**38** Garrity, S., Lee-Fowler, T., and Reinero, C. (2019). Feline asthma and heartworm disease: clinical features, diagnostics and therapeutics. *J. Feline Med. Surg.* 21 (9): 825–834.

**39** Venco, L., Genchi, M., Genchi, C. et al. (2011). Can heartworm prevalence in dogs be used as provisional data for assessing the prevalence of the infection in cats? *Vet. Parasitol.* 176 (4): 300–303.

**40** Traversa, D., Aste, G., Milillo, P. et al. (2010). Autochthonous foci of canine and feline infections by *Dirofilaria immitis* and *Dirofilaria repens* in central Italy. *Vet. Parasitol.* 169 (1-2): 128–132.

**41** Labarthe, N. and Guerrero, J. (2005). Epidemiology of heartworm: what is happening in South America and Mexico? *Vet. Parasitol.* 133 (2-3): 149–156.

**42** Maggi, R.G. and Krämer, F. (2019). A review on the occurrence of companion vector-borne diseases in pet animals in Latin America. *Parasite Vectors* 12 (1): 145.

**43** Dantas-Torres, F. and Otranto, D. (2020). Overview on *Dirofilaria immitis* in the Americas, with notes on other filarial worms infecting dogs. *Vet. Parasitol.* 109113.

**44** Lee, A.C., Montgomery, S.P., Theis, J.H. et al. (2010). Public health issues concerning the widespread distribution of canine heartworm disease. *Trends Parasitol.* 26 (4): 168–173.

**45** Little, S.E., Beall, M.J., Bowman, D.D. et al. (2014). Canine infection with *Dirofilaria immitis, Borrelia burgdorferi, Anaplasma* spp., and *Ehrlichia* spp. in the United States, 2010-2012. *Parasite Vectors* 7: 257.

**46** Little, S., Braff, J., Place, J. et al. (2021). Canine infection with *Dirofilaria immitis, Borrelia burgdorferi, Anaplasma* spp., and *Ehrlichia* spp. in the United States, 2013–2019. *Parasites Vectors* 14 (1): 10.

**47** Hays, K.M., Rodriguez, J.Y., Little, S.E. et al. (2020). Heartworm prevalence in dogs versus cats: multiple diagnostic modalities provide new insights. *Vet. Parasitol. X* 4: 100027.

**48** Bowman, D., Little, S.E., Lorentzen, L. et al. (2009). Prevalence and geographic distribution of *Dirofilaria immitis, Borrelia burgdorferi, Ehrlichia canis,* and *Anaplasma phagocytophilum* in dogs in the United States: results of a national clinic-based serological survey. *Vet. Parasitol.* 160 (1-2): 138–148.

**49** Slocombe, J.O. and Villeneuve, A. (1993). Heartworm in dogs in Canada in 1991. *Can. Vet. J.* 34 (10): 630–633.

**50** Jacobson, L.S., Ward, A.K., Lacaden, A.B., and Hornak, A.T. (2020). Prevalence of heartworm in relocated, local and outreach clinic dogs: a Canadian sheltering perspective. *Vet. Parasitol.* 109081.

**51** Bowman, D.D., Liu, Y., McMahan, C.S. et al. (2016). Forecasting United States heartworm *Dirofilaria immitis* prevalence in dogs. *Parasites Vectors* 9 (1): 540–540.

**52** Sałamatin, R.V., Pavlikovska, T.M., Sagach, O.S. et al. (2013). Human dirofilariasis due to *Dirofilaria repens* in Ukraine, an emergent zoonosis: epidemiological report of 1465 cases. *Acta Parasitol.* 58 (4): 592–598.

**53** Moskvina, T.V. and Ermolenko, A.V. (2018). Dirofilariasis in Russian Federation: a big problem with large distribution. *Russ. Open Med. J.* 7 (1): e0102.

**54** Capelli, G., Genchi, C., Baneth, G. et al. (2018). Recent advances on *Dirofilaria repens* in dogs and humans in Europe. *Parasites Vectors* 11 (1): 663.

**55** Széll, Z., Bacsadi, Á., Szeredi, L. et al. (2020). Rapid spread and emergence of heartworm resulting from climate and climate-driven ecological changes in Hungary. *Vet. Parasitol.* 280: 109067.

**56** Jokelainen, P., Mõtsküla, P.F., Heikkinen, P. et al. (2016). *Dirofilaria repens* microfilaremia in three dogs in estonia. *Vector Borne Zoonotic Dis.* 16 (2): 136–138.

**57** Pietikäinen, R., Nordling, S., Jokiranta, S. et al. (2017). *Dirofilaria repens* transmission in southeastern Finland. *Parasite Vectors* 10 (1): 561.

**58** Genchi, C., Bowman, D., and Drake, J. (2014). Canine heartworm disease (*Dirofilaria immitis*) in Western Europe: survey of veterinary awareness and perceptions. *Parasite Vectors* 7: 206.

**59** Fuehrer, H.P., Auer, H., Leschnik, M. et al. (2016). *Dirofilaria* in humans, dogs, and vectors in Austria (1978–2014) – from imported pathogens to the endemicity of *Dirofilaria repens*. *PLoS Negl.Trop. Dis.* 10 (5): e0004547.

**60** Welch, J.S., Dobson, C., and Freeman, C. (1979). Distribution and diagnosis of dirofilariasis and toxocariasis in Australia. *Aust. Vet. J.* 55 (6): 265–274.

**61** Nguyen, C., Koh, W.L., Casteriano, A. et al. (2016). Mosquito-borne heartworm *Dirofilaria immitis* in dogs from Australia. *Parasite Vectors* 9 (1): 535.

**62** Liu, C., Yang, N., He, J. et al. (2013). Prevalence of *Dirofilaria immitis* in dogs in Shenyang, Northeastern China. *Korean J. Parasitol.* 51 (3): 375–377.

**63** Yilmaz, E., Fritzenwanker, M., Pantchev, N. et al. (2016). The mitochondrial genomes of the zoonotic canine filarial parasites *Dirofilaria (Nochtiella) repens* and *Candidatus Dirofilaria* (*Nochtiella*) *Honkongensis* provide evidence for presence of cryptic species. *PLoS Negl.Trop. Dis.* 10 (10): e0005028.

**64** Yilmaz, E., Wongkamchai, S., Ramünke, S. et al. (2019). High genetic diversity in the *Dirofilaria repens* species complex revealed by mitochondrial genomes of feline microfilaria samples from Narathiwat, Thailand. *Transboundary Emerging. Dis.* 66 (1): 389–399.

**65** Pradeep, R.K., Nimisha, M., Pakideery, V. et al. (2019). Whether *Dirofilaria repens* parasites from South India belong to zoonotic *Candidatus Dirofilaria hongkongensis* (*Dirofilaria* sp. *hongkongensis*)? *Infect. Genet. Evol.* 67: 121–125.

**66** Burg, R.W., Miller, B.M., Baker, E.E. et al. (1979). Avermectins, new family of potent anthelmintic agents: producing organism and fermentation. *Antimicrob. Agents Chemother.* 15 (3): 361–367.

**67** Campbell, W.C. (2012). History of avermectin and ivermectin, with notes on the history of other macrocyclic lactone antiparasitic agents. *Curr. Pharm. Biotechnol.* 13 (6): 853–865.

**68** Van Voorhis, W.C., Hooft van Huijsduijnen, R., and Wells, T.N.C. (2015). Profile of William C. Campbell, Satoshi Ōmura, and Youyou Tu, 2015 Nobel Laureates in physiology or medicine. *Proc. Natl. Acad. Sci. U.S.A.* 112 (52): 15773–15776.

69 Shoop, W.L., Mrozik, H., and Fisher, M.H. (1995). Structure and activity of avermectins and milbemycins in animal health. *Vet. Parasitol.* 59 (2): 139–156.

70 Vercruysse, J. and Rew, R.S. (2002). *Macrocyclic Lactones in Antiparasitic Therapy*. CABI Pub.

71 Prichard, R.K. and Geary, T.G. (2019). Perspectives on the utility of moxidectin for the control of parasitic nematodes in the face of developing anthelmintic resistance. *Int. J. Parasitol. Drugs Drug Resist.* 10: 69–83.

72 Chabala, J.C., Mrozik, H., Tolman, R.L. et al. (1980). Ivermectin, a new broad-spectrum antiparasitic agent. *J. Med. Chem.* 23 (10): 1134–1136.

73 Ōmura, S. (2008). Ivermectin: 25 years and still going strong. *Int. J. Antimicrob. Agents* 31 (2): 91–98.

74 Ōmura, S. and Crump, A. (2014). Ivermectin: panacea for resource-poor communities? *Trends Parasitol.* 30 (9): 445–455.

75 Laing, R., Gillan, V., and Devaney, E. (2017). Ivermectin – old drug, new tricks? *Trends Parasitol.* 33 (6): 463–472.

76 Molyneux, D.H. and Ward, S.A. (2015). Reflections on the Nobel prize for medicine 2015 – the public health legacy and impact of avermectin and artemisinin. *Trends Parasitol.* 31 (12): 605–607.

77 Ashour, D.S. (2019). Ivermectin: from theory to clinical application. *Int. J. Antimicrob. Agents* 54 (2): 134–142.

78 World Health Organization (2019). World Health Organization model list of essential medicines: 21st list 2019. [cited 2020 June 18th]. Available from: https://apps.who.int/iris/bitstream/handle/10665/325771/WHO-MVP-EMP-IAU-2019.06-eng.pdf (accessed 18 June 2020).

79 Kim, Y.E., Remme, J.H., Steinmann, P. et al. (2015). Control, elimination, and eradication of river blindness: scenarios, timelines, and ivermectin treatment needs in Africa. *PLoS Negl.Trop. Dis.* 9 (4): e0003664.

80 Campbell, W.C. (1981). An introduction to the avermectins. *New Zealand Vet. J.* 29 (10): 174–178.

81 Campbell, W.C., Fisher, M.H., Stapley, E.O. et al. (1983). Ivermectin: a potent new antiparasitic agent. *Science (New York, NY)* 221 (4613): 823–828.

82 McKellar, Q.A. and Benchaoui, H.A. (1996). Avermectins and milbemycins. *J. Vet. Pharmacol. Ther.* 19 (5): 331–351.

83 Martin, R.J., Robertson, A.P., and Choudhary, S. (2020). Ivermectin: an anthelmintic, an insecticide, and much more. *Trends Parasitol.*

84 Soll, M.D., Carmichael, I.H., and Harvey, R.G. (1988). Prophylactic efficacy of sustained-release ivermectin against induced nematode infestations in cattle. *J. S. Afr. Vet. Assoc.* 59 (1): 9–11.

85 Prichard, R., Ménez, C., and Lespine, A. (2012). Moxidectin and the avermectins: consanguinity but not identity. *Int. J. Parasitol. Drugs Drug Resist.* 2: 134–153.

86 Shoop, W.L., DeMontigny, P., Fink, D.W. et al. (1996). Efficacy in sheep and pharmacokinetics in cattle that led to the selection of eprinomectin as a topical endectocide for cattle. *Intern. J. Parasitol.* 26 (11): 1227–1235.

**87** Shoop, W.L., Egerton, J.R., Eary, C.H. et al. (1996). Eprinomectin: a novel avermectin for use as a topical endectocide for cattle. *Int. J. Parasitol.* 26 (11): 1237–1242.

**88** Holste, J.E., Smith, L.L., Hair, J.A. et al. (1997). Eprinomectin: a novel avermectin for control of lice in all classes of cattle. *Vet. Parasitol.* 73 (1-2): 153–161.

**89** Pitt, S.R., Langholff, W.K., Eagleson, J.S., and Rehbein, S. (1997). The efficacy of eprinomectin against induced infections of immature (fourth larval stage) and adult nematode parasites in cattle. *Vet. Parasitol.* 73 (1-2): 119–128.

**90** Williams, J.C., Stuedemann, J.A., Bairden, K. et al. (1997). Efficacy of a pour-on formulation of eprinomectin (MK-397) against nematode parasites of cattle, with emphasis on inhibited early fourth-stage larvae of *Ostertagia* spp. *Am. J. Vet. Res.* 58 (4): 379–383.

**91** Holste, J.E., Colwell, D.D., Kumar, R. et al. (1998). Efficacy of eprinomectin against *Hypoderma* spp in cattle. *Am. J. Vet. Res.* 59 (1): 56–58.

**92** Rehbein, S.E., Winter, R.E., Visser, M.E. et al. (2005). Chorioptic mange in dairy cattle: treatment with eprinomectin pour-on. *Parasitol. Res.* 98 (1): 21–25.

**93** Kvaternick, V., Kellermann, M., Knaus, M. et al. (2014). Pharmacokinetics and metabolism of eprinomectin in cats when administered in a novel topical combination of fipronil, (S)-methoprene, eprinomectin and praziquantel. *Vet. Parasitol.* 202 (1-2): 2–9.

**94** Baker, C.F., Tielemans, E., Pollmeier, M.G. et al. (2014). Efficacy of a single dose of a novel topical combination product containing eprinomectin to prevent heartworm infection in cats. *Vet. Parasitol.* 202 (1-2): 49–53.

**95** Rehbein, S., Capári, B., Duscher, G. et al. (2014). Efficacy against nematode and cestode infections and safety of a novel topical fipronil, (S)-methoprene, eprinomectin and praziquantel combination product in domestic cats under field conditions in Europe. *Vet. Parasitol.* 202 (1-2): 10–17.

**96** Bai, S.H. and Ogbourne, S. (2016). Eco-toxicological effects of the avermectin family with a focus on abamectin and ivermectin. *Chemosphere* 154: 204–214.

**97** Dutton, C.J., Gibson, S.P., Goudie, A.C. et al. (1991). Novel avermectins produced by mutational biosynthesis. *J. Antibiot. (Tokyo)* 44 (3): 357–365.

**98** Goudie, A.C., Evans, N.A., Gration, K.A. et al. (1993). Doramectin – a potent novel endectocide. *Vet. Parasitol.* 49 (1): 5–15.

**99** Banks, B.J., Bishop, B.F., Evans, N.A. et al. (2000). Avermectins and flea control: structure-activity relationships and the selection of selamectin for development as an endectocide for companion animals. *Bioorg. Med. Chem.* 8 (8): 2017–2025.

**100** Bishop, B.F., Bruce, C.I., Evans, N.A. et al. (2000). Selamectin: a novel broad-spectrum endectocide for dogs and cats. *Vet. Parasitol.* 91 (3-4): 163–176.

**101** Takiguchi, Y., Mishima, H., Okuda, M. et al. (1980). Milbemycins, a new family of macrolide antibiotics: fermentation, isolation and physico-chemical properties. *J. Antibiot. (Tokyo)* 33 (10): 1120–1127.

**102** Takiguchi, Y., Ono, M., Muramatsu, S. et al. (1983). Milbemycins, a new family of macrolide antibiotics. Fermentation, isolation and physico-chemical properties of milbemycins D, E, F, G, and H. *J. Antibiot. (Tokyo)* 36 (5): 502–508.

**103** Garfield, R.A. and Reedy, L. (1992). The use of oral milbemycin oxime (Interceptor) in the treatment of chronic generalized canine demodicosis. *Vet. Dermatol.* 3: 231–235.

**104** Holm, B.R. (2003). Efficacy of milbemycin oxime in the treatment of canine generalized demodicosis: a retrospective study of 99 dogs (1995–2000). *Vet. Dermatol.* 14 (4): 189–195.

**105** Carter, G.T., Nietsche, J.A., Goodman, J.J. et al. (1987). LL-F42248 alpha, a novel chlorinated pyrrole antibiotic. *J. Antibiot. (Tokyo)* 40 (2): 233–236.

**106** Carter, G.T., Nietsche, J.A., Hertz, M.R. et al. (1988). LL-F28249 antibiotic complex: a new family of antiparasitic macrocyclic lactones. Isolation, characterization and structures of LL-F28249 alpha, beta, gamma, lambda. *J. Antibiot. (Tokyo)* 41 (4): 519–529.

**107** Ranjan, S., Trudeau, C., Prichard, R.K. et al. (1992). Efficacy of moxidectin against naturally acquired nematode infections in cattle. *Vet. Parasitol.* 41 (3): 227–231.

**108** Lespine, A., Martin, S., Dupuy, J. et al. (2007). Interaction of macrocyclic lactones with P-glycoprotein: structure-affinity relationship. *Eur. J. Pharm. Sci.* 30 (1): 84–94.

**109** McCall, J.W., Arther, R., Davis, W., and Settje, T. (2014). Safety and efficacy of 10% imidacloprid + 2.5% moxidectin for the treatment of *Dirofilaria immitis* circulating microfilariae in experimentally infected dogs. *Vet. Parasitol.* 206.

**110** Kryda, K., Six, R.H., Walsh, K.F. et al. (2019). Laboratory and field studies to investigate the efficacy of a novel, orally administered combination product containing moxidectin, sarolaner and pyrantel for the prevention of heartworm disease (*Dirofilaria immitis*) in dogs. *Parasites Vectors* 12 (1).

**111** Becskei, C., Thys, M., Doherty, P., and Mahabir, S.P. (2020). Efficacy of orally administered combination of moxidectin, sarolaner and pyrantel (Simparica Trio™) for the prevention of experimental *Angiostrongylus vasorum* infection in dogs. *Parasites Vectors* 13 (1): https://doi.org/10.1186/s13071-020-3948-z.

**112** Sutton, R.H., Atwell, R.B., and Boreham, P.F. (1985). Liver changes, following diethylcarbamazine administration, in microfilaremic dogs infected with *Dirofilaria immitis*. *Vet. Pathol.* 22 (2): 177–183.

**113** El-Shahawi, G.A., Abdel-Latif, M., Saad, A.H., and Bahgat, M. (2010). *Setaria equina*: in vivo effect of diethylcarbamazine citrate on microfilariae in albino rats. *Exp. Parasitol.* 126 (4): 603–610.

**114** Hawking, F. (1962). A review of progress in the chemotherapy and control of filariasis since 1955*. *Bull. World Health Organ.* 27 (4-5): 555–568.

**115** Pailet, A., Abadie, S.H., Smith, M.W., and Gonzalez, R.R. (1968). Chemotherapeutic heartworm control--the use of diethylcarbamazine in the control of *Dirofilaria immitis* infection in dogs in clinical trials. *Vet. Med.*, small animal clinician: VM, SAC 63 (7): 691–693.

**116** Prescott, C.W., O'Grady, A.S., and English, P.B. (1978). Dose rate of diethylcarbamazine for heartworm prophylaxis. *Aust. Vet. J.* 54 (8): 404–405.

**117** Chitkara, R.K. and Sarinas, P.S. (1997). *Dirofilaria*, visceral larva migrans, and tropical pulmonary eosinophilia. *Semin. Respir. Infect.* 12 (2): 138–148.

**118** Raynaud, J.P. (1992). Thiacetarsamide (adulticide) versus melarsomine (RM 340) developed as macrofilaricide (adulticide and larvicide) to cure canine heartworm infection in dogs. *Ann. Rech. Vet.* 23 (1): 1–25.

**119** Rawlings, C.A., Raynaud, J.P., Lewis, R.E. et al. (1993). Pulmonary thromboembolism and hypertension after thiacetarsamide vs melarsomine dihydrochloride treatment of *Dirofilaria immitis* infection in dogs. *Am. J. Vet. Res.* 54 (6): 920–925.

**120** McTier, T.L., McCall, J.W., Dzimianski, M.T. et al. (1994). Use of melarsomine dihydrochloride (RM 340) for adulticidal treatment of dogs with naturally acquired infections of *Dirofilaria immitis* and for clinical prophylaxis during reexposure for 1 year. *Vet. Parasitol.* 55 (3): 221–233.

**121** Maksimowich, D.S., Bell, T.G., Williams, J.F., and Kaiser, L. (1997). Effect of arsenical drugs on in vitro vascular responses of pulmonary artery from heartworm-infected dogs. *Am. J. Vet. Res.* 58 (4): 389–393.

**122** Bowman, D.D. and Drake, J. (2017). Examination of the "susceptibility gap" in the treatment of canine heartworm infection. *Parasites Vectors* 10 (Suppl 2): 513–513.

**123** American Heartworm Society (2020). Current canine guidelines for the prevention, diagnosis, and management of heartworm (*Dirofilaria immitis*) infection in dogs. [cited 2020 November 25th]. Available from: https://d3ft8sckhnqim2.cloudfront.net/images/pdf/2020_AHS_Canine_Guidelines_Summary_11_12.pdf?1605556516 (accessed 25 November 2020).

**124** European Society of Dirofilariosis and Angiostrongylosis (2017). Guidelines for Clinical Management of Canine Heartworm Disease. [cited 2020 November 25th]. Available from: https://www.esda.vet/wp-content/uploads/2017/11/GUIDELINES-FOR-CLINICAL-MANAGEMENT-OF-CANINE-HEARTWORM-DISEASE.pdf (accessed 25 November 2020).

**125** McCall, J.W. (2005). The safety-net story about macrocyclic lactone heartworm preventives: a review, an update, and recommendations. *Vet. Parasitol.* 133 (2-3): 197–206.

**126** McCall, J., Kramer, L., Genchi, C., et al. (2010) Effects of melarsomine dihydrochloride on two-month-old infections of *Dirofilaria immitis* and *Brugia pahangi* in dogs with dual infections. in *Proceedings of the American Heartworm Society Triennial Meeting*. Memphis: Poster.

**127** Pennisi, M.G., Tasker, S., Hartmann, K. et al. (2020). Dirofilarioses in cats: European guidelines from the ABCD on prevention and management. *J. Feline Med. Surg.* 22 (5): 442–451.

**128** Glaus, T.M., Jacobs, G.J., Rawlings, C.A. et al. (1995). Surgical removal of heartworms from a cat with caval syndrome. *J. Am. Vet. Med. Assoc.* 206 (5): 663–666.

**129** Lee, S.G., Moon, H.S., and Hyun, C. (2008). Percutaneous heartworm removal from dogs with severe heart worm (*Dirofilaria immitis*) infestation. *J. Vet. Sci.* 9 (2): 197–202.

**130** Bové, C.M., Gordon, S.G., Saunders, A.B. et al. (2010). Outcome of minimally invasive surgical treatment of heartworm caval syndrome in dogs: 42 cases (1999-2007). *J. Am. Vet. Med. Assoc.* 236 (2): 187–192.

**131** Turner, J.D., Marriott, A.E., Hong, D. et al. (2020). Novel anti-*Wolbachia* drugs, a new approach in the treatment and prevention of veterinary filariasis? *Vet. Parasitol.* 279: 109057.

**132** McCall, J.W., Kramer, L., Genchi, C. et al. (2011). Effects of doxycycline on early infections of *Dirofilaria immitis* in dogs. *Vet. Parasitol.* 176 (4): 361–367.

**133** Savadelis, M.D., Day, K.M., Bradner, J.L. et al. (2018). Efficacy and side effects of doxycycline versus minocycline in the three-dose melarsomine canine adulticidal heartworm treatment protocol. *Parasite Vectors* 11 (1): 671.

**134** Bazzocchi, C., Mortarino, M., Grandi, G. et al. (2008). Combined ivermectin and doxycycline treatment has microfilaricidal and adulticidal activity against *Dirofilaria immitis* in experimentally infected dogs. *Int. J. Parasitol.* 38 (12): 1401–1410.

**135** McCall, J.W., Genchi, C., Kramer, L. et al. (2008). Heartworm and *Wolbachia*: therapeutic implications. *Vet. Parasitol.* 158 (3): 204–214.

**136** Kramer, L., Grandi, G., Leoni, M. et al. (2008). Wolbachia and its influence on the pathology and immunology of *Dirofilaria immitis* infection. *Vet. Parasitol.* 158 (3): 191–195.

**137** Nelson, C.T., Myrick, E.S., and Nelson, T.A. (2017). Clinical benefits of incorporating doxycycline into a canine heartworm treatment protocol. *Parasite Vectors* 10 (Suppl 2): 515.

**138** Mealey, K.L., Bentjen, S.A., Gay, J.M., and Cantor, G.H. (2001). Ivermectin sensitivity in collies is associated with a deletion mutation of the mdr1 gene. *Pharmacogenetics* 11 (8): 727–733.

**139** Geyer, J., Döring, B., Godoy, J.R. et al. (2005). Frequency of the nt230 (del4) MDR1 mutation in Collies and related dog breeds in Germany. *J. Vet. Pharmacol. Ther.* 28 (6): 545–551.

**140** Neff, M.W., Robertson, K.R., Wong, A.K. et al. (2004). Breed distribution and history of canine mdr1-1Delta, a pharmacogenetic mutation that marks the emergence of breeds from the collie lineage. *Proc. Natl. Acad. Sci. U.S.A.* 101 (32): 11725–11730.

**141** Gramer, I., Leidolf, R., Döring, B. et al. (2011). Breed distribution of the nt230(del4) MDR1 mutation in dogs. *Vet. J.* 189 (1): 67–71.

**142** Dean, M., Rzhetsky, A., and Allikmets, R. (2001). The human ATP-binding cassette (ABC) transporter superfamily. *Genome Res.* 11 (7): 1156–1166.

**143** Mealey, K.L., Greene, S., Bagley, R. et al. (2008). P-glycoprotein contributes to the blood-brain, but not blood-cerebrospinal fluid, barrier in a spontaneous canine p-glycoprotein knockout model. *Drug Metab. Dispos.* 36 (6): 1073–1079.

**144** Juliano, R.L. and Ling, V. (1976). A surface glycoprotein modulating drug permeability in Chinese hamster ovary cell mutants. *Biochim. Biophys. Acta* 455 (1): 152–162.

**145** Schinkel, A.H., Smit, J.J.M., Tellingen, O.v. et al. (1994). Disruption of the mouse mdr1a P-glycoprotein gene leads to a deficiency in the blood-brain barrier and to increased sensitivity to drugs. *Cell (Cambridge)* 77 (4): 491–502.

**146** Schinkel, A.H., Wagenaar, E., Mol, C.A., and Deemter, L.v. (1996). P-glycoprotein in the blood-brain barrier of mice influences the brain penetration and pharmacological activity of many drugs. *J. Clin. Invest.* 97 (11): 2517–2524.

**147** Srikant, S. and Gaudet, R. (2019). Mechanics and pharmacology of substrate selection and transport by eukaryotic ABC exporters. *Nat. Struct. Mol. Biol.* 26 (9): 792–801.

**148** Merola, V.M. and Eubig, P.A. (2012). Toxicology of avermectins and milbemycins (macrocyclic lactones) and the role of P-glycoprotein in dogs and cats. *Vet. Clin. North Am. Small Anim. Pract.* 42: 313–vii.

**149** Geyer, J. and Janko, C. (2012). Treatment of MDR1 mutant dogs with macrocyclic lactones. *Curr. Pharm. Biotechnol.* 13 (6): 969–986.

**150** Stiedl, C.P. and Weber, K. (2017). Fast and simple detection methods for the 4-base pair deletion of canine MDR1/ ABCB1 gene by PCR and isothermal amplification. *J. Vet. Diagn. Invest.* 29 (2): 176–180.

**151** Tappin, S.W., Goodfellow, M.R., Peters, I.R. et al. (2012). Frequency of the mutant MDR1 allele in dogs in the UK. *Vet. Rec.* 171 (3): 72.

**152** Monobe, M.M., Junior, J.P.A., Lunsford, K.V. et al. (2015). Frequency of the MDR1 mutant allele associated with multidrug sensitivity in dogs from Brazil. *Vet. Med. (Auckl)* 6: 111–117.

**153** Marelli, S.P., Polli, M., Frattini, S. et al. (2020). Genotypic and allelic frequencies of MDR1 gene in dogs in Italy. *Vet. Rec. Open* 7 (1).

**154** Firdova, Z., Turnova, E., Bielikova, M. et al. (2016). The prevalence of ABCB1:c.227_230delATAG mutation in affected dog breeds from European countries. *Res. Vet. Sci.* 106: 89–92.

**155** Soussa, R.W., Woodward, A., Marty, M., and Cannon, C.M. (2020). Breed is associated with the ABCB1-1Δ mutation in Australian dogs. *Aust. Vet. J.* 98 (3): 79–83.

**156** Mealey, K.L., Munyard, K.A., and Bentjen, S.A. (2005). Frequency of the mutant MDR1 allele associated with multidrug sensitivity in a sample of herding breed dogs living in Australia. *Vet. Parasitol.* 131 (3-4): 193–196.

**157** Mizukami, K., Yabuki, A., Kohyama, M. et al. (2016). Molecular prevalence of multiple genetic disorders in Border collies in Japan and recommendations for genetic counselling. *Vet. J.* 214: 21–23.

**158** Dekel, Y., Machluf, Y., Stoler, A. et al. (2017). Frequency of canine nt230(del4) MDR1 mutation in prone pure breeds, their crosses and mongrels in Israel – insights from a worldwide comparative perspective. *BMC Vet. Res.* 13 (1): 333.

**159** Lee, J.-J., Lin, H.-Y., Chen, C.-A. et al. (2019). Development of an oligonucleotide microarray for simultaneous detection of two canine MDR1 genotypes and association between genotypes and chemotherapy side effects. *J. Vet. Sci.* 20 (1): 27–33.

**160** Silvestro, C.A., Soria, L.A., Conte, A., and Marrube, G. (2019). Two methods for genotyping a 4-base deletion in the canine ABCB1 gene. *J. Vet. Diagn. Investig.* 31 (6): 889–892.

**161** FDA (2018). The FDA is Seeking Input on the Evaluation of Approaches to Demonstrate Effectiveness of Heartworm Preventatives for Dogs. [cited 2020 November 23rd]; Available from: https://www.fda.gov/animal-veterinary/cvm-updates/fda-seeking-input-evaluation-approaches-demonstrate-effectiveness-heartworm-preventatives-dogs.

# 10

# Current Antifilarial Drugs – Mechanisms of Action

*Timothy G. Geary*[1,2,*], *Alan Long*[3], *and Lucienne Tritten*[4,5,6]

[1] *McGill University, Institute of Parasitology, 21,111 Lakeshore Road, Ste-Anne-de-Bellevue, H9X 3V9 QC, Canada*
[2] *Queen's University – Belfast, School of Biological Sciences, 19 Chlorine Gardens, Belfast BT9 5AJ, Northern Ireland*
[3] *Boehringer-Ingelheim, 3239 Satellite Blvd NW, Duluth, GA, 30096 USA*
[4] *University of Zurich, Institute of Parasitology, Winterthurerstrasse 266a, CH-8057 Zurich, Switzerland*
[5] *Swiss Tropical and Public Health Institute, Kreuzstrasse 2, CH-4123 Allschwil, Switzerland*
[6] *University of Basel, CH-4000 Basel, Switzerland*

**Abstract**

Filarial parasites of humans and animals represent a heterogeneous group of pathogenic nematodes, interacting throughout their life cycles with two hosts and, for most species, a bacterial endosymbiont. Various control strategies have been implemented since the mid-twentieth century, including vector control in endemic zones, mass drug administration campaigns, and preventive chemotherapy in companion animals. The challenges posed by the incompletely understood biology of these long-lived pathogens, their restricted accessibility, discrepant life stage-dependent sensitivities to drugs, as well as the potential for severe adverse events, have greatly complicated progress toward elimination of human filariases. This chapter chronicles the history of antifilarial medicines and reviews current understanding of their mechanisms of action, highlighting persisting knowledge gaps some 70 years after the first efforts were undertaken to control filarial infections.

## 10.1 Introduction

Relatively few filariid species are of major importance as parasites of humans and domesticated animals. Although generally not acutely lethal, these parasites cause significant morbidity that impairs productivity and quality of life and can lead to mortality in humans and their companion animals. The two-host life cycles of these pathogens include a definitive vertebrate host and an arthropod vector (mosquitoes, black flies; [1]); transmission is thus largely restricted to climactic zones that afford

*Corresponding author.

*Human and Animal Filariases: Landscape, Challenges, and Control*, First Edition.
Edited by Ronald Kaminsky and Timothy G. Geary.

facile vector development. The incidence and extent of pathology associated with filariases led to research devoted to the discovery and implementation of control measures targeted primarily at three diseases: onchocerciasis (or river blindness, due to *Onchocerca volvulus*) and lymphatic filariasis (LF; *Wuchereria bancrofti*, *Brugia timori* and *Brugia malayi*) in humans, and canine and feline heartworm disease (*Dirofilaria immitis*). Chemotherapy-based control measures (as opposed to targeted treatment of infected individuals) were introduced for LF and heartworm disease as early as the 1950s, following the discovery and development of diethylcarbamazine (DEC; [2]).

Vector control was the primary strategy for onchocerciasis until the introduction of ivermectin (IVM; [3, 4] (Figure 10.1)). A campaign of river-targeted insecticide treatments to reduce larval populations of black flies in the genus *Simulium*, which transmit *O. volvulus*, was implemented in highly onchocerciasis endemic regions in the 1970s and succeeded to some extent to reduce local incidence of infection and disease-related blindness [5]. Concurrent chemotherapeutic interventions for this disease were targeted to infected individuals but were of limited utility due to significant safety concerns. DEC was used to eliminate microfilariae in the skin but caused severe side effects, including blindness, due to drug-induced immune reactions to dying microfilariae in situ [6, 7]. Suramin and melarsoprol were used (along with surgical removal of nodules) as a macrofilaricidal strategy, but again led to severe side effects. These drugs are no longer used for onchocerciasis (Figure 10.2).

Filariae that cause LF are transmitted by many species of mosquito, and vector control strategies (primarily household spraying of DDT) were employed as a control measure for this disease with success in some circumstances [8]. The introduction of insecticide-treated bednets for malaria control has been beneficial to LF control in areas where the diseases are co-endemic [8]. The introduction of DEC

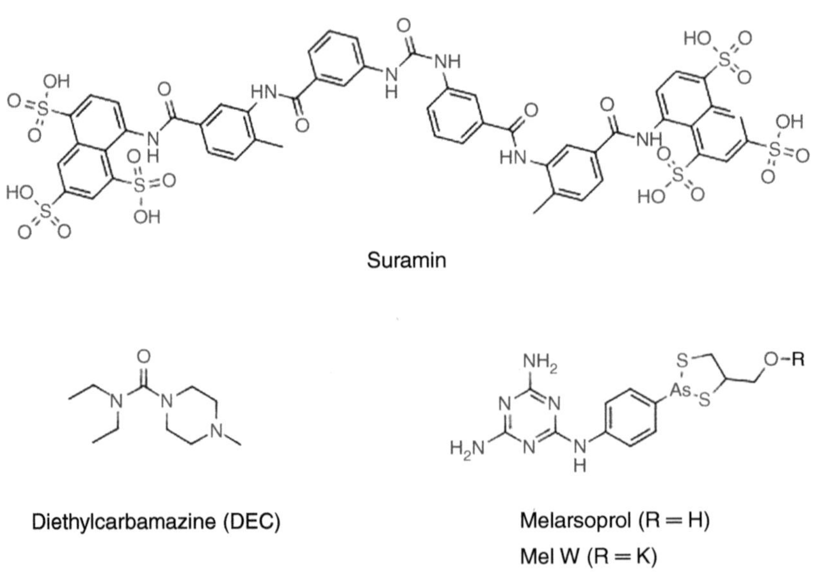


**Figure 10.1** Discontinued drugs for onchocerciasis.

Figure 10.2 Ivermectin is an ~80/20 mixture of the dihydroavermectin isomers $B_{1a}$ and $B_{1b}$.

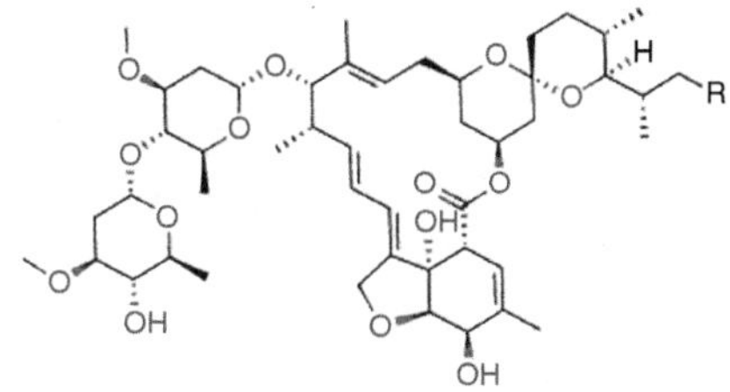


Ivermectin $B_{1a}$ (R=$CH_3$)
Ivermectin $B_{1b}$ (R=H)

treatment for LF began in the late 1940s; the drug was discovered in an effort to treat soldiers returning from LF endemic areas who were infected during WWII [2]. Monthly treatment with DEC safely suppressed microfilariae in blood of patients infected with *W. bancrofti* and had some macrofilaricidal efficacy; however, the drug caused more frequent adverse events in patients infected with *B. timori* or *B. malayi*, which restricted the use of intensive dosing schedules. Subsequent advances showed that once yearly doses of 6 mg/kg or the use of DEC-fortified salt were effective in suppressing microfilarial levels for prolonged periods but had less macrofilaricidal activity [2, 6, 9]. Importantly, this schedule caused acceptable levels of adverse events in patients infected with *Brugia* spp., enabling its wide introduction for control. Safe, simple macrofilaricidal options were not available, and thus chemotherapy was directed at reducing transmission rather than eliminating the pathology due to the presence of adult parasites in lymph vessels.

Like the parasites that cause LF, *D. immitis* is transmitted by many mosquito species. Vector control targeted specifically to prevent heartworm infections was not widely applied, although mosquito control programs to reduce malaria transmission in some areas likely had the ancillary benefit of reducing heartworm infections. Daily DEC treatment of dogs during mosquito season prevented the development of L3 to L4 stages but was ineffective after this transition occurred and was associated with severe side effects in microfilaremic dogs, necessitating diagnostic tests prior to starting treatment. Adult parasites are the pathogenic stage in heartworm disease and cannot be treated with DEC; these stages can be killed by arsenic-containing medicines, of which melarsomine is the only one widely available now.

The introduction of IVM in the 1980s [3], followed by several additional macrocyclic lactones (MLs) (Figure 10.3), revolutionized the ability to prevent infection by or transmission of *O. volvulus* and *D. immitis* by virtue of its prolonged activity against larval stages of filariid parasites after single low doses. The persistent microfilaricidal activity of IVM limits pathology in the skin and eye in onchocerciasis patients and limits transmission of *O. volvulus* and the species that cause LF [4]. In companion

Figure 10.3 The antibiotic doxycycline.

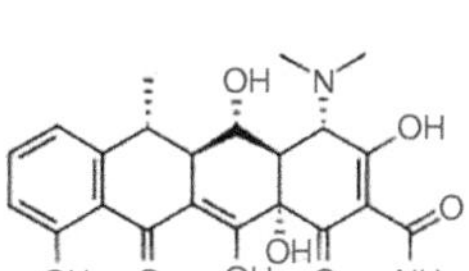


Doxycycline

animals, IVM and related drugs eliminate L4 stages when given monthly, preventing the development to the pathogenic adult stages [10]. IVM and other macrocyclic lactones changed the ability to control these infections to a remarkable degree, ushering in the modern era of antifilarial chemotherapy [4, 11].

Three important parasite life cycle stages of filarial parasites are targeted for treatment in mammalian hosts: microfilariae (L1), infective and developing larvae (L3 and L4), and adults. These stages are affected by chemotherapy to different extents in the major filarial diseases under review here.

## 10.2 Drugs Used in Onchocerciasis

Treatment of onchocerciasis prior to 1988 relied on DEC to remove microfilariae from the skin [2]; this remedy led to immune-dependent destruction of these stages in the skin and eye, causing significant inflammatory reactions in the skin known as the Mazzotti reaction, and in the eye, potentially causing blindness [6]. The intensity of these consequences hampered compliance, and DEC is no longer approved for use in onchocerciasis. Nodules containing adult stages could be removed by surgery, to the extent possible; many nodules are not palpable, and this procedure was rarely curative. Treatment with suramin or melarsoprol or its potassium salt, Mel W, provided some adulticidal efficacy but at a cost of significant, sometimes lethal, toxicity; these drugs are no longer used (see Ref. [7]).

As noted above, the introduction of IVM (as Mectizan) for human use for onchocerciasis in the late 1980s, and the subsequent decision by Merck & Co. to donate it for onchocerciasis control operations, revolutionized the field. The ability of single oral doses of 150 μg/kg to remove microfilariae from the skin and eye for many months with tolerable side effects had two beneficial effects in mass drug administration (MDA) programs: disease progression to blindness was effectively halted in treated populations, and the transmission of the parasite was markedly impeded in treated areas [12]. Twice-yearly dosing leads to better control of microfilariae in the skin and accelerated the progress toward the eradication of this parasite in the Americas [7, 13, 14].

An important exception to the otherwise impressive safety record of IVM (and also a serious complication of therapy with DEC) is the incidence of severe adverse central nervous system (CNS) events, including death, in patients harboring high levels of *Loa loa* microfilariae in the blood [15, 16]. The severity of the response retarded the introduction of IVM in regions in which *O. volvulus* and *L. loa* are co-endemic, and its use requires prescreening of target populations to identify heavily infected individuals, who are excluded from treatment. The severe adverse events appear to involve an inflammatory component, but we lack a full understanding of the pathophysiological basis for them in humans and the reason(s) for the apparent threshold of microfilaremia required for it to occur [17] (Table 10.1).

Extended (six week) regimens of daily doxycycline (Figure 10.4), an antibiotic that targets the *Wolbachia* symbiont essential for reproduction and survival of adult *O. volvulus*, has macrofilaricidal activity, although death of the parasite occurs slowly [18, 19]. Shorter regimens are suboptimally efficacious in this regard. A month-long course of doxycycline is feasible at the community level [20] but is not currently incorporated into control programs.

**Table 10.1** Mechanism of action of antifilarial drugs.

| Drugs | Parasite spp. | Stage killed | Efficacy | General mechanism of action | Alternative MoA | Comments |
|---|---|---|---|---|---|---|
| Diethylcarbamazine (DEC) | LF | mf | +++ | Likely indirect, stimulating host immune response | | |
| | LF | Adults | ++ | | | Only intensive regimens with resulting side effects |
| | *O. volvulus* | mf | +++ | | | Contraindicated |
| | *D. immitis* | L3 | +++ | | | Replaced by MLs; no efficacy against L4 |
| *Benzimidazoles* | | | | | | |
| Albendazole | LF | mf | +/− | Inhibits microtubule assembly in nematodes, impairing many cellular processes, with multiple phenotypic consequences | Protein secretion | Used annually or bi-annually, only in combination with other compounds. Repeated treatments may be more effective than infrequent single doses. Albendazole boosts anti-*Wolbachia* activity of antibiotics |
| | LF | Adults | + (++) | | | Single dose ineffective; intensive regimens show better efficacy but with side effects |
| | *L. loa* | mf | + | | | Modest efficacy in intensive regimens |
| | *O. volvulus* | mf | +/− | | | Evidence of efficacy lacking |

**Table 10.1** (Continued)

| Drugs | Parasite spp. | Stage killed | Efficacy | General mechanism of action | Alternative MoA | Comments |
|---|---|---|---|---|---|---|
| *Macrocyclic lactones* | | | | | | |
| Ivermectin | *O. volvulus*, LF | mf | +++ | Glutamate-gated chloride channel (GluCl) agonists. Long-term sterilization of adult filariae | Protein secretion; indirect effects on eliciting host immune responses | Mainstay for MDA programs for onchocerciasis; prevention of heartworm in dogs; used with caution in zones co-endemic for *L. loa* |
| | *O. volvulus*, LF | Adults | + | | | Prolonged sterilization; some macrofilaricidal efficacy after prolonged treatment |
| | *D. immitis* | L3/L4 | +++ | | | |
| | *D. immitis* | Adults | + | | | Requires prolonged regimens, not recommended |
| Selamectin | *D. immitis* | mf, L3/L4 | +++ | | | |
| | | Adults | + | | | Unknown efficacy |
| Moxidectin | *O. volvulus* | mf | +++ | | | Often administered in combination with other drugs |
| | *D. immitis* | mf, L3/L4 | +++ | | | |
| | | Adults | + | | | Unknown efficacy |

| | | | | | |
|---|---|---|---|---|---|
| Milbemycin oxime | *D. immitis* | mf, L3/L4 | +++ | | Often administered in combination with other drugs |
| | | mf | + | | Unknown efficacy |
| *Antibiotics* | | | | | |
| Doxycycline | LF | Adults | +++ | Targets the endosymbiont *Wolbachia*. Prevents translation by blocking interaction between tRNA and mRNA required for addition of amino acids at the ribosomal complex | Extended regimens sterilize and slowly kill |
| | *O. volvulus* | Adults | +++ | | Extended regimens sterilize and slowly kill |
| | *D. immitis* | L3/L4 | +++ | | Extended regimen prevents development of infective L3 in mosquitoes |
| | | Adults | + | | Reduces side effects of melarsomine treatment |
| *Arsenic-based* | | | | | |
| Melarsomine | *D. immitis* | Adults | +++ | Unclear; possible lethal oxidative stress may be induced through binding to sulfhydryl groups | |
| *Triple drug combination* | | | | | |
| DEC + ivermectin + albendazole | LF | mf | +++ | Unclear | Irreversible sterilization and slow killing of adult worms |

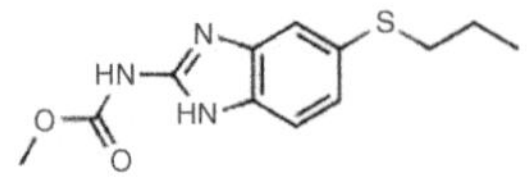

Albendazole

**Figure 10.4** Structure of albendazole.

## 10.3 Drugs Used in LF

DEC has been the mainstay for control of LF since its introduction some 70 years ago. It has been available as a medicated salt, and intensive dosing regimens have been used to treat infected individuals. In general, DEC has not been routinely used to kill adult parasites in lymph nests. Instead, MDA programs have been implemented to prevent the transmission of the parasites that cause LF. Until recently, these programs relied on annual treatment of people in endemic regions with a combination of DEC and albendazole (Figure 10.5) [14, 21], which has led to significant reductions in LF endemicity in many regions [22]. This regimen was replaced in regions co-endemic for LF and onchocerciasis with the combination of IVM plus albendazole to avoid the adverse events experienced after the administration of DEC to individuals infected with *O. volvulus*.

As discussed below, it has recently been reported that simultaneous administration of single standard doses of ivermectin + diethylcarbamazine + albendazole (IDA; triple drug therapy) produces very long-lasting suppression of microfilaremia

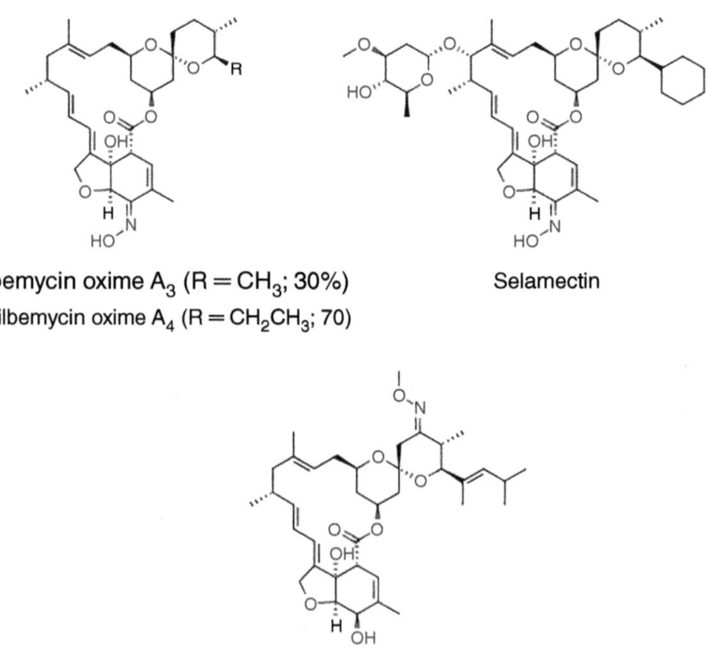


**Figure 10.5** Other macrocyclic lactones for heartworm treatment.

in LF patients, potentially due to macrofilaricidal efficacy [23–26]. More extensive, longer-term studies will reveal if this effect is sustained; reversion to a microfilariae-positive state in a recent study could also simply reflect re-infection [26]. A recent follow-up study five years after a single treatment showed almost a complete lack of microfilariae, although circulating filarial antigen was still present; it is possible that the effect of triple-drug therapy is irreversible sterilization rather than killing [27]. Permanent sterilization is equivalent to macrofilaricidal activity for the purposes of control. Importantly, this regimen has been shown to be safe [28, 29] and has been adopted as the treatment of choice for LF in non-onchocerciasis regions [30].

As is the case for onchocerciasis, extended-duration regimens of daily doxycycline doses are slowly macrofilaricidal in LF [19, 31]. However, this drug has not been incorporated into MDA programs, and, if the current results with the triple drug combination are consistently confirmed, it is unlikely to be needed.

## 10.4 Drugs Used in Heartworm Disease

Although DEC can prevent infection of dogs by *D. immitis*, daily administration during periods of mosquito presence was required for full efficacy, and overall adoption of daily DEC and compliance were low. The introduction of MLs as preventatives, beginning with IVM, revolutionized heartworm prophylaxis [10, 11, 32]. A single monthly dose of IVM was fully effective in preventing infection, assuming compliance, and encouraged widespread adoption of this practice by pet owners. Additional MLs, including milbemycin oxime, selamectin, and moxidectin, were subsequently introduced. Available products are formulated to enable oral, topical, or parenteral dosing, with the last strategy providing 6- or 12-month protection in a slow-release product [33]. The appearance of ML-resistant populations of *D. immitis* has compromised the efficacy of these drugs as heartworm preventatives [34]. The genotypic basis for this phenotype has not been fully defined. Evidence suggests that the phenotype affects all drugs in this class, but duration of exposure, which varies among the MLs, is a variable that may affect the degree of resistance [35].

Melarsomine (Figure 10.6) is used as an adulticide in infected dogs. A 30-day course of doxycycline typically precedes administration of melarsomine, which is given in three divided doses. The second and third dose are administered 30 and 31 days, respectively, following the first [32, 33]. The pathology associated with killing large worms in situ appears to be ameliorated by removal of the *Wolbachia* symbiont, especially when combined with pretreatment with IVM [36–38]. Doxycycline therapy has the added benefit of rendering circulating microfilariae incapable of developing to infectious L3 larvae in mosquitoes [36, 38]. This feature could help to reduce the transmission and spread of macrocyclic lactone-resistant strains from infected dogs.

No chemotherapy is approved for cats with patent heartworm infections; cats typically have few adult worms and transient microfilaremia. Supportive care is recommended with surgical removal of adult worms in some cases.

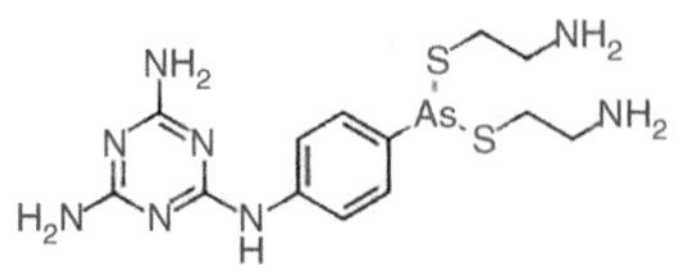


**Figure 10.6** The *D. immitis* adulticide melarsomine (injected as the dihydrochloride salt).

## 10.5 Drugs Used in Other Filariases

Two other filariid species have attracted chemotherapeutic attention, though to a lesser extent than the main foci of this chapter. The veterinary canine and feline parasite *Dirofilaria repens* is a concern in Europe, Asia, and Africa. The mosquito-borne *D. repens* can also infect humans but is not considered pathogenic in companion animals, and so has attracted less research than *D. immitis*. *D. repens* adults inhabit subcutaneous and intramuscular connective tissues, not the cardiopulmonary system, and microfilariae are found in the circulation. Prophylaxis can be achieved with monthly doses of macrocyclic lactones, and macrofilaricidal efficacy has been reported with long-term macrocyclic lactone therapy, which may be enhanced by the addition of doxycycline [39].

The human parasite *L. loa* is found in Africa and is co-endemic with LF and onchocerciasis in some areas [16]. As noted above, individuals bearing high levels of *L. loa* microfilaremia are at risk of serious adverse events following treatment with IVM (or DEC). Intensive dosing schedules of albendazole have shown some efficacy in reducing the microfilaremia, possibly due to sterilization of adult parasites, but are not of sufficient magnitude to entirely ameliorate the risk of IVM-MDA in endemic regions [40, 41].

## 10.6 Drug Effects Against Filariae are Species- and Host-Specific

Much of what we know about the pharmacology of antifilarial drugs comes from studies done in other kinds of nematodes, including the free-living species *Caenorhabditis elegans* [42]. No convenient laboratory animal can host the full development of *O. volvulus*, *W. bancrofti*, or *D. immitis*; even partial development requires some form of immunosuppression. An exception is *B. malayi* (and the closely related feline parasite *Brugia pahangi*), which can undergo full development in a number of small animals; these parasites have typically served as models for the other filariid species of interest. Furthermore, the efficacy and potency of antifilarial drugs vary greatly in the different host–parasite animal models that are amenable to laboratory operation [43] for reasons that are unclear. Samples of parasite material are very difficult or almost impossible to acquire, particularly adult stages of *O. volvulus* and *W. bancrofti*. It is also challenging to obtain samples of adult heartworms, which require the sacrifice of infected dogs or ferrets. Thus, the discussion that follows reflects the best available information, with appropriate caveats.

## 10.7 Mechanisms of Action and Basic Pharmacology of Antifilarial Drugs

Although hundreds of millions of people and companion animals have been treated with antifilarial chemotherapy for many decades, we know surprisingly, and depressingly, little about the mechanism of action of many of them. As has generally been the case for neglected tropical diseases, much greater investment has been made in the empirical search for efficacious drugs than in an effort to understand the basic biology of the pathogens and the pharmacology of the drugs that are available. It is important to recognize that all of the antifilarial drugs in use resulted from empirical drug discovery efforts primarily using infected animal models, almost all for indications in veterinary medicine.

### 10.7.1 DEC

#### 10.7.1.1 Background

The only antifilarial drug discovered explicitly for human filarial infections is DEC [2]. First identified in the late 1940s as part of an initiative to find safe treatments for soldiers who acquired LF during WWII, the drug remains a mainstay of LF control programs. In MDA programs, a single annual dose of DEC (given in combination with albendazole) rapidly removes microfilariae from the circulation and blocks development and release of new microfilariae from adult females for many months. Although previously used to treat onchocerciasis, DEC is now contraindicated because of the severity of side effects observed in the skin and eye due to immune-mediated killing of *O. volvulus* microfilariae in situ. It was also used to prevent heartworm infections in dogs; given daily during the transmission season, DEC targeted *D. immitis* L3 larvae. Diagnostic tests to preclude active infection with *D. immitis* were required before administration, as DEC was associated with severe side effects in microfilaremic dogs. The use of DEC for heartworm prophylaxis essentially ceased when the MLs were introduced.

#### 10.7.1.2 Mechanism of Action

DEC has little or no evident deleterious activity against any stage of many filarial nematodes in culture at pharmacologically relevant concentrations and durations of exposure ($C_{max}$ 7.5 μM, $t_{1/2}$ 10–12 hours; [26]). Instead, DEC has been thought to act indirectly by triggering immune responses against parasites in hosts [44]; these can include apparent activation of innate immune responses and attachment of host effector cells [45]. How the drug does this is not well understood, and little work has been done recently on this problem. The molecular target(s) of DEC in filariases are not fully defined; they may be present in both host and parasite. Therapeutic doses of the drug have a variety of immune-related and anti-inflammatory effects in animals [44, 45], but these are inconsistent and of uncertain relevance to the rapid action of the drug observed in human and canine filariases. Experimental evidence supports an interaction of DEC with the host cyclooxygenase pathway for the synthesis of prostaglandins, and a requirement for efficacy of host inducible nitric oxide synthase has been shown in a mouse model of filariasis [46], and prostaglandin synthesis in microfilariae is also inhibited by DEC [47]. How these systems are engaged by DEC and how they contribute to its efficacy in humans and dogs has not been determined, no evidence has been reported that DEC directly inhibits eicosanoid synthesis enzymes at pharmacologically relevant concentrations (or that recognizable homologs of such enzymes exist in nematodes). Recently, evidence has been published to document that *C. elegans* can synthesize prostaglandins by a non-cyclooxygenase-dependent pathway [48], which may explain previous reports of prostanoid synthesis by *B. malayi* microfilariae [47, 49]. DEC has also recently been shown to activate a family of TRP channels in microfilariae and adult *B. malayi* at concentrations in the range of reported $C_{max}$ values [50]. DEC caused a rapid but quickly resolved paralysis of these parasites in culture. Activation of TRP channels led to activation of SLO-1 $K^+$ channels, which are targeted by emodepside (see below); arachidonic acid metabolites may act similarly on TRP channels. How activation of TRP channels leads to immune-dependent clearance of microfilariae, whether these actions can explain the prolonged suppression of microfilaremia after a single DEC dose, and how the involvement of prostaglandins can be reconciled with the new observations should be research priorities. Considering that DEC is a mainstay of LF control programs and is given annually to many millions of humans, it is troubling that we still lack a comprehensive understanding of its mechanism(s) of antifilarial action.

#### 10.7.1.3 Pharmacodynamics

DEC acts very rapidly to clear microfilariae from the blood of LF patients and in some (but not all) animal models, and to rapidly cause the death of mf in the skin (and eyes) of onchocerciasis patients [2, 6, 7, 9]. What happens to L3 stages of *D. immitis* in DEC-treated dogs is unknown. Why the drug affects *D. immitis* L3 larvae and microfilariae, but not L4 larvae in dogs, is unknown. Intensive regimens of DEC in LF patients have some macrofilaricidal efficacy [2, 9] but are not used as part of the current elimination/eradication strategy. How DEC causes these effects is not known, and it is important to note that DEC has no apparent macrofilaricidal activity against *O. volvulus*. Importantly, a single dose of DEC in MDA programs causes significant and persistent diminution of microfilaremia for months (variable in duration among patients) [9], suggesting that the drug causes serious (but reversible) injury to or long-term sterilization of adult parasites. Although this persistent effect on microfilaremia following a single dose of DEC is the basis for its use in LF control programs, a pharmacological explanation for it remains elusive.

### 10.7.2 Albendazole

#### 10.7.2.1 Background

Albendazole is used primarily for the treatment of gastrointestinal nematode infections. The prototype of the benzimidazole class, thiabendazole, was introduced for this indication in the 1960s. Albendazole, a benzimidazole carbamate, has much better potency and efficacy than the progenitor [51]. Intensive regimens of albendazole (high and/or prolonged dosing) have some macrofilaricidal effects in LF patients but caused significant side effects thought to be associated with the death of adult worms. This observation led to a decision to include the drug at tolerable annual or biannual doses in combination with either DEC or IVM (in loiasis regions) [52] in LF MDA programs.

#### 10.7.2.2 Mechanism of Action

The mechanism of action of albendazole and other anthelmintic benzimidazoles is well understood: the drug destabilizes the tubulin–microtubule equilibrium [53]. Microtubules, which form essential dynamic structural networks in eukaryotic cells, are polymers of an α- and β-tubulin dimer. Microtubules develop and collapse by adding or losing tubulin dimers, and the concentration of tubulin in the cell is determined by the rate of dimerization of the α- and β-subunits. The benzimidazole binding site on tubulin, though incompletely solved, has been proposed to be centered on β-tubulin; residues 167, 198, and 200, all of which can be mutated to generate benzimidazole resistance, are thought to be key for drug binding [54, 55]. Binding to tubulin prevents the formation of microtubules, leading to the dissolution of microtubule networks in nematode cells. Benzimidazoles have higher affinity for nematode β-tubulin than for mammalian isoforms [53], which differ in the amino acid at position 200 (Phe in nematodes, Tyr in mammals; [56]). Loss of microtubule networks inhibits a plethora of cellular functions; in GI nematodes, a major consequence appears to be cessation of nutrient uptake across the gut epithelium. In contrast, the role of the filariid gut in nutrient acquisition is ill defined, as transcuticular uptake of small molecule nutrients is believed to predominate (see [57]). The formation and development of larvae in utero is also dependent on microtubule function during cell division, and the cessation of larval development and production in adult female filariids is a consequence of exposure to effective concentrations of benzimidazoles, an effect especially noted with flubendazole [58, 59].

#### 10.7.2.3 Pharmacodynamics

Albendazole is very poorly bioavailable after oral administration; its sulfoxide metabolite, which is an anthelmintic, reaches higher plasma concentrations and accounts for most of its systemic activity [51]. A recent revaluation of clinical data indicated that albendazole, given at single doses (400 or 800 mg, not adjusted for body weight), has at best limited antifilarial activity; its co-administration with DEC or IVM for treatment of LF provides no marked benefit over administration of DEC or IVM alone [60]. In contrast, albendazole has been reported to boost the activity of anti-*Wolbachia* antibiotics when given in multiple doses (see below); a pharmacological explanation for this effect remains obscure. Intensive and/or prolonged regimens of albendazole in loiasis patients had a modest but significant effect on microfilarial levels in blood, suggesting that prolonged or repeated exposure to the drug may be more effective than the dosing schedule employed in LF MDA programs [40, 41]. However, efficacy was insufficient to enable full deployment of current MDA strategies in loiasis regions. Importantly, the beneficial effects of treating GI nematode infections with albendazole may enhance community participation in the MDA programs.

Although the actions of albendazole that lead to its antifilarial effects are imprecisely known, it should be noted that this drug like other benzimidazoles inhibits protein secretion from parasitic nematodes (e.g. [61]). This effect is also seen in filarial parasites [62]. As discussed below, inhibition of the secretion of parasite-derived immunomodulatory molecules by the MLs may contribute to their antifilarial activity; whether this is also true for albendazole remains to be determined.

### 10.7.3 Macrocyclic Lactones (MLs)

#### 10.7.3.1 Background

MLs are the predominant anthelmintic drugs in veterinary medicine, and the use of IVM in human medicine opened the way to MDA for the control of onchocerciasis [11]. Importantly, the precedent set by Merck & Co. to donate IVM for onchocerciasis and LF control changed the way major pharmaceutical companies addressed neglected tropical diseases.

Four MLs are approved for the prevention or treatment of human and veterinary filariases. IVM is approved for use in LF, IVM and moxidectin are approved for the treatment of onchocerciasis, and these two drugs, milbemycin oxime and selamectin, are approved for the prevention of companion animal heartworm infections. These drugs are all derived from fermentations of *Streptomyces* spp. They are classified as avermectins (IVM and selamectin) or milbemycins (moxidectin and milbemycin oxime); the major difference between these classes is the presence of sugar residues attached at position C-13 (2 in IVM, 1 in selamectin, 0 in milbemycins) and the substituent at position C-25.

Although MLs are generally exceptionally safe drugs, overdoses can occur. Accumulation in the CNS is associated with severe toxicity. The drugs are normally excreted across the blood–brain barrier by a P-glycoprotein pump encoded on the *mdr1a* gene; loss-of-function mutations in this gene in dogs lead to IVM sensitivity [63]. Milbemycins are less affected by this pump and are generally safer in dogs.

IVM, the progenitor of the ML class (all commercially available members of which are semi-synthetic), is a product containing two isomers, 22,23-dihydroavermectin $B_{1a}$ and $B_{1b}$ in an 80 : 20 ratio [64]. It is given annually or biannually as a tablet in MDA programs for onchocerciasis, and monthly in various oral formulations for the prevention of heartworm disease in dogs. The other avermectin, selamectin [65], is available as a topical product. Milbemycin oxime is a product composed of the 5′ oxime of milbemycin $A_3$ and $A_4$ (30 : 70 ratio; [66]) and is administered as a tablet, often combined with other medicines. The milbemycin moxidectin [67] is available for companion animals in multiple dosage forms, again often in combination with other medicines, including oral tablets and topical formulations. It is also available as a slow-release preparation of drug-loaded microspheres, dosed every 6 or 12 months. It is approved for administration as a tablet for human onchocerciasis treatment [68] but has not yet been added to MDA programs for this indication.

Although the MLs share a common mechanism of action, they are not identical medicines [35, 63]. Among these differences, higher lipophilicity leads to the milbemycins generally having longer plasma residence times and more persistent activity than the avermectins. This is notable, for example, in the longer duration of suppression of microfilariae in the skin of onchocerciasis patients dosed with moxidectin compared to IVM [69].

#### 10.7.3.2 Mechanism of Action

Although not identical in pharmacology, these drugs share a mechanism of action: they are agonists at a family of ligand-gated ion channels (LGICs) [70] and the glutamate-gated chloride channels (GluCls) [71–74]. GluCls are restricted phylogenetically to invertebrates, most notably nematodes and arthropods. Glutamate binding causes opening of these channels, an event that is markedly potentiated and prolonged by the coincidental presence of an ML. Channel opening allows the passage of $Cl^-$ ions into the interior of nerve and muscle cells, lowering membrane potential and inhibiting the ability of the cell to contract or conduct a signal following exposure to an excitatory stimulus.

GluCls, like other LGICs, are composed of five subunits, which form homo- or heteropentamers, depending on the subunits and the cells in which they are expressed. Many distinct GluCl subunits are encoded in nematode genomes, although the complement varies to some degree [75]. GluCls composed of different subunits are located in tissues throughout the nematode body; the pharmacological actions of MLs in a given nematode species is presumed to be governed by this distribution [76].

Publication of a high-resolution crystal structure of a *C. elegans* GluCl (Cel-Glc-1) bound to ivermectin and glutamate enabled a molecular-level description of the presumed ML binding site [77]. This model predicts that ivermectin binds to an allosteric site in the plasma membrane near the extracellular domain at the interface between two subunits, inserting the cyclohexene moiety. Binding at this site, ivermectin induces a large-scale conformational change that extends to the neurotransmitter binding site, suggesting a mechanism for the effects of this drug on glutamate-gated currents [78].

The kinetics of activation of GluCls by IVM are compatible with a two-step process that leads to pseudo-irreversible binding [79]; binding is not covalent but is characterized by an extremely slow off-constant. It is thought that the exceptionally lipophilic drug partitions into the lipid bilayer of cell membranes, reaching a high local concentration which decreases very slowly as the drug dissolves in aqueous compartments outside the cell [78]. Fewer studies have been reported on other MLs, but the class is presumed to interact similarly with GluCls, although minor differences likely exist [63].

At higher concentrations, IVM and other MLs can affect the function of other LGICs, including those gated by gamma-aminobutyric acid (GABA), glycine, and histamine [76, 78]. The contribution of the effects of IVM on non-GluCl LGICs to its anthelmintic activity is uncertain, but these effects may account for at least some of the toxicity of IVM if it accumulates in the CNS.

#### 10.7.3.3 Pharmacodynamics

In GI nematodes and *C. elegans*, GluCls are prominently located in the somatic and pharyngeal neuromusculature, and ML-induced paralysis of these tissues leads to cessation of motion and feeding, respectively, both of which contribute to efficacy of these drugs [74]. The situation in filariid nematodes is different; MLs have little or no effect on motility of larval or adult stages in culture at pharmacologically relevant concentrations [73, 74]. Recent work suggests instead that IVM inhibits the release of parasite-derived immunodulatory molecules, primarily in extracellular vesicles, from the excretory–secretory pore [62, 80, 81]. Consequently, larvae are recognized and destroyed by the host immune response [74]. That the only GluCl detected in *B. malayi* microfilariae is localized to this pore is consistent with this hypothesis [62]. Additional experiments have shown that incubation with MLs stimulates the binding of canine leukocytes to *D. immitis* microfilariae [82, 83]. It is possible that this reflects the reduction in ML-induced release of parasite-derived immunomodulatory molecules; although parasite killing was not induced by cell attachment, this may lead to lethality in vivo.

In addition to larval stages of filarial parasites targeted by MLs in the host (L3, L4, and microfilaria), MLs induce long-term sterilization of adults [7, 84–86]. Killing of adult stages by MLs in vivo also occurs but is slow and requires repeated dosing [32, 87, 88]. It is extremely important to note that the activity of MLs against all stages varies with filariid species and is also host-dependent [43], for reasons that are not understood.

Activity against L3/L4 stages provides the basis for prophylaxis against *D. immitis* in companion animals; how these drugs prevent larvae in dogs and cats from maturing to adults has not been resolved experimentally. As the L4 stage last for more than a month during the *D. immitis* life cycle, treating companion animals on a monthly basis is an effective prophylaxis strategy. It is interesting to note that prophylactic activity against *O. volvulus* larvae in chimpanzees is not observed with IVM [89]; it is not clear if the drug has prophylactic activity against LF parasites in humans (it is not prophylactic against *Brugia* spp. in the jird; [43]).

Activity of IVM (and moxidectin) against microfilariae is the primary basis for its use in human filariases [4, 85]. Removing microfilariae from the blood or skin prevents transmission, a key pillar of control strategies, and the removal of these stages from the skin and eye of onchocerciasis patients is of pivotal therapeutic benefit in preventing blindness and dermal pathology [7]. Unlike DEC, adverse effects of MLs on microfilariae are generally mild and quickly reversible. MLs also remove *D. immitis* microfilariae in companion animals [32], generally without serious adverse events, and thus can prevent transmission of the parasite from infected animals. The activity of MLs on microfilariae is thought to be dependent on the host immune response, as noted above, but the precise mechanisms have not been resolved and may be species- and host-dependent.

More mature larval stages and young adults of *D. immitis* can be killed by prolonged monthly dosing with MLs, with lethal consequences even for most adult stages after long-term monthly treatment [87]; this effect is enhanced by simultaneous treatment with doxycycline [90]. The pharmacological basis for these effects is unknown. The use of MLs to treat adult stages of *D. immitis*, known as the "slow kill" protocol, is not currently recommended as a primary adulticidal strategy; concerns have been raised that this practice may have led to the selection of ML-resistant parasite strains [32]. As noted above, a degree of macrofilaricidal activity has been reported for *O. volvulus* after multiple rounds of annual treatment [88], but this action (the basis for which is unknown) is of insufficient magnitude and consistency to achieve eradication.

More importantly, MLs have profound nonlethal effects on adult filariids: single doses confer months-long sterility, marked by cessation of the production of new microfilariae and the accumulation of mal- or undeveloped larval stages in the uterus [84, 85, 91, 92]. It was shown that ivermectin dysregulates genes involved in meiosis and calcium signaling surrounding fertilization as well as embryo development in female *B. malayi* [93]. In addition, at least some of the cessation of microfilarial production appears to be due to effects on males, limiting insemination in *O. volvulus* [94] and *D. immitis* [86]. The pharmacology underlying these effects is not clearly understood.

One hypothesis is that MLs could inhibit pharyngeal function in adult filariids [74]. Although the acquisition of small nutrients seems to occur across the filarial cuticle [57], iron for the synthesis of new hemoproteins likely requires the ingestion of iron bound to host proteins. Prolonged inhibition of oral ingestion by MLs could lead to dysregulated sperm and larval development due to the inability to generate heme. A similar effect would result from the antibiotic-induced killing of *Wolbachia*, which provide the parasite with heme. However, evidence that adult filariae ingest host hemoproteins is lacking. An alternative explanation is based on the discovery of a GluCl (MbAVR-14) localized in reproductive tissues in adult male and female *B. malayi* [95]; similar tissue-level expression of the *C. elegans* AVR-14 ortholog has not been observed. The physiological function of BmAVR-14 has not been reported, but ML-induced hyperpolarization of the neuromuscular system associated with adult reproductive tissues could account for or contribute to the persistent sterilization phenotype.

### 10.7.4 Antibiotics

#### 10.7.4.1 Background

Almost all of the filariids that cause medically significant infections rely on symbiotic bacteria in the genus *Wolbachia* for reproduction and viability. Prolonged exposure to antibiotics reduces the population of these bacteria in filarial parasites, causing sterilization and gradual death of the adults [18, 19, 96]. *Wolbachia* contribute essential products of intermediary metabolism to the parasite, including heme [97–99]. Filariids cannot synthesize heme, and the absence of *Wolbachia*-derived heme likely prevents the formation of viable microfilariae, a developmental program that requires de novo hemoprotein synthesis. *Wolbachia* also can contribute vitamins and nucleic acid precursors to the host nematode, although it appears that the extent of this relationship may differ among filariid species.

#### 10.7.4.2 Mechanism of Action

The only antibiotic routinely used for filariasis is doxycycline, a semisynthetic macrolide in the tetracycline family originally derived from a *Streptomycin* fermentation. Tetracyclines are bacteriostatic protein synthesis inhibitors [100]. Although the details of its molecular target remain to be precisely defined, it is thought that tetracyclines interact with the 30S subunit of bacterial ribosomes, possibly through binding to the 16S rRNA component. Drug binding prevents translation by blocking the binding of aminoacyl-tRNA to the mRNA-ribosome complex. As noted below, other antibiotics with different mechanisms of action are also effective anti-*Wolbachia* agents.

#### 10.7.4.3 Pharmacodynamics

Prolonged exposure – typically, daily dosing for four to six weeks – is required for full macrofilaricidal efficacy in humans [18, 19, 96]. Sufficient durations of drug exposure in vitro and in vivo cause marked reductions in *Wolbachia* populations; these reductions can be irreversible. As the drug is bacteriostatic and not bactericidal, exposure must be long enough to prevent recovery of bacterial populations after treatment, a difficult outcome to confidently predict based on acute bacterial counts alone.

Antifilarial effects of doxycycline in at least some parasite species include sterilization of adult females, stunting of microfilariae, prevention of microfilarial maturation to infective larvae in mosquito vectors, and eventual death of adult stages [18, 19, 96]. The consequences of removal of *Wolbachia* may be related to loss of essential nutrients, including heme, but evidence has also been obtained that depletion of *Wolbachia* leads to apoptotic death of germline cells and in developing larvae in utero [101]. Apoptosis may also contribute to the death of adult worms following doxycycline treatment. Effects of doxycycline treatment on spermatogenesis in adult male filariids have not been reported in detail.

Doxycycline is administered to heartworm-infected dogs prior to treatment with the macrofilaricide melarsomine; reducing *Wolbachia* populations reduces the incidence and severity of side effects associated with the killing of adult parasites, presumably by minimizing the release of *Wolbachia* antigens upon parasite death [36, 38]. Doxycycline also prevents the development of L3 and L4 stages of *D. immitis* to adults, an effect also observed in some animal models of filariasis [102]. Although doxycycline does not kill *D. immitis* microfilariae, treatment prevents L3 stages that develop in mosquitoes from becoming competent for infecting a new host, blocking transmission [36, 38, 102]. This may be help limit the spread of ML-resistant strains of *D. immitis*.

### 10.7.5 Melarsomine

#### 10.7.5.1 Background

Arsenic-based compounds have been used for many decades for the treatment of parasitic infections; the only remaining valid indications are for the treatment of human African trypanosomiasis (melarsoprol) and for removal of adult heartworms (melarsomine).

#### 10.7.5.2 Mechanism of Action

Little information is available about the mechanism of action of melarsomine against adult heartworms. Studies on trypanosomes have shown that the active metabolite

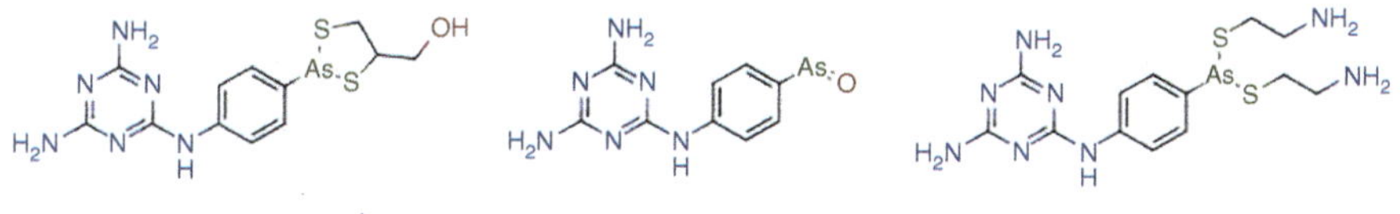


**Figure 10.7** Filaricidal arsenicals and the proposed active metabolite.

of melarsoprol, melarsen oxide (Figure 10.7) (with trivalent arsenic), binds covalently to sulfhydryl groups. A primary target appears to be trypanothione, resulting in the formation of a complex termed Mel T, which is thought to be toxic to these protozoan parasites through imprecisely defined mechanisms, although inhibition of trypanothione reductase may lead to lethal oxidative stress [103]. Inhibition of other sulfhydryl-dependent enzymes may contribute to the lethality. A key component of selective toxicity of melarsoprol for trypanosomes is its uptake into the parasite through three (or more) transporter systems, loss of which leads to drug resistance [104].

Whether and how melarsomine and/or melarsen oxide are selectively accumulated by adult heartworms has not been reported. Nematodes lack trypanothione and whether melarsen oxime generated from melarsomine can form toxic adducts with glutathione or thioredoxin in adult heartworms is unknown. Melarsomine does not kill *D. immitis* microfilariae in infected dogs, and any proposed mechanism of action must explain this stage specificity.

#### 10.7.5.3 Pharmacodynamics

Like melarsoprol, melarsomine is a prodrug which is converted to the active metabolite melarsen oxide [105]. Single doses of melarsomine are only partially efficacious against adult heartworms, and a three-dose regimen (a single dose followed one month later by two additional doses) is now standard practice in veterinary medicine [33]. This regimen reduces the incidence and severity of side effects associated with a more intensive treatment protocol, but the pharmacological basis for the improved macrofilaricidal efficacy achieved with this dosing schedule has not been resolved.

### 10.7.6 Triple Drug Combination

The use of DEC plus albendazole or IVM plus albendazole in annual dosing regimens is not associated with significant macrofilaricidal efficacy in LF, necessitating treatment of people living in endemic regions for many years to achieve eradication by suppression of transmission. As noted above, although initial evidence suggested that the addition of albendazole to these regimens increased efficacy in human filariases, recent evaluations show little benefit compared to treatment with DEC or IVM alone [60]. In contrast, combinations of DEC and IVM are additively more efficacious than either drug alone in suppressing microfilaremia in LF, although without marked macrofilaricidal effects [106, 107]. Recent observations of macrofilaricidal efficacy and safety of the triple combination of IVM + DEC + albendazole

(IDA) administered in a simultaneous single dose [23–26] is thus welcome but nonetheless somewhat surprising from a pharmacological perspective. The failure of combinations of any two of these drugs to reliably demonstrate macrofilaricidal efficacy and the lack of consistent evidence for positive contributions of albendazole as a micro- or macrofilaricide leave unresolved the mechanistic explanation for efficacy of IDA. Given the inability to examine this question in human subjects, it is important to analyze this interaction in animal models of filariasis. Longer-term studies will also be important to determine if macrofilaricidal activity is a general result of this regimen. Nevertheless, these results have changed the paradigm for LF control [30, 108] and, if more broadly confirmed, will unquestionably enhance prospects for eradication of this disease. Whether the combination can be safe and effective as a macrofilaricidal strategy for onchocerciasis remains to be seen.

### 10.7.7 Medicines in the Near-term Pipeline

The drive to discover, develop, and introduce new antifilarial drugs is driven by two main factors: the need for a macrofilaricide for onchocerciasis that is effective in short regimens, and the presence of macrocyclic lactone-resistant populations of *D. immitis*, which threaten the efficacy of current prophylaxis regimens. Significant investment has been made in these areas; the first is largely driven by academic, governmental, and nongovernmental organizations, while heartworm research has been the focus of the animal health industry.

Two types of molecules are in development for their ability to kill or permanently sterilize adult *O. volvulus*: (i) antibiotics and (ii) re-purposed drugs already approved for use in veterinary or human medicine.

#### 10.7.7.1 Antibiotics

Although doxycycline has proven macrofilaricidal activity in onchocerciasis, the requirement for a >30-day daily dosing regimen is unlikely to be compatible with MDA programs. The search for alternative antibiotics is focused on compounds that could deliver high efficacy in shorter regimens, preferably a week, but possibly two weeks [109]. The approved antibiotics minocycline and rifampicin (Figure 10.8) have attracted attention as potential macrofilaricides. Minocycline, which shares a mechanism of action with doxycycline, appears to act more quickly, although

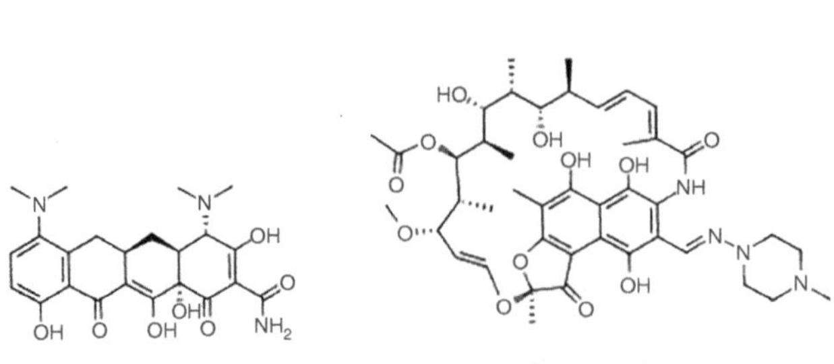


**Figure 10.8** Additional antibiotics for filaricidal usage.

still appears to require the same duration of treatment as doxycycline for full efficacy [110].

Rifampicin, an inhibitor of bacterial DNA-dependent RNA polymerase, exerts macrofilaricidal activity more rapidly than doxycycline in vitro and in animal models [111]. However, two- and four-week standard dosing regimens were not fully macrofilaricidal in onchocerciasis patients [112]. Experimental evidence and pharmacodynamic simulations suggest that a high-dose regimen of rifampicin could be safe and more effective as a macrofilaricide in an acceptably short course [113], but this has not yet been proven in human trials.

A few reports from humans and in animal models suggest that albendazole may contribute to efficacy if given in combination with antibiotics. Samples of *O. volvulus* obtained from patients treated with doxycycline for three weeks plus albendazole for three days showed modestly greater measures of efficacy than patients receiving doxycycline only, differences that seem to be unlikely to reflect full macrofilaricidal efficacy with the shorter course [113]. A study in an animal model showed that the addition of albendazole to courses of either minocycline or rifampicin led to faster macrofilaricidal efficacy than treatment with the antibiotics alone, suggestive of a possibly efficacious shorter course in humans [114]; how this finding can be reconciled with results from the study with doxycycline plus albendazole in humans [110] remains to be resolved. The mechanistic explanation for the interaction, if it is more than additive, remains to be determined.

Interest in this area has led to the identification of several new antibiotics that hold some promise for achieving high macrofilaricidal efficacy in acceptably short courses [111, 115, 116]. A derivative of tylosin (Figure 10.9), an antibiotic macrolide used in veterinary medicine, has advanced into development [117, 118]. Although mechanism of action studies have not been published, it is reasonable to assume that this derivative, TylaMac, acts like the prototype of the series as a bacteriostatic inhibitor of protein synthesis in *Wolbachia* through interactions with the 50S ribosomal subunit. Additional antibiotics with macrofilaricidal activity have been discovered in phenotypic screens; included are a number with unknown mechanisms of action as well as inhibitors of protein synthesis and of bacterial DNA-dependent RNA polymerase [111, 115, 116].

#### 10.7.7.2 Non-antibiotics

Three repurposed drugs are under investigation for direct macrofilaricidal activity. The most advanced is emodepside (Figure 10.10), a cyclooctadepsipeptide semi-

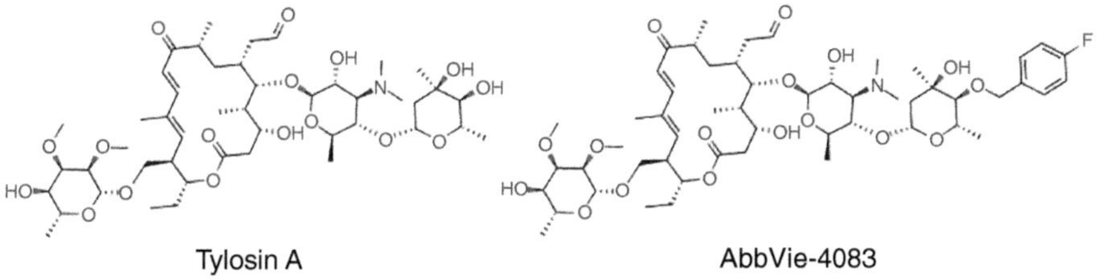


**Figure 10.9** Tylosin A and a semi-synthetic antibiotic (TylaMac).

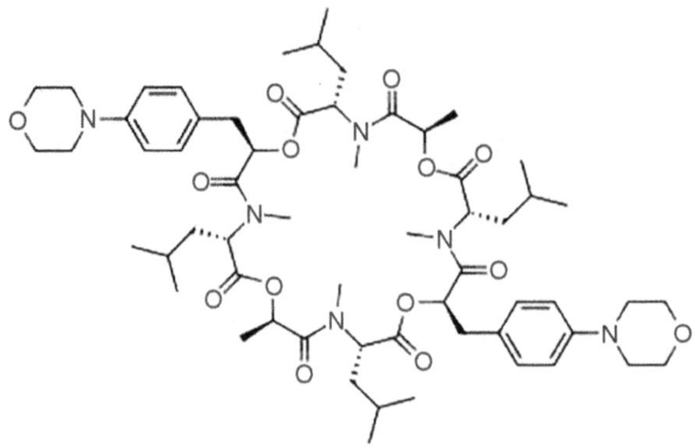


**Figure 10.10** The semi-synthetic anthelmintic cyclooctadepsipeptide emodepside.

synthetic natural product derived from a fungal fermentation. The primary anthelmintic target of emodepside is a nematode $Ca^{2+}$-activated $K^+$ channel, typified by the *C. elegans* channel encoded on the *slo-1* locus [119–122]. The channel is homologous to mammalian BK channels. Activation of this channel leads to the efflux of $K^+$ from cells in the nematode neuromuscular system, resulting in paralysis. Null *slo-1* mutants are highly resistant to emodepside, and expression of homologs from parasitic nematodes rescue emodepside sensitivity in these mutants [119].

Emodepside is used to treat gastrointestinal nematode infections in companion animals. Anthelmintics in this class exhibit potent, though species-dependent, antifilarial activity in animal models [123, 124], including activity against microfilariae and adult stages (macrofilaricidal and sterilizing effects). Emodepside is undergoing clinical development as a macrofilaricide for use in onchocerciasis through a partnership with DNDi, Bayer and Eisai. One concern in this regard is its activity against microfilariae; killing of these stages appears to cause significant adverse effects in dogs infected with *D. immitis* [125], and the drug is not licensed in the United States for use in dogs. Whether this activity will cause effects similar to DEC in humans infected with *O. volvulus* remains to be determined. Interestingly, medicinal chemistry efforts have led to the generation of emodepside analogs with considerably greater potency against L3 and L4 stages than microfilariae of *D. immitis* [125]; compounds with this property could have utility for heartworm prevention. The pharmacological basis for the stage-selective potency of the new derivatives is not known. That the prototype, emodepside, is equipotent against microfilariae and L4 larvae of *D. immitis* in culture suggests that neither differential SLO-1 abundance nor localization is responsible. The potency and efficacy of these compounds against adult filariae was not reported.

Auranofin (Figure 10.11), a gold salt, has been used for decades as a second-line drug for rheumatoid arthritis with a good safety record. It has many pharmacological activities in addition to immunomodulation and has attracted attention for several chemotherapeutic indications, including for viral, bacterial, and protozoal infections, and for cancer. Auranofin has macrofilaricidal activity in vitro and in

**Figure 10.11** The potential filaricide auranofin.

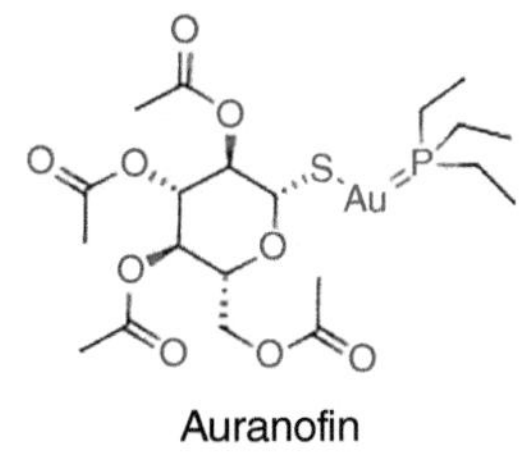


animal models of filariasis, although fairly prolonged dosing regimens are needed for full efficacy [126]. The parent drug has a very short plasma half-life, being essentially undetectable after oral administration and is present in an undefined form, possibly as the gold–phosphine complex bound to albumin. Measurements of gold indicate that this form has a very long residence time in the body [127]. The mechanism of antifilarial action of auranofin is imprecisely known; inhibition of the selenium-dependent enzyme thioredoxin reductase has been proposed as a general mechanism of action for multiple indications, and inhibition has been shown for *B. malayi* [126] but whether this is the only relevant target remains unresolved. A Phase I trial has been completed [127], with a focus on amebiasis, and the safety of the drug in this context has been confirmed. A Phase IIa follow-up study of auranofin is underway, again with the indication of GI protozoal infections.

Flubendazole and oxfendazole (Figure 10.12), which share a tubulin inhibition mechanism of action with albendazole, have been proposed as macrofilaricidal candidates. Flubendazole is approved for the treatment of human gastrointestinal nematode infections. This drug is very poorly bioavailable after oral dosing but has outstanding macrofilaricidal efficacy in multiple infection models, including human onchocerciasis patients, following parenteral administration, which produces a drug depot, leading to prolonged exposure to low concentrations [58]. Efforts to develop an orally bioavailable formulation with some efficacy succeeded, but the plasma concentrations reached had an unacceptable toxicity profile, and further development was halted [128]. Oxfendazole, which is approved for veterinary but not human use, is a prodrug of fenbendazole; oxfendazole has superior oral bioavailability and is a candidate under investigation for neurocysticercosis and other helminth infections in humans, including filariasis [129]. The drug has significant macrofilaricidal activity in some [130] but not all animal models of filariasis [131] following one to two weeks of oral dosing. A Phase I trial provided encouraging results for safety and pharmacokinetics in humans [132].

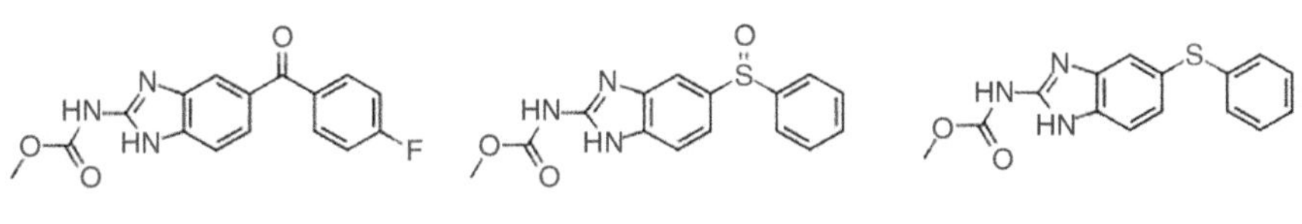


**Figure 10.12** Additional benzimidazoles under investigation as filaricides.

## 10.8 Conclusions and Priorities for Research

Despite the remarkable success of antifilarial chemotherapy in human and veterinary medicine, much remains to be learned about how these drugs work to prevent or eliminate filarial infections. As is the case for the human filarial pathogens, no convenient small animal model is available to support investigations into the basic pharmacology of heartworm preventatives or adulticides. The drugs used for filariases have enabled remarkable reductions in the incidence of these infections, and the potential of triple drug therapy to lead to elimination of LF as a public health concern, and its possible eradication, is highly encouraging. New macrofilaricides may be needed to achieve the same kind of results for onchocerciasis, and the worrisome appearance of ML-resistant populations of *D. immitis* may also require the discovery and development of new antifilarials. All such efforts can be bolstered by additional investment into the basic pharmacology of existing drugs used for these infections.

Among many important unresolved questions about antifilarial drugs are

1. How does DEC cause the removal of microfilariae from infected individuals? What are its targets in the host and/or parasite? Can the immunologically driven process be modulated to make DEC safe for use in onchocerciasis?
2. How does DEC sterilize adult LF parasites for prolonged periods after a single dose?
3. Can an animal model be developed to study the mechanism of macrofilaricidal activity of the triple drug combination? Does this effect, as is the case for many other therapies, vary with host and parasite species?
4. Does IVM paralyze the pharynx of adult filariids and does this contribute to its ability to sterilize these parasites?
5. What is the contribution of loss of sperm production in adult males to the sterilization observed in adult females in IVM-treated individuals?
6. Do the sterilizing effects of IVM and antibiotics converge on the inhibition of heme generation needed for de novo hemoprotein synthesis in developing embryos?
7. Does the anti-secretory effect of albendazole occur in filariae and does it contribute to the efficacy of the triple drug combination?
8. Does the ability of IVM to inhibit secretory processes in larval stages contribute to its efficacy as a preventative strategy for heartworms?
9. Lastly, can we develop an egg-to-egg culture system for a filarial parasite to enable genome-based research into drug targets and drug mechanisms of action?

## Acknowledgments

LT is holder of a grant from the Swiss National Science Foundation (PZ00P3_168080); TG is supported by a grant from the Natural Sciences and Engineering Council (NSERC) of Canada.

## References


1 Mäser, P. (2022). Filariae as organisms. In: *Advances in Control of Heartworm and Human Filariases* (ed. R. Kaminsky and T.G. Geary), Chapter 2. Weinheim: Wiley-VCH.

2 Hawking, F. (1978). Diethylcarbamazine: a review of the literature with special reference to its pharmacodynamics, toxicity, and use in the therapy of onchocerciasis and other filarial infections. World Health Organization. WHO/ONCHO/78.142. https://apps.who.int/iris/handle/10665/70735.

3 Campbell, W.C., Fisher, M.H., Stapley, E.O. et al. (1983). Ivermectin: a potent new antiparasitic agent. *Science* 221: 823–828.

4 Ottesen, E.A. and Campbell, W.C. (1994). Ivermectin in human medicine. *J. Antimicrob. Chemother.* 34: 195–203.

5 Boatin, B. (2008). The onchocerciasis control programme in West Africa (OCP). *Ann. Trop. Med. Parasitol.* 102 (suppl): 13–17.

6 Mackenzie, C.D. and Kron, M.A. (1985). Diethylcarbamazine: a review of its action in onchocerciasis, lymphatic filariasis, and inflammation. *Trop. Dis. Bull.* 83: R1–R34.

7 Higazi, T., Geary, T.G., and Mackenzie, C.D. (2014). Chemotherapy in the treatment, control and elimination of human onchocerciasis. *Res. Rep. Trop.l Med.* 5: 77–93.

8 Bockarie, M.J., Pederson, E.M., White, G.B., and Michael, E. (2009). Role of vector control in the Global Program to Eliminate Lymphatic Filariasis. *Annu. Rev. Entomol.* 54: 469–487.

9 Ottesen, E.A. (1985). Efficacy of diethylcarbamazine in eradicating infection with lymphatic-dwelling filariae in humans. *Rev. Infect. Dis.* 7: 341–356.

10 Guerrero, J., McCall, J.W., and Genchi, C. (2002). The use of macrocyclic lactones in the control and prevention of heartworm and other parasites in dogs and cats. In: *Macrocyclic Lactones in Antiparasitic Therapy* (ed. R.S. Rew and J. Vercruysse), 353–369. Oxon: CAB International.

11 Geary, T.G. (2005). Ivermectin 20 years on: maturation of a wonder drug. *Trends Parasitol.* 21: 530–532.

12 Cupp, E.W., Sauerbrey, M., and Richards, F. (2011). Elimination of human onchocerciasis: history of progress and current feasibility using ivermectin (Mectizan) monotherapy. *Acta Trop.* 120 (Suppl 1): S100–S108.

13 Sauerbrey, M., Rakers, L.J., and Richards, F.O. (2018). Progress toward elimination of onchocerciasis in the Americas. *Int. Health* 10 (suppl 1): i71–i78.

14 Specht, S., Kamgno, J., and Geary, T.G. (2022). Antifilarial chemotherapy: current options for humans. In: *Advances in Control of Heartworm and Human Filariases* (ed. R. Kaminsky and T.G. Geary), Chapter 7. Weinheim: Wiley-VCH.

15 Chippaux, J.P., Boussinesq, M., Gardon, J. et al. (1996). Severe adverse reaction risks during mass treatment with ivermectin in loiasis endemic areas. *Parasitol. Today* 12: 448–450.

16 Boussinesq, M. (2006). Loiasis. *Ann. Trop. Med. Parasitol.* 100: 715–731.

**17** Mackenzie, C.D., Geary, T.G., and Gerlach, J.A. (2003). Possible pathogenic pathways in the adverse clinical events seen following ivermectin administration to onchocerciasis patients. *Filarial J.* 2 (Suppl 1): 5–14.

**18** Hoerauf, A. (2006). New strategies to combat filariasis. *Expert Rev. Anti-Infective Ther.* 4: 211–222.

**19** Hoerauf, A. (2008). Filariasis: new drugs and new opportunities for lymphatic filariasis and onchocerciasis. *Curr. Opin. Infect. Dis.* 21: 673–681.

**20** Wanji, S., Tendongfor, N., Nji, T. et al. (2009). Community-directed delivery of doxycycline for the treatment of onchocerciasis in areas of co-endemicity with loiasis in Cameroon. *Parasites Vectors* 2: 39.

**21** Gyapong, J.O., Owusu, I.O., da Costa Vroom, F.B. et al. (2018). Elimination of lymphatic filariasis: current perspectives on mass drug administration. *Res. Rep. Trop.l Med.* 9: 25–33.

**22** Ichimori, K., King, J.D., Engels, D. et al. (2014). Global programme to eliminate lymphatic filariasis: the processes underlying programme success. *PLoS Negl.Trop. Dis.* 8: e0003328.

**23** Thomsen, E.K., Sanuku, N., Baea, M. et al. (2016). Efficacy, safety, and pharmacokinetics of coadministered diethylcarbamazine, albendazole, and ivermectin for treatment of Bancroftian filariasis. *Clin. Infect. Dis.* 62: 334–341.

**24** King, C.L., Suamani, J., Sanuku, N. et al. (2018). A trial of a triple-drug treatment for lymphatic filariasis. *N. Engl. J. Med.* 379: 1801–1810.

**25** Bjerum, C.M., Ouattara, A.F., Abdoulaye, M. et al. (2019). Efficacy and safety of a single dose of ivermectin, diethylcarbamazine and albendazole for treatment of lymphatic filariasis in Côte d'Ivoire: an open-label, randomized, controlled trial. *Clin. Infect. Dis.* pii: ciz1050. doi: https://doi.org/10.1093/cid/ciz1050.

**26** Edi, C., Bjerum, C.M., Ouattara, A.F. et al. (2019). Pharmacokinetics, safety, and efficacy of a single co-administered dose of diethylcarbamazine, albendazole and ivermectin in adults with and without *Wuchereria bancrofti* infection in Côte d'Ivoire. *PLoS Negl.Trop. Dis.* 13: e0007325.

**27** King, C.L., Weil, G.J., and Kazura, J.W. (2020). Single-dose triple-drug therapy for *Wuchereria bancrofti* - 5-year follow-up. *N. Engl. J. Med.* 382: 1956–1957.

**28** Weil, G.J., Bogus, J., Christian, M. et al. (2019). The safety of double- and triple-drug community mass drug administration for lymphatic filariasis: a multicenter, open-label, cluster-randomized study. *PLoS Med.* 16: e1002839.

**29** Hardy, M., Samuela, J., Kama, M. et al. (2020). The safety of combined triple drug therapy with ivermectin, diethylcarbamazine and albendazole in the neglected tropical diseases co-endemic setting of Fiji: a cluster randomised trial. *PLoS Negl.Trop. Dis.* 14: e0008106.

**30** WHO (2017). Guideline: Alternative mass drug administration regimens to eliminate lymphatic filariasis. Contract No.: WHO/HTM/NTD/PCT/2017.0. Geneva.

**31** Debrah, A.Y., Mand, S., Marfo-Debrekyei, Y. et al. (2007). Macrofilaricidal effect of 4 weeks of treatment with doxycycline on *Wuchereria bancrofti*. *Trop. Med. Int. Health* 12: 1433–1441.

**32** Bowman, D.D. and Atkins, C.E. (2009). Heartworm biology, treatment, and control. *Vet. Clin. N. Am.: Small Anim. Pract.* 39: 1127–1158.

**33** Ketsis, J. and Epe, C. (2022). Antifilarial chemotherapy: current options in veterinary medicine. In: *Advances in Control of Heartworm and Human Filariases* (ed. R. Kaminsky and T.G. Geary), Chapter 8. Weinheim: Wiley-VCH.

**34** Prichard, R.K. (2022). Drug resistance in filariae. In: *Advances in Control of Heartworm and Human Filariases* (ed. R. Kaminsky and T.G. Geary), Chapter 11. Weinheim: Wiley-VCH.

**35** Prichard, R.K. and Geary, T.G. (2019). Perspectives on the utility of moxidectin for the control of parasitic nematodes in the face of developing anthelmintic resistance. *Int. J. Parasitol. Drugs and Drug Resist.* 10: 69–83.

**36** McCall, J.W., Genchi, C., Kramer, L. et al. (2008). Heartworm and *Wolbachia*: therapeutic implications. *Vet. Parasitol.* 158: 204–214.

**37** Kramer, L., Grandi, G., Passeri, B. et al. (2011). Evaluation of lung pathology in *Dirofilaria immitis*-experimentally infected dogs treated with doxycycline or a combination of doxycycline and ivermectin before administration of melarsomine dihydrochloride. *Vet. Parasitol.* 176: 357–360.

**38** McCall, J.W., Kramer, L., Genchi, C. et al. (2014). Effects of doxycycline on heartworm embryogenesis, transmission, circulating microfilaria, and adult worms in microfilaremic dogs. *Vet. Parasitol.* 206: 5–13.

**39** Capelli, G., Genchi, C., Baneth, G. et al. (2018). Recent advances on *Dirofilaria repens* in dogs and humans in Europe. *Parasites Vectors* 11: 663.

**40** Klion, A.D., Massougbodji, A., Horton, J. et al. (1993). Albendazole in human loiasis: results of a double-blind, placebo-controlled trial. *J. Infect. Dis.* 168: 202–206.

**41** Kamgno, J., Nguipdop-Djomo, P., Gounoue, R. et al. (2016). Effect of two or six doses 800 mg of albendazole every two months on *Loa loa* microfilaraemia: A double blind, randomized, placebo-controlled trial. *PLoS Negl.Trop. Dis.* 10: e0004492.

**42** Geary, T.G. and Mackenzie, C.D. (2011). Progress and challenges in the discovery of macrofilaricidal drugs. *Expert Rev. Anti-infective Ther.* 9: 681–695.

**43** Campbell, W.C. (1982). Efficacy of the avermectins against filarial parasites: a short review. *Vet. Res. Commun.* 5: 251–262.

**44** Maizels, R. and Denham, D. (1992). Diethylcarbamazine (DEC): immunopharmacological interactions of an anti-filarial drug. *Parasitology* 105 (Suppl. 1): S49–S60.

**45** Peixoto, C.A. and Silva, B.S. (2014). Anti-inflammatory effects of diethylcarbamazine: a review. *Eur. J. Pharmacol.* 734: 35–41.

**46** McGarry, H.F., Plant, L.D., and Taylor, M.J. (2005). Diethylcarbamazine activity against *Brugia malayi* microfilariae is dependent on inducible nitric-oxide synthase and the cyclooxygenase pathway. *Filaria J.* 4: 4.

**47** Kanesa-thasan, N., Douglas, J.G., and Kazura, J.W. (1991). Diethylcarbamazine inhibits endothelial and microfilarial prostanoid metabolism in vitro. *Mol. Biochem. Parasitol.* 49: 11–19.

**48** Tiwary, E., Hu, M., Miller, M.A., and Prasain, J.K. (2019). Signature profile of cyclooxygenase-independent F2 series prostaglandins in *C. elegans* and their role in sperm motility. *Sci. Rep.* 9: 11750.

**49** Liu, L.X., Serhan, C.N., and Weller, P.F. (1990). Intravascular filarial parasites elaborate cyclooxygenase-derived eicosanoids. *J. Exp. Med.* 172: 993–996.

**50** Verma, S., Kashyap, S.S., Robertson, A.P., and Martin, R.J. (2020). Diethylcarbamazine activates TRP channels including TRP-2 in filaria, *Brugia malayi. Commun. Biol.* 3: 398.

**51** Horton, J. (2002). Albendazole: a broad spectrum anthelminthic for treatment of individuals and populations. *Curr. Opin. Infect. Dis.* 15: 599–608.

**52** Ottesen, E.A., Ismail, M.M., and Horton, J. (1999). The role of albendazole in programmes to eliminate lymphatic filariasis. *Parasitol. Today* 15: 382–386.

**53** Lacey, E. (1988). The role of the cytoskeletal protein, tubulin, in the mode of action and mechanism of drug resistance to benzimidazoles. *Int. J. Parasitol.* 18: 885–936.

**54** Robinson, M.W., McFerran, N., Trudgett, A. et al. (2004). A possible model of benzimidazole binding to beta-tubulin disclosed by invoking an inter-domain movement. *J. Mol. Graphics Modell.* 23: 275–284.

**55** Aguayo-Ortiz, R., Méndez-Lucio, O., Medina-Franco, J.L. et al. (2013). Towards the identification of the binding site of benzimidazoles to β-tubulin of *Trichinella spiralis*: insights from computational and experimental data. *J. Mol. Graphics Modell.* 41: 12–19.

**56** Geary, T.G., Nulf, S.C., Alexander-Bowman, S.J. et al. (1998). Cloning and characterization of cDNAs encoding beta-tubulin from *Dirofilaria immitis* and *Onchocerca volvulus. J. Parasitol.* 84: 356–360.

**57** Thompson, D.P. and Geary, T.G. (1995). The structure and function of helminth surfaces. In: *Biochemistry and Molecular Biology of Parasites* (ed. M. Müller and J.J. Marr), 203–232. London: Academic Press.

**58** Mackenzie, C.D. and Geary, T.G. (2011). Flubendazole, a potentially valuable macrofilaricide for lymphatic filariasis and onchocerciasis field programs. *Expert Rev. Anti-infective Ther.* 9: 497–501.

**59** Fischer, C., Ibiricu Urriza, I., Bulman, C.A. et al. (2019). Efficacy of subcutaneous doses and a new oral amorphous solid dispersion formulation of flubendazole on male jirds (*Meriones unguiculatus*) infected with the filarial nematode *Brugia pahangi. PLoS Negl.Trop. Dis.* 13: e0006787.

**60** Macfarlane, C.L., Budhathoki, S.S., Johnson, S. et al. (2019). Albendazole alone or in combination with microfilaricidal drugs for lymphatic filariasis. *Cochrane Database Syst. Rev.* (1): CD003753.

**61** Watts, S.D., Rapson, E.B., Atkins, A.M., and Lee, D.L. (1982). Inhibition of acetylcholinesterase secretion from *Nippostrongylus brasiliensis* by benzimidazole anthelmintics. *Biochem. Pharmacol.* 31: 3035–3040.

**62** Moreno, Y., Nabhan, J.F., Solomon, J. et al. (2010). Ivermectin disrupts the function of the excretory-secretory apparatus in microfilariae of *Brugia malayi. Proc. Natl. Acad. Sci. U.S.A.* 107: 20120–20125.

**63** Prichard, R., Ménez, C., and Lespine, A. (2012). Moxidectin and the avermectins: consanguinity but not identity. *Int. J. Parasitol.: Drugs and Drug Resist.* 2: 134–153.

**64** Shoop, W. and Soll, M. (2002). Ivermectin, abamectin and eprinomectin. In: *Macrocyclic Lactones in Antiparasitic Therapy* (ed. R.S. Rew and J. Vercruysse), 1–29. Oxon: CAB International.

**65** Conder, G.A. and Baker, W.J. (2002). Chemistry, pharmacology and safety: doramectin and selamectin. In: *Macrocyclic Lactones in Antiparasitic Therapy* (ed. R.S. Rew and J. Vercruysse), 30–50. Oxon: CAB International.

**66** Jung, M., Saito, A., Buescher, G. et al. (2002). Chemistry, pharmacology and safety: milbemycin oxime. In: *Macrocyclic Lactones in Antiparasitic Therapy* (ed. R.S. Rew and J. Vercruysse), 51–74. Oxon: CAB International.

**67** Rock, D.W., DeLay, R.L., and Gliddon, M.J. (2002). Chemistry, pharmacology and safety: moxidectin. In: *Macrocyclic Lactones in Antiparasitic Therapy* (ed. R.S. Rew and J. Vercruysse), 75–96. Oxon: CAB International.

**68** de Moraes, J. and Geary, T.G. (2020). FDA-approved antiparasitic drugs in the 21st century: a success for helminthiasis? *Trends Parasitol.* 36: 573–575.

**69** Opoku, N.O., Bakajika, D.K., Kanza, E.M. et al. (2018). Single dose moxidectin versus ivermectin for Onchocerca volvulus infection in Ghana, Liberia, and the Democratic Republic of the Congo: a randomised, controlled, double-blind phase 3 trial. *Lancet* 392: 1207–1216.

**70** Thompson, A.J., Lester, H.A., and Lummis, S.C.R. (2010). The structural basis of function in Cys-loop receptors. *Q. Rev. Biophys.* 43: 449–499.

**71** Cully, D.F., Wilkinson, H., Vassilatis, D.K. et al. (1996). Molecular biology and electrophysiology of glutamate-gated chloride channels of invertebrates. *Parasitology* 113 (Suppl): S191–S200.

**72** Martin, R.J., Robertson, A.P., and Wolstenholme, A.J. (2002). Mode of action of the macrocyclic lactones. In: *Macrocyclic Lactones in Antiparasitic Therapy* (ed. R.S. Rew and J. Vercruysse), 125–140. Oxon: CAB International.

**73** Wolstenholme, A.J. and Rogers, A.T. (2005). Glutamate-gated chloride channels and the mode of action of the avermectin/milbemycin anthelmintics. *Parasitology* 131 (Suppl): S85–S95.

**74** Geary, T.G. and Moreno, Y. (2012). Macrocyclic lactone anthelmintics: spectrum of activity and mechanism of action. *Curr. Pharm. Biotechnol.* 13: 866–872.

**75** Dent, J.A. (2010). The evolution of pentameric ligand-gated ion channels. *Adv. Exp. Med. Biol.* 683: 11–23.

**76** Laing, R., Gillan, V., and Devaney, E. (2017). Ivermectin - old drug, new tricks? *Trends Parasitol.* 33: 463–472.

**77** Hibbs, R.E. and Gouaux, E. (2011). Principles of activation and permeation in an anion-selective Cys-loop receptor. *Nature* 474: 54–60.

**78** Lynagh, T. and Lynch, J.W. (2012). Molecular mechanisms of Cys-loop ion channel receptor modulation by ivermectin. *Front. Mol. Neurosci.* 5: 60.

**79** Schaeffer, J.M. and Haines, H.W. (1989). Avermectin binding in *Caenorhabditis elegans*. A two-state model for the avermectin binding site. *Biochem. Pharmacol.* 38: 2329–2338.

**80** Wolstenholme, A.J., Maclean, M.J., Coates, R. et al. (2016). How do the macrocyclic lactones kill filarial nematode larvae? *Invertebr. Neurosci.* 16: 7.

**81** Harischandra, H., Yuan, W., Loghry, H.J. et al. (2018). Profiling extracellular vesicle release by the filarial nematode *Brugia malayi* reveals sex-specific differences in cargo and a sensitivity to ivermectin. *PLoS Negl.Trop. Dis.* 12: e0006438.

**82** Vatta, A.F., Dzimianski, M., Storey, B.E. et al. (2014). Ivermectin-dependent attachment of neutrophils and peripheral blood mononuclear cells to *Dirofilaria immitis* microfilariae in vitro. *Vet. Parasitol.* 206: 38–42.

**83** Berrafato, T., Coates, R., Reaves, B.J. et al. (2019). Macrocyclic lactone anthelmintic-induced leukocyte binding to *Dirofilaria immitis* microfilariae: influence of the drug resistance status of the parasite. *Int. J. Parasitol.: Drugs and Drug Resist.* 10: 45–50.

**84** Lok, J., Harpaz, T., and Knight, D. (1988). Abnormal patterns of embryogenesis in *Dirofilaria immitis* treated with ivermectin. *J. Helminthol.* 62: 175–180.

**85** Greene, B.M., Brown, K.R., and Taylor, H.R. (1989). Use of ivermectin in humans. In: *Ivermectin and Abamectin* (ed. W.C. Campbell), 311–323. New York: Springer-Verlag.

**86** Lok, J.B., Knight, D.H., Selavka, C.M. et al. (1995). Studies of reproductive competence in male *Dirofilaria immitis* treated with milbemycin oxime. *Trop. Med. Parasitol.* 46: 235–240.

**87** McCall, J.W. (2005). The safety-net story about macrocyclic lactone heartworm preventives: a review, an update, and recommendations. *Vet. Parasitol.* 133: 197–206.

**88** Walker, M., Pion, S.D.S., Fang, H. et al. (2017). Macrofilaricidal efficacy of repeated doses of ivermectin for the treatment of river blindness. *Clin. Infect. Dis.* 65: 2026–2034.

**89** Taylor, H.R., Trpis, M., Cupp, E.W. et al. (1988). Ivermectin prophylaxis against experimental *Onchocerca volvulus* infection in chimpanzees. *Am. J. Trop. Med. Hyg.* 39: 86–90.

**90** Bazzocchi, C., Mortarino, M., Grandi, G. et al. (2008). Combined ivermectin and doxycycline treatment has microfilaricidal and adulticidal activity against *Dirofilaria immitis* in experimentally infected dogs. *Int. J. Parasitol.* 38: 1401–1410.

**91** Albiez, E.J., Walter, G., Kaiser, A. et al. (1988). Histological examination of onchocercomata after therapy with ivermectin. *Trop. Med. Parasitol.* 39: 93–99.

**92** Mackenzie, C.D., Behan-Braman, A., Hauptman, J., and Geary, T. (2017). Assessing the viability and degeneration of the medically important filarial nematodes. In: *Nematology* (ed. M.M. Shah), 101–120. Rijeka: InTech Open Publishing.

**93** Ballesteros, C., Tritten, L., O'Neill, M. et al. (2016). Deciphering the loss of fertility in ivermectin-treated *Brugia malayi* females *in vitro*: a transcriptomic approach. *PLoS Negl.Trop. Dis.* 10: e0004929.

**94** Chavasse, D.C., Post, R.J., Davies, J.B., and Whitworth, J.A. (1993). Absence of sperm from the seminal receptacle of female *Onchocerca volvulus* following multiple doses of ivermectin. *Trop. Med. Parasitol.* 44: 155–158.

**95** Li, B.W., Rush, A.C., and Weil, G.J. (2014). High level expression of a glutamate-gated chloride channel gene in reproductive tissues of *Brugia malayi* may explain the sterilizing effect of ivermectin on filarial worms. *Int. J. Parasitol.: Drugs and Drug Resist.* 4: 71–76.
**96** Walker, M., Specht, S., Churcher, T.S. et al. (2015). Therapeutic efficacy and macrofilaricidal activity of doxycycline for the treatment of river blindness. *Clin. Infect. Dis.* 60: 1199–1207.
**97** Darby, A.C., Armstrong, S.D., Bah, G.S. et al. (2012). Analysis of gene expression from the *Wolbachia* genome of a filarial nematode supports both metabolic and defensive roles within the symbiosis. *Genome Res.* 22: 2467–2477.
**98** Bouchery, T., Lefoulon, E., Karadjian, G. et al. (2013). The symbiotic role of *Wolbachia* in Onchocercidae and its impact on filariasis. *Clin. Microbiol. Infect.* 19: 131–140.
**99** Luck, A.N., Anderson, K.G., McClung, C.M. et al. (2015). Tissue-specific transcriptomics and proteomics of a filarial nematode and its *Wolbachia* endosymbiont. *BMC Genomics* 16: 920.
**100** Chukwudi, C.U. (2016). rRNA binding sites and the molecular mechanism of action of the tetracyclines. *Antimicrob. Agents Chemother.* 60: 4433–4441.
**101** Landmann, F., Voronin, D., Sullivan, W., and Taylor, M.J. (2011). Anti-filarial activity of antibiotic therapy is due to extensive apoptosis after *Wolbachia* depletion from filarial nematodes. *PLoS Pathog.* 7: e1002351.
**102** McCall, J.W., Kramer, L., Genchi, C. et al. (2011). Effects of doxycycline on early infections of *Dirofilaria immitis* in dogs. *Vet. Parasitol.* 176: 361–367.
**103** Fairlamb, A.H., Henderson, G.B., and Cerami, A. (1989). Trypanothione is the primary target for arsenical drugs against African trypanosomes. *Proc. Natl. Acad. Sci. U.S.A.* 86: 2607–2611.
**104** Baker, N., de Koning, H.P., Mäser, P., and Horn, D. (2013). Drug resistance in African trypanosomiasis: the melarsoprol and pentamidine story. *Trends Parasitol.* 29: 110–118.
**105** Giordani, F., Morrison, L.J., Rowan, T.G. et al. (2016). The animal trypanosomiases and their chemotherapy: a review. *Parasitology* 143: 1862–1889.
**106** Moulia-Pelat, J.P., Glaziou, P., Weil, G.J. et al. (1995). Combination ivermectin plus diethylcarbamazine, a new effective tool for control of lymphatic filariasis. *Trop. Med. Parasitol.* 46: 9–12.
**107** Bockarie, M.J., Tisch, D.J., Kastens, W. et al. (2002). Mass treatment to eliminate filariasis in Papua New Guinea. *N. Engl. J. Med.* 347: 1841–1848.
**108** Fischer, P.U., King, C.L., Jacobson, J.A., and Weil, G.J. (2017). Potential value of triple drug therapy with ivermectin, diethylcarbamazine, and albendazole (IDA) to accelerate elimination of lymphatic filariasis and onchocerciasis in Africa. *PLoS Negl.Trop. Dis.* 11: e0005163.
**109** Taylor, M.J., Hoerauf, A., Townson, S. et al. (2014). Anti-*Wolbachia* drug discovery and development: safe macrofilaricides for onchocerciasis and lymphatic filariasis. *Parasitology* 141: 119–127.
**110** Klarmann-Schulz, U., Specht, S., Debrah, A.Y. et al. (2017). Comparison of doxycycline, minocycline, doxycycline plus albendazole and albendazole alone

in their efficacy against onchocerciasis in a randomized, open-label, pilot trial. *PLoS Negl.Trop. Dis.* 11: e0005156.

**111** Bakowski, M.A. and McNamara, C.W. (2019). Advances in antiwolbachial drug discovery for the treatment of parasitic filarial worm infections. *Trop. Med. Infect. Dis.* 4: 108.

**112** Specht, S., Mand, S., Marfo-Debrekyei, Y. et al. (2008). Efficacy of 2- and 4-week rifampicin treatment on the *Wolbachia* of *Onchocerca volvulus*. *Parasitol. Res.* 103: 1303–1309.

**113** Turner, J.D., Sharma, R., Al Jayoussi, G. et al. (2017). Albendazole and antibiotics synergize to deliver short-course anti-*Wolbachia* curative treatments in preclinical models of filariasis. *Proc. Natl. Acad. Sci. U.S.A.* 114: E9712–E9721.

**114** Aljayyoussi, G., Tyrer, H.E., Ford, L. et al. (2017). Short-course, high-dose rifampicin achieves *Wolbachia* depletion predictive of curative outcomes in preclinical models of lymphatic filariasis and onchocerciasis. *Sci. Rep.* 7: 210.

**115** Hawryluk, N. (2020). Macrofilaricides: an unmet medical need for filarial diseases. *ACS Infect. Dis.* 6: 662–671.

**116** Hawryluk, N. (2022). The antifilarial drug pipeline. In: *Advances in Control of Heartworm and Human Filariases* (ed. R. Kaminsky and T.G. Geary), Chapter 18. Weinheim: Wiley-VCH.

**117** von Geldern, T.W., Morton, H.E., Clark, R.F. et al. (2019). Discovery of ABBV-4083, a novel analog of Tylosin A that has potent anti-*Wolbachia* and anti-filarial activity. *PLoS Negl.Trop. Dis.* 13: e0007159.

**118** Taylor, M.J., von Geldern, T.W., Ford, L. et al. (2019). Preclinical development of an oral anti-*Wolbachia* macrolide drug for the treatment of lymphatic filariasis and onchocerciasis. *Sci. Transl. Med.* 11: eaau2086.

**119** Welz, C., Krüger, N., Schniederjans, M. et al. (2011). SLO-1-channels of parasitic nematodes reconstitute locomotor behaviour and emodepside sensitivity in *Caenorhabditis elegans slo-1* loss of function mutants. *PLoS Pathog.* 7: e1001330.

**120** Krücken, J., Harder, A., Jeschke, P. et al. (2012). Anthelmintic cyclooctadepsipeptides: complex in structure and mode of action. *Trends Parasitol.* 28: 385–394.

**121** Martin, R.J., Buxton, S.K., Neveu, C. et al. (2012). Emodepside and SLO-1 potassium channels: a review. *Exp. Parasitol.* 132: 40–46.

**122** Kulke, D., von Samson-Himmelstjerna, G., Miltsch, S.M. et al. (2014). Characterization of the Ca2+-gated and voltage-dependent K+-channel Slo-1 of nematodes and its interaction with emodepside. *PLoS Negl.Trop. Dis.* 8: e0003401.

**123** Zahner, H., Taubert, A., Harder, A., and von Samson-Himmelstjerna, G. (2001). Filaricidal efficacy of anthelmintically active cyclodepsipeptides. *Int. J. Parasitol.* 31: 1515–1522.

**124** Harder, A., Schmitt-Wrede, H.P., Krücken, J. et al. (2003). Cyclooctadepsipeptides–an anthelmintically active class of compounds exhibiting a novel mode of action. *Int. J. Antimicrob. Agents* 22: 318–331.

**125** Curtis, M.P., Sheehan, S.M., Kyne, G.M. et al. (2019). Endoparasitic depsipeptides. WO2019/108591 A1. Publication date 06 June 2019.

**126** Bulman, C.A., Bidlow, C.M., Lustigman, S. et al. (2015). Repurposing auranofin as a lead candidate for treatment of lymphatic filariasis and onchocerciasis. *PLoS Negl.Trop. Dis.* 9: e0003534.

**127** Capparelli, E.V., Bricker-Ford, R., Rogers, M.J. et al. (2017). Phase I clinical trial results of auranofin, a novel antiparasitic agent. *Antimicrob. Agents Chemother.* 61: e01947–e01916.

**128** Geary, T.G., Mackenzie, C.D., and Silber, S.A. (2019). Flubendazole as a macrofilaricide: history and background. *PLoS Negl.Trop. Dis.* 13: e0006436.

**129** Gonzalez, A.E., Codd, E.E., Horton, J. et al. (2019). Oxfendazole: a promising agent for the treatment and control of helminth infections in humans. *Expert Rev. Anti-infective Ther.* 17: 51–56.

**130** Zahner, H. and Schares, G. (1993). Experimental chemotherapy of filariasis: comparative evaluation of the efficacy of filaricidal compounds in *Mastomys coucha* infected with *Litomosoides carinii, Acanthocheilonema viteae, Brugia malayi* and *B. pahangi*. *Acta Trop.* 52: 221–266.

**131** Colella, V., Maia, C., Pereira, A. et al. (2018). Evaluation of oxfendazole in the treatment of zoonotic *Onchocerca lupi* infection in dogs. *PLoS Negl.Trop. Dis.* 12: e0006218.

**132** An, G., Murry, D.J., Gajurel, K. et al. (2019). Pharmacokinetics, safety, and tolerability of oxfendazole in healthy volunteers: a randomized, placebo-controlled first-in-human single-dose escalation study. *Antimicrob. Agents Chemother.* 63: e02255–e02218.

# 11

# Drug Resistance in Filariae

*Roger Prichard**

*Institute of Parasitology, McGill University, 21111 Lakeshore Road, Sainte Anne-de-Bellevue, H9X3V9, Canada*

**Abstract**

Macrocyclic lactones (MLs) are antifilarial drugs. Resistance occurs in *Dirofilaria immitis* to the MLs, ivermectin (IVM), milbemycin oxime, selamectin, and moxidectin (MOX). To date, resistance in *D. immitis* has arisen in the southern United States. Factors that may have contributed to selection of ML resistance are (i) high-intensity local transmission, (ii) a degree of inbreeding of the parasite, the fact that MLs act not only on the developing L3/L4 stages, but (iii) also have microfilaricidal properties and (iv) inhibit fecundity in adult parasites, (v) year-round monthly treatment, which (vi) may allow periods when drug concentrations are falling and insufficient, against less susceptible parasites to exert these antiparasitic effects. ML heartworm preventives have been used for over 30 years. This long-term selection pressure coupled with the heartworm life cycle make it not surprising that resistance has arisen in regions where transmission and drug treatments are highest. Tests to confirm ML resistance include a microfilariae reduction test and genetic markers. More research is needed to understand the mechanisms of resistance, factors that could reduce its spread or overcome resistance, as well as monitoring resistance.

IVM has been used to treat onchocerciasis for over 30 years, and in sub-Saharan Africa, to treat lymphatic filariasis (LF) in humans. Recently, MOX was registered for onchocerciasis. IVM and MOX remove microfilariae and suppress fecundity of adult *Onchocerca volvulus* for several months. There are a number of reports of a decrease in the anti-fecundity effect of IVM in some countries, and recently evidence of a reduction in the microfilaricidal effect of IVM. These reductions in the anti-fecundity and microfilaricidal effects of IVM indicate developing IVM resistance in *O. volvulus*.

Diethylcarbamazine (DEC) is widely used for LF. There have been some reports of poor responses to DEC treatment, but no concrete evidence of DEC resistance. Albendazole (ABZ) is used in combination with DEC or IVM, and more recently in a triple combination of these anthelmintics, to enhance the anti-fecundity effect on LF. There is no

*Corresponding author.

*Human and Animal Filariases: Landscape, Challenges, and Control*, First Edition.
Edited by Ronald Kaminsky and Timothy G. Geary.

evidence of resistance to ABZ in LF. However, single-nucleotide polymorphisms that confer ABZ resistance in other nematodes have been found in *Wuchereria bancrofti*. Doxycycline has been used against filariae in situations in which ML resistance is a concern or where there is a risk of severe adverse events after IVM use in people with high burdens of *Loa loa*. Doxycycline and related tetracyclines acts on the symbiotic *Wolbachia* in most filarial parasites, and the elimination of these bacteria impairs the viability of their filarial hosts. Antibiotic resistance to doxycycline can be selected, but so far has not been recorded when it is used as an anti-filarial treatment.

## 11.1 Introduction

Filarial nematode parasites are treated with a small number of anthelmintic drugs, which are reviewed in more detail in other chapters of this book [1–3]. This chapter deals more specifically with the evidence that their use in human and veterinary settings has led to the selection of drug-resistant populations of filarial parasites. Anthelmintic resistance is well-known phenotypically in the treatment of many nematode infections in veterinary medicine [4] and is of increasing concern for the treatment of human helminthiases [5]. For this chapter, "anthelmintic resistance" is defined as a heritable trait that enables persistent survival of a parasite population in the presence of normally effective drug concentrations or dosing regimens. Many explanations for inefficacy of treatments can be relevant in addition to resistance, the demonstration of which requires experimental evidence from the field or the laboratory.

## 11.2 Diethylcarbamazine

Diethylcarbamazine (DEC) was discovered in 1947 [6]. It is effective against microfilariae of *Wuchereria bancrofti*, *Brugia malayi*, and *Brugia timori*, which cause lymphatic filariasis (LF) in humans, and also has some adulticidal and embryo static effects against these parasites. While it shows similar effectiveness against *Onchocerca volvulus*, it should not be used in people infected with this parasite as it causes a severe pathology known as the Mazzotti reaction which can be life-threatening. DEC is also effective against *Loa loa* infection. In animals, DEC has been successfully used to prevent heartworm infection caused by *Dirofilaria immitis* but is little used these days for this indication because it requires daily treatment during the mosquito transmission season. DEC has been a mainstay of LF treatment and control programs for many years. While there are only a few reports of apparently reduced or variable efficacy of DEC against *W. bancrofti* [7, 8], there is no conclusive evidence that genetically determined resistance to DEC has affected the success of these programs.

## 11.3 Macrocyclic Lactones

A great advance was made in the treatment of filarial infections in animals and humans when it was discovered that ivermectin (IVM) was larvicidal and microfilaricidal against many filarial parasites. Initial studies using IVM in *O. volvulus*-infected people were conducted in 1982 [9, 10]; it was safe and effective in removing microfilariae and subsequently found to sterilize adult worms for three to six months [11]. In 1985, IVM was approved for use in humans at a dose of 150 μg/kg for onchocerciasis [12], as Mectizan® for oral administration. Approval was sought for registration of Heartgard-30® as an oral heartworm preventive in dogs in 1987, targeting developing L3/L4 larvae, with an IVM dose rate of 6–12 μg/kg (NADA 138-412, April 7, 1987). It should be noted that, consistent with its use in onchocerciasis against microfilarae, IVM at 50 μg/kg was markedly microfilaricidal in dogs with a patent *D. immitis* infection [13]. However, it was not registered for this indication in dogs.

Other anthelmintics in the macrocyclic lactone (ML) class were subsequently developed and registered for use. Milbemycin oxime, at 0.5 mg/kg orally, was approved in 1990, for use as an anthelmintic in dogs, including heartworm prevention (FDA NADA 140-915 Interceptor®, June 14, 1990). In some of the studies for registration, it was noted that in dogs already infected with adult *D. immitis*, this dose of milbemycin oxime was markedly microfilaricidal. However, it was not registered as a microfilaricide and was to be used as a preventive in dogs not already infected with heartworm.

Selamectin was approved as a topical treatment, at 6 mg/kg, for heartworm prevention, and treatment of fleas and some other parasites in dogs and cats in 2000 (NADA 141-152 Revolution®, June 13, 2000).

Moxidectin (MOX) was found to have high potency against microfilariae [14] and anti-reproductive effects on adult *Onchocerca* spp. [15]. The greater potency of MOX against the filarial nematode *D. immitis* compared with other MLs, was reported [16–18] and suggests that this may be generally true for other filariids. It was originally approved for heartworm prevention in dogs for up to two months, as ProHeart®, at an oral dose rate of 3 μg/kg in 1997 (NADA 141-051, ProHeart, 27 May 1997), but as with other heartworm preventives available at the time, treatment was recommended every 30 days. Subsequently, MOX was developed as injectable heartworm preventive formulations, at 6 monthly intervals (NADA 141-189, June 13, 2002, ProHeart 6) at 0.17 mg/kg, and 12 monthly intervals (ProHeart SR-12, 2001 in Australia; and ProHeart 12, in the United States, NADA 141-519, 2 July 2019) at 0.5 mg/kg. MOX was also developed as a topical formulation to be used each 30 days (NADA 141-251, Advantage Multi® [Advocate®], 20 December 2006), as a combination product with the ectoparasiticide, imidacloprid, and NADA 141-147 Coraxis® for heartworm prevention (without combination with an ectoparasiticide and without microfilaricide registration). Additional *D. immitis* microfilaricidal activity was approved for Advantage Multi (NADA 141-251, 24 October 2013). MOX was registered for use in humans for the treatment of onchocerciasis, as an 8-mg oral dose, by the US Food and Drug Administration on 13 June 2018 (https://www.Prnewswire.com/news-releases/us-fda-approves-moxidectin-for-thetreatment-of-river-blindness-300666114.html; 3). It was found to suppress microfilaridermia for more than 18 months after a single treatment [19].

## 11.4 Albendazole

Albendazole (ABZ) is not usually used by itself against filarial parasites, except in areas of Africa where LF is co-endemic with loaisis. ABZ resistance in gastrointestinal parasites in animals, caused by single-nucleotide polymorphisms (SNPs) in the β-tubulin gene, is reasonably common, and similar SNPs have been found in *W. bancrofti* [20]. ABZ is not particularly anti-filarial, but it does induce a very slow decline in microfilarial count [21] suggestive of an effect on adult worm reproduction. Its main benefit when used in combination with other antifilarial drugs is that it appears to enhance the anti-fecundity effect and it also removes soil-transmitted helminths as a collateral benefit. Other than the observation of the presence of SNPs in *W. bancrofti* and that the frequency of the SNPs appeared to increase after some rounds of ABZ + IVM combination treatment, there is no conclusive evidence of ABZ resistance in LF.

## 11.5 Combination Treatments used for Lymphatic Filariasis

The Global Program for the Elimination of Lymphatic Filariasis (GPELF) has recommended a combination of DEC plus ABZ in all LF endemic countries, except those in sub-Saharan Africa where onchocerciasis may be endemic. In sub-Saharan Africa, DEC is not recommended because of the risk of a Mazzotti reaction, and the combination of IVM plus ABZ is recommended for LF. The DEC + ABZ combination is highly microfilaricidal, removes a portion of the adult LF worm burden, and sterilizes the remaining adult worms for some months. ABZ in the combination appears to enhance the anti-fecundity effects and has collateral advantages of removing a high proportion of *Ascaris lumbricoides* and hookworms in the subjects and some of the *Trichuris trichiura* (soil transmitted helminths). The DEC + ABZ combination is thus very effective in interrupting transmission of lymphatic filariae for many months and also reducing the adult parasite burden with repeated treatments. The IVM + ABZ combination used in sub-Saharan Africa is also very good at removing microfilariae and suppressing reproduction in adult worms, thus having excellent anti-transmission properties. However, it has less anti-fecundity effect than DEC + ABZ, and very little adulticidal activity. In recent years, this has been superceded by the IDA triple drug combination (IVM + DEC + ABZ) in areas where co-infection with *O. volvulus* or *L. loa* is not a concern. IDA treatment appears to have significant macrofilaricidal activity after a single dose [22, 23] and is now the recommended treatment for LF [24].

## 11.6 Other Anthelmintics with Antifilarial Activity

Emodepside is another anthelmintic that has antifilarial activity and is being investigated as a candidate for development for this indication by a partnership between the Drugs for Neglected Diseases initiative (DNDi) and Bayer [3, 25, 26].

It is registered as a spot-on, Profender®, at 3 mg/kg, for the control of gastrointestinal nematodes in cats, and as an oral tablet given at 1 mg/kg for roundworm control in dogs, but is not so far, registered for anti-filarial use. Because it has a different mode of action from the MLs, acting on a calcium-activated potassium channel (SLO-1) [27], emodepside, or other anthelmintics which target this receptor, could prove useful in future for controlling filarial parasites resistant to MLs.

## 11.7 Doxycycline

Many filariae, including *D. immitis*, *O. volvulus*, and *W. bancrofi*, contain endosymbiotic *Wolbachia*. The *Wolbachia* are required for filarial larval development, the long-term viability of adult filariae, and embryogenesis. Enzymes involved in several biosynthetic pathways, such as for heme and riboflavin (vitamin B2), are absent from the genomes of filariae containing *Wolbachia*, such as *B. malayi*, but functional in *Wolbachia* [28–30]. Clinical trials in onchocerciasis and bancroftian filariasis patients showed that oral doxycycline (DOX) treatment mediated significant antifilarial curative activity [31, 32]. In both LF and onchocerciasis patients, macrofilaricidal activity was shown by loss of motility of adult filariae. A minimum four-week 100–200 mg/day dose of DOX was identified as minimally curative in both LF and onchocerciasis [32–37]. DOX treatment showed (i) *Wolbachia* depletion from filarial tissues after 2–4 months, (ii) blockade of embryogenesis, associated with degeneration and loss of interuterine developing embryos and a slow reduction of circulating or skin mf by 6–12 months, and (iii) adulticidal activity between 18 and 24 months posttreatment with the 4–6 weeks daily DOX regimen. DOX courses for 3 weeks or shorter periods did not result in significant macrofilaricidal activity in LF [36, 38, 39].

For heartworm-positive dogs, a 28-day regimen of DOX as part of a three-drug adulticidal regimen (DOX, an ML, and melarsomine) is now recommended [1]. Another aspect of DOX treatment of the mammalian host is that it can deplete microfilariae of *Wolbachia* and this can impair the infectivity of developing stages in the insect vector. In a *D. immitis* study, L3 derived from *Aedes aegypti* experimentally fed with microfilariae from DOX-treated dogs subsequently failed to develop into adults following inoculation into naïve dogs [40].

While long-term daily treatment with DOX is effective in removing infections of filarial parasites which contain endosymbiotic *Wolbachia*, the four to six weeks daily treatment required can prove a difficult regimen in some infected animals or humans, and there is an active search for other antibiotics with anti-*Wolbachia* activity which would be therapeutic with a much reduced treatment regimen [3, 26, 41, 42].

DOX resistance in *Wolbachia* has not been reported. However, bacterial drug resistance to DOX and other tetracyclines is common [43] and has been selected in dogs after DOX treatment for *D. immitis* [44]. The possibility of DOX resistance being selected in *Wolbachia* in filarial parasites, and thus reducing the effectiveness of DOX on filarial parasites, would presumably depend on ongoing selection pressure from extensive DOX usage and should be kept in mind.

## 11.8 Macrocyclic Lactone Action and Resistance

The most important pharmaceuticals used to control filarial parasites in animals and humans belong to the ML class of anthelmintics. In considering how resistance to these drugs may evolve, it is important to consider how they work and what stages of the filarial life cycle they target. The most sensitive targets for the MLs in nematodes, at nM concentrations, are considered to be glutamate-gated chloride channels (GluCls). The MLs have a pseudo-irreversible effect of opening these channels to the influx of $Cl^-$ which results in flaccid paralysis of muscle [45]. However, it is also important to note that MLs bind to, and activate, a wide range of other ligand-gated anion channels (e.g. GABA- and dopamine-, serotonin- and tyramine-gated chloride channels) [45], but effects on these ion channels require higher concentrations of the ML. In Clade 5 nematodes, such as *Caenorhabditis elegans*, *Haemonchus contortus*, hookworms, etc., the pharynx may be the most sensitive muscle paralyzed by MLs, resulting in the nematode being unable to take up nutrients. MLs also cause paralytic effects on body muscle, which in gastrointestinal nematode parasites may result in the parasites being unable to maintain their site of predilection in the gastrointestinal tract and leading to their physical removal. A more extensive discussion of possible mechanisms of ML resistance in nematodes, and of outstanding questions that remain about our understanding of ML resistance mechanisms, is available [46]. The effects of MLs seen in gastrointestinal nematodes may, however, be less important in filarial parasites as filariae are thought to be able to absorb nutrients through their cuticle and strong pharyngeal contraction is not required for feeding. Furthermore, some filariae, at some stages of their life cycle, such as adult *O. volvulus* in cutaneous nodules, and adult lymphatic filariae in lymph glands may be able to survive a temporary period with body muscles paralyzed without being physically removed from their "nests."

Although it is not conclusively proven, it has been hypothesized that GluCl expressed in the excretory pore cell may be paralyzed following ML exposure, resulting in filariae, particularly developing larval stages and microfilariae in the tissues or blood, being unable to adequately maintain the secretion of immunomodulatory chemicals and thus become much more susceptible to immune killing by the host [47].

In adult filariae, MLs can suppress reproduction for prolonged periods of time. In adult female *B. malayi*, strong GluCl expression signals were detected in the ovary, developing embryos and lateral hypodermal chords, with moderate expression in the uterus wall adjacent to stretched microfilariae [48]. GluCl genes were also highly expressed in adult male worms in spermatogonia, in the wall of the vas deferens, and in the lateral chords (but not in mature spermatozoa). In addition, the GluCl avr-14 was highly expressed in somatic muscles adjacent to the terminal end of the vas deferens. These results suggest that paralysis of muscles in these sites by MLs will impact gamete production and embryogenesis in filarial worms and explain the observed suppression of microfilaria production following ML treatment. Single ML treatment does not kill a significant proportion of adult filariae, but repeated exposure to ML may reduce the lifespan of adult worms [49].

To summarize, developing L3/L4 stages of filariae, such as *D. immitis*, and probably other filarial species are exquisitely sensitive to MLs (the basis of heartworm preventive control); microfilariae are sensitive to slightly higher concentrations of MLs for the microfilaricidal approach to reducing onchocerciasis and lymphatic filariasis transmission and morbidity (IVM and MOX) and the registered claim of microfilaricidal activity of Advantage Multi (MOX) against heartworm, and reproduction in adult filariae (the anti-fecundity effects of MLs that are critically important to reducing transmission of filariae and contributing to reducing morbidity). When ML resistance evolves in filariae, resistant L3/L4 can still develop into reproductively active adult parasites [50], resistant microfilariae survive ML treatment [51, 52] and can subsequently be transmitted by the relevant vector, and reproduction in the adult worms is suppressed for a shorter period of time, allowing more transmission and morbidity [53]. If the ML effect on any of these stages is diminished, the surviving parasites will be able to reproduce and pass on genes involved in the resistance phenotype. To evaluate the current situation of ML resistance and factors that may contribute to the evolution of resistance, it is necessary to consider different filarial infections separately.

## 11.9 ML Resistance in Heartworm

ML resistance in *D. immitis* is now well accepted. First indications that ML preventives may be failing were reported in 2005 [54], in an analysis of significant numbers of reports to the FDA of ineffectiveness of heartworm preventives, described as "adverse experiences of approved animal products" (commonly termed "LOE" for loss of effectiveness when compliance with heartworm prevention guidelines have been strictly followed). This paper reported that heartworm product complaints, relating to LOEs, began to be submitted to the Center for Veterinary Medicine (CVM) in 1998 and occurred rarely until 2000. The numbers increased dramatically between 2000 and 2003 and occurred mainly in the southern United States. Subsequently, most of these LOEs were ascribed to failure of compliance, because heartworm preventives were registered as being 100% effective based on initial registration trials.

Two reports then appeared of a lack of 100% effectiveness in dogs experimentally infected with an isolate called MP3; although the efficacy based on worm counts was high, respectively, 99.7% [55], and 94.5, 95.3, 95.3, and 100%, depending on which commercial heartworm preventive was used [18], these data raised the question of possible low-level ML resistance.

Clear examples of unequivocal resistance in two US *D. immitis* isolates, Td2008-2 and Jd2009-2, were obtained in three studies in which efficacies, based on worm counts, were respectively 31%, 71.4%, and 6.5%, along with evidence of genetic differences between these isolates and LOE samples on the one hand and susceptible isolates on the other hand [50]. Genetic evidence of resistance along with the phenotypic data clearly established that true drug resistance was responsible and not, for example, a problem for treatment of the dogs. This conclusion was strengthened by the demonstration that the resistance can be inherited between generations [56].

Subsequently, other studies [57, 58] have confirmed ML resistance to preventative treatment.

The microfilaricidal attributes of MLs used as heartworm preventives was not claimed in original registration approvals but had been shown previously ([13, 59–65]; FDA NADA 140-915, 1990). A clear case of a lack of expected efficacy *in vivo* in the microfilaricidal properties of IVM and milbemycin oxime, even when used repeatedly at high doses, was reported by Bourguinat et al. [51]; this study provided the first well-documented evidence for ML resistance in *D. immitis*. Microfilaricidal activity was approved in 2013 for Advantage Multi, which contains MOX for topical use at a recommended dose of 2.2 mg/kg (NADA 141-251, 24 October 2013) and is reported in a subsequent publication [66]. Based on the substantial body of work showing that MLs are microfilaricidal, Geary et al. proposed an *in vivo* microfilaraemia suppression test as a means of detecting heartworm infections containing ML-resistant *D. immitis* [67]. This method was used in dogs attending clinics in the United States having heartworm infections [52] and at the same time, the genetic profiles of the isolated microfilariae were evaluated using a panel of previously reported ML resistance genetic markers [50, 68]. The Ballesteros et al. study [50] indicated a number of very resistant infections coming from Louisiana, Tennessee, Arkansas, and Missouri, and other dogs with suspected resistant infections coming from these states plus Mississippi and Texas.

ML resistance in heartworm preventives has reached serious proportions. However, much is still to be learnt about whether it is occurring independently at different locations, whether it is being spread by movement of dogs or transmission by vectors, what is the general fitness cost of being resistant, and what we can expect in terms of its spread and our ability to control heartworm infections and disease in dogs and cats. What role, if any, will infections in untreated dogs and wild canines play in providing *refugia* from drug selection to mitigate against a worsening of the problem? Many other questions must be answered to provide guidance for maintaining our ability to prevent heartworm infection. Some of these questions are addressed below.

### 11.9.1 Selection for ML Resistance in Heartworm

If we assume, based on the report of Hampshire [54], that resistance first appeared around 1998, ML heartworm preventives had been used and recommended as the first line means to prevent heartworm disease for 10 years (Heartgard-30 was first approved in April 1987, see above). Given that heartworm preventives were registered as being 100% effective, can we assume that the genetic change(s) that causes ML resistance only arose and was selected in that 10–11 years period? Despite that fact that heartworm preventives were registered as being 100% effective and studies were submitted to the FDA showing 100% efficacy, it is very probable that they were never truly 100% effective. Vidyashankar et al. [69] have argued cogently that this may be the case. The arguments include the facts that early registration studies were performed with very few *D. immitis* isolates, which were maintained under experimental conditions, that severe "bottlenecking" occurs during the life cycle of

*D. immitis* and that very few dogs are used in treatment and control groups (typically 8–15 dogs), and each untreated dog has an infection with very few (breeding) adult male and female worms (typically 7 to 18 male or female worms/dog). Thus, efficacy studies used for registration purposes do not adequately reflect the likely genetic and phenotypic diversity that occurs in the wider *D. immitis* population; thousands of dogs infected with tens of thousands of worms from different isolates would need to be tested to adequately estimate the true variability in responsiveness of *D. immitis* to a heartworm preventive. Clearly, that is not practical and so a requirement for 100% effectiveness by registration authorities is a moot point. Given normal biological variation in organisms that are not propagated by cloning, very few if any pharmaceuticals are 100% effective in all circumstances. Nevertheless, a high level of efficacy should be expected when a preventive is registered.

A recent study [70] clearly demonstrates that against a very ML-resistant strain of *D. immitis* (JYD-34), the efficacy of MOX, used as a preventative in dogs, is dependent on the dose and the number of monthly treatments; when the dose is lower (e.g. at the original registered monthly rate of 3 μg/kg) and the challenged dogs are not treated at four-week intervals, more adult worms will establish and could subsequently breed and pass on their resistant genes. In contrast, when a high dose (24–60 μg/kg) was strictly repeated at four weekly intervals, efficacy against JYD-34 developing larvae was very high and approached 100%. These data are important for understanding how ML resistance in *D. immitis* likely was selected. ML heartworm preventives that were commercialized for monthly treatment were registered and used at the lowest dose rates that could provide apparent 100% efficacy in the limited studies that were required for registration, on the assumption that animal owners would treat their dogs at the recommended minimum dose rates and at the stipulated 30-day frequency. Studies of the pharmacokinetics of IVM heartworm preventive show that peak blood levels for the chewable oral formulation at 6 μg/kg occur at 8.5 hours (~3.7 ng/gm equivalent) and decay rapidly over the first three days and was less than 0.2 ng/gm equivalent by 10 days after treatment [71]. For the 6 mg/kg topical application of selamectin, the peak plasma concentration, occurring at around five days, reported in male dogs as 12.7 ng/ml and for female dogs it was 22.7 ng/ml. For both sexes, the mean residence time was 12.55 days [72]. Oral administration of 0.5 mg/kg of milbemycin oxime reaches peak plasma concentration by one to four days, with a $t_{1/2}$ of only 1.6 days (A3 form) and 3 days (A4 form) [73]. The pharmacokinetic studies show that these ML heartworm preventives reach peak concentrations not long after administration and then are cleared from the circulation within a few days. Thus, after the peak concentration is reached, decreasing levels are available to act on developing L3/L4 larvae of *D. immitis* until the next dose of preventive. If the next dose is delayed beyond 30 days, or is missed in some months, developing larvae are likely to be exposed to subtherapeutic levels of the ML. Any *D. immitis* with a genotype that makes them less susceptible to MLs will be more likely to survive and subsequently reproduce than a parasite that is fully susceptible. In this way, with selection pressure potentially every month, given that ML heartworm preventives have been recommended since 1987, selection for ML resistance in *D. immitis* is not surprising.

Other factors that are likely to have played a role in the development of ML resistance in *D. immitis* include the reasonable probability that inbreeding of *D. immitis* is common, the extent of *refugia* may not be high, and that MLs act on other life cycle stages, but at normal monthly dose rates are not lethal to some life cycle stages.

### 11.9.2 Inbreeding

A host with an adult breeding population of heartworm will often have only a few adult worms and any microfilariae taken up by a biting mosquito will be the progeny of those few female and male worms living in the mammalian host. Thus, a good proportion of microfilariae acquired by the transmitting mosquito are likely to be siblings or half-siblings. The small number of microfilariae acquired during a blood meal can develop to infectious L3 larvae in a few days and can then be transmitted to a new host when the mosquito takes another blood meal. Thus, the L3 larvae infecting the recipient mammalian host will also likely have a proportion of siblings and half-siblings similar to the microfilariae that the mosquito initially acquired. These related L3 will subsequently develop into breeding, genetically related, adult worms, unless the L3/L4 stages are truncated by effective chemoprophylaxis. Obviously, if the developing larvae are somewhat tolerant of the chemoprophylaxis, they may survive the intervention. Subsequent inbreeding will produce some progeny which may be even more resistant to ML prophylaxis than their parents. In this way, a moderately high propensity to inbreeding in filarial infections is likely to be more conducive to resistance selection [74] than would be parasite populations with little inbreeding, such as gastrointestinal nematodes that have very large and diverse populations on pasture or in the soil.

### 11.9.3 Refugia

In contrast to gastrointestinal nematodes where most of the nematode population will often be free-living stages on pasture/in soil, which are not subject to anthelmintic selection, filarial nematodes do not have any free-living stages; all of the filarial nematode population will be in a mammalian host (L4, adult, and microfilariae) or an insect vector (L1–L3). The vast majority of the population (>99.9%) overall will be the microfilariae in the mammalian host. These microfilariae can survive in a dog for two to three years and may total billions in a single dog. The proportion in the insect vector is miniscule. Thus, the proportion of the filarial population that is in *refugia* will be the miniscule fraction in vectors and those in infected mammalian hosts not subject to chemotherapy. In effect, the proportion of the domestic and wild canid population that is not under chemotherapy will be the main reservoir for *refugia*. In North America, and particularly in the southern United States where heartworm transmission is most intense, there have been very strong recommendations for many years to have all domesticated dogs on heartworm preventives 12 months of the year, because heartworm disease is very serious. To the extent that such campaigns have been successful, *refugia* will have diminished. It has been well established in other parasitic infections that, when *refugia* is low, resistance selection is likely to be high [75]. This study also provides evidence that "years to resistance" is likely to decrease in environments where parasite transmission is high. For heartworm infection, the Lower Mississippi River region is an environment where *D. immitis* transmission is high, which would be conducive to resistance development.

### 11.9.4 Effects of MLs on Different Life Cycle Stages of *D. immitis*

While ML heartworm preventives act primarily on L3/L4 stages in the mammalian host, they do have effects on other mammalian life cycle stages, such as microfilariae and adult worms. Indeed, killing microfilariae is the primary purpose of treatment of human filarial infections, such as in onchocerciasis (IVM or MOX) and LF (IVM). Against *D. immitis* microfilariae, the high MOX dose rate formulation, Advantage Multi has a registered claim as being microfilaricidal [67]. However, IVM and milbemycin oxime also have microfilaricidal effects ([13, 76]; FDA NADA 140-915 Interceptor, 14 June 14, 1990). The effectiveness of MLs as microfilaricides depends on the dose, frequency of treatment, and the degree of resistance of a population of *D. immitis*. This was illustrated by the observations of McTier et al. [77] who showed that microfilariae of the ZoeMO isolate, a relative of the highly ML-resistant JYD-34 isolate, were essentially eliminated over several months in dogs treated with MOX extended release products but were incompletely reduced by a single high oral dose of MOX (0.25 mg/kg). Thus, ML heartworm preventatives will more likely remove fully susceptible microfilariae, while microfilariae more tolerant to the ML treatment may survive, enriching the gene pool for resistance in subsequent generations.

A similar consideration is relevant to the effects of MLs on adult filariae. While MLs do not kill adult worms, they markedly interrupt their reproduction. This has been well documented for IVM [78] and MOX in *O. volvulus* [19], in which the suppression of fecundity by IVM lasts from 3 to 6 months, while that for MOX extended more than 15 months. Indeed, suboptimal response to IVM in onchocerciasis is manifested as a shortening of the anti-fecundity effect [79]. Blair and Campbell showed that a 50 μg/kg dose of IVM suppressed production of new *D. immitis* microfilariae for approximately two months in the dog [13]. A reduction of the anti-fecundity effect of MLs against *D. immitis* would manifest in adult heartworm showing a reduced susceptibility to the ML, as it does on *O. volvulus* [53]. Thus, for a period after ML heartworm preventative treatment, adult heartworms which were less affected by the treatment would be more likely to continue releasing microfilariae (replacing microfilariae eliminated by the treatment) than would more susceptible adult worms, and so more resistant adult worms would have a reproductive advantage over more susceptible worms, leading to selection for ML resistance. One could speculate that reliance on the "slow-kill" treatment method [80] to eliminate heartworm infections in dogs, in which repeated "subtherapeutic" administration of ML is used, would facilitate resistance selection acting on the adult worm fertility and also on the microfilaricidal effects of MLs.

To summarize the discussion so far on ML resistance selection in *D. immitis*, selection may occur at several points in the life cycle, particularly on L3/L4 larvae, the primary target for heartworm preventives, but be assisted by variable efficacy against microfilariae and any reduction in the anti-fecundity effect if MLs are used while adult worms are present. Selection for resistance is also likely to be greatly amplified by inbreeding in filariae and by high transmission conditions. On the other hand, as almost all of the filarial population exists in the mammalian host, recommendations to try to suppress heartworm infection by year-round monthly (or long acting) treatment, in as many dogs as possible, are likely to reduce *refugia*, which is mainly represented by infections in untreated dogs.

### 11.9.5 Other Factors in ML Resistance in *D. immitis*

While ML resistance in heartworm is mainly confined to the Lower Mississippi River region [52], it could be spread by the movement of dogs. Indeed, there is evidence of considerable migration of dogs in North America and of this migration increasing levels of heartworm infections [81]. Movement of dogs can introduce ML-resistant *D. immitis* to areas well away from the Lower Mississippi River region, as has already been reported [51]. So far, very little survey work has been done to provide information on the possible spread of ML resistance in heartworm.

The spread of ML resistance could also be facilitated by migration of mosquitoes. Although mosquitoes tend to remain in their local environment, vectors could play a role in spreading resistant strains locally, particularly in areas of high transmission.

Another factor that could affect the spread of ML resistance is if resistance confers some general fitness cost to resistant populations compared with susceptible populations, i.e. the greater the resistance fitness cost, the lower the tendency for resistance to spread. This has yet to be studied. However, a preliminary finding [58] suggests that there could be some fitness cost of ML resistance.

## 11.10 Detection, Diagnosis, and Monitoring for ML Resistance in Heartworm

Confirmation of ML resistance in heartworm infections is often confounded, because records of when treatments occurred may be incomplete or are difficult to confirm. Thus, a dog which has been mainly on regular monthly heartworm preventive may not have received a treatment in one or more months or the interval between monthly treatments is sometimes greater than 30 days. In these situations, if a dog becomes infected, it is likely to have been due to noncompliance with the treatment recommendations [82]. Thus, the existence of a heartworm infection in a dog whose owner believed that the dog had been on a preventative is not *de facto* proof of ML resistance.

The gold standard for showing that a *D. immitis* isolate contains ML-resistant parasites is by challenging a naïve dog which is on heartworm preventative therapy, under controlled conditions in which the treatment, timing, and infection challenge are all carefully monitored. This has been done in a number of studies [50, 56, 58] showing unequivocally that ML resistance exists. Furthermore, in these or associated studies, it was shown that resistance was inherited (had a genetic basis), and that changes at the DNA level were associated with the resistance trait [50, 56, 68, 83].

DNA-based tests of SNPs which correlate with ML resistance were initially developed by Bourguinat et al. [50, 84]. These SNP markers were further refined [68], and it was shown that a 2 SNP DNA test was very predictive of ML resistance as measured in a microfilariae suppression test carried out on field isolates in many states of the United States [52]. While these SNP markers do not tell us the mechanism of ML resistance in heartworm, they have been highly predictive in these and other unpublished analyses (Prichard, unpublished data). Further research is required to pinpoint the genetic changes which cause resistance.

Based on the early evidence that MLs have a microfilaricidal effect on *D. immitis* [13], Geary et al. proposed a microfilariae suppression test as a means of confirming a resistance diagnosis in heartworm infections [67]. The test involves undertaking a blood microfilaria count in a dog that is infected with heartworm, treating the dog with a dose of an ML that is usually microfilaricidal and then two to four weeks later having a second microfilarial count taken in order to determine the reduction in microfilariemia. In infections in which the heartworms are susceptible, the microfilarial count should be substantially reduced, while in cases of ML resistance the microfilarial count will not decrease significantly. This test was used and refined further with the microfilaricidal treatment Advantage Multi, which is registered as a microfilaricidal MOX formulation, and the follow-up blood count conducted three to four weeks after treatment [52]. This study, which used both the microfilariae reduction test and analysis of genetic markers for resistance, also provided some limited information on the distribution of ML resistance in the United States. However, there is a great need to expand on that limited study and monitor for the presence of ML resistance in *D. immitis* more widely and over time to better map the distribution of resistance and to see if it is spreading. Such a study would also possibly provide information on factors which are involved in the spread of resistance, as well as control practices which may delay or enhance the spread of the resistance and would be a service to help maintain animal health in heartworm endemic areas. It would also be prudent to monitor for ML heartworm preventative resistance in other parts of the world where *D. immitis* infection is common and dogs are regularly treated to prevent it.

## 11.11 ML Resistance in Human Filariae

Little work has been undertaken to assess whether there is evidence of ML resistance in human filariasis, such as in LF or onchocerciasis. This is partly because it is extremely difficult to monitor for possible resistance and even more difficult to unequivocally prove resistance in human filarial infections. The effects of IVM or MOX (the two MLs registered for human use) are to remove microfilariae (from the skin in onchocerciasis and from the blood in LF) and to suppress the adult worms from producing new microfilariae for several months. Because microfilaraemia is diurnal in *W. bancrofi* infections (the cause of over 90% of lymphatic filariasis), blood microfilarial counts are almost never done after IVM treatment for LF. In onchocerciasis, skin snip microfilarial counts have been done in a limited number of studies. However, some studies indicate that the anti-fecundity effect of IVM may be reduced in at least some areas where IVM treatment has been undertaken for many years. The anti-fecundity effect of IVM (and MOX) can be assessed in two ways, either by monitoring skin microfilarial counts at approximately 30, 90, 180, and 365 days after treatment in order to assess the repopulation of the skin of infected people with new microfilariae (after an initial peak drop by about 30 days) and/or by excising *Onchocerca* cutaneous nodules, digesting out the adult female worms and then conducting an embryogram (observing the state of development of embryos and developing microfilariae *in utero* in adult female worms at similar times after IVM treatment). Microfilarial repopulation and embryogram studies have been conducted in Ghana and Cameroon [53, 79, 85–87] and suggest decreased responsiveness of the anti-fecundity effect of IVM, and more recently a decrease in the microfilaricidal effect of IVM in some communities or regions where annual or more frequent IVM treatment had been conducted for many years. This is suggestive of a form of IVM resistance. Until recently, there was no evidence of a reduction in the microfilaricidal effect of IVM. However, using sensitive molecular tools (polymerase chain reaction (PCR) and loop-mediated amplification (LAMP) assays), Abong et al. have now shown a significant reduction in the ability of IVM to clear microfilariae 30 days after treatment of communities which had been under IVM treatment for over 20 years in Cameroon compared with other communities in Cameroon which have had less history of IVM treatment [87] and with historical data on the response of *O. volvulus* microfilariae to IVM [88]. It is very difficult to prove, in controlled experiments, resistance in human *O. volvulus* infections because, for ethical reasons, humans cannot be experimentally infected with this parasite and sampled repeatedly with nodulectomies and skin snips before and after treatment. Attempts to explain the apparent reduction in the anti-fecundity and now the microfilaricidal effects of IVM as a result of new infections do not stand up to close examination. The age of adult female worms recovered at nodulectomy 90 days after an IVM treatment was determined and embryograms undertaken on these worms [79]. The results showed that, in the populations showing a poor anti-fecundity effect of IVM, there were very few young adult female worms (possibly acquired since the IVM treatment) and of those few young worms, almost all were without microfilariae *in utero*. Therefore, new infections producing microfilariae could not have accounted for the more rapid rebound in microfilaridermia or for the continued presence of microfilariae 30 days after treatment in the communities/regions showing suboptimal responses to IVM. These suboptimally responding communities were compared with good/normally responding communities and with historical data using an individual-based onchocerciasis mathematical model [89]. The variability in the rate at which *O. volvulus* microfilariae repopulate the host's skin following IVM treatment was quantified. The model estimated a single skin repopulation rate for every host sampled, allowing reports of suboptimal responses to be statistically compared with responses from populations with no prior exposure to IVM. Statistically faster rates of skin repopulation, consistent with a suboptimal response, were observed in three Ghanaian villages (treated with IVM 12–17 times).

In addition to this phenotypic evidence of a change in the responsiveness of the anti-fecundity and microfilaricidal effects of IVM on *O. volvulus*, some studies have investigated genetic differences between suboptimally responding and normally responding *O. volvulus* [90, 91]. In the former study, significant differences in SNP frequencies in a β-tubulin gene were found between individual *O. volvulus* samples from communities showing a poor anti-fecundity response to IVM after many rounds of treatment compared to individual worm samples from communities that were either relatively naïve or responding well to IVM treatment. The work on β-tubulin followed other studies indicating allele frequency changes in this gene, when individual worms were genotyped, associated with repeated IVM treatment over many years [92, 93]. In another genomic analysis of *O. volvulus* worms from Ghana and Cameroon, next-generation whole genome sequencing was undertaken on pools of worms with each pool differentiated by country of origin, IVM treatment history, and response to treatment (duration of suppression of microfilaraemia) [91]. In addition, a subset of worms was analyzed individually at 130 SNP loci by Sequenom genotyping. The whole genome pooled sequencing revealed significant differentiation of good responder pools from poor responder pools, and the differences were not randomly distributed, but clustered in 31 quantitative trait loci. Single-worm sequencing at predetermined SNPs revealed geographical diversity and changes over time in the presence of drug pressure. Taken together, the data on genotyping indicate that IVM treatment pressure changes the genotypic profile of *O. volvulus*, which together with reduction in the anti-fecundity and microfilaricidal effects of IVM with repeated rounds of treatment, can be considered evidence of developing resistance to IVM in this human filarial parasite. Given the unequivocal evidence of ML resistance in the dog filarial parasite *D. immitis*, this should not be surprising and indicates that better monitoring for sustained efficacy of IVM in human filariae should be undertaken. Furthermore, because of the more potent effect of MOX in suppressing fecundity longer than IVM [19], this drug may have a role in hastening elimination of onchocerciasis (and perhaps LF) should ML resistance become more problematic for success in the control programs [94].

## 11.12 Conclusions and Future Directions

At present, MLs are the only practical and registered anthelmintics for the prevention of heartworm infections in animals. However, ML resistance occurs in some regions and isolates of *D. immitis* in the United States. It is important that the extent of this resistance is better understood, and that it is monitored for changes in its frequency and distribution over time. To do that, genotypic (e.g. resistance-associated SNPs) and phenotypic (e.g. microfilariae suppression test) monitoring need to occur. Ideally, organizations such as the American Heartworm Society (AHS), which brings together people from the animal health industry, veterinarians, and academics, could undertake such monitoring as AHS does with reports of heartworm infection incidence, from veterinary clinics across the United States.

Another need that arises with the development of ML resistance in heartworm is the development of a new class of heartworm preventatives/treatments. This is of considerable economic importance to animal health companies but is also important for veterinary clinics dealing with the problem of resistance in heartworms, and for animal welfare.

Further research to understand the mechanism(s) and genetics of ML resistance in *D. immitis* and *O. volvulus* is needed. Advances in this understanding will help improve diagnosis of resistance and may point to ways to reduce selection for resistance or to overcome it.

Because there may be a fitness cost associated with ML resistance, it would be interesting to attempt to quantitate this and to propose how this could be exploited to sustain the effectiveness of MLs against filariae. Other areas of research which could be productive include determining the efficacy of high doses of MLs against strains of parasites that are resistant to currently used doses and to establish whether resistance to ML heartworm preventives can be overcome with repeated monthly treatment with high doses of MLs, keeping in mind the need to retain target host safety. Now that there is good evidence of IVM resistance in *O. volvulus*, manifested as both reduction in the anti-fecundity and the microfilaricidal effects of IVM, it would be of interest to investigate whether MOX administration at three to six months intervals will overcome that resistance and drive onchocerciasis control toward elimination. Finally, it would be useful to have a better understanding of the role of duration of drug exposure on efficacy against ML-resistant filaria. The period of exposure of a parasite can be altered by the use of different anthelmintic formulations, for example in long-acting injectables [95]. Studies on ML-resistant *D. immitis* may provide insights into what is going on in terms of the development of IVM resistance in human filariae, such as *O. volvulus*.

## References

1 Ketzis, J. and Epe, C. (2022). Antifilarial chemotherapy: current options in veterinary medicine. In: *Advances in Control of Heartworm and Human Filariases* (ed. R. Kaminsky and T.G. Geary), Chapter 8. Weinheim, Germany: Wiley-VCH.

2 Specht, S., Kamgno, J., and Geary, T.G. (2022). Antifilarial chemotherapy: current options for humans. In: *Advances in Control of Heartworm and Human Filariases* (ed. R. Kaminsky and T.G. Geary), Chapter 7. Weinheim, Germany: Wiley-VCH.

3 Geary, T.G., Long, A., and Tritten, L. (2022). Current antifilarial drugs – mechanisms of action. In: *Advances in Control of Heartworm and Human Filariases* (ed. R. Kaminsky and T.G. Geary), Chapter 10. Weinheim, Germany: Wiley-VCH.

**4** Kaplan, R.M. and Vidyashankar, A.N. (2012). An inconvenient truth: global worming and anthelmintic resistance. *Vet. Parasitol.* 186: 70–78.

**5** King, C.H. (2019). Helminthiasis epidemiology and control: Scoring successes and meeting the remaining challenges. *Adv. Parasitol.* 103: 11–30.

**6** Busvine, J. (2012). *Disease transmission by insects: Its discovery and 90 years of effort to prevent it*, 260. Springer Science & Business Media ISBN 9783642457166. Archived from the original on 2017-09-10. From Wikipedia.

**7** Eberhard, M.L., Lowrie, R.C. Jr., and Lammie, P.J. (1988). Persistence of microfilaremia in bancroftian filariasis after diethylcarbamazine citrate therapy. *Trop. Med. Parasitol.* 39 (2): 128–130.

**8** Sankari, T., Subramanian, S., Hoti, S.L. et al. (2021). Heterogeneous response of *Wuchereria bancrofti*–infected persons to diethylcarbamazine (DEC) and its implications for the Global Programme to Eliminate Lymphatic Filariasis (GPELF). *Parasitol. Res.* 120: 311–319. https://doi.org/10.1007/s00436-020-06950-7.

**9** Aziz, M.A., Diallo, S., Diop, I.M. et al. (1982). Efficacy and tolerance of ivermectin in human onchocerciasis. *Lancet* 2 (8291): 171–173.

**10** Aziz, M.A., Diallo, S., Lariviere, M. et al. (1982). Ivermectin in onchocerciasis. *Lancet* 2 (8313): 1456–1457.

**11** Greene, B.M., Brown, K.R., and Taylor, H.R. (1989). Use of ivermectin in humans. In: *Ivermectin and Abamectin* (ed. W.C. Campbell), 311–323. New York: Springer-Verlag.

**12** Lindley, D. (1987). Merck's new drug free to WHO for river blindness programme. *Nature* 329 (6142): 752.

**13** Blair, L.S. and Campbell, W.C. (1979). Efficacy of avermectin B1a against microfilariae of *Dirofilaria immitis*. *Am. J. Vet. Res.* 40: 1031–1032.

**14** Tagboto, S.K. and Townson, S. (1996). *Onchocerca volvulus* and *O. lienalis*: the microfilaricidal activity of moxidectin compared with that of ivermectin *in vitro* and *in vivo*. *Ann. Trop. Med. Parasitol.* 90: 497–505.

**15** Trees, A.J., Graham, S.P., Renz, A. et al. (2000). *Onchocerca ochengi* infections in cattle as a model for human onchocerciasis: recent developments. *Parasitology* 120 (Suppl. 1): S133–S142.

**16** McCall, J.W., McTier, T.L., Holmes, R.A. et al. (1992). Prevention of naturally acquired heartworm infections in heartworm-naïve beagles by oral administration of moxidectin at an interval of either one or two months. In: *Proceedings of the Heartworm Symposium '92* (ed. M.D. Soll), 169–178. Batavia, IL: American Heartworm Society.

**17** McTier, T.L., McCall, J.W., Dzimianski, M.T. et al. (1992). Prevention of experimental heartworm infection in dogs with single oral doses of moxidectin. In: *Proceedings of the Heartworm Symposium '92* (ed. M.D. Soll), 165–168. Batavia, IL: American Heartworm Society.

**18** Blagburn, B.L., Dillon, A.R., Arther, R.G. et al. (2011). Comparative efficacy of four commercially available heartworm preventive products against the MP3 laboratory strain of *Dirofilaria immitis*. *Vet. Parasitol.* 176: 189–194.

**19** Opoku, N.O., Bakajika, D.K., Kanza, E.M. et al. (2018). Single dose moxidectin versus ivermectin for *Onchocerca volvulus* infection in Ghana, Liberia, and the Democratic Republic of the Congo: a randomised, controlled, double-blind phase 3 trial. *Lancet* 392: 1207–1216.

**20** Schwab, A.E., Boakye, D., Kyelem, D., and Prichard, R.K. (2005). Detection of benzimidazole-resistance associated mutations in the filarial nematode *Wuchereria bancrofti* and evidence for selection with albendazole and ivermectin treatment. *Am. J. Trop. Med. Hyg.* 73: 234–238.

**21** Pani, S., Subramanyam Reddy, G., Das, L. et al. (2002). Tolerability and efficacy of single dose albendazole, diethylcarbamazine citrate (DEC) or co-administration of albendazole with DEC in the clearance of *Wuchereria bancrofti* in asymptomatic microfilaraemic volunteers in Pondicherry, South India: a hospital-based study. *Filaria J.* 1 (1): 1.

**22** Thomsen, E.K., Sanuku, N., Baea, M. et al. (2016). Efficacy, safety, and pharmacokinetics of coadministered diethylcarbamazine, albendazole, and ivermectin for treatment of Bancroftian filariasis. *Clin. Infect. Dis.* 62: 334–341.

**23** King, C.L., Weil, G.J., and Kazura, J.W. (2020). Single-dose triple-drug therapy for *Wuchereria bancrofti* - 5-year follow-up. *N. Engl. J. Med.* 382: 1956–1957.

**24** WHO (2017) Guideline: Alternative mass drug administration regimens to eliminate lymphatic filariasis. Contract No.: WHO/HTM/NTD/PCT/2017.0. Geneva

**25** Kuesel, A. (2016). Research for new drugs for elimination of onchocerciasis in Africa. *Int. J. Parasitol. Drugs Drug Resist.* 6: 272–286.

**26** Hawryluk, N. (2022). The antifilarial drug pipeline. In: *Advances in Control of Heartworm and Human Filariases* (ed. R. Kaminsky and T.G. Geary), Chapter 18. Weinheim, Germany: Wiley-VCH.

**27** Abongwa, M., Martin, R.J., and Robertson, A.P. (2017). A brief review on the mode of action of antinematodal drugs. *Acta Vet. (Beogr.)* 67: 137–152.

**28** Foster, J., Ganatra, M., Kamal, I. et al. (2005). The *Wolbachia* genome of *Brugia malayi*: endosymbiont evolution within a human pathogenic nematode. *PLoS Biol.* 3 (4): e0030121.

**29** Li, Z. and Carlow, C.K. Characterization of transcription factors that regulate the type IV secretion system and riboflavin biosynthesis in *Wolbachia* of *Brugia malayi*. *PLoS One* 7 (12): e0051597.

**30** Wu, B., Novelli, J., Foster, J. et al. (2009). The heme biosynthetic pathway of the obligate *Wolbachia* endosymbiont of *Brugia malayi* as a potential anti-filarial drug target. *PLoS Negl.Trop. Dis.* 3 (7): e000475.

**31** Taylor, M.J., Makunde, W.H., McGarry, H.F. et al. (2005). Macrofilaricidal activity after doxycycline treatment of *Wuchereria bancrofti*: a double-blind, randomised placebo-controlled trial. *Lancet* 365: 2116–2121.

**32** Hoerauf, A., Specht, S., Buttner, M. et al. (2008). *Wolbachia* endobacteria depletion by doxycycline as antifilarial therapy has macrofilaricidal activity in onchocerciasis: a randomized placebo-controlled study. *Med. Microbiol. Immunol.* 197: 295–311.

**33** Debrah, A.Y., Mand, S., Specht, S. et al. (2006). Doxycycline reduces plasma VEGF-C/sVEGFR-3 and improves pathology in lymphatic filariasis. *PLoS Pathog.* 2: e92.

**34** Debrah, A.Y., Mand, S., Marfo-Debrekyei, Y. et al. (2007). Macrofilaricidal effect of 4 weeks of treatment with doxycycline on *Wuchereria bancrofti*. *Trop. Med. Int. Health* 12: 1433–1441.

**35** Hoerauf, A., Specht, S., Marfo-Debrekyei, Y. et al. (2009). Efficacy of 5-week doxycycline treatment on adult *Onchocerca volvulus*. *Parasitol. Res.* 104: 437–447.

**36** Debrah, A.Y., Mand, S., Marfo-Debrekyei, Y. et al. (2011). Macrofilaricidal activity in *Wuchereria bancrofti* after 2 weeks treatment with a combination of rifampicin plus doxycycline. *J. Parasitol. Res.* 2011: 201617.

**37** Turner, J.D., Tendongfor, N., Esum, M. et al. (2010). Macrofilaricidal activity after doxycycline only treatment of *Onchocerca volvulus* in an area of *Loa loa* co-endemicity: a randomized controlled trial. *PLoS Negl.Trop. Dis.* 4: e000660.

**38** Mand, S., Pfarr, K., Sahoo, P.K. et al. (2009). Macrofilaricidal activity and amelioration of lymphatic pathology in bancroftian filariasis after 3 weeks of doxycycline followed by single-dose diethylcarbamazine. *Am. J. Trop. Med. Hyg.* 81: 702–711.

**39** Turner, J.D., Mand, S., Debrah, A.Y. et al. (2006). A randomized, double-blind clinical trial of a 3-week course of doxycycline plus albendazole and ivermectin for the treatment of *Wuchereria bancrofti* infection. *Clin. Infect. Dis.* 42: 1081–1089.

**40** McCall, J.W., Kramer, L., Genchi, C. et al. (2014). Effects of doxycycline on heartworm embryogenesis, transmission, circulating microfilaria, and adult worms in microfilaremic dogs. *Vet. Parasitol.* 206: 5–13.

**41** Bakowski, M.A. and McNamara, C.W. (2019). Advances in antiwolbachial drug discovery for the treatment of parasitic filarial worm infections. *Trop. Med. Infect. Dis.* 4: 108.

**42** Turner, J.D., Marriott, A.E., David Hong, D. et al. (2020). Novel anti-*Wolbachia* drugs, a new approach in the treatment and prevention of veterinary filariasis? *Vet. Parasitol.* 279: 109057.

**43** Rudra, P., Hurst-Hess, K., Lappierre, P., and Ghosh, P. (2018). High levels of intrinsic tetracycline resistance in *Mycobacterium* abscesses are conferred by a tetracycline modifying monooxygenase. *Antimicrob. Agents Chemother.* 62: e00119-18.

**44** Tejedor-Junco, M.T., González-Martín, M., Bermeo-Garrido, E. et al. (2018). Doxycycline treatment for *Dirofilaria immitis* in dogs: impact on *Staphylococcus aureus* and *Enterococcus* antimicrobial resistance. *Vet. Res. Commun.* 42: 227–232.

**45** Prichard, R., Ménez, C., and Lespine, A. (2012). Moxidectin and the avermectins: Consanguinity but not identity. *Int. J. Parasitol. Drugs Drug Resist.* 2: 134–153.

**46** Prichard, R.K. and Geary, T.G. (2019). Perspectives on the utility of moxidectin for the control of parasitic nematodes in the face of developing anthelmintic resistance. *Int. J. Parasitol. Drugs Drug Resist.* 10: 69–83.

**47** Moreno, Y., Nabhan, J.F., Solomon, J. et al. (2010). Ivermectin disrupts the function of the excretory-secretory apparatus in microfilariae of *Brugia malayi*. *Proc. Natl. Acad. Sci. U.S.A.* 107: 20120–20125.

**48** Li, B.W., Rush, A.C., and Weil, G.J. (2014). High level expression of a glutamate-gated chloride channel gene in reproductive tissues of *Brugia malayi*

may explain the sterilizing effect of ivermectin on filarial worms. *Int. J. Parasitol. Drugs Drug Resist.* 4: 71–76.
**49** Walker, M., Pion, S.D.S., Fang, H. et al. (2018). Macrofilaricidal efficacy of repeated doses of ivermectin for the treatment of river blindness. *Clin. Infect. Dis.* 65: 2026–2034.
**50** Bourguinat, C., Lee, A.C.Y., Lizundia, R. et al. (2015). Macrocyclic lactone resistance in *Dirofilaria immitis*: Failure of heartworm preventives and investigation of genetic markers for resistance. *Vet. Parasitol.* 210: 167–178.
**51** Bourguinat, C., Keller, K., Bhan, A. et al. (2011). Macrocyclic lactone resistance in *Dirofilaria immitis*. *Vet. Parasitol.* 181: 388–392.
**52** Ballesteros, C., Pulaski, C.N., Bourguinat, C. et al. (2018). Clinical validation of molecular markers of macrocyclic lactone resistance in *Dirofilaria immitis*. *Int. J. Parasitol. Drugs Drug Resist.* 8: 596–606.
**53** Osei-Atweneboana, M.Y., Eng, J.K.L., Boakye, D.A. et al. (2007). Prevalence and intensity of *Onchocerca volvulus* infection and efficacy of ivermectin in endemic communities in Ghana: a two-phase epidemiological study. *Lancet* 369: 2021–2029.
**54** Hampshire, V.A. (2005). Evaluation of efficacy of heartworm preventive products at the FDA. *Vet. Parasitol.* 133: 191–195.
**55** Snyder, D.E., Wiseman, S., Cruthers, L.R., and Slone, R.L. (2011). Ivermectin and milbemycin oxime in experimental adult heartworm (*Dirofilaria immitis*) infection of dogs. *J. Vet. Int. Med.* 25: 61–64.
**56** Pulaski, C.N., Malone, J.B., Bourguinat, C. et al. (2014). Establishment of macrocyclic lactone resistant *Dirofilaria immitis* isolates in experimentally infected laboratory dogs. *Parasites Vectors* 7: 494.
**57** Blagburn, B.L., Arther, R.G., Dillon, A.R. et al. (2016). Efficacy of four commercially available heartworm preventative products against the JYD-34 laboratory strain of *Dirofilaria immitis*. *Parasites Vectors* 9: 191.
**58** McTier, T.L., Six, R.H., Pullins, A. et al. (2017). Efficacy of oral moxidectin against susceptible and resistant isolates of *Dirofilaria immitis* in dogs. *Parasites Vectors* 10 (Suppl 2): 482.
**59** McManus, E.C. and Pulliam, J.D. (1984). Histopathologic features of canine heartworm microfilarial infection after treatment with ivermectin. *Am. J. Vet. Res.* 45: 91–97.
**60** Bowman, D.D., Johnson, R.C., Ulrich, M.E. et al. (1992). Effects of long-term administration of ivermectin and milbemycin oxime on circulating microfilariae and parasite antigenemia in dogs with patent heartworm infections. In: *Proceedings of the Heartworm Symposium '92* (ed. M.D. Soll), 151–158. Batavia, IL: American Heartworm Society.
**61** Courtney, C.H., Zeng, Q.T., and Maler, M.M. (1998). The effect of chronic administration of milbemycin oxime and ivermectin on microfilaremias in heartworm-infected dogs. In: *Proceedings of the Heartworm Symposium '98* (ed. M.D. Soll), 193–199. Batavia, IL: American Heartworm Society.
**62** McCall, J.W., Ryan, W.G., Roberts, R.E., and Dzimianski, M.T. (1998). Heartworm adulticidal activity of prophylactic doses of ivermectin (6 ug/kg) plus pyrantel

administered monthly to dogs. In: *Proceedings of the Heartworm Symposium '98* (ed. M.D. Soll), 209–215. Batavia, IL: American Heartworm Society.

**63** Atkins, C. (2002). Canine heartworm disease: current treatment and prevention approaches. In: *Proceedings of the 26th Annual WALTHAM®Diets/OSU Symposium, Small Animal Cardiology* (ed. J.D. Bonagura, V.L. Fuentes and K.M. Muers). Waltham USA, Davis, CA.

**64** Venco, L., McCall, J.W., Guerrero, J., and Genchi, C. (2004). Efficacy of long-term monthly administration of ivermectin on the progress of naturally acquired heartworm infections in dogs. *Vet. Parasitol.* 124: 259–268.

**65** Bowman, D.D. and Torre, C.J. (2006). The effects of preventative dosages of macrolide treatments on circulating microfilariae in dogs with patent heartworm-*Dirofilaria immitis*-infections. *US Comp. Anim. Health* 9–11.

**66** Bowman, D.D., Charles, S.D., Arther, R., and Settje, T. (2015). Laboratory evaluation of the efficacy of 10% imidacloprid + 2.5% moxidectin topical solution (Advantage® Multi, Advocate®) for the treatment of *Dirofilaria immitis* circulating microfilariae in dogs. *Parasitol. Res.* 114 (Suppl. 1): S165–S174.

**67** Geary, T.G., Bourguinat, C., and Prichard, R.K. (2011). Evidence for macrocyclic lactone anthelmintic resistance in *Dirofilaria immitis*. *Top. Comp. Anim. Med.* 26: 186–192.

**68** Bourguinat, C., Keller, K., Xia, J. et al. (2017). Genetic profiles of ten *Dirofilaria immitis* isolates susceptible or resistant to macrocyclic lactone heartworm preventives. *Parasites Vectors* 10 (Suppl 2): 504.

**69** Vidyashankar, A.N., Jimenez Castro, P.D., and Kaplan, R.M. (2017). A statistical approach for evaluating the effectiveness of heartworm preventive drugs: what does 100% efficacy really mean? *Parasites Vectors* 10 (Suppl 2): 516.

**70** McTier, T.L., Six, R.H., Pullins, A. et al. (2019). Preventive efficacy of oral moxidectin at various doses and dosage regimens against macrocyclic lactone-resistant heartworm (*Dirofilaria immitis*) strains in dogs. *Parasites Vectors* 12: 444.

**71** Daurio, C.P., Cheung, E.N., Jeffcoat, A.R., and Skelly, B.J. (1992). Bioavailability of ivermectin administered orally to dogs. *Vet. Res. Commun.* 16: 125–130.

**72** Dupuy, J., Derlon, A.L., Sutra, J.F. et al. (2004). Pharmacokinetics of selamectin in dogs after topical application. *Vet. Res. Commun.* 28: 407–413.

**73** Letendre, L., Harriman, J., Drag, M. et al. (2017). The intravenous and oral pharmacokinetics of afoxolaner and milbemycin oxime when used as a combination chewable parasiticide for dogs. *J. Vet. Pharmacol. Therap.* 40: 35–43.

**74** Churcher, T.S., Schwab, A.E., Prichard, R.K., and Basáñez, M.-G. (2008). An analysis of genetic diversity and inbreeding in *Wuchereria bancrofti*: implications for the spread and detection of drug resistance. *PLoS Negl.Trop. Dis.* 2: e211.

**75** Leathwick, D.M., Sauermann, C.W., and Nielsen, M.K. (2019). Managing anthelmintic resistance in cyathostomin parasites: Investigating the benefits of refugia-based strategies. *Int. J. Parasitol. Drugs Drug Resist.* 10: 118–124.

**76** Sasaki, Y. and Kitagawa, H. (1993). Effects of milbemycin D on microfilarial number and reproduction of *Dirofilaria immitis* in dogs. *J. Vet. Med. Sci.* 55: 763–769.

**77** McTier, T.L., Pullins, A., Inskeep, G.A. et al. (2017). Microfilarial reduction following ProHeart® 6 and ProHeart® SR-12 treatment in dogs experimentally inoculated with a resistant isolate of *Dirofilaria immitis*. *Parasites Vectors* 10 (Suppl. 2): 485.

**78** Duke, B.O., Zea-Flores, G., and Munoz, B. (1991). The embryogenesis of *Onchocerca volvulus* over the first year after a single dose of ivermectin. *Trop. Med. Parasitol.* 42: 175–180.

**79** Osei-Atweneboana, M.Y., Awadzi, K., Attah, S.K. et al. (2011). Phenotypic evidence of emerging ivermectin resistance in *Onchocerca volvulus*. *PLoS Negl.Trop. Dis.* 5: e000998.

**80** Dixon-Jimenez, A.C., Coleman, A.E., Rapoport, G.S. et al. (2018). Approaches to canine heartworm disease treatment among alumni of a single College of Veterinary Medicine. *J. Am. Anim.Hosp. Assoc.* 54: 246–256.

**81** Drake, J. and Parrish, R.S. (2019). Dog importation and changes in heartworm prevalence in Colorado 2013-2017. *Parasites Vectors* 12: 207.

**82** Atkins, C.E., Murray, M.J., Olavessen, L.J. et al. (2014). Heartworm 'lack of effectiveness' claims in the Mississippi delta: computerized analysis of owner compliance--2004-2011. *Vet. Parasitol.* 206: 106–113.

**83** Bourguinat, C., Lefebvre, F., Sandoval, J. et al. (2017). *Dirofilaria immitis* JYD-34 isolate: whole genome analysis. *Parasites Vectors* 10 (Suppl 2): 494.

**84** Bourguinat, C., Keller, K., Blagburn, B. et al. (2011). Correlation between loss of efficacy of macrocyclic lactone heartworm preventatives and P-glycoprotein genotype. *Vet. Parasitol.* 176: 374–381.

**85** Pion, S.D., Nana-Djeunga, H., Kamgno, J. et al. (2013). Dynamics of *Onchocerca volvulus* microfilarial densities after ivermectin treatment in an ivermectin-naïve and a multi-treated population from Cameroon. *PLoS Negl.Trop. Dis.* 7: e0002084.

**86** Nana-Djeunga, H.C., Bourguinat, C., Pion, S.D. et al. (2014). Reproductive status of *Onchocerca volvulus* after ivermectin treatment in an ivermectin-naïve and a frequently treated population from Cameroon. *PLoS Negl.Trop. Dis.* 8: e0002824.

**87** Abong, R.A., Amambo, G.N., Chounna Ndongmo, P.W. et al. (2020). Differential susceptibility of *Onchocerca volvulus* microfilaria to ivermectin in two areas of contrasting history of mass drug administration in Cameroon: relevance of microscopy and molecular techniques for the monitoring of skin microfilarial repopulation within six months of direct observed treatment. *BMC Infect. Dis.* 20: 726.

**88** Basáñez, M.G., Pion, S.D., Boakes, E. et al. (2008). Effect of single-dose ivermectin on *Onchocerca volvulus*: a systematic review and meta-analysis. *Lancet Infect. Dis.* 8: 310–322.

**89** Churcher, T.S., Pion, S.D.S., Osei-Atweneboana, M.Y. et al. (2009). Identifying sub-optimal responses to ivermectin in the treatment of River Blindness. *Proc. Natl. Acad. Sci. U.S.A.* 106: 16716–16726.

**90** Osei-Atweneboana, M.Y., Boakye, D.A., Awadzi, K. et al. (2012). Genotypic analysis of β-tubulin in *Onchocerca volvulus* from communities and individuals showing poor parasitological response to ivermectin treatment. *Int. J. Parasitol. Drugs Drug Resist.* 2: 20–28.

**91** Doyle, S.R., Bourguinat, C., Nana-Djeunga, H.C. et al. (2017). Genome-wide analysis of ivermectin response by *Onchocerca volvulus* reveals that genetic drift and soft selective sweeps contribute to loss of drug sensitivity. *PLoS Negl.Trop. Dis.* 11: e0005816.

**92** Eng, J.K.L., Blackhall, W.J., Osei-Atweneboana, M.Y. et al. (2006). Ivermectin selection on β-tubulin: Evidence in *Onchocerca volvulus* and *Haemonchus contortus*. *Mol. Biochem. Parasitol.* 150: 229–235.

**93** Nana-Djeunga, H., Bourguinat, C., Pion, S.D.S. et al. (2012). Single Nucleotide Polymorphisms in β-tubulin selected in *Onchocerca volvulus* following repeated ivermectin treatment: possible indication of resistance selection. *Mol. Biochem. Parasitol.* 185: 10–18.

**94** Turner, H.C., Walker, M., Attah, S.K. et al. (2015). The potential impact of moxidectin on onchocerciasis elimination in Africa: an economic evaluation based on the Phase II clinical trial data. *Parasites Vectors* 19: 167.

**95** Bowman, D.D., Ohmes, C.M., Hostetler, J.A. et al. (2017). Efficacy of 10% imidacloprid + 2.5% moxidectin topical solution (Advantage Multi® for Dogs) for the prevention of heartworm disease and infection all month long. *Parasites Vectors* 10 (Suppl 2): 478.

# 12

## Elimination and Eradication of Human Filariases

*Boakye A. Boatin[1,2,3,*], Frank O. Richards Jr[4], Kapa D. Ramaiah[5], and John O. Gyapong[6]*

[1] *McGill University, Institute of Parasitology, Montreal, Quebec H2X 3V9, Canada*
[2] *University of Ghana, Noguchi Memorial Institute for Medical Research, Lymphatic Filariasis Support Centre for Africa, Legon, Ghana*
[3] *5569 Oakwood Drive Stone Mountain, GA 30087, USA*
[4] *453 Freedom Parkway, Atlanta, GA 30307, USA*
[5] *12, Bhaktavatsalam Street,Tagore Nagar, Lawspet, Puducherry, 605008, India*
[6] *Vice Chancellor's Office, University of Health & Allied Sciences AS, PMB 31 Ho VH-0194-8222 Volta Region, Ghana*

**Abstract**

Onchocerciasis, caused by *Onchocerca volvulus*, and lymphatic filariasis (LF), caused by *Wuchereria bancrofti*, *Brugia malayi*, and *Brugia timori*, are the most important causes of human filariases. These diseases primarily affect the poor, contribute to their poverty, and are a continuous obstacle to the socioeconomic development. Ninety-nine percent of onchocerciasis cases occurs in Africa, with the rest in Latin America and in Yemen. LF is more widespread, with 1.39 billion people living in areas at risk. These two diseases were targeted for elimination of the parasite and as public health problems, onchocerciasis by 2025 and LF by 2020.

Mass Drug Administration (MDA) is the strategy used in the elimination process for the two diseases. Ivermectin (IVM) is given once a year (in some cases more than once a year) to the at-risk population for decades for onchocerciasis, while MDA for LF has provided a combination of drugs once yearly for at least six years: albendazole (ALB) + IVM, ALB + diethylcarbamazine (DEC), or ALB twice a year in places where LF is coendemic with *Loa loa*. Recently, the triple drug combination of IVM + DEC + IVM (IDA) has shown evidence of macrofilaricidal activity in LF and is being introduced gradually in non-onchocerciasis LF endemic areas.

Progress toward elimination of onchocerciasis transmission has been slow in Africa but has been greatest in the Americas, where 11 of the 13 original foci in 4 of the 6 endemic countries have stopped MDA, and where completed posttreatment surveillance evaluations demonstrated no recrudescence of transmission.

Challenges faced by the programs include mapping of all transmission zones in Africa, resolution of the diagnostics platforms required to determine the threshold of

*Corresponding author.

*Human and Animal Filariases: Landscape, Challenges, and Control*, First Edition.
Edited by Ronald Kaminsky and Timothy G. Geary.

transmission interruption at the tail end of long-standing MDA programs, the need for a second-line drug suitable for MDA should there be resistance to IVM, and cross-border movements and migration that may result in re-establishment of transmission in areas where onchocerciasis has been interrupted or eliminated. Other challenges are lack of or poor political will, low priority, lack of monitoring and accountability, sub-optimal treatment coverage and monitoring and evaluation, inadequacy in data reporting, and difficulty of delivering effective MDA coverage in urban areas in the case of LF.

Nonetheless, results for LF are promising: 14 (19%) of the 73 countries targeted for control are in post-validation surveillance, whereas another 8 (11%) are in the post-MDA surveillance phase.

Onchocerciasis will be eliminated in the Americas in the expected time frame. Many countries in the Africa region, after the many rounds of MDA, are making onchocerciasis elimination a reality in the WHO-NTD 2021–2030 goals, but the *L. loa* issue and the need to complete mapping of hypoendemic areas remain challenges that need solutions in the near term. Progress has been made in LF elimination in all countries, and the introduction of IDA is likely to accelerate progress outside of Africa.

## 12.1 Introduction

The human filariases consist of a group of nematode parasites transmitted by arthropod vectors. These include onchocerciasis ("river blindness"), caused by *Onchocerca volvulus*, lymphatic filariasis (LF; "elephantiasis"), caused by *Wuchereria bancrofti, Brugia malayi, and B. timori,* loiasis ("eyeworm"), caused by *L. loa*, and mansonellosis, caused by *Mansonella perstans*, *Mansonella streptocerca,* or *Mansonella ozzardi*. Onchocerciasis and LF are of the greatest public health importance and will be covered in this chapter; loiasis will be mentioned in so far as its presence complicates the global efforts to eliminate onchocerciasis and LF.

Onchocerciasis and LF affect the poor, contribute to their poverty, and are an obstacle to the socioeconomic development. They belong to the group of Neglected Tropical Diseases (NTDs) that were targeted for elimination – onchocerciasis by 2025 [1] and LF by 2020 [1]. The largest number of persons afflicted with onchocerciasis (99% of all cases) occurs in Africa. The rest reside in two small foci, one in Latin America and the other in Yemen [2]. The World Health Organization (WHO) estimates that 205 million people live in onchocerciasis-endemic areas, although today there is little risk for these individuals to develop blindness or skin disease, as the prevalence and intensity of infections have been greatly reduced through regular administration of the microfilaricide ivermectin (IVM) [3]. The management of onchocerciasis in Africa has gone through several phases: administration of vector control alone from 1974 through the early nineties, followed by a phase in which management consisted of a combination of vector control and IVM treatment, and last, with the passage of the WHO resolution WHA 47.32 of 1994, through mass drug administration (MDA) of IVM (Mectizan®, donated by Merck & Co.). In the Americas, onchocerciasis was originally present in 6 countries, but 4 of these have now been verified as free of onchocerciasis transmission by WHO as a result of twice per year or four times per year (quarterly) IVM MDA [4].

LF is much more widely distributed than onchocerciasis. LF is endemic in 72 countries (in the Americas, Africa, South East Asia, the Middle East, and South East Asia), with an at-risk population of about 1.39 billion people [5]. Unlike onchocerciasis, which is primarily a rural disease, LF occurs in both urban and rural areas. In most endemic rural settings in Africa, LF is co-endemic with onchocerciasis. Given the global burden of LF, World Health Assembly Resolution WHA 50.29 of 1997 called for the elimination of LF as a public health problem [6]. The LF global program is based on two pillars: (i) to interrupt transmission of the parasite using MDA and (ii) to alleviate morbidities arising from LF infections, especially lymphedema/elephantiasis, male scrotal hydrocele, and acute attacks of lymphangitis [7].

The Onchocerciasis Control Programme for West Africa (OCP) was launched in 1974 in 9 (expanding to 11) countries. Its goal was to control onchocerciasis as a public health problem using vector control. In 1993, the Onchocerciasis Elimination Program for the Americas (OEPA) was launched with the goal of interrupting transmission throughout the region using a strategy of twice per year (six monthly) IVM MDA, reaching at least 85% overage of the eligible population in each treatment round. In 1996, the African Programme for Onchocerciasis Control (APOC) was launched in non-OCP-supported endemic countries in Africa with a strategy of annual MDA with IVM to control morbidity from onchocerciasis. In 2009, APOC changed its objective to elimination of onchocerciasis transmission in areas where this was practicable. APOC closed its operations in 2016 and was succeeded by the WHO Expanded Special Project for the Elimination of Neglected Tropical Diseases (ESPEN) in Africa. Among its broad and overarching roles in NTDs, ESPEN supports the transition to onchocerciasis elimination in Africa, including supporting country-led elimination programs that can garner domestic political and financial support as well as help with relevant capacity building.

The identification of LF as "eradicable" or "potentially eradicable" by the International Task force for Disease Eradication (ITFDE) [8] was followed by WHO's launch in 2000 of the Global Programme for the Elimination of Lymphatic Filariasis (GPELF), aiming to eliminate LF as a public health problem. MDA for the LF program has been based on a combination of 3 anthelmintics: albendazole (ALB, donated by GlaxoSmithKline) combined with diethylcarbamazine (DEC, donated by Essai) in LF but non-onchocerciasis endemic areas; and IVM (donated by Merck) in combination with ALB in areas coendemic for LF and onchocerciasis. Most recently, the triple therapy of IVM + DEC + ALB (IDA) has been introduced in LF but non-onchocerciasis endemic areas.

These regional and global initiatives fighting onchocerciasis and LF have run concurrently over the last two decades. This chapter sets out the conceptual framework, the methods in use, the progress, challenges, and the outlook arising from these initiatives for the elimination of onchocerciasis and LF in the global context. Some specific modalities within the various regions of the world are also presented. The emphasis will be on elimination of transmission and elimination as a public health problem, while the notion of global eradication of onchocerciasis and/or LF will only be mentioned briefly.

## 12.2 Definitions of Elimination of Transmission and of Public Health Problem and Eradication

It is important to make a clear distinction between these three terms. Elimination of transmission (EoT) is defined here as the reduction to zero of the incidence of infection caused by a specific pathogen in a defined geographic area, with minimal risk of reintroduction (based on careful post-intervention surveillance), as a result of deliberate efforts. Continued surveillance and potential action to prevent re-establishment of transmission may be required. WHO recognizes EoT through a process known as verification [9–11]. Elimination as a public health problem (EPHP) may result in interruption of transmission, but the primary goal is high-level control of morbidity resulting from the infection. Again, continued surveillance and potential action to prevent recurrent morbidity are required. WHO recognizes this accomplishment through a process known as validation. In both instances, MDA interventions may be stopped, but post-treatment surveillance (PTS) for a period after stopping MDA is required before EoT or EPHP may be verified or validated, respectively, and post-elimination surveillance follows indefinitely. Eradication, on the other hand, is "the permanent reduction to zero of a specific pathogen, as a result of deliberate efforts, with no more risk of reintroduction," documented by a WHO "certification process involving all countries on earth, regardless of whether they were previously known to be endemic" [11, 12]. In contrast to EoT and EPHP, post-certification surveillance is not required once global eradication is achieved.

## 12.3 Elimination of Onchocerciasis

Both OCP and APOC had the goal to reach a level of control sufficient to eliminate the disease as a public health problem. The idea of elimination of transmission and the parasite in Africa was not envisaged by either programs when first launched. OEPA was launched in response to a 1991 resolution of the Pan American Health Organization (PAHO) that called for the elimination of onchocerciasis from the Americas. The term "elimination" as used in OEPA is defined along the lines of the EoT definition provided above. Several observations have stoked current enthusiasm to embark on elimination of onchocerciasis transmission in Africa following proof-of-concept using sustained high coverage IVM MDA as the main intervention in OEPA. These include WHO verification of elimination of onchocerciasis transmission in four countries in Latin America, starting in 2012 in Colombia and followed in rapid succession by similar WHO verifications in Ecuador (2013), Mexico (2014), and Guatemala (2015) [14], announcement of elimination in the Abu Hamad focus by Sudan [13] and elimination in multiple foci in Uganda [14], findings that permitted IVM MDA to be stopped in two West African foci [15, 16], and growing evidence of progress toward interruption of transmission in several other foci in Africa.

### 12.3.1 Elimination of Onchocerciasis Transmission in Africa, Latin America, and Yemen

#### 12.3.1.1 Conceptual Framework

The OCP in West Africa largely achieved its objectives in the core 11 West African countries. Under OCP, the vector control effort interrupted transmission altogether in most areas of operation. The OCP EoT conceptual framework (CF) on which the vector control strategy was based is as depicted in Figure 12.1. The CF predicts that transmission is interrupted within two years of launching vector control, which then must be continued, in the ideal situation, for the 14-year lifespan of *O. volvulus* adult worms. At that point, vector control can be stopped but must be followed by surveillance. For the rest of Africa, i.e., the previously APOC-managed countries in which annual IVM MDA was the mode of intervention, the CF is profoundly different (Figure 12.2). The decline in transmission is much more gradual than seen with the OCP vector control model, since IVM is less effective in cutting transmission. IVM can suppress transmission of onchocerciasis for about six months after MDA. However, the reduction in fecundity of adult worms over time with repeated use of IVM [17] plays a role in reducing the reproductive lifespan of the adult worms, potentially lessening the 14-year period required for vector control alone. Thus, in areas treated with IVM MDA alone, the time to reach interruption of transmission (and subsequently stopping MDA) depends on the initial level of transmission, the level of treatment coverage, and the number of treatment rounds (Figures 12.3 and 12.4). It is generally assumed that annual MDA with good total population treatment coverage (>65%) will require 10–14 years to achieve this goal. Twice per year MDA has been estimated by modelers to reduce the time period to interruption by 40% [18], an approach used with great success in the Americas.

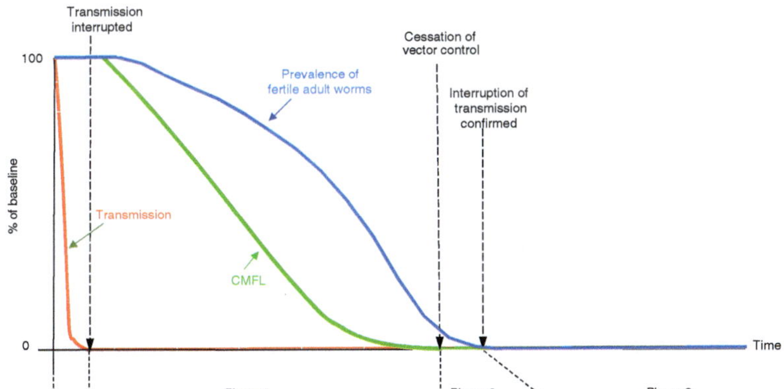


| | | | | |
|---|---|---|---|---|
| Intervention: | | Vector control | None | None |
| Duration: | 0–2 yr | 14 yr | 2 yr | Indefinite |
| Transmission: | Declining | Zero | Zero | Zero |
| Evaluation: | | Entomological and epidemiological evaluation | Post-control entomological evaluation | Routine surveillance |

**Figure 12.1** Conceptual Framework of elimination of onchocerciasis by ivermectin (OCP).

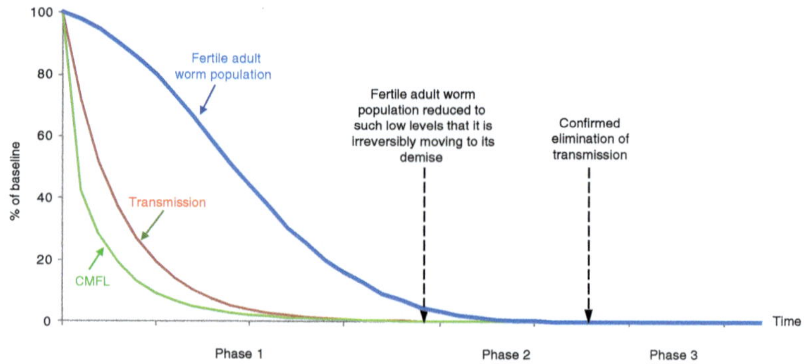


| | | | |
|---|---|---|---|
| Intervention: | Community-directed annual treatment withivermectin | None | None |
| Duration: | 6 to 20+ years | 3-5 years | Indefinite |
| Transmission: | Transmission progressively declining towards negligible levels | Irreversibly approaching zero | Zero |
| Evaluation: | Epidemiological evaluation of progresss towards elimination | Post treatment surveillance | Routine surveillance |

**Figure 12.2** Conceptual Framework of elimination of onchocerciasis by ivermectin (APOC).

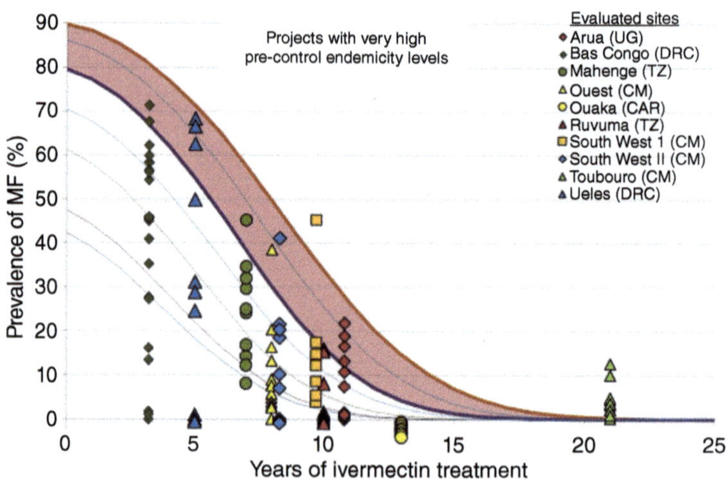


**Figure 12.3** Progress toward elimination in evaluated projects (projects with very high pre-control endemicity levels).

However, increasing the frequency of MDA per year could increase annual costs by around 60% [19]. In the MDA conceptual model, a post-MDA surveillance period of 3–5 years is required before elimination of transmission may be declared.

#### 12.3.1.2 Implementation of the Elimination Program for Onchocerciasis

IVM at a dose of 150 ug/body weight is the only agent used in the management of onchocerciasis to reduce morbidity and with the intention of stopping transmission of the parasite (elimination). IVM MDA in Africa is delivered primarily through a community-directed treatment with ivermectin (CDTi) strategy. It is important

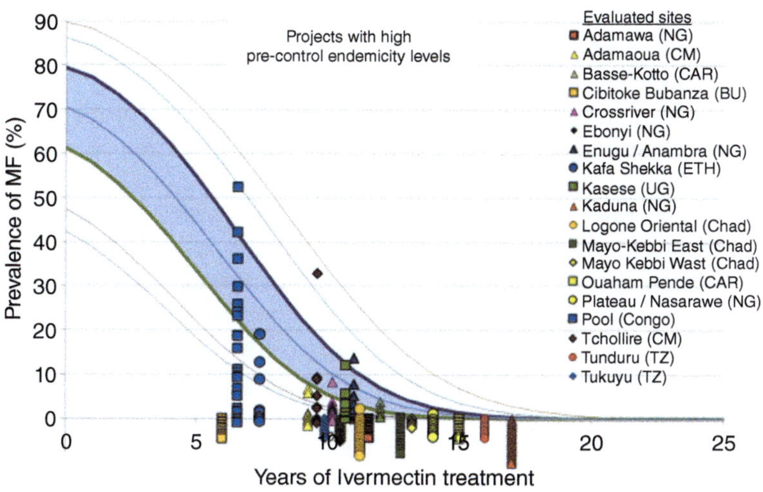


**Figure 12.4** Progress toward elimination in evaluated projects (projects with high pre-control endemicity levels).

to note that, unlike with vector control, the use of IVM rapidly results in the reduction of eye and skin morbidity in those already suffering from *O. volvulus* infections.

Several relevant modalities that are required must be considered if the march toward EoT of onchocerciasis is to be successful. These include the following:

#### 12.3.1.3 Mapping

Mapping is crucial to defining areas that need intervention. Most onchocerciasis-endemic areas have been mapped using either skin snip surveys or rapid epidemiological mapping of onchocerciasis (REMO), especially during the control period, when the focus was to discover areas with the greatest morbidity. The REMO maps provided optimum identification of areas that qualified for IVM distribution during the control era. For the EoT paradigm, it is essential to identify all areas that have sustained transmission, no matter if morbidity is a problem or not. IVM distribution programs must then be launched in these additional (hypoendemic) areas. This requires other diagnostics and approaches for mapping, such as remote sensing data, GIS spatial models, and OV16 testing in children and adults. Taken together, these approaches are called onchocerciasis elimination mapping (OEM), the purpose of which is to ensure that the new endemicity maps identify all communities that require treatment [2, 9, 20, 21].

#### 12.3.1.4 Coverage

Two important factors that make a positive or negative difference in an MDA intervention are "geographic coverage" and "therapeutic coverage." Geographic coverage describes the areas that have been identified to require IVM distribution. 100% geographic coverage indicates that the distribution program reaches all endemic communities in all transmission areas. Any endemic area not geographically covered during treatment remains a local source of transmission and will compromise the elimination agenda. Hypoendemic areas in the control era were excluded from treatment. All such low transmission areas need to be geographically covered for intervention in the elimination phase. A new protocol for OEM is being piloted for use in IVM-naïve areas in endemic countries to re-define required geographic coverage [2].

While geographic coverage focuses on communities, therapeutic coverage encompasses the coverage of all individuals in the population in each area that qualifies for IVM treatment. The minimum acceptable therapeutic coverage is 65%; the optimum level in light of the change of paradigm to onchocerciasis elimination by APOC is 80%. It should be noted that the therapeutic coverage calculation denominator includes persons ineligible for IVM (i.e., pregnant women, children under five years of age and the very sick) such that, ideally, programs are expected to approach close to 100% coverage of the eligible population in the quest for elimination of onchocerciasis. This is not always possible due to factors such as a high proportion of the populations composed of children under 5, pregnancy rates, absenteeism, illness, fatigue of multiple MDA programs, inadequate and poor social mobilization, decreased enthusiasm on the part of volunteer drug distributors, un-sustained political commitment, and abject refusals.

#### 12.3.1.5 Diagnostics

Diagnostics need to be more sensitive and specific to detect changes in low prevalence areas during elimination activities. As prevalence reaches zero, false positives (specificity), or the availability of confirmatory tests, become more critical. The skin snip method, hitherto the gold standard of OCP, APOC, and OEPA, is specific and quantitative (e.g., can be used to calculate intensity of infections expressed as the community microfilaria (mf) load-CMFL) but is unable to detect infections with very low microfilaridermias and is less and less acceptable to the population. Skin snips are now used mainly as a research tool for developing new diagnostics and testing candidate medicines for onchocerciasis.

WHO currently recommends the use of serology to detect IgG4 antibodies to the OV16 recombinant antigen, based on the sensitivity and specificity of the test when used at a population level to detect infection signals in low endemicity areas [22, 23]. The 4 countries in the Americas verified by WHO to have eliminated onchocerciasis relied in part on OV16 ELISA testing in children to indicate no recent transmission events in an area. The rapid diagnostic test (RDT) using OV16 has a high specificity (97–98%), is easy to use, and can provide rapid results [23, 24], although the ELISA method is more sensitive. An ideal diagnostic test, one that is field accessible, sufficiently specific and sensitive for detecting current infection, acceptable to the population and can be used to assess the elimination status in transmission zone, is urgently required to support the elimination process.

#### 12.3.1.6 Monitoring and Evaluation

Periodic monitoring to assess progress toward elimination and eventual evaluation exercises to meet WHO elimination criteria are essential parts of the EoT agenda.

Evaluations help detect shortcomings in process indicators of the intervention (therapeutic and geographic coverage of IVM MDA programs) and are important so that measures to correct such situations can be taken. In the African context, where interventions for elimination of onchocerciasis are undertaken through CDTi projects, procedures have been established for evaluation, analysis, and interpretation of results [25]. WHO/ESPEN recommends a first evaluation six years into intervention with IVM MDA, using skin snips to determine the decline in prevalence of infection as well as the CMFL in sentinel communities. The process is repeated every three years or so until the set elimination threshold (1/2000 polymerase chain reaction [PCR]-positive flies) level of infection is reached. Modeling is used to help interpret the results and to make a prediction of when the elimination threshold will be attained. This takes into account the pre-intervention endemicity level and the treatment coverage through the period under review [25].

#### 12.3.1.7 Modeling

Modeling has played an important role in onchocerciasis control and elimination programs. It was used as a prediction tool and helped developed forecasts that were used to make important OCP programmatic decisions as, for example, (i) when to stop larviciding, (ii) the likelihood of recrudescence of infection and how to control it, and (iii) possible outcomes/impact of different frequencies of IVM application based on levels of endemicity. Modeling remains extremely important in the elimination era. Since the original ONCHOSIM model was developed and used operationally by OCP and APOC, many other models for onchocerciasis have become available. Comparison of these models as well as their adaptation to the use of serology rather than skin snips in the onchocerciasis elimination era is still required, as real-life results become available to test their predictive values [18, 23].

#### 12.3.1.8 *Loa loa* Co-Endemicity

A major challenge in the quest to end onchocerciasis is the filarial parasite *L. loa*, which is often coendemic with onchocerciasis in central Africa. The vector of *L. loa* is *Chrysops* spp. deerflies that reside in forested areas. Clinically, *L. loa* infections are known for the tendency of adult filarial worms to move across the conjunctiva of the eye. The mf of this parasite appear in the blood, often in massive numbers exceeding 30 000 per ml. These mf are sensitive to IVM, which causes their rapid death, which can provoke central nervous system adverse events such as altered consciousness and (infrequently) coma [26, 27]. The Rapid Assessment Procedure for *Loa loa* (RAPLOA), which uses eyeworm self-reports as a proxy for *L. loa* infection, has been utilized to identify likely areas of high-density infection where IVM treatment expansion in hypoendemic onchocerciasis communities has up to now been avoided [28, 29].

As noted earlier, IVM MDA treatment of hypoendemic areas is necessary under the new EoT onchocerciasis paradigm, since untreated hypoendemic areas may constitute a source of ongoing transmission. As yet, a solution has not been found to this challenge, resulting in many countries (such as Gabon and Democratic Republic of Congo) being left out of the 2025 elimination goal.

#### 12.3.1.9 Challenges and Alternative Approaches to Intervention

Broad areas in Africa have been mapped and provided MDA for some years; these may reach EoT by 2025. However, mapping of all transmission zones in Africa, which is needed to determine where IVM treatment still must be administered, is not yet complete. This unknown treatment need is a major factor that makes it unlikely that Africa will reach 80% EoT by 2025.

The diagnostics required to determine the threshold of transmission interruption at the tail end of long-standing MDA programs still need to be resolved. Although model predictions and previous empirical information indicate that infection and transmission do not have to be completely zero to trigger halting MDA [9], optimal diagnostics to confirm trends are essential.

Annual MDA has been largely satisfactory and adequate in many endemic African countries. Some countries/projects have used twice a year treatment in an attempt to reduce the time to elimination (according to modeling by about 40%). In problematic areas where coverage has not been optimum or new areas starting treatment for the first time, twice a year or some other frequency of treatment may be warranted. The decision to change to twice a year or another frequency of treatment should take into consideration the cost implications.

IVM is the drug of choice for MDA. Although resistance has not been explicitly demonstrated, there have been reports of suboptimal IVM response [30]. The presence of a second-line drug suitable for MDA, preferably a macrofilaricide that could sterilize or kill adult worms, would be useful. Moxidectin (MOXI) is in the same class as IVM but has a considerably longer half-life. A single annual dose of MOXI has the same effect at suppressing mf in skin as twice per year IVM, and modeling suggests a similar reduction in time to elimination by about 40% [31]. It may be instructive to note, however, that despite a similar mode of action as IVM, theoretically MOXI may retain efficacy against IVM-resistant strains, although intestinal parasites resistant to IVM show some degree but not complete cross-resistance to MOXI [32]. Doxycycline, which kills the *Wolbachia* endosymbionts necessary for adult worm health, is macrofilaricidal against *O. volvulus* [33], but the prolonged daily treatment course is not suitable for MDA except in specific situations.

Cross-border movements and migration may result in re-establishment of transmission in areas where it has been interrupted or eliminated. Additionally, migration of infected flies may be assisted (especially in West Africa) by seasonal wind currents that may carry them hundreds of kilometers to reinstate transmission. Conflicts and difficult to access regions tend to disrupt the frequency with which communities receive IVM. All these phenomena affect coverage and can increase the duration of treatment required to attain elimination.

#### 12.3.1.10 Current Status and Progress toward Elimination of Onchocerciasis

OCP and APOC supported onchocerciasis-endemic countries to set up CDTi projects for onchocerciasis control. The CDTi approach is still valid for implementing IVM distribution in the elimination phase. CDTi is usually carried out once a year, but it has been used for twice a year distribution in Africa in former OCP (Togo, Ghana,

Benin, etc.) and APOC countries (Uganda, Sudan, Nigeria, and Ethiopia). The OEPA countries in the Americas have used twice or four times per year treatment to accelerate interruption of transmission [4]. Through annual MDA with IVM, elimination of infection has been achieved in a few focal areas in Africa [34]. It is important to recognize that the duration of MDA required to achieve elimination depends on pretreatment endemicity level, frequency of treatment, and treatment coverage. The lower the endemicity level, the fewer the number of rounds of treatment needed to attain elimination [9]. WHO Weekly epidemiological record No. 47 [2] for the period 2017–2018 on onchocerciasis reports that (i)142.5 million received IVM, and (ii) there was 88.5% geographical coverage, ranging from a low of 69.4% to 17 countries achieving 100% coverage. Table 12.1 summarizes the global MDA status for onchocerciasis for 2017.

Although the geographic coverage is high in several countries, onchocerciasis elimination requires 100% coverage in all areas concerned. The aim, therefore, is to strive to attain this level of coverage, with minimum 65% therapeutic coverage. Several countries have reported focal interruption of transmission of onchocerciasis, and stoppage of MDA has been reported in 3 countries (Ethiopia, Nigeria, and Uganda) that have started PTS after meeting the criteria for stopping MDA according to the 2016 WHO guidelines [10]. In Uganda, several sub-national areas have completed the required three years of PTS, and as a result, more than 1.15 million

**Table 12.1** Global MDA for onchocerciasis in 2017.

| MDA/WHO region | Africa | Americas | Eastern Mediterranean | Global |
|---|---|---|---|---|
| Total population requiring MDA in 2017 | 204 135 056 | 30 561 | 1 058 579 | 205 151 196 |
| Population no longer requiring MDA | 1 157 303 | 538 517 | 120 000 | 1 815 820 |
| No. of districts requiring MDA | 1 635 | 2 | 37 | 1 674 |
| No. of districts delivering MDA | 1 447 | 2 | 4 | 1 453 |
| No. of districts no longer requiring MDA | 8 | 11 | 3 | 22 |
| Proportion of districts achieving effective coverage | 88.9 | 100 | 100 | 88.9 |
| Reported population treated in 2017 | 142 037 943 | 23 141 | 419 037 | 142 480 |
| Geographical coverage (%) | 88.5 | 100 | 10.8 | 86.8 |
| National coverage (%) | 69.6 | 75.7 | 38.6 | 69.4 |

Source: Adapted from WHO [2].

Ugandans are now considered to live in an area where transmission has been eliminated [2].

Progress toward elimination of onchocerciasis transmission has been greatest in the Americas, where 11 of the 13 original foci have stopped MDA and completed PTS evaluations that demonstrated no recrudescence of transmission. The 4 countries of the Americas that have been verified by WHO to be free of transmission are the only countries in the world that have achieved that status. 96% of MDA in the region has been stopped compared to peak MDA delivery in 2008 (Figure 12.5), and 94% of the original population at risk now reside in areas that have completed PTS (Map 12.1). The remaining MDA programs in the Americas are now only directed at about 35,000

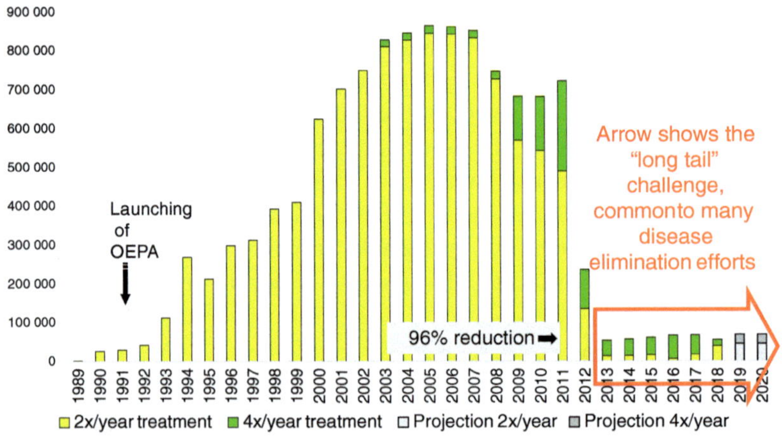


**Figure 12.5** 1989–2018 History of Ivermectin Treatment in the Americas.

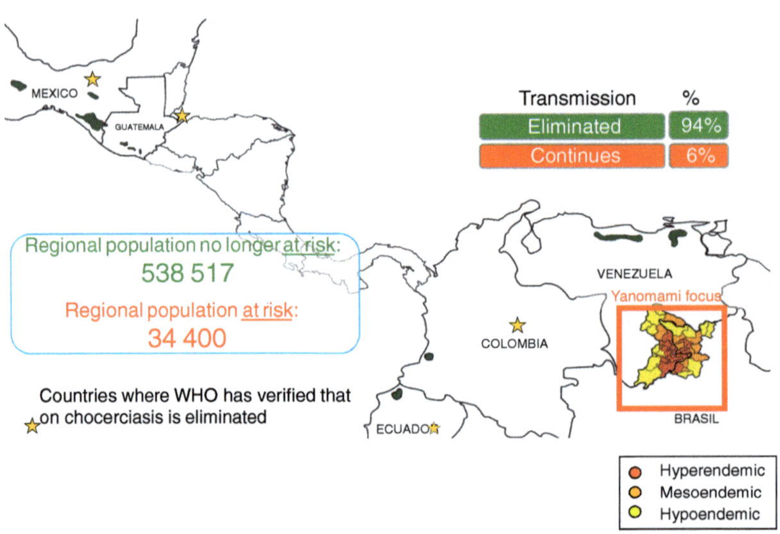


**Map 12.1** Onchocerciasis transmission status in the Americas, and range of endemicity in the remaining transmission focus.

indigenous people residing in an extremely remote area in the Amazon jungle on the border between Brazil and Venezuela, accessible only by air or boat.

#### 12.3.1.11 Outlook for the Future

The elimination of onchocerciasis will be attained by 2025 in the OEPA countries and most likely Yemen. Indeed, the challenge for the Americas in completing the "final inch" is well-reflected in the required "long-tail" of continued interventions in remote and difficult areas. Undoubtedly, there will be similar "long tail" areas that will remain to be treated in Africa as the final goal of transmission elimination is approached near the end of this decade. The establishment of independent national onchocerciasis elimination committees by ministries of Health has been recommended by WHO/ESPEN to facilitate the drive for the elimination program activities. Optimistically and realistically, one would expect a large proportion of Africa to have attained elimination of onchocerciasis transmission by 2025 with the operationally problematic areas following by 2030.

#### 12.3.1.12 Research

Operations and basic research remain very important to the elimination agenda. Priorities include: improved diagnostics; optimization of the use of current intervention tools such as IVM, MOXI, doxycycline, and vector control; a continued search for novel (especially macrofilaricidal) agents; monitoring for the emergence of IVM resistance; standardization of methods for mapping, monitoring and evaluation and surveillance; modeling to support monitoring, evaluation, and PTS decisions; development of approaches to stop onchocerciasis transmission in coendemic *L. loa* areas; and development of a strong evidence base to advise and, if necessary, revise current WHO onchocerciasis elimination guidelines.

## 12.4 Elimination of Lymphatic Filariasis

Compared to onchocerciasis, LF is a global disease, with over one billion people at risk in over 80 countries; more than 120 million are infected in both rural and urban areas. Over 56% of the global burden of LF is in Asia (especially in the Indian subcontinent), an estimated 37% in Africa and the remainder (6%) in the Western Pacific, Americas, and Middle East. Africa's total at-risk population is estimated to be 432 million, residing in 36 of 49 sub-Saharan countries [35, 36].

The Global Programme for the Elimination of Lymphatic Filariasis (GPELF) was established in early 2000 with the goal of eliminating LF as a public health problem by 2020. Several advances and factors provided the impetus to the WHA, international communities, national governments, and pharmaceutical entities to pursue the elimination of LF. These include: (i) the declaration of the ITFDE in 1993 that LF could be eradicated [8]; (ii) subsequent passage of WHA Resolution 50.29; (iii) new understanding about the severity and impact of the disease; (iv) new, practical, and reliable point-of-care diagnostic and monitoring tools [37, 38]; (v) the prospect that LF could be eliminated through a five to six year annual

MDA program with drugs donated by pharmaceutical companies; (vi) new disability management approaches; (vii) development of partnerships among academia, researchers, and donors; and (viii) proof-of-principle examples of successful elimination programs from countries such as China, Korea, Japan, and the Solomon Islands [33, 39, 40].

Unlike onchocerciasis programs in Africa, the GPELF was set up to be an elimination rather than a control program. One of the most important differences was the much lower prevalence required for launching LF MDA; during the disease mapping stage of the LF program, 1% circulating filarial antigen was set as the MDA threshold (compared to 30–40% infection threshold for onchocerciasis in APOC). Another important difference was that the LF program, unlike the onchocerciasis program, quickly moved away from complicated microscopic (parasitological) tests to rapid finger-prick-based serological diagnostics. Hitherto, night blood testing for mf was the standard test for LF. This test, although highly specific, was insensitive at low prevalence rates, cumbersome for teams (as they had to be present in villages at night in most LF areas where mf circulate after 10 p.m.), and inconvenient to the population. The advent of a rapid format filarial antigen-based diagnostic test that could be done during the day [38] revolutionized the ability to rapidly map areas for MDA via a method known as RAGFIL (rapid assessment of geographical distribution of filariasis) [41].

## 12.5 Framework/Steps toward Elimination of Transmission

The framework/steps used in the implementation of the LF program include mapping, at least five years of MDA, a stop MDA assessment, and then five years of post-MDA surveillance (Figure 12.6). The critical decision of when to stop MDA is based on a transmission assessment survey (TAS) that evaluates antigen levels in children six to seven years of age [42]. It is assumed that if antigen levels in these children are below 2%, transmission in the larger community has been interrupted and MDA may be stopped. Post-MDA surveillance consists of repeating TAS surveys in children three and five years following the end of MDA. When an entire country

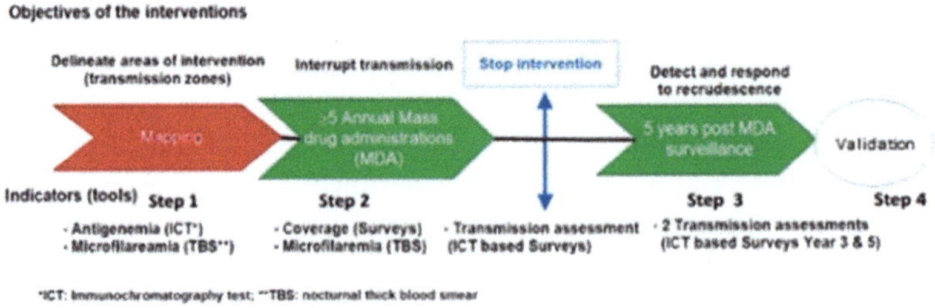


**Figure 12.6** Steps towards elimination of transmission of lymphatic filariasis. Source: Adapted from Sodahon et al. [5].

has completed these steps, it may request WHO validation of the elimination of LF as a public health problem.

In contrast to the REMO approach (using nodule rates) to support the development of CDTI projects as the base unit for implementation of MDA, the LF program uses immunochromatographic test (ICT) antigen-based test and the RAGFIL to identify Implementation Units (IUs) for MDA. These IUs are usually the entire district in which a village with >1% antigenemia is identified. MDA is carried out annually for five to six years if the ICT threshold is 1% or higher.

### 12.5.1 Implementation of the LF Elimination Program

#### 12.5.1.1 MDA for LF

The entire at-risk population in LF endemic areas is targeted for MDA to interrupt parasite transmission. MDA, which is usually carried out by health workers or/and volunteer community drug distributors (CDD), is generally conducted door-to-door, and the treatment is ideally directly observed by the CDD (various combinations to this accepted strategy, however, are in use). The recommended drugs used in MDA are administered mostly in once-yearly, single doses of two drug combinations: 6 mg/kg DEC + 400 mg ALB; or 150 μg/kg IVM + 400 mg ALB (in areas that are also endemic for onchocerciasis); 400 mg ALB is also used twice per year in areas that are coendemic for *L. loa* [43]. As in onchocerciasis MDA, dosing poles are used for IVM treatment to approximate weight, thus avoiding the use of scales. An alternative and equally effective community-wide regimen used in a few settings is the use of common table salt or cooking salt fortified with DEC. The successful Chinese LF elimination program was largely based on a combination of MDA and DEC fortified salt [44].

The primary aim is to administer MDA once a year for a minimum of five years, with coverage in each round reaching a minimum of 65% of the population. The objective of MDA is to reduce the level (prevalence and density) of microfilaremia in infected individuals in the communities to levels at which transmission can no longer be sustained by the vector.

Triple drug therapy (IDA) is a new and more efficacious regimen that combines all three current medicines used in double combination LF MDA [45, 46]. In clinical studies, one dose is capable of permanently eliminating microfilaraemia. IDA theoretically can interrupt transmission of LF in two annual cycles if good coverage is achieved. It was recently successfully implemented in four districts of India and in the entire country of Timor-Leste with extensive advocacy and social mobilization inputs. It may be expanded to more districts in coming years in India, Indonesia, and Myanmar.

#### 12.5.1.2 Vector Control

Other interventions may be considered as adjuncts to the main regimen of MDA. Integrated vector management is encouraged to accelerate and sustain the elimination process. Long-lasting insecticidal bed nets (LLLIN) distributed by the malaria control programs have an important synergistic effect with MDA, especially in areas where LF is transmitted by *Anopheles* mosquitoes.

#### 12.5.1.3 Morbidity Management and Disability Prevention (MMDP)

Despite the benefits of an effective MDA program, individuals who already have damage from past LF infections still suffer because the injury to the lymphatic system is permanent and not reversed by antifilarial medicines. Those who live with these disease manifestations need knowledge, support, and in some cases of hydrocele, surgical intervention. MMDP aims to: (i) reduce the misery of acute attacks of adenolymphangitis (fever, pain, and inflammation of swollen limbs related to secondary bacterial or fungal infections due to poor hygiene); (ii) halt and, if possible, reverse the progression of swollen limbs, and (iii) provide surgical intervention for hydrocele. In addition, more focus is being given to relieving the mental health struggles brought on by the suffering, stigma, and hopelessness of this disease. This work will improve productivity and enhance quality of life.

MMDP is the "second pillar" of WHO's requirements for validation of elimination of LF as a public health problem. Unfortunately, funding for this second pillar is lacking compared to the support that the MDA pillar enjoys. The focus is to enable health facilities to help treat persons afflicted by elephantiasis or acute attacks and create a referral program to facilities equipped to perform hydrocele surgeries. Overall, the program relies on ministries of health to strengthen their delivery of these services within the existing peripheral health care infrastructure.

### 12.5.2 Current Status and Progress toward Elimination of Lymphatic Filariasis

The LF elimination program has made remarkable progress since it was established in 2000. The number of treatments in 2000 was around 2 million. In the space of 17 years (2000–2017), a cumulative total of 7.1 billion treatments have been delivered to about 890 million people. In 2017, 465.4 million of the targeted 585.8 million people that required MDA from 37 reporting countries received treatment. The overall program coverage was 79.4%, while geographic and national coverage were 70.7% and 52.4%, respectively (Table 12.1) [47]. The desired 100% geographical coverage was not achieved. During the same period, an estimated 157.1 million children 2–14 years of age (preschool and school-aged) children received MDA.

In Africa, the number of people estimated to require MDA was >343.2 million. More than 202.1 million received treatment. The reported number that received MDA was >202.1 million, and the geographic and national coverages were 84.0% and 58.9%, respectively.

As of 2015, 18 LF endemic countries had reduced infection prevalence to levels at which transmission is unsustainable.

The highest proportion of the burden of LF globally is found in the south-east Asia region (SEAR); treatment results in this region are partially summarized in Table 12.2. During the last 18 years, national programs were gradually expanded and a staggering 5.38 billion treatments were reportedly administered to 853 million people. MDA programs cured or prevented an estimated 56.75 million LF cases as of 2013 [48].

**Table 12.2** Progress and current status of program to eliminate LF in SEAR countries as of 2017.

| Country | Number of IUs | Population of IUs (millions) | Number of treatments delivered (millions) | Number of treatments consumed (millions) | % Population no longer requiring MDA | Current program status |
|---|---|---|---|---|---|---|
| India | 256 | 630 | 6377.61 | 4449.5 | 39 | MDA and surveillance |
| Indonesia | 236 | 102 | 453.05 | 280.05 | 43 | MDA and surveillance |
| Myanmar | 45 | 40 | 289.06 | 258.41 | 25 | MDA and surveillance |
| Nepal | 63 | 25 | 166.77 | 124.39 | 60 | MDA and surveillance |
| Timor-Leste | 13 | 12.8 | 4.63 | 3.47 | 100 | MDA likely to be stopped |
| Bangladesh | 19 | 33 | 264.51 | 220.24 | 100 | Surveillance |
| Thailand | 350 | 0.17 | 1.21 | 1.07 | 100 | LF elimination acknowledged |
| Sri Lanka | 8 | 10.46 | 52.77 | 44.79 | 100 | LF elimination acknowledged |
| Maldives | 1 | 0.00 | 0.007 | 0.006 | 100 | LF elimination acknowledged |

Source: Data from Molyneux et al. [39] and WHO [49].

Maldives, Thailand, and Sri Lanka, with an endemic population of 10.63 million, have eliminated LF. Bangladesh is expected to complete surveillance soon and submit the dossier to WHO to obtain LF elimination validation. Timor-Leste, with a population of 12.8 million, became the first country in the world to implement nation-wide triple drug therapy in 2019 and thus earned quicker eligibility to stop MDA [49].

The three large endemic countries in SEAR, India, Indonesia, and Myanmar, along with Nepal, include 600 endemic IUs and a population of 797 million. The enormity, diversity, accessibility and logistics issues, and weaker health systems in some districts pose a challenge to LF elimination efforts. Nonetheless, the national programs achieved steady progress during the last 18 years. MDA geographic coverage is 100%, and MDA was completed and no longer required in 248 districts, with a population of about 380 million. MDA is in progress in the remaining 352 districts (population 417 million). Among the districts, 1 to >12 rounds of MDA were implemented and an average of 6.41 treatments reportedly consumed as of 2017.

GPELF's progress for 2017, represented by the status of MDA in countries, is shown in Figure 12.7. The countries are at various stages of MDA; 3 countries have yet to start MDA and 14 countries are in the post-validation and surveillance stage. Almost half (49%) of the countries in the GPELF are undertaking MDA at the stage in which MDA is scaled to all endemic IUs.

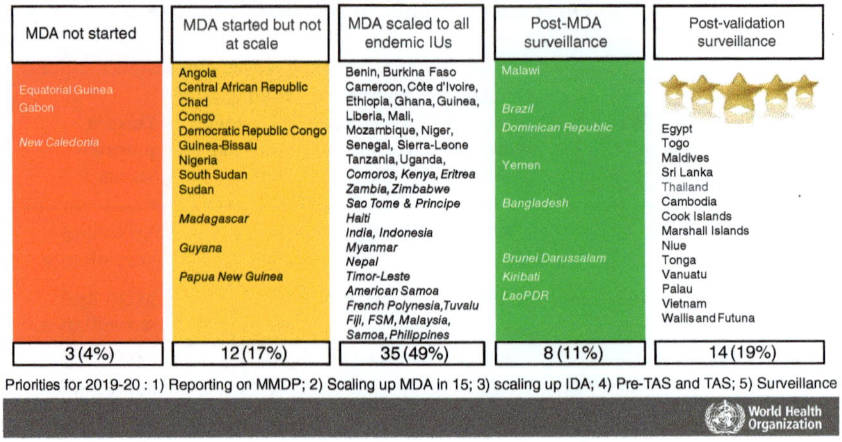


**Figure 12.7** GPELF Progress: MDA status of countries 2018.

### 12.5.3 Outlook for the Future

The original target set for the elimination of LF by 2020 could not be met. There is, however, optimism that, given the rapid and considerable progress that has been made in virtually all LF endemic regions, there is considerable momentum to move toward realization of the elimination goal. In addition, the new three drug IDA combination should accelerate progress toward elimination outside of African countries where DEC cannot be used because of *L. loa* and onchocerciasis.

### 12.5.4 Challenges

Challenges faced by the LF elimination program are common to many of the national programs. They include: (i) lack of or poor political will, which calls for concerted efforts at advocacy; (ii) low priority and lack of monitoring and accountability, resulting in neglect of the program at various levels; (iii) sub-optimal treatment coverage and monitoring and evaluation (M&E), leading to too many rounds of MDAs; (iv) inadequacy in data reporting; and (v) difficulty of delivering effective MDA coverage in urban areas.

## 12.6 Conclusion

The next WHO NTD roadmap for the period 2021–2030 is expected to set ambitious goals for 2030 with respect to elimination of onchocerciasis and LF. The current situation is consistent with the elimination of onchocerciasis from the Americas. Many countries in the Africa region are expected to have profited from the many rounds of MDA, making its elimination a reality in the time frame of 2021–2030, but the *L. loa* issue and the need to complete mapping of hypoendemic areas will remain challenges that need solutions in the near term. LF has made progress toward elimination in countries in all regions and the introduction of IDA [50] is likely to accelerate this progress outside of Africa. Support for MMDP as the second pillar of LF EPHP will remain a challenge to be met as long as outreach health delivery services are weak.

## References


**1** World Health Organization (2010). First WHO report on neglected tropical diseases: working to overcome the global impactworking to overcome the global impact of neglected tropical diseases. World Health Organization, Geneva. 171 pp.

**2** World Health Organization (2018). Progress report on the elimination of human onchocerciasis, 2017–2018. *Weekly Epidemiol. Rec.* 93 (47): 633–648.

**3** Boatin, B.A. and Amazigo, U.V. (2016). Onchocerciasis. In: *Neglected Tropical Diseases- Sub-Saharan Africa* (ed. J. Gyapong and B. Boatin), 159–181. Springer International Publishing Switzerland.

**4** Sauerbrey, M.S., Rakers, L.J., and Richards, F.O. (2018). Progress toward elimination of onchocerciasis in the Americas. *Int. Health* 10: i71–i78.

**5** Sodahon, Y., Malecela, M., and Gyapong, J.O. (2016). Lymphatic filariasis (elephantiasis). In: *Neglected Tropical Diseases – Sub-Saharan Africa* (ed. J. Gyapong and B. Boatin), 187–230. Springer International Publishing Switzerland.

**6** Gyapong, J.O. (2016). An overview of neglected tropical diseases in Sub-Saharan Africa. In: *Neglected Tropical Diseases- Sub-Saharan Africa* (ed. J. Gyapong and B. Boatin), 1–14. Springer International Publishing Switzerland.

**7** Ottesen, E.A. (2000). The global programme to eliminate lymphatic filariasis. *Trop. Med. Int. Health* 5 (9): 161–167.

**8** Centres for Disease Control and Prevention (1993). Recommendation of the international task force for disease eradication. *Morb. Mortal. Wkly. Rep.* 42 (RR-16): 1–27.

**9** Dadzie, Y., Amazigo, U.V., Boatin, B.A., and Seketeli, A. (2018). Is onchocerciasis elimination in Africa feasible by 2025: a perspective based on lessons learnt from the African control programmes. *Infect. Dis. Poverty* 7: 63.

**10** World Health Organization (2016). *Guidelines for Stopping Mass Drug Administration and Verifying Elimination of Human Onchocerciasis*. Geneva, Switzerland: World Health Organization http://apps.who.int/iris/bitstream/10665/204180/1/9789241510011.

**11** Dowdle, W.R. (1998). The principles of disease elimination and eradication. *Bull. World Health Org.* 76 (Suppl 2): 22–25.

**12** Cochi, S.L. and Dowdle, W.R. (2011). *Disease Eradication in the 21st Century: Implications for Global Health*. Cambridge, MA: The MIT Press.

**13** Zarroug, I.M.A., Hashim, K., El Mubark, W.A. et al. (2016). The first successful confirmed elimination of an onchocerciasis focus in Africa: Abu Hamed, Sudan. *Am. J. Trop. Med. Hyg.* 95: 1037–1040.

14 Katabarwa, M., Lakwo, T., Habomugisha, P. et al. (2018). After 70 years of fighting an age-old scourge, onchocerciasis in Uganda, the end is in sight. *Int. Health* 10: i79–i88.

15 Diawara, L., Traore, M.O., Badji, A. et al. (2009). Feasibility of onchocerciasis elimination with ivermectin treatment in endemic foci in Africa: first evidence from studies in Mali and Senegal. *PLoS Negl.Trop. Dis.* 3: e497.

16 Traore, M.O., Sarr, M.D., Badji, A. et al. (2012). Proof-of-principle of onchocerciasis elimination with ivermectin treatment in endemic foci in Africa: final results of a study in Mali and Senegal. *PLoS Negl.Trop. Dis.* 6 (9): e1825.

17 Plaisier, A.P., Alley, E.S., Boatin, B.A., and Van Oortmarssen, G.J. (1995). Irreversible effects of ivermectin on adult parasites in onchocerciasis patients in the onchocerciasis control programme in West Africa. *J. Infect. Dis.* 172: 204–210.

18 Stolk, W.A., Walker, M., Coffeng, L.E. et al. (2015). Required duration of mass ivermectin treatment for onchocerciasis elimination in Africa: a comparative modelling analysis. *Parasites Vectors* 8: 552.

19 Turner, H.C., Osei-Atweneboana, M.Y., Walker, M. et al. (2013). The cost of annual versus biannual community-directed treatment of onchocerciasis with ivermectin: Ghana as a case study. *PLoS Negl.Trop. Dis.* 7 (9): e0002452.

20 O'Hannon, S.J., Slater, H.A.C., Cheke, R.A. et al. (2016). Model-based geostatistical mapping of the prevalence of *Onchocerca volvulus* in West Africa. *PLoS Negl.Trop. Dis.* 10: e0004328.

21 Jacob, B.G., Novak, R.J., Toe, L.D. et al. (2013). Validation of a remote sensing model to identify *Simulium damnosum* s.l breeding sites in sub-Saharan Africa. *PLoS Negl.Trop. Dis.* 7: e0002342.

22 Richards, F.O. Jr., Katabarwa, M., Bekele, F.M. et al. (2018). Operational performance of the *Onchocerca volvulus* 'OEPA' Ov16 ELISA serological assay in mapping, guiding decisions to stop mass drug administration, and posttreatment surveillance surveys. *Am. J. Trop. Med. Hyg.* 99: 749–752.

23 Gass, K.M. (2018). Rethinking the serological threshold for onchocerciasis elimination. *PLoS Negl. Trop. Dis.* 12: e0006249.

24 Unnasch, T.R., Golden, A., Cama, V., and Cantey, P.T. (2018). Diagnostics for onchocerciasis in the era of elimination. *Int. Health* 10 (suppl. 1): i20–i26.

25 Tekle, A.H., Zoure, H.G., Noma, M. et al. (2016). Progress towards onchocerciaisi elimination in the participating countries of the African programme for onchocerciasis control: epiemiological evalution results. *Infect. Dis. Poverty* 5: 66.

26 Boussinesq, M., Gardon, J., Gardon-Wendel, N. et al. (1998). Three probable cases of *Loa loa* encephalopathy following ivermectin treatment for onchocerciasis. *Am. J. Trop. Med. Hyg.* 58: 461–469.

27 Twum-Danso, N.A. (2003). *Loa loa* encephalopathy temporally related to ivermectin administration reported from onchocerciasis mass treatment programs from 1989 to 2001: implications for the future. *Filaria J.* 2 (Suppl. 1): S7.

28 Zoure, H.G., Wanji, S., Noma, M. et al. (2011). The geographic distribution of *Loa loa* in Africa: results of large-scale implementation of the rapid assessment procedure for Loiasis (RAPLOA). *PLoS Negl.Trop. Dis.* 5: e0001210.

**29** Kamgno, J., Nana-Djeunga, H.C., Pion, S.D. et al. (2018). Operationalization of the test and not treat strategy to accelerate the elimination of onchocerciasis and lymphatic filariasis in Central Africa. *Int. Health* 10: i49–i53.

**30** Awadzi, K., Attah, S.K., Addy, E.T. et al. (2004). Thirty month follow-up of suboptimal responders to multiple treatments with ivermectin, in two onchocerciasis-endemic foci in Ghana. *Ann. Trop. Med. Parasitol.* 98 (4): 231–249.

**31** Opoku, N.O., Bakajika, D.K., Kanza, E.M. et al. (2018). Single dose moxidectin versus ivemectin for *Onchocerca volvulus* infection in Ghana, Liberia, and Democratic Republic of the Congo: a randomised, controlled, double-blind phase 3 trial. *Lancet* 392 (10154): 1207–1216.

**32** Bygarski, E.E., Prichard, R.K., and Ardelli, B.F. (2014). Resistance to the macrocyclic lactone moxidectin is mediated in part by membrane transporter P-glycoproteins: implications for control of drug resistant parasitic nematodes. *Int. J. Parasitol. Drugs Drug Res.* 4: 143–151.

**33** Hoerauf, A., Specht, S., Marfo-Debrekyei, Y. et al. (2009). Efficacy of 5-week doxycycline treatment on adult *Onchocerca volvulus*. *Parasitol. Res.* 104: 437–447.

**34** Tekle, A.H., Elhassan, E., Isiyaku, S., Amazigo, U.V. (2009) Impact of long-term treatment of onchocerciasis with ivermectin in Kaduna. World Health Organisation informal consultation on elimination of onchocerciasis transmission with current tools in Africa-shrinking the map. Ouagadougou: African programme for onchocerciasis control: 2009.

**35** World Health Organization (2012). Global programme to eliminate lymphatic filariasis: progress report 2011. *Weekly Epidemiol. Rec.* 87 (37): 346–356.

**36** Addis, D. and Global Alliance to Eliminate lymphatic Filariasis (2010). The 6th meeting of the global Alliance to eliminate filariasis: a half-time reviews of lymphatic filariasis elimination and integration with the control of other neglected tropical diseases. *Parasites Vectors* 3: 100.

**37** Ottesen, E.A., Duke, B.O., Karam, M., and Behbehani, K. (1997). Strategies and tools for the control/elimination of lymphatic filariasis. *Bull. World Health Org.* 75 (6): 491–503.

**38** Weil, G.J., Lammie, P.J., and Weiss, N. (1997). The ICT filariasis test: a rapid-format antigen test for diagnosis of bancroftian filariasis. *Parasitol. Today* 13: 401–404.

**39** Molyneux, D.H., Neira, M., Liese, B., and Heymann, D. (2000). Lymphatic filariasis: setting the scene for elimination. *Trans. R. Soc. Trop. Med. Hyg.* 94: 589–591.

**40** Molyneux, D.H. and Zagaria, N. (2002). Lymphatic filariasis elimination: progress in global programme development. *Ann. Trop. Med. Parasitol.* 96 (Suppl 2): S15–S40.

**41** Gyapong, J.O. and Remme, J.H. (2001). The use of grid sampling methodology for rapid assessment of the distribution of bancroftian filariasis. *Trans. R. Soc. Trop. Med. Hyg.* 95 (6): 681–686.

**42** World Health Organization (2006) *Preventive Chemotherapy in Human Helminthiasis: A Manual for Health Professionals and Programme Managers*. WHO, Geneva. http://www.who.int/iris/handle/10665/43545.

**43** World Health Organisation (2012). *Provisional Strategy for Interrupting Lymphatic Filariasis Transmission in Loiasis-Endemic Countries: Country Report of the Meeting on Lymphatic Filariasis, Malaria and Integrated Vector Management*. Geneva: WHO.

**44** De-jian, S., Xu-li, D., and Ji-hui, D. (2013). The history of the elimination of lymphatic filariasis in China. *Infect. Dis. Poverty* 2: 30.

**45** King, C.L., Suamani, J., Sanuku, N. et al. (2018). A trial of triple-drug treatment for lymphatic filariasis. *N. Engl. J. Med.* 379 (19): 1801–1810.

**46** Weil, G.J., Bogus, J., Christian, M. et al. (2019). The safety of double- and triple-drug community mass drug administration for lymphatic filariasis: a multicentre, open-label, cluster-randomized study. *PLoS Med.* 16 (6): e100283946.

**47** World Health Organisation (2018). Global programme to eliminate lymphatic filariasis: progress report, 2017. *Weekly Epidemiol. Rec.* 93: 589–604.

**48** Ramaiah, K.D. and Ottesen, E.A. (2014). Progress and impact of 13 years of the global programme to eliminate lymphatic filariasis on reducing the burden of filarial disease. *PLoS Negl.Trop. Dis.* 8 (11): e0003319.

**49** World Health Organization. (2017). Guideline – Alternative mass drug administration regimens to eliminate lymphatic filariasis. World Health Organization. Geneva. WHO/HTM/NTD/PCT/2017.07.

**50** King, C.L., Suamani, J., Sanuku, N. et al. (2018). A trial of triple-drug treatment for lymphatic filariasis. *N. Engl. J. Med.* 379 (19): 1801–1810.

# Part II

# Drug Discovery for Novel Antifilarials

# 13

# Global Economics of Heartworm Disease

*Darrell Klug*[1,*] *and Jason Drake*[2]

[1] *MBA, Darrell Klug Consulting, 5 Waldron Ct., Greensboro, NC 27408, USA*
[2] *DVM, Dip. ACVM-Parasitology, Global Technical Marketer – Pet Health Parasiticides, Elanco Animal Health, 2500 Innovation Way, Greenfield, IN 46140, USA*

**Abstract**

Heartworm disease caused by infection with *Dirofilaria immitis* causes a pathological impact on dogs and an economic impact on pet owners. This chapter outlines the worldwide economic impact of heartworm disease on the pet owner. The cost is calculated by separating cost components into the cost of prevention (heartworm medication), the cost of treatment, and the opportunity cost of treatment. From a geographic standpoint, separate estimates on the impact within the key heartworm countries (United States, Australia, Japan, Italy, Spain, and Canada) and the rest of the world in aggregate are provided. These calculations provide an estimated total global cost of heartworm disease of US$ 2.47 billion dollars. The cost of prevention makes up 93%, or $2.30 billion of this total, with the cost of treatment representing 6%, or $146 million, and the opportunity cost to the pet owner of 1% or $24.6 million. From a geographic standpoint, the United States accounts for 65% of the costs ($1.6 billion), Japan 10% ($257 million), Canada 7% ($174 million), Italy 6% ($150 million), Spain 4% ($101 million), Australia 3% ($80 million), and the rest of the world 4% ($106 million). The global costs of heartworm will likely increase in the future as the disease spreads geographically, the prevalence of heartworm increases in countries where it is currently found, and the standard of care for pets continues to increase throughout the world.

## 13.1 Introduction

Heartworm disease caused by *Dirofilaria immitis* not only carries a pathologic impact on pet dogs and cats but also brings significant emotional and economic impacts on pet owners and provides economic opportunities for practicing veterinarians and pharmaceutical manufacturers. Dogs and cats can suffer from clinical signs and pathology related to heartworm infection, while pet owners can struggle

*Corresponding author.

*Human and Animal Filariases: Landscape, Challenges, and Control*, First Edition.
Edited by Ronald Kaminsky and Timothy G. Geary.

with the emotional impacts of concern for their pet while infected and during treatment, plus the additional stress related to the costs of testing, treatment, and prevention. Practicing veterinarians and veterinary health care teams try to balance making the best medical recommendations for at-risk patients while also helping pet owners manage costs related to treatment and prevention of heartworms. Additionally, sales of diagnostics, treatments, and preventives represent a significant source of revenue for veterinary hospitals, which often underwrites the costs of other veterinary services. Similarly, manufacturers of heartworm diagnostics, treatments, and preventives can profit from increased awareness of the risks of heartworm disease through increasing sales of heartworm-related products. In this chapter, we explore the financial implications related to diagnosis, treatment, and prevention of heartworm disease in pets.

## 13.2 Background and Current Situation

*Dirofilaria immitis* infections in pets are found in many countries around the world. *D. immitis* is commonly diagnosed in pet dogs and cats in tropical and subtropical regions around the world [1–3]. Currently, the United States is the largest "heartworm market" based upon total sales of heartworm preventives, heartworm treatments, and diagnostic tests. The most detailed data sets for analysis related to heartworm prevalence, usage of heartworm prevention, and other economic factors related to heartworm disease also come from this market. For areas outside United States, many of the economic impacts we discuss have been extrapolated from the US market, while also considering prevalence data and pet populations from a variety of countries. Globally, the most well-developed heartworm markets, based upon routine testing and usage of prevention, include United States, Canada, southwestern Europe, Japan, and Australia. Emerging heartworm markets, where heartworm is endemic but testing and preventive use is lower, includes much of Asia, Mexico, Central America, South America, and Africa.

A recent study analyzing heartworm prevalence in the United States and the usage of preventatives in dogs reported an overall prevalence of 1.28%, with approximately 2/3 of 77 million dogs receiving no heartworm prevention. In addition, this study showed results from over 9 million heartworm tests performed in 2016 and reported via maps created by the Companion Animal Parasite Council (CAPC). According to CAPC, these data represent fewer than 30% of all heartworm tests run in the United States, meaning more than 50% of dogs in the United States were not tested for heartworms in 2016. The dispensing data from this study showed that, in 2016, 20 million dogs were dispensed an average of 8.6 doses of heartworm preventative, or a total of 172 million doses [4].

Prevalence rates around the world vary widely, and we are currently limited in many areas to small studies or local dog populations, typically either pet dog studies or shelter dog studies [5–9]. A few studies have also looked at the prevalence of heartworms in wild canids like foxes, jackals, and coyotes [7, 10, 11].

While local prevalence studies are helpful for determining historic risk of infection, we are seeing a trend involving animal rescue groups and animal shelter networks, which have started moving large numbers of dogs for the purpose of adoption and re-homing. Testing and treatment of parasites is not a requirement prior to movement of dogs in some areas. A survey of animal rescue organizations reported that 2/3 of the organizations provided no medical services related to heartworms, whether that be testing, treatment, or prevention, prior to dog shipments [12]. As animals are relocated, it is possible for them to also bring parasitic infections into local populations. For example, a recent study revealed an increase in heartworm prevalence in Colorado over a period of time when over 114,000 dogs were brought into the state by animal welfare organizations [13]. This animal movement phenomenon is not limited to the United States. International dog movement has also raised concerns with increasing infectious diseases and the spread of parasites related to pets for sale and adoption in the United Kingdom and in western Europe which originate from eastern European sources [14, 15]. As animals are transported into new locations, the risks of parasitic infections may arise in areas not previously thought to be endemic. With no foreseeable end to animal movement and relocation, the future heartworm risk and associated economic impacts seem to be on the rise.

## 13.3 The Global Economic Cost of Heartworm Disease

Economic considerations include a number of factors, including at-risk pet populations, diagnostic testing, pretreatment patient assessments, adulticide treatments, heartworm disease preventatives, hospital visits, and time investments by pet owners and hospital teams. In this section, we examine these elements of cost, calculate a value for each, and estimate the total economic cost of heartworm globally.

Worldwide pet spending is growing significantly as pet owners in developed countries spend more on high-quality food, technology (e.g. pet tracking), and services like boarding and grooming. Pet owners in developing countries are following the trend of humanization of pets as it relates to care. Total spending on pet care globally is expected to be $223 billion in 2019 [16], with the United States accounting for $95.7 billion or 43% of that total, and veterinary care making up 30% of the spend in the United States [17] (Table 13.1). According to the American Pet Products Association (APPA), spending is expected to continue growing as lower prices make care more accessible to a broader market.

Spending on heartworm makes up a significant portion of companion animal veterinary care spending, as seen in the following analysis. We calculate overall out-of-pocket costs on prevention and treatment and also build in an economic estimate of the cost of a pet owner's time when it comes to both prevention and treatment. Most cost calculations focus on out-of-pocket costs only. However, by including the pet owners' cost of time, we provide a more comprehensive and truer estimate of the economic impact for the total cost of the disease.

**Table 13.1** Veterinary spending by category in the United States.

| Category | Spend (billions $s) | % of total spend |
|---|---|---|
| Food | $36.9 | 39% |
| Supplies/over the counter (OTC) medicine | $19.2 | 20% |
| Veterinary care | $29.3 | 31% |
| Other services | $10.3 | 11% |
| Total | $95.7 | 100% |

Our calculations only include the cost to the pet owner. From the standpoint of a veterinarian, there are no unrecovered out-of-pocket costs for the treatment or prevention of heartworm. The veterinarian gets paid for his or her time, and for the diagnostic testing, treatment, and/or preventative products. There is the possibility of an opportunity cost whereby the veterinarian could be spending their time in a more efficient way, or in a more lucrative activity; however, this is beyond the scope of this chapter. So, for the purposes of this chapter, we focus on the complete economic cost to the pet owner, including both out-of-pocket expenses and opportunity costs of time.

The calculations are separated into three different sections, as there are different levels of data available and different levels of heartworm significance in these areas.

(1) the United States
(2) Key Countries of Canada, Australia, Japan, Italy, and Spain
(3) Rest of the World (ROW)

There is more information available about the United States, less available for Key Countries, and even less in the Rest of the World category. Therefore, our calculations require more extrapolation for Key Countries compared to the United States, and even more extrapolation for Rest of the World. However, the United States accounts for a large percentage of the global heartworm spend, so we are comfortable with the global estimates. The formula for the calculation is below:

$$\text{Economic cost of heartworm(EC)} = \text{Cost of prevention } (P) + \text{Cost of treatment } (T) + \text{Opportunity cost of treatment (OC)}$$

where:

$P$ = number of dogs in the country × % HW (heartworm) prevention use × average compliance × monthly cost of heartworm prevention
$T$ = pet owner cost of treatment (charged by the veterinarian) × estimated number of heartworm cases per year × % treated (not slow-kill method)
OC of treatment = ((time driving to veterinarian × trips to the veterinarian + time at veterinarian × number of appointment/treatment visits) × cost per hour) + (number of veterinary visits × estimated travel cost per visit) × number of heartworm cases treated per year.

**Table 13.2** Economic cost of heartworm calculations summary.

| | US | Canada | Japan | Australia | Italy | Spain | ROW | Total World |
|---|---|---|---|---|---|---|---|---|
| Number of dogs (millions) | 69.9 | 8.2 | 12.0 | 3.7 | 7.0 | 4.7 | 365.5 | |
| % treated with HW Preventative | 33% | 33% | 33% | 33% | 33% | 33% | 0.5% | |
| Avg doses per year | 8.6 | 8.6 | 8.6 | 8.6 | 8.6 | 8.6 | 6 | |
| Avg Cost of HW Preventative Monthly Dose | $ 7.42 | $ 7.42 | $ 7.42 | $ 7.42 | $ 7.42 | $ 7.42 | $ 7.42 | |
| **Cost of Prevention ($ millions)** | **$ 1,471.95** | **$ 172.68** | **$ 252.70** | **$ 77.91** | **$ 147.41** | **$ 98.97** | **$ 81.36** | **$ 2,302.98** |
| Treatment of HW Cost to Vet | $ 1,000.00 | $ 1,000.00 | $ 1,000.00 | $ 1,000.00 | $ 1,000.00 | $ 1,000.00 | $ 1,000.00 | |
| Prevalence of HW | 1.28% | 0.16% | 0.30% | 0.30% | 0.30% | 0.30% | 0.05% | |
| Number of dogs test positive (millions) | 0.89 | 0.013 | 0.036 | 0.011 | 0.021 | 0.014 | 0.183 | |
| % treated with Slow Kill | 75% | 75% | 75% | 75% | 75% | 75% | 75% | |
| % treated with Fast Kill | 12.5% | 12.5% | 12.5% | 12.5% | 12.5% | 12.5% | 12.5% | |
| % not treated | 12.5% | 12.5% | 12.5% | 12.5% | 12.5% | 12.5% | 12.5% | |
| **Cost of Treatment ($ millions)** | **$ 111.84** | **$ 1.64** | **$ 4.50** | **$ 1.39** | **$ 2.63** | **$ 1.76** | **$ 22.84** | **$ 146.98** |
| Time Driving to vet (hours) | 0.5 | 0.5 | 0.5 | 0.5 | 0.5 | 0.5 | 0.5 | |
| Trips to the vet | 6 | 6 | 6 | 6 | 6 | 6 | 6 | |
| Time at vet (hours) | 0.5 | 0.5 | 0.5 | 0.5 | 0.5 | 0.5 | 0.5 | |
| Number of Appt/Treatment Visits | 5 | 5 | 5 | 5 | 5 | 5 | 5 | |
| Cost per hour | $ 27.98 | $ 17.31 | $ 18.12 | $ 20.72 | $ 16.80 | $ 12.87 | $ 9.25 | |
| Miles per visit | 10 | 10 | 10 | 10 | 10 | 10 | 10 | |
| Cost per mile | $ 0.58 | $ 0.58 | $ 0.58 | $ 0.58 | $ 0.58 | $ 0.58 | $ 0.58 | |
| Cost of time | $ 153.89 | $ 95.21 | $ 99.66 | $ 113.96 | $ 92.40 | $ 70.79 | $ 50.88 | |
| Cost of travel | $ 34.80 | $ 34.80 | $ 34.80 | $ 34.80 | $ 34.80 | $ 34.80 | $ 34.80 | |
| Total cost per visit | $ 188.69 | $ 130.01 | $ 134.46 | $ 148.76 | $ 127.20 | $ 105.59 | $ 85.68 | |
| **Total Cost for All Visits ($ millions)** | **$ 21.10** | **$ 0.21** | **$ 0.61** | **$ 0.21** | **$ 0.33** | **$ 0.19** | **$ 1.96** | **$ 24.60** |
| **Total Cost $ Millions** | **$ 1,604.89** | **$ 174.53** | **$ 257.80** | **$ 79.51** | **$ 150.36** | **$ 100.92** | **$ 106.16** | **$ 2,474.18** |

There is an opportunity cost of time for prevention as well. For the purposes of these calculations, we assume that the trip to the veterinarian, during which a pet owner purchases or refills their heartworm preventative, is a trip that is of dual use for a health check as well and so do not include this cost for heartworm *per se*.

We go into greater detail for US calculations but follow the same procedure for the Key Country and Rest of the World sections. Table 13.2 provides a summary of the data used in the calculations.

## 13.4 Economic Cost of Heartworm in the United States

The total economic cost of heartworm in the United States can be calculated by adding estimates for the current costs of prevention, treatment, and the opportunity cost for pet owners' time and associated travel costs.

### 13.4.1 Cost of Prevention: The Data and Calculation

According to PetSecure [18], there were 69.9 million dogs in the United States in 2017 (see Table 13.3). Of these dogs, 1/3 is treated with a heartworm preventative, averaging 8.6 monthly doses per year [4].

The average cost to the pet owner for monthly heartworm preventative depends on several variables:

(1) The size of the dog. Typically, monthly preventative for larger dogs costs more than for smaller dogs. For example, according to July 2019 prices on petmeds

**Table 13.3** Pet populations in various heartworm endemic regions.

| Country | Estimated dog population (millions) |
|---|---|
| United States | 69.9 |
| Brazil | 35.8 |
| China | 27.4 |
| Russia | 12.5 |
| Japan | 12.0 |
| Philippines | 11.6 |
| India | 10.2 |
| Canada | 8.2 |
| France | 7.2 |
| Italy | 7.0 |
| Spain | 4.7 |
| Australia | 3.7 |
| Total World estimate | 471 |

.com [19], the cost for a leading heartworm/endoparasiticide product (six-month package) is $5.25 per monthly dose for a dog under 25 lb, $6.75 per monthly dose for a dog between 26 and 50 lb, and $8.25 per monthly dose for a dog over 50 lb.

(2) Location of purchase. Online sites are frequently less expensive than buying at the veterinary clinic. While many online sites appear to have close to the same price, veterinary clinic prices can be 20–30% higher. Veterinarians and pharmaceutical companies sometimes offer incentives and rebates to the consumer to offset this difference in price. We do not include these incentives in our cost estimates.

(3) Formulation (oral, injectable, or topical), and whether or not the product is bought as an endoparasite product or a combination endo-ectoparasite product. If the product purchased is a combination endo-ectoparasite product, there is some subjectivity in the percentage value of the heartworm component of the product.

As a result of these variables, there is a wide range for what a pet owner pays for heartworm prevention. For the purpose of this calculation, we must make some assumptions to derive a realistic average price paid for heartworm prevention by month, as we want to determine a number, not a range. It would be possible to calculate a very granular price per month, but the lack of publicly available data and the subjectivity that is necessary even if we had the data would still make the number an educated guess (i.e. how do you put a value on the heartworm component in a combination endo-ectoparasite product?). To keep it somewhat simple and conservative, we use the six-month package price of a leading branded heartworm preventative for the middle weight range, 25–50 lb, from petmeds.com as of July 2019 [19], and increase that price by 10% to compensate for the portion of pet owners that may purchase at veterinary clinics at a higher price. This gives an average monthly price of $6.75 plus 10% for a total of $7.42 per month on average for what a pet owner pays for heartworm prevention. Several online resources were reviewed, including 4Paws4Rescue [20], which quoted an average monthly price between $3.75 and $8.33, and DVM360, which quoted an average price between $5.00 and $20.83 per month [21]. In all the reviewed sites, a wide range was given and our estimate was in the middle of the ranges. Utilizing these data, we can calculate the cost of heartworm prevention in the United States.

Cost of prevention = number of dogs in the country (69 900 000)
× %HW prevention use (33%) × average compliance (8.6 months)
× average monthly cost ($7.42) = **$1.62 billion**.

So, the total cost to the pet owner for heartworm prevention in the United States is **$1.62 billion** per year.

#### 13.4.1.1 Cost of Treatment – The Data and Calculation

There are three conventional methods for treatment of a dog that has tested positive for heartworm. The "slow kill" method is typically for milder cases and consists of giving the dog a heartworm preventative over an extended period of time. This method is not recommended by either CAPC or the American Heartworm

Society (AHS). The heartworm preventative medication (often combined with doxycycline and restricted activity) reportedly kills the adult worms over a period of one to two years. This method is controversial as it may also allow continued disease progression and may select for resistance to the heartworm preventative. The "fast kill" method involves using a heartworm preventative, doxycycline, and either two or three injections of melarsomine, which is Food and Drug Administration (FDA)-approved for the treatment of adult *D. immitis*. The third option for severe cases (caval syndrome) is surgical removal of the nearly foot-long worms.

The cost of treatment for heartworm is calculated as follows: estimated average veterinary treatment price to the pet owner × number of heartworm cases per year × the estimated percent of cases treated.

Veterinarians typically charge one fee to the pet owner for the entire treatment for heartworms. According to the websites Costhelper Pets and Petcare, the price a veterinarian typically charges a pet owner for treatment of heartworm is around $1,000 [22]. Searching various online sources reveals pricing estimates ranging from $400 to $1,800 to treat adult heartworm infections, with most estimates centering around $1,000. We use this number for our economic cost calculation.

Based on the study by Drake and Wiseman, the prevalence of heartworm in the United States in 2016 was 1.28% [4], meaning that about 985 600 dogs tested positive for heartworm that year. We use this number in our calculation for the cost of treatment. A broad range appears when we search for the number of dogs actually treated after testing positive for heartworm. If we use as the basis for our number a study that investigated management choices for heartworm treatment in Mississippi [23], we can estimate that about 12.5% of heartworm-positive dogs are treated with fast kill, 12.5% of dogs are not treated, and 75% of dogs are treated with slow kill. At this point, we make the assumption that the slow kill dogs are "on heartworm prevention" and are counted in the previous prevention calculation and therefore are not included in this cost of treatment calculation. Thus, about 123 200 dogs are treated with the fast kill method by veterinarians for heartworms.

The calculation for cost of treatment for heartworms in the United States is:

$$\text{Veterinarian charge for treatment (\$1,000)} \times \text{number of HW case per year (985 600)} \\ \times \text{percent treated (12.5\%)} = \mathbf{\$123\,200\,000}.$$

Therefore, the total cost to the pet owner for treatment of heartworm (money paid to the vet) is **$123 200 000** per year.

### 13.4.2 Opportunity Cost of Pet Owner Time for Treatment

The pet owner incurs non-veterinary costs for the treatment of heartworm that need to be included when calculating a total cost of heartworm. The pet owner spends travel time to and from the veterinarian, appointment time with the veterinarian, and incurs the expense of transportation to and from the veterinarian. To determine a specific number, we again make a set of general assumptions for the variables. For this calculation, we assume the pet owner cost calculation for only the fast kill treatment method, as the slow kill involves minimal time and few veterinary visits from the pet owner. The calculation for non-veterinary cost that we label as "opportunity cost" is as follows:

OC of treatment = (((time driving to veterinary hospital
× number of trips to the veterinary hospital + time at veterinary hospital
× number of appointment/treatment visits) × cost per hour)
+ (number of veterinary visits × estimated travel cost per visit))
× number of heartworm cases treated per year).

To calculate these costs, we assume a treatment protocol as outlined below (we assume 3 melarsomine injections):

(1) Pet owner takes the pet to the veterinarian for initial diagnosis. Dog is tested and confirmed at clinic, then an additional test is performed and sent to the diagnostic laboratory for confirmation.
(2) Veterinarian then calls the pet owner confirming heartworm positive results.
(3) Pet owner visits veterinarian for diagnostic work-up for staging the disease, including radiographs, ultrasound, blood work.
(4) Pet owner picks up heartworm preventative and doxycycline.
(5) Thirty days later, the pet owner brings the pet in to the clinic for the 1st injection of melarsomine.
(6) As this is the three-dose protocol, the pet owner brings the pet in again 30 days later for the second injection of melarsomine.
(7) The pet owner then brings the pet back to the veterinarian for the 3rd injection 24 hours after the second injection.

We assume that each visit to the veterinarian (except for step 4) includes travel, travel time, and 30 minutes in the veterinarian clinic. Average travel distance we set at 10 mi (to the clinic and back) and 30 minutes of travel time. The Internal Revenue Service (IRS) used 0.58 cents per mile for standard deductible costs for operating an automobile in 2019 [24], and we use the same figure to estimate the pet owner cost of driving back and forth to the clinic. Based on the protocol above, there are six trips to the vet, with five of these visits including some "appointment or treatment" time. Finally, according to the US Bureau of Labor Statistics, the average hourly earnings of all employees in the United States in 2019 is $27.98 [25]. Based on these assumptions and data, the calculation for the opportunity cost of treatment to the pet owner is as follows:

(Time driving to veterinarian and back (30 minutes)
× number of trips driving to the veterinary hospital (6))
+ (time at the veterinary hospital (30 minutes)
× number of of appointment/treatment visits (5))
× cost per hour ($27.98) + (number of veterinary visits (6)
× estimated travel cost per visit (10 mi at 0.58 per mi = $5.80))
× number of of heartworm cases treated per year (123 200).

This leads to the following calculation:

$$(((30 \text{ minutes} \times 6 \text{ visits}) + (30 \text{ minutes} \times 5 \text{ appointments}) \times \$27.98 \text{ per hour}) + (6 \text{ visits} \times 10 \text{ miles per visit} \times 0.58 \text{ per mile})) \times 123\,200 =$$

$((5.5 \text{ hours} \times \$27.98) + (60 \text{ mi} \times 0.58)) \times 123\,200 =$ **\$23 246 608** opportunity/time cost to the pet owner in the United States per year.

### 13.4.3 Total Cost of Heartworm in the United States

Therefore, the total cost of heartworm to pet owners in the United States is the cost of prevention (\$1620 million) + the cost of treatment (\$123 million) + the opportunity cost to the pet owner (\$23 million) = **\$1.766 billion**. It is interesting to note that 92% of the costs to the pet owner are from purchase of heartworm preventative.

## 13.5 Economic Cost of Heartworm in Key Countries (Australia, Japan, Italy, Spain, and Canada)

We use the same calculations to estimate the cost of heartworm in the Key Countries of Australia, Japan, Italy, Canada, and Spain. Few data are available on which to base many of the necessary calculations, and so estimates have been made when no source can be found for the needed data. The key data that change for each of these countries were the number of dogs and the average salary per hour, which are built into the opportunity cost of treatment.

The data used for the number of dogs in each of the countries can be found in Table 13.2, with all numbers except Canada and the total world estimate coming from Pet Secure [18]. The number of dogs in Canada was obtained online at the Canadian Animal Health Institute [26], and the total world numbers were obtained online at Statista [27]. Reported prevalence of heartworm for Australia was 0.3% [28] and 0.16% for Canada [29]. Calculations for Italy, Spain, and Japan were performed using the same prevalence reported for Australia [28]. The income per hour cost for the opportunity cost of time for pet owners for Canada, Japan, Australia, Italy, and Spain was found through the following sources [30–33]. Estimates and calculations are in US Dollars to facilitate comparisons and calculation of a global number.

At the highest level, costs are driven by the number of dogs and the percent administered heartworm prevention in each country. These data were not readily available for these countries, so the US numbers were used for estimation purposes.

## 13.6 Total Cost of Heartworm in Key Countries

The total annual cost of heartworm in these countries combined is estimated to be \$763 million. We find that the cost of prevention in these countries is slightly under \$750 million, the cost of treatment is \$12 million, and the opportunity cost of time is \$1.54 million. The cost of prevention contributes the significant majority of the cost, making up 98% of the total. The lower estimated prevalence of heartworm in these countries versus the United States drives the cost of treatment and the opportunity cost down.

### 13.6.1 Economic Cost of Heartworm in the Rest of the World

We again use the same calculation process to estimate the economic cost of heartworm in the remaining countries of the world. Very few data are available for these countries, as heartworm is not seen as a major issue in most of the world. Key variables for this calculation are the number of dogs in the remaining countries throughout the world, an estimated percent treated with heartworm preventative (we estimated 0.5%, as we know most of the areas either do not have heartworm and/or do not treat), average doses per year when treated, and heartworm prevalence (estimated at 0.05%). The numbers (except for numbers of animals) are estimated using logical deduction comparing to the numbers that we know, and based on industry experience.

### 13.6.2 Total Cost of Heartworm Rest of World

Using the numbers summarized in Table 13.2, we estimate that the total economic cost of heartworm in the "Rest of World" is $106 million per year. Around 77% of this total, or $81.4 million, is cost of prescribed heartworm preventative. About 22% ($22.8 million) is the treatment cost of heartworm, and the remaining $2 million, or approximately 2%, comes from the opportunity cost of the pet owner.

## 13.7 Conclusions

The economic cost (EC) of heartworm disease caused by *D. immitis* in dogs is a significant portion of estimated overall pet care costs globally, considering it is only one health issue for one species. The estimated annual global costs total $2.47 billion, with approximately 65% of those costs coming from the United States, an additional 31% coming from the key countries of Canada, Japan, Australia, Italy, and Spain, and the remaining 4% coming from the rest of the world. Most of the spend (approximately 93%) comes from spending on heartworm preventatives, with an additional 6% coming from the cost of treatment and 1% the opportunity cost of time for the pet owner. It is expected that these costs will increase as the number of dogs treated with preventatives rises, and the incidence of heartworm increases in countries throughout the world which, historically, have not had significant heartworm prevalence.

## References


1 Companion Animal Parasite Council (CAPC) Guidelines (2020). Heartworm. https://capcvet.org/guidelines/heartworm/ (accessed 21 September 2020).

2 European Scientific Counsel Companion Animal Parasitology (ESCCAP) (2019). GL5: Control of vector-borne diseases in dogs and cats. https://www.esccap.org/page/GL5+Control+of+VectorBorne+Diseases+in+Dogs+and+Cats/29/ (accessed 23 July 2019).

3 TrocCAP Canine Guidelines – Heartworm. http://www.troccap.com/canine-guidelines/other-systems/heartworm/ (accessed 23 July 2019).

**4** Drake, J. and Wiseman, S. (2018). Increasing incidence of *Dirofilaria immitis* in dogs in USA with focus on the southeast region 2013–2016. *Parasites Vectors* 11 (1): 39.

**5** Labarth, N.V., Paiva, J.P., Reifur, L. et al. (2014). Updated canine infection rates for *Dirofilaria immitis* in areas of Brazil previously identified as having a high incidence of heartworm-infected dogs. *Parasites Vectors* 7 (1): 493.

**6** Vieira, A.L., Viera, M.J., Oliveira, J.M. et al. (2014). Prevalence of canine heartworm (*Dirofilaria immitis*) disease in dogs of central Portugal. *Parasites* 21: 5.

**7** Tolnai, Z., Szell, Z., Sproch, A. et al. (2014). *Dirofilaria immitis*: an emerging parasite in dogs, red foxes, and golden jackals in Hungary. *Vet. Parasitol.* 203 (3-4): 339–342.

**8** Lu, T., Wong, J., Tan, T., and Hung, Y.-W. (2017). Prevalence and epidemiology of canine and feline heartworm infection in Taiwan. *Parasites Vectors* 10 (2): 484.

**9** Morchón, R., Carretón, E., Gonzalez-Miguel, J., and Mellado-Hernández, I. (2012). Heartworm disease (*Dirofilaria immitis*) and their vectors in Europe – New distribution trends. *Front. Physiol.* 3: 196.

**10** Sacks, B.N. and Blejwas, K. (2000). Effects of canine heartworm (*Dirofilaria immitis*) on body condition and activity of free-ranging coyotes (*Canis latrans*). *Can. J. Zool.* 78 (6): 1042–1051.

**11** Marks, C.A. and Bloomfield, T.E. (1998). Canine heartworm (*Dirofilaria immitis*) detected in red foxes (*Vulpes vulpes*) in urban Melbourne. *Vet. Parasitol.* 78 (2): 147–154.

**12** Simmons, K. and Hoffman, C. (2016). Dogs on the move: factors impacting animal shelter and rescue organizations' decisions to accept dogs from distant locations. *Animals* 6 (2): 11.

**13** Drake, J. and Parrish, R.S. (2019). Dog importation and changes in heartworm prevalence in Colorado 2013–2017. *Parasites Vectors* 12 (1): 207.

**14** Anonymous (2017). Call for government action as illegal puppy sales fuel canine welfare crisis. *Vet. Rec.* 181 (4): 77.

**15** Norman, C., Stavisky, J., and Westgarth, C. (2020). Importing rescue dogs into the UK: reasons, methods and welfare considerations. *Vet. Rec.* 186 (8): 248.

**16** Global Market Insights. https://www.gminsights.com/industry-analysis/pet-care-market (accessed 5 October 2020).

**17** American Pet Products Association. Pet Industry Size and Ownership Statistics. https://www.americanpetproducts.org/press_industrytrends.asp (accessed 11 October 2020).

**18** Pet Secure. A Guide to Worldwide Pet Ownership. https://www.petsecure.com.au/pet-care/a-guide-to-worldwide-pet-ownership/ (accessed 11 October 2020).

**19** Petmeds. https://www.1800petmeds.com/?MID=1435&CID=21846 (accessed 15 July 2019).

**20** 4paws4rescue. https://www.4paws4rescue.com/heartworm-information (accessed 15 July 2019).

**21** DVM 360 (2018). Handle Concerns about the Price of Heartworm Prevention. http://veterinarymedicine.dvm360.com/handle-concerns-about-price-heartworm-prevention (accessed 15 July 2019).

**22** Costhelper pets and petcare. How Much Does Heartworm Treatment. https://pets.costhelper.com/heartworm-treatment.html (accessed 15 July 2019).

**23** Ku, T.N. (2016). Investigating management choices for canine heartworm disease in northern Mississippi. *Parasites Vectors* 10 (2): 474.

**24** IRS.gov. IRS issues Standard Mileage Rates for 2019. https://www.irs.gov/newsroom/irs-issues-standard-mileage-rates-for-2019 (accessed 15 July 2019).

**25** U.S. Bureau of Labor Statistics. Economic News Release. https://www.bls.gov/news.release/empsit.t19.htm (accessed 15 July 2019).

**26** Canadian Animal Health Institute (2019). Latest Canadian Pet Population Figures Released. https://www.cahi-icsa.ca/press-releases/latest-canadian-pet-population-figures-released (accessed 15 July 2020).

**27** Statista (2018). Number of Dogs and Cats kept as pets Worldwide. https://www.statista.com/statistics/1044386/dog-and-cat-pet-population-worldwide/ (accessed 15 July 2020).

**28** Nguyen, C.K., Wei, L., Casteriano, A. et al. (2016). Mosquito-borne heartworm *Dirofilaria immitis* in dogs from Australia. *Parasites Vectors* 9 (1): 535.

**29** Klotins, K.C., Martin, S.W., Bonnett, B.N., and Peregrine, A.S. (2000). Canine heartworm testing in Canada: are we being effective? *Can. Vet. J.* 41 (12): 929–937.

**30** Statista (2019). Median Average Wage Rate of All Employees in Canada in 2018. https://www.statista.com/statistics/439934/median-hourly-wage-by-type-of-job-canada/ (accessed 15 July 2019).

**31** Japan Times. https://www.japantimes.co.jp/news/2020/02/07/business/economy-business/japan-part-time-workers-rise-wages-drop/#.XsQw4xNKhp8 (accessed 15 July 2019).

**32** Australian Bureau of Statistics. https://www.abs.gov.au/statistics/labour/earnings-and-work-hours/average-weekly-earnings-australia/latest-release (accessed 15 July 2019).

**33** (2019). https://www.thenewbarcelonapost.com/el-salario-medio-mensual-bruto-por-comunidad-autonoma/ (accessed 15 July 2019).

# 14

# Product Profiles for New Drugs Against Human and Animal Filariasis

*Sabine Specht*[1,*] *and Ronald Kaminsky*[2,*]

[1] *Drugs for Neglected Diseases initiative, Filarial Program, 15 Chemin Camille-Vidart, 1202, Geneva, Switzerland*
[2] *ParaConsulting, Altenstein 13, 79685, Häg-Ehrsberg, Germany*

**Abstract**

For a drug project to succeed in delivering the right medicine for the right patient, there must be a clear understanding of the critical features of the final therapeutic product from the very beginning. These features are defined in the Target Product Profile (TPP), which is a list of the desired and essential attributes required for a specific drug to be a successful product. The TPP is an important strategic planning and decision-making tool that is used to define: the target population, the levels of efficacy and safety, the route of administration, the dosing schedule, the properties of the formulated drug, and financial parameters of the formulated drug. In an optimal setting, the TPP guides the complete drug discovery and development process from the start of target identification, through clinical study design, and to final drug registration. In human and animal health, TPPs for filarial diseases can help focus the drug discovery process and reduce the high attrition rates typically associated with drug discovery and development. In addition, this approach helps more quickly identify targets and compound series or candidate characteristics, which can never achieve the parameters defined in the TPP. This facilitates the rapid termination of such projects and allows realignment of valuable resources into more promising projects. We focus in this chapter on the most important characteristics of TPPs against human onchocerciasis and canine filariasis and discuss fundamental differences, similarities, and potential synergies.

## 14.1 Introduction

Whether to invest in a Research and Development (R&D) program for products against filarial diseases is initially a strategic decision. Neglected tropical diseases (NTDs), including human filariases, are closely associated with poverty and thus no high-value commercial market exists; decisions about control efforts are steered by the global health community rather than patients, in contrast to the situation

*Corresponding authors.

*Human and Animal Filariases: Landscape, Challenges, and Control*, First Edition.
Edited by Ronald Kaminsky and Timothy G. Geary.

in Animal Health, for which decisions on R&D investments depend largely on economic considerations.

Product profiles are set guidelines to facilitate decisions within R&D programs on success or failure of a project. The ultimate goal of such an R&D program is the development of a new therapeutic with sufficient benefits over existing therapies to justify introduction to the market. To ensure that the requirements for introduction of a new drug are established clearly and are useful for guiding the drug discovery and development process, a target product profile (TPP) is established at the beginning of the program. The TPP is a list of characteristics that defines and prioritizes the key attributes of the intended new agent and formalizes what constitutes success. In Animal Health R&D programs, TPPs are often decisive for which projects are pursued and when they are terminated. In contrast, in public health, TPPs help guide strategic decisions of stakeholders and are critical to allow flexibility and adjustment to changing landscapes. Therefore, TPPs for human filariasis evolve and mature as a project advances and need to be reassessed periodically to ensure the continuing validity of ideal and minimally acceptable criteria.

Here we propose TPPs for both veterinary and human filarial diseases. During drug development, it is frequently the case that certain criteria in an accepted TPP can't be met, and decision-makers must determine if this failure is critical for the project or can be overcome or compensated by additional beneficial criteria. Therefore, TPPs are often combined with Minimum Product Profiles in Animal Health, which define the lowest values for criteria that must be met in order to pursue the project. Similarly, for NTDs, TPPs are separated into ideal TPPs and acceptable TPPs.

One-sixth of the world's population – approximately 1 billion people – is infected with pathogens that cause NTDs, including the vector-borne parasitic diseases, African sleeping sickness, Chagas' disease, leishmaniasis, schistosomiasis, and filariasis, including onchocerciasis. Although drugs currently used to treat many NTDs have serious limitations, only 1% of all new drugs to reach the market in the past 25 years were for neglected diseases [1, 2]. Drug discovery and development for NTDs are largely driven by the unmet medical needs put forward by the global health community. Since in this area no high return of investment is possible, minimizing waste of money and resources is even more important in this under-funded arena. To ensure any drug discovery project is addressing the complex requirements of patients and health care providers and can deliver a benefit over existing therapies, the ideal and acceptable attributes of a novel drug need to be predefined. In particular, the TPP needs to take into account existing therapies and define those attributes required for the new therapy to present a significant advantage over them.

In contrast, the strategic decision for investment in drug discovery in the Animal Health industry is based on needs and market opportunities. Heartworm in dogs and cats, caused by *Dirofilaria immitis*, is a priority due to its impact on the health and well-being of these companion animals, a substantial economic value (reviewed by Klug and Drake in Chapter 13 of this volume). Thus, this market segment is highly competitive and has recently drawn even more attention due to the occurrence of drug-resistant *D. immitis* isolates [3–5].

We summarize major differences among TPPs and point to opportunities to synergize efforts in both areas in line with the One Health concept. While there is a good chance that antifilarial drugs will have useful activity against important human and veterinary filariae, it cannot be taken for granted that the same chemical will be developed into a drug for both human and veterinary medicine. Indeed, such development will depend on how well the characteristics of a particular drug candidate will fit the partially overlapping but still distinct TPPs. An additional barrier for human drug development is the fact that currently, no intensive and dedicated drug discovery program for antifilarials is ongoing, while in Animal Health, competitive research is pursued and pipelines are likely to be populated with diverse candidates at various levels of development. Discovery of human antifilarials could instead be based on transfer from R&D activities in Animal Health. Against that background, it is important to define and compare TPPs for new human and animal antifilarial drugs to investigate potential differences and to identify product characteristics that enable synergies early in any project.

## 14.2 Target Product Profile for Human Filariasis

The NTD concept was developed to draw attention to overlooked patient groups and was embedded in the millennium development goals (MDGs) [6]. Twenty diseases are recognized as NTDs by World Health Assembly resolutions, among them human filarial diseases. Progressing to the end of NTDs is now firmly embedded within the sustainable development goals (SDGs) for 2030, under target 3.3 (https://unstats.un.org/sdgs/metadata/files/Metadata-03-03-05.pdf), reflecting the promise to “leave no one behind.” In 2020, the World Health Organization (WHO) renewed specific targets for control, elimination, and eradication (https://www.who.int/neglected_diseases/WHONTD-roadmap-2030/en), including the need for a macrofilaricide for onchocerciasis. The 2012 London declaration (https://unitingtocombatntds.org/london-declaration-neglected-tropical-diseases) united pharmaceutical companies, donors, endemic countries, and nongovernment organizations in the recognition that new drugs need to be developed to achieve the elimination goals.

The WHO is currently updating the NTD road map that identifies critical goals and the actions required to reach the targets set for 2030, established through global consultation with the NTD community (https://www.who.int/neglected_diseases/WHONTD-roadmap-2030/en). To reach the program goals, it is recognized that new drugs or drug regimens that kill or permanently sterilize adult filarial worms (without rapidly killing microfilariae) would improve chances for elimination and are likely to accelerate accomplishment of that goal. Current treatment of river blindness and lymphatic filariasis relies on preventive chemotherapy (PC), administered as a single annual or biannual dose, either alone or in combination, to control morbidity in populations at risk of infection or illness (https://apps.who.int/iris/bitstream/handle/10665/83962/9789241505499_eng.pdf). For lymphatic filariasis, recent development of the triple therapy combination (IDA: ivermectin-diethylcarbamazine-albendazole) has provided an effective macrofilaricidal strategy for elimination [7], and so we focus in this chapter on the TPP for a macrofilaricide for onchocerciasis. With the recent registration of moxidectin for use in humans, it is considered that new microfilaricides are not needed for human filarial disease, and so TPPs for that indication will not be addressed here.

## 14.3 Target Population/Use Case

Important for generating a TPP is definition of the target population and how the product will be used (use case). In onchocerciasis, a new macrofilaricide or long-term sterilizing treatment would be beneficial for:

- Mass drug administration (MDA) campaigns, if the safety and tolerability profile is suitable, to drastically reduce the number of MDA cycles (lasting 10–15 years as currently required).
- Test-and-Treat strategies (TNT) for therapy of patients in endemic areas outside MDA campaigns when diagnostic tools are available, especially in "mop-up" campaigns after the disease burden has been reduced by MDA programs, rendering them no longer cost-effective, or in areas where regular ivermectin (IVM) distribution is difficult.
- Test-and-not-Treat (TaNT) campaigns in areas where *Loa loa* is co-endemic when the macrofilaricidal drug also has rapid microfilaricidal activity.
- Individual case management.

### 14.3.1 MDA

MDA is a means of delivering safe and inexpensive essential medicines based on the principles of preventive chemotherapy, in which populations or subpopulations are offered treatment without individual diagnosis. MDA in endemic areas aims to prevent and alleviate symptoms and morbidity on the one hand and reduce transmission on the other, together improving global health [8]. Insufficient coverage, however, is one of the main factors hindering progress toward elimination. Reasons for this situation are many, including difficult to treat areas (for instance, due to ongoing conflicts), lack of financial resources, and inadequate political engagement. Additionally, program fatigue in MDA has been reported after years of implementation, and an individual in an endemic area may find repeated MDA inconvenient or may lose confidence in the MDA campaign. As the population is treated as a whole, irrespective of individual infection status, an additional aspect must be considered. Unlike a vaccine, MDA for onchocerciasis does not provide an immediate benefit for an uninfected person. With effective MDA, the benefits for the entire population clearly exceed the risk, but when prevalence has been reduced, the risk/benefit ratio may narrow considerably. The balance between making MDA compulsory, which potentially limits the autonomy of an individual, and allowing individuals to opt out, potentially jeopardizing the entire effort of the campaign if onchocerciasis is reintroduced to the treated community, must be constantly evaluated, a topic also discussed with regard to malaria [9].

A curative macrofilaricide in a single-dose regimen would be ideal to simplify elimination campaigns on a large scale. Most importantly, such a medicine must be highly safe and sufficiently well-tolerated to be given across the age spectrum, including children, adolescents, and pregnant women. This requires testing of a new drug on thousands of patients for safety, as currently planned for moxidectin to allow treatment without prescription or health worker supervision (clinicaltrial .gov NCT04311671) An ideal MDA drug would be safe for use in *Loa*-coendemic areas, i.e. does not target microfilariae but only adult worms (see below).

### 14.3.2 Test-and-Treat (TNT)

A TNT approach for filariasis requires diagnosis of infection and/or considerations of any contraindications before a decision is made to treat and with what regimen, assuming the availability of sufficiently specific and sensitive diagnostic tests. This approach can increase confidence in the benefit provided by treatment as opposed to MDA, particularly in asymptomatic individuals, and secondly, can address safety, particularly in areas with prior experience of severe adverse events. TNT can be carried out at the individual level (case management) – this would most likely require the availability of the registered product in pharmacies throughout endemic areas at a reasonable price. A programmatic TNT approach requires participation of the community during each TNT campaign and would need funding for such activities. Cost-effectiveness at the operational level remains to be fully evaluated by national and international programs; however, modeling of costs associated with switching from MDA to more focused TNT approaches suggests that it may be cheaper than the current MDA approach, especially when overall disease prevalence declines [10].

### 14.3.3 Test-and-Not-Treat (TaNT)

This strategy applies to *L. loa–Onchocerca volvulus* co-endemic areas. IVM causes serious adverse events in patients with high microfilarial densities of *L. loa*, another filarial nematode common in central Africa [11]. The expected number of onchocerciasis-infected individuals in co-endemic areas is predicted to be 3.6 million in 2025. Of these, 25.000 people are estimated to be coinfected with *L. loa* with microfilarial densities >20.000/ml blood. Such individuals need to be identified before enrolment in an MDA campaign to minimize the risk of severe adverse events. In some areas in Cameroon where onchocerciasis is hypo-endemic, the TaNT strategy is currently being evaluated with screening for loiasis using the “LoaScope” [12]. Subjects with loiasis microfilaremia rates >20 000 mf/ml are excluded from IVM treatment. Patients excluded but positive for onchocerciasis receive doxycycline for four weeks (ongoing DND*i* program, unpublished). This approach allows a focused treatment plan for onchocerciasis in *Loa* co-endemic areas.

## 14.4 Target Product Profile for Onchocerciasis

TPPs exist for human parasitic diseases such as malaria [13] and Chagas’ Disease [14]. Currently, only two TPPs for onchocerciasis are publicly available, one generated by the WHO (https://www.who.int/tdr/publications/documents/rna-drug .pdf) and one by the Drugs for Deglected Diseases *initiative* (DND*i*) (Table 14.1).

**Table 14.1** Target product profile for onchocerciasis.

| Variable | Acceptable | Ideal |
|---|---|---|
| Indication | For the treatment of onchocerciasis | For the treatment of onchocerciasis |
| Product description | Oral dosage form, intramuscular or subcutaneous injection | Oral dosage form |
| Target population | All individuals with the exception of pregnant women and children younger than 5 yr | All individuals who are at risk for onchocerciasis |
| Treatment regimen | • Oral dose, once or twice a day<br>• Duration of treatment up to 14 d<br>• One single im or sc injection or repeated after a week (2 injections)<br>• One dose for adults and weight/age-adjusted or height-based dosing for children | • Oral dose, once a day, up to 3 d<br>• One dosage for all ages |
| Efficacy | Superior to comparator in eliminating skin microfilariae at 24 mo with evidence of impacting adult worms (killing adults or sterilization) | Superior to comparator in eliminating skin microfilariae at 24 mo with evidence of impact adult worms (killing adults or sterilization) |
| Safety | **Adverse events**<br>Minor and manageable side effects<br>Monitoring for AE manageable at local healthcare post<br>Moderate impact on activities of daily living<br>No severe Mazzoti reaction<br>No severe adverse ocular reaction<br><br>**Population for restricted use at registration**<br>Pregnant women<br>Lactating woman (duration according to PK of the drug)<br>**Precaution/Warnings**<br>Concomitant infections (e.g. loiasis)<br>Acute illness (e.g. fever, bacterial infection)<br><br>**Use in specific populations**:<br>Pretreatment assessment and careful posttreatment follow-up should be available for patients with *Loa loa* coinfection.<br>Exclusion of high *L. loa* mf/ml co-infected patients | **Adverse events**<br>• No monitoring for AE required<br>• No impact on activities of daily living<br>• No Mazzoti reaction<br>• No adverse ocular reaction<br><br>**Population for restricted use at registration**<br>• None<br><br>**Precaution/Warnings**<br>• None<br><br>**Use in specific populations:**<br>• Safe for use in patients co-infected with *L. loa*<br>No monitoring needed (no rapid microfilaricidal activity) |

*(Continued)*

**Table 14.1** (Continued)

| Variable | Acceptable | Ideal |
|---|---|---|
| | **Drug–drug interactions:**<br>Manageable for individual case treatment | **Drug–drug interactions:**<br>No clinically significant drug–drug interaction with commonly used antiparasitic and anti-infective drugs<br>No evidence for clinically significant adverse interactions with long-term/chronic use drugs (e.g. anti-tuberculosis drugs, anti-retrovirals, contraceptives)<br>No evidence for clinically significant adverse interactions with commonly administered MDA drugs (e.g. ivermectin, praziquantel, benzimidazoles, azithromycin), and antimalarial drugs |
| Shelf life | 3 yr in zone IVb | More than 3 yr in zone IVb |

The DND*i* is a nonprofit R&D organization based in Geneva that is committed to developing new treatments for patients with NTDs, including onchocerciasis. Whereas the WHO TPP was formulated within a working group focusing on RNA-based drugs, DND*i*'s TPP targets drugs that, independent of the mode of action, lead to death or long-term sterilization of adult filarial parasites (https://www.dndi.org/diseases-projects/filarial-diseases/target-product-profile-filarial-diseases). The first TPP targets applicability for MDA, whereas the second also addresses other treatment strategies, such as multiple daily treatments for up to 14 days for case management and TNT scenarios. Key properties also include safety, heat stability, palatability, and acceptable production costs.

## 14.5 Challenges

The development of pharmaceutical products for NTD patients in low- and middle-income countries (LMIC) presents additional challenges in addition to those mentioned in the discussion section.

### 14.5.1 Formulation Development

There is a critical need to develop appropriate formulations of medications that are suitable across weight bands and age groups, including children and adolescents. In addition to pharmacokinetics and safety, suitable formulations are important to ensure correct dosing and compliance with treatment. Factors that influence compliance include tablet, capsule, or pill size and number needed per dose, frequency of dosing, volume of solution, palatability, food requirement, and side effects attributed to medications. Innovative alternatives, such as long-acting formulations, could be considered and developed as alternatives to repeated daily oral medications.

### 14.5.2 Supply Chain

Medicines for filarial infections need to be shipped to and stored in LMIC, where climate conditions can be hot and dry or hot and humid. Product stability must consider those conditions to attain the intended shelf life in the target countries. Appropriate packaging with greatly reduced weight and volume reduces shipment and storage costs. The ideal product should be easily transported, not require refrigeration, and be readily available to people who need it.

## 14.6 Product Profiles for Animal Health

Although only one macrofilaricide against the adult *D. immitis* is available [15], almost all R&D efforts in Animal Health for new antifilarial drugs are focused on novel heartworm preventatives (https://www.grandviewresearch.com/industry-analysis/flea-tick-heartworm-products-market). We propose two TPPs, one for a pure heartworm preventative (Table 14.2) and one for an endo-parasiticide (Table 14.3), and highlight key features of novel antifilarial drug candidates.

### 14.6.1 Spectrum of Preventative Drugs

Due to the spectrum of the actives and customer preferences, these can be categorized as pure antifilarial drugs, as endo-parasiticides for control and prevention of heartworm and other pathogenic nematodes, and as endecto-parasiticides covering additionally the control of ectoparasites such as fleas and ticks. Accordingly, the TPPs differ: for a pure antifilarial preventative drug, the targeted parasite species (primary therapeutic area) is *D. immitis* and, as an added benefit, *Dirofilaria repens*. In contrast, for a competitive endo-parasiticide, the target parasite species include numerous pathogenic nematodes such as hookworms, roundworms, whipworms in addition to major tapeworms (Table 14.3). At present, all preventative products contain a macrocyclic lactone (ML) such as IVM, moxidectin, selamectin, or milbemycin oxime as the active ingredient. Beyond targeting L3 and L4 stages of *Dirofilaria* spp., MLs are also active against gastrointestinal nematodes. Therefore, the same active can be employed not only as a heartworm preventative but also as a gastrointestinal dewormer for companion animals. However, that profile is not identical for all actives, and it is possible to conceive of chemicals or vaccines, which target only larval stages of *Dirofilaria* spp. However, pet owners prefer a "once for all solution," i.e. one treatment should cover as many pathogens as possible, hence the importance of endecto-parasiticides.

**Table 14.2** Target product profile for a heartworm chemotherapeutic preventative for animal health.

| Product features | TPP (target product profile) | MPP (minimum product profile) |
|---|---|---|
| Primary therapeutic area | Prevention of dirofilariasis caused by *D. immitis* (heartworm disease) or subcutaneous dirofilariasis caused by *D. repens* | Prevention of dirofilariasis caused by *D. immitis* (heartworm disease) |
| Primary species | Dogs | Dogs |
| Secondary species | Cats | N/A |
| Age of species | Puppies and dogs, kittens, and cats: 6 wk and older | Puppies and dogs: 12 wk and older |
| Mechanism of action | New chemical class, new mode of action, a potential ML resistance breaker | New mode of action |
| Clinical description of effects | Indicated for the prevention of heartworm disease caused by *D. immitis*, and prevention and control of subcutaneous *D. repens* | Indicated for the prevention of heartworm disease caused by *D. immitis* |
| Comparative efficacy | 100% against *D. immitis* infection, proven efficacy against ML-resistant *D. immitis* isolates | 100% against *D. immitis* infection |
| Comparative safety | Equivalent safety to leading heartworm products; free from neurological or collie-related side effects | Equivalent safety to leading heartworm products |
| Site of action | Systemic | Systemic |
| Route(s) of administration | • Oral if it has 1 mo duration, longer duration is a plus<br>• Injectable if it has 6–12 mo duration<br>• Topical if it has 1 mo duration | Oral or topical |
| Onset of action | Class-leading onset of action | Comparable to leading heartworm products |
| Duration of action | • 1 mo for an oral<br>• 6–12 mo for an injectable<br>• 1 mo for a topical | 1 mo for an oral |
| Dosing regimen | According to route of administration | Once per month minimum |
| Dosage form | Oral chewie or gel, injectable, thin strip, drops, granules or liquid – convenience is key | Oral or topical |
| Antigens included (vaccines only) | Vaccine as stand alone | Vaccine in combination with a ML |
| Adjuvant system (vaccines only) | TBD | TBD |
| Shelf life | 3 yr | 2 yr |

(Continued)

**Table 14.2** (Continued)

| Product features | TPP (target product profile) | MPP (minimum product profile) |
|---|---|---|
| Precautions/ contraindications (or lack thereof) | Similar to leading HW/IP products | Similar to leading HW/IP products |
| Patentability | TBD | TBD |
| *Registration goal* | TBD, in 20XX | TBD, in 20XX |
| Pricing (specify to veterinarian or to consumer/farmer) | USD XX /dose (avg.) to veterinarian | USD XX plus/dose (avg.) to veterinarian |
| Launch countries | USA, Canada, Australia, New Zealand, Europe, Brazil, Japan plus all other HW markets | USA, Canada |
| Estimated peak sales (in USD) | XX $ mio USD | XX $ mio USD |
| Estimated gross margin % | TBD % | TBD % |

### 14.6.2 Efficacy and Resistance Breaking

Resistance of *D. immitis* isolates to MLs has been demonstrated (3–5) and has finally been accepted by the American Heartworm Society: "Every compound currently marketed in every form of administration (oral, topical, and parenteral) has been shown to be less than perfect in at least one study" (https://www.heartwormsociety.org/veterinary-resources). Against this background, the mode of action of a new antifilarial drug is critical in the TPP. This includes the equipotency of a new chemical against ML-resistant and -sensitive *D. immitis* isolates. Meeting the regulatory requirement for a new heartworm antifilarial remains a challenge, because the FDA continues to set the required efficacy at 100%, apparently including efficacy against contemporaneous field strains, which will possibly include ML-resistant isolates.

### 14.6.3 Routes of Administration and Duration of Activity

Convenience of drug administration is a key driver for innovation [16] and crucial for a new heartworm preventative. Currently, three possible routes of administration exist: oral, topical, and injectable, each with advantages and disadvantages. Oral administration for dogs offers the opportunity to combine drug administration with a tasty chewie to almost make this a treat for the dog, fostering, beyond health care, the owner–dog relationship. Oral application is not generally a favorable treatment route for cats. Injectables are usually more costly and typically require a visit to a veterinary clinic. However, they are accepted as trade-off for prolonged duration of activity, if, for example, a product prevents disease for up to one year.

**Table 14.3** Target product profile for an endo-parasiticide for dogs and cats.

| Product features | TPP (target product profile) | MPP (minimum product profile) |
|---|---|---|
| Primary therapeutic area | Prevention of dirofilariasis caused by *D. immitis* (heartworm disease) and subcutaneous dirofilariasis caused by *D. repens*, additionally for the treatment of parasitic infections due to<br>*Dogs:*<br>Larval, pre-adult and adult stages of hookworms (*Ancylostoma caninum*, *Ancylostoma braziliense*, *Uncinaria stenocephala*), roundworms (*Toxocara canis*, *Toxascaris leonina*), whipworms (*Trichuris vulpis*), and tapeworms (*Dipylidium caninum*, *Taenia* spp., and *Echinococcus multilocularis*, *E. granulosus*, *Mesocestoides* spp.);<br>Prevention and treatment of disease caused by *Angiostrongylus vasorum*, *Crenosoma vulpis*<br>*Cats:*<br>Larval, pre-adult and adult stages of hookworms (*Ancylostoma tubaeforme*, *Uncinaria stenocephala*), roundworms (*Toxocara cati*, *Toxascaris leonina*) and tapeworms (*Dipylidium caninum*, *Taenia* spp., and *Echinococcus* spp.) | Indicated for the prevention of heartworm disease caused by *Dirofilaria immitis*, and for the treatment and control of parasitic infections due to<br>*Dogs*:<br>Adult stages of hookworms (*Ancylostoma caninum*, *Uncinaria stenocephala*), roundworms (*Toxocara canis* and *leonina*), whipworms (*Trichuris vulpis)* and tapeworms (*Dipylidium caninum*, *Echinococcus multilocularis*);<br>Treatment of disease caused by *Angiostrongylus vasorum* and *Crenosoma vulpis*<br>*Cats*:<br>Adult and immature stages of hookworms (*Ancylostoma tubaeforme*, *Uncinaria stenocephala*), roundworms (*Toxocara cati*, *Toxacara leonina*) and tapeworms (*Dipylidium caninum*, *Taenia* spps., and *Echinococcus* spp.);<br>The minimum should always be at parity with the spectrum of competition products |
| Primary species | Dogs and cats | Dogs and cats |
| Secondary species | N/A | N/A |
| Age of species | Dogs 2 wk and older, and cats 4 wk and older | Dogs 6 wk and cats 6 wk and older, only if one can claim superiority in spectrum, efficacy, or convenience over competition products |
| Mechanism of action | New chemical class, new mode of action, a potential ML resistance breaker | New chemical class |
| Clinical description of effects | Indicated for the prevention of heartworm infection, and prevention and control of subcutaneous *D. repens* infection, plus prevention and/or control of intestinal parasites | Prevention of heartworm infection, prevention and/or control of intestinal parasites |

*(Continued)*

**Table 14.3** (Continued)

| Product features | TPP (target product profile) | MPP (minimum product profile) |
|---|---|---|
| Comparative efficacy | 100% against *D. immitis* infection, proven efficacy against ML-resistant *D. immitis* isolates, plus superior efficacy to competition products against common parasites in dogs and cats | 100% against *D. immitis* infection, plus equivalent efficacy to competition products against common parasites in dogs and cats |
| Comparative safety | Superior safety to leading heartworm products; free from neurological or collie-related side effects | Equivalent safety to competition product |
| Site of action | Systemic and/or topical (cat) | Systemic |
| Route(s) of administration | Flexible route of administration: oral and topical | Oral or topical |
| Onset of action/immunity | Comparable to competition product | At parity with competition product |
| Duration of action/immunity | 3 mo or longer | At parity with competition product |
| Dosing regimen | Once every 3 mo | Monthly |
| Dosage form | Palatable oral with high acceptability in dogs and cats with high scoring on "spontaneous uptake" (e.g. novel formulation, thin strip, chewie or paste) | Oral application with comparable score as competition product on "spontaneous uptake" in dogs and cats |
| | Ideally large dose band allowing only 1 tablet/spot-on per animal treatment regardless of weight | Dosing band and number of tabs, etc., comparable to competition product |
| Antigens included (vaccines only) | N/A | N/A |
| Adjuvant system (vaccines only) | N/A | N/A |
| Shelf life | 5 yr | 3 yr |
| Precautions/ contraindications (or lack thereof) | Superior to competition product | Similar to competition product |
| Patentability | TBD | TBD |
| Registration goal | 20XX | 20XX or later |
| Pricing (specify to veterinarian or to consumer/farmer) | Premium in the CAB endo category | At parity with competitor |
| Launch countries | Europe, USA, Canada, Australia, Japan, ASIA, South America | Europe, USA, Canada, Australia, Japan |
| Estimated peak sales (in USD) | XY $mm USD | Less than XY $ mm USD |
| Estimated GM% | TBD | TBD |

### 14.6.4 Financial Parameters

Beyond the need for novel drugs, investment in R&D programs is based on financial opportunities. Consequently, the desired gross margin is critical in the TPP, which is impacted by cost of goods, the estimated selling price of the new product, and its potential market penetration. Naturally, the qualities of technical features compared with available competitive products will impact those financial parameters. In addition, time needed for development and registration and the intended launch date are important and part of TPPs. These parameters can't be generalized and depend on the individual project, but will influence the gross profit margin of the product. The desired gross profit margin percentage is set by decision-makers within an Animal Health company and, although usually not publicly accessible, is decisive for the project.

## 14.7 Discussion

In Animal Health, product profiles are almost contract-like, in that not meeting standards defined in the minimum product profile (MPP) is a cutoff point for a project. The termination or continuation of programs in Animal Health is finally based on economic viability of the individual project. TPPs are fairly stable during a project but may change depending on market parameters or on market changes forced by changes in the landscape of competitor products.

In contrast, TPPs in human filariasis are guidelines, likely to be adjusted according to changing landscapes and feasibilities in public health, making it difficult to act via clear cutoff points. With drug development timelines of 10–15 years, current approaches try to anticipate what the disease prevalence may be at the time of registration of a new drug. This exercise will help to formulate the use case for a macrofilaricide. Funding environments are constantly challenged, and therefore control and elimination of NTDs require persistent international cooperation to make sustainable progress over a long period of time.

### 14.7.1 Similarities, Differences, and Opportunities for Synergy of Human and AH TPPs

The obvious difference between potential human and animal antifilarial products is the target populations. In Animal Health, the major target species is dogs, with a secondary species of cats. Species-specific parameters such as the route of administration and interspecies differences in pharmacology will influence bioavailability, dosage, and efficacy of any agent. While several routes of administration are possible, a noninvasive formulation is preferred for humans and animals, particularly oral formulations with acceptable taste and size. On the other hand, drug characteristics such as safety and efficacy are not negotiable and are independent of the target species. Acceptable tolerability and safety limits are set by drug approval authorities, i.e. FDA, EMEA, and country regulatory agencies. Importantly, some of the required studies, including environmental safety, preclinical and reproductive toxicology, and manufacturing, are similar or identical in drug development for NTDs and Animal Health.

Drug development is handicapped by high attrition rates, and many promising molecules fail during preclinical development or in subsequent toxicological, safety, and efficacy testing; thus, R&D costs in aggregate are very high. The level of investment into R&D for new products for NTDs, as reported in the annual Global Funding of Innovation for Neglected Diseases (G-FINDER) surveys, shows that few NTD areas receive anywhere near the level of funding required; and that funding, when it is available, is rarely allocated in a manner likely to move products through the pipeline to patients [16]. Therefore, no dedicated drug development pipeline for human filariasis is in place, and it is essential that stakeholders, funders, industry, academics, and NGOs adopt a cooperative approach and share responsibility to reduce risks and overcome these obstacles. Joint efforts are being made to cut the cost of R&D for new drugs for NTDs and increase attractiveness of this sector to funders and investors. Supportive programs by the FDA (priority voucher program: https://www.fda.gov/about-fda/center-drug-evaluation-and-research-cder/tropical-disease-priority-review-voucher-program) and EMA (article 58: https://www.ema.europa.eu/en/human-regulatory/marketing-authorisation/medicines-use-outside-european-union) aim to increase incentives for companies to engage in drug development for NTDs.

Commercially available and late pipeline Animal Health products should be evaluated for human indications to shorten drug development timelines and enhance affordability of drug discovery and development. Repurposing is a common strategy in other therapeutic areas, and drugs are typically transferred from human to animal health, such as for cancer, nonparasitic infectious diseases or allergies. It is the opposite case in parasitology. While the de novo development of anthelmintics is commercially not attractive for human use, development of new drugs for animal health is (relatively) financially rewarding and therefore much better supported and further advanced. Furthermore, drug repurposing typically has a higher chance of success with an already proven drug target in nematodes of veterinary importance. Some impressive examples demonstrate successful repurposing of veterinary drugs for human use, including benzimidazoles, IVM, praziquantel, moxidectin, and triclabendazole [17]. This approach has also been adopted by the DND*i*, which is currently investigating emodepside (in collaboration with Bayer AG) and ABBV-4083 (a tylosin derivative, jointly developed in collaboration with the Anti-Wolbachia [AWOL] consortium, and the pharma partner AbbVie). The other lead compound is the off-patent veterinary product oxfendazole for potential human use (https://www.dndi.org/diseases-projects/filarial-diseases).

#### 14.7.1.1 Sales and Access

In both cases, a new drug should be as inexpensive as possible, albeit for different reasons. In Animal Health, low cost of goods might enable a sufficiently high gross profit margin, depending on the selling price and market share, to justify the initiation and pursuit of an anti-filarial R&D project. However, cost of goods for products for companion animals, such as heartworm preventatives, is not as critical as for products for livestock. For farmers, costs for Animal Health must be integrated in their business expenses for producing animals, meat, milk, wool, and other animal products and are therefore price-sensitive. This is in contrast to the majority of dog or cat owners, who have built up more or less strong, often family type, relationships with their pets and are prepared to pay a premium for superiority over products of the competition. As noted above, the resulting Animal Health financial opportunity is crucial and defined in the TPP.

In global health, successful development and securing regulatory approval do not guarantee that a product will be available to people who need it. Challenges to access in LMIC include guaranteeing high distribution coverage, ensuring affordability, and adoption of medicines, at both provider and end-user levels. Developing countries often lack the financial resources and infrastructure needed to ensure access to medicines. Therefore, affordable cost, which is impacted by the cost of goods and distribution, is a vital part of decision-making in NTD treatment and control programs. This is ultimately measured by achieving the SDGs [18] and/or costs per averted disability-adjusted life year (DALY); the estimated cost per averted DALY of annual MDA varies between US$ 3 and US$ 30 [19, 20]. The economic success of IVM distribution is largely based on the fact that the drug is donated free of charge by Merck & Co. If the costs of the drug were included (calculated at US$ 1.5 per tablet), the economic balance would be highly unfavorable [20] and IVM may not have been taken up by the community to eliminate filarial diseases. Additional costs include international shipping and in-country distribution costs. These costs can be relatively high, reflecting in part a limited ability of governments and individuals in affected countries to pay for the efforts needed. The exact cost of goods is thus difficult to estimate and highly depends on the use case and thus the target population.

Therefore, it is important to acquire accurate information on the target population to estimate acceptable levels of cost of the formulated drug, based on data provided by health care workers and physicians, health regulators, and policymakers, especially from disease endemic countries. This estimate further depends on the use case of a macrofilaricidal drug, i.e. in MDA programs delivering billions of treatments to the population at risk or in a more targeted approach. For TNT approaches, higher cost of goods may be acceptable, but must be as low as possible for MDA programs. Mathematical models have been developed and are constantly updated to provide optimal information for key stakeholders [21, 22]. Economic analyses are currently being conducted [19], comparing the cost-effectiveness of current elimination strategies with other potential interventions, such as T(a)NT approaches with hypothetical macrofilaricidal drugs. It must be noted that many obstacles affect the accuracy of assumptions used in modeling, such as political instability, development of resistance, or MDA program fatigue. In the absence of consolidated and clear global forecasts and demand planning, there may be no definitive incentive for companies or partnerships to initiate development of such drugs. Facing these challenges, the signatories of the London Declaration made a clear statement that it is important to take the initiative now for the development of new drugs to avoid a scenario in 2030, when elimination targets may not be achieved and valuable time wasted [22].

#### 14.7.1.2 Efficacy

A substantial difference in the TPPs for new antifilarial drugs is the required efficacy. In Animal Health, the FDA requires 100% efficacy for new products against contemporaneous isolates of *D. immitis*. The 100% requirement is unique for Animal Health approvals and scientifically questionable, but nonetheless it remains established for heartworm preventatives, although for other anthelmintics for ruminants or companion animal gastrointestinal nematodes, "effectiveness should be 90% or higher calculated using transformed (geometric means) data" (https://www.fda.gov/regulatory-information/search-fda-guidance-documents/cvm-gfi-111-vich-gl19-effectiveness-anthelmintics-specific-recommendations-canine).

In contrast, for a novel antifilarial human drug, the minimal requirement for efficacy is not yet decided. However, a minimum standard would be non-inferiority to current drugs (i.e. IVM or moxidectin for a microfilaricide). For onchocerciasis, a new drug would be used to prevent disease transmission and disease-associated morbidity and multiple treatment rounds are accepted, while in Animal Health, a new drug will be used for preventive chemotherapy. Therefore, the major difference lies in the therapeutic strategies.

A new mode of action is envisioned in both areas. In Animal Health, it is included in the TPP: adding a requirement for full efficacy against ML-resistant isolates. Among the MLs, moxidectin at use doses has been shown to retain efficacy against at least some IVM-resistant isolates of *D. immitis* [23, 24]. Although all MLs are thought to share the same mechanism of action, the mechanisms of resistance developed by the parasites may differentially affect individual MLs. In addition, moxidectin seems to have higher intrinsic potency against filarial nematodes than other MLs [25]. While moxidectin and IVM both bind to the same transmembrane domain of glutamate-gated chloride channels, their interaction with them differs [26]. Furthermore, moxidectin has different pharmacokinetic parameters than other MLs in dogs, including a longer half-life and larger volume of distribution [27]. Therefore, it was not surprising that moxidectin showed (at least for a time) full efficacy against ML-resistant GI-nematodes in sheep [28], cattle [29], and dogs [24], independent of the occurrence of moxidectin-resistant isolates [25].

In human filariasis, resistance to IVM has been observed [30], but has not been proven on a large scale. On the historic background in Animal Health, it would not be unexpected after more than 20 years of IVM distribution that resistance would develop in onchocerciasis as well. Moxidectin, recently repurposed from animal health and registered for human use, has shown much longer suppression of microfilaridermia than IVM after a single dose. If production costs and delivery hurdles are successfully passed, implementation of moxidectin may result in fewer MDA cycles. In addition, the significantly longer duration of action may make moxidectin a preferred choice over IVM in areas in which regular annual or biannual treatments cannot be guaranteed. Nevertheless, moxidectin does not kill the adult parasite and microfilariae reappear after some time; as is the case for dirofilariasis, drugs with new modes of actions would be a major step ahead, albeit not presently fixed in the human TPP. The most desirable aim is to develop a drug with a different mode of action and with long-term sterilizing or macrofilaricidal activity, but the feature of "resistance breaker" may not be decisive.

Five NTDs are currently controlled through MDA using safe, single-dose, or combination medicines: LF, onchocerciasis, schistosomiasis, soil-transmitted helminths, and trachoma. For these, no combination products are available in a single formulation. However, the global health community is moving toward integrated approaches for NTD control and elimination activities. Such integrated approaches may include the simultaneous delivery of multiple drugs for the treatment of several NTDs, including filarial infections. Such combination products are already standard in Animal Health. Heartworm preventatives are rarely limited to use for prevention of dirofilariasis, but are often included in products for other canine of feline helminth infections, and thus are classified as endo-parasiticides. In addition, endecto-parasiticides, products containing additional actives against most common ectoparasites, are commercially available and widely used.

### 14.7.2 Vaccines as Future Tools for Filarial Control

With the research activities of the Onchocerciasis Vaccine for Africa (TOVA) initiative (https://www.riverblindnessvaccinetova.org), a new vaccine may enter Phase I trials by 2023. However, its efficacy in humans remains to be proven (see chapter on Development of a vaccine against *O. volvulus* by Abraham et al.). If this or any other vaccine demonstrates safety and efficacy at a minimum of 50% prevention of establishment of inoculated worms and > 90% reduction of microfilaridermia levels (TOVA TPP; [31]), it may be used alone or in combination with anthelmintics in programs of "vaccine-linked chemotherapy" to prevent reinfection following MDA.

A safe and efficacious canine vaccine against *D. immitis* has the potential to disrupt the current heartworm market. It would provide veterinarians and pet owners with a nonchemical alternative to the only current option of prevention with regularly administered MLs. However, with the exception of Dictol® for cattle against *Dictyocaulus viviparus* and Barbervax® for sheep against *Haemonchus contortus*, no anthelmintic vaccines have been commercialized. In contrast to vaccines against ruminant pathogenic nematodes, a heartworm vaccine for pets would likely have to be 100% efficacious or close to that, given current FDA requirements for preventatives. Additional characteristics of a canine vaccine that will affect its acceptance by veterinarians and pet owners now using MLs are the number and frequency of immunization boosters and the period of full protection, which would need to be longer than what is possible with MLs. These parameters are also important for a human antifilarial vaccine. Current technology advancements will certainly enable identification of potential vaccine targets. The tricky part will be selection of best antigens and the subsequent engineering of a vaccine that provides safe and long-duration protection.

### 14.7.3 Game Changer That Would Change TPPs

For human filarial diseases, the game changer would be a safe, cheap, and easy to administer drug with a curative single dose that kills adult worms safely. If such a drug acts only on adult worms but not microfilariae, it could enable its use for MDA in loiasis co-endemic areas, which have not yet been included in MDA programs due to the risk of severe adverse events [32]. Ultimately, the cost of goods must be manageable by the global health community. Such a drug could be implemented for MDA, T(a)NT, and case management. Advocating such a game-changer drug profile ("best case") for NTDs bears the danger that less ideal macrofilaricidal projects may be dropped too early; in this context, one should be aware that no drug was initially developed for MDA purposes. If a given drug candidate is considered to show macrofilaricidal potential, the minimally acceptable frequency of treatment remains to be elucidated.

In Animal Health, game-changing superiority over competitor products could be achieved on the basis of convenience of administration, duration of protection, and spectrum of parasites. A candidate product with a one year or longer period of protection clearly would have advantages over monthly or bimonthly medication, regardless if the product was a vaccine or a slow-release drug depot created by injection. A novel and superior convenient route of administration could be achieved if such a medication could be given as a non-injectable treat. Finally, protection over a long period not only against canine filariasis but also against other pathogenic helminths and most common ectoparasites, including disease-transmitting vectors, would be clearly superior to present options.

## 14.8 Conclusions

A TPP guides the development of an antiparasitic compound through its life, starting with drug discovery and development until access to the patient population. A TPP sets specific requirements that a compound must meet for initiation and continuing its development to finally become a registered and available medicine. Establishing a TPP includes careful consideration of the impact of the disease, the target population, the mode of action and route of administration of the active, efficacy and safety levels, and economic value. This exercise allows rational decisions to be made on continuation or termination of a project and thus enables focusing limited investments on the most promising projects.

NTDs in humans are generally nonprofitable and drug development is driven by unmet medical needs. Investments into these areas are therefore limited, requiring the need to avoid high-risk development programs and to minimize development costs to enable new drugs to reach patients. In contrast, it is a strategic decision in the Animal Health industry to enter into a disease area based on needs and market opportunities. The veterinary market, especially for companion animals, offers a potential return on investments, which is a strong driver for drug discovery and development. For filarial diseases, the Animal Health pipeline is much more advanced compared with human medicine. Therefore, the potential of repurposing actives should be exploited, particularly for antifilarials, to enable drug development for human use. Further efforts should be made in line with the One Health concept and available animal health products and promising candidates should be exploited for human indications to make drug development affordable.

A comparison of filaricide TPPs in veterinary and human medicine shows that similarities and clear differences exist between these areas. Similarities are linked to the parasitic organisms, since closely related species in the same biological family are targeted. Chances are high that the intrinsic activity of compound is expressed against animal and human pathogenic filarial species. Safety and toxicology requirements are similar or even identical. In addition, a new mode of action is desirable to add superior efficacy and to either avoid the development of drug resistance in humans or to break it in animal pathogens such as *D. immitis*. Low cost of goods for final products is desirable, although for different reasons. In Animal Health, low cost of goods enables a sufficient gross margin to meet economic requirements of the TPP, while in human NTDs, affordable costs are required to enable product distribution to the target populations, mostly located in resource-limited settings.

Key differences are related to the host species, the parasitic stage targeted, required efficacy, and the sociological environment the drugs are used in. The focus in Animal Health, mainly for canine heartworm disease, is on preventative drugs targeting two larval lifestages (L3 and L4), while in the NTD onchocerciasis, the search is for a macrofilaricide with efficacy against adults. Furthermore, in Animal Health, the required efficacy is set by regulatory authorities at 100%, while in human medicine, against the background of limited available candidates, efficacy is currently driven by need for additive effects, i.e. macrofilaricidal or long-term inhibition of embryogenesis, to reach superiority over current treatments. These must demonstrate sufficient efficacy to substantially shorten the time needed for parasite elimination, thus to contribute to healthy lives and promote well-being for all at all ages.

## References


1 Trouiller, P., Olliaro, P., Torreele, E. et al. (2002). Drug development for neglected diseases: a deficient market and a public-health policy failure. *Lancet* 359: 2188–2194.

2 Pedrique, B., Strub-Wourgaft, N., Some, C. et al. (2013). The drug and vaccine landscape for neglected diseases (2000-11): a systematic assessment. *Lancet Glob. Health* 1: e371–9.1.

3 Bourguinat, C., Lee, A.C., Lizundia, R. et al. (2015). Macrocyclic lactone resistance in *Dirofilaria immitis*: failure of heartworm preventives and investigation of genetic markers for resistance. *Vet. Parasitol.* 210: 167–178.

4 Bourguinat, C., Keller, K., Bhan, A. et al. (2011). Macrocyclic lactone resistance in *Dirofilaria immitis*. *Vet. Parasitol.* 181: 388–392.

5 Geary, T.G., Bourguinat, C., and Prichard, R.K. (2011). Evidence for macrocyclic lactone anthelmintic resistance in *Dirofilaria immitis*. *Top Companion Anim. Med.* 26: 186–192.

6 Smith, J. and Taylor, E.M. (2016). What is next for NTDs in the era of the sustainable development goals? *PLoS Negl. Trop. Dis.* 10: e0004719.

7 Fischer, P.U., King, C.L., Jacobson, J.A., and Weil, G.J. (2017). Potential value of triple drug therapy with ivermectin, diethylcarbamazine, and albendazole (IDA) to accelerate elimination of lymphatic filariasis and onchocerciasis in Africa. *PLoS Negl.Trop. Dis.* 11: e0005163.

8 Webster, J.P., Molyneux, D.H., Hotez, P.J., and Fenwick, A. (2014). The contribution of mass drug administration to global health: past, present and future. *Philos. Trans. R. Soc. Lond. B Biol. Sci.* 369: 20130434.

9 Cheah, P.Y. and White, N.J. (2016). Antimalarial mass drug administration: ethical considerations. *Int. Health* 8: 235–238.

10 Kim, Y.E., Stolk, W.A., Tanner, M., and Tediosi, F. (2017). Modelling the health and economic impacts of the elimination of river blindness (onchocerciasis) in Africa. *BMJ Glob. Health* 2: e000158.

11 Vinkeles Melchers, N.V.S., Coffeng, L.E., Boussinesq, M. et al. (2020). Projected number of people with onchocerciasis-loiasis co-infection in Africa, 1995 to 2025. *Clin. Infect. Dis.* 70: 2281–2289.

12 Boussinesq, M., Gardon, J., Kamgno, J. et al. (2001). Relationships between the prevalence and intensity of *Loa loa* infection in the central province of Cameroon. *Ann. Trop. Med. Parasitol.* 95: 495–507.

13 Burrows, J.N., Duparc, S., Gutteridge, W.E. et al. (2017). New developments in anti-malarial target candidate and product profiles. *Malar. J.* 16: 26.

14 Porras, A.I., Yadon, Z.E., Altcheh, J. et al. (2015). Target product profile (TPP) for Chagas disease point-of-care diagnosis and assessment of response to treatment. *PLoS Negl. Trop. Dis.* 9: e0003697.

15 Bowman, D.D. and Drake, J. (2017). Examination of the "susceptibility gap" in the treatment of canine heartworm infection. *Parasites Vectors* 10: 513.

16 Kaminsky, R. (2018). Where is the breakthrough innovation for parasite control? *Trends Parasitol.* 34: 99–101.

17 de Moraes, J. and Geary, T.G. (2020). FDA-approved antiparasitic drugs in the 21st century: a success for helminthiasis? *Trends Parasitol.* 36: 573–575.

18 Engels, D. (2016). Neglected tropical diseases in the sustainable development goals. *Lancet* 387: 223–224.

19 Turner, H.C., Walker, M., Pion, D.D.S. et al. (2019). Economic evaluations of onchocerciasis interventions: a systematic review and research needs. *Trop. Med. Int. Health* 24: 788–816.

20 Water, H.R., Rehwinkel, J.A., and Burham, G. (2004). Economic evaluation of Mectizan distribution. *Trop. Med. Int. Health* 9: A16–A25.

21 Behrend, M.R., Basanez, M.G., Hamley, J.I.D. et al. (2020). Modelling for policy: the five principles of the neglected tropical diseases modelling consortium. *PLoS Negl.Trop. Dis.* 14: e0008033.

22 Group NTDMCO (2019). The World Health Organization 2030 goals for onchocerciasis: insights and perspectives from mathematical modelling: NTD modelling consortium onchocerciasis group. *Gates Open Res.* 3: 1545.

23 Blagburn, B.L., Dillon, A.R., Arther, R.G. et al. (2011). Comparative efficacy of four commercially available heartworm preventive products against the MP3 laboratory strain of *Dirofilaria immitis*. *Vet. Parasitol.* 176: 189–194.

**24** McTier, T.L., Six, R.H., Pullins, A. et al. (2017). Efficacy of oral moxidectin against susceptible and resistant isolates of *Dirofilaria immitis* in dogs. *Parasites Vectors* 10: 482.

**25** Prichard, R.K. and Geary, T.G. (2019). Perspectives on the utility of moxidectin for the control of parasitic nematodes in the face of developing anthelmintic resistance. *Int. J. Parasitol. Drugs Drug. Resist.* 10: 69–83.

**26** Forrester, S.G., Prichard, R.K., and Beech, R.N. (2002). A glutamate-gated chloride channel subunit from *Haemonchus contortus*: expression in a mammalian cell line, ligand binding, and modulation of anthelmintic binding by glutamate. *Biochem. Pharmacol.* 63: 1061–1068.

**27** Al-Azzam, S.I., Fleckenstein, L., Cheng, K.J. et al. (2007). Comparison of the pharmacokinetics of moxidectin and ivermectin after oral administration to beagle dogs. *Biopharm. Drug Dispos.* 28: 431–438.

**28** Lloberas, M., Alvarez, L., Entrocasso, C. et al. (2015). Comparative pharmacokinetic and pharmacodynamic response of single and double intraruminal doses of ivermectin and moxidectin in nematode-infected lambs. *N. Z. Vet. J.* 63: 227–234.

**29** Waghorn, T.S., Miller, C.M., and Leathwick, D.M. (2016). Confirmation of ivermectin resistance in *Ostertagia ostertagi* in cattle in New Zealand. *Vet. Parasitol.* 229: 139–143.

**30** Osei-Atweneboana, M.Y., Awadzi, K., Attah, S.K. et al. (2011). Phenotypic evidence of emerging ivermectin resistance in *Onchocerca volvulus*. *PLoS Negl.Trop. Dis.* 5: e000998.

**31** Hotez, P.J., Bottazzi, M.E., Zhan, B. et al. (2015). The onchocerciasis vaccine for Africa – TOVA – initiative. *PLoS Negl.Trop. Dis.* 9: e0003422.

**32** Kamgno, J., Pion, S.D., Mackenzie, C.D. et al. (2009). *Loa loa* microfilarial periodicity in ivermectin-treated patients: comparison between those developing and those free of serious adverse events. *Am. J. Trop. Med. Hyg.* 81: 1056–1061.

# 15

# Discovery and Development of New Antifilarial Drugs (*In Vitro* Assays)

*Lucien Rufener[1,*], Alexandre Vernudachi[2], Ronald Kaminsky[3], and Thomas Duguet[1]*

[1] *INVENesis Sàrl, Rue de Neuchâtel 15A, CH-2072, St-Blaise, Switzerland*
[2] *INVENesis France, Bâtiment 311, Route de Crotelles, 37380, Nouzilly, France*
[3] *paraC Consulting, Altenstein 13, 79685, Haeg-Ehrsberg, Germany*

**Abstract**

The worldwide threat represented by filarial species has imposed challenging research initiatives among academic and private contributors toward identifying novel or complementary antiparasitic treatments. As a result, past decades have seen the emergence of a variety of nematode-applied drug screening methods. Some of them consisted in revisiting classical assays directly inherited from model organisms such as the free-living nematode *Caenorhabditis elegans* or the strongyle parasite *Haemonchus contortus*. Such methods often imply phenotypic observation of hit compound effect on the motility/locomotion or the development of different nematode instars. Applied to filarial species, the use of the dog heartworm model *Dirofilaria immitis* has highlighted supplementary challenges imposed by the species life cycle and maintenance requirements, which imposes a reflection on alternative assays to ease screening processes and further *in vivo* testing. The era of microfluidic technologies represents a source of powerful tools, which would enable fine characterization of phenotypic effects induced by any active ingredient in development. The motility trap assay, currently adapted to strongyle species, is one of the most recent accomplishments in combining motility observation and automated recording in a nematode-applied fluidic device. As a result, the MTA enables fine anthelmintic sensitivity assessment and fills the gaps of classical methods as it dramatically reduces the duration of the assay, the number of required animals, and enable the detection of relevant and more representative effective concentrations. In summary, the preliminary data presented in this chapter reflect a novel standardized approach that has the potential to be applied to filariid species, a prospective that is hoped to stimulate a new dynamic in anthelmintic research.

*Corresponding author.

*Human and Animal Filariases: Landscape, Challenges, and Control*, First Edition.
Edited by Ronald Kaminsky and Timothy G. Geary.

## 15.1 Introduction

### 15.1.1 *In Vitro* Tests in General

Most available antiparasitic drugs, or at least the lead molecules for all classes, were characterized in low-throughput screens that employ parasites (either target or alternative model species) in infected host animals. In the anthelmintic area, almost all of the work that led to marketed drugs over the past eight decades was performed by the animal health industry [1, 2]. Indeed, anthelmintics used in livestock or companion animal species can bring considerable financial revenue due to markets and cost recovery much bigger than for human helminth diseases. In the advancement of drug discovery, a screening-based approach remains an intensively explored strategy, employing infected-animal screening or alternatively whole-organism *in vitro* screens which utilize various life stages of target or model parasites. Furthermore, model organisms, particularly the free-living nematode species *C. elegans*, were used for both routine and specialized screening research initiatives. Indeed, in an industrial setup, screening can be pursued on a regular basis or can be conducted in time- or number of available compounds-limited campaigns. In all the latter assays, the parasites are maintained in a microtiter plate in the presence of candidate compounds and phenotypic parameters such as the viability, motility, and/or development of the worms are recorded [3–7]. When a phenotypic effect is observed with a given molecule, it implies that the compound possesses physicochemical properties that allow it to enter the parasite, to penetrate membrane barriers, and to avoid the detoxification mechanisms of the parasite, and finally to reach its molecular target.

For the discovery of new anthelmintics, the classical egg hatch assay (EHA) was often used as a primary screen [8, 9] as well as free-living larval stages of parasitic species like the sheep pathogenic *H. contortus* [10, 11]. Larvae can be fairly easy obtained in large quantities, and thus, this strategy appears to be for *H. contortus* a good compromise between throughput, amount of compound required, and laboratory feasibility. For historical and pragmatic reasons, the pharmaceutical industry mostly uses whenever possible the original target parasite species for screening compound libraries and evaluating the efficacy of lead molecules. However, some parasitic nematodes, including the canine heartworm *Dirofilaria immitis*, have more complex life cycle, making it challenging to examine the impact of small molecules in high-throughput, quick read-out paradigms preferably used to identify molecules with activity against this filarial species in large compound libraries.

The free-living nematode *C. elegans* offers a convenient alternative model system to search for new compounds with anthelmintic properties [12–15]. The majority, if not all, of marketed anthelmintics express some degree of activity against *C. elegans* [16]. However, some exceptions highlight the caution required with this model organism. For example, the nAChR antagonist derquantel appears to be active on adult worms only if the cuticle integrity is artificially disrupted [17]. Also, the insensitivity of *C. elegans* to morantel is explained by the absence of orthologs of the parasite-specific nAChR genes *acr-26* and *acr-27* [18]. These latter findings reflect the ability to employ techniques for the molecular manipulation of *C. elegans*, which has been instrumental in understanding the mechanism of action (MoA) of anthelmintic compounds [12, 15, 19–23]. Nevertheless, regardless of the nematode species used in an *in vitro* screen, the activity of any identified hit compound has to be verified by *in vivo* testing independent of knowledge of a molecular target.

One of the main challenges in the search for new active compounds with the potential to become a drug remains the identification of hits that maintain antiparasitic activity during the transition from *in vitro* assays to *in vivo* studies [24]. Screening of large compound libraries against infectious agents such as bacteria, viruses, and parasites has yielded thousands of hits in a given time, resulting in an acute need for rational criteria to select the best candidates for further testing and development [25–27]. Often, only a small number of the *in vitro* identified hits are active in the animal model, which can rapidly increase research costs as only a minority of hits are validated for further development.

However, these lead molecules might lack essential pharmacokinetic (PK) characteristics needed to detect efficacy *in vivo*: action governed either by total drug exposure or exposure to a peak concentration (AUC/$C_{MAX}$), or by duration of drug levels above the minimal effective concentration. A major limitation of current *in vitro* assays is that compound selection is mostly done with static concentrations; target proteins, cells, or intact organisms are exposed to nominal concentrations of active compounds for long periods of time (up to a week in some cases). The true concentration experienced by the parasite can vary in an uncontrolled way by precipitation of the active ingredient, adsorption to the labware, or through metabolism by the parasite or by associated microorganisms, which are almost always present in *in vitro* assays employing parasites, particular in those which use parasites obtained from manure. *In vivo*, exposure profiles are typically very different than those used in culture; anthelmintics are absorbed, distributed, metabolized, and excreted [28] by the host, leading to dynamic and unstable concentrations in blood and tissues. Drug metabolism and pharmacokinetics (DMPKs) is an important part of studies often referred to as ADME (absorption, distribution, metabolism, and excretion; [29]). In mammalian models, many compounds disappear quickly from the body, thanks to very efficient detoxification systems. In the authors' experience, most failures in the early stages of *in vivo* testing are due to lack of efficacy often due to an unfavorable DMPK profile.

Currently, several phenotypic low- to high-throughput screening assays (LTS and HTS, respectively) have been adapted for evaluating anthelmintic activity (Table 15.1): the EHA [40] reviewed and harmonized by Samson-Himmelstjerna et al. [8], the larval development assay (LDA) [41, 42], the larval migration inhibition assay (LMIA) [43], the larval motility assay (LMA) [44], the larval exsheathment inhibition assay (LEIA) [36], the larval feeding inhibition assay (LFIA) [37, 45], and more recently the automated larval migration assay (ALMA) [39] (Table 15.1).

The LMA utilizes motility as an indicator of larval fitness and drug efficacy and thus has the fewest limitations with regard to parasite life stage and species. LMIA and LMA assays have not been well standardized across laboratories, and despite

**Table 15.1** Overview of key model *in vitro* screening assays developed for nematodes.

| Assay | Nematode species | Life stage | Endpoint | Duration of assay | References |
|---|---|---|---|---|---|
| EHA | *C. elegans* | Eggs | Number of hatched eggs and L1 | 15 h | [30] |
| | *H. contortus* | | | 48 h | [8] |
| LDA | *C. elegans* | Eggs to L1/adults | Number of developed larvae/adults | 24 h (L1) to 96 h (adults) | [19] |
| | *H. contortus* | Eggs to L3 | Number of developed L3 | 6–8 d | [31, 32] |
| LMA/LMIA | *H. contortus* | L3 | Number of migrated larvae | 24 h incubation 24 h migration | [33, 34] |
| | *D. immitis* | L3 | Number of migrated larvae | 48 h incubation 2 h migration | [35] |
| LEIA | *H. contortus* | Ensheathed L3 | Number of exsheathed larvae | 3 h | [36] |
| LFIA | *H. contortus* | L1 | Number of fluorescent fed larvae | 4 h | [37, 38] |
| ALMA | *H. contortus* | L2 | Number of migrated larvae | 25 min | [39] |

numerous publications that describe the LMIA, no consistent protocol has been established yet [46]. The LMIA is based on the ability of larvae to migrate through a fine mesh screen; larvae that are negatively affected by the test compound are less able to migrate through the mesh. Obviously, a major limiting factor to this assay is the differing size of larval stages of different parasite species, requiring custom selection of mesh size for every species [33–35]. Additionally, a single time point is usually measured to estimate migration capability, thus requiring numerous assays to optimize incubation time for each nematode species and drug. The LMA requires an observer to assign a motility score for each larva on a 0–3 scale or to classify it as motile or non-motile [31]. Consequently, the LMA is highly subjective, vulnerable to reader bias, poorly quantitative, and extremely low throughput. The ALMA takes advantage of the natural auto-fluorescence of most gastro-intestinal nematodes larval stages by measuring real-time fluorescence correlated with the migration rate of worms through a 20 μM sieve. Blanchard and collaborators have shown a highly significant correlation between fluorescence and the number of larvae that migrated through the sieve ($R^2 = 0.9935$) after 25 minutes [39]. As for the LMIA, the size of the sieve pores has to be adapted and experimentally selected according to the diameter of the parasite species under investigation. Another drawback of this new technology is the very low throughput (one single read at a time) and the high number of larvae required (7500 per concentration to be tested).

Technological developments have accelerated the adaptation of computer image processing coupled with high definition video for studying and characterizing the movement of nematodes [47]. WormAssay is a specific application of this type of technology developed at the University of California, San Francisco (UCSF) to assess motility of macro-parasites (i.e. visible to the naked eye) in cell culture plates. This can be applied for high-throughput screening of potential anthelmintic compounds [5, 48]. The program analyzes differences in worm position in successive video frames to determine the rate of movement, which contrasts with more classical approaches that measure dead or paralyzed worms as an endpoint. With computerized movement recording systems, the lower the movement, the lower the motility number and more effective the compound is against the parasite. Recently, based on this technology, the Worminator was developed to provide a relatively low-cost, easy to assemble platform for developing genera- and stage-specific assays for measuring sensitivity to anthelmintics, as well as screening new compounds for anthelmintic properties [49]. Interestingly, a recent application of similar technology [48] identified auranofin, a repurposed FDA-approved drug, as a macrofilaricide against the human filariid species *Brugia malayi* and *Onchocerca volvulus* [50]. In addition to improving screening throughput, sensitivity, and accuracy, this new generation of algorithm-based processes [14] brings novel tools for the fine detection of compound-induced behaviors and the identification of poorly explored MoAs.

New innovative tools have taken advantage of microfluidics to assess worm movement [14]. Microfluidic devices have emerged as powerful tools for high-throughput analysis of *C. elegans*, going beyond the possibilities offered by standard assay formats [51–55]. Microfluidic devices permit accurate spatiotemporal application of compounds, so that concentration–response curves and data sets with enriched information can be obtained [56]. Microfluidics enables the manipulation of very small (nanoliter to picoliter) volumes of fluids inside very small channels (micrometer sized). Microfluidic technology is precise and automated and, therefore, applicable to drug discovery for cost-efficient and ultrahigh-throughput assays using biological materials that are in limited supply [57, 58]. Because microfluidic devices require small volumes, liquid pharmaceutical libraries of compounds can be screened conveniently. These systems may be particularly amenable for small worms and for assessing drug effects on various phenotypes, including neurophysiological and behavioral responses [59–62]. Overall, microfluidic technologies have demonstrated the potential to considerably improve many aspects of worm-based research by providing higher performance when applied to protocols ranging from observations of worm behavior, embryo selection and age synchronization, reversible worm immobilization, imaging and drug exposure [56].

In most initiatives implementing HTS assays to discover new nematicide classes, an LDA is usually performed with worms developing in the presence of static nominal compound concentrations. The analysis is done using machine vision software, yielding objective data that are not subject to scorer bias. This automated LDA often leads to high hit rates (depending on the starting concentration used and the artificially set threshold), which must be correlated with compound concentration used throughout the primary screening process. A current alternative is the use of single stage worms (mainly non-feeding L3-larvae), measuring the effect of static levels of a drug on their dorsoventral oscillations. In this HTS assay, the worms are not required to do anything more than oscillate. However, this test fails to identify the main classes of nematicides (e.g. macrocyclic lactones; MLs), biasing the set of identified compounds to favor those with severe toxicity. In animal health, some drugs are also tested with *in vitro* DMPK models borrowed from human pharma research. This strategy results in a large number of *in vivo* studies with a very low success rate. In other words, this approach reflects the need of technologies enabling a better mimicking of the *in vivo* situation with the use of additional parameters than attainable with only motility as a readout and the challenge of modifying drug concentration over time in an HTS-compatible system.

### 15.1.2 Screening Assays for Heartworms

The filarial nematode *D. immitis* can be found globally [63–65]. The endemic areas are expanding due to climate changes that enhance propagation of infectious larvae in mosquitoes [66, 67]. Parasite-associated pathologies are broad, but most serious manifestation are lung damages and heart failure [63, 68]. Diagnosis of *D. immitis* infection is based on adult or antigen detections (by echocardiography and ELISA, respectively) or circulating microfilariae (mf) [69–72]. Because treatment of infected dogs to eliminate adult worms is challenging and risky [73], control efforts are focused on preventative treatments. The American Heart Worm Society recommends preventative treatment with MLs such as ivermectin, milbemycin, or moxidectin to eliminate the tissue-migrating 3$^{rd}$ and 4$^{th}$-stage larvae (L3-larvae and L4-larvae) before they enter into the blood stream [68, 73]. Dogs in endemic areas generally need to be treated on a monthly basis to ensure protection from adult heartworm infection. The commercial market for the prevention of filarial infection of dogs is estimated to be around US$ 8 billion per year in the USA, the largest companion animal health market [74].

Anthelmintic resistance (AR) in gastrointestinal nematode (GIN) parasites constitutes a major problem for livestock health and productivity worldwide [75, 76]. The common practice of intensive farming methods coupled with a heavy dependence on anthelmintics has resulted in a serious escalation in the prevalence and distribution of AR in many of the most important GIN parasite species, leading to the emergence of isolates resistant to all commercially available anthelmintics [77, 78]. In contrast to the situation with GIN, filarial parasites such as *D. immitis* were considered to be at a low risk for developing AR [79]. Despite the constant widespread use of MLs for prevention, no treatment failures were documented for almost three decades. However, since 2005, an increased number of loss of efficacy (LOE) cases have been reported, in which dogs develop mature heartworm infections despite receiving monthly doses of MLs [80–84]. The fact that only one class of anthelmintics is usefully effective for this indication, coupled with the rise of resistance, stresses the urgent need for discovery and development of new anthelmintic groups with, ideally, confirmed *in vivo* activity against ML-resistant *D. immitis* larvae.

A well-known constraint for chemotherapy of filariae is imposed by their biology. For preventative-based control, anthelmintics need to reach L3/L4 larvae as they migrate through the body of the host, where they are difficult to locate. In addition, *D. immitis* cannot be expelled from the body like GINs from the intestinal tract, so treatments that induce temporary immobilization are not sufficient. Last but not least, *D. immitis* development (L3-larvae to adult) in the tissues of the host is a slow process (>five months). Hitting the filariae at the right place, at the right time, with a sufficient concentration of anthelmintic remains challenging and stresses the difficulties faced in the discovery of new preventive and also curative (macrofilaricidal) treatments. Nevertheless, filarial parasites have a few weak points while in their vertebrate host, which can be used to our advantage when designing the next generation of filaricidal compounds: (i) they need to feed; (ii) they need to molt; and (iii) they cannot temporarily leave the host to avoid drug exposure. Ideally, to maximize the chance to hit the worms, a new filaricide compound should have an irreversible MoA from which worms cannot recover or which express activity over a sufficiently long duration. In the first case, $C_{max}$ is the most critical DMPK parameter, while in the second case, the compound half live ($t_{1/2}$) becomes the most critical DMPK factor in the treatment scheme.

### 15.1.3 Which Nematode Species/Model Is Best?

The first decision in establishing a screening concept for the identification of *D. immitis* active compounds resides in the choice of the parasite species to be employed. As outlined earlier, the final goal will require testing in an infected animal. With filarial nematodes, an additional difficulty resides in the insect vector intermediate host. In addition, the long migration of the intermediate life stages in the final host is radically different than the rather straightforward location of maturing GINs. For those reasons, it remains important to consider the advantages and disadvantages of different parasitic or non-parasitic nematode models available before setting up a filaricide screening platform.

#### 15.1.3.1 Filarial Models: Introduction

*Dirofilaria immitis* is found in over 30 mammalian species, including foxes, cats, wolves, coyotes, and other wild carnivores, but is mainly a concern in dogs [65, 85, 86] . Dogs are the most important definitive hosts and serve as the main reservoir of infection. The parasite is named "heartworm" due to the location of the adults in the arteries of the lungs and occasionally in the right ventricle of the heart. Very common in numerous tropical countries, it is widespread throughout the Far West, Equatorial Africa (multiple heartworm species), and in the Pacific and also occurs in North/South America, Australia, North Africa, and South/East Europe [63, 64, 68, 87, 88], depending on the distribution and seasonal presence of competent mosquito species. The life cycle of this parasite involves two hosts: a mosquito (of the genera *Culex, Aedes, Psorophora, Mansonia,* or *Anopheles*) as an intermediate host and typically a dog (*Canis lupus familiaris*) as the final host. The pre-patency period in a newly infected dog is six to nine months. During this time, the larval stages migrate through the host body until they reach the target tissues, the pulmonary arteries and the right ventricle. Migrating larvae pass several host tissues, and the larvae undergo two molts while escaping the host immune system. Details of the migration path are largely unknown [89, 90] along with the time required for this migration. Knowing the localization of larvae would enhance our understanding of the PK parameters that should be met to select the optimal compound and treatment scheme.

#### 15.1.3.2 Use of *D. immitis* as a Screening Model

The *D. immitis* mf model used for *in vitro* screens has been implemented in several public institutions, including the College of Veterinary Medicine of Auburn University, AL and the University of Georgia, among others, as well as in animal health-related industrial laboratories. When infected dogs are available, detection of mf in the dog blood enables the harvest of sufficient number of parasites for HTS screening campaigns.

#### 15.1.3.3 Disadvantages

The *D. immitis* model is complex and has a long-life cycle that requires dogs and mosquitoes. Drug discovery efforts against *D. immitis* are hampered by the difficulty of accessing the target larval stages (L3 and L4) for high-throughput phenotypic screens. Indeed, even obtaining L3-larval stages from infected mosquitoes is rather time- and energy-consuming. Given the challenges faced to sustain a high-throughput screen with L3- or L4-larval stages of *D. immitis*, an alternative for whole organism-based anti-filarial screening is to use mf. Such screens presume that different developmental stages share the same functional biology of essential drug targets, as observed with the major classes of anthelmintics in *C. elegans* [16, 91]. This hypothesis is supported by the example of monepantel (MPTL) activity on *C. elegans*. Indeed, MPTL was shown to target L2 larvae to adult worms, which coincides with expression of the MPTL-sensitive *acr-23* gene across these life stages [22]. Interestingly, in our experience, mf-based assays often reveal questionable results, such as a very high hit rate (5–10% with compounds tested at 10 μM, most of which are toxic to the host), necessitating counter-screens to minimize the number of compounds to be advanced. In addition, and most importantly, MLs are hardly detectable in mf-based assays because they do not paralyze or kill this stage at concentrations that are relevant *in vivo* [92, 93].

#### 15.1.3.4 Conclusions

The use of L3-larvae and other life stages *in vitro* in multiple types of screens faces challenging limitations that have not generally been more productive for the identification of new classes of anthelmintics as opposed to simpler screens using *C. elegans* [15, 19]. A potential solution for antifilarial screening would be to adapt migration assays with mf under dynamic compound exposure conditions. Indeed, the use of screening assays under static concentrations and with motility only as a readout could explain the lack of compounds showing efficacy when tested *in vivo*, further stressing the need for the development of new assays. Based on the difficulties and the costs required for sustainable access to a constant production of good quality *D. immitis* L3-larvae, other nematodes, such as *H. contortus* or *C. elegans*, offer interesting alternatives.

### 15.1.4 *Haemonchus contortus*

#### 15.1.4.1 Introduction

*Haemonchus contortus* belongs to the phylum Nematoda (roundworms), more precisely to the family of Trichostrongylidae. *Haemonchus* species are the largest nematodes found in the abomasum of ruminants. They range from 10 to 20 mm in length for males and 18 to 30 mm in length for females. They are reddish when freshly obtained because they are blood feeders. They can pierce the epithelial membrane of the abomasum with the help of a tiny lancet found in their small buccal capsule. *H. contortus* is also called the barber pole worm because the white ovaries of female wind around the red blood-filled intestine. Males are easily recognizable by their copulatory bursa and spicules at the posterior end. The *H. contortus* life cycle is monoxenic and direct. The predilection site is the abomasum of its host.

#### 15.1.4.2 Advantages

Many reasons make *H. contortus* an excellent parasite to use in screens for new anthelmintics, including new filaricides. First, it is one of the parasitic nematode species for which AR is a common problem and has been intensively studied [94, 95]. Second, it is closely related to many other important GIN species of ruminants for which AR is emerging [96]. Third, as *H. contortus* is a parasite of sheep, it is possible to undertake experimental work in its natural host rather than in laboratory model hosts. This is not the case for human or cattle parasites for ethical and cost considerations, respectively. Fourth, female *H. contortus* are extremely prolific, with a female producing up to 10 000 eggs per day. This allow for the collection of high number of eggs and the ready production of L3-larvae for DNA extraction or drug selection *in vitro*. Furthermore, as the adults are quite large, DNA extraction from a single individual is possible, permitting genotyping assays. L3-larvae can also be cryopreserved, allowing stable access to isolates or intermediate resistant lines [97]. In addition, *H. contortus* is closely related phylogenetically to *C. elegans* [98–101]. Finally, all nematicides on the market today are active against *H. contortus*, with very low $EC_{50s}$ for the MLs, making this parasite a relevant and cost-effective model for filaricide screening.

#### 15.1.4.3 Disadvantages

An obvious disadvantage to the use of *H. contortus* to screen for new filaricides is that it is an obligatory parasite of sheep, requiring animals to maintain it in a laboratory. In addition, this parasite does not have *Wolbachia* endosymbionts like *D. immitis* and most medically important filarial parasites. Also, it belongs to nematode clade V (versus clade III for filariae), meaning that, from an evolutionary point of view, *H. contortus* is closer to *C. elegans* than to *D. immitis*. Finally, most screening assays using *H. contortus*, including the LDA and the EHA, are still labor-intensive, despite increasing efforts to refine handling and readouts.

#### 15.1.4.4 Conclusions

Even though *H. contortus* is not in the same clade as filarial parasites and is a parasite of sheep, not dogs, it remains an excellent parasitic nematode model to screen for new filaricides. Most anthelmintic classes that affect *D. immitis* also control *H. contortus in vitro* and *in vivo*. Of course, compounds with a MoA specific to *D. immitis* will not be discovered, but such compounds are expected to be the overwhelming minority. Indeed, more than 80% of the anthelmintics on the market target the parasite nervous system. The *D. immitis* receptome contains considerably fewer ligand-gated ion channel subunits than *H. contortus* or *C. elegans*, potentially explained by the fact that *D. immitis* parasites live in a much more controlled environment (e.g. mosquitoes and mammals) and are never exposed to the environment (meaning that less neurological plasticity is required). As a consequence, chances for false positives are certainly higher than false negatives when screening with *H. contortus* for new filaricides.

### 15.1.5 *C. elegans*

#### 15.1.5.1 Introduction

The free-living nematode *C. elegans* lives in soil and feeds on microbes. To emulate their natural environment, *C. elegans* is typically cultured in the laboratory on agar plates seeded with a lawn of *Escherichia coli*. *C. elegans* has two sexes: hermaphrodite and male. In 1977, Platzer and colleagues used, for the first time, *C. elegans* to evaluate the anthelmintic activities of nine benzimidazoles and suggested that it could be used in a test for anthelmintic screening [102]. Recent advances in screens using *C. elegans*, including in handling, liquid workflow, read-out, and data analysis, have facilitated high-throughput drug screens, including for anthelmintics [5].

#### 15.1.5.2 Advantages

As noted, parasitic nematodes are difficult to work with, requiring passage through their host(s) for maintenance of their life cycle. This greatly complicates quantitative experiments in their natural habitat. Methods for conducting forward and reverse genetics are also at best primitive. Thus, *C. elegans* has routinely been exploited as a more "user-friendly" model system, one that is also highly tractable to molecular genetic techniques. Since its first introduction to biology in the early 1960s, *C. elegans* has played a pivotal role in elucidating genetic pathways controlling important cellular processes (e.g. development, cell death, aging, RNAi). In the past decade, *C. elegans* has emerged as a tool for drug discovery: its small size (~1 mm in length), short generation time (~three days), ability to produce ~300 offspring in ~three days, genetic amenability and conservation of cellular processes across species, make *C. elegans* an excellent tool for whole organism-based HTS.

The majority of marketed anthelmintics are active against *C. elegans*, and the use of this model system has been instrumental in improving the understanding of the mechanism of action of several anthelmintic compounds, including levamisole [103, 104], benzimidazole [105], and the amino-acetonitrile derivatives [21, 22]. Its relatively short lifespan of two to three weeks allows the drug discovery process to be studied over the whole life of the organism and in a time-specific manner. A large collection of mutant strains is readily available at the Caenorhabditis Genetics Center (CGC; https://cgc.umn.edu) to easily study a significant portion of its genome, and many genetic tools have been developed that allow manipulation of single genes or groups of genes, such as ENU (*N*-ethyl-*N*-nitrosourea) or EMS (ethyl methanesulfonate) mutagenesis, genetic interference by double-stranded RNA (RNAi) and more recently the CRISPR technology [15, 106, 107].

For parasites that are difficult to obtain in large numbers or too resource demanding to maintain in the laboratory (e.g. *D. immitis*), transgenic *C. elegans* strains offer an interesting alternative to express parasite drug targets. This model has been exploited to validate anthelmintic targets from parasitic nematodes as well as drug resistance mechanisms for benzimidazoles, MLs, monepantel, and emodepside [22, 108–112]. Recently, Courtot et al. used *C. elegans* as a functional tool to express a novel class of acetylcholine receptors for which no ortholog could be identified in its genome [18]. They confirmed the functional expression of a receptor made of *H. contortus* or *Parascaris equorum* ACR-26 and ACR-27 subunits, rendering *C. elegans* sensitive to morantel and pyrantel, to which wild-type *C. elegans* is insensitive. Last, *C. elegans* offers the ability to simultaneously evaluate drug efficacy and absorption, distribution, metabolism, excretion, or toxicity (ADMET) characteristics at the initial stages of the drug discovery pipeline [113]. Those examples highlight the value of *C. elegans* expressing molecular targets from parasitic nematodes as a relevant screening tool for the discovery of novel anthelmintics and filaricides in particular. Indeed, as mentioned above, *C. elegans* is easily amenable to efficient motility assays with a large repertoire of scorable phenotypes available.

#### 15.1.5.3 Disadvantages

Although the utility of the worm as a versatile genetic tool is certain, there are distinct limitations for using *C. elegans* in drug discovery:

1. It is not a parasite and therefore lacks many of the adaptations required for parasitism and the potential anthelmintic targets associated with those processes [12].
2. *Caenorhabditis elegans* has been thought to be a poor candidate for screening due to the relatively inefficient drug accumulation caused by the impermeability of the cuticle to non-water-soluble compounds [16, 114] and the selective uptake of drugs by the intestine.
3. *Escherichia coli* (OP50) is typically added to the culture as the primary food source. The use of live bacteria could lower the effective concentration delivered to the worms due to metabolism or degradation by the bacteria. For these reasons, higher initial compound concentrations (25–100 μM) are usually used in *C. elegans*-based drug screens.

4. Despite its similarity to parasitic nematodes, with many of them grouped in the same phylogenetic clade (clade V), the correlation of activity against *C. elegans* to parasitic nematodes is not universal. Two key human anthelmintics, albendazole and pyrantel, show no activity against *C. elegans* larvae [115].
5. Compounds active against larval stages are not necessarily active against adult worms, the traditional targets of anthelmintic chemotherapy, given stage-dependent differences in target expression [116].
6. Drug effects can vary significantly even among closely related nematode species.
7. *in vitro* – *in vivo* correlation is poor for many standard anthelmintics, since these drugs do not induce death of worms but rather impair motility or cause other subtle alterations (difficult to detect and quantitate by eye in an *in vitro* system) that still result in the expulsion of the parasitic worms [117, 118]. Ivermectin, for example, has excellent trichuricidal properties in the *Trichuris muris* mouse model despite having no detectable *in vitro* activity. A newly developed assay, the Motility Trap Assay (MTA), might offer a solution to this specific issue (see next section).

#### 15.1.5.4 Conclusions

*Caenorhabditis elegans* was used for many years as a substitute organism for parasites in whole-organism screens [119]. These efforts were largely unsuccessful; no new anthelmintic class was brought to market based on the activity against this organism [12]. Recent data show that the accumulation of xenobiotics in this worm is much more limited than expected based on physical chemical characteristics [114], suggesting that it may not be a faithful mimic of drug permeation and accumulation in adult nematode parasites. This finding also suggests that use of *C. elegans* as a primary anthelmintic screen may have missed important actives; there is every reason to develop new and improved screening methods for this purpose. It is promising in this regard that the use of worms with defects in cuticle structure demonstrated enhanced sensitivity to chemicals due to increased accumulation, improving the drug sensitivity correlation between *C. elegans* and parasitic nematodes [118, 119].

It is undeniable that drug discovery using *C. elegans* offers opportunities that are currently not possible with parasitic species. Keiser proposed the use of *C. elegans* for the screening of large libraries with unknown anthelmintic properties, while smaller, better characterized libraries (e.g. from animal health companies) should be directly tested on models of human and veterinary nematode infections, ideally both larval and adult stages [120]. The recent advancement made with genome editing technologies (e.g. CRISPR-Cas9) makes it possible to express parasitic targets in *C. elegans*, offering an unprecedent, powerful and affordable screening tool.

### 15.1.6 Motility Trap Assay (MTA) with *H. contortus*

#### 15.1.6.1 Introduction

Measuring motility of nematodes in an accurate and reproducible way has long been challenging but necessary to identify active compounds in a high-throughput manner. Nevertheless, the use of motility only as a readout under static drug concentrations does not mimic properly the *in vivo* situation. In addition, being able to test the potential reversibility of the compound-induced phenotype would help to determine the proper *in vivo* treatment regimen. The MTA was designed to resolve these challenges and was developed in collaboration with the INRAE (Institut National de Recherche pour l'Agriculture, l'Alimentation et l'Environnement, Nouzilly, France) and the CSEM (Centre Suisse d'Electronique et de Microtechnique, Neuchâtel, Switzerland). This new innovative and proprietary worm migration device was inspired by the ALMA technology [39, 121] but with HTS in mind. Briefly, 100 *H. contortus* L3-larvae are deposited in a dedicated well and allowed to migrate toward a trap area, where the worm number increases over a 21-minute time window. The migration process is monitored by an automatized reading system, which measures the surface of worms at the destination trap every three minutes, generating a migration curve over time. Thanks to the automatized reading system, this assay is totally objective, as it removes the reader bias that can occur in other assays, such as the LMA. The number of L3-larvae exposed to test compounds reaching the trap area is then compared to drug-free controls.

#### 15.1.6.2 Advantages

The MTA proof of concept was revealed by testing MLs with *H. contortus* L3-larvae (non-developing worms). While classical *in vitro* assays (e.g. L3-larvae immersion assay; Table 15.2) yield micromolar $EC_{50s}$ with MLs [34, 122], the ALMA platform and the MTA provide $EC_{50s}$ in the nanomolar range. This is more in agreement with the *in vivo* situation (Figure 15.1 and Table 15.2). The need for few parasites (only 100 L3-larvae compared to 7500 with the ALMA) and its simple recording analysis protocol makes it suitable for relatively HTS with a throughput of 3200

**Table 15.2** $EC_{50}$ values for a set of anthelmintic compounds measured with three different screening assays (MTA; immersion assay; ALMA).

| | MTA (4 h) | | Immersion assay[a)] (6 h) | | Immersion assay[a)] (24 h) | | ALMA[b)] (4 h) | |
|---|---|---|---|---|---|---|---|---|
| **Compounds** | **$EC_{50}$ (nM)** | **Std. Err.** | **$EC_{50}$ (nM)** | **Std. Err.** | **$EC_{50}$ (nM)** | **Std. Err.** | **$EC_{50}$ (nM)** | **Std. Err.** |
| Emodepside | 275.0 | 67.3 | 7 900.0 | 4 100.0 | 3 500 | 800.0 | NA | NA |
| Monepantel | 906.0 | 312.6 | 78 500.0 | 41 500.0 | 2 500.0 | 1 300.0 | NA | NA |
| Ivermectin | 26.5 | 7.4 | 78 500.0 | 41 500.0 | 2 500.0 | 1 300.0 | 6.6 | 0.2 |
| Moxidectin | 245.0 | 49.3 | NA | NA | 4 830.0[c)] | NA | NA | NA |
| Pyrantel | 338.0 | 117.9 | NA | NA | NA | NA | 770 | 10 |
| Levamisole | 1 257.0 | 299.0 | NA | NA | NA | NA | 1 140 | 30 |

a) The INVENesis *H. contortus* L3-larvae assay measures the effect of compounds on L3-larvae of *H. contortus*. Approximately 100 L3-larvae are deposited in a 384-well plate and the treatment formulated in DMSO. Plates are incubated at 25 °C for 24 hour. The effect of compounds is measured as motility reduction using an automated data acquisition system. Efficacy is expressed in % motility reduction compared to negative controls.
b) ALMA assay, *H. contortus* L3-larvae Weybridge strain [121].
c) Worminator assay, *H. contortus* L3-larvae, UGA-SUSC strain [122].

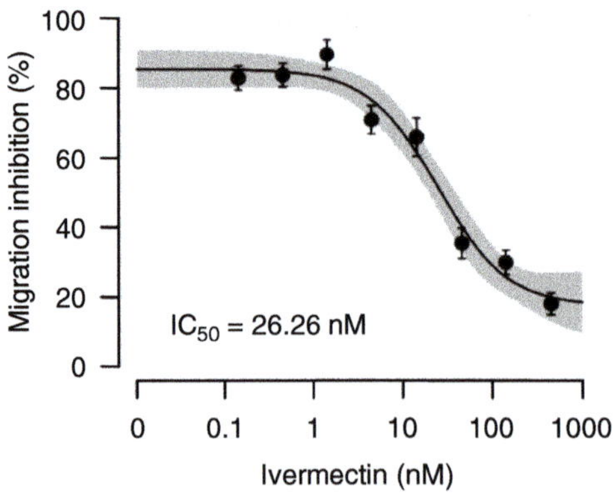


**Figure 15.1 Ivermectin concentration–response curve obtained in the MTA assay using *H. contortus* L3-larvae.** Inhibition of migration is expressed in percent compared to placebo. Each point represents the average of 4 independent experiments, each in triplicate. Standard deviation is shown as well as 95% confidence intervals (gray).

data points in triplicates per month. A key aspect of the MTA technology relies on its high discriminating power, as it enables shorter incubation times than other available or published tests (four hours versus eight hours or more generally 72 hours) [4, 7, 123, 124].

Another interesting aspect of the MTA resides in the fact that incubation of L3-larvae in test compounds and the migration are two distinct processes that take place in two different plates. Thus, the incubation time can be adapted according to the optimization phase of the screening: 24 hours incubation allows the detection of hits and a shorter and more discriminative four hours incubation detects hit compounds with fast action (Figure 15.2). In addition, the fact that a compound effect is irreversible or not should be considered when selecting compounds for *in vivo* studies and designing treatment schemes for these studies. Here again,

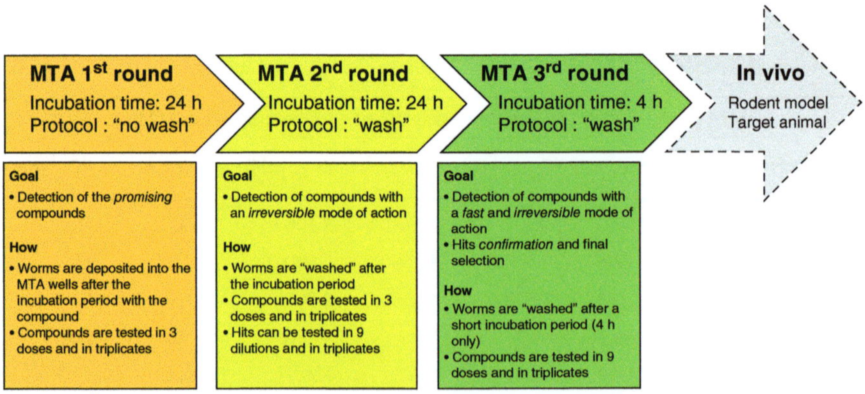


**Figure 15.2 Proposed screening flowchart using the MTA for the discovery of new filaricide compounds.** 1st round: incubation time 24 hour along with "no wash" method to detect promising hits. 2nd round: incubation time 24 hour along with "wash" method to detect compounds with an irreversible mode of action. Such compounds have a higher chance to remain efficacious when applied *in vivo*. 3rd round: incubation time four hours along with "wash" method to discriminate among hit compounds that act as quickly as ivermectin and remain effective after wash-out, mimicking the declining concentration of the drug *in vivo* due to the metabolism by and excretion from the host.

the design feature which makes incubation and migration two distinct processes in the MTA allows analysis of worm recovery to compounds, as it is possible to "wash" away the compounds before migration takes place. The MTA encompasses this parameter with two versions of the test: the "wash" and "no-wash" methods (Figure 15.2).

As the principle of the MTA is not based on the use of a mesh, it avoids the major limiting factor of the LMIA (e.g. custom selection of mesh size; see section above). There is no need to modify the MTA plate to adapt to different sizes of larval stages of different parasite species (although this remains to be proven with species other than *H. contortus*).

In the AR context, worm motility is not a suitable phenotype to assess the emergence of resistance in parasite populations. Available assays that detect AR face numerous limitations, such as manually counting parasite, operator-dependent interpretation of results, and typically low throughput [122]. The MTA, being exempt of the limitations cited above, could lead to fast and efficient detection of AR even from a mixture of different parasitic species. Last but not least, since the MTA allows recollection of tested worms, it is possible to perform subsequent molecular analysis, for instance to determine the proportion of a resistance marker in worms that reach the destination trap.

#### 15.1.6.3 Disadvantages

Screening for new filaricides using *H. contortus* L3-larvae might be seen as a disadvantage as it is not the target species. However, even though *H. contortus* does not belong to the same clade as filarial parasites and is a parasite of sheep and not dog, it remains an excellent parasitic nematode model to screen for new filaricides. As previously discussed, chances for false positives are certainly higher than false negatives when screening with *H. contortus* for new filaricides.

Another well-known and common weakness of most larval-based assays resides in the use of non-target stages (e.g. free-living L3-larvae of trichostrongyloid species), the results of which may impair reliable conclusions about compound efficacy against target stages. For heartworm preventatives, which are intended to act against larval stages, this is less of a disadvantage. The use of CRISPR-engineered *C. elegans* strains expressing relevant parasitic targets (see above) could compensate for this disadvantage and offer an interesting, HTS-compatible alternative.

#### 15.1.6.4 Conclusions

The MTA is an *in vitro* assay that can detect MLs in the nanomolar range using non-developing L3-larvae. It appears particularly interesting to screen repurposing libraries of compounds, as numerous anthelmintic candidates may have been missed in previous screening campaigns due to the poor *in vivo* predictability of other tests. The different versions of the MTA which allow variation of (i) incubation time and (ii) the possibility to wash compounds make this *in vitro* assay extremely polyvalent and a tool of choice to study numerous aspects of compounds prior to *in vivo* studies. A potential screening cascade for new filaricides using the MTA as a selection tool is outlined in Figure 15.2.

**Table 15.3** Summary of model species key points.

| Species | Requirements | Advantages | Disadvantages |
|---|---|---|---|
| *D. immitis* | Dogs<br>Mosquitoes | – High microfilaremia<br>– Standardized mf preparation<br>– L4 filariae: the target life stage<br>– The right parasite | – Mf: not the target life stage<br>– Costly and challenging<br>– Access to high number of L4-larvae |
| *H. contortus* | Sheep | – High production of eggs/L3-larvae<br>– Standardized egg test<br>– Exsheathment of L3-larvae<br>– Microplate test<br>– Automated | – Poor prediction for other clades of nematodes<br>– Challenging development to L4-larvae |
| *C. elegans* | n.a. | – Plenty of mutants available<br>– Production is easy and standardized | – Non-parasitic<br>– Detoxification enzymes |

### 15.1.7 Summary/Conclusion

Despite numerous descriptions of new lead molecules with promising efficacy against nematodes, including filarial worms, there has been relatively limited progress in the development and marketing of new drugs. Geary et al. [125] pointed out the major challenges associated with development and translation to the market of new anthelmintics. In addition to the hurdles of achieving efficacy with an acceptable therapeutic index and the ability to develop formulations that deliver the PK profile necessary for efficacy, the drug has to demonstrate acceptable human safety, in addition to low cost-of-goods. Regarding heartworm, the market supports the cost of development for an efficacious drug for this application, but the limitations imposed by the parasite itself (e.g. complex life cycle, slow development within the host, rise of AR) dictate the search for alternative molecules and the development of novel compound evaluation methods (Table 15.3). The MTA, currently developed with *H. contortus* L3-larvae, is a response to these challenges, with the aim of refining the characterization of active compounds in a more representative manner. This is also encouraged by the most recently collected data demonstrating the effectiveness of standard anthelmintics in both *H. contortus* and *D. immitis* [92]. Preliminary data presented in this chapter reflect a novel standardized approach that has the potential to be applied to filariid species, a prospective that is hoped to stimulate a new dynamic in anthelmintic research.

## Acknowledgments

We sincerely thank Jacques Bouvier for critical review of the manuscript.

## References


1 Geary, T.G., Woo, K., McCarthy, J.S. et al. (2010). Unresolved issues in anthelmintic pharmacology for helminthiases of humans. *Int. J. Parasitol.* 40: 1–13.

2 Woods, D.J., Vaillancourt, V.A., Wendt, J.A., and Meeus, P.F. (2011). Discovery and development of veterinary antiparasitic drugs: past, present and future. *Future Med. Chem.* 3: 887–896.

3 Smout, M.J., Kotze, A.C., McCarthy, J.S., and Loukas, A. (2010). A novel high throughput assay for anthelmintic drug screening and resistance diagnosis by real-time monitoring of parasite motility. *PLoS Negl.Trop. Dis.* 4: e885.

4 Preston, S., Jabbar, A., Nowell, C. et al. (2015). Low cost whole-organism screening of compounds for anthelmintic activity. *Int. J. Parasitol.* 45: 333–343.

5 Partridge, F.A., Brown, A.E., Buckingham, S.D. et al. (2018). An automated high-throughput system for phenotypic screening of chemical libraries on *C. elegans* and parasitic nematodes. *Int. J. Parasitol.: Drugs Drug Resist.* 8: 8–21.

6 Preston, S., Jabbar, A., Nowell, C. et al. (2016). Practical and low-cost whole-organism motility assay: a step-by-step protocol. *Mol. Cell. Probes* 30: 13–17.

7 Liu, M., Landuyt, B., Klaassen, H. et al. (2019). Screening of a drug repurposing library with a nematode motility assay identifies promising anthelmintic hits against *Cooperia oncophora* and other ruminant parasites. *Vet. Parasitol.* 265: 15–18.

8 von Samson-Himmelstjerna, G., Coles, G.C., Jackson, F. et al. (2009). Standardization of the egg hatch test for the detection of benzimidazole resistance in parasitic nematodes. *Parasitol. Res.* 105: 825.

9 Abriola, L., Hoyer, D., Caffrey, C.R. et al. (2019). Development and optimization of a high-throughput screening method utilizing *Ancylostoma ceylanicum* egg hatching to identify novel anthelmintics. *PLoS One* 14: e0217019.

10 Geary, T.G. (2016). *Haemonchus contortus*: applications in drug discovery. *Adv. Parasitol.* 93: 429–463.

11 Jiao, Y., Preston, S., Hofmann, A. et al. (2020). A perspective on the discovery of selected compounds with anthelmintic activity against the barber's pole worm-where to from here? *Adv. Parasitol.* 108: 1–45.

12 Geary, T.G. and Thompson, D.P. (2001). *Caenorhabditis elegans*: how good a model for veterinary parasites? *Vet. Parasitol.* 101: 371–386.

13 Burns, A.R., Luciani, G.M., Musso, G. et al. (2015). *Caenorhabditis elegans* is a useful model for anthelmintic discovery. *Nat. Commun.* 6: 7485.

14 Buckingham, S.D., Partridge, F.A., and Sattelle, D.B. (2014). Automated, high-throughput, motility analysis in *Caenorhabditis elegans* and parasitic nematodes: applications in the search for new anthelmintics. *Int. J. Parasitol.: Drugs Drug Resist.* 4: 226–232.

**15** Carretero, M., Solis, G.M., and Petrascheck, M. (2017). *C. elegans* as model for drug discovery. *Curr. Top. Med. Chem.* 17: 2067–2076.

**16** Holden-Dye, L. and Walker, R.J. (2014). Anthelmintic drugs and nematicides: studies in *Caenorhabditis elegans*. *WormBook* 1: 1–29.

**17** Ruiz-Lancheros, E., Viau, C., Walter, T.N. et al. (2011). Activity of novel nicotinic anthelmintics in cut preparations of *Caenorhabditis elegans*. *Int. J. Parasitol.* 41: 455–461.

**18** Courtot, E., Charvet, C.L., Beech, R.N. et al. (2015). Functional characterization of a novel class of morantel-sensitive acetylcholine receptors in nematodes. *PLoS Pathog.* 11: e1005267.

**19** Simpkin, K.G. and Coles, G.C. (1981). The use of *Caenorhabditis elegans* for anthelmintic screening. *J. Chem. Technol. Biotechnol.* 31: 66–69.

**20** Gilleard, J.S. (2004). The use of *Caenorhabditis elegans* in parasitic nematode research. *Parasitology* 128: S49–S70.

**21** Kaminsky, R., Ducray, P., Jung, M. et al. (2008). A new class of anthelmintics effective against drug-resistant nematodes. *Nature* 452: 176–180.

**22** Rufener, L., Bedoni, N., Baur, R. et al. (2013). *acr-23* encodes a monepantel-sensitive channel in *Caenorhabditis elegans*. *PLoS Pathog.* 9: e1003524–e1003524.

**23** Sepulveda-Crespo, D., Reguera, R.M., Rojo-Vazquez, F. et al. (2020). Drug discovery technologies: *Caenorhabditis elegans* as a model for anthelmintic therapeutics. *Med. Res. Rev.* 40: 1715–1753.

**24** Mackenzie, C.D. and Geary, T.G. (2013). Addressing the current challenges to finding new anthelminthic drugs. *Expert Rev. Anti-infective Ther.* 11: 539–541.

**25** Mayr, L.M. and Fuerst, P. (2008). The future of high-throughput screening. *J. Biomol. Screening* 13: 443–448.

**26** Al-Ali, H. (2016). The evolution of drug discovery: from phenotypes to targets, and back. *Med. Chem. Commun.* 7: 788–798.

**27** Loso, M.R., Garizi, N., Hegde, V.B. et al. (2017). Lead generation in crop protection research: a portfolio approach to agrochemical discovery. *Pest Manage. Sci.* 73: 678–685.

**28** Lifschitz, A., Lanusse, C., and Alvarez, L. (2017). Host pharmacokinetics and drug accumulation of anthelmintics within target helminth parasites of ruminants. *N. Z. Vet. J.* 65: 176–184.

**29** Wan, H. (2013). What ADME tests should be conducted for preclinical studies? *ADMET & DMPK* 1: 19–28.

**30** Sant'anna, V., Vommaro, R.C., and de Souza, W. (2013). *Caenorhabditis elegans* as a model for the screening of anthelminthic compounds: ultrastructural study of the effects of albendazole. *Exp. Parasitol.* 135: 1–8.

**31** Gill, J.H., Redwin, J.M., van Wyk, J.A., and Lacey, E. (1991). Detection of resistance to ivermectin in *Haemonchus contortus*. *Int. J. Parasitol.* 21: 771–776.

**32** Kotze, A.C., O'Grady, J., Emms, J. et al. (2009). Exploring the anthelmintic properties of Australian native shrubs with respect to their potential role in livestock grazing systems. *Parasitology* 136: 1065–1080.

**33** Kotze, A.C., Le Jambre, L.F., and O'Grady, J. (2006). A modified larval migration assay for detection of resistance to macrocyclic lactones in *Haemonchus contortus*, and drug screening with Trichostrongylidae parasites. *Vet. Parasitol.* 137: 294–305.

**34** Demeler, J., Kuttler, U., El-Abdellati, A. et al. (2010). Standardization of the larval migration inhibition test for the detection of resistance to ivermectin in gastro intestinal nematodes of ruminants. *Vet. Parasitol.* 174: 58–64.

**35** Evans, C.C., Moorhead, A.R., Storey, B.E. et al. (2013). Development of an *in vitro* bioassay for measuring susceptibility to macrocyclic lactone anthelmintics in *Dirofilaria immitis*. *Int. J. Parasitol.: Drugs Drug Resist.* 3: 102–108.

**36** Bahuaud, D., Martinez-Ortiz de Montellano, C., Chauveau, S. et al. (2006). Effects of four tanniferous plant extracts on the *in vitro* exsheathment of third-stage larvae of parasitic nematodes. *Parasitology* 132: 545–554.

**37** Jackson, F. and Coop, R.L. (2000). The development of anthelmintic resistance in sheep nematodes. *Parasitology* 120 (Suppl): S95–S107.

**38** Álvarez-Sánchez, M.A., Pérez García, J., Bartley, D. et al. (2005). The larval feeding inhibition assay for the diagnosis of nematode anthelmintic resistance. *Exp. Parasitol.* 110: 56–61.

**39** Blanchard, A., Guégnard, F., Charvet, C.L. et al. (2018). Deciphering the molecular determinants of cholinergic anthelmintic sensitivity in nematodes: when novel functional validation approaches highlight major differences between the model *Caenorhabditis elegans* and parasitic species. *PLoS Pathog.* 14: e1006996.

**40** Le Jambre, L.F. (1976). Egg hatch as an *in vitro* assay of thiabendazole resistance in nematodes. *Vet. Parasitol.* 2: 385–391.

**41** Taylor, M.A. (1990). A larval development test for the detection of anthelmintic resistance in nematodes of sheep. *Res. Vet. Sci.* 49: 198–202.

**42** Coles, G.C., Jackson, F., Pomroy, W.E. et al. (2006). The detection of anthelmintic resistance in nematodes of veterinary importance. *Vet. Parasitol.* 136: 167–185.

**43** Wagland, B.M., Jones, W.O., Hribar, L. et al. (1992). A new simplified assay for larval migration inhibition. *Int. J. Parasitol.* 22: 1183–1185.

**44** Martin, P.J. and Le Jambre, L.F. (1979). Larval paralysis as an *in vitro* assay of levamisole and morantel tartrate resistance in *Ostertagia*. *Vet. Sci. Commun.* 3: 159–164.

**45** Taylor, M.A., Hunt, K.R., and Goodyear, K.L. (2002). Anthelmintic resistance detection methods. *Vet. Parasitol.* 103: 183–194.

**46** Matthews, J.B., McArthur, C., Robinson, A., and Jackson, F. (2012). The *in vitro* diagnosis of anthelmintic resistance in cyathostomins. *Vet. Parasitol.* 185: 25–31.

**47** Krajacic, P., Shen, X., Purohit, P.K. et al. (2012). Biomechanical profiling of *Caenorhabditis elegans* motility. *Genetics* 191: 1015–1021.

**48** Marcellino, C., Gut, J., Lim, K.C. et al. (2012). WormAssay: a novel computer application for whole-plate motion-based screening of macroscopic parasites. *PLoS Negl.Trop. Dis.* 6: e0001494.

**49** Storey, B., Marcellino, C., Miller, M. et al. (2014). Utilization of computer processed high definition video imaging for measuring motility of microscopic nematode stages on a quantitative scale: "The Worminator". *Int. J. Parasitol.: Drugs Drug Resist.* 4: 233–243.

**50** Bulman, C.A., Bidlow, C.M., Lustigman, S. et al. (2015). Repurposing auranofin as a lead candidate for treatment of lymphatic filariasis and onchocerciasis. *PLoS Negl.Trop. Dis.* 9: e0003534.

**51** Chronis, N., Zimmer, M., and Bargmann, C.I. (2007). Microfluidics for *in vivo* imaging of neuronal and behavioral activity in *Caenorhabditis elegans*. *Nat. Methods* 4: 727–731.

**52** Hulme, S.E., Shevkoplyas, S.S., Apfeld, J. et al. (2007). A microfabricated array of clamps for immobilizing and imaging *C. elegans*. *Lab Chip* 7: 1515–1523.

**53** Chung, K., Crane, M.M., and Lu, H. (2008). Automated on-chip rapid microscopy, phenotyping and sorting of *C. elegans*. *Nat. Methods* 5: 637–643.

**54** Boyd, W.A., McBride, S.J., Rice, J.R. et al. (2010). A high-throughput method for assessing chemical toxicity using a *Caenorhabditis elegans* reproduction assay. *Toxicol. Appl. Pharmacol.* 245: 153–159.

**55** Xian, B., Shen, J., Chen, W. et al. (2013). WormFarm: a quantitative control and measurement device toward automated *Caenorhabditis elegans* aging analysis. *Aging Cell* 12: 398–409.

**56** Cornaglia, M., Lehnert, T., and Gijs, M.A.M. (2017). Microfluidic systems for high-throughput and high-content screening using the nematode *Caenorhabditis elegans*. *Lab Chip* 17: 3736–3759.

**57** Dittrich, P.S. and Manz, A. (2006). Lab-on-a-chip: microfluidics in drug discovery. *Nat. Rev. Drug Discovery* 5: 210–218.

**58** Whitesides, G.M. (2006). The origins and the future of microfluidics. *Nature* 442: 368–373.

**59** Carr, J.A., Parashar, A., Gibson, R. et al. (2011). A microfluidic platform for high-sensitivity, real-time drug screening on *C. elegans* and parasitic nematodes. *Lab Chip* 11: 2385–2396.

**60** Chung, K., Zhan, M., Srinivasan, J. et al. (2011). Microfluidic chamber arrays for whole-organism behavior-based chemical screening. *Lab Chip* 11: 3689–3697.

**61** Lockery, S.R., Hulme, S.E., Roberts, W.M. et al. (2012). A microfluidic device for whole-animal drug screening using electrophysiological measures in the nematode *C. elegans*. *Lab Chip* 12: 2211–2220.

**62** Lycke, R., Parashar, A., and Pandey, S. (2013). Microfluidics-enabled method to identify modes of *Caenorhabditis elegans* paralysis in four anthelmintics. *Biomicrofluidics* 7: 64103–64103.

**63** Simón, F., Siles-Lucas, M., Morchón, R. et al. (2012). Human and animal dirofilariasis: the emergence of a zoonotic mosaic. *Clin. Microbiol. Rev.* 25: 507–544.

**64** Genchi, C. and Kramer, L.H. (2020). The prevalence of *Dirofilaria immitis* and *D. repens* in the Old World. *Vet. Parasitol.* 280: 108995.

**65** Dantas-Torres, F. and Otranto, D. (2020). Overview on *Dirofilaria immitis* in the Americas, with notes on other filarial worms infecting dogs. *Vet. Parasitol.* 282: 109113.

**66** Montarsi, F., Ciocchetta, S., Devine, G. et al. (2015). Development of *Dirofilaria immitis* within the mosquito *Aedes* (Finlaya) *koreicus*, a new invasive species for Europe. *Parasites Vectors* 8: 177.

**67** Hertig, E. (2019). Distribution of *Anopheles* vectors and potential malaria transmission stability in Europe and the Mediterranean area under future climate change. *Parasites Vectors* 12: 18.

**68** Bowman, D.D. and Atkins, C.E. (2009). Heartworm biology, treatment, and control. *Vet. Clin. N. Am.: Small Anim. Pract.* 39: 1127–1158.

**69** Henry, L.G., Brunson, K.J., Walden, H.S. et al. (2018). Comparison of six commercial antigen kits for detection of *Dirofilaria immitis* infections in canines with necropsy-confirmed heartworm status. *Vet. Parasitol.* 254: 178–182.

**70** Little, S., Saleh, M., Wohltjen, M., and Nagamori, Y. (2018). Prime detection of *Dirofilaria immitis*: understanding the influence of blocked antigen on heartworm test performance. *Parasites Vectors* 11: 186.

**71** Liu, J., Drexel, J., Andrews, B. et al. (2018). Comparative evaluation of 2 in-clinic assays for vector-borne disease testing in dogs. *Top. Companion Anim. Med.* 33: 114–118.

**72** Evans, C., Pilotte, N., Williams, S., and Moorhead, A. (2022). Veterinary diagnosis of filarial infection. In: *Advances in Control of Heartworm and Human Filariases* (eds. R. Kaminsky and T.G. Geary), Chapter 6. Weinheim: Wiley-VCH.

**73** Ketzis, J. and Epe, C. (2022). Antifilarial chemotherapy: current options in veterinary medicine. In: *Advances in Control of Heartworm and Human Filariases* (eds. R. Kaminsky and T.G. Geary), Chapter 8. Weinheim: Wiley-VCH.

**74** Klug, D. and Drake, J. (2022). Global economics of heartworm disease. In: *Advances in Control of Heartworm and Human Filariases* (eds. R. Kaminsky and T.G. Geary), Chapter 13. Weinheim: Wiley-VCH.

**75** Fleming, S.A., Craig, T., Kaplan, R.M. et al. (2006). Anthelmintic resistance of gastrointestinal parasites in small ruminants. *J. Vet. Internal Med.* 20: 435–444.

**76** Kaplan, R.M. and Vidyashankar, A.N. (2012). An inconvenient truth: global worming and anthelmintic resistance. *Vet. Parasitol.* 186: 70–78.

**77** Kaminsky, R., Bapst, B., Stein, P.A. et al. (2011). Differences in efficacy of monepantel, derquantel and abamectin against multi-resistant nematodes of sheep. *Parasitol. Res.* 109: 19–23.

**78** Van den Brom, R., Moll, L., Kappert, C., and Vellema, P. (2015). *Haemonchus contortus* resistance to monepantel in sheep. *Vet. Parasitol.* 209: 278–280.

**79** Prichard, R.K. (2005). Is anthelmintic resistance a concern for heartworm control? What can we learn from the human filariasis control programs? *Vet. Parasitol.* 133: 243–253.

**80** Hampshire, V.A. (2005). Evaluation of efficacy of heartworm preventive products at the FDA. *Vet. Parasitol.* 133: 191–195.

**81** Bourguinat, C., Keller, K., Bhan, A. et al. (2011). Macrocyclic lactone resistance in *Dirofilaria immitis*. *Vet. Parasitol.* 181: 388–392.

**82** Bourguinat, C., Lee, A.C., Lizundia, R. et al. (2015). Macrocyclic lactone resistance in *Dirofilaria immitis*: failure of heartworm preventives and investigation of genetic markers for resistance. *Vet. Parasitol.* 210: 167–178.

**83** Pulaski, C.N., Malone, J.B., Bourguinat, C. et al. (2014). Establishment of macrocyclic lactone resistant *Dirofilaria immitis* isolates in experimentally infected laboratory dogs. *Parasites Vectors* 7: 494.

**84** Prichard, R. (2022). Drug resistance in filariae. In: *Advances in Control of Heartworm and Human Filariases* (eds. R. Kaminsky and T.G. Geary), Chapter 11. Weinheim: Wiley-VCH.

**85** Knight, D.H. (1987). Heartworm infection. *Vet. Clin. N. Am.: Small Anim. Pract.* 17: 1463–1518.

**86** Moroni, B., Rossi, L., Meneguz, P.G. et al. (2020). *Dirofilaria immitis* in wolves recolonizing northern Italy: are wolves competent hosts? *Parasites Vectors* 13: 482.

**87** Genchi, C., Kramer, L.H., and Prieto, G. (2001). Epidemiology of canine and feline dirofilariasis: a global view. In: *Heartworm Infection in Humans and Animals* (eds. F. Simón and C. Genchi), 121–134. Salamanca: Ediciones Universidad de Salamanca.

**88** Genchi, C., Bowman, D., and Drake, J. (2014). Canine heartworm disease (*Dirofilaria immitis*) in Western Europe: survey of veterinary awareness and perceptions. *Parasites Vectors* 7: 206.

**89** Kotani, T. and Powers, K.G. (1982). Developmental stages of *Dirofilaria immitis* in the dog. *Am. J. Vet. Res.* 43: 2199–2206.

**90** Supakorndej, P., McCall, J.W., and Jun, J.J. (1994). Early migration and development of *Dirofilaria immitis* in the ferret, *Mustela putorius furo*. *J. Parasitol.* 80: 237–244.

**91** Holden-Dye, L. and Walker, R.J. (2007). Anthelmintic drugs. *WormBook* 1: 1–13.

**92** Godel, C., Kumar, S., Koutsovoulos, G. et al. (2012). The genome of the heartworm, *Dirofilaria immitis*, reveals drug and vaccine targets. *FASEB J.* 26: 4650–4661.

**93** Maclean, M.J., Savadelis, M.D., Coates, R. et al. (2017). Does evaluation of in vitro microfilarial motility reflect the resistance status of *Dirofilaria immitis* isolates to macrocyclic lactones? *Parasites Vectors* 10 (Suppl 2): 480.

**94** Prichard, R. (2001). Genetic variability following selection of *Haemonchus contortus* with anthelmintics. *Trends Parasitol.* 17: 445–453.

**95** Wolstenholme, A.J. and Kaplan, R.M. (2012). Resistance to macrocyclic lactones. *Current Pharm. Biotechnol.* 13: 873–887.

**96** Kaplan, R.M. (2004). Drug resistance in nematodes of veterinary importance: a status report. *Trends Parasitol.* 20: 477–481.

**97** Gill, J.H. and Redwin, J.M. (1995). Cryopreservation of the first-stage larvae of trichostrongylid nematode parasites. *Int. J. Parasitol.* 25: 1421–1426.

**98** Blaxter, M. (1998). *Caenorhabditis elegans* is a nematode. *Science* 282: 2041–2046.

**99** Blaxter, M.L., De Ley, P., Garey, J.R. et al. (1998). A molecular evolutionary framework for the phylum Nematoda. *Nature* 392: 71–75.

**100** Bürglin, T.R., Lobos, E., and Blaxter, M.L. (1998). *Caenorhabditis elegans* as a model for parasitic nematodes. *Int. J. Parasitol.* 28: 395–411.

**101** Blaxter, M. and Koutsovoulos, G. (2015). The evolution of parasitism in Nematoda. *Parasitology* 142: S26–S39.

**102** Platzer, E.G. and Friedman, P.A. (1977). Growth inhibition of *Caenorhabditis elegans* with benzimidazoles. *J. Nematol.* 9: 280.

**103** Lewis, J.A., Wu, C.H., Berg, H., and Levine, J.H. (1980). The genetics of levamisole resistance in the nematode *Caenorhabditis elegans*. *Genetics* 95: 905–928.

**104** Boulin, T., Gielen, M., Richmond, J.E. et al. (2008). Eight genes are required for functional reconstitution of the *Caenorhabditis elegans* levamisole-sensitive acetylcholine receptor. *Proc. Natl. Acad. Sci. U.S.A.* 105: 18590–18595.

**105** Driscoll, M., Dean, E., Reilly, E. et al. (1989). Genetic and molecular analysis of a *Caenorhabditis elegans* beta-tubulin that conveys benzimidazole sensitivity. *J. Cell Biol.* 109: 2993–3003.

**106** Boulin, T. and Hobert, O. (2012). From genes to function: the *C. elegans* genetic toolbox. *Wiley Interdiscip. Rev. Dev. Biol.* 1: 114–137.

**107** Chen, C., Fenk, L.A., and de Bono, M. (2013). Efficient genome editing in *Caenorhabditis elegans* by CRISPR-targeted homologous recombination. *Nucleic Acids Res.* 41: e193.

**108** Kwa, M.S., Veenstra, J.G., Van Dijk, M., and Roos, M.H. (1995). Beta-tubulin genes from the parasitic nematode *Haemonchus contortus* modulate drug resistance in *Caenorhabditis elegans*. *J. Mol. Biol.* 246: 500–510.

**109** Cook, A., Aptel, N., Portillo, V. et al. (2006). *Caenorhabditis elegans* ivermectin receptors regulate locomotor behaviour and are functional orthologues of *Haemonchus contortus* receptors. *Mol. Biochem. Parasitol.* 147: 118–125.

**110** Guest, M., Bull, K., Walker, R.J. et al. (2007). The calcium-activated potassium channel, SLO-1, is required for the action of the novel cyclo-octadepsipeptide anthelmintic, emodepside, in *Caenorhabditis elegans*. *Int. J. Parasitol.* 37: 1577–1588.

**111** Glendinning, S.K., Buckingham, S.D., Sattelle, D.B. et al. (2011). Glutamate-gated chloride channels of *Haemonchus contortus* restore drug sensitivity to ivermectin resistant *Caenorhabditis elegans*. *PLoS One* 6: e22390.

**112** Welz, C., Krüger, N., Schniederjans, M. et al. (2011). SLO-1-channels of parasitic nematodes reconstitute locomotor behaviour and emodepside sensitivity in *Caenorhabditis elegans slo-1* loss of function mutants. *PLoS Pathog.* 7: e1001330.

**113** O'Reilly, L.P., Luke, C.J., Perlmutter, D.H. et al. (2014). *C. elegans* in high-throughput drug discovery. *Adv. Drug Delivery Rev.* 69-70: 247–253.

**114** Burns, A.R., Wallace, I.M., Wildenhain, J. et al. (2010). A predictive model for drug bioaccumulation and bioactivity in *Caenorhabditis elegans*. *Nat. Chem. Biol.* 6: 549–557.

**115** Hu, Y., Ellis, B.L., Yiu, Y.Y. et al. (2013). An extensive comparison of the effect of anthelmintic classes on diverse nematodes. *PLoS One* 8: e70702.

**116** Geary, T.G., Sangster, N.C., and Thompson, D.P. (1999). Frontiers in anthelmintic pharmacology. *Vet. Parasitol.* 84: 275–295.

**117** Thompson, D.P., Klein, R.D., and Geary, T.G. (1996). Prospects for rational approaches to anthelmintic discovery. *Parasitology* 113: S217–S238.

**118** Kudelska, M.M., Holden-Dye, L., O'Connor, V., and Doyle, D.A. (2017). Concentration-dependent effects of acute and chronic neonicotinoid exposure on the behaviour and development of the nematode *Caenorhabditis elegans. Pest Manage. Sci.* 73: 1345–1351.

**119** Xiong, H., Pears, C., and Woollard, A. (2017). An enhanced *C. elegans* based platform for toxicity assessment. *Sci. Rep.* 7: 9839.

**120** Keiser, J. (2015). Is *Caenorhabditis elegans* the magic bullet for anthelminthic drug discovery? *Trends Parasitol.* 31: 455–456.

**121** Charvet, C.L., Guégnard, F., Courtot, E. et al. (2018). Nicotine-sensitive acetylcholine receptors are relevant pharmacological targets for the control of multidrug resistant parasitic nematodes. *Int. J. Parasitol.: Drugs Drug Resist.* 8: 540–549.

**122** George, M.M., Lopez-Soberal, L., Storey, B.E. et al. (2018). Motility in the L3 stage is a poor phenotype for detecting and measuring resistance to avermectin/milbemycin drugs in gastrointestinal nematodes of livestock. *Int. J. Parasitol.: Drugs Drug Resist.* 8: 22–30.

**123** Panic, G., Duthaler, U., Speich, B., and Keiser, J. (2014). Repurposing drugs for the treatment and control of helminth infections. *Int. J. Parasitol.: Drugs Drug Resist.* 4: 185–200.

**124** Jiao, Y., Preston, S., Song, H. et al. (2017). Assessing the anthelmintic activity of pyrazole-5-carboxamide derivatives against *Haemonchus contortus. Parasites Vectors* 10: 272.

**125** Geary, T.G., Sakanari, J.A., and Caffrey, C.R. (2015). Anthelmintic drug discovery: into the future. *J. Parasitol.* 101: 125–133.

# 16

# *In Vivo* Models for the Discovery of New Antifilarial Drugs

*Sandra Schorderet-Weber*[1,*] *and Sabine Specht*[2]

[1] *Consultant Parasitology, Neuchâtel, Switzerland*
[2] *Drugs for Neglected Diseases initiative, 15 Chemin Louis-Dunant, 1202 Geneva, Switzerland*

**Abstract**

The search for novel, safe, and effective antifilarials has led to the development of various screening and evaluation tools. Among them, many rodent models have been investigated and optimized over the last decades. As filarial nematodes are specific to their hosts, establishing representative models in rodents or other animals remains very difficult. The worm species, its development and location in the model host, and the resulting pathology never completely mimic the target filarial worm in its natural host. Experience over the last decades has shown that different approaches and models need to be investigated to get a good overview of the potential of a drug. This process is, however, time-consuming and not always adapted to high-throughput screens and chemical optimization. So far, most efforts have been focused on re-purposing known antiparasitics (or antibiotics in the case of *Wolbachia* elimination). Characterizing and comparing well-known fully optimized drugs in models has at least allowed a better understanding of the suitability of these models. Identifying novel classes and running chemical optimization in models, as is done for other parasites, may be a much bigger challenge, but this may be helped by the recent development of transgenic, often immuno-compromised, models that can harbor the target worm parasite, sometimes up to the adult stage. It is, however, too early to predict the added value of these models, which are still far from the reality in humans. Here, we list current existing models and their use for evaluating known and potentially new antifilarial compounds in human and veterinary medicine and highlight important aspects to consider when using these models.

## 16.1 Introduction

Since the 1940s, researchers have used animal models in their research and development efforts to discover drugs against filarial parasites. Depending on the

Sandra Schorderet-Weber and Sabine Specht are contributed equally to the manuscript.
*Corresponding author.

*Human and Animal Filariases: Landscape, Challenges, and Control*, First Edition.
Edited by Ronald Kaminsky and Timothy G. Geary.

veterinary or human needs, different goals have been prioritized, such as sterilizing adult females or directly killing microfilariae to block transmission, killing the adult worms, or preventing the development of infective larvae into adult worms.

Every host–parasite combination has its own strengths and limitations. The larval biology of filarial worms presents not only common but also highly distinct features, which appear to be fundamental and need to be considered when choosing or interpreting an animal model for research and development of antifilarial drugs. The difficulty of finding good representative *in vivo* models lies in the specificity of filarial worms for their host. Transposing, for example, human filarial worms or dog heartworms into rodent species is hardly possible due to the size of the adult worms, the body compartment they may occupy, or rodent immune responses. A few recent attempts to overcome this problem by using transgenic rodent hosts, immunologically transformed to tolerate human or dog filarial species, have been described and are discussed in this chapter. Thus, the search for new antifilarial drugs has mostly been performed with a different worm species in a different host than the final targets. In this chapter, we describe the life cycles of the rodent filarial worms most frequently used for drug testing and the *in vivo* models employing them that are most relevant for human and animal health.

To illustrate the complexity of the problem, we present a set of standard antifilarial drugs and their performance in various animal models (Table 16.1). Many known drugs give different outcomes depending on the rodent model chosen. Many publications report the parallel testing of a drug in different rodent models to check for consistency of activity (efficacy and potency) across models. This approach, including results from rodent models infected with other nematode species, such as the gastrointestinal parasites *Haemonchus contortus* and *Trichostrongylus colubriformis* [40], could be the best way to identify and prioritize drug candidates coming out of *in vitro* screens.

However, besides macrocyclic lactones (MLs) and benzimidazoles, and more recently also cyclodepsipeptides, no new antifilarial chemical class has yet emerged from the screening efforts of many research groups, highlighting a big gap between activity in *in vitro* and *in vivo* models, and confirmation of this activity against the target filarial parasite in the final host. Filarial worms are embedded deep in the host's body, and adequate drug pharmacokinetics is necessary to reach them. This aspect may be impossible to address with a model using a surrogate host and a different parasite species and needs careful consideration.

In human filariasis, for which the main objective has been to stop transmission, the minimum requirement besides safety for new drugs is microfilaricidal activity. It has become clear that additional drugs, ideally both safe and macrofilaricidal, are needed to achieve elimination. In dog heartworm, the focus continues to be on killing the infective larvae before they reach the blood circulation and pulmonary arteries. For both, the challenge remains to identify novel, safe, and effective drugs that serve the purpose of prevention and/or elimination, providing additional tools for difficult to treat areas.

**Table 16.1** Activity of drugs in rodent models.

| Drug | Dosage (mg/kg) | Frequency | Treatment mode | Treatment time (day post-infection) | Model worm species | Rodent host | Microfilariae (mf) reduction (%) | | Adult worms (Mf) reduction (%) | Notes | References |
|---|---|---|---|---|---|---|---|---|---|---|---|
| | | | | | | | Day 3/7 post treatment | At necropsy (from d42) | | | |
| *Macrocyclic lactones* | | | | | | | | | | | |
| Ivermectin | 0.05 | 5, Q24h | s.c. or p.o. | 65–69 | *A. viteae* | *M. natalensis* | 100 | 100 | n.a. | | [1] |
| | 0.1 | 1 | p.o. | 65 | *A. viteae* | *M. natalensis* | 100 | 100 | n.a. | | |
| | 6.25 | 5, Q24h | p.o. | 65–69 | *A. viteae* | *M. natalensis* | 100 | 100 | 73 | | |
| | 6.25 | 5, Q24h | s.c. | 65–69 | *A. viteae* | *M. natalensis* | 100 | 100 | 84 | | |
| | 0.05 | 5, Q24h | s.c. | 63–67 | *A. viteae* | *M. coucha* | 100 | 100 | 80 | Mf activity at 5 × 6.25 mg/kg | [2] |
| | 0.5 | 1 | s.c. | 65 | *A. viteae* | *M. coucha* | 100 | 100 | 64 | | [3] |
| | 0.05 | 1 | s.c. | 65 | *A. viteae* | *M. coucha* | 100 | 100 | 57 | | |
| | 0.005 | 1 | s.c. | 65 | *A. viteae* | *M. coucha* | 100 | 100 | 57 | | |
| | 0.0005 | 1 | s.c. | 65 | *A. viteae* | *M. coucha* | 70 | 70 | 35 | | |
| | 0.2 | 1 | p.o. | 6 | *A. viteae* | *M. unguiculatus* | n.d. | >50 | 99 | Treatment on L3/early L4 | Sager and Pautrat, personnal communication |
| | 0.2 | 1 | i.p. | 6 | *A. viteae* | *M. unguiculatus* | n.d. | >50 | 92 | Treatment on L3/early L4 | |
| | 0.1 | 1 | s.c. | 6 | *A. viteae* | *M. unguiculatus* | n.d. | >50 | 97 | Treatment on L3/early L4 | |

*(Continued)*

**Table 16.1** (Continued)

| Drug | Dosage (mg/kg) | Frequency | Treatment mode | Treatment time (day post-infection) | Model worm species | Rodent host | Microfilariae (mf) reduction (%) | | Adult worms (Mf) reduction (%) | Notes | References |
|---|---|---|---|---|---|---|---|---|---|---|---|
| | | | | | | | Day 3/7 post treatment | At necropsy (from d42) | | | |
| | 6.25 | 5, Q24h | s.c. | 115–119 | *B. malayi* | *M. coucha* | 40 | 90 | n.a. | | [2] |
| | 5 | 1 | s.c. | 115 | *B. malayi* | *M. coucha* | 80 | 92 | n.a. | | [3] |
| | 0.5 | 1 | s.c. | 115 | *B. malayi* | *M. coucha* | 60 | 42 | n.a. | | |
| | 0.2 | 1 | p.o. | 2 | *B. malayi* | Balb/c mice SCID | n.d. | 81 | n.d. | mf i.v. infusion | [4] |
| | 15 | 7, Q24h | p.o. | 1–7 | *B. malayi* | Balb/c mice SCID | n.d. | n.d. | 57 | Count on day 8, efficacy on L3-L4 | [5] |
| | 15 | 7, Q24h | p.o. | 1–7 | *B. malayi* | Balb/c mice | n.d. | n.d. | 40 | Count on day 8, efficacy on L3-L4 | |
| | 5 | 1 | i.p. | 2 | *B. malayi* | Balb/c mice SCID | 76 | n.d. | n.d. | mf i.v. infusion | |
| | 5 | 1 | i.p. | 2 | *B. malayi* | Balb/c mice | 79 | n.d. | n.d. | mf i.v. infusion | |
| | 0.5 | 5, Q24h | i.p. | 4–8 | *B. pahangi* | Balb/c mice | n.d. | n.d. | 8–82 | Larval recovery d29–32 | [6] |
| | 12.5 | 5, Q24h | s.c. | 115–119 | *B. pahangi* | *M. coucha* | 85 | 100 | n.a. | | [2] |
| | 5 | 1 | s.c. | 115 | *B. pahangi* | *M. coucha* | 10 | 92 | n.a. | | [3] |
| | 1.56 | 5, Q24h | s.c or p.o. | 85–89 | *L. carinii* | *M. natalensis* | 100 | 100 | n.a. | | [1] |

| | | | | | | | | | | |
|---|---|---|---|---|---|---|---|---|---|---|
| 0.2 | 1 | p.o. | 85 | *L. carinii* | *M. natalensis* | 100 | 100 | n.a. | | |
| 6.25 | 5, Q24h | s.c. or p.o. | 85–89 | *L. carinii* | *M. natalensis* | 100 | 100 | n.a. | | |
| 0.03 | 5, Q24h | s.c | 85–89 | *L. carinii* | *M. coucha* | 100 | >95 | n.a.* | *No Mf activity at 5 × 0.05 mg/kg | [2] |
| 5 | 1 | s.c. | 85 | *L. carinii* | *M. coucha* | 100 | 100 | n.a. | | [3] |
| 0.5 | 1 | s.c. | 85 | *L. carinii* | *M. coucha* | 100 | 98 | n.a. | | |
| 0.05 | 1 | s.c. | 85 | *L. carinii* | *M. coucha* | 100 | 90 | n.a. | | |
| 0.005 | 1 | s.c. | 85 | *L. carinii* | *M. coucha* | 100 | 65 | n.a. | | |
| 0.2 | 3, Q24h | p.o. | 0–2 | *L. sigmodontis* | Balb/c mice | n.d. | n.d. | 94 | Treatment on L3 | Sager and Pautrat, personnal communication |
| 0.2 | 5, Q24h | s.c. | 4–8 | *M. dessetae* | *P. oris* | n.d. | n.d. | 100 | Treatment on L3 | [7] |
| 0.2 | 5, Q24h | s.c. | 47–51 | *M. dessetae* | *P. oris* | n.d. | n.d. | 77 | Treatment on pre-adult worms | |
| 0.2 | 5, Q24h | s.c. | 189–193 | *M. dessetae* | *P. oris* | 70 | 60 | 71 | Treatment on adult worms | |
| 0.15 | 1 | p.o. | 3 | *O. ochengi* | *M. auratus* | n.d. | 100 | n.d. | mf s.c. infusion | [8] |
| 0.6 | 1 | p.o. | 3 | *O. ochengi* | *M. auratus* | n.d. | 100 | n.d. | mf s.c. infusion | |

(Continued)

**Table 16.1** (Continued)

| Drug | Dosage (mg/kg) | Frequency | Treatment mode | Treatment time (day post-infection) | Model worm species | Rodent host | Microfilariae (mf) reduction (%) | | Adult worms (Mf) reduction (%) | Notes | References |
|---|---|---|---|---|---|---|---|---|---|---|---|
| | | | | | | | Day 3/7 post treatment | At necropsy (from d42) | | | |
| | 0.006 | 1 | s.c. | 30 | *D. immitis* | *Rat (Sprague–Dawley IGS)* | n.d. | n.d. | 100 | Immuno-suppressor in diet | [9] |
| | 0.003 | 1 | s.c. | 30 | *D. immitis* | *Rat (Sprague–Dawley IGS)* | n.d. | n.d. | 96.5 | Immuno-suppressor in diet | |
| | 0.001 | 1 | s.c. | 30 | *D. immitis* | *Rat (Sprague–Dawley IGS)* | n.d. | n.d. | 89.3 | Immuno-suppressor in diet | |
| Milbemicin A4 oxime | 0.5 | 1 | s.c. | 65 | *A. viteae* | *M. coucha* | 82 | 75 | 56 | | [3] |
| | 0.05 | 1 | s.c. | 65 | *A. viteae* | *M. coucha* | 778 | 0 | 31 | | |
| | 0.005 | 1 | s.c. | 65 | *A. viteae* | *M. coucha* | 68 | 30 | 47 | | |
| | 0.0005 | 1 | s.c. | 65 | *A. viteae* | *M. coucha* | 58 | 35 | 35 | | |
| | 5 | 1 | s.c. | 115 | *B. malayi* | *M. coucha* | 48 | 57 | n.a. | | |
| | 0.5 | 1 | s.c. | 115 | *B. malayi* | *M. coucha* | 30 | 0 | n.a. | | |
| | 5 | 1 | s.c. | 115 | *B. pahangi* | *M. coucha* | 82 | 92 | n.a. | | |
| | 5 | 1 | s.c. | 85 | *L. carinii* | *M. coucha* | 100 | 1000 | n.a. | | |
| | 0.5 | 1 | s.c. | 85 | *L. carinii* | *M. coucha* | 100 | 70 | n.a. | | |
| | 0.05 | 1 | s.c. | 85 | *L. carinii* | *M. coucha* | 90 | 55 | n.a. | | |

| | | | | | | | | | | | |
|---|---|---|---|---|---|---|---|---|---|---|---|
| | 0.005 | 1 | s.c. | 85 | *L. carinii* | *M. coucha* | 90 | 48 | n.a. | | |
| Milbemycin oxime | 5 | 1 | p.o. | 6 | *A. viteae* | *M. unguiculatus* | n.d. | >50 | 95 | Treatment on L3/early L4 | Sager and Pautrat, personnal communication |
| | 3.2 | 1 | p.o. | 6 | *A. viteae* | *M. unguiculatus* | n.d. | >50 | 89 | Treatment on L3/early L4 | |
| | 2.5 | 1 | p.o. | 6 | *A. viteae* | *M. unguiculatus* | n.d. | >50 | 100 | Treatment on L3/early L4 | |
| | 1 | 1 | p.o. | 6 | *A. viteae* | *M. unguiculatus* | n.d. | >50 | 77 | Treatment on L3/early L4 | |
| | 1 | 1 | p.o. | 13 | *A. viteae* | *M. unguiculatus* | n.d. | >50 | 96 | Treatment on L4 | |
| | 0.5 | 1 | p.o. | 6 | *A. viteae* | *M. unguiculatus* | n.d. | n.a. | n.a. | Treatment on L3/early L4 | |
| | 0.5 | 5, Q24h | p.o. | 5–9 | *A. viteae* | *M. unguiculatus* | n.d. | >50 | 98 | Treatment on L3/early L4 | |
| | 0.5 | 5, Q168h | p.o. | 7, 14, 21, 28, 35 | *A. viteae* | *M. unguiculatus* | n.d. | >50 | 95 | Treatment on L4/preadults | |
| | 3 | 5, Q24h | p.o. | 0–4 | *L. sigmodontis* | *Balb/c mice* | n.d. | n.d. | 80 | Treatment on L3 | |

*(Continued)*

**Table 16.1** (Continued)

| Drug | Dosage (mg/kg) | Frequency | Treatment mode | Treatment time (day post-infection) | Model worm species | Rodent host | Microfilariae (mf) reduction (%) | | Adult worms (Mf) reduction (%) | Notes | References |
|---|---|---|---|---|---|---|---|---|---|---|---|
| | | | | | | | Day 3/7 post treatment | At necropsy (from d42) | | | |
| Milbemycin | 5 | 5, Q24h | s.c. | 63–67 | *A. viteae* | *M. coucha* | >95 | >95 | n.a.* | *No Mf activity at 5 × 5 mg/kg | [2] |
| | 5 | 5, Q24h | s.c. | 115–119 | *B. malayi* | *M. coucha* | <95 | <95 | n.a.* | *No Mf activity at 5 × 5 mg/kg | |
| | 5 | 5, Q24h | s.c. | 115–119 | *B. pahangi* | *M. coucha* | <95 | <95 | n.a.* | *No Mf activity at 5 × 5 mg/kg | |
| | 0.5 | 5, Q24h | s.c. | 85–89 | *L. carinii* | *M. coucha* | >95 | >95 | n.a.* | *No Mf activity at 5 × 5 mg/kg | |
| Moxidectin | 0.5 | 1 | s.c. | 65 | *A. viteae* | *M. coucha* | 100 | 100 | 74 | | [3] |
| | 0.05 | 1 | s.c. | 65 | *A. viteae* | *M. coucha* | 100 | 100 | 44 | | |
| | 0.005 | 1 | s.c. | 65 | *A. viteae* | *M. coucha* | 0 | 100 | 34 | | |
| | 0.0005 | 1 | s.c. | 65 | *A. viteae* | *M. coucha* | 50 | 57 | 74 | | |
| | 0.5 | 1 | p.o. | 6 | *A. viteae* | *M. unguiculatus* | n.d. | >50 | 94 | Treatment on L3/early L4 | Sager and Pautrat, personnal communication |
| | 1 | 1 | p.o. | 6 | *A. viteae* | *M. unguiculatus* | n.d. | >50 | 100 | Treatment on L3/early L4 | |

| | | | | | | | | | |
|---|---|---|---|---|---|---|---|---|---|
| 20 | 1 | s.c. | 5–8 mo | *B. malayi* | *M. coucha* | 18 | 71 | 49 | [10] |
| 5 | 1 | s.c. | 115 | *B. malayi* | *M. coucha* | 62 | 95 | n.a. | [3] |
| 0.5 | 1 | s.c. | 115 | *B. malayi* | *M. coucha* | 62 | 97 | n.a. | |
| 10 | 1 | p.o. | Trans-planted adult worms | *B. malayi* | *M. unguiculatus* | n.d. | 51 | 49 | [10] |
| 10 | 1 | s.c. | Trans-planted adult worms | *B. malayi* | *M. unguiculatus* | n.d. | 54 | 56 | |
| 20 | 1 | p.o. | Trans-planted adult worms | *B. malayi* | *M. unguiculatus* | n.d. | 60 | 63 | |
| 20 | 1 | s.c. | Trans-planted adult worms | *B. malayi* | *M. unguiculatus* | n.d. | 63 | 73 | |
| 40 | 1 | p.o. | Trans-planted adult worms | *B. malayi* | *M. unguiculatus* | n.d. | 63 | 66 | |
| 40 | 1 | s.c. | Trans-planted adult worms | *B. malayi* | *M. unguiculatus* | n.d. | 66 | 71 | |

(Continued)

**Table 16.1** (Continued)

| Drug | Dosage (mg/kg) | Frequency | Treatment mode | Treatment time (day post-infection) | Model worm species | Rodent host | Microfilariae (mf) reduction (%) Day 3/7 post treatment | Microfilariae (mf) reduction (%) At necropsy (from d42) | Adult worms (Mf) reduction (%) | Notes | References |
|---|---|---|---|---|---|---|---|---|---|---|---|
| | 5 | 1 | s.c. | 115 | *B. pahangi* | *M. coucha* | 70 | 100 | n.a. | | [3] |
| | 5 | 1 | s.c. | 85 | *L. carinii* | *M. coucha* | 100 | 100 | n.a. | | |
| | 0.5 | 1 | s.c. | 85 | *L. carinii* | *M. coucha* | 100 | 100 | n.a. | | |
| | 0.05 | 1 | s.c. | 85 | *L. carinii* | *M. coucha* | 100 | 100 | n.a. | | |
| | 0.005 | 1 | s.c. | 85 | *L. carinii* | *M. coucha* | n.a. | 64 | n.a. | | |
| | 0.003 | 1 | s.c. | 30 | *D. immitis* | *Rat (Sprague–Dawley IGS)* | n.d. | n.d. | 100 | Immuno-suppressor in diet | [9] |
| | 0.001 | 1 | s.c. | 30 | *D. immitis* | *Rat (Sprague–Dawley IGS)* | n.d. | n.d. | 91.7 | Immuno-suppressor in diet | |
| Doramectin | 0.5 | 1 | s.c. | 65 | *A. viteae* | *M. coucha* | 100 | 100 | 75 | | [3] |
| | 0.05 | 1 | s.c. | 65 | *A. viteae* | *M. coucha* | 100 | 100 | 65 | | |
| | 0.005 | 1 | s.c. | 65 | *A. viteae* | *M. coucha* | 100 | 100 | 63 | | |
| | 0.0005 | 1 | s.c. | 65 | *A. viteae* | *M. coucha* | 80 | 45 | 6 | | |
| | 5 | 1 | s.c. | 115 | *B. malayi* | *M. coucha* | 80 | 93 | n.a. | | |
| | 0.5 | 1 | s.c. | 115 | *B. malayi* | *M. coucha* | 55 | 90 | n.a. | | |
| | 5 | 1 | s.c. | 115 | *B. pahangi* | *M. coucha* | 0 | 96 | n.a. | | |

| | | | | | | | | | | | |
|---|---|---|---|---|---|---|---|---|---|---|---|
| | 5 | 1 | s.c. | 85 | *L. carinii* | *M. coucha* | 100 | 100 | n.a. | | |
| | 0.5 | 1 | s.c. | 85 | *L. carinii* | *M. coucha* | 100 | 100 | n.a. | | |
| | 0.05 | 1 | s.c. | 85 | *L. carinii* | *M. coucha* | 100 | 90 | n.a. | | |
| | 0.005 | 1 | s.c. | 85 | *L. carinii* | *M. coucha* | 100 | 70 | n.a. | | |
| *Carbamates* | | | | | | | | | | | |
| Diethyl carbamazine (DEC) | 250 | 5, Q24h | p.o. | 2–6 | *A. viteae* | *M. natalensis* | n.d. | 100 | 97 | | [11] |
| | 500 | 5, Q24h | p.o. | 28–32 | *A. viteae* | *M. natalensis* | n.d. | 98 | 55 | | [11] |
| | 100 | 5, Q24h | p.o. | 63–67 | *A. viteae* | *M. coucha* | >95 | 85 | 66 | | [2] |
| | 50 | 5, Q24h | i.p. | 65–69 | *A. viteae* | *M. coucha* | n.d. | 90 | n.a. | | [12] |
| | 250 | 5, Q24h | p.o. | 115–119 | *B. malayi* | *M. coucha* | >95 | 93 | 50 | | [2] |
| | 50 | 7, Q24h | p.o. | 1–7 | *B. malayi* | Balb/c mice SCID | n.d. | n.d. | 100 | Count on day 8, efficacy on L3-L4 | [5] |
| | 50 | 7, Q24h | p.o. | 1–7 | *B. malayi* | Balb/c mice | n.d. | n.d. | >99 | Count on day 8, efficacy on L3-L4 | [5] |
| | 100 | 1 | p.o. | trans-planted adult worms | *B. malayi* | *M. unguiculatus* | n.d. | 25 | 13 | | [10] |
| | 100 | 1 | s.c. | trans-planted adult worms | *B. malayi* | *M. unguiculatus* | n.d. | 25 | 41 | | [10] |

(Continued)

**Table 16.1** (Continued)

| Drug | Dosage (mg/kg) | Frequency | Treatment mode | Treatment time (day post-infection) | Model worm species | Rodent host | Microfilariae (mf) reduction (%) | | Adult worms (Mf) reduction (%) | Notes | References |
|---|---|---|---|---|---|---|---|---|---|---|---|
| | | | | | | | Day 3/7 post treatment | At necropsy (from d42) | | | |
| | 50 | 5, Q24h | p.o. | 5–8 mo | *B. malayi* | *M. coucha* | 70 | n.a. | 40 | | [10] |
| | 25 | 5, Q24h | i.p. | 4–8 | *B. pahangi* | Balb/c mice | n.d. | n.d. | n.a.-50 | larval recovery d29–32 | [6] |
| | 250 | 5, Q24h | p.o. | 115–119 | *B. pahangi* | *M. coucha* | 80 | 100 | 90 | | [2] |
| | 200 | 5, Q24h | i.p. | 140–144 | *B. pahangi* | *M. unguiculatus* | n.d. | n.d. | 67 | Delayed effect, 8 weeks post-treatment | [13] |
| | 25 | 5, Q24h | s.c. | trans-planted adult worms | *B. pahangi* | *M. unguiculatus* | n.d. | n.a. | 15 | | [14] |
| | 50 | 5, Q24h | i.p. | >150 | *B. pahangi* | *M. coucha* | 92 | 17 | n.a. | | [14] |
| | 100 | 5, Q24h | p.o. | 85–89 | *L. carinii* | *M. coucha* | >95 | 0 | n.a.* | *No Mf activity at 5 × 250 mg/kg | [2] |
| | 50 | 5, Q24h | i.p. | 85–89 | *L. carinii* | *S. hispidus* | n.d. | n.a. | n.a. | | [12] |
| | 400 | 5, Q24h | p.o. | 4–8 | *M. dessetae* | *P. oris* | n.d. | n.d. | 100 | Treatment on L3 | [7] |

| | | | | | | | | | | | |
|---|---|---|---|---|---|---|---|---|---|---|---|
| | 400 | 5, Q24h | p.o. | 47–51 | *M. dessetae* | *P. oris* | n.d. | n.d. | 95 | Treatment on pre-adult worms | [7] |
| | 400 | 5, Q24h | p.o. | 189–193 | *M. dessetae* | *P. oris* | 100 | 10 | 94 | Treatment on adult worms | [7] |
| DEC citrate | 350 | 5, Q24h | i.p. | Not specified | *A. viteae* | *M. natalensis* | n.d. | >90 | n.a. | | [15] |
| | 125 | 5, Q24h | p.o. | 90–94 | *L. carinii* | *M. natalensis* | 99 | 74 | n.a. | | [16] |
| | 6 | 5, Q24h | i.p. | Not specified | *L. carinii* | *S. hispidus* | n.d. | >90 | n.a. | | [15] |
| *Benzimidasoles* | | | | | | | | | | | |
| Albendazole | 50 | 5, Q24h | p.o. | 2–6 | *A. viteae* | *M. natalensis* | n.d. | 100 | 97 | | [11] |
| | 300 | 5, Q24h | p.o. | 28–32 | *A. viteae* | *M. natalensis* | n.d. | 98 | 74 | | |
| | 25 | 5, Q24h | s.c. | 63–67 | *A. viteae* | *M. coucha* | n.a. | >95 | >95 | | [2] |
| | 50 | 5, Q24h | s.c. | 115–119 | *B. malayi* | *M. coucha* | n.a. | >95 | >95 | | |
| | 50 | 7, Q24h | p.o. | 1–7 | *B. malayi* | Balb/c mice SCID | n.d. | n.d. | 98 | Count on day 8, efficacy on L3-L4 | [5] |
| | 50 | 7, Q24h | p.o. | 1–7 | *B. malayi* | Balb/c mice | n.d. | n.d. | 100 | Count on day 8, efficacy on L3-L4 | |
| | 50 | 5, Q24h | s.c. | 115–119 | *B. pahangi* | *M. coucha* | n.a. | >95 | >95 | | [2] |
| | 6.25 | 5, Q24h | s.c. | 85–89 | *L. carinii* | *M. coucha* | n.a. | >95 | >95* | *Mf activity at 5 × 50 mg/kg | |

(Continued)

**Table 16.1** (Continued)

| Drug | Dosage (mg/kg) | Frequency | Treatment mode | Treatment time (day post-infection) | Model worm species | Rodent host | Microfilariae (mf) reduction (%) Day 3/7 post treatment | Microfilariae (mf) reduction (%) At necropsy (from d42) | Adult worms (Mf) reduction (%) | Notes | References |
|---|---|---|---|---|---|---|---|---|---|---|---|
| | 50 | 5, Q24h | s.c. | 4–8 | *M. dessetae* | *P. oris* | n.d. | n.d. | 100 | Treatment on L3 | [7] |
| | 50 | 5, Q24h | s.c. | 47–51 | *M. dessetae* | *P. oris* | n.d. | n.d. | 100 | Treatment on pre-adult worms | |
| | 50 | 5, Q24h | s.c. | 189–193 | *M. dessetae* | *P. oris* | 60 | 100 | 100 | Treatment on adult worms | |
| Albendazole/ ivermectin | 10/0.04 | 5, Q24h | s.c. | 190–194 | *M. dessetae* | *P. oris* | n.d. | n.d. | 37 | | [17] |
| | 10/0.04 | 5, Q24h | s.c. | 4–8 | *M. dessetae* | *P. oris* | n.d. | n.d. | 100 | Treatment on L3 | [7] |
| | 10/0.04 | 5, Q24h | s.c. | 47–51 | *M. dessetae* | *P. oris* | n.d. | n.d. | 85 | Treatment on pre-adult worms | |
| | 10/0.04 | 5, Q24h | s.c. | 189–193 | *M. dessetae* | *P. oris* | 0 | 0 | 37 | Treatment on adult worms | |
| Cambenda-zole | 50 | 5, Q24h | s.c. | 63–67 | *A. viteae* | *M. coucha* | n.a. | >95 | n.a. | | [2] |
| | 50 | 5, Q24h | s.c. | 115–119 | *B. malayi* | *M. coucha* | n.a. | n.a. | n.a. | | |
| | 25 | 5, Q24h | s.c. | 85–89 | *L. carinii* | *M. coucha* | n.a. | >95 | >95* | *Mf activity at 5 × 50 mg/kg | |

| | | | | | | | | | | | |
|---|---|---|---|---|---|---|---|---|---|---|---|
| Ciclobendazole | 50 | 5, Q24h | s.c. | 63–67 | *A. viteae* | *M. coucha* | n.a. | >95 | >95 | | [2] |
| | 6,25 | 5, Q24h | s.c. | 85–89 | *L. carinii* | *M. coucha* | n.a. | >95 | >95* | *Mf activity at 5 × 25 mg/kg | |
| Fenbendazole | 200 | 5, Q24h | p.o. | 2–6 | *A. viteae* | *M. natalensis* | n.d. | 15 | 20 | | [11] |
| | 200 | 5, Q24h | p.o. | 28–32 | *A. viteae* | *M. natalensis* | n.d. | 85 | 25 | | |
| | 50 | 5, Q24h | s.c. | 63–67 | *A. viteae* | *M. coucha* | n.a. | n.a. | n.a. | | [2] |
| | 50 | 5, Q24h | s.c. | 115–119 | *B. malayi* | *M. coucha* | n.a. | n.a. | n.a. | | |
| | 10 | 5, Q24h | i.p. | 4–8 | *B. pahangi* | *Balb/c mice* | n.d. | n.d. | 100 | Larval recovery d29–32 | [6] |
| | 50 | 5, Q24h | s.c. | 115–119 | *B. pahangi* | *M. coucha* | n.a. | n.a. | n.a. | | [2] |
| | 25 | 5, Q24h | s.c. | 85–89 | *L. carinii* | *M. coucha* | n.a. | >95 | >95* | *Mf activity at 5 × 50 mg/kg | |
| Flubendazole | 50 | 5, Q24h | p.o. | 2–6 | *A. viteae* | *M. natalensis* | n.d. | 100 | 99 | | [11] |
| | 100 | 5, Q24h | p.o. | 28–32 | *A. viteae* | *M. natalensis* | n.d. | 100 | 99 | | |
| | 1.56 | 5, Q24h | s.c. | Trans-planted adult worms | *A. viteae* | *M. unguiculatus* | n.d. | n.d. | 100 | | [18] |
| | 100 | 5, Q24h | s.c. | Trans-planted adult worms | *A. viteae* | *M. unguiculatus* | n.d. | n.d. | 100 | | [19] |

*(Continued)*

Table 16.1 (Continued)

| Drug | Dosage (mg/kg) | Frequency | Treatment mode | Treatment time (day post-infection) | Model worm species | Rodent host | Microfilariae (mf) reduction (%) Day 3/7 post treatment | At necropsy (from d42) | Adult worms (Mf) reduction (%) | Notes | References |
|---|---|---|---|---|---|---|---|---|---|---|---|
| | 6.25 | 5, Q24h | s.c. | 63–67 | *A. viteae* | *M. coucha* | 0 | 100 | 100 | | [2] |
| | 100 | 5, Q24h | s.c. | Trans-planted adult worms | *A. viteae* | Balb/c mice | n.d. | n.d. | 88 | | [20] |
| | 50 | 5, Q24h | s.c. | Trans-planted adult worms | *A. viteae* | Balb/c mice | n.d. | n.d. | 100 | | |
| | 100 | 5, Q24h | s.c. | Trans-planted adult worms | *A. viteae* | *M. unguiculatus* | n.d. | n.d. | 100 | | |
| | 50 | 5, Q24h | s.c. | Trans-planted adult worms | *A. viteae* | *M. unguiculatus* | n.d. | n.d. | 55 | | |
| | 12.5 | 5, Q24h | s.c. | 115–119 | *B. malayi* | *M. coucha* | 0 | 100 | 96 | | [2] |
| | 10 | 5, Q24h | s.c. | Trans-planted adult worms | *B. malayi* | Balb/c mice SCID | n.d. | n.d. | 100 | | [5] |
| | 10 | 5, Q24h | s.c. | Trans-planted adult worms | *B. malayi* | Balb/c mice | n.d. | n.d. | 100 | | |

| | | | | | | | | | | |
|---|---|---|---|---|---|---|---|---|---|---|
| 2 | 1 | p.o. | 2 | *B. malayi* | Balb/c mice SCID | n.d. | 12 | n.d. | mf i.v. infusion | [4] |
| 40 | 1 | p.o. | 2 | *B. malayi* | Balb/c mice SCID | n.d. | 49 | n.d. | mf i.v. infusion | |
| 10 | 5, Q24h | i.p. | 4–8 | *B. pahangi* | Balb/c mice | n.d. | n.d. | 100 | Larval recovery d29–32 | [6] |
| 100 | 5, Q24h | s.c. | Trans-planted adult worms | *B. pahangi* | *M. unguiculatus* | n.d. | n.d. | 100 | | [19] |
| 6.25 | 5, Q24h | s.c. | 115–119 | *B. pahangi* | *M. coucha* | 0 | 100 | 97* | *Mf activity at 5 × 25 mg/kg | [2] |
| 100 | 5, Q24h | s.c. | Trans-planted adult worms | *B. pahangi* | Balb/c mice | n.d. | n.d. | 100 | | [20] |
| 50 | 5, Q24h | s.c. | Trans-planted adult worms | *B. pahangi* | Balb/c mice | n.d. | n.d. | 100 | | |
| 100 | 5, Q24h | s.c. | Trans-planted adult worms | *B. pahangi* | *M. unguiculatus* | n.d. | n.d. | 100 | | |
| 50 | 5, Q24h | s.c. | Trans-planted adult worms | *B. pahangi* | *M. unguiculatus* | n.d. | n.d. | 55 | | |

(Continued)

**Table 16.1** (Continued)

| Drug | Dosage (mg/kg) | Frequency | Treatment mode | Treatment time (day post-infection) | Model worm species | Rodent host | Microfilariae (mf) reduction (%) | | Adult worms (Mf) reduction (%) | Notes | References |
|---|---|---|---|---|---|---|---|---|---|---|---|
| | | | | | | | Day 3/7 post treatment | At necropsy (from d42) | | | |
| | 1.56 | 5, Q24h | s.c. | Trans-planted adult worms | *Brugia pahangi* | *M. unguiculatus* | n.d. | n.d. | 100 | | [18] |
| | 1.6 | 5, Q24h | s.c. | 85–89 | *L. carinii* | *M. coucha* | 0 | 100 | 100* | *Mf activity at 5 × 12.5 mg/kg | [2] |
| | 40 | 1 | p.o. | 76 or 90 | *L. sigmodontis* | *M. unguiculatus* | n.d. | 91 | 85 | | [21] |
| | 2 | 5, Q24h | p.o. | 76–80, or 90–94 | *L. sigmodontis* | *M. unguiculatus* | n.d. | 95 | 79 | | |
| | 6 | 5, Q24h | p.o. | 76–80, or 90–94 | *L. sigmodontis* | *M. unguiculatus* | n.d. | 95 | 71 | | |
| | 15 | 5, Q24h | p.o. | 76–80, or 90–94 | *L. sigmodontis* | *M. unguiculatus* | n.d. | 99 | 84 | | |
| | 2 | 10, Q24h | p.o. | 76–85, or 90–99 | *L. sigmodontis* | *M. unguiculatus* | n.d. | 99.5 | 59 | | |
| | 6 | 10, Q24h | p.o. | 76–85, or 90–99 | *L. sigmodontis* | *M. unguiculatus* | n.d. | 99.7 | 77 | | |
| | 15 | 10, Q24h | p.o. | 76–85, or 90–99 | *L. sigmodontis* | *M. unguiculatus* | n.d. | 99.9 | 97 | | |
| | 2 | 1 | s.c. | 76 or 90 | *L. sigmodontis* | *M. unguiculatus* | n.d. | 100 | 98 | | |

| | | | | | | | | | | |
|---|---|---|---|---|---|---|---|---|---|---|
| 10 | 1 | s.c. | 76 or 90 | *L. sigmodontis* | *M. unguiculatus* | n.d. | 100 | 100 | | |
| 10 | 5, Q24h | s.c. | 76–80, or 90–94 | *L. sigmodontis* | *M. unguiculatus* | n.d. | 100 | 100 | | |
| 10 | 5, Q24h | s.c. | Transplanted adult worms | *O. ochengi* | Balb/c mice SCID | n.d. | n.d. | 93 | | [4] |
| 10 | 1 | s.c. | Transplanted adult worms | *O. ochengi* | Balb/c mice SCID | n.d. | n.d. | 82 | | |
| 0.2 | 5, Q24h | p.o. | Transplanted adult worms | *O. ochengi* | Balb/c mice SCID | n.d. | n.d. | n.a. | Solid dispersion | |
| 1.5 | 5, Q24h | p.o. | Transplanted adult worms | *O. ochengi* | Balb/c mice SCID | n.d. | n.d. | 19 | Solid dispersion | |
| 15 | 5, Q24h | p.o. | Transplanted adult worms | *O. ochengi* | Balb/c mice SCID | n.d. | n.d. | 30 | Solid dispersion | |
| 10 | 5, Q24h | s.c. | Transplanted adult worms | *O. ochengi* | Balb/c mice SCID | n.d. | n.d. | 97 | | [5] |

*(Continued)*

**Table 16.1** (Continued)

| Drug | Dosage (mg/kg) | Frequency | Treatment mode | Treatment time (day post-infection) | Model worm species | Rodent host | Microfilariae (mf) reduction (%) | | Adult worms (Mf) reduction (%) | Notes | References |
|---|---|---|---|---|---|---|---|---|---|---|---|
| | | | | | | | Day 3/7 post treatment | At necropsy (from d42) | | | |
| Mebendazole | 25 | 5, Q24h | p.o. | 2–6 | *A. viteae* | *M. natalensis* | n.d. | 100 | 92 | Sterile females | [11] |
| | 100 | 5, Q24h | p.o. | 28–32 | *A. viteae* | *M. natalensis* | n.d. | 100 | 94 | No dose dependence | |
| | 200 | 5, Q24h | p.o. | 63–67 | *A. viteae* | *M. coucha* | n.a. | 80 | 36 | | [2] |
| | 12.5 | 1 | s.c. | 63 | *A. viteae* | *M. coucha* | n.a. | 75 | 73 | | |
| | 200 | 5, Q24h | p.o. | 115–119 | *B. malayi* | *M. coucha* | n.a. | 73 | 26 | | |
| | 6.25 | 5, Q24h | s.c. | 115–119 | *B. malayi* | *M. coucha* | n.a. | 74 | 87 | | |
| | 100 | 5, Q24h | p.o. | 115–119 | *B. pahangi* | *M. coucha* | n.a. | 73 | 11 | | |
| | 6.25 | 5, Q24h | s.c. | 115–119 | *B. pahangi* | *M. coucha* | n.a. | 81 | 87 | | |
| | 10 | 5, Q24h | i.p. | 4–8 | *B. pahangi* | Balb/c mice | n.d. | n.d. | 100 | Larval recovery d29–32 | [6] |
| | 25 | 5, Q24h | p.o. | 85–89 | *L. carinii* | *M. coucha* | n.a. | 88 | n.a. | | [2] |
| | 0.8 | 5, Q24h | s.c. | 85–89 | *L. carinii* | *M. coucha* | n.a. | 94 | n.a. | | |
| Oxfendazole | 200 | 5, Q24h | p.o. | 2–6 | *A. viteae* | *M. natalensis* | n.d. | 100 | 94 | | [11] |
| | 100 | 5, Q24h | p.o. | 28–32 | *A. viteae* | *M. natalensis* | n.d. | 99 | 80 | No dose dependence | |

| | | | | | | | | | | | |
|---|---|---|---|---|---|---|---|---|---|---|---|
| | 25 | 5, Q24h | s.c. | 63–67 | *A. viteae* | *M. coucha* | n.a. | >95 | >95* | *Mf activity at 5 × 25 mg/kg | [2] |
| | 50 | 5, Q24h | s.c. | 115–119 | *B. malayi* | *M. coucha* | n.a. | >95 | >95 | | |
| | 6.25 | 5, Q24h | s.c. | 85–89 | *L. carinii* | *M. coucha* | n.a. | >95 | >95* | *Mf activity at 5 × 25 mg/kg | |
| Parbendazole | 200 | 5, Q24h | p.o. | 90–94 | *L. carinii* | *M. natalensis* | 34 | 97 | n.a. | | [22] |
| Thiabendazole | 200 | 5, Q24h | p.o. | 2–6 | *A. viteae* | *M. natalensis* | n.d. | 29 | 9 | | [11] |
| | 200 | 5, Q24h | p.o. | 28–32 | *A. viteae* | *M. natalensis* | n.d. | 81 | 30 | | |
| *Imidazothiazoles* | | | | | | | | | | | |
| Levamisole | 3.1 | 5, Q24h | p.o. | 63–67 | *A. viteae* | *M. coucha* | 100 | 0 | n.a. | | [2] |
| | 6.25 | 5, Q24h | p.o. | 2–6 | *A. viteae* | *M. natalensis* | n.d. | 100 | 98 | | [11] |
| | 12.5 | 5, Q24h | p.o. | 28–32 | *A. viteae* | *M. natalensis* | n.d. | 100 | 100 | | |
| | 50 | 5, Q24h | p.o. | 63–67 | *A. viteae* | *M. coucha* | 100 | >95 | 97 | | [2] |
| | 50 | 5, Q24h | p.o. | Trans-planted adult worms | *A. viteae* | Balb/c mice | n.d. | n.d. | 61 | | [20] |
| | 50 | 5, Q24h | p.o. | Trans-planted adult worms | *A. viteae* | *M. unguiculatus* | n.d. | n.d. | 100 | | |
| | 25 | 5, Q24h | p.o. | 115–119 | *B. malayi* | *M. coucha* | 97 | 93 | 24* | *Mf activity at 5 × 75 mg/kg | [2] |
| | 10 | 5, Q24h | i.p. | 4–8 | *B. pahangi* | Balb/c mice | n.d. | n.d. | 100 | Larval recovery d29–32 | [6] |

(Continued)

**Table 16.1** (Continued)

| Drug | Dosage (mg/kg) | Frequency | Treatment mode | Treatment time (day post-infection) | Model worm species | Rodent host | Microfilariae (mf) reduction (%) | | Adult worms (Mf) reduction (%) | Notes | References |
|---|---|---|---|---|---|---|---|---|---|---|---|
| | | | | | | | Day 3/7 post treatment | At necropsy (from d42) | | | |
| | 25 | 5, Q24h | p.o. | 115–119 | *B. pahangi* | *M. coucha* | 100 | 83 | n.a. | | [2] |
| | 50 | 5, Q24h | p.o. | 115–119 | *B. pahangi* | *M. coucha* | 100 | >95 | 81* | *Mf activity at 5 × 75 mg/kg | |
| | 50 | 5, Q24h | p.o. | Trans-planted adult worms | *B. pahangi* | Balb/c mice | n.d. | n.d. | 0 | | [20] |
| | 50 | 5, Q24h | p.o. | Trans-planted adult worms | *B. pahangi* | *M. unguiculatus* | n.d. | n.d. | 71 | | |
| | 25 | 5, Q24h | p.o. | 90–94 | *L. carinii* | *M. natalensis* | 98 | 37 | 69 | | [22] |
| | 12.5 | 5, Q24h | p.o. | 85–89 | *L. carinii* | *M. coucha* | 100 | 60 | n.a.* | *No Mf activity at 5 × 100 mg/kg | [2] |
| Tetramisole | 25 | 5, Q24h | p.o. | 90–94 | *L. carinii* | *M. natalensis* | 98 | 78 | n.a. | | [22] |

| | | | | | | | | | | | |
|---|---|---|---|---|---|---|---|---|---|---|---|
| *Amidines* | | | | | | | | | | | |
| Pyrantel tartrate | 40 | 5, Q24h | p.o. | 90–94 | *L. carinii* | *M. natalensis* | n.a. | n.a. | n.a. | | [22] |
| Methyl-pyrantel tartrate | 40 | 5, Q24h | p.o. | 90–94 | *L. carinii* | *M. natalensis* | n.a. | n.a. | n.a. | | [22] |
| *Anilines* | | | | | | | | | | | |
| Amidantel | 100 | 5, Q24h | p.o. | 63–67 | *A. viteae* | *M. coucha* | 100 | >95 | 95* | *Mf activity at 5 × 200 mg/kg | [2] |
| | 250 | 5, Q24h | p.o. | 115–119 | *B. malayi* | *M. coucha* | 88 | 50 | n.a. | | |
| | 250 | 5, Q24h | p.o. | 115–119 | *B. pahangi* | *M. coucha* | 85 | 70 | n.a. | | |
| | 250 | 5, Q24h | p.o. | 85–89 | *L. carinii* | *M. coucha* | 94 | 40 | n.a. | | |
| *Cyclodepsi-peptides* | | | | | | | | | | | |
| PF1022A | 100 | 5, Q24h | p.o. | 115–119 | *B. malayi* | *M. coucha* | 100 | 50 | n.a. | | [23] |
| PF1022A | 100 | 5, Q24h | p.o. | 85–89 | *L. sig-modontis* | *M. coucha* | 100 | 42 | n.a. | | |
| Emodepside | 100 | 1 | spot on | 3 | *A. viteae* | *M. coucha* | n.d. | 98 | 90 | Treatment on L3 | [24] |
| | 100 | 1 | spot on | 14 | *A. viteae* | *M. coucha* | n.d. | 100 | 100 | Treatment on L4 | |
| | 100 | 1 | spot on | 28 | *A. viteae* | *M. coucha* | n.d. | 95 | 86 | Treatment on preadults | |
| | 100 | 1 | p.o. | 65 | *A. viteae* | *M. coucha* | n.d. | n.d. | 96 | | [23] |

*(Continued)*

Table 16.1 (Continued)

| Drug | Dosage (mg/kg) | Frequency | Treatment mode | Treatment time (day post-infection) | Model worm species | Rodent host | Microfilariae (mf) reduction (%) | | Adult worms (Mf) reduction (%) | Notes | References |
|---|---|---|---|---|---|---|---|---|---|---|---|
| | | | | | | | Day 3/7 post treatment | At necropsy (from d42) | | | |
| | 12.5 | 1 | Spot on | 65 | *A. viteae* | *M. coucha* | 100 | 99 | 96* | *Mf activity at 1 × 100 mg/kg | |
| | 100 | 1 | Spot on | 3 | *B. malayi* | *M. coucha* | n.d. | 0 | 21 | Treatment on L3 | [24] |
| | 100 | 1 | Spot on | 12 | *B. malayi* | *M. coucha* | n.d. | 65 | 42 | Treatment on L4 | |
| | 100 | 1 | Spot on | 38 | *B. malayi* | *M. coucha* | n.d. | 72 | 18 | Treatment on preadults | |
| | 50 | 1 | p.o. | 115 | *B. malayi* | *M. coucha* | <95 | <95 | n.a.* | *No Mf activity at 5 × 100 mg/kg | [23] |
| | 6.25 | 1 | s.c. | 115 | *B. malayi* | *M. coucha* | 98 | 75 | n.a.* | *No Mf activity at 5 × 100 mg/kg | |
| | 12.5 | 1 | Spot on | 115 | *B. malayi* | *M. coucha* | 95 | 20 | n.a.* | *No Mf activity at 5 × 100 mg/kg | |
| | 100 | 1 | Spot on | 14 | *L. sigmodontis* | *M. coucha* | n.d. | 20 | 31 | Treatment on L4 | [24] |
| | 100 | 1 | Spot on | 28 | *L. sigmodontis* | *M. coucha* | n.d. | 85 | 36 | Treatment on preadults | |

| | | | | | | | | | | | |
|---|---|---|---|---|---|---|---|---|---|---|---|
| | 3.1 | 1 | p.o. | 85 | *L. sigmodontis* | *M. coucha* | 95 | 60 | 81* | *Mf activity at 1 × 100 mg/kg | [23] |
| | 12.5 | 1 | p.o. | 85 | *L. sigmodontis* | *M. coucha* | 100 | 92 | 100* | *Mf activity at 5 × 100 mg/kg | |
| | 3.1 | 1 | s.c. | 85 | *L. sigmodontis* | *M. coucha* | 100 | 90 | n.a. | | |
| | 12.5 | 1 | s.c. | 85 | *L. sigmodontis* | *M. coucha* | 100 | 98 | 100* | *Mf activity at 5 × 100 mg/kg | |
| | 12.5 | 1 | Spot on | 85 | *L. sigmodontis* | *M. coucha* | 100 | 98 | n.a. | | |
| | 100 | 1 | Spot on | 3 | *L. sigmodontis* | *M. coucha* | n.d. | 50 | 43 | Treatment on L3 | [24] |
| | 5 | 1 | s.c. | 30 | *D. immitis* | *Rat (Sprague–Dawley IGS)* | n.d. | n.d. | 100 | Immuno-suppressor in diet | [9] |
| | 1 | 1 | s.c. | 30 | *D. immitis* | *Rat (Sprague–Dawley IGS)* | n.d. | n.d. | 75.6 | Immuno-suppressor in diet | |
| Depsi-1 | 10 | 3, Q96h | s.c. | 30, 34, 38 | *D. immitis* | *Rat (Sprague–Dawley IGS)* | n.d. | n.d. | 100 | Immuno-suppressor in diet | |
| Depsi-2 | 10 | 3, Q96h | s.c. | 30, 34, 38 | *D. immitis* | *Rat (Sprague–Dawley IGS)* | n.d. | n.d. | 100 | Immuno-suppressor in diet | |
| Depsi-3 | 10 | 3, Q96h | s.c. | 30, 34, 38 | *D. immitis* | *Rat (Sprague–Dawley IGS)* | n.d. | n.d. | 100 | Immuno-suppressor in diet | |

*(Continued)*

**Table 16.1** (Continued)

| Drug | Dosage (mg/kg) | Frequency | Treatment mode | Treatment time (day post-infection) | Model worm species | Rodent host | Microfilariae (mf) reduction (%) | | Adult worms (Mf) reduction (%) | Notes | References |
|---|---|---|---|---|---|---|---|---|---|---|---|
| | | | | | | | Day 3/7 post treatment | At necropsy (from d42) | | | |
| *Phenylurea* | | | | | | | | | | | |
| Suramin | 40 | 5, Q24h | s.c. | 28–32 | *A. viteae* | *M. natalensis* | n.d. | 86 | 53 | | [11] |
| | 40 | 5, Q24h | s.c. | 63–67 | *A. viteae* | *M. coucha* | <95 | <95 | >95 | | [2] |
| | 40 | 5, Q24h | s.c. | 2–6 | *A. viteae* | *M. natalensis* | n.d. | 99 | 100 | | [11] |
| | 40 | 5, Q24h | s.c. | 115–119 | *B. malayi* | *M. coucha* | n.a. | n.a. | n.a. | | [2] |
| | 40 | 5, Q24h | i.p. | 4–8 | *B. pahangi* | Balb/c mice | n.d. | n.d. | 7–87 | larval recovery d29–32 | [6] |
| | 40 | 5, Q24h | s.c. | 115–119 | *B. pahangi* | *M. coucha* | n.a. | n.a. | n.a. | | [2] |
| | 50 | 5, Q24h | s.c. | 90–94 | *L. carinii* | *M. natalensis* | 0 | 50 | 100 | | [25] |
| | 40 | 5, Q24h | s.c. | 85–89 | *L. carinii* | *M. coucha* | n.a. | n.a. | n.a. | | [2] |
| | 40 | 5, Q24h | s.c. | 4–8 | *M. dessetae* | *P. oris* | n.d. | n.d. | 100 | Treatment on L3 | [7] |
| | 40 | 5, Q24h | s.c. | 47–51 | *M. dessetae* | *P. oris* | n.d. | n.d. | 70 | Treatment on pre-adult worms | |
| | 40 | 5, Q24h | s.c. | 189–193 | *M. dessetae* | *P. oris* | 20 | 100 | 93 | Treatment on adult worms | |

| | | | | | | | | | | | |
|---|---|---|---|---|---|---|---|---|---|---|---|
| *Thiocyanates* | | | | | | | | | | | |
| Amoscanate | 50 | 5, Q24h | p.o. | 2–6 | *A. viteae* | *M. natalensis* | n.d. | 100 | 97 | No males | [11] |
| | 50 | 5, Q24h | p.o. | 28–32 | *A. viteae* | *M. natalensis* | n.d. | 98 | 75 | Females shorter | |
| | 100 | 5, Q24h | p.o. | 63–67 | *A. viteae* | *M. coucha* | >95 | >95 | 100 | | [2] |
| | 100 | 5, Q24h | s.c. | Trans-planted adult worms | *A. viteae* | *M. unguiculatus* | n.d. | n.d. | 33 | Anti-schistosome | [18] |
| | 100 | 5, Q24h | s.c. | Trans-planted adult worms | *A. viteae* | *M. unguiculatus* | n.d. | n.d. | 100 | | [19] |
| | 100 | 5, Q24h | s.c. | Trans-planted adult worms | *A. viteae* | Balb/c mice | n.d. | n.d. | 54 | | [20] |
| | 50 | 5, Q24h | s.c. | Trans-planted adult worms | *A. viteae* | Balb/c mice | n.d. | n.d. | 54 | | |
| | 100 | 5, Q24h | s.c. | Trans-planted adult worms | *A. viteae* | *M. unguiculatus* | n.d. | n.d. | 100 | | |
| | 50 | 5, Q24h | s.c. | Trans-planted adult worms | *A. viteae* | *M. unguiculatus* | n.d. | n.d. | 78 | | |

(Continued)

**Table 16.1** (Continued)

| Drug | Dosage (mg/kg) | Frequency | Treatment mode | Treatment time (day post-infection) | Model worm species | Rodent host | Microfilariae (mf) reduction (%) | | Adult worms (Mf) reduction (%) | Notes | References |
|---|---|---|---|---|---|---|---|---|---|---|---|
| | | | | | | | Day 3/7 post treatment | At necropsy (from d42) | | | |
| | 25 | 5, Q24h | p.o. | 115–119 | *B. malayi* | *M. coucha* | 100 | 100 | >95 | | [2] |
| | 100 | 5, Q24h | s.c. | Trans-planted adult worms | *B. pahangi* | *M. unguiculatus* | n.d. | n.d. | 17 | Anti-schistosome | [18] |
| | 200 | 5, Q24h | i.p. | 4–8 | *B. pahangi* | Balb/c mice | n.d. | n.d. | n.a. | Larval recovery d29–32 | [6] |
| | 25 | 5, Q24h | p.o. | 115–119 | *B. pahangi* | *M. coucha* | 92 | 90 | >95 | | [2] |
| | 100 | 5, Q24h | s.c. | Trans-planted adult worms | *B. pahangi* | *M. unguiculatus* | n.d. | n.d. | 0 | | [19] |
| | 100 | 5, Q24h | s.c. | Trans-planted adult worms | *B. pahangi* | Balb/c mice | n.d. | n.d. | 74 | | [20] |
| | 50 | 5, Q24h | s.c. | Trans-planted adult worms | *B. pahangi* | Balb/c mice | n.d. | n.d. | 61 | | |
| | 100 | 5, Q24h | s.c. | Trans-planted adult worms | *B. pahangi* | *M. unguiculatus* | n.d. | n.d. | 0 | | |

| | | | | | | | | | | |
|---|---|---|---|---|---|---|---|---|---|---|
| | 50 | 5, Q24h | s.c. | Transplanted adult worms | *B. pahangi* | *M. unguiculatus* | n.d. | n.d. | 29 | |
| | 25 | 5, Q24h | p.o. | 85–89 | *L. carinii* | *M. coucha* | 100 | >95 | 100 | [2] |
| CGP 6140 (amoscanate derivative) | 100 | 5, Q24h | s.c. | Transplanted adult worms | *A. viteae* | *M. unguiculatus* | n.d. | n.d. | 100 | [19] |
| | 100 | 5, Q24h | s.c. | Transplanted adult worms | *A. viteae* | Balb/c mice | n.d. | n.d. | 18 | [20] |
| | 100 | 5, Q24h | s.c. | Transplanted adult worms | *A. viteae* | *M. unguiculatus* | n.d. | n.d. | 49 | |
| | 100 | 5, Q24h | s.c. | Transplanted adult worms | *B. pahangi* | *M. unguiculatus* | n.d. | n.d. | 0 | [19] |
| | 100 | 5, Q24h | s.c. | Transplanted adult worms | *B. pahangi* | Balb/c mice | n.d. | n.d. | 38 | [20] |
| | 100 | 5, Q24h | s.c. | Transplanted adult worms | *B. pahangi* | *M. unguiculatus* | n.d. | n.d. | 2 | |
| Nitroscanate | 50 | 5, Q24h | p.o. | 2–6 | *A. viteae* | *M. natalensis* | n.d. | 21 | 14 | [11] |
| | 50 | 5, Q24h | p.o. | 28–32 | *A. viteae* | *M. natalensis* | n.d. | 41 | 0 | |

*(Continued)*

**Table 16.1** (Continued)

| Drug | Dosage (mg/kg) | Frequency | Treatment mode | Treatment time (day post-infection) | Model worm species | Rodent host | Microfilariae (mf) reduction (%) | | Adult worms (Mf) reduction (%) | Notes | References |
|---|---|---|---|---|---|---|---|---|---|---|---|
| | | | | | | | Day 3/7 post treatment | At necropsy (from d42) | | | |
| *Aminoquinolines* | | | | | | | | | | | |
| Amodiaquine | 100 | 5, Q24h | p.o. | 2–6 | *A. viteae* | *M. natalensis* | n.d. | 59 | 24 | | [11] |
| | 100 | 5, Q24h | p.o. | 28–32 | *A. viteae* | *M. natalensis* | n.d. | 82 | 64 | | |
| | 25 | 5, Q24h | p.o. | 90–94 | *L. carinii* | *M. natalensis* | 32 | 97 | 84 | | [25] |
| *Organo-phosphates* | | | | | | | | | | | |
| Metrifonate | 100 | 5, Q24h | p.o. | 63–67 | *A. viteae* | *M. coucha* | 60 | 0 | n.a. | | [2] |
| | 100 | 5, Q24h | p.o. | 115–119 | *B. malayi* | *M. coucha* | 10 | 0 | n.a. | | |
| | 100 | 5, Q24h | p.o. | 115–119 | *B. pahangi* | *M. coucha* | 0 | 0 | n.a. | | |
| | 25 | 5, Q24h | i.p. | 4–8 | *B. pahangi* | Balb/c mice | n.d. | n.d. | n.a. | Larval recovery d29–32 | [6] |
| | 200 | 5, Q24h | p.o. | 90–94 | *L. carinii* | *M. natalensis* | 99 | 0 | n.a. | | [22] |
| | 100 | 5, Q24h | p.o. | 85–89 | *L. carinii* | *M. coucha* | 100 | 0 | n.a. | | [2] |
| Fenthion | 10 | 5, Q24h | p.o. | 2–6 | *A. viteae* | *M. natalensis* | n.d. | 0 | 0 | | [11] |
| | 10 | 5, Q24h | p.o. | 28–32 | *A. viteae* | *M. natalensis* | n.d. | 0 | 1 | | |
| *Nitrofuranes* | | | | | | | | | | | |
| Nitrofurantoin | 50 | 5, Q24h | p.o. | 2–6 | *A. viteae* | *M. natalensis* | n.d. | 99 | 80 | | [11] |
| | 100 | 5, Q24h | p.o. | 2–6 | *A. viteae* | *M. natalensis* | n.d. | 100 | 100 | | |

| | | | | | | | | | |
|---|---|---|---|---|---|---|---|---|---|
| 100 | 5, Q24h | p.o. | 28–32 | *A. viteae* | *M. natalensis* | n.d. | 41 | n.a. | |
| 100 | 5, Q24h | s.c. | Trans-planted adult worms | *A. viteae* | *M. unguiculatus* | n.d. | n.d. | 81 | [19] |
| 100 | 5, Q24h | s.c. | Trans-planted adult worms | *A. viteae* | *M. unguiculatus* | n.d. | n.d. | 81 | [20] |
| 50 | 5, Q24h | s.c. | Trans-planted adult worms | *A. viteae* | *M. unguiculatus* | n.d. | n.d. | 0 | |
| 100 | 5, Q24h | s.c. | Trans-planted adult worms | *A. viteae* | Balb/c mice | n.d. | n.d. | 42 | |
| 50 | 5, Q24h | s.c. | Trans-planted adult worms | *A. viteae* | Balb/c mice | n.d. | n.d. | 0 | |
| 100 | 5, Q24h | p.o. | 63–67 | *A. viteae* | *M. coucha* | 0 | 70 | n.a. | [2] |
| 100 | 5, Q24h | p.o. | 115–119 | *B. malayi* | *M. coucha* | 92 | 80 | 70 | |
| 100 | 5, Q24h | s.c. | Trans-planted adult worms | *B. pahangi* | *M. unguiculatus* | n.d. | n.d. | 100 | [20] |
| 100 | 5, Q24h | s.c. | Trans-planted adult worms | *B. pahangi* | *M. unguiculatus* | n.d. | n.d. | 100 | |

*(Continued)*

**Table 16.1** (Continued)

| Drug | Dosage (mg/kg) | Frequency | Treatment mode | Treatment time (day post-infection) | Model worm species | Rodent host | Microfilariae (mf) reduction (%) Day 3/7 post treatment | At necropsy (from d42) | Adult worms (Mf) reduction (%) | Notes | References |
|---|---|---|---|---|---|---|---|---|---|---|---|
| | 50 | 5, Q24h | s.c. | Trans-planted adult worms | *B. pahangi* | *M. unguiculatus* | n.d. | n.d. | 0 | | |
| | 100 | 5, Q24h | s.c. | Trans-planted adult worms | *B. pahangi* | Balb/c mice | n.d. | n.d. | 55 | | |
| | 50 | 5, Q24h | s.c. | Trans-planted adult worms | *B. pahangi* | Balb/c mice | n.d. | n.d. | 13 | | |
| | 100 | 5, Q24h | p.o. | 115–119 | *B. pahangi* | *M. coucha* | 88 | 100 | 60 | | [2] |
| | 100 | 5, Q24h | p.o. | 85–89 | *L. carinii* | *M. coucha* | 85 | 99 | 100 | | |
| Hydroxymethyl nitrofurantoin | 150 | 5, Q24h | p.o. | 63–67 | *A. viteae* | *M. coucha* | 100 | 70 | n.a. | | [2] |
| | 150 | 5, Q24h | p.o. | 115–119 | *B. malayi* | *M. coucha* | 97 | 96 | 56 | | |
| | 150 | 5, Q24h | p.o. | 85–89 | *L. carinii* | *M. coucha* | 99 | 99 | 100 | | |
| Nifurtimox | 150 | 5, Q24h | p.o. | 2–6 | *A. viteae* | *M. natalensis* | n.d. | 0 | 14 | | [11] |
| | 150 | 5, Q24h | p.o. | 28–32 | *A. viteae* | *M. natalensis* | n.d. | 86 | 47 | | |
| | 100 | 5, Q24h | p.o. | 63–67 | *A. viteae* | *M. coucha* | 0 | 50 | n.a. | | [2] |

| | | | | | | | | | | | |
|---|---|---|---|---|---|---|---|---|---|---|---|
| | 100 | 5, Q24h | p.o. | 115–119 | *B. malayi* | *M. coucha* | 20 | 0 | n.a. | | |
| | 100 | 5, Q24h | p.o. | 85–89 | *L. carinii* | *M. coucha* | 0 | 100 | 100 | | |
| Furazolidone | 100 | 5, Q24h | p.o. | 63–67 | *A. viteae* | *M. coucha* | 75 | 75 | 33 | | [2] |
| | 75 | 5, Q24h | p.o. | 115–119 | *B. malayi* | *M. coucha* | 92 | 99 | 18 | | |
| | 100 | 5, Q24h | p.o. | 85–89 | *L. carinii* | *M. coucha* | 0 | 100 | 100 | | |
| Furapyrimidone | 100 | 5, Q24h | s.c. | Trans-planted adult worms | *A. viteae* | *M. unguiculatus* | n.d. | n.d. | 36 | | [19] |
| | 100 | 5, Q24h | p.o. | 63–67 | *A. viteae* | *M. coucha* | 100 | 96 | 12 | | [2] |
| | 100 | 5, Q24h | p.o. | 115–119 | *B. malayi* | *M. coucha* | 100 | 98 | 82 | | |
| | 100 | 5, Q24h | s.c. | Trans-planted adult worms | *B. pahangi* | *M. unguiculatus* | n.d. | n.d. | 38 | | [19] |
| | 100 | 5, Q24h | p.o. | 115–119 | *B. pahangi* | *M. coucha* | 100 | 98 | 81 | | [2] |
| | 50 | 5, Q24h | p.o. | 85–89 | *L. carinii* | *M. coucha* | 0 | >95 | 100 | | |
| *Furanes* | | | | | | | | | | | |
| Nigericin | 6.25 | 5, Q24h | s.c. | Trans-planted adult worms | *A. viteae* | *M. unguiculatus* | n.d. | n.d. | 91 | Antibiotic | [18] |
| *Arsenicals* | | | | | | | | | | | |
| Thiacertas-amide | 15 | 5, Q24h | s.c. | 63–67 | *A. viteae* | *M. coucha* | 70 | 90 | >95 | | [2] |
| | 20 | 5, Q24h | s.c. | 115–119 | *B. malayi* | *M. coucha* | <95 | <95 | >95* | *Mf activity at 5 × 2.5 mg/kg | |

(Continued)

**Table 16.1** (Continued)

| Drug | Dosage (mg/kg) | Frequency | Treatment mode | Treatment time (day post-infection) | Model worm species | Rodent host | Microfilariae (mf) reduction (%) | | Adult worms (Mf) reduction (%) | Notes | References |
|---|---|---|---|---|---|---|---|---|---|---|---|
| | | | | | | | Day 3/7 post treatment | At necropsy (from d42) | | | |
| | 20 | 5, Q24h | s.c. | 115–119 | *B. pahangi* | *M. coucha* | <95 | <95 | >95* | *Mf activity at 5 × 2.5 mg/kg | |
| | 20 | 5, Q24h | s.c. | 85–89 | *L. carinii* | *M. coucha* | 70 | >95 | >95 | | |
| *Flavonoids* | | | | | | | | | | | |
| Rutine | 25 | 5, Q24h | s.c. | Trans-planted adult worms | *B. pahangi* | *M. unguiculatus* | n.d. | n.a. | n.a. | | [14] |
| Flavone | 25 | 5, Q24h | s.c. | Trans-planted adult worms | *B. pahangi* | *M. unguiculatus* | n.d. | n.a. | 38 | | |
| Hesperetin | 25 | 5, Q24h | s.c. | Trans-planted adult worms | *B. pahangi* | *M. unguiculatus* | n.d. | n.a. | n.a. | | |
| Chysin | 25 | 5, Q24h | s.c. | Trans-planted adult worms | *B. pahangi* | *M. unguiculatus* | n.d. | n.a. | n.a. | | |
| Naringin | 25 | 5, Q24h | s.c. | Trans-planted adult worms | *B. pahangi* | *M. unguiculatus* | n.d. | n.a. | n.a. | | |

| | | | | | | | | | | |
|---|---|---|---|---|---|---|---|---|---|---|
| Naringenin | 25 | 5, Q24h | s.c. | Transplanted adult worms | *B. pahangi* | *M. unguiculatus* | n.d. | n.a. | 73 | |
| | 100 | 5, Q24h | i.p. | >150 | *B. pahangi* | *M. coucha* | 20 | 53 | 51 | |
| *Substituted azines compounds* | | | | | | | | | | |
| Compound 2 | 3 | 5, Q24h | p.o. | 5–9 | *A. viteae* | *M. unguiculatus* | n.d. | >50 | >90 | [26] |
| Compounds 18, 90 | 10 | 5, Q24h | p.o. | 5–9 | *A. viteae* | *M. unguiculatus* | n.d. | >50 | >90 | |
| Compound 104 | 23 | 5, Q24h | s.c. | 5–9 | *A. viteae* | *M. unguiculatus* | n.d. | >50 | >90 | |
| Compounds 6, 57, 65, 80 | 32 | 5, Q24h | p.o. | 5–9 | *A. viteae* | *M. unguiculatus* | n.d. | >50 | >90 | |
| *Pyridinyl/pyrimidinyl compounds* | | | | | | | | | | |
| Compound 1 | 3 | 5, Q24h | p.o. | 5–9 | *A. viteae* | *M. unguiculatus* | n.d. | >50 | >80 | [27] |
| *Sulfonyl-aminobenzamide compounds* | | | | | | | | | | |
| Compounds 1.5, 1.10, 1.13, 1.17, 2.1, 2.21, 2.26, 2.51 | 10 | 5, Q24h | p.o. | 5–9 | *A. viteae* | *M. unguiculatus* | n.d. | >50 | >80 | [28] |
| Compounds 1.6, 1.7, 1.8, 1.25, 1.30, 1.32, 2.24, 4.13, 4.24 | 10 | 5, Q24h | p.o. | 5–9 | *A. viteae* | *M. unguiculatus* | n.d. | >50 | >80 | [29] |

*(Continued)*

**Table 16.1** (Continued)

| Drug | Dosage (mg/kg) | Frequency | Treatment mode | Treatment time (day post-infection) | Model worm species | Rodent host | Microfilariae (mf) reduction (%) | | Adult worms (Mf) reduction (%) | Notes | References |
|---|---|---|---|---|---|---|---|---|---|---|---|
| | | | | | | | Day 3/7 post treatment | At necropsy (from d42) | | | |
| *Benzazole derivatives* | | | | | | | | | | | |
| Group 1 (benzothiazoles) | 25–100 | 5, Q24h | p.o. | 65–69 | *A. viteae* | *M. natalensis* | >95 | n.d. | >95 | | [30, 31] |
| | 12.5–100 | 5, Q24h | p.o. | 115–119 | *B. malayi* | *M. natalensis* | >95 | n.d. | >95 | | |
| | 12.5–50 | 5, Q24h | p.o. | 115–119 | *B. pahangi* | *M. natalensis* | >95 | n.d. | >95 | | |
| | 12.5–50 | 5, Q24h | p.o. | 85–89 | *L. carinii* | *M. natalensis* | >95 | n.d. | >95 | | |
| Group 2 (benzothiazoles) | 12.5–50 | 5, Q24h | p.o. | 65–69 | *A. viteae* | *M. natalensis* | >95 | n.d. | >95 | Best derivative of all groups, CGP 20376: mf and Mf activity at 5 × 6.25 mg/kg | |
| | 6.25–50 | 5, Q24h | p.o. | 115–119 | *B. malayi* | *M. natalensis* | >95 | n.d. | >95 | | |
| | 6.25–25 | 5, Q24h | p.o. | 115–119 | *B. pahangi* | *M. natalensis* | >95 | n.d. | >95 | | |
| | 6.25–25 | 5, Q24h | p.o. | 85–89 | *L. carinii* | *M. natalensis* | >95 | n.d. | >95 | | |
| Group 3 (benzoxazoles) | 100 | 5, Q24h | p.o. | 65–69 | *A. viteae* | *M. natalensis* | >97 | n.d. | n.a. | | |
| | 50–100 | 5, Q24h | p.o. | 115–119 | *B. malayi* | *M. natalensis* | >95 | n.d. | >95 | | |
| | 50–100 | 5, Q24h | p.o. | 115–119 | *B. pahangi* | *M. natalensis* | >95 | n.d. | >95 | | |
| | 50–100 | 5, Q24h | p.o. | 85–89 | *L. carinii* | *M. natalensis* | >95 | n.d. | >95 | | |

| | | | | | | | | | | |
|---|---|---|---|---|---|---|---|---|---|---|
| Group 4 (benzoxazoles) | 50–100 | 5, Q24h | p.o. | 65–69 | *A. viteae* | *M. natalensis* | >95 | n.d. | >95 | |
| | 25–100 | 5, Q24h | p.o. | 115–119 | *B. malayi* | *M. natalensis* | >95 | n.d. | >95 | |
| | 25–100 | 5, Q24h | p.o. | 115–119 | *B. pahangi* | *M. natalensis* | >95 | n.d. | >95 | |
| | 25–100 | 5, Q24h | p.o. | 85–89 | *L. carinii* | *M. natalensis* | >95 | n.d. | >95 | |
| *Benzothiazole derivatives* | | | | | | | | | | |
| CGP 21306 | 25 | 1 | p.o. | 3 | *A. viteae* | *M. natalensis* | n.d. | n.d. | >95 | [32] |
| | 50 | 1 | p.o. | 28 | *A. viteae* | *M. natalensis* | n.d. | n.d. | >95 | |
| | 25 | 1 | p.o. | 3 | *B. malayi* | *M. natalensis* | n.d. | n.d. | >95 | |
| | 50 | 1 | p.o. | 38 | *B. malayi* | *M. natalensis* | n.d. | n.d. | >95 | |
| | 25 | 1 | p.o. | 3 | *B. pahangi* | *M. natalensis* | n.d. | n.d. | >95 | |
| | 25 | 1 | p.o. | 32 | *B. pahangi* | *M. natalensis* | n.d. | n.d. | >95 | |
| CGP 21835 | 50 | 1 | p.o. | 3 | *A. viteae* | *M. natalensis* | n.d. | n.d. | >95 | |
| | 50 | 1 | p.o. | 28 | *A. viteae* | *M. natalensis* | n.d. | n.d. | >95 | |
| | 25 | 1 | p.o. | 3 | *B. malayi* | *M. natalensis* | n.d. | n.d. | >95 | |
| | 25 | 1 | p.o. | 38 | *B. malayi* | *M. natalensis* | n.d. | n.d. | >95 | |
| | 25 | 1 | p.o. | 3 | *B. pahangi* | *M. natalensis* | n.d. | n.d. | >95 | |
| | 25 | 1 | p.o. | 32 | *B. pahangi* | *M. natalensis* | n.d. | n.d. | >95 | |

(Continued)

**Table 16.1** (Continued)

| Drug | Dosage (mg/kg) | Frequency | Treatment mode | Treatment time (day post-infection) | Model worm species | Rodent host | Microfilariae (mf) reduction (%) | | Adult worms (Mf) reduction (%) | Notes | References |
|---|---|---|---|---|---|---|---|---|---|---|---|
| | | | | | | | Day 3/7 post treatment | At necropsy (from d42) | | | |
| CGP 26701 | 50 | 1 | p.o. | 3 | *A. viteae* | *M. natalensis* | n.d. | n.d. | >95 | | |
| | 50 | 1 | p.o. | 28 | *A. viteae* | *M. natalensis* | n.d. | n.d. | >95 | | |
| | 50 | 1 | p.o. | 38 | *B. malayi* | *M. natalensis* | n.d. | n.d. | >95 | | |
| | 50 | 1 | p.o. | 3 | *B. pahangi* | *M. natalensis* | n.d. | n.d. | >95 | | |
| | 100 | 1 | p.o. | 32 | *B. pahangi* | *M. natalensis* | n.d. | n.d. | >95 | | |
| CGP 21833 | 100 | 1 | p.o. | 3 | *A. viteae* | *M. natalensis* | n.d. | n.d. | <95 (89) | | |
| | 100 | 1 | p.o. | 28 | *A. viteae* | *M. natalensis* | n.d. | n.d. | <95 | | |
| | 100 | 1 | p.o. | 3 | *B. malayi* | *M. natalensis* | n.d. | n.d. | >95 | | |
| | 100 | 1 | p.o. | 38 | *B. malayi* | *M. natalensis* | n.d. | n.d. | >95 | | |
| | 100 | 1 | p.o. | 3 | *B. pahangi* | *M. natalensis* | n.d. | n.d. | >95 | | |
| | 100 | 1 | p.o. | 32 | *B. pahangi* | *M. natalensis* | n.d. | n.d. | >95 | | |
| CGP 26702 | 100 | 1 | p.o. | 3 | *A. viteae* | *M. natalensis* | n.d. | n.d. | <95 | | |
| | 100 | 1 | p.o. | 28 | *A. viteae* | *M. natalensis* | n.d. | n.d. | >95 | | |
| | 100 | 1 | p.o. | 38 | *B. malayi* | *M. natalensis* | n.d. | n.d. | >95 | | |
| | 100 | 1 | p.o. | 3 | *B. pahangi* | *M. natalensis* | n.d. | n.d. | >95 | | |
| | 100 | 1 | p.o. | 32 | *B. pahangi* | *M. natalensis* | n.d. | n.d. | >95 | | |

| | | | | | | | | | |
|---|---|---|---|---|---|---|---|---|---|
| CGP 20308 | 25 | 1 | p.o. | 3 | *A. viteae* | *M. natalensis* | n.d. | n.d. | >95 |
| | 50 | 1 | p.o. | 28 | *A. viteae* | *M. natalensis* | n.d. | n.d. | >95 |
| | 25 | 1 | p.o. | 3 | *B. malayi* | *M. natalensis* | n.d. | n.d. | >95 |
| | 25 | 1 | p.o. | 38 | *B. malayi* | *M. natalensis* | n.d. | n.d. | >95 |
| | 12.5 | 1 | p.o. | 3 | *B. pahangi* | *M. natalensis* | n.d. | n.d. | >95 |
| | 12.5 | 1 | p.o. | 32 | *B. pahangi* | *M. natalensis* | n.d. | n.d. | >95 |
| CGP 20376 | 25 | 1 | p.o. | 3 | *A. viteae* | *M. natalensis* | n.d. | n.d. | >95 |
| | 50 | 1 | p.o. | 28 | *A. viteae* | *M. natalensis* | n.d. | n.d. | >95 |
| | 12.5 | 1 | p.o. | 3 | *B. malayi* | *M. natalensis* | n.d. | n.d. | >95 |
| | 25 | 1 | p.o. | 38 | *B. malayi* | *M. natalensis* | n.d. | n.d. | >95 |
| | 25 | 1 | p.o. | 3 | *B. pahangi* | *M. natalensis* | n.d. | n.d. | >95 |
| | 25 | 1 | p.o. | 32 | *B. pahangi* | *M. natalensis* | n.d. | n.d. | >95 |
| CGP 24588 | 50 | 1 | p.o. | 3 | *A. viteae* | *M. natalensis* | n.d. | n.d. | >95 |
| | 100 | 1 | p.o. | 28 | *A. viteae* | *M. natalensis* | n.d. | n.d. | >95 |
| | 12.5 | 1 | p.o. | 3 | *B. malayi* | *M. natalensis* | n.d. | n.d. | >95 |
| | 25 | 1 | p.o. | 38 | *B. malayi* | *M. natalensis* | n.d. | n.d. | >95 |
| | 25 | 1 | p.o. | 3 | *B. pahangi* | *M. natalensis* | n.d. | n.d. | >95 |
| | 50 | 1 | p.o. | 32 | *B. pahangi* | *M. natalensis* | n.d. | n.d. | >95 |

*(Continued)*

**Table 16.1** (Continued)

| Drug | Dosage (mg/kg) | Frequency | Treatment mode | Treatment time (day post-infection) | Model worm species | Rodent host | Microfilariae (mf) reduction (%) | | Adult worms (Mf) reduction (%) | Notes | References |
|---|---|---|---|---|---|---|---|---|---|---|---|
| | | | | | | | Day 3/7 post treatment | At necropsy (from d42) | | | |
| CGP 20309 | 100 | 1 | p.o. | 3 | *A. viteae* | *M. natalensis* | n.d. | n.d. | >95 | | |
| | 100 | 1 | p.o. | 28 | *A. viteae* | *M. natalensis* | n.d. | n.d. | <95 | | |
| | 50 | 1 | p.o. | 3 | *B. malayi* | *M. natalensis* | n.d. | n.d. | >95 | | |
| | 50 | 1 | p.o. | 38 | *B. malayi* | *M. natalensis* | n.d. | n.d. | >95 | | |
| | 50 | 1 | p.o. | 3 | *B. pahangi* | *M. natalensis* | n.d. | n.d. | >95 | | |
| | 50 | 1 | p.o. | 32 | *B. pahangi* | *M. natalensis* | n.d. | n.d. | >95 | | |
| CGP 24589 | 100 | 1 | p.o. | 3 | *A. viteae* | *M. natalensis* | n.d. | n.d. | <95 | | |
| | 100 | 1 | p.o. | 28 | *A. viteae* | *M. natalensis* | n.d. | n.d. | >95 | | |
| | 50 | 1 | p.o. | 3 | *B. pahangi* | *M. natalensis* | n.d. | n.d. | >95 | | |
| | 100 | 1 | p.o. | 32 | *B. pahangi* | *M. natalensis* | n.d. | n.d. | >95 | | |
| *Quinolones (Quinol-4(1H)-one-3-carboxamide derivatives)* | | | | | | | | | | | |
| Compound 4a | 200 | 5, Q24h | p.o. | Mature infection | *A. viteae* | *M. coucha* | n.d. | 90 | 100 | | [33] |
| Compound 4e | 200 | 5, Q24h | p.o. | Mature infection | *A. viteae* | *M. coucha* | n.d. | 24 | 80 | | |

| | | | | | | | | | | |
|---|---|---|---|---|---|---|---|---|---|---|
| *Thiosemicarbazone derivatives* | | | | | | | | | | |
| Compound 1 | 50 | 5, Q24h | s.c. | Trans-planted adult worms | *A. viteae* | *M. unguiculatus* | n.d. | n.d. | 74 | [34] |
| Compound 2 | 6.25 | 5, Q24h | s.c. | Trans-planted adult worms | *A. viteae* | *M. unguiculatus* | n.d. | n.d. | 70 | |
| Compound 10 | 50 | 5, Q24h | s.c. | Trans-planted adult worms | *A. viteae* | *M. unguiculatus* | n.d. | n.d. | 76 | |
| Compound 12 | 100 | 5, Q24h | s.c. | Trans-planted adult worms | *A. viteae* | *M. unguiculatus* | n.d. | n.d. | 80 | |
| Compound 6 | 100 | 5, Q24h | s.c. | Trans-planted adult worms | *B. pahangi* | *M. unguiculatus* | n.d. | n.d. | 80 | |
| Compound 7 | 12.5 | 5, Q24h | s.c. | Trans-planted adult worms | *B. pahangi* | *M. unguiculatus* | n.d. | n.d. | 76 | |
| Compound 14 | 100 | 5, Q24h | s.c. | Trans-planted adult worms | *B. pahangi* | *M. unguiculatus* | n.d. | n.d. | 100 | |
| Cyclohexanol compound 2b | 50 | 5, Q24h | i.p. | 65–69 | *A. viteae* | *M. coucha* | n.d. | n.a. | 89 | [12] |
| | 30 | 5, Q24h | i.p. | 85–89 | *L. carinii* | *S. hispidus* | n.d. | n.a. | n.a. | |

*(Continued)*

**Table 16.1** (Continued)

| Drug | Dosage (mg/kg) | Frequency | Treatment mode | Treatment time (day post-infection) | Model worm species | Rodent host | Microfilariae (mf) reduction (%) | | Adult worms (Mf) reduction (%) | Notes | References |
|---|---|---|---|---|---|---|---|---|---|---|---|
| | | | | | | | Day 3/7 post treatment | At necropsy (from d42) | | | |
| Cyclooctanol compound 2f | 50 | 5, Q24h | i.p. | 65–69 | *A. viteae* | *M. coucha* | n.d. | n.a. | 100 | | |
| | 30 | 5, Q24h | i.p. | 85–89 | *L. carinii* | *S. hispidus* | n.d. | n.a. | n.a. | | |
| *Thioxanthenes* | | | | | | | | | | | |
| Flupentixol | 12.5 | 5, Q24h | s.c. | Trans-planted adult worms | *A. viteae* | *M. unguiculatus* | n.d. | n.d. | 96 | Neuroleptic | [18] |
| | 12.5 | 5, Q24h | s.c. | Trans-planted adult worms | *B. pahangi* | *M. unguiculatus* | n.d. | n.d. | 15 | | |
| *Cellular dyes* | | | | | | | | | | | |
| Ethidium | 12.5 | 5, Q24h | s.c. | Trans-planted adult worms | *A. viteae* | *M. unguiculatus* | n.d. | n.d. | 89 | Anti-African *Try-panosoma* | [18] |
| | 12.5 | 5, Q24h | s.c. | Trans-planted adult worms | *B. pahangi* | *M. unguiculatus* | n.d. | n.d. | 7 | | |
| Gentian violet | 2 | 5, Q24h | s.c. | 90–94 | *L. carinii* | *M. natalensis* | n.a. | n.a. | n.a. | Dyes | [22] |

| | | | | | | | | | | | |
|---|---|---|---|---|---|---|---|---|---|---|---|
| Hoechst 28637 | 125 | 5, Q24h | p.o. | 90–94 | *L. carinii* | *M. natalensis* | 97 | 34 | n.a. | | |
| Hoechst 29691 | 100 | 5, Q24h | p.o. | 90–94 | *L. carinii* | *M. natalensis* | 97 | 34 | 63 | | |
| Hoechst 33258 free base | 40 | 5, Q24h | s.c. | 90–94 | *L. carinii* | *M. natalensis* | 99 | 90 | n.a. | | [16] |
| Hoechst 33258 diphosphate | 20 | 5, Q24h | s.c. | 90–94 | *L. carinii* | *M. natalensis* | 100 | 95 | n.a. | | [16] |
| | 10 | 5, Q24h | s.c. | 90–94 | *L. carinii* | *M. natalensis* | 99.5 | 63 | 60 | | [22] |
| Hoechst 33258 × 3 HCL | 40 | 5, Q24h | s.c. | 90–94 | *L. carinii* | *M. natalensis* | 93 | 43 | n.a. | | [16] |
| | 2.5 | 5, Q24h | s.c. | 90–94 | *L. carinii* | *M. natalensis* | 99 | 91 | n.a. | | [22] |
| Methylen violet | 4 | 5, Q24h | s.c. | 90–94 | *L. carinii* | *M. natalensis* | n.a. | n.a. | n.a. | | |
| *Thiazines* | | | | | | | | | | | |
| Isothipendyl | 6.25 | 5, Q24h | s.c. | Transplanted adult worms | *A. viteae* | *M. unguiculatus* | n.d. | n.d. | 96 | Anti-*Trypanosoma cruzi* | [18] |
| | 6.25 | 5, Q24h | s.c. | Transplanted adult worms | *B. pahangi* | *M. unguiculatus* | n.d. | n.d. | 24 | | |

(Continued)

**Table 16.1** (Continued)

| Drug | Dosage (mg/kg) | Frequency | Treatment mode | Treatment time (day post-infection) | Model worm species | Rodent host | Microfilariae (mf) reduction (%) | | Adult worms (Mf) reduction (%) | Notes | References |
|---|---|---|---|---|---|---|---|---|---|---|---|
| | | | | | | | Day 3/7 post treatment | At necropsy (from d42) | | | |
| *Alkaloids* | | | | | | | | | | | |
| Neoeudistomin analog 1d | 50 | 5, Q24h | i.p. | Not specified | *A. viteae* | *M. natalensis* | n.d. | 73 | 100 | | [15] |
| | 30 | 5, Q24h | i.p. | Not specified | *L. carinii* | *S. hispidus* | n.d. | 0 | 0 | | |
| Neoeudistomin analog 2a | 50 | 5, Q24h | i.p. | Not specified | *A. viteae* | *M. natalensis* | n.d. | 0 | 0 | | |
| | 30 | 5, Q24h | i.p. | Not specified | *L. carinii* | *S. hispidus* | n.d. | 0 | 98 | | |
| Neoeudistomin analog 2c | 50 | 5, Q24h | i.p. | Not specified | *A. viteae* | *M. natalensis* | n.d. | 33 | 0 | | |
| | 30 | 5, Q24h | i.p. | Not specified | *L. carinii* | *S. hispidus* | n.d. | 73 | 68 | | |
| *Antimonate compounds* | | | | | | | | | | | |
| Stibocaptate acid | 400 | 5, Q24h | s.c. | 90–94 | *L. carinii* | *M. natalensis* | n.a. | n.a. | 64 | | [25] |
| Stibophen | 315 | 5, Q24h | s.c. | 90–94 | *L. carinii* | *M. natalensis* | 48 | 100 | n.a. | | |
| Tartar emetic | 30 | 5, Q24h | s.c. | 90–94 | *L. carinii* | *M. natalensis* | 22 | 100 | n.a. | | |
| *Compounds from Trachyspernum ammi* | | | | | | | | | | | |
| 2-Isopropyl-5-methylphenol | 50 | 5, Q24h | i.p. | 5–6 mo old infection | *B. malayi* | *M. coucha* | 46 | 0 | 59 | | [35] |

| | | | | | | | | | | |
|---|---|---|---|---|---|---|---|---|---|---|
| *Diarylheptanoids compounds from Alnus nepalensis* | | | | | | | | | | |
| Methanolic fraction | 200 | 5, Q24h | s.c. | Trans-planted adult worms | *B. malayi* | *M. coucha* | n.d. | n.d. | 38 | | [36] |
| Chloroformic fraction | 100 | 5, Q24h | s.c. | Trans-planted adult worms | *B. malayi* | *M. coucha* | n.d. | n.d. | 51 | | |
| Butanol extract | 100 | 5, Q24h | s.c. | Trans-planted adult worms | *B. malayi* | *M. coucha* | n.d. | n.d. | 40 | | |
| Methanolic fraction | 100 | 5, Q24h | s.c. | Trans-planted adult worms | *B. malayi* | *M. unguiculatus* | n.a. | n.d. | 49 | | |
| Chloroformic fraction | 100 | 5, Q24h | s.c. | Trans-planted adult worms | *B. malayi* | *M. unguiculatus* | n.a. | n.d. | 37 | | |
| Butanol extract | 100 | 5, Q24h | s.c. | Trans-planted adult worms | *B. malayi* | *M. unguiculatus* | n.a. | n.d. | 38 | | |
| *Others* | | | | | | | | | | | |
| WR215498 | 100 | 5, Q24h | s.c. | Trans-planted adult worms | *A. viteae* | *M. unguiculatus* | n.d. | n.d. | 82 | Antifilarial | [18] |

*(Continued)*

**Table 16.1** (Continued)

| Drug | Dosage (mg/kg) | Frequency | Treatment mode | Treatment time (day post-infection) | Model worm species | Rodent host | Microfilariae (mf) reduction (%) Day 3/7 post treatment | At necropsy (from d42) | Adult worms (Mf) reduction (%) | Notes | References |
|---|---|---|---|---|---|---|---|---|---|---|---|
| | 100 | 5, Q24h | s.c. | Trans-planted adult worms | *B. pahangi* | *M. unguiculatus* | n.d. | n.d. | 97 | | |
| WR229428 | 100 | 5, Q24h | s.c. | Trans-planted adult worms | *A. viteae* | *M. unguiculatus* | n.d. | n.d. | 66 | Antifilarial | |
| | 100 | 5, Q24h | s.c. | Trans-planted adult worms | *B. pahangi* | *M. unguiculatus* | n.d. | n.d. | 100 | | |
| WR237379 | 100 | 5, Q24h | s.c. | Trans-planted adult worms | *A. viteae* | *M. unguiculatus* | n.d. | n.d. | 95 | Antifilarial | |
| | 100 | 5, Q24h | s.c. | Trans-planted adult worms | *B. pahangi* | *M. unguiculatus* | n.d. | n.d. | 52 | | |

| | | | | | | | | | | | |
|---|---|---|---|---|---|---|---|---|---|---|---|
| WR24199 | 12.5 | 5, Q24h | s.c. | Transplanted adult worms | *A. viteae* | *M. unguiculatus* | n.d. | n.d. | 95 | Anti-African *Trypanosoma* | |
| | 12.5 | 5, Q24h | s.c. | Transplanted adult worms | *B. pahangi* | *M. unguiculatus* | n.d. | n.d. | 26 | | |
| WR7396 | 25 | 5, Q24h | s.c. | Transplanted adult worms | *A. viteae* | *M. unguiculatus* | n.d. | n.d. | 95 | Antimalarial | |
| | 25 | 5, Q24h | s.c. | Transplanted adult worms | *B. pahangi* | *M. unguiculatus* | n.d. | n.d. | 32 | | |
| Bisamide | 10 | 1 | s.c. | 30 | *D. immitis* | *Rat (Sprague–Dawley IGS)* | n.d. | n.d. | 24.6 | Immunosuppressor in diet | [9] |
| Isoxazoline | 30 | 1 | s.c. | 30 | *D. immitis* | *Rat (Sprague–Dawley IGS)* | n.d. | n.d. | 28.9 | Immunosuppressor in diet | |
| *Wolbachia control* | | | | | | | | | | | |
| Doxycyclin | 50 | 14, Q24h | i.p. | 1–14 | *L. sigmodontis* | Balb/c mice | n.d. | n.d. | >99 | *Wolbachia* reduction in female worms | [37] |
| | 200 | 7, Q24h | p.o. | 1–7 | *L. sigmodontis* | Balb/c mice | n.d. | n.d. | 100 | *Wolbachia* growth inhibition | |

*(Continued)*

**Table 16.1** (Continued)

| Drug | Dosage (mg/kg) | Frequency | Treatment mode | Treatment time (day post-infection) | Model worm species | Rodent host | Microfilariae (mf) reduction (%) | | Adult worms (Mf) reduction (%) | Notes | References |
|---|---|---|---|---|---|---|---|---|---|---|---|
| | | | | | | | Day 3/7 post treatment | At necropsy (from d42) | | | |
| | 100 | 14, Q24h | p.o. | 1–14 | *L. sigmodontis* | Balb/c mice | n.d. | n.d. | 100 | *Wolbachia* growth inhibition | |
| Minocyclin | 25 | 10, Q24h | i.p. | 1–10 | *L. sigmodontis* | Balb/c mice | n.d. | n.d. | >99 | *Wolbachia* reduction in female worms | |
| | 25 | 56, Q12 | s.c. | Start 6 w postinfection | *B. malayi* | Balb/c mice SCID | n.d. | n.d. | 98 | *Wolbachia* reduction in female worms | [5] |
| Tigecyclin | 25 | 10, Q24h | i.p. | 1–10 | *L. sigmodontis* | Balb/c mice | n.d. | n.d. | >99 | *Wolbachia* reduction in female worms | [37] |
| Rifapentine | 15 | 14, Q24h | s.c. | Transplanted adult worms | *O. ochengi* | Balb/c mice SCID | n.d. | n.d. | 99 | *Wolbachia* reduction in female worms | [5] |
| Rifapentin/ Moxifloxacin | 50/200 | 7, Q24h/14, Q12h | i.p. | 1–7 | *L. sigmodontis* | Balb/c mice | n.d. | n.d. | >99 | *Wolbachia* reduction in female worms | [37] |

| | | | | | | | | | | | |
|---|---|---|---|---|---|---|---|---|---|---|---|
| Doxycyclin/ Rifapentin/ Moxifloxacin | 50/50/200 | 6/6, Q24h/12, Q12h | p.o. | 1–6 | *L. sigmodontis* | Balb/c mice | n.d. | n.d. | >99 | *Wolbachia* reduction in female worms | |
| Doxycyclin/ Rifapentin/ Moxifloxacin | 50/15/200 | 3/3, Q24h/6, Q12h | i.p. | 1–3 | *L. sigmodontis* | Balb/c mice | n.d. | n.d. | >99.9 | *Wolbachia* reduction in female worms | |
| Minocyclin/ Rifapentin/ Moxifloxacin | 50/15/200 | 4/4, Q24h/8, Q12h | p.o. | 1–4 | *L. sigmodontis* | Balb/c mice | n.d. | n.d. | >99.9 | *Wolbachia* reduction in female worms | |
| *Quinazolines* | | | | | | | | | | | |
| CBR417 | 50 | 4, Q24h | p.o. | 13 wk | *L. sigmodontis* | *M. unguiculatus* | n.d. | 100 | >99.8 | *Wolbachia* reduction in female worms; no significant reduction in adult worm numbers | [38, 39] |
| CBR417 | 50 | 7, Q24h | p.o. | 16 wk | *L. sigmodontis* | *M. unguiculatus* | n.d. | 100 | >97.7 | | |
| CBR417 | 20 | 7, Q24h | p.o. | 16 wk | *L. sigmodontis* | *M. unguiculatus* | n.d. | <90 | <90 | | |
| CBR417 | 10 | 7, Q24h | p.o. | 16 wk | *L. sigmodontis* | *M. unguiculatus* | n.d. | <90 | <90 | | |
| CBR490 | 75 | 7, Q12h | p.o. | 13 wk | *L. sigmodontis* | *M. unguiculatus* | n.d. | 100 | >99.9 | | |
| CBR490 | 25 | 7, Q12h | p.o. | 13 wk | *L. sigmodontis* | *M. unguiculatus* | n.d. | 100 | >99 | | |

*(Continued)*

**Table 16.1** (Continued)

| Drug | Dosage (mg/kg) | Frequency | Treatment mode | Treatment time (day post-infection) | Model worm species | Rodent host | Microfilariae (mf) reduction (%) | | Adult worms (Mf) reduction (%) | Notes | References |
|---|---|---|---|---|---|---|---|---|---|---|---|
| | | | | | | | Day 3/7 post treatment | At necropsy (from d42) | | | |
| CBR490 | 20 | 7, Q24h | p.o. | 16 wk | *L. sigmodontis* | *M. unguiculatus* | n.d. | <90 | <90 | | |
| CBR490 | 10 | 7, Q24h | p.o. | 16 wk | *L. sigmodontis* | *M. unguiculatus* | n.d. | <90 | <90 | | |
| CBR490 | 25 | 7, Q12h | p.o. | 16 wk | *L. sigmodontis* | *M. unguiculatus* | n.d. | 100 | >99.6 | | |
| CBR417 | 40 | 7, Q24h | p.o. | 23 wk | *B. pahangi* | *M. unguiculatus* | n.d. | >90 | 100 | *Wolbachia* reduction in female worms | [38] |
| CBR490 | 40 | 7, Q24h | p.o. | 23 wk | *B. pahangi* | *M. unguiculatus* | n.d. | <90 | >99.9 | | |
| CBR490 | 20 | 7, Q24h | p.o. | 23 wk | *B. pahangi* | *M. unguiculatus* | n.d. | >90 | >90 | | |
| CBR490 | 10 | 7, Q24h | p.o. | 23 wk | *B. pahangi* | *M. unguiculatus* | n.d. | >90 | >90 | | |

Abbreviations: i.p., intraperitoneal; i.v., intravenous; Mf, macrofilariae; mf, microfilariae; n.a., not active; n.d., not done; p.o., per oral; s.c., subcutaneous; wk, weeks.

## 16.2 Major Rodent Filariae Life Cycles

### 16.2.1 *Litomosoides sigmodontis*

*Litomosoides sigmodontis* was discovered and described by Chandler in 1931 in the cotton rat (*Sigmodon hispidus*) found in Texas, USA. It was later reclassified as *Litomosoides carinii*, a very similar filaria of the genus *Litomosoides* recovered from a squirrel, as it was considered identical. It was not until the 1980s that Odile Bain confirmed the distinct morphological characteristics of *Litomosoides* spp. from squirrels and from cotton rats. Consequently, the species now being maintained in laboratory rodents has reverted to its original name, *L. sigmodontis*. The experimental hosts are jirds, mice, albino rats, and *Mastomys* spp. The natural host, however, is the cotton rat (Figure 16.1) [41]. In cotton rats, inoculation of as few as 5 third-stage infective larvae (L3) is sufficient to establish a patent

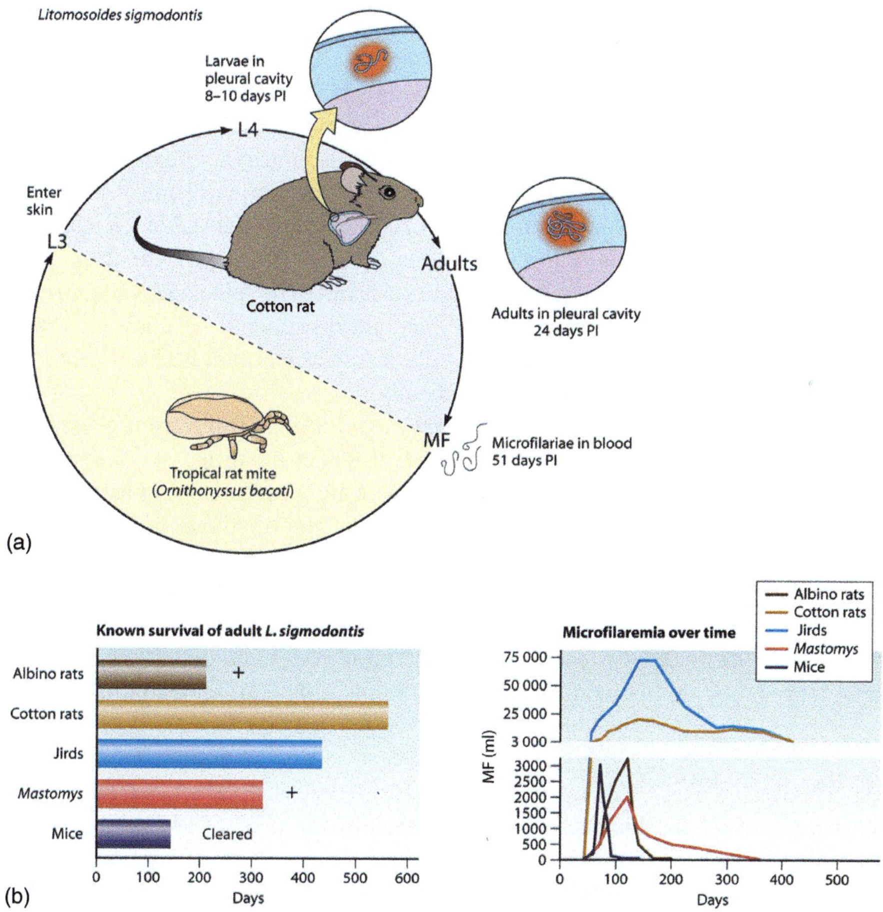


**Figure 16.1** Life cycle of *Litomosoides sigmodontis* in its natural host, the cotton rat. Source: Figure 5 from Morris et al. [41]/with permission from American Society for Microbiology.

infection [42]. After introduction by the bite of a tropical rat mite (*Ornythonyssus bacoti*), L3s travel preferentially to the pleural and pericardial cavities of the cotton rat. Experimental infections suggest that approximately 21% of inoculated larvae survive to adulthood [43]. When parasite burdens reach approximately 400 worms, the peritoneal cavity of the cotton rat also becomes parasitized [44]. Cotton rats and jirds are the preferred animals to maintain the parasitic cycle in the laboratory due to their continuously high microfilaria (mf) levels. However, for screening purposes, small rodents, especially mice, are preferred, as they are easy to breed and handle, and up-to-date immunological and genetic tools are available.

In BALB/c mice, L3s enter small-vessel lymphatics shortly after inoculation [45] and later localize preferentially to the pleural cavity by four days postinfection (p-i). A few adult worms can occasionally be found in the peritoneal cavity. Two molts occur within the pleural cavity at 8 to 12 days, and 25 to 30 days p-i, and patency commences at 50 days p-i. Adult worm numbers start to decline much earlier than in the natural host. This decline begins at around 70 days p-i, and most worms are cleared by 16 weeks p-i [46]. This is important to note, as depending on the mode of action of a drug, macrofilaricidal efficacy may not be seen in this short time window.

BALB/c, BALB/k, and BALB/b mice are transiently permissive, with BALB/c mice sustaining the longest period of microfilaremia [47]. Female BALB/c mice are more susceptible to infection than male BALB/c mice, as measured by both adult worm burden and microfilaremia, but in other strains of mice, males are more susceptible [47, 48]. Between 30% and 100% of infected BALB/c mice become microfilaremic, depending on the inoculation protocol [49, 50]. In the CBA, C3H, and DBA strains, worms develop to the adult stage but male spiculae are malformed, preventing microfilaremia [47]. All B10 mice are resistant to infection, including those with H-2d MHC [47], and 129/SvJ mice are semi-resistant [51]. BALB/c mice are susceptible and can produce patent infections.

*Litomosoides sigmodontis* is a widely used model of filarial infection in mice for discovery of direct-acting anthelmintics and anti-*Wolbachia* drugs (see Chapter 22). It further represents a spectrum of parasitological and immunological features that mimic some of those seen in human infections. Three methods of infection are commonly used: exposure to infected mites, subcutaneous inoculation of L3 larvae obtained from mite dissection, and subcutaneous inoculation of L3 larvae obtained from the pleural cavity of recently infected jirds [49, 51, 52].

Recently, this model was used to identify a number of candidates that have now been moved forward into clinical trials in humans [2, 49, 53] (see also Chapter 17).

### 16.2.2 *Acanthocheilonema viteae*

Since the early 1950s, this filarial parasite of the jird, *Meriones lybicus*, has been considered a valuable laboratory model for the study of the biology of filarial worms [54] and the search for novel antifilarial drugs. This filarial nematode was first described under the name *Dipetalonema vite* or *witei*, then *D. viteae* [55], and finally *A. viteae*.

Adult *A. viteae* mainly dwell in the subcutaneous tissues and external muscle layers of their rodent host, with preferred sites on the back, axilla, and groin [54].

The worm only successfully reproduces in a limited range of experimental hosts, such as jirds (*Meriones unguiculatus*) [56], multimammate rats (*Mastomys natalensis*) [57], and golden hamster (*Mesocricetus auratus*) [55]. *A. viteae* is among the few filarial worms lacking *Wolbachia* [58]. This particular feature should be taken into account when screening for drugs targeting adult worm reproduction and survival via *Wolbachia* inhibition.

Unsheathed *A. viteae* mf circulating in rodent blood are taken up by an Argasid tick of the *Ornithodoros* genius (Figure 16.2). In the wild, they develop into L3s in *Ornithodoros tartakovskyi* ticks that commonly live in rodent burrows [54, 59], but *Ornithodoros moubata* is most often used in the laboratory as intermediate vector host. This tick species can be easily reared and infected with *A. viteae* mf by using artificial membrane feeding. The tick infection rate can be controlled, and high numbers of L3s can be produced when needed with limited laboratory resources

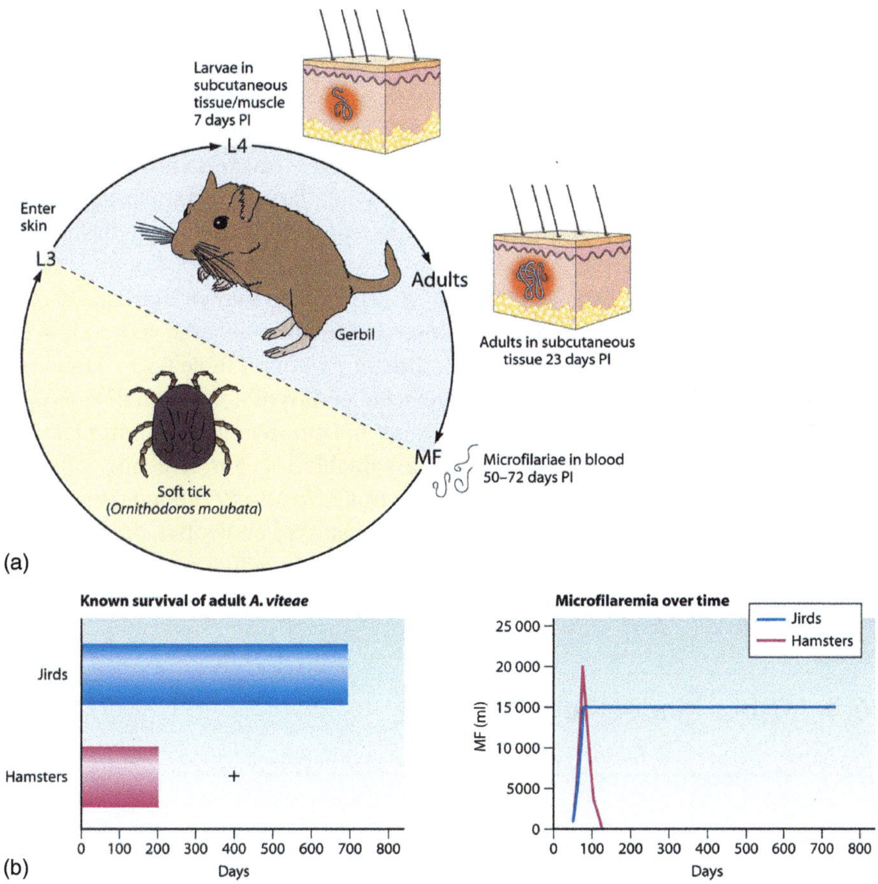


**Figure 16.2** (a) Life cycle of *Acanthocheilonema viteae* in its natural host, the jird. (b), Left: Known survival of worms after infection in jirds and hamsters (+ most likely the worms survive longer). Right: Rough outline of the course of microfilaremia over time after infection with 20 L3s in jirds or 160 L3 in hamsters. Source: Figure 1 from Morris et al. [41]/with permission from American Society for Microbiology.

[60, 61]. The mf ingested by ticks in a blood meal actively cross the wall of the tick midgut and enter muscle cells within a few days. In muscle cells, they molt to the L2 stage, acquire a functioning gut, go through a rapid growth phase, and molt to L3 [54, 55]. Mature infective larvae leave the muscle cells and accumulate in the tick body cavity. Full development takes about 27 days in ticks maintained at 27 °C [11] and 20–25 days in ticks maintained at 29 °C [59]. L3 larvae can survive for a very long time in their arthropod host, for periods of up to one year. (In the laboratory, *O. moubata* is unable to transmit the larvae directly to the rodent [61]. Infective larvae are collected from dissected ticks and injected subcutaneously into the mammalian host.) Twenty-four hours after infection, most larvae can be found in host muscles around the biting site [62]. They begin a migration phase through subcutaneous tissues and musculature and can be found in all parts of the rodent body three days later. Five to seven days postinfection, larvae moult to the L4 stage [11] and grow rapidly. They reach the L5 stage after around 23 days p-i [63], but oocyte production has already started [56]. Circulating mf appears in the jird blood seven to nine weeks p-i, and a positive microfilaremia can be observed for up to two years, providing that individuals of both sexes survive that long [41, 63]. In jirds, microfilaremia can reach very high levels (>1000 mf/μl blood) [61, 64], with no impact on host health status or lifespan. However, severe neurophysiological problems can occur between 30 and 90 days p-i when larvae migrate to the central nervous system. Ataxia and paralysis, often requiring euthanasia, have been reported [65].

*Acanthocheilonema viteae* is generally more sensitive to known filaricide classes (MLs, benzimidazoles) compared with other rodent filariae used as models for drug screening, such as *L. sigmodontis* or *Brugia pahangi* (Table 16.1). However, its longer life cycle and absence of the *Wolbachia* endosymbiont often disqualify it for use in the search for novel antifilarial drugs and compound class optimization. Nonetheless, the *A. viteae* model remains a valuable tool for screening of compounds against heartworm. Both *A. vitae* and *Dirofilaria immitis* undergo migration through host muscular tissues during L3 and L4 stages; preventive drugs against *D. immitis* target these stages, and molecules known to be active against heartworm are also active against *A. viteae* larval stages.

## 16.3 Other Models for Filariae

### 16.3.1 *Onchocerca ochengi*

*Onchocerca ochengi* is a nodule-dwelling bovine parasite closely related to *O. volvulus*. Although with limited veterinary importance, *O. ochengi* has become a natural model of human onchocerciasis due to the similar relationship it has with its host. Since its discovery, many insights into the natural history of the infection have been obtained, including into reproductive biology, pre-patency periods, and differential susceptibility to infection among cattle [40]. Selected as a tertiary drug screen by the World Health Organization (WHO), several standard drugs have been tested in this natural infection, suggesting similar drug susceptibility of *O. ochengi* and *O. volvulus*. Ivermectin (IVM) at 200 μg/kg, for example, exhibited microfilaricidal activity, but no significant macrofilaricidal activity [66]. However, the same study also revealed a lack of macrofilaricidal activity of suramin, which contrasts with the effect observed in humans [67, 68]. While one of the key attributes of this model is the ability to easily perform several *ex vivo* assays at the same time (MTT, embryogenesis, motility observation, histology) due to the abundance of nodules, certain differences need to be taken into account. These include, but are not limited to, the absence of clinical symptoms in cattle, the location of the nodules (intradermal), and the fact that there is only one female per nodule. Furthermore, infection timelines are like those in human onchocerciasis and are not feasible for rapid drug screening efforts. This model may be useful for gaining additional data for a lead candidate, but refined parameters need to be measured in humans.

### 16.3.2 *Brugia malayi* and *Brugia pahangi*

Although the domestic cat, a natural host of both *B. malayi* and *B. pahangi* [69–71], has been experimentally infected and used for the evaluation of diethylcarbamazine (DEC) [72], a suitable model for rapid screening of potential filaricides was not available until Ash and Riley [73] showed *M. unguiculatus* to be a good experimental host for *B. pahangi*. Suswillo and Denham [74] transplanted adult *B. pahangi* worms into the peritoneal cavity of jirds and used this model to screen potential antifilarial compounds. With respect to experimental infection of the *B. malayi* subperiodic strain, jirds and multimammate rats are good animal models [75, 76]. After inoculation of infective larvae, the third molt occurs within 7–8 days and the final one by 29–35 days. The prepatent period in jirds lasts 93 days, and 107 days in *Mastomys coucha*. Although these animal models harbor patent infections for periods beyond six months, they do not fully mimic human lymphatic filariasis: in rodents, despite the lack of obvious pathology, most of the worms are localized in different organs in the animals, while filarial worms are limited to the lymphatic system in human [77].

Leaf monkeys (*Presbytis* spp.) have been extensively studied to determine their suitability as hosts for lymphatic filarial parasites [78]. However, due to ethical concerns as well as costs and maintenance efforts, this model is not preferred for drug screening. This also applies to the use of dogs and cats. Nonetheless, these models may be chosen to address specific questions, especially those with a focus on pathology.

Although wild-type mice are refractory to the full developmental cycle of *Brugia* spp., each stage can survive for limited periods of time. Adult male and female worms can be implanted into the peritoneal cavity and survive for around 90 days [79]. The female worms continue to produce mf under these conditions. In addition, mf alone, injected intraperitoneally, will survive at least 28 days [80], and if injected intravenously, they will circulate in the bloodstream for approximately 65 days [81]. Although L3 of *B. malayi* (injected either intraperitoneally or subcutaneously) will not normally survive longer than 10 days [82], athymic and severe combined immunodeficiency (SCID) mice are susceptible to full infection with the development of mf-producing adults following L3 injection by either route [83]. Adult worms implanted into athymic mice will also survive longer than in intact mice.

## 16.4 Immuno-Compromised Models

A significant hurdle for studying the biology of parasites and for the development of new therapeutics and diagnostics has been the absence of predictive small animal models. One reason is the potential host's ability to prevent infection by non-coevolved species, making it a challenge to introduce filarial parasites of human and veterinary importance into small rodents. To overcome this obstacle, immunodeficient mouse strains that have been extremely useful for immunology, infectious diseases, cancer, stem cell biology, and other research areas have also enabled the introduction of parasites into otherwise resistant hosts. Several immunodeficient mouse strains are available, of which those with SCID [84] are the most widely used in filarial research (Table 16.2). In the absence of T and B cells, SCID mice are fully susceptible to *B. malayi* [86]. Within 6–10 weeks after injection of infective Bm L3 larvae, both male and female worms were found in 90% of the mice and peripheral mf were detected [91]. Studies comparing the efficacy of standard anthelmintics in immunodeficient mice, however, are scarce (Table 16.1). Results show strong larvicidal activity of albendazole (ABZ) and DEC, and partial activity for IVM, in both wild-type (WT) and SCID mice. To further assess responses of mf to these drugs, *B. malayi* mf were inoculated via the tail vein into SCID mice or respective controls. As opposed to the experimental setup *in vitro*, IVM induced

**Table 16.2** Transgenic models in rodents.

| Worm species | Parasitic stage used | Parasite survival | References |
|---|---|---|---|
| *O. volvulus* | Nodule implant | Survival >20 wk | [85] |
| *O. ochengi* | Nodule implant | | [5] |
| *B. malayi* | L3 injection s.c. | Full patency | [86] |
| *O. volvulus* | L3 injection s.c. | L4 (not investigated beyond 8 wk) | [87] |
| *O. volvulus* | L3 injection s.c. | L4 (not investigated beyond 8 wk) | |
| *O. volvulus* | L3 injection s.c. | L4 (not investigated beyond 8 wk) | |
| *O. volvulus* | L3 injection s.c. | L4 (not investigated beyond 8 wk) | |
| *Loa loa* | L3 injection s.c. | Young adults (70 d) | [88] |
| *B. malayi* | L3 injection i.p | 12 wk | [89] |
| *O. ochengi* | implant into pleural cavity | 38 d | |
| *L. loa* | L3 injection s.c. | Full patency | [90] |

Abbreviations: s.c, subcutaneous; i.p., intraperitoneal; wk, weeks.

rapid reductions, suggesting involvement of the host immune system in mf removal. As SCID mice are devoid of T and B cells, other mechanisms, as yet unidentified, play a role in rapid mf removal.

A first attempt to maintain the human infectious species *O. volvulus* outside of a human or primate host was made by implanting onchocercomata (skin nodules containing adult worms) subcutaneously into SCID mice. After up to 20 weeks in the mouse, onchocercomata contained both viable adult worms and mf [85]. This model has been recently used by Halliday [5] with *O. ochengi* nodules extracted from cattle. Being the closest species to *O. volvulus*, this could be an alternative to implantation of *O. volvulus* nodules. However, the functionality of nodule implantation as a screening model remains to be validated. It is unknown how parasites in nodules are affected by living in a foreign environment, whether nodules will be fully re-vascularized, and how rejection mechanisms by the host impact viability of the worms.

Recently, a highly immune-compromised mouse strain that has profound defects in adaptive and innate immune responses has entered the scene. NSG mice can be replenished with human immune cells to build humanized mouse models. Interestingly, NSG mice can support the complete life cycle of the human nematode *Strongyloides stercoralis*, whereas immunologically competent mice cannot. NSG and humanized NSG mice were infected with 100 *O. volvulus* infective larvae (L3). In each of the different humanized mouse models, worms advanced to the L4 stage and there was a trend for higher parasite recovery in mice with cellular engraftment compared to unengrafted NSG mice, albeit it was not statistically significant. Although this model is not yet used for drug discovery, it is a helpful tool to identify parasite-derived biomarkers measurable in both urine and serum [87]. Similar approaches are being investigated for heartworm research. A model using caninized SCID mice bearing *D. immitis* worms in the peritoneal cavity has been recently patented [92]. Whether these models can be established for high-throughput screening remains to be evaluated.

Drugs that are developed against onchocerciasis will eventually have to be counter-screened against *Loa loa* because of safety concerns about microfilaricidal drugs in co-endemic areas. Development of a fully patent infection was established in immunodeficient lymphopenic mice lacking the common gamma chain ($\gamma$c) cytokine signaling pathway, providing an alternative to splenectomised non-human primates. In these mice, a single dose of IVM, but not benzimidazoles, induced rapid mf clearance >90% [90], allowing this model to be used as a counter screen in human filarial drug development.

Rodents immunosuppressed by administration of glucocorticoids incorporated in their diet, already described in the case of gastrointestinal nematodes models in jirds and rats [93, 94], could be an interesting alternative to the use of transgenic immuno-compromised animals. A recently published patent application [9] describes the development to full patency of *D. immitis* in hydrocortisone-fed rats. In this model, all standard endoparasitic drugs were successful, opening new perspectives in drug screening against heartworm (Table 16.1).

## 16.5 Models for Human Filariae

The trickiest question in drug development for filarial parasites has always been: which is the best model to use to prioritize and select drug candidates by accurately predicting efficacy in humans? The simple answer is that there is not one. The choice of model needs to be guided by the programmatic question being addressed, for example, breaking of infection transmission or identifying and understanding the direct effect of a chemotherapeutic agent on the worm's reproductive capability and/or its viability. The complexity of the nematode life cycle requires, therefore, that attention be given not only to species/host combinations and larval stages but also to the anatomical location and the presence or absence of the immune system. Therefore, results from model infections always need to be interpreted with caution.

IVM, DEC, and the recently registered moxidectin are potent inhibitors of microfilaremia. In the past, IVM and DEC have been excellent drugs to address the most important issue: elimination of filarial diseases as a public health problem. These drugs have been remarkably successful in reducing transmission and clinical symptoms. However, it is important to note that these drugs would have been missed if a stringent drug screening cascade had been applied with *in vitro* prior to *in vivo* testing. This sort of disconnect unfortunately hinders the setup of a straightforward "one-fits-all" screening cascade.

As mass drug administration with IVM has successfully reduced infection intensity over the last two decades, it is currently debated whether additional tools should be developed, and if so, which. Without doubt, a drug that either kills the adult worms or permanently sterilizes females would be a useful tool for tackling the remaining filarial infections on the road to elimination.

For drugs acting via removal of *Wolbachia*, the assessment is relatively easy, since *Wolbachia* measurement is a reliable surrogate marker, leading to long-term inhibition of embryogenesis and ultimately a slow death of adult parasites (see Chapter 24), as shown in studies in humans treated with long-term regimens of doxycycline.

The situation is more complex for direct-acting drugs. Drug candidates with a clear effect across the models may act on mf directly and/or attack adult worms quickly. The amount of antigen released by large numbers of dying worms may lead to development of adverse events, and thus there is a cost to safety or feasibility. Other drugs may act more subtly and only show effects in a few host/parasite combinations, but eventually with the same final outcome, i.e., death of the parasite. This requires careful investigation of the host/parasite environment chosen, so that potentially slowly acting but highly effective drugs are not missed in the development cascade. The investigator must be guided by the totality of the data when moving a potential drug forward. Ultimately, only parasite-infected humans contain all components needed to evaluate the efficacy of a drug.

PK/PD modeling is a helpful tool to predict the human efficacious dose, and *in vivo* models have progressed for this purpose over recent years. These should also be applied to drug development for human filarial infections. However, none of the models reflects all the necessary parameters that are present in human infections. Whereas *L. sigmodontis* has been developed as a high-throughput model for drug discovery in onchocerciasis, the parasites are in the pleural cavity, rather than in subcutaneous nodules. *O. ochengi* is the closest relative to *O. volvulus* with many similarities to it, but differences between them make confident predictions rather difficult. At best, drug candidates that have shown efficacy in more than one model and have been proven to be safe in humans should be tested for proof-of-concept prior any final selection is made. Only then can a fully informed decision be taken whether or not this compound should be taken further along the development pipeline.

## 16.6 Models for Heartworm

The discovery of IVM in the late 1970s, with its exceptional spectrum of activity against nematodes, including heartworm [95, 96], opened an era of effective heartworm prevention. Other MLs (milbemycin oxime, moxidectin, and selamectin) with different modes of administration or protection periods were later developed. For a long time, the effectiveness of MLs at preventing the development of *D. immitis* in dogs and cats, along with the difficulty of the dog model itself as an experimental system (involving long and expensive studies), did not motivate animal health companies to specifically look for new anti-heartworm drugs with alternative modes of action. The situation changed in the last decade with the emergence of *D. immitis* strains resistant to MLs in the Mississippi basin [97]. The development of tools for screening new chemical entities in an *in vitro* high-throughput format (see Chapter 15) and the establishment of complete screening streams with *in vivo* rodent models and dog heartworm are recent, and little has been published. Finding the right rodent model that mimics the fate of heartworm in dogs is challenging. There is no rodent filarial worm with a life cycle resembling that of *D. immitis* in dogs. Models in ferrets exist [96] but are not suitable for a discovery screening platform when the amount of compound available is limited. Ferrets are not necessary easier to maintain than dogs, and *D. immitis* life cycle is not shorter in these animals. Recently, Mills et al. [9] were able to obtain a *D. immitis* patent infection in male Sprague–Dawley rats fed on a glucocorticoid diet. The treatment started with 200 ppm hydrocortisone acetate eight days prior infection with 50 *D. immitis* larvae. The immunosuppressive treatment was pursued for another 13 days and then reduced to 50 ppm until necropsy that took place 112–120 days p-i. Infection rates in rats were close to 100%, and an average of 5–7 fully mature worms were recovered from the heart and pulmonary arteries. No particular mortality was observed in rats due to the hydrocortisone treatment or the filarial worms. The experiment was extended without increase in rat mortality up to 260 days p-i to obtain circulating mf. Standard preventive antifilarials (MLs, but also emodepside) were active in the rat model after treatment 30 days p-i (Table 16.1). This new rat model, although allowing efficacy assessment one month earlier than with dogs or ferrets, is still taking about 17 weeks. These timelines remain suboptimal when thinking about lead chemical optimization and understanding of structure activity relationship. Nonetheless, it clearly opens new perspectives for the search of novel heartworm preventives.

Other filarial species have also been used as surrogates to *D. immitis* for the identification of new chemical entities potentially active against heartworm. Gauvry and her team [26–29] described the use of a model with *A. viteae* in male jirds for chemical optimization. Rodents are infected by subcutaneous injection and treated by oral gavage daily between days 5 and 9 p-i, targeting L3 and early L4 migrating stages. This migration phase is also present in *D. immitis* L3 and L4 stages in a similar compartment, giving some hope that drugs positive in the jird model may also show activity against heartworm in dogs. Major antifilarials, such as MLs and benzimidazoles, have indeed been found to be positive in *A. viteae* rodent models at very low doses (Table 16.1). Like the rat model cited earlier, the model described by Gauvry et al. takes time. Necropsy is performed only after 84 days, and only animals with clearly reduced microfilaremia are selected for complete dissection and counting of adult worms. Here again, the model duration may be a significant hurdle for chemical optimization, since clear structure–activity relationships can only be established every 12 weeks. Parallel hypotheses need to be pursued to optimize the process.

*Litomosoides sigmodontis* is another model of choice because of its shorter life cycle and the ease of larval or adult recovery from the pleural cavity [50]. The worm is typically less sensitive to anthelmintics than *A. viteae* but detects the known anthelmintic classes (Table 16.1). However, little has been disclosed about how representative this model is for heartworm in dogs.

Another option followed by Köhler et al. [40] was to perform the *in vitro* screening step against *D. immitis* microfilariae and L3–L4 larvae and then test the best compounds in jirds infected with *H. contortus* and *T. colubriformis* gastrointestinal nematodes to generate an *in vivo* proof-of-concept.

A model using caninized SCID mice that permit development of *D. immitis* in the peritoneal cavity up to the adult stage has been recently described [92]. The advantages of having the right worm species in a rodent might be limited by the different final location of the worm and the use of genetically transformed animals that lack effective immune responses.

The future will tell whether one of these models can be more successful than the others at identifying novel chemical entities as antifilarials. The pharmacokinetic behavior of a drug is critical for efficacy against worms that migrate through the body of their host for weeks before reaching the target organ. There is a risk that using rodent models for chemical optimization will result in identifying the best drug for the rodent but not necessarily for the target host. As is the case for research into drugs for the elimination of human filariae, research into drugs to prevent heartworm requires collection of information from different sources, *in vitro* and *in vivo* in different models, and should include, at a very early stage, some PK/PD evaluation in the dog. Although the ideal PK profile requirements for a drug to be successful at controlling heartworm are still unclear, in addition to data from models, data generated in the target host are essential to taking informed decisions and selecting the most promising candidates for efficacy testing in the long and tedious dog/heartworm model.

## 16.7 Conclusions

It is important to note that, in the context of drug development, simple numerical quantification is used to assess the most commonly used parameters. With this approach, there is a risk that potentially useful drugs with particular characteristics may be discarded. Therefore, a more detailed assessment is warranted that involves functional (biochemistry, fertility) and histological parameters. We have seen in this chapter that no model is ideal and that the accumulation of information (activity *in vitro* against different worm stages, activity in different *in vivo* models, pharmacokinetic parameters in the final hosts, etc.) may be needed to identify novel and effective antifilarials.

To understand the irreversible points of parasite degeneration induced by new drugs, it is necessary to investigate changes at the level of the worm itself and to evaluate drug-induced pathological events. For example, depletion of *Wolbachia* with doxycycline results in long-term inhibition of embryogenesis and eventually killing of the adult parasites. Whereas this long-term effect has been described in humans, surrogate models, such as *L. sigmodontis*, are too short-lived for it to be observed. But since the results of *Wolbachia* depletion are very well described (see Chapter 24), candidate selection for anti-wolbachial drugs is relatively straightforward. It becomes more complicated for direct acting drugs, and particularly for the discovery of novel classes of antifilarials for which chemical optimization is required. With new technical developments, it may become feasible to implement new methods to reduce the use of animals, refine endpoints, and bypass invasive techniques for assessing antifilarial drug efficacy. The evaluation of filarial parasites and responses to drug exposure still require improvement and harmonization.

## Acknowledgments

The authors thank Dr. François Pautrat for his support in the patent search, and Dr. Heinz Sager and François Pautrat for kindly providing efficacy data of MLs in the *A. viteae* model in jirds.

## References

**1** Zahner, H., Sänger, I., Lämmler, G., and Müller, H.A. (1987). Effect of ivermectin in *Dipetalonema viteae* and *Litomosoides carinii* infections of *Mastomys natalensis*. *Trop. Med. Parasitol.* 38: 117–122.

**2** Zahner, H. and Schares, G. (1993). Experimental chemotherapy of filariasis: comparative evaluation of the efficacy of filaricidal compounds in *Mastomys coucha* infected with *Litomosoides carinii*, *Acanthocheilonema viteae*, *Brugia malayi* and *B. pahangi*. *Acta Trop.* 52: 221–266.

**3** Schares, G., Hofmann, B., and Zahner, H. (1994). Antifilarial activity of macrocyclic lactones: comparative studies with ivermectin doramectin, milbemycin A4 oxime, and moxidectin in *Litomosoides carinii*, *Acanthocheilonema viteae*, *Brugia malayi*, and *B. pahangi* infection in *Mastomys coucha*. *Trop. Med. Parasitol.* 45: 97–106.

**4** Sjoberg, A., Pionnier, N., Alijayyoussi, G. et al. (2018). Short-course oral flubendazole does not mediate significant efficacy against *Onchocerca* adult male worms of *Brugia* microfilariae in murine infection models. *PLoS Negl.Trop. Dis.* 13: e0006356.

**5** Halliday, A., Guimaraes, A.F., Tyrer, H.E. et al. (2014). A murine macrofilaricide pre-clinical screening model for onchocerciasis and lymphatic filariasis. *Parasit. Vectors* 7: 472.

**6** Devaney, E., Howells, R.E., and Smith, G. (1985). *Brugia pahangi* in the BALB/c mouse: a model for testing filaricidal compounds. *J. Helminthol.* 59: 95–99.

**7** Duarte, Z., Gantier, J.C., and Gayral, P. (1994). Activity of albendazole-ivermectin combination and other filaricidal drugs against infective larvae, preadult, microfilariae and adult worms of *Molinema dessetae* in the rodent *Proechimys oris*. *Parasite* 1: 57–64.

**8** Mbah, G.E., Ayiseh, R.B., and Cho-Ngwa, F. (2016). Development and validation of an *Onchocerca ochengi* microfilarial hamster model for onchocerciasis drugs screens. *BMC Infect. Dis.* 16: 404.

**9** Mills, B.J., McTier, T.L., Knauer, C.S., Woods, D.J. (2020) Anthelmintic laboratory animal model for heartworm. *International Patent Application* PCT/US2020/012523.

**10** Verma, M., Pathak, M., Shahab, M. et al. (2014). Moxidectin causes adult worm mortality of human lymphatic filarial parasite *Brugia malayi* in rodent models. *Folia Parasitol.* 61: 561–570.

**11** Hartmann, S. (1979). Zur Chemoprophylaxe der Filariose: Experimentelle Untersuchungen and der *Dipetalonema viteae*-Infektion der *Mastomys natalensis*. PhD Thesis. Justus-Liebig University. Giessen, Germany.

**12** Agarwal, A., Awasthi, S.K., and Murthy, P.K. (2011). *In vivo* antifilarial activity of some cyclic and acyclic alcohols. *Med. Chem. Res.* 20: 430–434.

**13** Fujimaki, Y., Sithithaworn, P., Mitsui, Y., and Aoki, Y. (2004). Delayed macrofilaricidal activity of diethylcarbamazine against *Brugia pahangi* in Mongolian jirds. *J. Helminthol.* 78: 293–295.

**14** Lakshmi, V., Joseph, S.K., Srivastava, S. et al. (2010). Antifilarial activity *in vitro* and *in vivo* of some flavonoids tested against *Brugia malayi*. *Acta Trop.* 116: 127–133.

**15** Agarwal, A., Agarwal, S., Singh, S.N. et al. (1995). Synthesis of neoeudistomin analogs – as potential filaricides. *Bioorg. Med. Chem. Lett.* 5: 1545–1548.

**16** Lämmler, G., Herzog, H., Saupe, E., and Schütze, H.R. (1971). Chemotherapeutic studies on *Litomosoides carinii* infection of *Mastomys natalensis*. 1. The filaricidal action of 2,6-benzimidazoles. *Bull. WHO* 44: 751–756.

**17** Duarte, Z., Gantier, J.C., and Gayral, P. (1994). Macrofilaricidal activity of albendazole-ivermectin combination: histopathological evaluation of adult *Molinema dessetae*. *Trop. Med. Parasitol.* 45: 209–213.

**18** Kinnamon, K.E., Klayman, D.L., Poon, B.T. et al. (1994). Filariasis testing in a jird model: new drug leads from some old standbys. *Am. J. Trop. Med. Hyg.* 51: 791–796.

**19** Court, J.P., Tables, J.N., Lees, G.M. et al. (1988). *Dipetalonema viteae* and *Brugia pahangi* transplant infections in gerbils for use in antifilarial screening. *J. Helminthol.* 62: 1–9.

**20** Stables, J.N., Lees, G.M., and Rankin, R. (1988). The potential of mice as animal models for antifilarial screening. *Trop. Med. Parasitol.* 39: 25–28.

**21** Hübner, M.P., Ehrens, A., Koschel, M. et al. (2019). Macrofilaricidal efficacy of single and repeated oral and subcutaneous doses of flubendazole in *Litomosoides sigmodontis* infected jirds. *PLoS Neg. Trop. Dis.* 13: e0006320.

**22** Lämmler, G., Herzog, H., and Schütze, H.R. (1971). Chemotherapeutic studies on *Litomosoides carinii* infection of *Mastomys natalensis*. 2. The activity of drugs against microfilariae. *Bull. WHO* 44: 757–763.

**23** Zahner, H., Taubert, A., Harder, A., and von Samson-Himmelstjerna, G. (2001). Filaricidal efficacy of anthelmintically active cyclodepsipeptides. *Int. J. Parasitol.* 31: 1515–1522.

**24** Zahner, H., Taubert, A., Harder, A., and von Sanson-Himmelstjerna, G. (2001). Effects of Bay 44-4400, a new cyclodepsipeptide, on developing stages of filariae (*Acanthocheilonema viteae, Brugia malayi, Litomosoides sigmodontis*) in the rodent *Mastomys coucha*. *Acta Trop.* 80: 19–28.

**25** Lämmler, G., Herzog, H., and Schütze, H.R. (1971). Chemotherapeutic studies on *Litomosoides carinii* infection of *Mastomys natalensis*. 3. The activity of drugs against adult parasites. *Bull. WHO* 44: 765–770.

**26** Gauvry, N., Pautra, F. (2014). Substituted azines as pesticides. *International Patent Publication* WO 2014/023723 A1.

**27** Gauvry, N., Pautrat, F. (2015). New compounds. *International Patent Publication* WO 2015/071417 A1.

**28** Gauvry, N., Pautrat, F., Perret, J-L., Tahtaoui, C. (2015). New sulfonylaminobenzamide compounds. *International Patent Publication* WO 2015/195423 A1.

**29** Gauvry, N., Tahtaoui, C. (2016). Sulfonylaminobenzamide compounds as anthelmintics. *International Patent Publication* WO 2016/033341 A1.

**30** Zahner, H., Striebel, H.P., Schütze, H.R. et al. (1988). Antifilarial activities of benzazole derivatives. 1. Macrofilaricidal effects against *Litomosoides carinii, Dipetalonema viteae, Brugia malayi* and *B. pahangi* in *Mastomys natalensis*. *Trop. Med. Parasitol.* 39: 14–18.

**31** Zahner, H., Striebel, H.P., Schütze, H.R. et al. (1988). Antifilarial activities of benzazole derivatives. 2. Microfilaricidal effects against *Litomosoides carinii, Acanthocheilonema viteae, Brugia malayi* and *B. pahangi* in *Mastomys natalensis*. *Trop. Med. Parasitol.* 39: 284–290.

**32** Zahner, H., Striebel, H.P., Sänger, I., and Schütze, H.R. (1990). Antifilarial activities of benzazole derivatives. 3 effects of benzothiazoles on third stage larvae and preadult worms of *Acanthocheilonema viteae, Brugia malayi* and *B. pahangi* in *Mastomys natalensis. Trop. Med. Parasitol.* 41: 407–410.

**33** Srivastava, S.K., Chauhan, P.M.S., Bhaduri, A.P. et al. (2000). Quinolones: novel probes in antifilarial chemotherapy. *J. Med. Chem.* 43: 2275–2279.

**34** Kinnamon, K.E., McCall, J.W., Engle, R.R. et al. (1994). A new class of antifilariasis compounds: a preliminary look. *Mil. Med.* 159: 368–372.

**35** Mathew, N., Misra-Bhattacharya, S., Perumal, V., and Muthuswamy, K. (2008). Antifilarial lead molecules isolated from *Trachyspermum ammi. Molecules* 13: 2156–2168.

**36** Yadav, D., Kushwaha, V., Saxena, K. et al. (2013). Diarylheptanoid compounds from *Alnus nepalensis* expressed *in vitro* and *in vivo* antifilarial activity. *Acta Trop.* 128: 509–517.

**37** Specht, S., Pfarr, K.M., Arriens, S. et al. (2018). Combinations of registered drugs reduce treatment times required to deplete *Wolbachia* in the *Litomosoides sigmodontis* mouse model. *PLoS Negl.Trop. Dis.* 12: e0006116.

**38** Hübner, M.P., Gunderson, E., Vogel, I. et al. (2020). Short-course quinazoline drug treatments are effective in the *Litomosoides sigmodontis* and *Brugia pahangi* jird models. *Int. J. Parasitol. Drugs Drug Res.* 12: 18–27.

**39** Bakowski, M.A., Shiroodi, R.K., Lui, R. et al. (2019). Discovery of short-course antiwolbachial quinazolines for elimination of filarial worm infections. *Sci. Transl. Med.* 11: 491.

**40** Köhler, A., Welz, C., Börngen, K., Ilg, T. et al. (2017). Pyrazolopyrimidine derivatives. *International patent publication* WO 2017/178416 A1.

**41** Morris, C.P., Evans, H., Larsen, S.E., and Mitre, E. (2013). A comprehensive, model-based review of vaccine and repeat infection trials for filariasis. *Clin. Microbiol. Rev.* 26: 381–421.

**42** Wenk, P. and Mossinger, J. (1991). Recovery of adult stages and microfilaraemia after low dose inoculation of third stage larvae of *Litomosoides carinii* in *Sigmodon hispidus. J. Helminthol.* 65: 219–225.

**43** Jaquet, C. (1980) Comparison of the infection in relation to the immune response. PhD Thesis. Department of Entomology, London School of Hygiene and Tropical Medicine, London, England. Neuchatel, CH: 113.

**44** Bertram, D.S. (1966). Dynamics of parasitic equilibrium in cotton rat filariasis. *Adv. Parasitol.* 4: 255–319.

**45** Bain, O., Wanji, S., Vuong, P.N. et al. (1994). Larval biology of six filariae of the sub-family Onchocercinae in a vertebrate host. *Parasite* 1: 241–254.

**46** Marechal, P., Le Goff, L., Hoffman, W. et al. (1997). Immune response to the filaria *Litomosoides sigmodontis* in susceptible and resistant mice. *Parasite Immunol.* 19: 273–279.

**47** Petit, G., Diagne, M., Marechal, P. et al. (1992). Maturation of the filaria *Litomosoides sigmodontis* in BALB/c mice; comparative susceptibility of nine other inbred strains. *Ann. Parasitol. Hum. Comp.* 67: 144–150.

**48** Graham, A.L., Taylor, M.D., Le Goff, L. et al. (2005). Quantitative appraisal of murine filariasis confirms host strain differences but reveals that BALB/c females are more susceptible than males to *Litomosoides sigmodontis*. *Microbes Infect.* 7: 612–618.

**49** Hübner, M.P., Torrero, M.N., McCall, J.W., and Mitre, E. (2009). *Litomosoides sigmodontis*: a simple method to infect mice with L3 larvae obtained from the pleural space of recently infected jirds (*Meriones unguiculatus*). *Exp. Parasitol.* 123: 95–98.

**50** Marechal, P., Le Goff, L., Petit, G. et al. (1996). The fate of the filaria *Litomosoides sigmodontis* in susceptible and naturally resistant mice. *Parasite* 3: 25–31.

**51** Specht, S., Saeftel, M., Arndt, M. et al. (2006). Lack of eosinophil peroxidase or major basic protein impairs defense against murine filarial infection. *Infect. Immun.* 74: 5236–5243.

**52** Al-Qaoud, K.M., Fleischer, B., and Hoerauf, A. (1998). The Xid defect imparts susceptibility to experimental murine filariosis--association with a lack of antibody and IL-10 production by B cells in response to phosphorylcholine. *Int. Immunol.* 10: 17–25.

**53** Taylor, M.J., von Geldern, T.W., Ford, L. et al. (2019). Preclinical development of an oral anti-*Wolbachia* macrolide drug for the treatment of lymphatic filariasis and onchocerciasis. *Sci. Transl. Med.* 11: eaau2086.

**54** Terry, A., Tery, R.J., and Worms, M.J. (1961). *Dipetalonema witei*, filarial parasite of the jird, *Meriones libycus*. II. The reproductive system, gametogenesis and development of the microfilaria. *J. Parasitol.* 47: 703–711.

**55** Bain, O. (1967). Larval biology and mechanism of transmission of the filarian *Dipetalonema viteae*. *Ann. Parasitol. Hum. Comp.* 42: 211–267.

**56** Johnson, M.H., Orihel, T.C., and Beaver, P.C. (1974). *Dipetalonema viteae* in the experimentally infected jird, *Meriones unguiculatus*. I. Insemination, development from egg to microfilaria, reinsemination, and longevity of mated and unmated worms. *J. Parasitol.* 60: 302–309.

**57** Sanger, I. and Lammler, G. (1979). On *Dipetalonema viteae* infection of *Mastomys natalensis*. *Tropenmed. Parasitol.* 30: 81–87.

**58** Bandi, C., Anderson, T.J., Genchi, C., and Blaxter, M.L. (1998). Phylogeny of *Wolbachia* in filarial nematodes. *Proc. Biol. Sci.* 265: 2407–2413.

**59** Hawking, F. and Worms, M. (1961). Transmission of filarioid nematodes. *Annu. Rev. Entomol.* 6: 413–432.

**60** Wirtz, H.P. and Raybould, J.N. (1986). Artificial feeding of West African *Simulium damnosum* Theobald s.l. (Diptera: Simuliidae) through membranes and their subsequent fecundity. *Trop. Med. Parasitol.* 37: 143–148.

**61** Lucius, R. and Textor, G. (1995). *Acanthocheilonema viteae*: rational design of the life cycle to increase production of parasite material using less experimental animals. *Appl. Parasitol.* 36: 22–33.

**62** Eisenbeiss, W.F., Apfel, H., and Meyer, T.F. (1991). Recovery, distribution, and development of *Acanthocheilonema viteae* third- and early fourth-stage larvae in adult jirds. *J. Parasitol.* 77: 580–586.

63 Beaver, P.C., Orihel, T.C., and Johnson, M.H. (1974). *Dipetalonema viteae* in the experimentally infected jird, *Meriones unguiculatus*. II. Microfilaremia in relation to worm burden. *J. Parasitol.* 60: 310–315.

64 Schrempf-Eppstein, B., Kern, A., Textor, G., and Lucius, R. (1997). *Acanthocheilonema viteae*: vaccination with irradiated L3 induces resistance in three species of rodents (*Meriones unguiculatus*, *Mastomys coucha*, *Mesocricetus auratus*). *Trop. Med. Int. Health* 2: 104–110.

65 Weinstein, P.P. and Highman, B. (1974). Infection of the jird, *Meriones unguiculatus*, with the filarial worm, *Dipetalonema viteae*: central nervous system invasion and pathology. *J. Parasitol.* 60: 138–148.

66 Renz, A., Trees, A.J., Achu-Kwi, D. et al. (1995). Evaluation of suramin, ivermectin and CGP 20376 in a new macrofilaricidal drug screen, *Onchocerca ochengi* in African cattle. *Trop. Med. Parasitol.* 46: 31–37.

67 Duke, B.O. (1968). The effects of drugs on *Onchocerca volvulus*. 3. Trials of suramin at different dosages and a comparison of the brands Antrypol, Moranyl and Naganol. *Bull. World Health Organ.* 39: 157–167.

68 Awadzi, K., Hero, M., Opoku, N.O. et al. (1995). The chemotherapy of onchocerciasis XVIII. Aspects of treatment with suramin. *Trop. Med. Parasitol.* 46: 19–26.

69 Buckley, J.J. and Edeson, J.F. (1956). On the adult morphology of *Wuchereria* sp. (*malayi*?) from a monkey (*Macaca irus*) and from cats in Malaya, and on *Wuchereria pahangi* n.sp. from a dog and a cat. *J. Helminthol.* 30: 1–20.

70 Edeson, J.F. and Wharton, R.H. (1957). The transmission of *Wuchereria malayi* from man to the domestic cat. *Trans. R. Soc. Trop. Med. Hyg.* 51: 366–370.

71 Edeson, J.F. (1959). Studies on filariasis in Malaya: the periodicity of the microfilariae of *Brugia malayi* and *B. pahangi* in animals. *Ann. Trop. Med. Parasitol.* 53: 381–387.

72 Edeson, J.F. and Laing, A.B. (1959). Studies on filariasis in Malaya: the effect of diethylcarbamazine on *Brugia malayi* and *B. pahangi* in domestic cats. *Ann. Trop. Med. Parasitol.* 53: 394–399.

73 Ash, L.R. and Riley, J.M. (1970). Development of *Brugia pahangi* in the jird, *Meriones unguiculatus*, with notes on infections in other rodents. *J. Parasitol.* 56: 962–968.

74 Suswillo, R.R. and Denham, D.A. (1977). A new system of testing for filaricidal activity using transplanted adult brugia in the jird. *J. Parasitol.* 63: 591–592.

75 Ash, L.R. and Riley, J.M. (1970). Development of subperiodic *Brugia malayi* in the jird, *Meriones unguiculatus*, with notes on infections in other rodents. *J. Parasitol.* 56: 969–973.

76 Sanger, I., Lammler, G., and Kimmig, P. (1981). Filarial infections of *Mastomys natalensis* and their relevance for experimental chemotherapy. *Acta Trop.* 38: 277–288.

77 Murthy, P.K., Tyagi, K., Roy Chowdhury, T.K., and Sen, A.B. (1983). Susceptibility of *Mastomys natalensis* (GRA strain) to a subperiodic strain of human *Brugia malayi*. *Indian. J. Med. Res.* 77: 623–630.

**78** Misra, S., Tyagi, K., and Chatterjee, R.K. (1997). Experimental transmission of nocturnally periodic *Wuchereria bancrofti* to Indian leaf monkey (*Presbytis entellus*). *Exp. Parasitol.* 86: 155–157.

**79** Suswillo, R.R., Owen, D.G., and Denham, D.A. (1980). Infections of *Brugia pahangi* in conventional and nude (athymic) mice. *Acta Trop.* 37: 327–335.

**80** Rajasekariah, G.R., Puri, P.M., Chandrashekar, R., and Subrahmanyam, D. (1988). Clearance of *Brugia pahangi* microfilariae in immunized mice. *Immunol. Cell. Biol.* 66 (Pt 5–6): 331–336.

**81** Fanning, M.M. and Kazura, J.W. (1983). Genetic association of murine susceptibility to *Brugia malayi* microfilaraemia. *Parasite Immunol.* 5: 305–316.

**82** Carlow, C.K. and Philipp, M. (1987). Protective immunity to *Brugia malayi* larvae in BALB/c mice: potential of this model for the identification of protective antigens. *Am. J. Trop. Med. Hyg.* 37: 597–604.

**83** Vincent, A.L., Sodeman, W.A., and Winters, A. (1980). Development of *Brugia pahangi* in normal and nude mice. *J. Parasitol.* 66: 448.

**84** Bosma, M.J. and Carroll, A.M. (1991). The SCID mouse mutant: definition, characterization, and potential uses. *Annu. Rev. Immunol.* 9: 323–350.

**85** Rajan, T.V., Nelson, F.K., Cupp, E. et al. (1992). Survival of *Onchocerca volvulus* in nodules implanted in immunodeficient rodents. *J. Parasitol.* 78: 160–163.

**86** Nelson, F.K., Greiner, D.L., Shultz, L.D., and Rajan, T.V. (1991). The immunodeficient scid mouse as a model for human lymphatic filariasis. *J. Exp. Med.* 173: 659–663.

**87** Patton, J.B., Bennuru, S., Eberhard, M.L. et al. (2018). Development of *Onchocerca volvulus* in humanized NSG mice and detection of parasite biomarkers in urine and serum. *PLoS Negl.Trop. Dis.* 12: e0006977.

**88** Tendongfor, N., Wanji, S., Ngwa, J.C. et al. (2012). The human parasite *Loa loa* in cytokine and cytokine receptor gene knock out BALB/c mice: survival, development and localization. *Parasit. Vectors* 5: 43.

**89** Aljayyoussi, G., Tyrer, H.E., Ford, L. et al. (2017). Short-course, high-dose rifampicin achieves *Wolbachia* depletion predictive of curative outcomes in preclinical models of lymphatic filariasis and onchocerciasis. *Sci. Rep.* 7: 210.

**90** Pionnier, N.P., Sjoberg, H., Chunda, V.C. et al. (2019). Mouse models of *Loa loa*. *Nat. Commun.* 10: 1429.

**91** Rajan, T.V., Babu, S., Sardinha, D. et al. (1999). Life and death of *Brugia malayi* in the mammalian host: passive death vs active killing. *Exp. Parasitol.* 93: 120–122.

**92** Abraham, D., Hess, J., Bondesen, B.A., Harrington, J.M. (2018). *In vivo* model for parasitic worm infection and methods for evaluating antiparasitic compounds, including compounds active against canine heartworm. *International patent publication WO 2018/148392 A1.*

**93** Conder, G.A., Jen, L.W., Marbury, K.S. et al. (1990). A novel anthelmintic model utilizing jirds, *Meriones unguiculatus*, infected with *Haemonchus contortus*. *J. Parasitol.* 76: 168–170.

**94** Gration, K.A., Bishop, B.F., Martin-Short, M.R., and Herbert, A. (1992). A new anthelmintic assay using rats infected with *Trichostrongylus colubriformis*. *Vet. Parasitol.* 42: 273–279.

**95** Chabala, J.C., Mrozik, H., Tolman, R.L. et al. (1980). Ivermectin, a new broad-spectrum antiparasitic agent. *J. Med. Chem.* 23: 1134–1136.

**96** Blair, L.S., Williams, E., and Ewanciw, D.V. (1982). Efficacy of ivermectin against third-stage *Dirofilaria immitis* larvae in ferrets and dogs. *Res. Vet. Sci.* 33: 386–387.

**97** Bourguinat, C., Lee, A.C., Lizundia, R. et al. (2015). Macrocyclic lactone resistance in *Dirofilaria immitis*: failure of heartworm preventives and investigation of genetic markers for resistance. *Vet. Parasitol.* 210: 167–178.

# 17

# *In Vivo* Assays – Discovery and Development of New Antifilarial Drugs in Companion Animals

*Heinz Sager[1,*], Regina Lizundia[1], and William H. White[2]*

[1] *Elanco Tiergesundheit AG, Mattenstrasse 24A, 4058, Basel, Switzerland*
[2] *Elanco Animal Health, Alfred-Nobel-Str. 50, 40789, Monheim, Germany*

**Abstract**

Heartworm disease or dirofilariasis is a globally distributed, life-threatening condition primarily afflicting canines, and to a lesser extent cats, and is caused by the mosquito-vectored parasitic filarial nematode, *Dirofilaria immitis*. In dogs, untreated heartworm disease can result in extensive damage to the pulmonary arterial system and heart, potentially leading to congestive heart failure and even death. Unlike dogs, cats are generally able to mount an effective immune response to *D. immitis*, but the unlucky minority that develop dirofilariasis often die from feline-specific heartworm-associated respiratory disease. Effective treatment, therefore, is critical for the health and well-being of our pets. For almost 40 years, the primary means of controlling dirofilariasis has been through the widespread use of a single class of drug, the macrocyclic lactones, to eliminate larval stages of the parasite, thereby preventing development of adult heartworm disease in the host. As with any type of anti-infective therapy and irrespective of the mechanism of action, natural selection has, over time, invariably led to an increase in the incidence of treatment failure due to macrocyclic lactone resistance. To address the threat posed by resistance, it will be necessary to identify new anthelmintic drugs with unique mechanisms of action, that are able to convey a high level of preventive efficacy in regions of the world where the treatment failure due to macrocyclic lactone resistance is growing. In this chapter, we describe and reduce to practice, detailed methodologies that are inherent with, and unique to, the discovery of new preventive or curative treatments for dirofilariasis due to *D. immitis*, including mosquito breeding and parasite propagation, experimental infection of animals and laboratory and field trials for safety and effectiveness. The current regulatory environment, as well as requirements for the approval of new preventive treatments, is also discussed, with an emphasis on challenges that may arise from the inconsistent application and/or interpretation of globally harmonized guidelines.

*Corresponding author.

*Human and Animal Filariases: Landscape, Challenges, and Control*, First Edition.
Edited by Ronald Kaminsky and Timothy G. Geary.

## 17.1 Introduction

Dirofilariasis, or heartworm disease caused by infection with *Dirofilaria immitis*, is predominantly described for dogs but may also occur in cats. The fact that it is a life-threatening disease and that resistance issues against existing prophylactic treatments are of concern increases the pressure to identify new drugs. *In vivo* models are of high importance for the evaluation of new agents effective against *D. immitis*. Despite the fact that *in vitro* models may help to identify small molecules with a high degree of intrinsic activity against *D. immitis*, they cannot replace – or simulate only in a very preliminary way – the complex host–parasite interaction. As mentioned in the previous chapter, rodent models allow a first evaluation of identified hits in filaria-infected mammals. However, these models have limitations, as noted in Chapter 15. Finally, testing in the target animal remains the only way to reliably demonstrate efficacy against this parasite.

Considering the long duration of the life cycle of *D. immitis* and respecting efforts to follow animal welfare 3R recommendations (reduce, refine, and replace), it is obvious that it is not possible to screen hundreds or thousands of compounds for their activity in dogs. Such studies require large amounts of compounds and end in the euthanasia of the dogs, followed by necropsy and worm counting. It takes at least six months after infection until patency is reached [1]. Although such studies may be shortened by aiming for immature stages in the right heart and pulmonary arteries, they do not allow rapid evaluation of compounds of interest.

For the evaluation of heartworm efficacy, two approaches are of most interest: (i) control of larvae and (ii) elimination of adults and/or microfilariae. Although the two treatment strategies aim for different life cycle stages, the endpoint of evaluation remains the late stages in the right heart and in the pulmonary artery. The focus of current research lies in the former, i.e. preventive treatment with compounds that eliminate early larval stages migrating in tissues. In that regard, it is not optimal to wait several months after experimental infection until adult stages have developed. It would be much more attractive to examine survival of larvae during the first two to three months after infection, or even earlier if possible. The difficulty lies in the identification of early infection biomarkers that diminish in case of successful treatment. Many efforts are ongoing in that direction; however, no real breakthrough has been achieved.

The second approach, aiming to clear adult infection or eliminating microfilariae, can be simplified by the use of donor dogs. In such studies, microfilariae or adult *D. immitis* are harvested from dogs with existing infections and transplanted to the study dogs. By doing so, the study duration can be considerably shortened, compared to a standard protocol using third-stage-larvae for experimental infection.

In addition to direct treatment against *D. immitis*, alternative approaches aim for blocking parasite transmission from mosquitoes to dogs [2]. In these cases, it will not be sufficient to simply demonstrate repellency or rapid kill of mosquitos. The final proof of prevention of disease transmission will still rely on the absence of adult *D. immitis* in exposed dogs.

In the following sections, several aspects of the heartworm model in dogs (and to a lesser extent in cats) will be highlighted. We focus on experimental infection models for treatment against larvae, but other setups, like treatment against adults or natural infection models, will also be discussed. Although other filarial nematodes, such as *Dirofilaria repens*, have obtained more attention recently, little information is available on experimental infection models and they will be discussed only marginally.

## 17.2 Requirements, Infrastructure and Safety Measures for Experimental Infections and Mosquito Breeding

The life cycle of *D. immitis* is closely linked to the presence of suitable vectors and climatic conditions that allow full development of infective larvae in the intermediate host. To maintain the parasite in an experimental environment, certain conditions for the respective life stages must be met. The breeding of mosquitos and the development of infective third-stage larvae are described below (mosquito-breeding and production of L3s).

Studies that include animals must respect local animal welfare requirements. These normally define the housing conditions (e.g. surface area per dog, natural light exposure, access to outdoor infrastructure, interaction with other dogs, etc.) and manipulations (blood sampling, physical examinations, etc.). In addition, regulations may be made for biosafety reasons. In non-endemic areas, work with microfilaremic dogs may only be allowed in mosquito-proof facilities. The use of mosquito nets may be a possibility, but single layers may not prevent exposure if dogs are in direct contact with the net. In contrast to measures for malaria prophylaxis, it is not recommended to impregnate the nets with insecticides or repellents, as this may impact the outcome of mosquito breeding or efficacy studies.

The development of *D. immitis* larvae in mosquitos is strongly dependent on environmental temperature [3], while further development in the dog is thought to show only low variation due to the constant body temperature. However, the infection rate after experimental injection of *D. immitis* L3 seems better when the average temperatures are at 20 °C or above (internal communication). Furthermore, microfilaremia is highly seasonal and seems to depend on day length and temperature changes [4]. This leads to the requirement for indoor housing for dogs in areas with moderate or cold climatic conditions, especially for donor dogs, which should remain highly microfilaremic throughout the year.

The mosquito-proof housing of dogs is an important measure for non-endemic areas to prevent introduction of *D. immitis* to the local mosquito population. In endemic areas, prevention of exposure the other way is also important, as study dogs may be exposed to pathogen-carrying mosquitos. Such exposure may result in unplanned (patent) infections of dogs, which impacts the outcome of efficacy studies and may contaminate characterized heartworm isolates.

Dirofilariasis is a zoonotic disease, which increases requirements on the qualification of the personnel working with the parasites, the mosquitos and the dogs. The most critical is the handling of L3-carrying mosquitos. Accidental release of

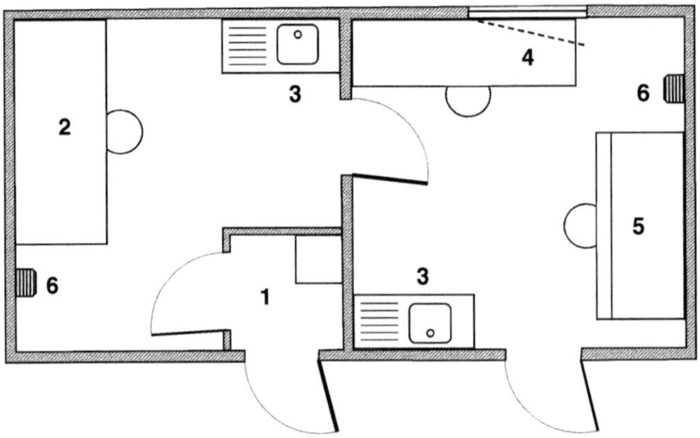


**Figure 17.1** Example of a mosquito breeding unit and lab for isolation of third-stage larvae of *Dirofilaria* spp. The lab on the left is climate-controlled to maintain a temperature of 28 °C and relative humidity of 80%. It is equipped with a two-door entry (1) to prevent escape of mosquitos. Lab coats are available at the entry. The feeding of mosquitos can be done on the work station (2). For isolation of *Dirofilaria* spp. L3, mosquito cages are transferred to the isolation lab on the right. The window is equipped with mosquito nets (4). Isolation of L3 is done in the laminar flow (5). Both labs have water supply (3) and mosquito traps (6). All air tubes (climate system, laminar flow, etc.) have mosquito nets or filters to prevent mosquito escape.

*D. immitis*-infected mosquitos (especially when using isolates with demonstrated anthelmintic resistance) must be avoided by all means. Access to mosquito breeding locations should therefore be limited to trained personnel only, and the entry should be equipped with two doors (or mosquito-proof nets; see Figure 17.1). Furthermore, consideration should be given to appropriate disposal of material. Mosquito larvae and pupae should be immersed in hot water at approximately 80 °C until they do not move, while adult mosquitoes and eggs should be frozen at −20 °C for at least 12 hours.

## 17.3 Isolation of *D. immitis* from the Field

*Dirofilaria immitis* can be maintained under laboratory conditions. Nevertheless, this is costly and requires extensive infrastructure to cover all life stages in mosquitos as well as in the definitive host. The starting point may be an infected dog coming from the field. In some areas, it may be possible to introduce a patent *D. immitis*-infected dog into a laboratory environment and to use it as source of microfilaremic blood. In many countries, it will be difficult to get animal welfare approval for use of a dog not bred for laboratory purposes. An additional challenge may be the incomplete history of such a dog (vaccination, exposure to pathogens, etc.), which increases the risk of importing unrecognized diseases into the laboratory environment. Thus, it may be more attractive to simply get a sample of peripheral blood from a *D. immitis*-infected dog from a clinic or shelter to establish a new isolate. Anticoagulated blood samples should be sent without delay to the laboratory. While no specific cooling procedure is required, the samples must not be exposed to temperatures below 0 °C. When evaluating the density of microfilariae, their viability has to be considered. This is normally based on motility.

Mosquitos feeding on microfilaremic blood can harbor infective third-stage larvae within two to three weeks. It is possible to ship such mosquitos for infection of a dog. The number of larvae per mosquito may range between 2.6 and 6.8 [5], but one has to consider that a proportion of the mosquitos remains uninfected. Thus, a sufficient number of mosquitos will be required to infect one or several dogs. For shipment, it is also important to respect all biosafety requirements, while using packaging material that allows airflow and ideally feeding of the mosquitos.

Getting access to new field isolates may be facilitated by contacts with veterinary practices in endemic areas or through institutions working with *D. immitis*. The more information one can gather on the infected dog, the better. Heartworm isolates from dogs that underwent regular prophylactic treatment have a high risk of being resistant to macrocyclic lactones. Thus, the characterization of a new isolate is recommended. This can be done by genetic profiling, *in vitro* assays and *in vivo* exposure to drugs. However, some authorities specifically request the use of recent, non-characterized field isolates for pivotal heartworm efficacy studies.

Dogs infected with *D. immitis* can be kept for several years. Our experience is that dogs can remain without clinical symptoms over a period of at least five years while maintaining an adequate microfilaremia. Normally, the first year of patency results in lower microfilariae concentrations than the following years [4].

Microfilariae of *D. immitis* exhibit seasonal periodicity, resulting in higher counts in peripheral blood during summer (northern hemisphere) compared to winter. Even under standardized climatic conditions, seasonality can be observed. Some experiments indicate that heartworm isolates may adapt to laboratory conditions and lose seasonality after a few passages [4]. Whether this also impacts other characteristics of the respective isolates remains an open question, but it may be one of the arguments to maintain the requirement for acquiring new isolates from the field for pivotal efficacy studies.

## 17.4 Mosquito Breeding and Production of L3

An impressive number of mosquito species belonging to the genera *Aedes*, *Anopheles*, and *Culex* are competent vectors for *D. immitis* and *D. repens*, as long as climatic conditions allow development of the larvae. In a standardized laboratory environment, climatic conditions can be optimized for larval development (Figure 17.2a). When it comes to the selection of the best mosquito species for the production of L3, one has to evaluate the availability of isolates, the ease of handling, and the expected yield of infective larvae. Our experience is based on *Aedes aegypti*, but other species may provide equivalent or even better results. Examples of research protocols for rearing mosquitos, collecting infective larvae, and experimentally infecting dogs or cats can be found on the FR3 homepage [6].

Mosquitos become infected by ingesting microfilaremic blood (Figure 17.2b). Under laboratory conditions, this will only exceptionally be done directly on the dog.

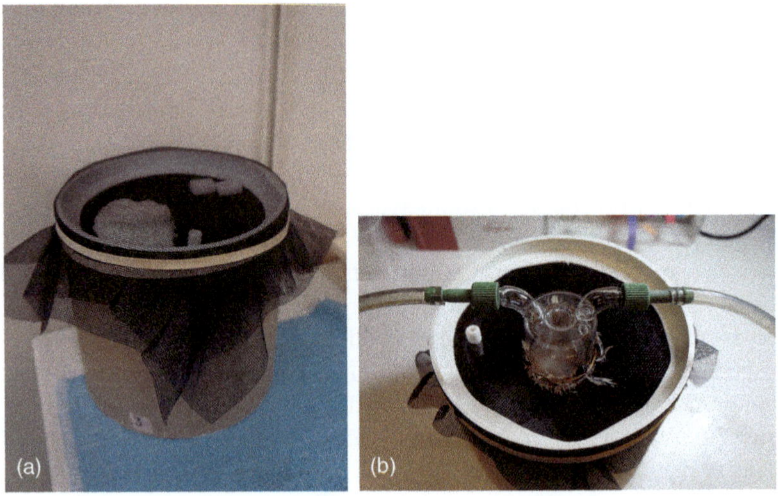


**Figure 17.2** (a) Mosquito cage. The cardboard box is closed with a mosquito net and sealed with a plastic lid whose central part has been cut. A rubber band holds the mosquito net tightly. A wet cotton square and several pieces of sugar are placed on top of the mosquito net and serve as water and food supply. The second mosquito net is placed above the cardboard box and secured with another rubber band. (b) Mosquito feeding with blood. For mosquito feeding, the upper net and the wet cotton are removed from the cardboard box. Adult mosquitoes inside the cardboard box are fed with dog blood using a glass feeding system composed of a mouse skin and glass chamber that remains open, so that blood can be poured into the chamber. The chamber is connected with plastic tubes to a water bath set at 37 °C to warm the system.

Much more standardized is artificial feeding. Blood samples from microfilaremic dogs are typically diluted with blood from non-infected dogs to avoid too high concentrations of microfilariae. Values of 3000–5000 microfilariae per mL blood seem suitable.

After a period of 13–21 days in a climate-controlled room at 28 °C/80% relative humidity with access to 5% glucose and water *ad libitum*, third-stage larvae are fully developed and can be harvested from the mosquitos. To do so, mosquitos are crushed and viable *Dirofilaria* larvae purified by active migration and washing steps in culture media, such as Hank's balanced salt solution (HBSS). The number of third-stage larvae harvested from mosquitos can vary considerably. It may therefore be useful to include backups in case of insufficient numbers of the primary production. Our experiments resulted in average production of 450 L3 per mosquito cage. However, in some cases, the harvest of L3 went as low as 50 per cage.

## 17.5 Experimental Infection of Dogs and Cats

Experimental infection of target animals is currently the only way to assess the efficacy of a test compound. The following sections describe the steps for experimental infection of dogs. Particularities for infection of cats will be highlighted at the end of this section.

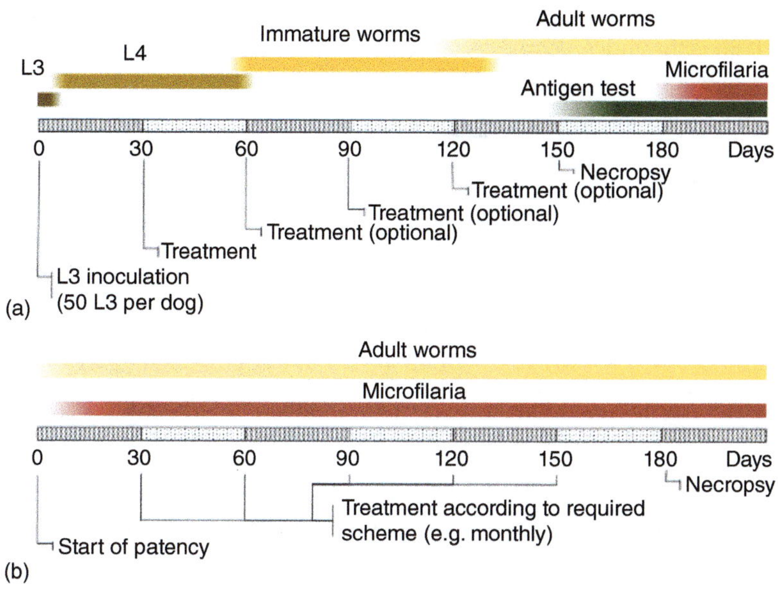


**Figure 17.3** Experimental study setup for preventive treatment (a) and for treatment of adult *D. immitis* (b). The period between experimental infection with third-stage larvae and the appearance of circulating microfilariae in peripheral blood is approximately six months. In a standard study design for preventive treatment (a), the first administration is done 30 days after infection. Additional treatments may follow, e.g. in monthly intervals. The evaluation of efficacy is based on worm counts five months or later after infection. When treating adult heartworm (b), a patent infection has to be established, either by infection with third-stage larvae or by transplantation of adult males and females. The treatment scheme and the time of necropsy depend on the characteristics of the test compound and formulation. The calculation of efficacy will be based, as for preventive treatment, on the counts of adult worms in the treated group compared to the non-treated control group.

As discussed earlier, two approaches are typically used to evaluate efficacy of an experimental compound in dogs: to control larvae (preventive treatment) or to eliminate adult worms (adulticide treatment). Each approach targets different parasite stages and requires different study designs (see Figure 17.3). Here, we focus mainly on the description of the study design for efficacy evaluation of preventive treatment. The main activities include (i) *D. immitis* inoculation, (ii) treatment, (iii) monitoring of infection status and onset of patency, (iv) worm count at necropsy, and (v) efficacy calculation.

### 17.5.1 *D. immitis* Inoculation

Each inoculum generally consists of 50–100 infective *D. immitis* L3 suspended in HBSS. Animals are infected by subcutaneous injection in the inguinal area or in the region of the neck (between the shoulder blades). The inoculation dose and the number of inoculation sites may vary to attempt to increase the infection rate. In addition, if the number of L3 larvae obtained following the initial harvest of mosquitoes is inadequate to inoculate all animals on the same day, it is possible to infect dogs twice, one week apart (i.e. 25 L3 on day 0 and 25 L3 on day 7). Spreading the number of inoculations through several days might be helpful to better mimic the natural exposure of dog by infected mosquitoes.

### 17.5.2 Treatment

Generally, one month after parasite infection (day $30 \pm 3$), a single dose of the preventive compound under test is administered to the dog. The group size for treatment and non-treated controls is typically four to eight animals. The route of administration can be oral, topical, or parenteral. Additional treatments can be done at monthly intervals (~day 60 and ~day 90) to try to eliminate remaining larvae. Other treatment schemes are conceivable, depending on the intended product claim (Figure 17.3a). A non-treated (vehicle) control group is used to confirm adequacy of infection of the heartworm isolate used.

Dogs are observed at different time points after the treatment to ensure that the entire treatment is given (i.e. to detect vomiting in case of oral administration via gelatin capsules) and to detect possible adverse events related to the test compound.

Blood samples are usually taken pre-dosing and at different time points after the treatment for analysis to better understand the pharmacokinetic profile of the test compound.

### 17.5.3 Monitoring Infection Status and Onset of Patency (Antigen Tests and Microfilaremia)

Currently available heartworm antigen tests detect proteins secreted mainly by adult female *D. immitis* [7]. Although the test has limitations and may provide false-negative results, this is normally not an issue in experimentally infected dogs: the expected number of females and the total number of adult worms are considerably above the detection level [8]. More challenging is the fact that it takes at least five months until positive results can be obtained (Figure 17.3a). Many study designs schedule necropsy of the dogs five to six months after infection. Therefore, the value of this assay in the experimental protocol is relatively low. Many efforts have been made to identify markers that allow confirmation of infection at an earlier stage, i.e. before the (pre-) adult worms reach the heart and pulmonary artery. An antigen that seemed promising was DiT33. Antibodies against this protein could be detected between 9 and 11 weeks after experimental infection. Dogs given a drug-abbreviated infection, and therefore only exposed to infective larvae and early fourth-stage larvae, did not contain detectable levels of DiT33 [9]. Although such an assay would allow early detection of *D. immitis* and of successful prophylactic treatment, it never attained wide distribution, nor was it commercialized.

The presence of microfilariae is monitored using a direct blood smear and the modified Knott's technique [10, 11]. The earliest that microfilariae can be detected is about six months postinfection, but almost always by seven months (day 210) post-infection. Generally, in experimental efficacy studies evaluating preventive compounds, microfilaremia is not evaluated, as necropsy is performed before patency is reached.

It can be concluded that the antigen test and the determination of microfilaremia in peripheral blood provide limited value to studies with dogs experimentally infected with *D. immitis*. These assays do not provide early information on the infection status of the dogs, nor can they be considered as endpoints, i.e. even in cases of clear negative or positive results, necropsy of the dogs for the determination of the adult worm burden is required. Nevertheless, the tests can be helpful tools when establishing a new infection protocol, to estimate when (pre-) adult stages reach the vascular system. Finally, monitoring microfilaremia in donor dogs is essential, when precise information on microfilaremia in blood fed to mosquitos is needed.

### 17.5.4 Necropsy and Worm Counts

Between five and six months after infection (~day 150–180), dogs are sedated, euthanized via intravenous injection of an approved euthanasia solution, and then necropsied. The necropsy procedure involves removal of the heart and lungs, dissection, and examination of the anterior vena cava, right atrium, right ventricle, and pulmonary arteries (Figure 17.4). All adult *D. immitis* worms found in the cardiopulmonary system are collected, identified, sexed, and counted (Figure 17.5). Only worms that are normal in both appearance and motility are considered live. All other worms are considered dead. The WAAVP guidelines mention an expected average of 28 worms per non-treated control dog when infecting with 50 larvae [12].

Personnel performing the necropsy should be blinded to the treatment group to which the animals have been allocated.

### 17.5.5 Efficacy Calculation (Criteria to Grant a Claim for Registration)

The US Food and Drug Administration Center for Veterinary Medicine requires 100% efficacy for laboratory studies following VICH GL19 [13] to support the approval of heartworm disease preventive products.

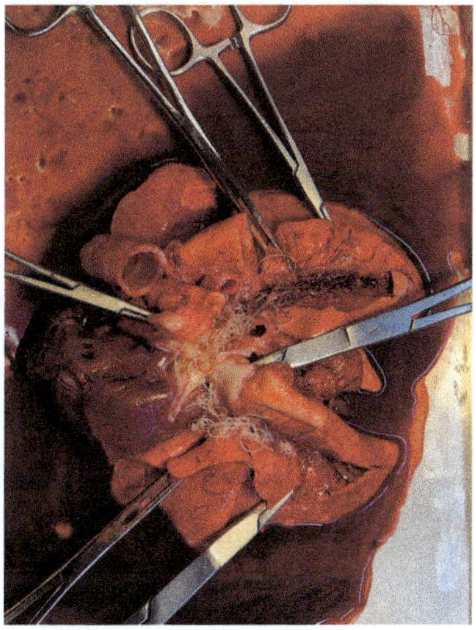

**Figure 17.4** Dissection of heart and pulmonary vessels approximately five months after experimental infection with *D. immitis*. A large number of adult stages was identified and macroscopically evaluated for viability and gender.

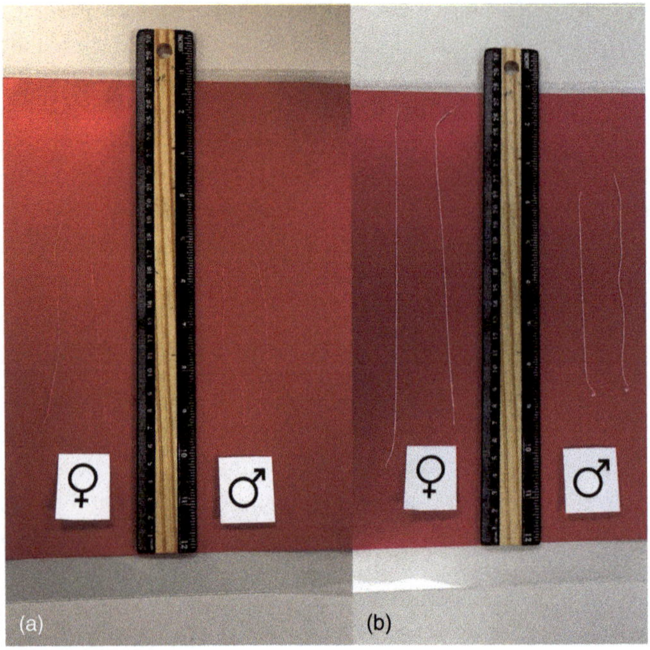


**Figure 17.5** Male and female adult worm pairs collected 153 days post-infection from dogs in a laboratory study. Female worms are usually longer than male worms. The tail of the male worm is spirally coiled. (a) Worms from a dog treated with an experimental research compound are thinner and show reduced size compared to (b) worms collected from a non-treated control dog.

According to VICH GL19 and GL20 guidelines [13, 14], requirements for approval of an anthelmintic product are based on statistically significant differences between the treated and control groups and on calculated percent effectiveness (which must be 100% for preventatives) using the Abbott formula:

$$\text{Efficacy (\%)} = 100 \times \frac{M_c - M_t}{M_c}$$

$M_t$ = arithmetic/geometric mean of adult worms in the treated group
$M_c$ = arithmetic/geometric mean of adult worms in the control group

Efficacy can be calculated using the arithmetic or the geometric mean. Differences in efficacy might be observed depending on whether arithmetic or geometric means are used. Log-transformed parasite counts tend to follow a normal distribution more closely than do non-transformed parasite counts. Therefore, the geometric mean is a more appropriate estimate of central tendency and has less potential for misinterpretation than the arithmetic mean.

### 17.5.6 Experimental Infection of Cats

The infection methodology for and life cycle of *D. immitis* in cats is generally similar to dogs, though cats are not a natural definitive host for *D. immitis*. As with dogs, cats harboring adult heartworms can develop caval syndrome [15], but that is where similarities between the two species end. Compared to dogs, the pre-patency interval in cats is approximately one to two months longer, adult worm burdens are lower, microfilaremia is often absent or lower in magnitude if present, and the rate of successful infection following experimental inoculation with L3s is lower [16–18]. For the clinician, the diagnosis, clinical presentation, and subsequent treatment of dirofilariasis in cats are also markedly different from dogs in several respects [19–21]. Cats mount an aggressive immune response to *D. immitis* infection, that while resulting in self-cure in the majority of cases, can also lead to feline-specific heartworm-associated respiratory disease (HARD) as well as anaphylactic shock and death [21–26]. The only sound approach for the treatment of heartworm disease in cats, therefore, is regular monthly treatment with a heartworm preventive.

As already mentioned, the experimental infection methodology for cats is essentially the same as described for dogs, with the caveat that successful infection rates are lower. This necessitates inoculation of a larger number of animals to secure the required number of heartworm-positive cats and appropriate statistical power for evaluating effectiveness of treatment. Unlike dogs, assessment of drug safety in heartworm-positive animals is contraindicated for cats [12]. Treatments are administered in the same fashion and interval as for dogs, if designed to gain a heartworm prevention claim. Monitoring of infection status is trickier in cats, as antigen tests do not work as well [19]. Necropsy and worm count protocols are similar to those for dogs, but one should expect fewer worms overall and more animals in the control group with zero worms. According to WAAVP guidelines, 70% of cats given 100 larvae will harbor an average of four to five adult worms. The recommendation is to aim for a group size of nine cats to ensure six infected animals [12]. Adult worms have a shorter life span and are smaller in cats than in dogs [27, 28]. Similar rules apply for gaining an effectiveness claim, via VICH GL20 [14].

## 17.6 Transplantation of Adult Worms

Surgical transplantation of adult *D. immitis* has been described in dogs [29–31] as well as in cats [32]. Refinement of the technique by Rawlings and McCall [33] resulted in a reliable and relatively safe method of establishing well-controlled infections in dogs that are suitable for supporting a variety of studies, including research into the pathophysiology of dirofilariasis, evaluating the safety of heartworm preventives in heartworm-positive animals, and evaluating adulticidal or microfilaricidal activity of antifilarial drugs. The transplantation method reduces the number and magnitude of variables arising in naturally or experimentally infected animals, such as unpredictable and often excessive worm burdens, unknown worm gender ratios, and highly variable degrees of microfilaremia. Surgical transplantation of adult worms ultimately results in greater predictability and uniformity of heartworm disease progression between animals, as well as significant reductions in undesirable sequelae in the host due to excessive worm burden (for example, caval syndrome) [15, 33, 34]. Adult heartworms may be surgically transplanted into cats for similar purposes; in fact, transplantation could be viewed as more critical for cats despite a higher post-implantation death rate [35], given that this species exhibits even greater variability than dogs in terms of predictable infection rate and clinical manifestations arising from heartworm infection. For the sake of brevity, we limit discussion to dogs unless otherwise specified, recognizing that the general methodology of transplantation into cats is similar (though fewer worms are transplanted, for obvious reasons of host physical size).

The first step in surgical transplantation is the generation of a suitable number of heartworm-positive donor dogs. Early work utilized donor dogs that had been inoculated via subcutaneous injection with large numbers of infective larvae, often up to 400 L3s [33]. Such heavily infected dogs tended to develop caval syndrome, and given the inherent value of these animals and animal welfare concerns, inoculation with fewer larvae, usually in the range of 50–100 L3s [23], reduces the potential for complications such as caval syndrome. As early as five months after inoculation, screening of infected donor dogs via heartworm antigen test may be initiated. Echocardiography may also be useful in helping to exclude donor dogs with an insufficient worm burden. Worms may be harvested from antigen-positive dogs prior to becoming sexually mature, though waiting until donor dogs become microfilaremic (usually about six to seven months post-infection) can reduce the amount of time required for recipient dogs to become microfilaremic following transplantation surgery.

Typically, 5–15 adult female–male pairs of worms are transplanted into a heartworm-naïve dog via jugular venotomy [13, 33]. Strict worm gender balance does not seem to be as critical as ensuring that each dog receives a sufficient number of females and harbors the same total worm burden. In the current state of the art, dog recovery rates are very high, with few if any post-surgical complications. Following successful transplantation, these dogs are suitable for use in a variety of different research or drug development activities, most frequently involving studies to evaluate drug safety in heartworm-positive dogs; studies to evaluate potential microfilaricidal activity of drugs; and studies to evaluate potential adulticidal activity of drugs. Potential adulticidal and microfilaricidal activities are pertinent to drug safety. Generation of a patent infection via surgical implantation of adult worms into healthy, naïve recipient dogs provides some advantages over the more traditional L3 infection method (subcutaneous injection). First, the interval of time between transplantation of worms and evaluation of drug safety in dogs with a patent infection can be as short as two weeks, resulting in studies of significantly shorter duration. Second, since the life cycle from injection of L3s to worm maturation and microfilaria production is bypassed with the surgical method, the potential for heartworm-associated pathology (i.e. vascular inflammation, caval syndrome) will likely be reduced, thereby minimizing potentially confounding factors that could make interpretation of safety data more difficult. Third, the transplantation method tends to reduce animal-to-animal variability in microfilaria counts, which improves overall confidence that mean group changes in microfilaria burden are reflective of the true changes to be expected in a larger patient population.

## 17.7 Field Trials

Due to the complexity and severity of dirofilariasis, heartworm preventive field trials represent an integral part of a drug development program, primarily to obtain a more complete picture of the effects and potential adverse events of treatment under real-world conditions. While the mandate for such a field trial may differ between regulatory agencies, completion of a trial is an absolute requirement for approval of any new canine heartworm preventive in the United States. While agencies in some countries may rely exclusively on data from a large US field trial, others may stipulate that smaller-scale field trials be conducted in the country to confirm drug effectiveness and safety in local populations. In cats, controlled laboratory effectiveness studies are considered adequate to support approval of a claim, and heartworm preventive field trials are not mandated. As with all field studies, safety is always a primary objective, though increasing emphasis has been placed on effectiveness over the past decade. This would seem to be a logical response on the part of regulatory agencies, at least in part, to an increase in LOE (lack of efficacy) reports with commercial heartworm preventives [36, 37], as well as the fact that an increasing number of strains obtained from the field exhibit reduced susceptibility to macrocyclic lactone treatment in laboratory studies. Field trials also facilitate collection of information from pet owners about user-friendliness of the product, for example, ease of treatment, palatability (if oral), or visual appearance of skin and hair coat at the application site (if topical).

The typical heartworm field trial is of relatively straightforward design, though a number of noteworthy requirements are unique to this type of study. The investigational veterinary product (IVP) under evaluation should be in a near-final or final presentation form and manufactured under current Good Manufacturing Practice standards. For ethical reasons, a negative control (placebo) group is never utilized in a field trial, and the safety and efficacy of the IVP are compared to the safety and efficacy of a commercially available control veterinary product (CVP) that is tested in parallel. Preferably, the CVP selected for comparison should employ a similar route of administration and treatment frequency (e.g. both monthly oral treatments, but different active ingredients), or alternatively, the same active ingredient with different routes of administration (e.g. monthly oral versus monthly topical). Only a limited number of individuals involved in a trial should be unmasked; to maintain impartiality, it is critical that the clinical investigators and veterinary staff remain unaware of treatment group assignment for enrolled patients. Veterinary staff involved in administration or dispensing of treatments should be blinded, as well as owners and any individuals responsible for third-party monitoring and quality oversight of the study. A minimum of 100 patients, or evaluable cases in each treatment group, is considered necessary to generate data that offer significant inferential value to support conclusions about safety and effectiveness. The size of a typical trial and the required number of patients necessitate the use of multiple clinics, preferably spread across a relatively broad geographic area that is representative of heartworm disease incidence. While there is no specific guidance around the number of veterinary clinics that may participate in the study, care should be taken to ensure that no single clinic accounts for more than one-third to two-fifths of the total number of evaluable cases. That being said, it is currently recommended that at least half of the evaluable cases originate from regions with the highest prevalence of heartworm disease, which in the United States is the southeastern portion of the country that includes the Mississippi Delta and Mississippi River Valley.

As a general rule, dogs from single- and multi-animal households are eligible for recruitment, though only one dog in any given household should be enrolled in the study. For multi-pet households, consideration must be given to ensure that non-enrolled animals on anthelmintic and/or endectocide treatment do not present the potential for inadvertent cross-contamination of dogs receiving the IVP or CVP; for example, non-enrolled pets should not receive a topical treatment with a parasiticide product having antifilarial activity. General good health, a complete medical history, and negative heartworm antigen and modified Knott's tests [10, 11] are the primary criteria used to establish patient eligibility. Dog age and weight are not usually considered to be of critical importance with the exception that they should match the intended age and weight restrictions for the IVP under evaluation. Concomitant treatment, with commonly used prescription medications other than anthelmintic or antifilarial drugs, for example a nonsteroidal anti-inflammatory, is frequently permissible. Exclusionary criteria include stray, feral and/or shelter dogs (due to an inability to establish an accurate medical history), or dogs intended for breeding purposes during the study, as well as females that are pregnant or lactating. Once suitability for enrollment is confirmed, a dog should be randomly allocated to one of the two treatment groups at a target ratio that is specified in the trial protocol.

The timing and duration for a field trial are critical. It is recommended that the vast majority of dogs be enrolled and receive their first treatment during peak heartworm transmission season, which in North America falls between June 1 and July 31 [28]. Additional cases enrolled outside this window may be justifiable, particularly in geographic regions where peak heartworm transmission season begins before 1 June. American Heartworm Society and Companion Animal Parasite Council guidelines promote year-round protection from heartworm infection, so for each dog, a trial duration of approximately one year should be anticipated, from the time of enrollment to the final clinic visit. Dogs should receive the appropriate number of treatments that are designed to provide heartworm protection for the entire period. For example, a monthly heartworm preventive would be administered approximately every 28–32 days, whereas a dog receiving a long-acting heartworm preventive that is intended to provide six months of protection would receive two treatments, the first administered at the time of enrollment and the second administered six months later. The first treatment should always be administered by the veterinary staff at the clinic, but subsequent treatments may be given by the pet owner at home, unless the intended product concept requires veterinary administration (i.e. injection). During the course of the trial, a thorough assessment of each dog should be conducted by clinic personnel at routine intervals, beginning at enrollment and with follow-up visits at least every three months. At each clinic visit, a complete physical examination and heartworm antigen test should be performed, with blood drawn for a modified Knott's test to detect circulating microfilaria. The antigen tests conducted through the first eight months of the trial are critical for determining continued eligibility of a dog for measurement of preventive efficacy; it is during this window of time that a positive antigen test would confirm that the dog had a preexisting, pre-patent heartworm infection that was not detectable at the time of enrollment. These dogs should be removed immediately from the study, and owners should be counseled by the veterinarian about treatment options to address adult heartworm infection. Additionally, owners should receive training to support daily assessment and documentation of any abnormal clinical signs or behavior in their dogs, particularly following administration of treatments.

The effectiveness of treatments is assessed via heartworm antigen and modified Knott's tests and reflected numerically as the percentage of eligible evaluable cases that remain antigen negative with no evidence of circulating microfilaria for the duration of the study. Regulatory agency hurdles remain quite stringent for a field trial, with the anticipated and desired outcome being the same as with pivotal laboratory effectiveness studies, in which all dogs remain free from detectable heartworm infection. However, as pointed out by Vidyashankar and colleagues [38], from a statistical standpoint, an observed value of 100% preventive efficacy in laboratory studies that utilize a limited number of dogs should not be construed to infer that the treatment in question will be 100% efficacious under all circumstances. In the case of a field trial, regulatory agencies may determine that it is acceptable for a very small number of eligible evaluable cases in the IVP group to be positive for heartworm infection (i.e. a treatment failure) provided that a comparable or greater number of failures are observed in the CVP group and non-inferiority of the IVP is demonstrated.

One of the most obvious limitations of a field trial, in terms of evaluating preventive efficacy of a drug, is the uncertainty around *D. immitis* exposure. While mosquitos are ubiquitous and there is a plethora of historical mosquito burden data for North America, direct methods for assessing *D. immitis* burden in a given mosquito population are not amenable to implementation at a field trial scale. In theory, it is assumed that a significant number of dogs will be exposed to *D. immitis*-infected mosquitos at least once during the course of a trial, particularly in geographic areas with very high mosquito burdens and historically high incidence rates of reported heartworm-positive dogs. However, these are indirect measurements that, in reality, can only infer likely exposure without providing conclusive evidence for it. Nevertheless, field trials provide critical information about drug safety, and the large number of dogs in a trial provides key inferential value, that when taken together with laboratory data, enables reliable assessment of drug effectiveness under real-world conditions.

## 17.8 Characteristics of Experimental and Natural Infection Models

| | Rodent models | Experimental infection with L3 | Transplantation of adult worms | Natural infection |
|---|---|---|---|---|
| Indication | Early research: First *in vivo* testing of compounds identified *in vitro* | Research and development: Proof-of-concept and pivotal studies. Efficacy against larval and adult stages of *D. immitis* (and *D. repens*) | Research and development: Proof-of-concept and pivotal studies. Efficacy against adult stages and microfilariae of *D. immitis*, safety of heartworm preventives in heartworm-positive animals | Development: Field studies. Efficacy against larval and adult stages of *D. immitis* (and *D. repens*) |
| Infrastructure and resources | Rodent facility, rooms for parasite breeding, rooms for necropsy and worm counts | Dog facility, mosquito breeding (or access to L3), rooms for necropsy and worm counts | Dog facility, mosquito breeding or access to donor dogs, rooms for surgery (transplantation) necropsy and worm counts | Veterinary clinics for client-owned dogs, rooms for clinical examination, and access to diagnostic tests (antigen, microfilaremia) |
| Number of animals needed | 4–10 mice or gerbils per group | 4–8 dogs per group<br>8–12 cats per group | 4–8 dogs per group<br>8–12 cats per group | ≥100 dogs per group (client owned) |
| API quantities | Several milligrams | Several grams | Several grams | Several kilograms |
| Time | ~3 mo | ~6 mo | 2–12 mo | ≥1 yr |
| Costs per study | ~10–50 k USD | ~500 k USD | 200–800 k USD | ≥1 M USD |
| Predictability | Low to medium | High | High | High |
| Risks (infection failure, non-compliance, etc.) | Standardized laboratory setting reduces the risk of variability | Standardized laboratory setting reduces the risk of variability | Standardized laboratory setting reduces the risk of variability | Increased risk of incomplete history of dogs and noncompliance of dog owners |

## 17.9 Requirements of Authorities (CVM, EMA)

*D. immitis* has been identified across a very large region of the world that includes North and South America, Australia, Japan, and both central and Mediterranean coastal areas of Europe [3, 23, 39–44] and is considered by global regulatory agencies as a serious disease of concern to dogs and cats. In North America and specifically the United States, *D. immitis* is prevalent across large regions of the country, particularly in areas where the climate is conducive to prolific mosquito populations and a relatively long mosquito season. These factors, along with the serious nature of dirofilariasis, contribute in part to what is the most stringent possible requirement for disease treatment or prevention in humans or animals, that of 100% efficacy. To gain an approval for a heartworm prevention claim in the United States, at least two pivotal dose confirmation laboratory studies and a field trial are necessary. Some regulatory agencies may also require that at least one laboratory study be conducted against a local heartworm isolate, so there is a high likelihood that more than two dose confirmation studies will be required to support investigational drugs intended for commercialization on a global scale. Some agencies have greater evidentiary requirements around how a proposed minimum dosage was determined, and if *D. immitis* is the dose-limiting parasite, the need for comprehensive dose escalation work should be anticipated.

Prior to approval of the first macrocyclic lactone, ivermectin, in the United States in 1987, heartworm disease was prevented through daily administration of flavored treats containing diethylcarbamazine. The introduction of ivermectin revolutionized the landscape of heartworm disease, primarily because it conveyed complete protection (100% preventive efficacy) against the development of heartworm disease in a more convenient once-monthly treatment regimen for the pet owner [45]. Since the advent of ivermectin, 100% efficacy has been the requirement set forth by multiple regulatory authorities, including the CVM, to support approval of a heartworm prevention claim for an investigational animal drug. However, there is increasing recognition within the heartworm community that the concept of "100% efficacy" is an artificial and misleading benchmark, and results from limited numbers of dogs in laboratory studies should not be construed to indicate that a particular treatment can provide complete protection against development of heartworm infection under all circumstances [38]. In the United States, the current regulatory environment is such that the requirement for a drug to be granted approval as a heartworm preventive is significantly more stringent than comparable regulations governing minimum protection rates (i.e. disease prevention) for vaccines against deadly canine diseases, including parvovirus, distemper, hepatitis, and even rabies. Furthermore, an increasing tempo in the identification and characterization of heartworm isolates with reduced susceptibility (or resistance) to macrocyclic lactones has resulted in a conundrum, whereby expectations of regulatory agencies do not always align with internationally accepted guidelines stating that dose confirmation work should not be conducted against known drug-resistant strains of parasites [46]. This leads to the question of whether a result of <100% efficacy against an isolate confirmed *a priori* or *ex post facto* to be resistant to the drug or drug class under evaluation should be classified as a treatment failure or an invalid study. While this distinction may seem to be semantic in nature, the practical outcome of whether or not an investigational drug is deemed approvable for a heartworm prevention claim currently hinges on a regulatory agency's interpretation of language in the harmonized guideline. VICH GL7 [46] guidance would suggest that this type of scenario should result in the study being categorized as invalid for purposes of determining drug efficacy; at the time of writing this article, the European Medicines Agency seems to have adopted this line of thinking for at least one commercial heartworm preventive [47].

For over 30 years, macrocyclic lactone class anthelmintics have been the mainstay for the prevention of heartworm infection and have provided pets and their owners with an unprecedented level of protection against heartworm disease. While most LOE cases are believed to be caused by poor owner compliance with manufacturer treatment recommendations, the number of heartworm isolates obtained from the field and characterized in the laboratory as being less susceptible *in vivo* to one or more macrocyclic lactone class drugs is growing. The mechanisms behind the high level of efficacy historically observed with macrocyclic lactones are complex, but evidence of macrocyclic lactone resistance is now undeniable and regulatory agencies must work with academic and industry researchers to develop reasonable policies that address this threat.

## References


**1** Kotani, T. and Powers, K.G. (1982). Developmental stages of *Dirofilaria immitis* in the dog. *Am. J. Vet. Res.* 43: 2199–2206.

**2** McCall, J.W., Varloud, M., Hodgkins, E. et al. (2017). Shifting the paradigm in *Dirofilaria immitis* prevention: blocking transmission from mosquitoes to dogs using repellents/insecticides and macrocyclic lactone prevention as part of a multimodal approach. *Parasites Vectors* 10 (Suppl 2): 525.

**3** Genchi, C., Mortarino, M., Rinaldi, L. et al. (2011). Changing climate and changing vector-borne disease distribution: the example of *Dirofilaria* infection in Europe. *Vet. Parasitol.* 176: 295–299.

**4** Lovis, L., Grandjean, M., Overney, L. et al. (2017). Seasonality and circadian variation of microfilaremia in dogs experimentally infected with *Dirofilaria immitis*. *Vet. Parasitol.* 243: 235–241.

**5** Bartholomay, L.C., Farid, H.A., El Kordy, E., and Christensen, B.M. (2001). Short report: A practical technique for the cryopreservation of *Dirofilaria immitis*, *Brugia malayi*, and *Wuchereria bancrofti* microfilariae. *Am. J. Trop. Med. Hyg.* 65: 162–163.

**6** FR3: http://www.filariasiscenter.org/protocols/Protocols (accessed 13 December 2021).

**7** Courtney, C.H. and Cornell, J.A. (1990). Evaluation of heartworm immunodiagnostic tests. *J. Am. Vet. Med. Assoc.* 197 (6): 724–729.

**8** Atkins, C.E. (2003). Comparison of results of three commercial heartworm antigen test kits in dogs with low heartworm burdens. *J. Am. Vet. Med. Assoc.* 222 (9): 1221–1223.

**9** Hong, X.Q., Santiago Mejia, J., Kumar, S. et al. (1996). Cloning and expression of DiT33 from *Dirofilaria immitis*: a specific and early marker of heartworm infection. *Parasitology* 112: 331–338.

**10** Georgi, J.R. and Georgi, M.E. (1992). Heartworms and other filariids. In: *Canine Clinical Parasitology*, 192–198. Philadelphia, PA: Lea & Febiger.

**11** Knott, J. (1939). A method for making microfilarial surveys on day blood. *Trans. R. Soc. Trop. Med. Hyg.* 33: 191–196.

**12** Jacobs, D.E., Arakawa, A., Courtney, C.H. et al. (1994). World Association for the Advancement of Veterinary Parasitology (W.A.A.V.P.) guidelines for evaluating the efficacy of anthelmintics for dogs and cats. *Vet. Parasitol.* 52: 179–202.

**13** European Agency for the Evaluation of Medicinal Products, Veterinary Medicines and Information Technology. (2001). Efficacy of anthelmintics: specific recommendations for canines. *VICH Topic GL19*, CVMP/VICH/835/99-FINAL (London, 30 July 2001). https://www.ema.europa.eu/en/documents/scientific-guideline/vich-gl19-efficacy-anthelmintics-specific-recommendations-canines-step-7_en.pdf (accessed 13 December 2021).

**14** European Agency for the Evaluation of Medicinal Products, Veterinary medicines and Information Technology. (2001). Efficacy of anthelmintics: specific recommendations for feline. *VICH Topic GL20*, CVMP/VICH/545/00-FINAL (London, 30 July 2001). https://www.ema.europa.eu/en/documents/scientific-guideline/vich-gl20-efficacy-anthelmintics-specific-recommendations-feline-step-7_en.pdf (accessed 13 December 2021).

**15** Strickland, K.N. (1998). Canine and feline caval syndrome. *Clin. Tech. Small Anim. Pract.* 13: 88–95.

**16** Venco, L., Genchi, C., Genchi, M. et al. (2008). Clinical evolution and radiographic findings of feline heartworm infection in asymptomatic cats. *Vet. Parasitol.* 158: 232–237.

**17** Bowman, D.D. and Atkins, C.E. (2009). Heartworm biology, treatment, and control. *Vet. Clin. North Am. Small Anim. Pract.* 39: 1127–1158.

**18** McCall, J.W., Dzimianski, M.T., McTier, T.L. et al. (1992). Biology of experimental heartworm infection in cats. In: *Proceedings of the American Heartworm Symposium '92, Austin, Texas (27–29 March 1992)* (ed. M.D. Soll), 71–79. Batavia, IL: American Heartworm Society.

**19** Litster, A.L. and Atwell, R.B. (2008). Feline heartworm diseases: a clinical review. *J. Feline Med. Surg.* 10: 137–144.

**20** Nelson, C.T. (2008). *Dirofilaria immitis* in cats: diagnosis and management. *Compend. Contin. Educ. Vet.* 30: 393–400.

**21** Venco, L., Marchesotti, F., and Manzocchi, S. (2015). Feline heartworm disease: a 'Rubik's-cube-like' diagnostic and therapeutic challenge. *J. Vet. Cardiol.* 17: S190–S201.

**22** Simón, F., Kramer, L.H., Román, A. et al. (2007). Immunopathology of *Dirofilaria immitis* infection. *Vet. Res. Comm.* 31: 161–171.

**23** McCall, J.W., Genchi, C., Kramer, L.H. et al. (2008). Heartworm disease in animals and humans. *Adv. Parasitol.* 66: 193–285.

**24** Lee, A.C.Y. and Atkins, C.A. (2010). Understanding feline heartworm infection: disease, diagnosis, and treatment. *Top Comp. Anim. Med.* 25: 224–230.

**25** Dillon, A.R., Blagburn, B.L., Tillson, M. et al. (2017). Heartworm-associated respiratory disease (HARD) induced by immature adult *Dirofilaria immitis* in cats. *Parasites Vectors* 10 (Suppl 2): 514.

**26** Dillon, A.R., Blagburn, B.L., Tillson, M. et al. (2017). The progression of heartworm associated respiratory disease (HARD) in SPF cats 18 months after *Dirofilaria immitis* infection. *Parasites Vectors* 10 (Suppl 2): 533.

**27** Payne, P., Rehm, C., Nelson, C.T. et al. (2020). Current feline guidelines for the prevention, diagnosis, and management of heartworm (*Dirofilaria immitis*) infection in Cats. American Heartworm Soceity. https://heartwormsociety.org/veterinary-resources/american-heartworm-society-guidelines (accessed 13 December 2021).

**28** Nelson, C.T., McCall, J.W., Jones, S., and Moorhead, A. (2020). Current canine guidelines for the prevention, diagnosis, and management of heartworm (Dirofilaria immitis) infection in dogs. American Heartworm Soceity. https://heartwormsociety.org/veterinary-resources/american-heartworm-society-guidelines (accessed 13 December 2021).

**29** Mann, P.H. and Fratta, I. (1953). Transplantation of adult heartworms, *Dirofilaria immitis*, into dogs and cats. *J. Parasitol.* 39: 139–144.

**30** Hawking, P., Sawyer, T.K., and Worms, M.J. (1963). Transport of *Dirofilaria immitis* across the Atlantic and successful reimplantation into dogs. *J. Parasitol.* 49: 1035–1036.

**31** Carlisle, C.H. (1970). The experimental production of heartworm disease in the dog. *Aust. Vet. J.* 46: 190–194.

**32** Donahoe, J.M. (1975). Experimental infection of cats with *Dirofilaria immitis*. *J. Parasitol.* 61: 599–605.

**33** Rawlings, C.A. and McCall, J.W. (1985). Surgical transplantation of adult *Dirofilaria immitis* to study heartworm infection and disease in dogs. *Am. J. Vet. Res.* 46: 221–224.

**34** Atkins, C.E., Keene, B.W., and McGuirk, S.M. (1988). Pathophysiological mechanism of cardiac dysfunction in experimentally induced heartworm caval syndrome in dogs: an echocardiographic study. *Am. J. Vet. Res.* 49: 403–410.

**35** Atkins, C.E., Arther, R.G., Ciszewski, D.K. et al. (2008). Echocardiographic quantification of *Dirofilaria immitis* in experimentally infected cats. *Vet. Parasitol.* 158: 164–170.

**36** Drake, J., Heinz-Loomer, C., Rotenberry, A., and Wiseman, S. (2018). Increasing incidence of *Dirofilaria immitis* in dogs in USA 2013–2017. In: *Proceedings of the 63rd Annual AAVP Meeting*, July 14–17, 2018, Denver, CO, USA.

**37** Drake, J. and Wiseman, S. (2018). Increasing incidence of *Dirofilaria immitis* in dogs in the USA with focus on the southeast region 2013–2016. *Parasites Vectors* 11: 39.

**38** Vidyashankar, A.N., Jimenez Castro, P.D., and Kaplan, R.M. (2017). A statistical approach for evaluating the effectiveness of heartworm preventive drugs: what does 100% efficacy really mean? *Parasites Vectors* 10 (Suppl 2): 516.

**39** Bidgood, A. and Collins, G.H. (1996). The prevalence of *Dirofilaria immitis* in dogs in Sydney. *Aust. Vet. J.* 73: 103–104.

**40** Bowman, D.D., Little, S.E., Lorentzen, L. et al. (2009). Prevalence and geographic distribution of *Dirofilaria immitis, Borrelia burgdorferi, Ehrlichia canis*, and

*Anaplasma phagocytophilum* in dogs in the United States: results of a national clinic-based serologic survey. *Vet. Parasitol.* 160: 138–148.

**41** Vezzani, D., Carbajo, A.E., Fontanarrosa, M.F. et al. (2011). Epidemiology of canine heartworm in its southern distribution limit in South America: Risk factors, inter-annual trend and spatial patterns. *Vet. Parasitol.* 176: 240–249.

**42** Otranto, D., Dantas-Torres, F., Brianti, E. et al. (2013). Vector-borne helminths of dogs and humans in Europe. *Parasites Vectors* 6: 16.

**43** Wang, D., Bowman, D.D., Brown, H.E. et al. (2014). Factors influencing U.S. canine heartworm (*Dirofilaria immitis*) prevalence. *Parasites Vectors* 7: 264.

**44** Nguyen, C., Koh, W.L., Casteriano, A. et al. (2016). Mosquito-borne heartworm *Dirofilaria immitis* in dogs from Australia. *Parasites Vectors* 9: 535.

**45** Plue, R.E., Acre, K.E., Coleman, M.W. et al. (1989, 1989). Efficacy and field safety of two new ivermectin formulations for the prevention of canine heartworm disease. In: *Proceedings of the Heartworm Symposium '89*, March 17–19, 1989, Charleston, South Carolina (ed. G.F. Otto, R.F. Jackson, D.H. Knight, et al.). American Heartworm Society.

**46** European Agency for the Evaluation of Medicinal Products, Veterinary Medicines and Information Technology. (2000). Efficacy requirements for anthelmintics: overall guidelines. *VICH Topic GL7*, CVMP/VICH/832/99-corr (London, 7 December 2000). https://www.ema.europa.eu/en/vich-gl7-efficacy-anthelmintics-general-requirements (accessed 13 December 2021).

**47** European Medicines Agency: EMEA/V/C/003842/000. (2014). Committee for Medicinal Products for Veterinary Use (CVMP), Assessment Report for NEXGARD SPECTRA, 6 November 2014. https://www.ema.europa.eu/en/documents/assessment-report/nexgard-spectra-epar-public-assessment-report_en.pdf (accessed 13 December 2021).

# 18

# The Antifilarial Drug Pipeline

*Natalie A. Hawryluk**

*Bristol Myers Squibb, Global Health Research & Early Development, 10300 Campus Point Drive, Suite 100, San Diego, CA 92121, USA*

**Abstract**

The current therapies for managing human filarial infections fall short in efficient treatment of onchocerciasis and lymphatic filariasis. Current approved treatments, including ivermectin (IVM), diethylcarbamazine (DEC), and albendazole (ABZ), require repeat administration over prolonged periods of time in order to target and eliminate the L1 larval stage (microfilariae). Both single and combinatorial drug therapies for onchocerciasis have drawn concerns over adverse events and increasing emergence of IVM-resistant strains, therefore creating a great need for compounds that exhibit adult stage selectivity (macrofilaricidal).

The reposing of veterinary anthelmintics such as moxidectin (MOX), flubendazole, oxfendazole (OXF), and emodepside has proven to be lucrative with MOX being approved in 2018 for treatment of onchocerciasis and both OXF and emodepside moving into Phase II clinical trials for onchocerciasis. Along with medicinal chemistry efforts, multiple compounds have shown significant progress toward the treatment of filarial diseases; however, more information and dedicated discovery efforts are needed to determine the future of antifilarial treatments.

## 18.1 Current Therapies for Control of Filarial Diseases

The current therapies (Table 18.1) for managing human filarial infections target the elimination of microfilariae (mf), the L1 larval stage, via repeated administration of ivermectin (IVM) or diethylcarbamazine (DEC), in combination with albendazole (ABZ) for lymphatic filariasis (LF). In addition, these drugs provide a prolonged block of the production of new mf, leading to sustained reductions in transmission and, for onchocerciasis, pathology [1]. IVM is one of most extensively used antiparasitic drugs in human and veterinary medicine. To date, the impact of IVM on human filarial diseases has been astonishing. Owing to widespread distribution of IVM,

*Corresponding author.

*Human and Animal Filariases: Landscape, Challenges, and Control*, First Edition.
Edited by Ronald Kaminsky and Timothy G. Geary.

**Table 18.1** Currently approved therapies.

| Molecule | Source/affiliation | Mechanism of action |
|---|---|---|
| ABZ<br>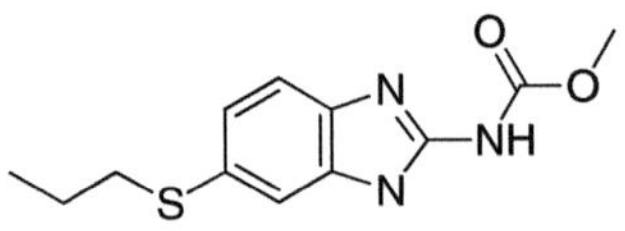 | Originally from Smith Kline Corporation | β-tubulin inhibitor |
| DCE<br>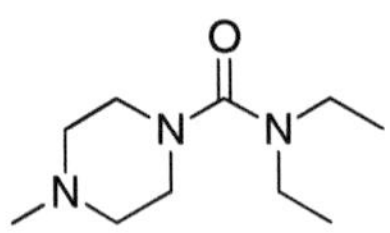 | American Cyanamid Co | Interferes with arachidonic acid and nitric oxide metabolism |
| IVM<br>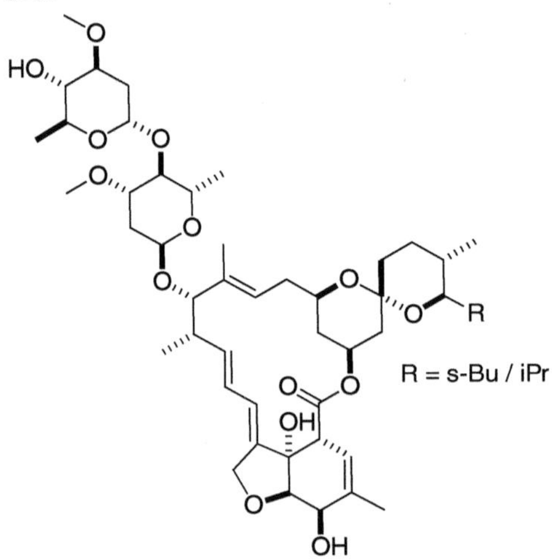 | Merck & Co | Chloride channel modulators |
| MOX<br>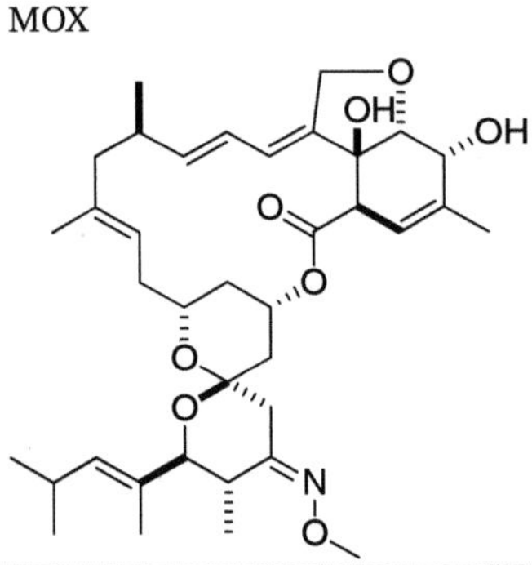 | American Cyanamid Co/Wyeth/Wyeth Holdings Corp<br>Medicines Development for Global Health | Chloride channel modulators |

the Mectizan Donation Program run by Merck & Co. has reduced the occurrence of onchocerciasis and LF for millions of people in the poorest countries in the world [2]. While DEC has shown evidence of macrofilaricidal activity and has been used extensively to treat LF since 1947, it is used as a microfilaricide and long-term sterilant. The mechanism of action of DEC remains unclear, though it has been suggested that it interferes with arachidonic acid and nitric oxide metabolism in microfilariae [3, 4]. Lastly, ABZ, part of the benzimidazole class of broad-spectrum anthelmintic agents, is generally used to treat enteric helminthiasis; however, more recently it has been used in combination for treatment of LF. ABZ functions via inhibition of tubulin polymerization [1, 5].

### 18.1.1 Ivermectin, Diethylcarbamazine, Albendazole

Annual mass drug administration (MDA) programs for the treatment and control of LF rely on three agents, DEC, IVM and ABZ, as either single drugs or in combination. For over 30 years, MDA of IVM has been used as a single agent for treatment of onchocerciasis in endemic areas [6]. IVM, as a single oral dose of 150 μg/kg every 6–12 months, prevents disease transmission by decreasing the number of skin mf (mf per mg of skin) for up to 30 days post treatment and causes long-term sterility of female worms. IVM, however, is not a macrofilaricide, as it does not kill the adult worms. Neurological events consisting of coma, some of which resulted in encephalopathy, Parkinson's disease, or death, have been observed in a limited patient population with very high blood levels of *Loa loa* mf treated with IVM in loiasis and onchocerciasis co-endemic areas of Cameroon and the Democratic Republic of Congo (DRC) [7]. DEC is effective against loaisis; however, it is not recommended in areas of high levels of *L. loa* microfilaremia due to severe adverse side effects, including encephalitis, retinal hemorrhage, and Mazotti reactions, all caused by the rapid immune-mediated killing of mf [8]. Concerns of serious *L. loa*-related post-IVM adverse events and the possible emergence of IVM-resistant *O. volvulus* have prompted the need for compounds that exhibit adult stage selectivity (macrofilaricidal) or longer-lasting (preferably permanent) sterilizing effects.

### 18.1.2 Combination Therapies

Utilization of two drug combinations, IA (IVM and ABZ) or DA (DEC and ABZ), is highly dependent on geography and co-endemicity of LF with either onchocerciasis or loiasis. The WHO recently amended its guidelines for LF MDA to allow for triple-drug therapy, IDA (IVM, DEC, and ABZ) in onchocerciasis or loiasis non-endemic areas. IDA was shown to be the most effective regimen and was equally well-tolerated compared with the current DEC + ABZ MDA regimen [9–12].

### 18.1.3 Moxidectin

In 2018, MOX was approved for onchocerciasis for people over the age of 11 in the United States (https://www.prnewswire.com/news-releases/us-fda-approves-moxidectin-for-the-treatment-of-river-blindness-300666114.html). Similar to IVM, MOX is a repurposed veterinary medicine; the macrocyclic lactone (ML) class of anthelmintics acts by binding to glutamate-gated chloride channels (GluCls) [1]. However, MOX has a considerably longer half-life in plasma, 20 days compared with IVM (two days), allowing for either less frequent treatment or improved efficacy with similar frequency of treatment. MOX has also demonstrated effectiveness in IVM-resistant animal helminth infections [13, 14]. Development of MOX as an oral treatment of onchocerciasis was supported by Medicines Development for Global Health (MDGH) through the Global Health Investment Fund (GHIF) program [15]. In multiple clinical trials comparing the efficacy, safety, and tolerability of MOX and IVM in *O. volvulus* infected patients, the data suggested that MOX was efficacious with a favorable safety profile for MDA [16]. MOX showed superiority to IVM at 12 months posttreatment using skin mf density as the primary efficacy outcome; the MOX cohort (adjusted geometric mean 0·6 [95% CI 0·3–1·0]) was lower than the IVM group (adjusted geometric mean 4·5 [95% CI 3·5–5·9]; a skin mf density difference of 3·9 [95% CI 3·2–4·9], $p < 0.0001$; treatment difference 86%) [17].

While these current treatments are successful in reducing the prevalence of the diseases, administration of the therapies requires repeated treatments for 10–12 years, as they do not kill adult worms.

## 18.2 Experimental Therapies and Drug Discovery Approaches

The IDA (triple-drug) regimen recently adopted for the treatment and control of LF appears to be macrofilaricidal [1, 9–12], reducing the urgency of developing new drugs for this indication. However, current treatments used for elimination of onchocerciasis have limitations, in that they are ineffective against the macrofilaria, or adult worms, thus requiring repeated and prolonged treatments. The need for a novel macrofilaricide is imperative to achieve eradication of this disease. This is further complicated by the threat of emerging resistance. Recently, there has been a focus on finding well-tolerated macrofilaricidal drug regimens either through re-evaluating known molecules or dedicated drug discovery efforts (Table 18.2). Another approach has been new antibacterial drugs that target the *Wolbachia* symbiont of the major human filarial pathogens and are discussed elsewhere in this book [1, 21]

### 18.2.1 Flubendazole

A member of the benzimidazole (BZD) class, flubendazole (FBZ), is reported to be the best macrofilaricidal molecule within this class, having demonstrated lethal effects on multiple filarial species in animal models [22]. Initially developed as a paste by Janssen Pharmaceutical N.V. for treatment of gastrointestinal nematode infections in companion animals, FBZ is approved for human use in Europe and a good candidate for therapy against filariasis [23]. Unfortunately, similar to other

**Table 18.2** Current experimental therapies.

| Molecule | Source/ affiliation | Mechanism of action | Current status |
|---|---|---|---|
| FBZ 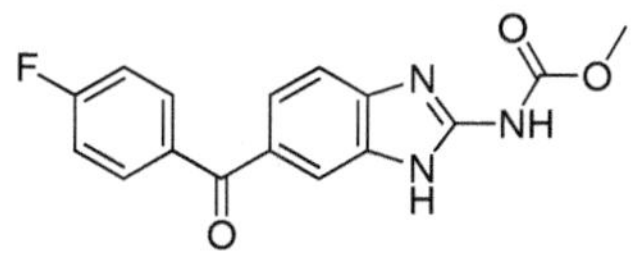 | Janssen Pharmaceutica/DNDi | β-tubulin inhibitor | Clinical evaluation halted |
| OXF 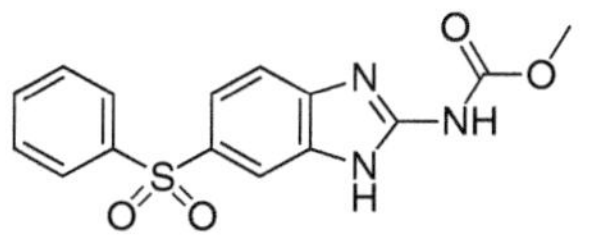 | Originally from Syntex/DNDi | β-tubulin inhibitor | Phase I MAD completed-2019 |
| Emodepside 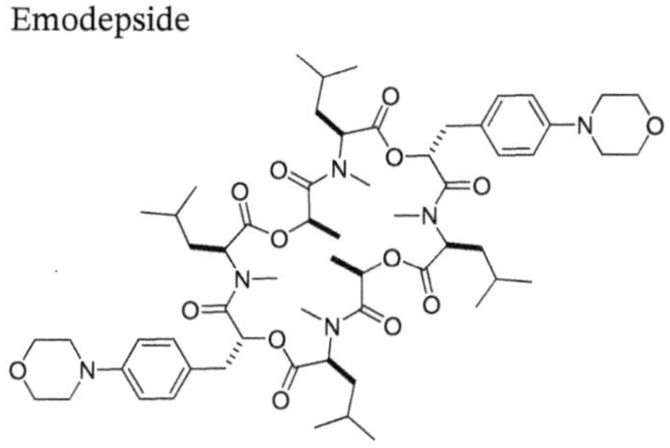 | Bayer HealthCare/DNDi | Modulator of the BK/SLO-1 $Ca^{2+}$ -activated $K^+$ channel | Phase I FIH completed 2017<br>Phase I MAD completed 2018<br>Phase II planned 2019 |
| Auranofin [18] 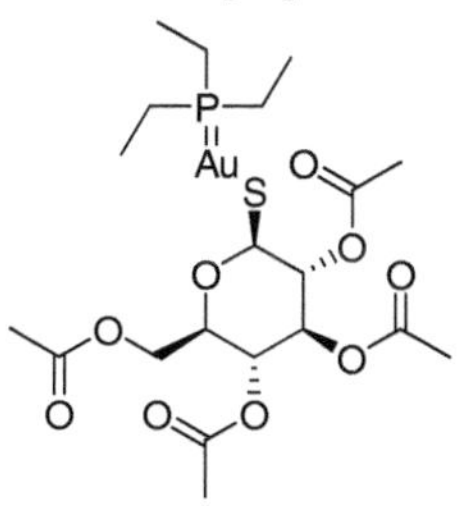 | Originally from Smith Kline Corporation/UCSF | Thioredoxin reductase (TrxR) inhibitor | Unknown |
| Anacor-AN8799 [19, 20] 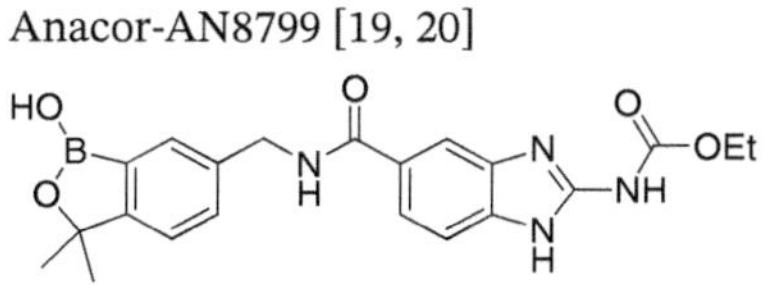 | Anacor (currently Pfizer) | β-tubulin inhibitor | POC achieved in preclinical models |
| Anacor-AN-15470 [20] 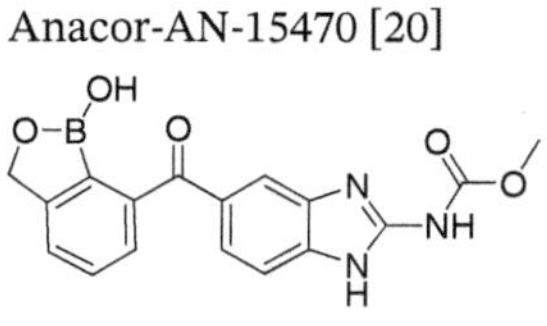 | Anacor (currently Pfizer) | β-tubulin inhibitor | POC achieved in preclinical models |

FIH: First in Human; MAD: Multiple Ascending Dose; POC: Proof of Concept

BZD anthelmintics, FBZ has limited water solubility and standard oral human dosage forms as tablets or suspensions displayed low systemic bioavailability [24]. An improved formulation of FBZ was reported by Janssen and several partners, including DND*i* and AbbVie, with enhanced bioavailability for use as an oral macrofilaricide for the treatment of LF and onchocerciasis (https://www.jnj.com/media-center/press-releases/johnson-johnson-joins-public-and-private-partners-in-the-largest-coordinated-action-to-date-to-eliminate-or-control-neglected-tropical-diseases). An amorphous solid dispersion formulation (ASD FBZ) was developed by Janssen and partners, which exhibits good oral bioavailability in animals [25]. ASD FBZ provided improved systemic absorption for efficacy and pharmacokinetic evaluation and allowed for preclinical evaluation. ASD FBZ dosed orally (0.2–15 mg/kg) for five days showed no significant decrease in worm burden in jirds (*Meriones unguiculatus*) infected with the filarial nematode *Brugia pahangi*, in contrast to five days of single subcutaneous (SC) injections of 10 mg/kg of FBZ. At doses as low as 1.5 mg/kg, extensive ultrastructural damage in developing embryos and mf was observed [25]. Dose and treatment duration dependent reduction of adult worm burden was observed with ASD FBZ in the *L. sigmodontis* infected jird model. Oral dosing with ASD FBZ for 10 days at 15 mg/kg gave a 95% reduction in worm burden and a 25% cure rate. Comparatively, SC administration of FBZ (2 mg/kg single dose) was more effective, providing 98.2% reduction in the adult worm burden and a cure rate of 58%. Additionally, oral ASD FBZ treatment, 15 mg/kg for five days, significantly reduced the number of later embryonal stages within female adult worms, with 43% of analyzed female worms being patent compared with 91% in the control group. Oral ASD FBZ treatment caused degeneration of filarial uterine tissue, preventing filarial embryogenesis and the release of mf; when compared with SC administration of FBZ, oral ASD FBZ did not prove to be as effective in reducing adult worm burden [26].

While being effective, FBZ also interferes with microtubules and has reproductive toxicity [27]. It is reported that embryotoxicity had been observed at concentrations above 0.25 μg/ml *in vivo* [28–30]. It has been suggested that exposure duration in patients should not exceed one day, due to carcinogenic risk associated with a positive effect in the *in vivo* micronucleus test. Utilization of ASD FBZ allowed re-evaluation of the toxicity of FBZ, including potential genotoxicity. The teratogenicity and FBZ-induced toxicity risk were deemed monitorable and controllable in patients if stringent precautions were applied with respect to testicular toxicity [31]. However, it was concluded that the risk/benefit associated with the use of orally bioavailable FBZ as a macrofilaricide did not substantiate further development [27].

### 18.2.2 Oxfendazole

As an improved profile was desired, a BZD related to FBZ, oxfendazole (OXF), was identified as a drug candidate for the treatment of onchocerciasis with improved oral bioavailability compared with FBZ. OXF has a moderate effect on *O. gutturosa* adult worm motility *in vitro*; only 16% reduction was observed after 24 hours at the highest concentration of 50 μM. When tested against *O. volvulus* pre-adults (L5 stage), motility inhibition after 14 days of exposure to OXF was marginal ($IC_{50}$ = 7.59 μM); however, after washout and up to 33 additional days in culture, almost complete inhibition of motility was observed ($IC_{50}$ = 0.1 μM). OXF dosed at 12.5 and 25 mg/kg BID PO or 5 and 25 mg/kg QD SC for five days achieved sterile cure (100% adult worm reduction) in *L. sigmodontis* infected mice. Oral OXF treatments (10 days, 5 mg/kg, BID) in patently infected jirds reduced peripheral microfilaremia by >99% within three weeks post treatment along with complete inhibition of embryogenesis in the remaining female adult worms at necropsy 16 weeks after treatment [32]. OXF does not cause rapid mf clearance, which is deemed critical with regard to potential microfilaricidal toxicity associated with rapid mf death [32]. Compared with oral FBZ, oral OXF is more efficacious, achieving a complete reduction of *L. sigmodontis* adult worm burden in infected mice when administered as a conventional suspension. Additionally, OXF has yet to exhibit analogous toxicity to FBZ in cell-based toxicity assays. OXF is currently under development for the treatment of neurocysticercosis (NCT02234570). Based on encouraging Phase I study data in a multiple ascending dose evaluation of the safety and PK of OXF (3, 7.5, or 15 mg/kg) given orally daily to healthy volunteers, oral administration for five consecutive days at the predicted human efficacious dose, 1.5–4.1 mg/kg, is considered achievable. Based on preclinical efficacy and Phase I data, DND*i* is repurposing OXF as a macrofilaricidal treatment for filarial indications along with a planned proof-of-concept Phase II clinical study for onchocerciasis [33].

### 18.2.3 Emodepside

Emodepside (BAY 44-4400), a potent macrofilaricide, is being investigated in clinical trials as a potential treatment for onchocerciasis. Emodepside, a cyclooctadepsipeptide, is registered in Europe and the United States as a combination product for the treatment of gastrointestinal nematode infections in companion animals. Emodepside is a novel modulator of BK/SLO-1 $Ca^{2+}$ -activated $K^{+}$ channels in nematodes, insects, and humans [34], which paralyzes adult filarial worms via a mode of action distinct from previous anthelmintics [1, 15]. In 2014, Bayer HealthCare entered into a legal agreement with DND*i* for collaboration on development of emodepside. Favorable preliminary results were observed in a first-in-human double-blind, placebo-controlled study of single ascending doses in healthy male volunteers (NCT02661178) and in a repeat dose study evaluating pharmacokinetic, safety, and tolerability of a liquid formulation given over 10 days (NCT03383614) [35]. Emodepside was found to be safe up to 40 mg and well tolerated up to 20 mg, with rapid absorption, dose proportionality, and a long terminal half-life [36]. The evaluation of safety and efficacy of emodepside in a Phase II clinical trial in Ghana is planned [37, 38].

### 18.2.4 Auranofin

Auranofin, a registered antirheumatoid agent, was identified in a high-throughput screen of over 2000 FDA-approved drugs on adult *Brugia* species as a potential macrofilaricide by scientists from the University of California, San Francisco [18]. Auranofin is a gold complex targeting thioredoxin reductase (TrxR), a key enzyme involved in defense against oxidative damage due to oxygen metabolism. It is suggested that TrxR is likely the target of auranofin in *Brugia*. Auranofin is highly effective in killing both male and female *B. malayi* and *Onchocerca* adult worms and inhibiting molting of *O. volvulus* L3 to the L4 stage with $IC_{50}$ values <1 μM. Auranofin shows a 58–91% overall reduction of *B. pahangi* adult worm burden *in vivo* in infected jird models (5 mg/kg BID weekdays and SID weekends for 28 days for a total of 48 doses). Additionally, it is not effective in killing *O. ochengi* and *L. loa* microfilariae, an important consideration for treatment in *L. loa* endemic areas. Since auranofin is already an FDA-approved drug, the path to clinical trials could be efficient, particularly since rheumatoid arthritis patients who were treated with auranofin for an average of six months exhibited few side effects, with the most common being diarrhea [39].

### 18.2.5 AN8799 and AN15470

In order to overcome the limited solubility and embryotoxicity associated with FBZ, Anacor (currently Pfizer) prepared benzoxaborole-benzimidazole analogs of FBZ [40, 41]. Initially, several small-molecule boron-containing molecules for the treatment of river blindness demonstrated killing of *B. malayi* adult worms in one to two days at 10 μM *in vitro* [40]. More recently, a series of benzimidazole–benzoxaborole amide linked hybrid molecules were described that exhibited good *in vitro* activity against *O. volvulus*. In particular, AN8799 showed acceptable pharmacokinetic properties and when dosed s.c. as a suspension at 100 mg/kg/day for 14 days demonstrated effective killing of *B. malayi*, *B. pahangi*, and *L. sigmodontis* in infected gerbils [40]. Although the proof-of-concept efficacy demonstrated with AN8799 was encouraging, the lack of oral bioavailability was of concern, prompting further modification of the series. Further optimization of a benzimidazole–benzoxaborole series resulted in hybrid molecules linked via a ketone versus the amide. This resulted in the identification of AN15470, which maintained acceptable pharmacokinetic properties and enabled evaluation by oral dosing in *in vivo* filarial models. When dosed for 14 or 28 days, AN15470 was 99% effective in killing female *B. pahangi* worms in infected gerbils [41]. Further evaluation of AN15470 in other *in vivo* efficacy models, including the examination of worm damage using transmission electron microscopy (TEM), is ongoing and to be reported.

### 18.2.6 1,2,4-Thiadiazol-5-amines

Recently, Celgene Global Health (currently Bristol Myers Squibb) disclosed a series of substituted di(pyridin-2-yl)-1,2,4-thiadiazol-5-amines as potential macrofilaricides. This series of compounds are effective at *ex vivo* killing of adult *O. gutturosa*, *B. malayi*, *B. pahangi*, and *L. sigmodontis* [42]. Further evaluation of these compounds is ongoing and will be reported in due course.

## 18.3 Conclusion

Significant progress has been made toward the goals of elimination and control of filarial diseases through repurposing veterinary drugs and screening of FDA-approved compounds. While IVM has been vital for the treatment of filarial diseases, it has been implied that IVM treatments will not lead to the elimination of onchocerciasis in Africa [19]. The effectiveness of IVM has been compromised due to growing concerns of resistance development, which was initially observed in ruminants. With the recent registration of MOX for human use, it has been suggested that MOX selects less rapidly for resistance; though there is evidence that MOX-resistant parasite populations have evolved, and that MOX remains more potent than the other avermectins against nematodes exhibiting ML resistance [20, 43]. Similar concerns for emerging resistance with ABZ and DEC have been expressed with regard to LF [44] and for ABZ in soil-transmitted helminths [45].

Use of combination IDA treatment in LF has been shown to be beneficial and should be examined further in onchocerciasis, after depleting mf levels to avoid DEC-induced toxicity. Regarding macrofilaricides, recent modeling data suggest that a macrofilaricidal-only drug regimen may not be sufficient to reach elimination goals and a combination of macrofilaricidal and microfilaricidal treatments would provide the ideal outcome in a timely manner [46]. The therapies currently in the clinic will provide critical information for future dedicated discovery efforts for antifilarial drugs.

## References


1. Geary, T.G., Long, A., and Tritten, L. (2022). Current antifilarial drugs – mechanisms of action. In: *Advances in Control of Heartworm and Human Filariases* (ed. R. Kaminsky and T.G. Geary), Chapter 10. Weinheim, Germany: Wiley-VCH.
2. Laing, R., Gillan, V., and Devaney (2017). Ivermectin – old drug, new tricks? *Trends Parasitol.* 33 (6): 463–472.
3. Maizels, R.M. and Denham, D.A. (1992). Diethylcarbamazine (DEC): immunopharmacological interactions of an anti-filarial drug. *Parasitology* 105 (Suppl): S49–S60.
4. McGarry, H.F., Plant, L.D., and Taylor, M.J. (2005). Diethylcarbamazine activity against *Brugia malayi* microfilariae is dependent on inducible nitric-oxide synthase and the cyclooxygenase pathway. *Filaria J.* 4: 4.
5. Taman, A. and Azab, M. (2014). Present-day anthelmintics and perspectives on future new targets. *Parasitol. Res.* 113 (7): 2425–2433.

6 Ashour, D.S. (2019). Ivermectin: From theory to clinical application. *Int. J. Antimicrob. Agents* 54 (2): 134–142.

7 Holmes, D. (2013). *Loa loa*: neglected neurology and nematodes. *Lancet Neurol.* 12 (7): 631–632.

8 Townson, S., Ramirez, B., Fakorede, F. et al. (2007). Challenges in drug discovery for novel antifilarials. *Expert Opin. Drug Discovery* 2 (s1): S63–S73.

9 Fischer, P.U., King, C.L., Jacobson, J.A., and Weil, G.J. (2017). Potential value of triple drug therapy with ivermectin, diethylcarbamazine, and albendazole (IDA) to accelerate elimination of lymphatic filariasis and onchocerciasis in Africa. *PLoS Negl.Trop. Dis.* 11 (1): e0005163.

10 Irvine, M.A., Stolk, W.A., Smith, M.E. et al. (2017). Effectiveness of a triple-drug regimen for global elimination of lymphatic filariasis: a modelling study. *Lancet Infect. Dis.* 17 (4): 451–458.

11 King, C.L., Suamani, J., Sanuku, N. et al. (2018). A trial of a triple-drug treatment for lymphatic filariasis. *N. Engl. J. Med.* 379 (19): 1801–1810.

12 Weil, G.J., Bogus, J., Fischer, P.U. et al. (2019). The safety of double- and triple-drug community mass drug administration for lymphatic filariasis: A multicenter, open-label, cluster-randomized study. *PLoS Med.* 16 (6): e1002839.

13 Prichard, R.K. (2022). Drug resistance in filariae. In: *Advances in Control of Heartworm and Human Filariases* (ed. R. Kaminsky and T.G. Geary), Chapter 11. Weinheim, Germany: Wiley-VCH.

14 Bowman, D.D., McTier, T.L., Mahabir, S.P. et al. (2017). Evaluation of the efficacy of ProHeart(®) 6 (moxidectin) against a resistant isolate of *Dirofilaria immitis* (JYD-34) in dogs. *Parasites Vectors* 10 (Suppl 2): 502.

15 Kuesel, A.C. (2016). Research for new drugs for elimination of onchocerciasis in Africa. *Int. J. Parasitol. Drugs Drug Resist.* 6 (3): 272–286.

16 Milton, P., Hamley, J.I.D., Walker, M., and Basáñez, M.G. (2020). Moxidectin: an oral treatment for human onchocerciasis. *Expert Rev. Anti. Infect. Ther.* 18 (11): 1067–1081.

17 Opoku, N.O., Bakajika, D.K., Kanza, E.M. et al. (2018). Single dose moxidectin versus ivermectin for *Onchocerca volvulus* infection in Ghana, Liberia, and the Democratic Republic of the Congo: a randomised, controlled, double-blind phase 3 trial. *Lancet* 392 (10154): 1207–1216.

18 Bulman, C.A., Bidlow, C.M., Lustigman, S. et al. (2015). Repurposing auranofin as a lead candidate for treatment of lymphatic filariasis and onchocerciasis. *PLoS Negl.Trop. Dis.* 9 (2): e0003534.

19 Boussinesq, M. (2008). Onchocerciasis control: biological research is still needed. *Parasite* 15 (3): 510–514.

20 Prichard, R., Menez, C., and Lespine, A. (2012). Moxidectin and the avermectins: Consanguinity but not identity. *Int. J. Parasitol. Drugs Drug Resist.* 2: 134–153.

21 Hübner, M.P., Pfarr, K., and Hoerauf, A. (2022). *Wolbachia* endosymbionts as treatment targets for filarial diseases. In: *Advances in Control of Heartworm and Human Filariases* (ed. R. Kaminsky and T.G. Geary), Chapter 24. Weinheim, Germany: Wiley-VCH.

**22** Ceballos, L., Mackenzie, C., Geary, T. et al. (2014). Exploring the potential of flubendazole in filariasis control: evaluation of the systemic exposure for different pharmaceutical preparations. *PLoS Negl.Trop. Dis.* 8 (5): e0002838.

**23** Mackenzie, C.D. and Geary, T.G. (2011). Flubendazole: a candidate macrofilaricide for lymphatic filariasis and onchocerciasis field programs. *Expert Rev. Anti Infect. Ther.* 9 (5): 497–501.

**24** Moreno, L., Alvarez, L., Mottier, L. et al. (2004). Integrated pharmacological assessment of flubendazole potential for use in sheep: disposition kinetics, liver metabolism and parasite diffusion ability. *J. Vet. Pharmacol. Ther.* 27 (5): 299–308.

**25** Fischer, C., Ibiricu Urriza, I., Bulman, C.A. et al. (2019). Efficacy of subcutaneous doses and a new oral amorphous solid dispersion formulation of flubendazole on male jirds (*Meriones unguiculatus*) infected with the filarial nematode *Brugia pahangi. PLoS Negl. Trop. Dis.* 13 (1): e0006787.

**26** Hübner, M.P., Ehrens, A., Koschel, M. et al. (2019). Macrofilaricidal efficacy of single and repeated oral and subcutaneous doses of flubendazole in *Litomosoides sigmodontis* infected jirds. *PLoS Negl.Trop. Dis.* 13 (1): e0006320.

**27** Geary, T.G., Mackenzie, C.D., and Silber, S.A. (2019). Flubendazole as a macrofilaricide: History and background. *PLoS Negl.Trop. Dis.* 13 (1): e0006436.

**28** Longo, M., Zanoncelli, S., Colombo, P.A. et al. (2013). Effects of the benzimidazole anthelmintic drug flubendazole on rat embryos in vitro. *Reprod. Toxicol.* 36: 78–87.

**29** Longo, M., Zanoncelli, S., Messina, M. et al. (2014). In vivo preliminary investigations of the effects of the benzimidazole anthelmintic drug flubendazole on rat embryos and fetuses. *Reprod. Toxicol.* 49: 33–42.

**30** Tweats, D.J., Johnson, G.E., Scandale, I. et al. (2016). Genotoxicity of flubendazole and its metabolites in vitro and the impact of a new formulation on in vivo aneugenicity. *Mutagenesis* 31 (3): 309–321.

**31** Lachau-Durand, S., Lammens, L., van der Leede, B.-j. et al. (2019). Preclinical toxicity and pharmacokinetics of a new orally bioavailable flubendazole formulation and the impact for clinical trials and risk/benefit to patients. *PLoS Negl.Trop. Dis.* 13 (1): e0007026.

**32** Hubner, M.P., Martin, C., Specht, S. et al. (2020). Oxfendazole mediates macrofilaricidal efficacy against the filarial nematode *Litomosoides sigmodontis in vivo* and inhibits *Onchocerca spec.* motility *in vitro. PLoS Negl.Trop. Dis.* 14 (7): e0008427.

**33** DNDi. (2018). Oxfendazole. Retrieved 24 February 2022, from https://www.dndi.org/diseases-projects/portfolio/oxfendazole/

**34** Crisford, A., Ndukwe, K., O'Connor, V. et al. (2015). The cyclooctadepsipeptide anthelmintic emodepside differentially modulates nematode, insect and human calcium-activated potassium (SLO) channel alpha subunits. *PLoS Negl.Trop. Dis.* 9 (10): e0004062.

**35** DNDi. (2016). *First in man clinical trial of emodepside (BAY 44-4400)*, NCT02661178. clinicaltrials.gov.

**36** Gillon, J.Y.A., van den Berg, F., Dequatre Cheeseman, K., Hopchet, N. et al. (2018) A single-center, First-in-Human, randomized, double-blind, placebo-controlled, parallel-group study to investigate the safety, tolerability and pharmacokinetics of escalating single doses of emodepside (BAY44-4400) in healthy male subjects Annual Society of Tropical Medicine and Hygiene Annual Meeting, New Orleans, LA.

**37** DNDi. (2018). Emodepside. Retrieved 24 February 2022, from https://www.dndi.org/diseases-projects/portfolio/emodepside/.

**38** DNDi. (2019). Emodepside-Phase II. Retrieved 24 February 2022, from https://dndi.org/research-development/portfolio/emodepside/.

**39** Furst, D.E. (1983). Mechanism of action, pharmacology, clinical efficacy and side effects of auranofin. An orally administered organic gold compound for the treatment of rheumatoid arthritis. *Pharmacotherapy* 3 (5): 284–298.

**40** Akama, T., Freund, Y.T., Berry, P. et al. (2019). Macrofilaricidal benzimidazole-benzoxaborole hybrids as an approach to the treatment of river blindness. Part 1. Amide linked analogs. *ACS Infect. Dis.* 6 (2): 173–179.

**41** Carter, D.S., Jacobs, R.T., Freund, Y.R. et al. (2020). Macrofilaricidal benzimidazole-benzoxaborole hybrids as an approach to the treatment of river blindness: Part 2. Ketone linked analogs. *ACS Infect. Dis.* 6 (2): 180–185.

**42** Hawryluk, N., Canan, S.S., Condroski, K.R., et al. (2020). Heterocyclic compounds and their use for treatment of helminthic infections and diseases. US Patent application #20200339559, published 29 October.

**43** Prichard, R.K., Geary, T.G., and T.G. (2019). Perspectives on the utility of moxidectin for the control of parasitic nematodes in the face of developing anthelmintic resistance. *Int. J. Parasitol. Drugs Drug Resist.* 10: 69–83.

**44** Schwab, A.E., Boakye, D.A., Kyelem, D., and Prichard, R.K. (2005). Detection of benzimidazole resistance-associated mutations in the filarial nematode *Wuchereria bancrofti* and evidence for selection by albendazole and ivermectin combination treatment. *Am. J. Trop. Med. Hyg.* 73 (2): 234–238.

**45** Prichard, R.K. (2007). Markers for benzimidazole resistance in human parasitic nematodes? *Parasitology* 134 (8): 1087–1092.

**46** Walker, M., Hamley, J.I.D., Milton, P. et al. (2020). Designing antifilarial drug trials using clinical trial simulators. *Nat. Commun.* 11 (1): 2685.

# Part III

# New Frontiers for Control of Antifilarial Diseases

# 19

# The Host–Helminth Interface as a Rich Resource for Novel Drug Targets

*Thomas B. Duguet*[1] *and Lucienne Tritten*[2,3,4,*]

[1] *INVENesis Sàrl, rue de Neuchâtel 15A, CH-2072 Saint-Blaise, Switzerland*
[2] *University of Zurich, Institute of Parasitology, Winterthurerstrasse 266a, CH-8057 Zurich, Switzerland*
[3] *Swiss Tropical and Public Health Institute, Kreuzstrasse 2, CH-4123 Allschwil, Switzerland*
[4] *University of Basel, CH-4000 Basel, Switzerland*

**Abstract**

The evolutionary arms races underlying all host–helminth associations have resulted in intricate negotiations enabling a successful infection. At the host–parasite interface, a continuous dialogue takes place, where a plethora of molecules are exchanged, with crucial consequences on the outcome of the infection. This molecular exchange appears to be a rich source of potential therapeutic targets. Here, we present what recent research has brought to light on this topic for helminths, focusing on filarial nematodes.

## 19.1 Helminth–Host Interactions

Among the highly abundant and diverse helminth species, parasitism is a prevalent lifestyle. It is estimated that between 75 000 and 300 000 helminth species are parasites of vertebrates [1]. In phylum Nematoda alone, parasitism is thought to have evolved at least 15 times independently [2, 3]. Hence, even in humans (until recently), being parasitized by nematodes represented the normal situation rather than an exception. Being important pathogens of plants, humans, and animals, there is considerable interest in understanding the molecular and genomic adaptations of nematodes that enable parasitism [4]. In addition, the interactive interface between host and parasite, shaped by the constant exchange of mediators in a molecular dialogue, may reveal key molecules that support successful infection. Parasitic helminths have evolved various strategies to thrive in their vertebrate hosts, in a context of activated immune defenses. These strategies encompass evasion from, and modulation or subversion of, each component of the immune response that clears or prevents infection by nonadapted parasite species [5]. An evolutionary arms race underlies all host–parasite associations in which hosts

*Corresponding author.

*Human and Animal Filariases: Landscape, Challenges, and Control*, First Edition.
Edited by Ronald Kaminsky and Timothy G. Geary.

have constrained the detrimental impact of parasites, while the pathogens have learned how to overcome host responses to survive and propagate. Hence, the intricate negotiations between hosts and parasites have led to the evolution of highly elaborate relationships, characterized by complex molecular dialogues in precisely delimited conditions that permit a successful and sustained infection. In that context, much attention has been given to molecules released by parasites that influence host immune responses. The best characterized molecules are soluble excretory/secretory (E/S) proteins. Their immunomodulatory properties have been shown in many instances to facilitate the establishment of infection and result in the creation of a permissive environment for parasites to thrive [6]. Among well-studied immunomodulatory E/S proteins from filarial nematodes are cytokine homologs and protease inhibitors (e.g., MIF-1 and serpin-2 of *Brugia malayi*), or the strong anti-inflammatory phosphorylcholine-containing glycoprotein ES-62 from the filarial nematode *Acanthocheilonema viteae* [7, 8]).

Filarial nematodes have complex life cycles, as all are transmitted by various arthropod vectors to a restricted set of permissive vertebrate hosts. Some species rely on the endosymbiotic bacterium *Wolbachia*, some are endosymbiont-free. Filarial infections are not acutely life-threatening; they may persist for many years and can cause long-term suffering and tissue damage. The adult worms reside in (or migrate through) specific tissues where they reproduce and release microfilariae, the characteristic first-stage larvae [9]. These circulate in blood or localize to the skin of their hosts. Hence, different life histories are expected to be reflected in the different molecules released at the host–parasite interface, since the challenges parasites experience in different hosts and localizations differ.

Current control of filarial infections mainly relies on chemotherapeutic strategies, predominantly directed at microfilariae in humans and developing larvae in companion animals. Several gaps remain in the spectrum of action of our available drug panel, which may compromise elimination goals. Few treatments used in mass drug administration campaigns show direct adulticidal (macrofilaricidal) effects, which are predicted to accelerate progress toward global elimination of human filariases [10]. Since 2017, the World Health Organization recommends a triple-drug therapy for lymphatic filariasis, which appears to be macrofilaricidal and is now being broadly implemented [11, 12]. In addition, the modes of action of these compounds have not always been fully resolved, a lack of knowledge that may hamper their optimal use. Despite the millions of people (and even more animals) affected by filarial infections, the pipeline of new antifilarial candidates is limited. Therefore, efforts to discover novel drug targets are still needed [13].

In addition to killing the parasite, finding ways to render permissive hosts nonpermissive represents an alternative approach. The idea behind vaccines is that a well-rounded immune response is elicited, engaging both innate and adaptive arms of immunity, conferring long-lasting protection against a pathogen [14]. However, our understanding of how components of host immunity interact, and how hosts and parasites function together, is limited, impeding the implementation of a rational antigen targeting strategy. Capturing global changes induced by an infection at the systems-, cellular and molecular levels has become possible through recently developed analytical tools and shows promise for the development of targeted strategies that prevent infection.

Genomes, transcriptomes, and proteomes of filarial parasites are becoming increasingly available [15] and will open new avenues to illuminate new targets for infection control and diagnostics. Research aimed at identifying the molecules exchanged by parasites and hosts will build our understanding of the interactive interface between them, enabling us to decipher the mechanisms used to influence each other and allowing us to deduce the important molecular events that result in an established infection. This knowledge can ultimately support identification of vulnerable nematode pathways that can be targeted by novel interventions. This chapter provides an overview of current research advancements in this field and provides a perspective on their potential as therapeutics.

## 19.2 Filarial Nematodes Release Proteins

E/S products refer to molecules released into the host environment by parasites either through excretion (e.g., waste products or uterine fluid) or by an active secretory process; at least some of these molecules are expected to exert functional roles at the host–parasite interface. Analysis using sensitive methods such as mass spectrometry and genome mining approaches allowed the generation of large E/S protein lists. These represent various types of molecules, including a large number of enzymes, parasite-derived antigens, etc. [16]. Because they circulate in the extracellular compartment, E/S proteins have captured substantial interest for the development of therapeutic and diagnostic applications, since they are more accessible to drugs compared with somatic proteins [17]. E/S proteins are involved in various essential biological processes, including proteolysis, cell–cell communication, and nutrient acquisition [18]. Roles of parasite E/S proteins in the modulation of host immune functions are also broadly recognized; they may interfere with complement activation, chemotaxis, lymphocyte responses, etc. [7, 18]. Table 19.1 summarizes briefly the available filarial "secretomes."

The contribution of E/S proteins to immune evasion is believed to be very important. Indeed, the fact that ivermectin-treated *B. malayi* microfilariae lose their capacity to release E/S proteins *in vitro* has been proposed as mode of action of the drug [28–31]. By impairing the contractility of muscles around the secretory pore, ivermectin-reduced E/S protein release may lead to the inability of the parasite to evade the normally effective host immune response.

Discovery of E/S proteins can be done by proteomics (i.e., analysis by mass spectrometry) based on material produced in the laboratory or using computational approaches to search expressed sequence tags (ESTs) databases [32, 33]. Producing sufficient material from *in vitro* helminth cultures may be difficult; some stages are challenging or even impossible to maintain *in vitro* for the required time. Therefore, genome-wide identification workflows may represent an efficient way to prioritize downstream experimental investigations of potentially important E/S proteins as new drug targets. A recent study identified and functionally annotated proteins

**Table 19.1** E/S proteins of filarial nematodes.

| Species | Number of proteins | Dominant proteins/ families | References |
|---|---|---|---|
| *Acanthocheilonema viteae* | NA | ES-62 (leucyl-aminopeptidase) | [8] |
| *Brugia malayi* | Up to 852 | Triose-phosphate isomerase, phosphatidylethanolamine-binding protein 1, galectin 1 | [19–22] |
| *Dirofilaria immitis* | Up to 110 | Phosphatidylethanolamine-binding protein, transthyretin-like proteins | [23, 76] |
| *Loa loa* | 5 | Av33 | [24] |
| *Litomosoides sigmodontis* | 302 | Cysteine protease inhibitor, glutathione S-transferase, major sperm proteins, ES-62, Av33 | [25] |
| *Onchocerca ochengi* | 94 | transthyretin-like proteins, von Willebrand factor type-d domain proteins, cysteine protease inhibitor, Av33 | [26] |
| *Onchocerca volvulus* | NA | secreted larval acidic proteins, retinol-binding protein, cysteine protease inhibitor, superoxide dismutase, glutathione S-transferase | Reviewed in [27] |

Data were derived from experiments on different stages, and proteins and protein families highly represented in secretomes depend on the stage examined. The stages selected, number of organisms cultivated, and the methods of analysis are likely responsible for the large discrepancies in numbers and types of proteins identified. NA: not available

encoded in the genomes of 73 nematode species. Following matching to the DrugBank target database and RNA interference (RNAi) phenotypes in *Caenorhabditis elegans*, the authors reported 62 potential targets for human parasitic species and 390 targets for plant parasitic species [33].

Glycoproteins (and lipids) are abundant in helminth E/S preparations and likely to interact with the host innate immune system via pattern recognition receptors [34]. The extent and type of posttranslational modifications of E/S proteins remain difficult to assess. However, in some cases, immunomodulatory properties may be due primarily to their levels of glycosylation, etc. The best-characterized helminth E/S protein, ES-62 from *A. viteae*, is a potent anti-inflammatory molecule. The multiple immunomodulatory effects of the tetrameric glycoprotein ES-62 depend on phosphorylcholine moieties, attached via an N-type glycan [35]; this moiety is effective when added to other proteins.

Helminth protein secretomes are a potentially rich source of targets for vaccines and novel anthelmintics. Intriguingly, these proteins may also lead to the development of novel therapies for inflammatory and autoimmune disorders; awareness of the fact that some worm infections show protective effects is increasing [36]. Clearly, recombinant proteins would almost certainly be better accepted by patients than inoculation with live parasite ova or larvae.

While a thorough characterization of the protein candidate is necessary, there are a number of potential pitfalls, from the design of an immunogenic molecule to its manufacture. For instance, ES-62 is a very potent anti-inflammatory molecule, but is too large to be produced on a large scale at acceptable costs. Therefore, its structural features have inspired the design of small-molecule analogues to treat inflammatory disorders [8, 37]. It was pointed out that structurally closely related compounds may vary with respect to immunomodulatory activity and hence require careful analysis at the individual level, the effects being more selective than those observed with the native ES-62 molecule [37].

## 19.3 Extracellular Vesicles Contribute to Establishing a Permissive Infection

The past few years have seen parasite-derived extracellular vesicles (EVs) become a main focus of helminth research as newly discovered key elements of the molecular dialogue with the host. EVs comprise both exosomes (50–150 nm in diameter) and microvesicles (up to 1 μm in diameter) [38] and are released by virtually all eukaryotic cells into their environment and may be taken up by and affect the function of distant cells. As such, EV trafficking is emerging as a central mechanism in intercellular communication in mammals [39]. EVs express markers of the parent cells and are specifically enriched in molecules reflecting their subcellular origin or that were selectively packaged into them. The presence of bioactive proteins, lipids, nucleic acids, and metabolites has been demonstrated; their relative abundance is cell-type-specific and may depend on the physiological (and pathological) status.

It is also hypothesized that the molecular composition of EVs and their cargo is specific to each host–parasite association, thus reflecting the biological requirements associated with localization of the parasite in the host at different stages in the life cycle. This is nicely illustrated by the differing effects observed upon administration of EVs from *Trichuris muris* and *Nippostrongylus brasiliensis* to their common host, the mouse [40]. At the time of writing, EVs and their cargo have been characterized in E/S products from 20 parasitic helminth species, including filarial parasites. Experimental evidence supports important functions for EV cargo in host–parasite relationships, with consequences on host cells that may determine pathogenicity, invasion, and longevity of infections [34, 41–44].

Indeed, parasite EVs have been shown to be taken up by host cells. They attach to recipient cells through surface proteins such as tetraspanins and integrins, among others [45], after which membranes fuse and vesicle contents are delivered into the cell. This phenomenon appears to be facilitated by various endocytic internalization pathways [34]. Helminth EVs impact the activation profile of the recipient cell: upon internalization by intestinal cells, *Heligmosomoides polygyrus* EVs suppressed the expression of key genes in innate immunity and inflammation pathways, including dual specificity phosphatase 1 (*Dusp-1*; preferentially downregulating interleukin (IL)-6 while upregulating levels of immunosuppressive IL-10) and interleukin-1 receptor-like 1 (IL-33 receptor) [46, 47]. Similarly, EVs from the cestode *Echinococcus multilocularis* suppressed the levels of nitric oxide produced by activated macrophages via downregulation of inducible nitric oxide synthase expression [48]. Supported by several other reports, e.g., [40, 49–52], the various regulatory effects observed seem to be relevant for the establishment and survival of the parasites in their hosts. EVs from filarial parasites have also been characterized [53, 54]. Protein and nucleic-acid-filled EVs are released *in vitro* by all *B. malayi* mammalian stages (microfilariae, L3, L4, adult males and females) [29, 54]; the E/S pore has been proposed as a candidate site for EV release in microfilariae [54]. *B. malayi* EVs are enriched in effectors such as a cathepsin-L like cysteine protease, reflecting transition between stages involving molting, but whose intriguing presence in the extracellular compartment points to some manipulation of the host–parasite interface, as suggested for other parasites [55, 56]. *B. malayi* EVs transport quantities of small noncoding RNAs among which, microRNAs (miRNAs) play essential roles in regulating gene expression. Far larger amounts of miRNAs were detected in L3 cultures than in adult cultures, despite the greater biomass adult worms represent [54]. Interestingly, several miRNAs displayed homology to host miRNAs; this finding suggests the possibility that parasite-derived miRNAs tap into existing host gene regulatory pathways [54]. *B. malayi* EVs are internalized by murine macrophages *in vitro* and elicit a classically activated phenotype [54]. In total, 18% of female-derived EV proteins were predicted to be immunomodulators or pathogenesis-related effectors (e.g., cytokine macrophage migration inhibitory factor homolog (MIF-1/BmMIF-1), glycan-binding proteins (galectins 1 and 2), heat shock proteins, thioredoxin peroxidases, and GAPDH) [29].

Helminth EVs are enriched in vaccine candidates [57]: for instance, 31% of *S. mansoni* proteins contained in EVs are homologues of vaccine candidates or therapeutic targets [56], many of them predicted to have potential against multiple stages of the parasite. In addition, four of the five most abundant EV proteins are known vaccine candidates or protective antigens from other trematodes [56], including a saposin-like protein [58], a tetraspanin (currently in clinical development) [59], the tegumental protein Sm-29 [60], and a component of cytoplasmic dynein [61].

Vaccination with helminth EVs has been studied in experimental animal models, with encouraging results for *H. polygyrus* [46, 47], *N. brasiliensis* [40], *Echinostoma caproni* [50], and *Fasciola hepatica* [62]. The presence of relevant proteins in or on EVs that are internalized by host cells may explain how the host has access to them to mount an efficient immune response, blocking uptake by host cells. Hence, if EV internalization and cargo delivery to host cells play a pivotal role in the establishment of an infection, disruption of this process may explain why vaccines directed against EV membrane proteins show efficacy [63].

EV surface proteins are hypothesized to be even more accessible to antibody binding and possible blockade of EV uptake by host cells [63]. EVs from *F. hepatica* display 380 individual proteins on their surface; this complex profile suggests various modes of biogenesis, and cellular origins, of different EV subpopulations [64].

EV exposure to anti-tetraspanin antibodies reduced the uptake of *Opisthorchis viverrini* EVs by host cells [65]; however, pretreatment of *F. hepatica* EVs with host serum reactive against these surface proteins instead facilitated uptake by host cells [47, 64]. In fact, exposing proteins on the EV surface for targeting by host antibodies may represent a benefit to the parasite, increasing its immunomodulatory capabilities. Consequently, target specificity is likely to be required to identify valid vaccine candidates.

## 19.4 MicroRNAs, the New Immunomodulators

The regulation of gene expression is a fundamental mechanism driving biological processes and is governed by different modes of action. The action of transcription factors on gene promoter sites is one of the most important regulatory systems. The role of epigenetics, and processes involving DNA structural changes, has also gained interest as it finds applications in cancer research and other diseases [66–68]. Since the identification of microRNAs (miRNAs) in the free-living nematode *C. elegans* [69, 70], posttranscriptional gene regulation has opened new fields of investigation enforced by the current era of complex genome sequencing. Being central to gene regulation in all organisms, the importance of miRNAs is particularly crucial in parasitic helminth species as key regulators of host–parasite relationships, and they are therefore increasingly investigated as platforms for the identification of new therapeutic targets [71, 72].

miRNAs represent a major subset of small (21–25 nucleotides) noncoding RNAs initially functionally characterized in *C. elegans lin-4* and *let-7*; both regulate worm development [69, 70]. In animals, miRNA synthesis has been described in detail [73]. It involves multiple conformational changes and associations with intermediate proteins to enable a partial base complementarity with any region of targeted messenger RNAs (mRNAs). RNA-induced silencing complexes (RISCs) are assembled in the cytosol and directly downregulate protein translation by destabilizing transcript structure [74]. Hence, the interaction between an miRNA and a target transcript typically results in gene repression.

The increasing availability of helminth genomes and transcriptomes has dramatically expanded the collections of miRNAs, and the constant input of current sequencing projects continuously enriches our knowledge of helminth-derived noncoding RNAs. Moreover, miRNA databases [75, 76] enable downstream analyses focusing on functional characteristics, evolution, diversity, conservation, and gene expression modulation across a large variety of species. The last decade has seen a steadily increasing number of newly sequenced "miRNAomes" from a variety of model species, sometimes across life stages, or directly from host biofluids. These species include parasitic nematodes of plants [77] and animals, including filarial nematode species [53, 78] such as *Dirofilaria immitis* [79–81], *B. malayi* [29, 82, 83], *Brugia pahangi* [84], *O. volvulus* [85], *O. ochengi* [85, 86], *Loa loa* [86], and *Litomosoides sigmodontis* [87]. Pan-species comparative analyses have highlighted conserved groups of miRNAs, such as the miR-36 family shared by all screened helminth species, whereas others may be clade specific, such as miR-2 and miR-71 [83, 88–91], or novel miRNAs, being the product of different evolutionary mechanisms based on gene duplication, mutation, and arm switching [73].

Immunomodulatory roles played by helminth miRNAs can be investigated through the development and interrogation of bioinformatic tools that enable prediction of target mRNAs and networks based on miRNA structural features and miRNA–protein interactions, among others [92]. The roles of miRNAs in nematode development, modulation of physiological functions such as reproduction, and others have been particularly well described in *C. elegans* and highlight the complexity of miRNA biology. The hypothesis that parasite-derived miRNAs target host genes involved in immune responses has been well documented in a variety of host–parasite relationships [41, 71, 72]. Indeed, detection of parasite miRNAs in host fluid circulation supports the hypothesis that parasite-derived miRNAs modulate expression of host genes. Based on their presence in plasma, parasite-derived miRNAs have gained interest as biomarkers, diagnostic tools, and potential therapeutics [71, 72, 93–99]. With some exceptions [100], miRNA release by parasites seems to mostly take place through the production of EVs, and most recent studies have developed host target prediction tools supported by experimental validations, as illustrated with filariid species [46, 53].

miRNAs appear essential in the development of parasites and for their successful establishment within their respective host. Often exploited as diagnostic markers, miRNAs have, in some cases, also been considered as potential therapeutic targets [101]. Classical approaches such as knocking-down miRNA genes or implementing exogenous transcripts, also known as miRNA sponges to form inactive complexes [102, 103], remain rarely applied to helminth species [104], while impairing miRNA functions through the synthesis of anti-miRNAs and/or miRNA inhibitors (also known as "antagomirs") has been more seriously considered [105, 106]. They both consist of blocking specific miRNA action, which has sometimes proven efficacious (but not always) in a variety of applications against hepatitis C virus [107], cancers (oncomiRs) [108], inflammatory/metabolic diseases [109–111], and others reviewed elsewhere [107, 112].

In nematodes, miRNAs as therapeutic targets remain a poorly investigated concept despite recent identification of candidate miRNAs and their respective antagomiRs or mimics. For example, *hco-miR-228* and *hco-miR-235* from the gastrointestinal parasite of small ruminants *Haemonchus contortus* are abundantly expressed in L3 (infective stage larvae) and may maintain development arrest until ingestion by the host. Interestingly, the *in vitro* use of specific miRNA inhibitors significantly increased the number of L3 to L4 transitioning animals while miRNA mimics exhibited no apparent change on L4 numbers [113]. In filarial nematodes, the "miRNAome" of key model species would theoretically enable the identification of potentially targetable essential miRNAs at the center of the host–parasite relationship. However, in *B. pahangi*, although miR-5864, a member of the *let-7* family, was found to be highly upregulated during the mosquito to mammalian host transition, the use of anti-miR sequences failed to induce significant changes in host target mRNA expression levels. The authors of this study report promiscuous activity of miRNAs and the overlapping functions of other untargeted miRNAs, which would lead research toward a multi-miRNA targeting approach [114]. As indicated by Poole and colleagues, further analysis of filariid miRNAs and their respective targets will provide insights on targetable miRNA-related pathways [83]. Analysis of the *B. malayi* "miRNAome" has revealed up to nine filarial-specific miRNA families, such as the large miR-2 family or the microfilariae subfamily miR-71 [83]. Despite being potential targets for antagomir-based therapies, this option has poorly been developed in filariids, as the off-target risk remains a strong limitation and requires adjustments such as targeting whole miRNA families [115].

Host-derived miRNAs released in response to parasite invasion could also be investigated as therapeutic pathways. Indeed, in *B. malayi* infected mice, miR-125b-5p, miR-146a-5p, miR-199b-5p, and miR-378-3p appear differentially expressed in parasite-exposed macrophages [116]. The use of miR-378-3p inhibitors or mimics negatively regulates macrophage proliferation by interfering with the PI3K/Akt-signaling pathway, highlighting potential parasite targeting pathways.

The expansion of the set of filariid miRNAomes offers a rich reservoir of potentially targetable miRNAs, which, along with extended studies on pathway analysis, may reveal key host response actions that are required for prevention or rejection of an infection.

## 19.5 DNA Vaccines

The era of genome sequencing has greatly influenced the search for novel targeting pathways and the development of vaccine strategies. The last decades have seen intensive development of DNAs encoding key parasitic antigens as a vaccine strategy, administered to activate protective host immune response against a variety of parasitic diseases such as protozoan and helminth infections, including filariases. At a close contact point between host and parasites, several parasite-derived proteins have been proposed as candidate vaccines and the era of -omics is expected to expand the current library of targetable antigens. Identifying parasite proteins that are required for immune evasion would lead to a more rational selection of candidate vaccine antigens; new strategies for the generation of transgenic filarial nematodes may soon enable this platform [117]. The concept here is to turn a permissive into a nonpermissive host by disabling the parasite immunomodulation strategy.

The quest for novel antiparasitic drugs remains a major axis of research. However, the history of vaccine development against infectious and noninfectious diseases (i.e., cancer) supports renewed attention to parasitic helminths, including filariids. Considerable recent efforts have been made to develop innovative vaccine strategies as illustrated by DNA vaccines. Briefly, the host receives an exogenous DNA encoding a candidate antigen protein of parasitic origin that will educate the host immune system to generate immune-based protection. Since their first application against influenza [118], DNA vaccines have been investigated against some cancers and infectious diseases. A large variety of immune responses can be generated [119], and research has also focused on increasing the robustness of immune response while developing multi-antigen vaccines and adjuvants [120–123]. As the genomes of many parasitic species have become available, DNA vaccines have gained particular interest against protozoan parasites (malaria, leishmaniasis, trypanosomiasis, etc.) [124]; the approach is now being investigated against complex parasitic helminth species. For example, research on schistosomiasis vaccine strategies has developed several DNA-based vaccine candidates with the objective of reducing mortality, morbidity, and/or transmission, while potentially complementing current chemotherapy [125]. Several candidate DNA vaccines, mostly encoding tegument-derived proteins, have been advanced [126, 127], and efforts are being made to improve efficacy by targeting antigen presenting cells, developing adjuvants, and testing different routes of immunization [128].

Control of human filariases relies almost exclusively on chemotherapy and vector-control strategies, which, despite being undeniably successful, remain time- and labor-demanding and require prolonged sustainable funding. Vaccination represents a novel cost-effective complementary approach that has been initiated by immunizing mice or jirds with attenuated L3 larvae of *O. volvulus* [129] or *B. malayi* [130]. A few surface or secreted proteins have been proposed as candidate antigens. For example, the *O. volvulus* L3-specific chitinase induced a decrease in parasite survival in mice to a maximum of 53% when injected as a plasmid DNA vaccine [131]. Several *B. malayi* antigens, such as Bm5 (paramyosin), BmHSP-70 (heat shock protein), BmIF (intermediate filament), and Bm14 (serodiagnostic antigen), were tested in mice and helped measuring the amplitude of antibody responses, the result of different routes of immunization as well as the relevance of using multivalent vaccine strategies [132, 133]. Since then, the increase in genome sequencing projects continues to provide new DNA-based vaccine strategies, and a variety of candidate antigens have been proposed as monovalent or multivalent solutions.

Among these candidate targets, *B. malayi* ALT-1 and ALT-2 (abundant larval transcripts) are particularly abundant in L3s and jird immunizations with ALT-1 reduced parasite survival by 76% [134]. The function of ALT proteins remains poorly understood, but some evidence implies a role in protective immunity, possibly participating in parasite establishment [135–137]. Since the first effective monovalent rodent vaccines based on BmALT-2 [138] and BmALT-1 [135], a variety of multivalent combinations have been proposed to boost immune responses and improve protection in rodent models. For example, BmALT-2 in combination with thioredoxin peroxidase conferred 78% protection against L3 larvae in mice [139]. Protection was 80% with the venom allergen BmVAH [140] and up to 85% against adult stages with the immunoreactive vespid venom homolog BmVAL-1 [141] if complemented with a heterologous protein boost regimen [142]. Finally, a synergistic enhancement of protective immune response was obtained with a multivalent combination of BmALT-2 and a small heat shock protein that induced 90% larval death when given as a prime boost regimen [143].

To date, different vaccine formulas (whole worm, subunits, DNA-based, etc.) have been tested in animal models. However, translation for use in humans has not yet been undertaken as safety and logistics remain challenging. Novel approaches switching from empirical to more rational antigen-selection strategies have already been explored by developing multi-epitope-based vaccines against several infectious diseases [144–146]. Such a strategy has recently been applied to filariids, which aimed at combining B-cell and T-cell epitopes to generate a chimeric vaccine with gained antigenicity and a potential for cross-protection between multiple parasitic nematode species [147].

## 19.6 Conclusion

To eliminate helminth infections, it is likely that new drugs or vaccines will be required. In recent years, many new players in the molecular dialogue between parasite and host have been discovered, and we expect even more will be found in the future. The roles of sugar moieties, lipids, metabolites, and other noncoding RNA species remain to be characterized, and much more work needs to be done to identify the host pathways targeted by parasite-derived miRNAs that enable a sustained infection.

Identifying key parasite-derived proteins that make hosts permissive to infection should allow us to rationally target immune responses that will convert them to nonpermissive, at least for the target parasite. Unlike anthelmintics, vaccination is not likely to be broad-spectrum; however, filarial diseases, including onchocerciasis, lymphatic filariasis, and heartworm disease, are amenable to species-specific vaccination for control [147–150].

Developing an in-depth mechanistic understanding of host–parasite communication is an essential step for informed antigen discovery as it pertains to EVs and vaccine research in general. Indeed, a shift in practice is likely required, from empirical to more rational antigen-selection strategies. Evaluating the full impact of helminth mediators on the infective process (e.g., beneficial to the pathogen, to the host, or both) is required to understand physiological functions and may offer the perspective for vaccination and other therapeutic approaches [63].

## Acknowledgments

LT is holder of a grant by the Swiss National Science Foundation (PZ00P3_168080). TBD was supported by a postdoctoral award from Merck KGaA.

## References

1 Dobson, A., Lafferty, K.D., Kuris, A.M. et al. (2008). Homage to Linnaeus: How many parasites? How many hosts? *Proc. Natl. Acad. Sci. U.S.A.* 105: 11482–11489.

2 Weinstein, S.B. and Kuris, A.M. (2016). Independent origins of parasitism in Animalia. *Biol. Lett.* 12: 20160324.

3 Blaxter, M. and Koutsovoulos, G. (2015). The evolution of parasitism in *Nematoda. Parasitology* 142 (Suppl 1): S26–S39.
4 Viney, M. (2018). The genomic basis of nematode parasitism. *Brief. Funct. Genomics* 17: 8–14.
5 Maizels, R.M. and McSorley, H.J. (2016). Regulation of the host immune system by helminth parasites. *J. Allergy Clin. Immunol.* 138: 666–675.
6 Allen, J.E. and Maizels, R.M. (2011). Diversity and dialogue in immunity to helminths. *Nat. Rev. Immunol.* 11: 375–388.
7 Hewitson, J.P., Grainger, J.R., and Maizels, R.M. (2009). Helminth immunoregulation: The role of parasite secreted proteins in modulating host immunity. *Mol. Biochem. Parasitol.* 167: 1–11.
8 Pineda, M.A., Lumb, F., Harnett, M.M., and Harnett, W. (2014). ES-62, a therapeutic anti-inflammatory agent evolved by the filarial nematode *Acanthocheilonema viteae. Mol. Biochem. Parasitol.* 194: 1–8.
9 Cross, J.H. (1996). Filarial nematodes. In: *Medical Microbiology*, 4e (ed. S. Baron). Galveston, TX: University of Texas Medical Branch at Galveston. Available from: https://www.ncbi.nlm.nih.gov/books/NBK7844/.
10 Geary, T.G. (2012). Are new anthelmintics needed to eliminate human helminthiases? *Curr. Opin. Infect. Dis.* 25: 709–717.
11 WHO (2021). Global programme to eliminate lymphatic filariasis: progress report, 2020. *Wkly. Epidemiol. Rec.* 96: 589–604.
12 Fischer, P.U., King, C.L., Jacobson, J.A., and Weil, G.J. (2017). Potential value of triple drug therapy with ivermectin, diethylcarbamazine, and albendazole (IDA) to accelerate elimination of lymphatic filariasis and onchocerciasis in Africa. *PLoS Negl.Trop. Dis.* 11: e0005163.
13 Boatin, B.A., Basáñez, M.-G., Prichard, R.K. et al. (2012). A research agenda for helminth diseases of humans: towards control and elimination. *PLoS Negl. Trop. Dis.* 6: e0001547.
14 Sharma, M., Krammer, F., García-Sastre, A., and Tripathi, S. (2019). Moving from empirical to rational vaccine design in the "Omics" era. *Vaccines* 7: 89.
15 Lustigman, S., Grote, A., and Ghedin, E. (2017). The role of "omics" in the quest to eliminate human filariasis. *PLoS Negl. Trop. Dis.* 11: e0005464.
16 Harnett, W. (2014). Secretory products of helminth parasites as immunomodulators. *Mol. Biochem. Parasitol.* 195: 130–136.
17 Bonin-Debs, A.L., Boche, I., Gille, H., and Brinkmann, U. (2004). Development of secreted proteins as biotherapeutic agents. *Expert Opin. Biol. Ther.* 4: 551–558.
18 Maizels, R.M. and Yazdanbakhsh, M. (2003). Immune regulation by helminth parasites: cellular and molecular mechanisms. *Nat. Rev. Immunol.* 3: 733–744.
19 Moreno, Y. and Geary, T.G. (2008). Stage- and gender-specific proteomic analysis of *Brugia malayi* excretory-secretory products. *PLoS Negl. Trop. Dis.* 2: e000326.
20 Hewitson, J.P., Harcus, Y.M., Curwen, R.S. et al. (2008). The secretome of the filarial parasite, *Brugia malayi*: proteomic profile of adult excretory-secretory products. *Mol. Biochem. Parasitol.* 160: 8–21.

**21** Bennuru, S., Semnani, R., Meng, Z. et al. (2009). *Brugia malayi* excreted/secreted proteins at the host/parasite interface: stage- and gender-specific proteomic profiling. *PLoS Negl. Trop. Dis.* 3: e000410.

**22** Bennuru, S., Meng, Z., Ribeiro, J.M.C. et al. (2011). Stage-specific proteomic expression patterns of the human filarial parasite *Brugia malayi* and its endosymbiont *Wolbachia*. *Proc. Natl. Acad. Sci. U.S.A.* 108: 9649–9654.

**23** Geary, J., Satti, M., Moreno, Y. et al. (2012). First analysis of the secretome of the canine heartworm, *Dirofilaria immitis*. *Parasites Vectors* 5: 140.

**24** Hertz, M.I., Nana-Djeunga, H., Kamgno, J. et al. (2018). Identification and characterization of *Loa loa* antigens responsible for cross-reactivity with rapid diagnostic tests for lymphatic filariasis. *PLoS Negl. Trop. Dis.* 12: e0006963.

**25** Armstrong, S.D., Babayan, S.A., Lhermitte-Vallarino, N. et al. (2014). Comparative analysis of the secretome from a model filarial nematode (*Litomosoides sigmodontis*) reveals maximal diversity in gravid female parasites. *Mol. Cell. Proteomics* 13: 2527–2544.

**26** Armstrong, S.D., Xia, D., Bah, G.S. et al. (2016). Stage-specific proteomes from *Onchocerca ochengi*, sister species of the human river blindness parasite, uncover adaptations to a nodular lifestyle. *Mol. Cell. Proteomics* 15: 2554–2575.

**27** Gomez-Fuentes, S., Morales-Ruiz, V., López-Recinos, D. et al. (2019). Biological role of excretory-secretory proteins in endemic parasites of Latin America and the Caribbean. *J. Helminthol.* 94: e53.

**28** Moreno, Y., Nabhan, J.F., Solomon, J. et al. (2010). Ivermectin disrupts the function of the excretory-secretory apparatus in microfilariae of *Brugia malayi*. *Proc. Natl. Acad. Sci. U.S.A.* 107: 20120–20125.

**29** Harischandra, H., Yuan, W., Loghry, H.J. et al. (2018). Profiling extracellular vesicle release by the filarial nematode *Brugia malayi* reveals sex-specific differences in cargo and a sensitivity to ivermectin. *PLoS Negl. Trop. Dis.* 12: e0006438.

**30** Loghry, H.J., Yuan, W., Zamanian, M. et al. (2020). Ivermectin inhibits extracellular vesicle secretion from parasitic nematodes. *J. Extracell. Vesicles* 10: e12036.

**31** Moreno, Y., Geary, T.G., and Tritten, L. (2021). When secretomes meet anthelmintics: Lessons for therapeutic interventions. *Trends Parasitol.* 37: 468–475.

**32** Nagaraj, S.H., Gasser, R.B., and Ranganathan, S. (2008). Needles in the EST haystack: large-scale identification and analysis of excretory-secretory (ES) proteins in parasitic nematodes using expressed sequence tags (ESTs). *PLoS Negl.Trop. Dis.* 2: e000301.

**33** Gahoi, S., Singh, S., and Gautam, B. (2019). Genome-wide identification and comprehensive analysis of Excretory/Secretory proteins in nematodes provide potential drug targets for parasite control. *Genomics* 111: 297–309.

**34** Zakeri, A., Hansen, E.P., Andersen, S.D. et al. (2018). Immunomodulation by helminths: intracellular pathways and extracellular vesicles. *Front. Immunol.* 9: 2349.

35 Harnett, W., Harnett, M., and Byron, O. (2003). Structural / functional aspects of ES-62 - a secreted immunomodulatory phosphorylcholine-containing filarial nematode glycoprotein. *Curr. Protein Pept. Sci.* 4: 59–71.

36 Helmby, H. (2015). Human helminth therapy to treat inflammatory disorders - where do we stand? *BMC Immunol.* 16: 12.

37 Al-Riyami, L., Pineda, M.A., Rzepecka, J. et al. (2013). Designing anti-inflammatory drugs from parasitic worms: a synthetic small molecule analogue of the *Acanthocheilonema viteae* product ES-62 prevents development of collagen-induced arthritis. *J. Med. Chem.* 56: 9982–10002.

38 van Niel, G., D'Angelo, G., and Raposo, G. (2018). Shedding light on the cell biology of extracellular vesicles. *Nat. Rev. Mol. Cell Biol.* 19: 213–228.

39 Turchinovich, A., Samatov, T.R., Tonevitsky, A.G., and Burwinkel, B. (2013). Circulating miRNAs: cell-cell communication function? *Front. Genet.* 4: 119.

40 Eichenberger, R.M., Ryan, S., Jones, L. et al. (2018). Hookworm secreted extracellular vesicles interact with host cells and prevent inducible colitis in mice. *Front. Immunol.* 9: 850.

41 Tritten, L. and Geary, T.G. (2018). Helminth extracellular vesicles in host-parasite interactions. *Curr. Opin. Microbiol.* 46: 73–79.

42 Coakley, G., Maizels, R.M., and Buck, A.H. (2015). Exosomes and other extracellular vesicles: the new communicators in parasite infections. *Trends Parasitol.* 31: 477–489.

43 Eichenberger, R.M., Sotillo, J., and Loukas, A. (2018). Immunobiology of parasitic worm extracellular vesicles. *Immunol. Cell Biol.* 96: 704–713.

44 Marcilla, A., Trelis, M., Cortés, A. et al. (2012). Extracellular vesicles from parasitic helminths contain specific excretory/secretory proteins and are internalized in intestinal host cells. *PLoS One* 7: e45974.

45 Mulcahy, L.A., Pink, R.C., and Carter, D.R.F. (2014). Routes and mechanisms of extracellular vesicle uptake. *J. Extracell. Vesicles* 4: 3.

46 Buck, A.H., Coakley, G., Simbari, F. et al. (2014). Exosomes secreted by nematode parasites transfer small RNAs to mammalian cells and modulate innate immunity. *Nat. Commun.* 5: 5488.

47 Coakley, G., McCaskill, J.L., Borger, J.G. et al. (2017). Extracellular vesicles from a helminth parasite suppress macrophage activation and constitute an effective vaccine for protective immunity. *Cell Rep.* 19: 1545–1557.

48 Zheng, Y., Guo, X., Su, M. et al. (2017). Regulatory effects of *Echinococcus multilocularis* extracellular vesicles on RAW264.7 macrophages. *Vet. Parasitol.* 235: 29–36.

49 Wang, L., Li, Z., Shen, J. et al. (2015). Exosome-like vesicles derived by *Schistosoma japonicum* adult worms mediate M1 type immune- activity of macrophages. *Parasitol. Res.* 114: 1865–1873.

50 Trelis, M., Galiano, A., Bolado, A. et al. (2016). Subcutaneous injection of exosomes reduces symptom severity and mortality induced by *Echinostoma caproni* infection in BALB/c mice. *Int. J. Parasitol.* 46: 799–808.

**51** Zhu, L., Liu, J., Dao, J. et al. (2016). Molecular characterization of *S. japonicum* exosome-like vesicles reveals their regulatory roles in parasite-host interactions. *Sci. Rep.* 6: 25885.

**52** Meningher, T., Barsheshet, Y., Ofir-Birin, Y. (2020). Schistosomal extracellular vesicle-enclosed miRNAs modulate host T helper cell differentiation. *EMBO Rep.* 21: e47882.

**53** Quintana, J.F., Babayan, S.A., and Buck, A.H. (2017). Small RNAs and extracellular vesicles in filarial nematodes: From nematode development to diagnostics. *Parasite Immunol.* 39: e12395.

**54** Zamanian, M., Fraser, L.M., Agbedanu, P.N. et al. (2015). Release of small RNA-containing exosome-like vesicles from the human filarial parasite *Brugia malayi*. *PLoS Negl. Trop. Dis.* 9: e0004069.

**55** Dalton, J.P., Neill, S.O., Stack, C. et al. (2003). *Fasciola hepatica* cathepsin L-like proteases: biology, function, and potential in the development of first generation liver fluke vaccines. *Int. J. Parasitol.* 33: 1173–1181.

**56** Sotillo, J., Pearson, M., Potriquet, J. et al. (2016). Extracellular vesicles secreted by *Schistosoma mansoni* contain protein vaccine candidates. *Int. J. Parasitol.* 46: 1–5.

**57** Drurey, C., Coakley, G., and Maizels, R.M. (2020). Extracellular vesicles: new targets for vaccines against helminth parasites. *Int. J. Parasitol.* 50: 623–633.

**58** Espino, A.M. and Hillyer, G.V. (2003). Molecular cloning of a member of the *Fasciola hepatica* saposin-like protein family. *J. Parasitol.* 89: 545–552.

**59** Tran, M.H., Pearson, M.S., Bethony, J.M. et al. (2006). Tetraspanins on the surface of *Schistosoma mansoni* are protective antigens against schistosomiasis. *Nat. Med.* 12: 835–840.

**60** Cardoso, F.C., Pacífico, R.N.A., Mortara, R.A., and Oliveira, S.C. (2006). Human antibody responses of patients living in endemic areas for schistosomiasis to the tegumental protein Sm29 identified through genomic studies. *Clin. Exp. Immunol.* 144: 382–391.

**61** Rezende, C.M.F., Silva, M.R., Santos, I.G.D. et al. (2011). Immunization with rP22 induces protective immunity against *Schistosoma mansoni*: effects on granuloma down-modulation and cytokine production. *Immunol. Lett.* 141: 123–133.

**62** Roig, J., Saiz, M.L., Galiano, A. et al. (2018). Extracellular vesicles from the helminth *Fasciola hepatica* prevent DSS-induced acute ulcerative colitis in a T-lymphocyte independent mode. *Front. Microbiol.* 9: 1036.

**63** Kifle, D.W., Sotillo, J., Pearson, M.S., and Loukas, A. (2017). Extracellular vesicles as a target for the development of anti-helminth vaccines. *Emerg. Top. Life Sci.* 1: 659–665.

**64** de la Torre-Escudero, E., Gerlach, J.Q., Bennett, A.P.S. et al. (2019). Surface molecules of extracellular vesicles secreted by the helminth pathogen *Fasciola hepatica* direct their internalisation by host cells. *PLoS Negl.Trop. Dis.* 13: e0007087.

**65** Chaiyadet, S., Sotillo, J., Smout, M. et al. (2015). Carcinogenic liver fluke secretes extracellular vesicles that promote cholangiocytes to adopt a tumorigenic phenotype. *J. Infect. Dis.* 212: 1636–1645.

**66** Tarayrah, L. and Chen, X. (2013). Epigenetic regulation in adult stem cells and cancers. *Cell Biosci.* 3: 41.

**67** Moutinho, C. and Esteller, M. (2017). MicroRNAs and epigenetics. *Adv. Cancer Res.* 135: 189–220.

**68** Yao, Q., Chen, Y., and Zhou, X. (2019). The roles of microRNAs in epigenetic regulation. *Curr. Opin. Chem. Biol.* 51: 11–17.

**69** Ambros, V. and Horvitz, H.R. (1984). Heterochronic mutants of the nematode *Caenorhabditis elegans*. *Science* 226: 409–416.

**70** Ruvkun, G. and Giusto, J. (1989). The *Caenorhabditis elegans* heterochronic gene *lin-14* encodes a nuclear protein that forms a temporal developmental switch. *Nature* 338: 313–319.

**71** Britton, C., Winter, A.D., Gillan, V., and Devaney, E. (2014). microRNAs of parasitic helminths - Identification, characterization and potential as drug targets. *Int. J. Parasitol. Drugs Drug Resist.* 4: 85–94.

**72** Cai, P., Gobert, G.N., and McManus, D.P. (2016). MicroRNAs in parasitic helminthiases: current status and future perspectives. *Trends Parasitol.* 32: 71–86.

**73** Wahid, F., Shehzad, A., Khan, T., and Kim, Y.Y. (2010). MicroRNAs: synthesis, mechanism, function, and recent clinical trials. *Biochim. Biophys. Acta* 1803: 1231–1243.

**74** Chekulaeva, M. and Filipowicz, W. (2009). Mechanisms of miRNA-mediated post-transcriptional regulation in animal cells. *Curr. Opin. Cell Biol.* 21: 452–460.

**75** Moore, A.C., Winkjer, J.S., and Tseng, T.-T. (2015). Bioinformatics resources for microRNA discovery. *Biomark. Insights* 10: 53–58.

**76** Griffiths-Jones, S., Grocock, R.J., van Dongen, S. et al. (2006). miRBase: microRNA sequences, targets and gene nomenclature. *Nucleic Acids Res.* 34: D140–D144.

**77** Jaubert-Possamai, S., Noureddine, Y., and Favery, B. (2019). MicroRNAs, new players in the plant-nematode interaction. *Front. Plant Sci.* 17: 1180.

**78** Soichot, J., Endriss, Y., Marti, H., and Tritten, L. (2020). Secreted microRNA data from the parasitic filarial nematode *Acanthocheilonema viteae*. *Data Brief* 29: 105334.

**79** Fu, Y., Lan, J., Wu, X. et al. (2013). Identification of *Dirofilaria immitis* miRNA using Illumina deep sequencing. *Vet. Res.* 44: 3.

**80** Tritten, L., Clarke, D., Timmins, S. et al. (2016). *Dirofilaria immitis* exhibits sex- and stage-specific differences in excretory/secretory miRNA and protein profiles. *Vet. Parasitol.* 232: 1–7.

**81** Braman, A., Weber, P.S., Tritten, L. et al. (2018). Further characterization of molecular markers in canine *Dirofilaria immitis* infection. *J. Parasitol.* 104: 697–701.

**82** Poole, C.B., Davis, P.J., Jin, J., and McReynolds, L.A. (2010). Cloning and bioinformatic identification of small RNAs in the filarial nematode, *Brugia malayi*. *Mol. Biochem. Parasitol.* 169: 87–94.

**83** Poole, C.B., Gu, W., Kumar, S. et al. (2014). Diversity and expression of microRNAs in the filarial parasite, *Brugia malayi*. *PLoS One* 9: e96498.

**84** Winter, A.D., Weir, W., Hunt, M. et al. (2012). Diversity in parasitic nematode genomes: the microRNAs of *Brugia pahangi* and *Haemonchus contortus* are largely novel. *BMC Genomics* 13: 4.

**85** Quintana, J.F., Makepeace, B.L., Babayan, S.A. et al. (2015). Extracellular *Onchocerca*-derived small RNAs in host nodules and blood. *Parasites Vectors* 8: 58.

**86** Tritten, L., O'Neill, M., Nutting, C. et al. (2014). *Loa loa* and *Onchocerca ochengi* miRNAs detected in host circulation. *Mol. Biochem. Parasitol.* 198: 14–17.

**87** Quintana, J.F., Kumar, S., Ivens, A. et al. (2019). Comparative analysis of small RNAs released by the filarial nematode *Litomosoides sigmodontis in vitro* and *in vivo*. *PLoS Negl. Trop. Dis.* 13: e0007811.

**88** Palakodeti, D., Smielewska, M., and Graveley, B.R. (2006). MicroRNAs from the planarian *Schmidtea mediterranea*: a model system for stem cell biology. *RNA* 12: 1640–1649.

**89** Huang, J., Hao, P., Chen, H. et al. (2009). Genome-wide identification of *Schistosoma japonicum* microRNAs using a deep-sequencing approach. *PLoS One* 4: e8206.

**90** de Souza Gomes, M., Muniyappa, M.K., Carvalho, S.G. et al. (2011). Genome-wide identification of novel microRNAs and their target genes in the human parasite *Schistosoma mansoni*. *Genomics* 98: 96–111.

**91** Sotillo, J., Robinson, M.W., Kimber, M.J. et al. (2020). The protein and microRNA cargo of extracellular vesicles from parasitic helminths – current status and research priorities. *Int. J. Parasitol.* 50: 635–645.

**92** Akhtar, M.M., Micolucci, L., Islam, M.S. et al. (2016). Bioinformatic tools for microRNA dissection. *Nucleic Acids Res.* 44: 24–44.

**93** Manzano-Román, R. and Siles-Lucas, M. (2012). MicroRNAs in parasitic diseases: potential for diagnosis and targeting. *Mol. Biochem. Parasitol.* 186: 81–86.

**94** Hoy, A.M., Lundie, R.J., Ivens, A. et al. (2014). Parasite-derived microRNAs in host serum as novel biomarkers of helminth infection. *PLoS Negl.Trop. Dis.* 8: e0002701.

**95** Tritten, L., Burkman, E., Moorhead, A. et al. (2014). Detection of circulating parasite-derived microRNAs in filarial infections. *PLoS Negl.Trop. Dis.* 8: e0002971.

**96** Siles-Lucas, M., Morchon, R., Simon, F., and Manzano-Roman, R. (2015). Exosome-transported microRNAs of helminth origin: new tools for allergic and autoimmune diseases therapy? *Parasite Immunol.* 37: 208–214.

**97** Tritten, L., Burkman, E.J., Clark, T., and Verocai, G.G. (2021). Secretory microRNA profiles of third- and fourth-stage *Dirofilaria immitis* larvae with different macrocyclic lactone susceptibility: In search of biomarkers for early detection of infection. *Pathogens* 10: 786.

**98** Mu, Y., McManus, D.P., Gordon, C.A., and Car, P. (2021). Parasitic helminth-derived microRNAs and extracellular vesicle cargos as biomarkers for helminthic infections. *Front. Cell Infect. Microbiol.* 11: 708952.

**99** Macfarlane, C.L., Quek, S., Pionnier, N. et al. (2020). The insufficiency of circulating miRNA and DNA as diagnostic tools or as biomarkers of treatment efficacy for *Onchocerca volvulus*. *Sci. Rep.* 10: 6672.

**100** Taylor, P.J., Hagen, J., Faruqu, F.N. et al. (2020). *Trichinella spiralis* secretes abundant unencapsulated small RNAs with potential effects on host gene expression. *Int. J. Parasitol.* 50: 697–705.

**101** Baumann, V. and Winkler, J. (2014). miRNA-based therapies: strategies and delivery platforms for oligonucleotide and non-oligonucleotide agents. *Future Med. Chem.* 6: 1967–1984.

**102** Esau, C.C. (2008). Inhibition of microRNA with antisense oligonucleotides. *Methods* 44: 55–60.

**103** Ebert, M.S. and Sharp, P.A. (2010). MicroRNA sponges: progress and possibilities. *RNA* 16: 2043–2050.

**104** He, X., Sun, Y., Lei, N. et al. (2018). MicroRNA-351 promotes schistosomiasis-induced hepatic fibrosis by targeting the vitamin D receptor. *Proc. Natl. Acad. Sci. U.S.A.* 115: 180–185.

**105** van Rooij, E., Purcell, A.L., and Levin, A.A. (2012). Developing microRNA therapeutics. *Circ. Res.* 110: 496–507.

**106** Krützfeldt, J., Rajewsky, N., Braich, R. et al. (2005). Silencing of microRNAs in vivo with "antagomirs.". *Nature* 438: 685–689.

**107** Drury, R.E., O'Connor, D., and Pollard, A.J. (2017). The clinical application of microRNAs in infectious disease. *Front. Immunol.* 8: 1182.

**108** Balacescu, O., Visan, S., Baldasici, O., Balacescu, L. et al. (2019) MiRNA-based therapeutics in oncology, realities, and challenges, in *Antisense Therapy* (eds S. Sharad and S. Kapur), IntechOpen., London, UK, https://doi.org/10.5772/intechopen.81847.

**109** Thum, T., Gross, C., Fiedler, J. et al. (2008). MicroRNA-21 contributes to myocardial disease by stimulating MAP kinase signalling in fibroblasts. *Nature* 456: 980–984.

**110** Worm, J., Stenvang, J., Petri, A. et al. (2009). Silencing of microRNA-155 in mice during acute inflammatory response leads to derepression of c/ebp Beta and down-regulation of G-CSF. *Nucleic Acids Res.* 37: 5784–5792.

**111** Trajkovski, M., Hausser, J., Soutschek, J. et al. (2011). MicroRNAs 103 and 107 regulate insulin sensitivity. *Nature* 474: 649–653.

**112** Rupaimoole, R. and Slack, F.J. (2017). MicroRNA therapeutics: towards a new era for the management of cancer and other diseases. *Nat. Rev. Drug Discovery* 16: 203–222.

**113** Marks, N.D., Winter, A.D., Gu, H.Y. et al. (2019). Profiling microRNAs through development of the parasitic nematode *Haemonchus* identifies nematode-specific miRNAs that suppress larval development. *Sci. Rep.* 9: 17594.

**114** Winter, A.D., Gillan, V., Maitland, K. et al. (2015). A novel member of the *let-7* microRNA family is associated with developmental transitions in filarial nematode parasites. *BMC Genomics* 16: 331.

**115** Obad, S., dos Santos, C.O., Petri, A. et al. (2011). Silencing of microRNA families by seed-targeting tiny LNAs. *Nat. Genet.* 43: 371–378.

**116** Rückerl, D., Jenkins, S.J., Laqtom, N.N. et al. (2012). Induction of IL-4Rα-dependent microRNAs identifies PI3K/Akt signaling as essential for IL-4-driven murine macrophage proliferation *in vivo*. *Blood* 120: 2307–2316.

**117** Liu, C., Mhashilkar, A.S., Chabanon, J. et al. (2018). Development of a toolkit for piggyBac-mediated integrative transfection of the human filarial parasite *Brugia malayi*. *PLoS Negl.Trop. Dis.* 12: e0006509.

**118** Ulmer, J.B., Donnelly, J.J., Parker, S.E. et al. (1993). Heterologous protection against influenza by injection of DNA encoding a viral protein. *Science* 259: 1745–1749.

**119** Liu, M.A. (2003). DNA vaccines: a review. *J. Intern. Med.* 253: 402–410.

**120** Krieg, A.M., Yi, A.K., Matson, S. et al. (1995). CpG motifs in bacterial DNA trigger direct B-cell activation. *Nature* 374: 546–549.

**121** Hemmi, H., Takeuchi, O., Kawai, T. et al. (2000). A Toll-like receptor recognizes bacterial DNA. *Nature* 408: 740–745.

**122** Ramshaw, I.A. and Ramsay, A.J. (2000). The prime-boost strategy: exciting prospects for improved vaccination. *Immunol. Today* 21: 163–165.

**123** Ito, K., Ito, K., Shinohara, N., and Kato, S. (2003). DNA immunization via intramuscular and intradermal routes using a gene gun provides different magnitudes and durations on immune response. *Mol. Immunol.* 39: 847–854.

**124** Dumonteil, E. (2007). DNA Vaccines against protozoan parasites: advances and challenges. *J. Biomed. Biotechnol.* 2007: 90520.

**125** Cherfas, J. (1991). New hope for vaccine against schistosomiasis. *Science* 251: 630–631.

**126** McManus, D.P. and Loukas, A. (2008). Current status of vaccines for schistosomiasis. *Clin. Microbiol. Rev.* 21: 225–242.

**127** Tebeje, B.M., Harvie, M., You, H. et al. (2016). Schistosomiasis vaccines: where do we stand? *Parasites Vectors* 9: 528.

**128** Da'dara, A.A., Harn, D.A. (2005). DNA vaccines against tropical parasitic diseases. *Expert Rev. Vaccines* 4: 575–589.

**129** Lange, A.M., Yutanawiboonchai, W., Lok, J.B. et al. (1993). Induction of protective immunity against larval *Onchocerca volvulus* in a mouse model. *Am. J. Trop. Med. Hyg.* 49: 783–788.

**130** Yates, J.A. and Higashi, G.I. (1985). *Brugia malayi*: vaccination of jirds with 60 cobalt-attenuated infective stage larvae protects against homologous challenge. *Am. J. Trop. Med. Hyg.* 34: 1132–1137.

**131** Harrison, R.A., Wu, Y., Egerton, G., and Bianco, A.E. (1999). DNA immunisation with *Onchocerca volvulus* chitinase induces partial protection against challenge infection with L3 larvae in mice. *Vaccine* 18: 647–655.

**132** Li, B.W., Zhang, S., Curtis, K.C., and Weil, G.J. (1999). Immune responses to *Brugia malayi* paramyosin in rodents after DNA vaccination. *Vaccine* 18: 76–81.

**133** Li, B.-W., Rush, A., Zhang, S.R. et al. (2004). Antibody responses to *Brugia malayi* antigens induced by DNA vaccination. *Filaria J.* 3: 1.

**134** Gregory, W.F., Atmadja, A.K., Allen, J.E., and Maizels, R.M. (2000). The abundant larval transcript-1 and -2 genes of *Brugia malayi* encode stage-specific candidate vaccine antigens for filariasis. *Infect. Immun.* 68: 4174–4179.

**135** Ramachandran, S., Kumar, M.P., Rami, R.M.V. et al. (2004). The larval specific lymphatic filarial ALT-2: induction of protection using protein or DNA vaccination. *Microbiol. Immunol.* 48: 945–955.

**136** Gomez-Escobar, N., Bennett, C., Prieto-Lafuente, L. et al. (2005). Heterologous expression of the filarial nematode *alt* gene products reveals their potential to inhibit immune function. *BMC Biol.* 3: 8.

**137** Hoerauf, A., Satoguina, J., Saeftel, M., and Specht, S. (2005). Immunomodulation by filarial nematodes. *Parasite Immunol.* 27: 417–429.

**138** Thirugnanam, S., Pandiaraja, P., Ramaswamy, K. et al. (2007). *Brugia malayi*: comparison of protective immune responses induced by *Bm-alt-2* DNA, recombinant Bm-ALT-2 protein and prime-boost vaccine regimens in a jird model. *Exp. Parasitol.* 116: 483–491.

**139** Anand, S.B., Murugan, V., Prabhu, P.R. et al. (2008). Comparison of immunogenicity, protective efficacy of single and cocktail DNA vaccine of *Brugia malayi* abundant larval transcript (ALT-2) and thioredoxin peroxidase (TPX) in mice. *Acta Trop.* 107: 106–112.

**140** Anand, S.B., Kodumudi, K.N., Reddy, M.V., and Kaliraj, P. (2011). A combination of two *Brugia malayi* filarial vaccine candidate antigens (BmALT-2 and BmVAH) enhances immune responses and protection in jirds. *J. Helminthol.* 85: 442–452.

**141** Murray, J., Gregory, W.F., Gomez-Escobar, N. et al. (2001). Expression and immune recognition of *Brugia malayi* VAL-1, a homologue of vespid venom allergens and *Ancylostoma* secreted proteins. *Mol. Biochem. Parasitol.* 118: 89–96.

**142** Kalyanasundaram, R. and Balumuri, P. (2011). Multivalent vaccine formulation with BmVAL-1 and BmALT-2 confer significant protection against challenge infections with *Brugia malayi* in mice and jirds. *Res. Rep. Trop. Med.* 2011: 45–56.

**143** Samykutty, A., Dakshinamoorthy, G., and Kalyanasundaram, R. (2010). Multivalent vaccine for lymphatic filariasis. *Procedia Vaccinol.* 3: 12–18.

**144** Nezafat, N., Karimi, Z., Eslami, M. et al. (2016). Designing an efficient multi-epitope peptide vaccine against *Vibrio cholerae* via combined immunoinformatics and protein interaction based approaches. *Comput. Biol. Chem.* 62: 82–95.

**145** Ali, M., Pandey, R.K., Khatoon, N. et al. (2017). Exploring dengue genome to construct a multi-epitope based subunit vaccine by utilizing immunoinformatics approach to battle against dengue infection. *Sci. Rep.* 7: 9232.

**146** Meza, B., Ascencio, F., Sierra-Beltrán, A.P. et al. (2017). A novel design of a multi-antigenic, multistage and multi-epitope vaccine against *Helicobacter pylori*: an *in silico* approach. *Infect. Genet. Evol.* 49: 309–317.

**147** Shey, R.A., Ghogomu, S.M., Esoh, K.K. et al. (2019). *In-silico* design of a multi-epitope vaccine candidate against onchocerciasis and related filarial diseases. *Sci. Rep.* 9: 4409.

**148** Madhumathi, J., Prince, P.R., Anugraha, G. et al. (2014). Multi-epitope peptide vaccines for human lymphatic filariasis. *Int. J. Infect. Dis.* 21: 19.

**149** Madhumathi, J., Prince, P.R., Rao, D.N. et al. (2017). Epitope mapping of *Brugia malayi* ALT-2 and the development of a multi-epitope vaccine for lymphatic filariasis. *J. Helminthol.* 91: 43–54.

**150** Anugraha, G., Madhumathi, J., Prince, P.R. et al. (2015). Chimeric epitope vaccine from multistage antigens for lymphatic filariasis. *Scand. J. Immunol.* 82: 380–389.

# 20

# Functional Genomics of Filariae

*Eileen Devaney* and Collette Britton**

*University of Glasgow, Institute of Biodiversity, Animal Health and Comparative Medicine, College of Medical, Veterinary and Life Sciences, Bearsden Road, Glasgow G61 1QH, Scotland, UK*

**Abstract**

The genome of the filarial worm *Brugia malayi* was one of the first parasitic nematode genomes to be sequenced, over a decade ago. Since then, improved technology has facilitated sequencing of the genomes and transcriptomes of many more filarial species and of their *Wolbachia* endosymbiont. It is now possible to identify genes and gene families that are present in filariae, to compare across species to identify gene gain and loss and to identify in which developmental stages specific genes are expressed. However, functional studies have lagged behind progress in genome sequencing. Some genes are conserved across the nematode phylum, and functional information can be gained from the model nematode *Caenorhabditis elegans*. However, many genes in filarial nematodes are novel, and direct manipulation in the parasite is required to verify their function. Here we review successes to date with RNA interference (RNAi) technology and the phenotypes observed in a variety of filarial species and life cycle stages following RNAi by soaking. In addition to success with gene knock-down, significant progress has been made in transfection methodology allowing the expression of transgenes in filarial worms, holding promise for determining spatial and temporal gene expression patterns. CRISPR-Cas9 knockout has proved effective in *C. elegans* and in the parasitic nematode *Strongyloides* and, building upon the success of transfection, has the potential to revolutionize functional genomics in filarial nematodes. Complementing these technologies are improvements in the *in vitro* culture of filarial species to enhance development, which is important for phenotypic screening. Together, the application of these technologies holds promise for advancing knowledge of filarial gene function.

## 20.1 Introduction

Filarial worms are among the most fascinating of parasitic nematodes. Their potential for longevity, related to their extraordinary ability to modulate host immune

*Corresponding authors.

*Human and Animal Filariases: Landscape, Challenges, and Control*, First Edition.
Edited by Ronald Kaminsky and Timothy G. Geary.

responses, and the intimate relationship of many pathogenic species with their *Wolbachia* endosymbiont, mark them out from most other parasitic nematodes. However, these are challenging worms to manipulate in the laboratory; they are obligate parasites with lengthy developmental cycles in the vertebrate host and require an arthropod vector to complete the life cycle. Currently, there are no methods for propagating the whole life cycle *in vitro,* although certain life cycle stages can undergo some development in culture (see below). In recent years, the publication of a number of filarial genomes and transcriptomes has facilitated the identification of multiple genes with potentially important functions in worm development and survival ([1–3] and WormBase ParaSite [https://parasite.wormbase.org/index.html]). However, many filarial genes lack annotation (hypothetical or unknown genes), and thus methods for identifying likely function are important, particularly for predicting targets that could form the basis of novel therapies. At present, functional genomics tools are limited for filarial nematodes, and the model nematode *Caenorhabditis elegans* remains the "go-to" system for further analysis. However, success with gene silencing and transfection of *Brugia malayi* holds promise for the future. In this chapter, we discuss how to identify genes of interest in filarial worms and review the tools available for functional genomics.

## 20.2 Exploiting Filarial Omics Data

### 20.2.1 Mining the Genome

Numerous parasitic nematode genomes have now been sequenced to varying depths and with varying degrees of annotation [3]. These provide an important source of information for identifying gene families conserved throughout the filariae, as well as those that are greatly expanded or, conversely, have been lost in filarial worms. WormBase ParaSite currently lists 11 filarial genomes, three species of *Brugia* (*malayi, pahangi,* and *timori*), three species of *Onchocerca* (*volvulus, ochengi,* and *flexuosa*), the human parasites *Wuchereria bancrofti* and *Loa loa*, the dog heartworm, *Dirofilaria immitis*, and the laboratory model species *Acanthocheilonema viteae* and *Litomosoides sigmondontis.* A draft *B. malayi* genome was first published in 2007 [4], and both the *B. malayi* and *O. volvulus* genomes are now assembled into chromosomes [3, 5, 6]. In addition to providing abundant information on genome structure, content, and diversity, these studies enable the understanding of filarial genome evolution, as observed by the identification of a recently evolved Y chromosome in *O. volvulus* [6].

The genomes of filarial nematodes encode fewer proteins than those of many other nematodes, with approximately 11 000–15 000 protein coding genes, depending on species, compared with approximately 20 000 in *C. elegans.* While there has been extensive gene loss, expansion of specific genes in some filarial nematodes can help indicate important functions within the parasite and/or in host–parasite interactions. For example, *O. volvulus* is unique among the filariae in containing genes encoding all of the G-protein-coupled receptor (GPCR) families present in *C. elegans* and shows remarkable expansion of some GPCR families, possibly required for sensing host signals [6]. As *O. volvulus* adult worms are not localized to a particular organ of the body, the intriguing possibility also exists that the expansion of GPCRs may reflect the requirement for males to find females for mating.

### 20.2.2 The Filarial Transcriptome

As well as genome sequencing, RNA sequencing (RNA-Seq) has been widely used to describe the transcriptome of various filarial species at different stages of development (reviewed in [2]). Using these data, it is possible to build up a picture of transcripts that are highly expressed at key points in the life cycle and that may play important roles in the development and/or survival of the parasite. In addition, many RNA-Seq projects have also co-sequenced transcripts expressed by the *Wolbachia* endosymbiont in an attempt to define the role that *Wolbachia* play in parasite survival (for example [7]). By identifying key metabolic genes absent from filarial worms that are present in *Wolbachia*, it may be possible to better understand the nature of this symbiotic relationship. Examples of pathways missing or partially missing in filarial worms are: purine/pyrimidine salvage pathways and a heme biosynthesis pathway [8, 9]. The heme pathway is absent in nematodes, implying that heme must be supplied from the environment or perhaps via the endosymbiont, which possesses a full heme biosynthetic pathway. In support of this hypothesis, knockdown of *hrg-1*, a key gene in the heme acquisition pathway, in adult *B. malayi* inhibited microfilariae (Mf) production [10]. More sophisticated tools, such as flux balance analysis, have been used to reconstruct metabolic pathways in filarial worms and their *Wolbachia* and indicate that the endosymbiont may provide a source of purines to the worm. Using a combination of imaging and cell biology methods, Foray et al. [11] described the role of the *Wolbachia* in germline stem cell development in adult *Brugia* female worms. These elegant studies demonstrated that *Wolbachia* are essential for germline proliferation and maintaining stem cell quiescence and are unlikely to act (at least with respect to the germline) through nucleotide supplementation. In contrast, it was suggested that metabolic supplementation may be provided by *Wolbachia* present in hypodermal cells of the worm and may promote longevity [11]. More recent studies have shed further light on the mutualistic relationship between worm and *Wolbachia*, building upon the observation that the bacteria lack key genes required for glycolysis [12]. It was demonstrated that supplying exogenous pyruvate, the end point of glycolysis, stimulated growth of the *Wolbachia* within the worm, providing further evidence of a true symbiosis.

### 20.2.3 Tissue-Specific Gene/Protein Expression

The large size of some adult filarial worms has allowed dissection and analysis of tissue-specific gene expression. Luck et al. [13] dissected the body wall, intestine, and reproductive tissues from adult male and female *D. immitis* and processed these for both transcriptomic and proteomic profiling. That study highlighted the contribution of the reproductive tissue to the overall transcriptome of adult worms and showed a reasonably good correlation between the transcriptome and the proteome (~35% of tissue-specific transcripts having a peptide "hit"). A similar proteomic analysis of adult *B. malayi* identified proteins enriched in each tissue [14], with particular emphasis on those in the digestive tract, which might constitute "hidden antigens" and therefore be interesting vaccine candidates. In addition to tissue-specific transcriptomic profiling, understanding where and when a gene is expressed can yield important information. For this purpose, *in situ* hybridization has been used to study the localization of specific transcripts in filarial worms. Li et al. [15] localized a likely target of ivermectin, the glutamate-gated chloride channel (GluCl) gene *avr-14*, in adults and Mf of *B. malayi*. The expression of this GluCl subunit in the reproductive system of adult *B. malayi*, as well as in developing embryos, helps explain the mechanisms by which ivermectin sterilizes filarial nematodes.

Rather than sequencing a whole worm or a particular tissue, methods are now available for single-cell RNA-Seq that could be applied to filarial worms. Single-cell RNA-Seq of *C. elegans* L2 stages resolved 27 different cell types, including neuronal cells that were sparsely represented in the worm [16]. The transcriptomic data from this analysis correlated well with whole animal RNA-Seq and resolved tissue-specific expression patterns. Moreover, the sensitivity of the method was demonstrated when examining neuronal gene expression, where genes expressed in individual neurons were identified. Data such as these have been used to configure whole worm gene expression atlases for different *C. elegans* life cycle stages (e.g., [16, 17]), providing an extremely useful (and searchable) resource. While such technology has yet to be applied to filarial worms, the significant contribution of reproductive tissue to the total transcriptome of adult worms may provide additional challenges and suggest that single-cell RNA-Seq of filariae might be better focused on sexually immature stages, at least initially.

### 20.2.4 Using miRNA Profiling to Identify Genes of Interest

In recent years, improved understanding of genome and transcriptome content has resulted in the identification and characterization of regulatory noncoding RNA species, including microRNAs (miRNAs), long noncoding RNAs, and small interfering RNAs (siRNA), each of which can have distinct roles in eukaryotes [18]. miRNAs have been sequenced from a range of filarial species, including *B. pahangi* [19], *B. malayi* [20], and *D. immitis* [21]. Work in our laboratory focused on *B. pahangi* miRNAs that were differentially expressed in various life cycle stages. Many miRNAs identified had no homology to miRNAs in miRBase (ftp://mirbase.org/pub/mirbase21), while others belonged to well-characterized miRNA families [19]. These data were used to examine both miRNA and gene expression in parallel as the L3 of *B. pahangi* moves from the mosquito vector to the mammalian host, with the aim of better understanding how the parasite adapts to a novel environment at this key transition in the life cycle. From five miRNAs that were differentially expressed (up- or downregulated) upon infection of the mammalian host, we focused on a novel member of the *let-7* family, *Bpa-miR-5364* [22]. Using RNA-Seq data from corresponding life cycle stages and a variety of miRNA target prediction tools, we identified several mRNAs, the expression of which correlated negatively with *Bpa-mir-5364* expression. While an interaction between *Bpa-miR-5364* and three selected target mRNAs could be demonstrated *in vitro* in a heterologous transfection system, it was not possible to inhibit *Bpa-miR-5364* in the worm using antisense oligonucleotides, highlighting the challenges involved in defining gene function in the parasite itself.

### 20.2.5 The Filarial Secretome and Immunomodulation

Detailing the proteome and miRNAome of filarial secreted products is important in progressing understanding of host–parasite interactions and immunomodulation. The majority of studies on the excretory-secretory (ES) products of filarial worms have focused on protein constituents as the major source of immune-modulators (reviewed in [23]). Many potential immune regulators have been identified in ES products and their activity and mode of action confirmed. One of the best-characterized filarial ES products is ES-62, a phosphorylcholine containing molecule that interferes with immune cell signaling and mediates a range of anti-inflammatory responses [24]. Other molecules identified in filarial ES include a cystatin that blocks antigen processing in the MHC class II pathway [25]. Interestingly, this effect was specific to filarial cystatin, as the *C. elegans* homologue showed no such activity [26].

More recent studies have considered the role of extracellular vesicles (EV) containing miRNAs in modulating immune responses to nematode infections. This is based on the original observations of Buck et al. [27], who demonstrated the transfer of miRNAs in EV from the gastrointestinal nematode *Heligmosomoides polygyrus* to mammalian cells and their ability to target immune genes. Filarial parasites also release EV containing miRNAs, and these have been shown to enter murine macrophages and affect expression of immune genes [28]. Additionally, *Onchocerca*-specific miRNAs have been detected in human serum from *O. volvulus* infected patients and serum from cattle infected with *O. ochengi* [29], confirming that parasite miRNAs are released *in vivo* and raising the possibility of using miRNAs as biomarkers of infection.

## 20.3 Gene Silencing and Editing in Filariae

### 20.3.1 RNA Interference and CRISPR/Cas

While genes of interest can be identified from the genome and their degree of conservation and homology ascertained *in silico*, it is important to confirm function by modulating levels of gene expression and investigating resulting phenotypes. RNA interference (RNAi) is a reverse genetic technique developed for the model nematode *C. elegans*, in which it can be mediated by feeding bacteria expressing dsRNA, by soaking worms in dsRNA, or by microinjection of dsRNA. In filarial worms, which do not feed on bacteria and for which microinjection is difficult, most efforts have concentrated on RNAi by soaking. Aboobaker and Blaxter in 2003 [30] were the first to demonstrate knock-down of two housekeeping genes (β-tubulin and RNA polymerase II) in adult female *B. malayi* and the subsequent death of the worms. RNAi knock-down of a gene encoding a microfilarial sheath protein (specifically expressed by adult female worms and incorporated into the Mf sheath) was also successful, with a phenotype observed only in Mf [30]. Other functionally relevant targets for knock-down in filarial worms include genes involved in molting, such as those encoding cathepsin-like cysteine proteases, which were shown to be required for the L3 to L4 molt in *O. volvulus* [31]. Landmann et al. [32] successfully targeted a number of structural genes in adult *B. malayi* that have severe phenotypes in *C. elegans*, using a mixture of short overlapping RNAs generated by enzymatic digestion of dsRNA. Adult worms were subjected to soaking for two to five days, and embryos from treated worms phenocopied RNAi results in *C. elegans*. More recent studies aimed at elucidating drug mode of action have used RNAi to knock down predicted target genes of anthelmintic drugs. For example, Verma et al. [33] demonstrated the effect of various cholinergic anthelmintics on adult *B. malayi*. By using RNAi targeting specific subunits of the nicotinic acetylcholine receptor (nAChR), the role of each subunit in the response to different drugs was demonstrated. A summary of RNAi studies in filarial species is presented in Table 20.1.

While most attempts at RNAi in filarial worms have focused on soaking accessible stages maintained *in vitro,* such as L3s or adult worms, Song et al. [38] injected dsRNA into the thorax of *B. malayi*-infected mosquitoes containing L2 or L3 stages and showed good levels of knock-down of a cathepsin-L gene. This novel approach resulted in parasites with a range of motility phenotypes, which were unable to migrate from the thorax to the head and mouthparts [38].

Genome editing using CRISPR-Cas9 is now widely used experimentally to precisely mutate DNA sequences and has been successfully applied in *Strongyloides stercoralis* [41], a parasitic nematode with the advantage of a free-living stage in the life cycle. Building upon the success of transgenesis in *Strongyloides* species [42], a Cas-9 construct targeting CRISPR sites within the muscle gene *unc-22* was micro-injected into free-living adult female worms. The expected twitcher phenotype was apparent in a proportion of F1 offspring and, when passaged through a gerbil host, F2 mutants were recovered. This impressive finding was complemented by additional experiments in which the same technology was used to introduce a transgene into the related species, *S. ratti*. These studies are important because they mark a new frontier in parasitic nematode biology and hopefully will encourage the application of similar methods to less tractable species, such as filarial worms.

### 20.3.2 Transgenesis: *C. elegans* and Beyond

As alluded to previously, the most widely used functional genomics tool for analyzing filarial genes relies upon the use of *C. elegans*. The free-living nematode provides a "tool-box" of reagents and methods, including a range of plasmid

**Table 20.1** Summary of RNAi studies in filarial parasitic nematodes.

| Species | Stage | Gene | Method | Effect | References |
|---|---|---|---|---|---|
| *B. malayi* | Adult female | RNA polymerase, Beta-tubulin, Sheath protein | Soaking in dsRNA | Adult death Adult death Short mf | [30] |
| | Adult female | γ-tubulin Polarity determinant Protein PAR-1 Cell junction protein AJM-1 | Soaking in heterogenous dsRNA | Emb cytokinesis Polarity defects | [32] |
| | Adult female | Cathepsin L & Z | Soaking in dsRNA or siRNA | Reduced mf release and embryo viability | [34] |
| | Adult female L3 | trehalose 6-phosphate phosphatase | Soaking in siRNA | Reduction in Mf impaired viability | [35] |
| | Adult female | Collagen proyl-4 hydroxylase subunits | Soaking in heterogenous siRNAs | mf morphology defects | [36] |
| | Adult L3 | UDP-Galactopyranose mutase | Soaking in siRNAs | Reduced motility, Fewer mf and embryos Reduced motility or death | [37] |
| | Adult | nACh Receptors unc-29+unc-38 | Soaking in dsRNA | Reduced motility | [33] |
| | L3 | Cathepsin L | Heterogenous siRNAs injected into mosquito thorax | Reduced motility Reduced migration | [38] |
| *L. sigmodontis* | Adult female | Actin | Soaking in dsRNA | Reduced motility Reduced mf release | [39] |
| *O. volvulus* | L3 | cathepsin L & Z | Soaking in dsRNA | Incomplete molting | [31] |
| | L3 | Serine protease inhibitor | Soaking in dsRNA | Reduced molting and viability | [40] |

vectors, selectable markers, and reporter constructs that facilitate its application in determining spatial and temporal gene expression and as a heterologous expression system for parasite genes (reviewed in [43]). Transgenesis of *C. elegans* with promoter–reporter constructs has identified the expression pattern of filarial genes [44] and of *C. elegans* homologues of filarial genes of interest [45]. In addition, the availability of large numbers of mutant strains from the *C. elegans* Genetics Center

(CGC) provides an excellent resource to determine whether a particular phenotype can be rescued by expression of a filarial gene, indicating functional conservation. The results have proved to be somewhat unpredictable, but potentially valuable in revealing novel functions or structures of filarial proteins. For example, attempts to rescue a *C. elegans daf-21* (*hsp90* mutant) with the *Brugia* orthologue (84% amino acid identity) were not successful, while partial rescue was achieved with a construct containing the *Haemonchus contortus* open reading frame (88% amino acid identity to *C. elegans*) [46]. A similar approach was used to examine prolyl 4-hydroxylase (C-P4H) enzymes, required for cuticle collagen formation. Winter et al. [36] demonstrated functional conservation of the C-P4H β subunit, with complete rescue of a *C. elegans* mutant (*pdi-2*) with the homologous *B. malayi* gene. In contrast, no rescue was observed following transgenesis with genes encoding the α-subunits, PHY-1 or PHY-2, singly or in combination. However, human C-P4H α-subunit genes can rescue the *C. elegans* mutant phenotypes. This work indicated that the surprising result with *B. malayi* homologues was not due to evolutionary distance, but may result from a novel structure required for *B. malayi* C-P4H activity, which has potential relevance for drug targeting.

Rather than using *C. elegans* as a transgenic expression system, recent efforts have concentrated upon transformation of filarial parasites. Building upon early successes with biolistic bombardment, in which luciferase reporters were transfected into *B. malayi* embryos [47], the Unnasch laboratory recently reported transfection and stable integration in *B. malayi* using *piggyBac* vectors [48]. In their original work, embryos, L3, and adult females were all biolistically transformed, with optimal activity of the luciferase reporter in embryos [49]. One of the challenges of attempting to transfect adult female filariae is their size. In *C. elegans,* constructs are injected into the gonad, while the long length of the filarial reproductive tract makes it difficult to know where to inject. Higazi et al. [49] micro-injected into three separate regions of the female reproductive tract and recorded most luciferase activity from the medial region. However, transfected embryos were unable to develop further.

In subsequent studies, the L3 stage of *B. malayi* was transfected via calcium phosphate precipitation with a luciferase reporter under the control of the *Bm-hsp-70* promoter both *in vitro* and *in vivo* within the peritoneal cavity of jirds [50]. In these studies, maintenance of the L3 *in vitro*, under conditions that promoted molting to the L4, resulted in successful transformation, but these parasites were not capable of survival *in vivo*. By injecting L3 into jirds along with the transfection mix, it was possible to recover adult worms and Mf that were transformed (i.e. able to secrete luciferase into the medium). In these studies, an impressive 67% of adult females and 60% of adult males were transformed and, moreover, a small number of Mf derived from these adult worms inherited the transgene. Finally, it was shown that transformed Mf could infect mosquitoes, resulting in L3 carrying the transgene. In addition to demonstrating the feasibility of establishing transgenic lines of filariae, these experiments are important for a number of reasons: they defined the necessary promoter and 3′UTR sequences required for expression, highlighted the importance of including intronic sequence in the construct, developed the use of a secreted luciferase reporter gene, and indicated that molting parasites might be more susceptible to transfection.

While calcium phosphate precipitation was successful, it was much less efficient than biolistic bombardment [50], and additionally, it was not clear how transgenes were maintained in the filarial genome. Further studies from this group used the *piggyBac* system and lipofectamine to stably transform *B. malayi* L3 in an updated culture system containing Bovine Embryo Skeletal Muscle (BESM) cells in which high levels of L3 molt to the L4 [48]. These transfected L4 were competent to develop when injected into jirds and produced small numbers of transgenic Mf in which the transgene was integrated, mostly into intergenic regions of the genome.

### 20.3.3 Cryopreservation and *In Vitro* Culture of Filarial Worms

While impressive progress has been made in transgenesis, the complex and lengthy life cycle of filarial worms provides additional challenges. It is possible to cryopreserve larval stages (Mf and L3) efficiently and to recover them from liquid nitrogen while maintaining their infectivity for their respective hosts [51–53]. This is an important consideration for enabling long-term storage of transformed worms. However, there are presently no culture systems that permit the development of the whole life cycle *in vitro*, meaning that Mf must be fed to mosquitoes to generate L3. The Mf of *Brugia* are intracellular parasites within the mosquito host, and this presumably restricts their ability to develop *in vitro* [54]. However, better success has been reported with L3 stages, which can be cultured to L4 and beyond, depending upon species. The L3 of *D. immitis* molt quite rapidly, after 48–72 hours *in vitro* [55], while *O. volvulus*, requires four to five days in culture [56], and *Brugia* spp. require a longer period of culture (six to eight days). This observation may relate to specific differences in the life cycle: both *D. immitis* and *O. volvulus* L3 develop to the L4 in the tissues surrounding the bite site, while *Brugia* spp. migrate rapidly to the lymphatic system [57]. Elegant *in vivo* studies using labeled *L. sigmodontis* recently detailed the unidirectional movement of L3 within the lymphatics [58]. Determining whether the observed difference in behavior and/or *in vivo* environment is reflected in differences *in vitro* will require additional studies.

More recently, a method was published for obtaining early adults of *O. volvulus in vitro* [59]; this technique involved allowing L3 to molt to L4 in the standard culture system, which includes human PBMC, and then switching the L4 to a different system containing a mammalian cell line. The best results were obtained by containing *O. volvulus* L4 within a transwell overlaying either human dermal fibroblasts (HDF) or human umbilical vein endothelial cells (HUVEC) in medium supplemented with 25% fetal bovine serum, 1% nonessential amino acids, and 0.1% lipid mixture. Under these conditions, the L4 molted to the L5 and underwent significant growth. These experiments underpin the importance of attempting to mimic the *in vivo* environment of filarial nematodes; these are highly adapted organisms that tend to be extremely host-specific within both vector and mammalian hosts and in general will only develop in specific anatomical sites. Progress in mammalian culture systems, including the development of organoids [60], may provide clues as to how best to optimize filarial culture. These advances are highly relevant to *in vitro* application of gene knockdown technology and phenotypic analysis and screening for new therapeutics.

## 20.4 Conclusions

While considerable progress has been made in developing tools for functional genomic analysis in filarial worms, several obstacles remain, perhaps the most significant being the obligate parasitic life cycles, requiring two hosts for development. Great strides have been made in the control of human filarial infection, using ivermectin/moxidectin for onchocerciasis and diethylcarbamazine/albendazole or ivermectin/albendazole for the lymphatic species (see Chapter 6). In small foci of onchocerciasis, such as in Latin America, control has been achieved, but in the major African foci, an estimated 21 million people remain infected, despite >30 years of ivermectin distribution. Moreover, the specter of ivermectin resistance is now a reality [61]. Thus much remains to be learned as to how best to combat these important infections; an improved understanding of their genomes and the development of additional functional genomic tools will help accelerate control strategies.

## Acknowledgments

The authors thank Dr. Roz Laing, University of Glasgow for comments on the manuscript.

## References

1 Grote, A., Lustigman, S., and Ghedin, E. (2017). Lessons from the genomes and transcriptomes of filarial nematodes. *Mol. Biochem. Parasitol.* 215: 23–29.
2 Bennuru, S., O'Connell, E.M., Drame, P.M., and Nutman, T.B. (2018). Mining filarial genomes for diagnostic and therapeutic targets. *Trends Parasitol.* 34: 80–90.
3 International Helminth Genomes Consortium (2019). Comparative genomics of the major parasitic worms. *Nat. Genet.* 51: 163–174.
4 Ghedin, E., Wang, S., Spiro, D. et al. (2007). Draft genome of the filarial nematode *Brugia malayi*. *Science* 317: 1756–1760.
5 Choi, Y.J., Tyagi, R., McNulty, S.N. et al. (2016). Genomic diversity in *Onchocerca volvulus* and its *Wolbachia* endosymbiont. *Nature Microbiol.* 2: 16207.
6 Cotton, J.A., Bennuru, S., Grote, A. et al. (2016). The genome of *Onchocerca volvulus*, agent of river blindness. *Nature Microbiol.* 2: 16216.
7 Grote, A., Voronin, D., Ding, T. et al. (2017). Defining *Brugia malayi* and *Wolbachia* symbiosis by stage-specific dual RNA-Seq. *PLoS Negl. Trop. Dis.* 11 (3): e0005357.

**8** Foster, J., Ganatra, M., Kamal, I. et al. (2005). The *Wolbachia* genome of *Brugia malayi*: endosymbiont evolution within a human pathogenic nematode. *PLoS Biol.* 3: e21.

**9** Slatko, B.E., Taylor, M.J., and Foster, J.M. (2010). The *Wolbachia* endosymbiont as an antifilarial nematode target. *Symbiosis* 51: 55–65.

**10** Luck, A.N., Yuan, X., Voronin, D. et al. (2016). Heme acquisition in the parasitic filarial nematode *Brugia malayi*. *FASEB J.* 30 (10): 3501–3514.

**11** Foray, V., Perez-Jimenez, M.M., Fattouh, N., and Landmann, F. (2018). *Wolbachia* control stem cell behavior and stimulate germline proliferation in filarial nematodes. *Dev. Cell* 45: 198–211.

**12** Voronin, D., Schnall, E., Grote, A. et al. (2019). Pyruvate produced by *Brugia* spp via glycolysis is essential for maintaining the mutualistic relationship between the parasite and its endosymbiont, *Wolbachia*. *PLoS Pathog.* 15 (9): e1008085.

**13** Luck, A.N., Anderson, K.G., McClung, C.M. et al. (2015). Tissue-specific transcriptomics and proteomics of a filarial nematode and its *Wolbachia* endosymbiont. *BMC Genomics* 16: 920.

**14** Morris, C.P., Bennuru, S., Kropp, L.E. et al. (2015). A proteomic analysis of the body wall, digestive tract and reproductive tract of *Brugia malayi*. *PLoS Negl. Trop. Dis.* 9: e0004054.

**15** Li, B.W., Rush, A.C., and Weil, G.J. (2014). High level expression of a glutamate gated chloride channel gene in reproductive tissues of *Brugia malayi* may explain the sterilizing effect of ivermectin on filarial worms. *Int. J. Parasitol. Drugs Drug Resist.* 4: 71–76.

**16** Cao, J., Packer, J.S., Ramani, V. et al. (2017). Comprehensive single-cell transcriptional profiling of a multicellular organism. *Science* 357: 661–667.

**17** Packer, J.S., Zhu, Q., Huynh, C. et al. (2019). A lineage-resolved molecular atlas of *C. elegans* embryogenesis at single cell resolution. *Science* 365 (6459): eaax1971.

**18** Aalto, A.P. and Pasquinelli, A.E. (2012). Small non-coding RNAs mount a silent revolution in gene expression. *Curr. Opin. Cell Biol.* 24: 333–340.

**19** Winter, A.D., Weir, W., Hunt, M. et al. (2012). Diversity in parasitic nematode genomes: the microRNAs of *Brugia pahangi* and *Haemonchus contortus* are largely novel. *BMC Genomics* 13: 4.

**20** Poole, C.B., Gu, W., Kumar, S. et al. (2014). Diversity and expression of microRNAs in the filarial parasite *Brugia malayi*. *PLoS One* 9 (5): e96498.

**21** Fu, Y., Lan, J., Wu, X. et al. (2013). Identification of *Dirofilaria immitis* miRNA using Illumina deep sequencing. *Vet. Res.* 44: 3.

**22** Winter, A.D., Gillan, V., Maitland, K. et al. (2015). A novel member of the let-7 microRNA family is associated with developmental transitions in filarial nematode parasites. *BMC Genomics* 16: 331.

**23** Hewitson, J.P., Harcus, Y.M., Curwen, R.S. et al. (2008). The secretome of the filarial parasite *Brugia malayi*: proteomic profile of adult excretory-secretory products. *Mol. Biochem. Parasitol.* 160: 8–21.

**24** Harnett, M.M. and Harnett, W. (2017). Can parasitic worms cure the modern world's ills. *Trends Parasitol.* 33: 694–705.

**25** Manoury, B., Gregory, W.F., Maizels, R.M., and Watts, C. (2001). Bm-CPI-2, a cystatin homolog secreted by the filarial parasite *Brugia malayi*, inhibits class II MHC-restricted antigen processing. *Curr. Biol.* 11: 447–451.

**26** Gregory, W.F. and Maizels, R.M. (2008). Cystatins from filarial parasites: evolution, adaptation and function in the host-parasite relationship. *Int. J. Biochem. Cell Biol.* 40 (6-7): 1389–1398.

**27** Buck, A.H., Coakley, G., Simbari, F. et al. (2014). Exosomes secreted by nematode parasites transfer small RNAs to mammalian cells and modulate innate immunity. *Nat. Commun.* 5: 5488.

**28** Zamanian, M., Fraser, L.M., Agbedanu, P.N. et al. (2015). Release of small RNA-containing exosome-like vesicles from the human filarial parasite *Brugia malayi*. *PLoS Negl. Trop. Dis.* 9: e0004069.

**29** Quintana, J.F., Makepeace, B.L., Babayan, S.A. et al. (2015). Extracellular *Onchocerca*-derived small RNAs in host nodules and blood. *Parasites Vectors* 8: 58.

**30** Aboobaker, A.A. and Blaxter, M.L. (2003). Use of RNA interference to investigate gene function in the human filarial nematode parasite *Brugia malayi*. *Mol. Biochem. Parasitol.* 129: 41–51.

**31** Lustigman, S., Zhang, J., Liu, J. et al. (2004). RNA interference targeting cathepsin L and cathepsin Z-like cysteine proteases of *Onchocerca volvulus* confirmed their essential function during L3 molting. *Mol. Biochem. Parasitol.* 138: 165–170.

**32** Landmann, F., Foster, J.M., Slatko, B.E., and Sullivan, W. (2012). Efficient *in vitro* RNA interference and immunofluorescence-based phenotype analysis in a human parasitic nematode, *Brugia malayi*. *Parasites Vectors* 5: 16.

**33** Verma, S., Kashyap, S.S., Robertson, A.P., and Martin, R.J. (2017). Functional genomics in *Brugia malayi* reveal diverse muscle nAChRs and differences between cholinergic anthelmintics. *Proc. Natl. Acad. Sci. U.S.A.* 114: 5539–5544.

**34** Ford, L., Zhang, J., Liu, J. et al. (2009). Functional analysis of the cathepsin-like cysteine protease genes in adult *Brugia malayi* using RNA interference. *PLoS Negl. Trop. Dis.* 3 (2): e377.

**35** Kushwaha, S., Singh, P.K., Shahab, M. et al. (2012). *In vitro* silencing of *Brugia malayi* trehalose-6-phospahte phosphatase impairs embryogenesis and *in vivo* development of infective larvae in jirds. *PLoS Negl. Trop. Dis.* 6: e1770.

**36** Winter, A.D., McCormack, G., Myllyharju, J., and Page, A.P. (2014). Prolyl 4-hydroxlase activity is essential for development and cuticle formation in the human infective parasitic nematode *Brugia malayi*. *J. Biol. Chem.* 288: 1750–1761.

**37** Misra, S., Gupta, J., and Misra-Bhattacharya, S. (2017). RNA interference mediated knockdown of *Brugia malayi* UDP-Galactopyranose mutase severely affects parasite viability, embryogenesis and *in vivo* development of infective larvae. *Parasites Vectors* 10: 34.

**38** Song, C., Gallup, J.M., Day, T.A. et al. (2010). Development of an *in vivo* RNAi protocol to investigate gene function in the filarial nematode, *Brugia malayi*. *PLoS Pathog.* 6 (12): e1001239.

**39** Pfarr, K., Heider, U., and Hoerauf, A. (2006). RNAi mediated silencing of actin expression in adult *Litomosoides sigmodontis* is specific, persistent and results in a phenotype. *Int. J. Parasitol.* 36: 661–669.

**40** Ford, L., Guiliano, D.B., Oksov, Y. et al. (2005). Characterization of a novel filarial serine protease inhibitor, Ov-SPI-1, from *Onchocerca volvulus*, with potential multifunctional roles during development of the parasite. *J. Biol. Chem.* 280: 40845–40856.

**41** Gang, S.S., Castelletto, M.L., Bryant, A.S. et al. (2017). Targeted mutagenesis in a human-parasitic nematode. *PLoS Pathog.* 13: e1006675.

**42** Junio, A.B., Li, X., Massey, H.C. Jr., et al. (2008). *Strongyloides stercoralis*: cell- and tissue-specific transgene expression and co-transformation with vector constructs incorporating a common multifunctional 3′ UTR. *Exp. Parasitol.* 118: 253–265.

**43** Ward, J.D. (2015). Rendering the intractable more tractable: tools from *Caenorhabditis elegans* ripe for import into parasitic nematodes. *Genetics* 201: 1279–1294.

**44** Hashmi, S., Britton, C., Liu, J. et al. (2002). Cathepsin L is essential for embryogenesis and development of *Caenorhabditis elegans*. *J. Biol. Chem.* 277: 3477–3486.

**45** Thompson, F.J., Britton, C., Wheatley, I. et al. (2002). Biochemical and molecular characterization of two cytidine deaminases in the nematode *Caenorhabditis elegans*. *Biochem. J.* 365: 99–107.

**46** Gillan, V., Maitland, K., McCormack, G. et al. (2009). Functional genomics of *hsp-90* in parasitic and free-living nematodes. *Int. J. Parasitol.* 39 (10): 1071–1081.

**47** Higazi, T.B. and Unnasch, T.R. (2013). Biolistic transformation of *Brugia malayi*. *Methods Mol. Biol.* 940: 103–115.

**48** Liu, C., Mhashilkar, A.S., Chabanon, J. et al. (2018). Development of a toolkit for piggyBac-mediated integrative transfection of the human filarial parasite *Brugia malayi*. *PLoS Negl. Trop. Dis.* 12 (5): e0006509.

**49** Higazi, T.B., Merriweather, A., Shu, L. et al. (2002). *Brugia malayi*: transient transfection by microinjection and particle bombardment. *Exp. Parasitol.* 100 (2): 95–102.

**50** Xu, S., Liu, C., Tzertzinis, G. et al. (2011). *In vivo* transfection of developmentally competent *Brugia malayi* infective larvae. *Int. J. Parasitol.* 41: 355–362.

**51** Ham, P., James, E., and Bianco, A. (1979). *Onchocerca* spp: cryopreservation of microfilariae and subsequent development in the insect host. *Exp. Parasitol.* 47 (3): 384–391.

**52** Lowrie, R. (1983). Cryopreservation of the microfilariae of *Brugia malayi*, *Dirofilaria corynodes*, and *Wuchereria bancrofti*. *Am. J. Trop. Med. Hyg.* 32: 138–145.

**53** Trpis, M., Scoles, G., and Struble, R. (1993). Cryopreservation of infective larvae of *Onchocerca volvulus* (Filarioidea: Onchocercidae). *J. Parasitol.* 79 (5): 695.

**54** Devaney, E. (1981). The development of *Dirofilaria immitis* in cultured Malpighian tubules. *Acta Trop.* 38: 251–260.

**55** Devaney, E. (1985). *Dirofilaria immitis*: the molting of the infective larva *in vitro*. *J. Helminthol.* 59: 47–50.

**56** Lustigman, S., Huima, T., Brotman, B., and Prince, A.M. (1990). *Onchocerca volvulus*: biochemical and morphological characteristics of the surface of third- and fourth-stage larvae. *Exp. Parasitol.* 71: 489–495.

**57** Schacher, J.F. (1962). Developmental stages of *Brugia pahangi* in the final host. *J. Parasitol.* 48: 693–706.

**58** Kilarski, W.W., Martin, C., Pisano, M. et al. (2019). Inherent biochemical traits enable infective filariae to disseminate through collecting lymphatic vessels. *Nat. Commun.* 10: 2985.

**59** Voronin, D., Tricoche, N., Jawahar, S. et al. (2019). Development of a preliminary *in vitro* drug screening assay based on a newly established culturing system for pre-adult fifth-stage *Onchocerca volvulus* worms. *PLoS Negl. Trop. Dis.* 13 (1): e0007108.

**60** Yin, Y.B., de Jonge, H.R., Wu, X., and Yin, Y.L. (2019). Mini-gut: a promising model for drug development. *Drug Discovery Today* 24: 1784–1794.

**61** Doyle, S.R., Bourguinat, C., Nana-Djeunga, H.C. et al. (2017). Genome-wide analysis of ivermectin response by *Onchocerca volvulus* reveals that genetic drift and soft selective sweeps contribute to loss of drug sensitivity. *PLoS Negl. Trop. Dis.* 11 (7): e0005816.

# 21

## Development of a Vaccine Against *Onchocerca volvulus*

*David Abraham*[1], *Ben Makepeace*[2], *and Sara Lustigman*[3,*]

[1] *Thomas Jefferson University, Sidney Kimmel Medical College, Department of Microbiology and Immunology, Philadelphia, PA 19107, USA*
[2] *University of Liverpool, Institute of Infection & Global Health, Department of Infection Biology, Liverpool L3 5RF, UK*
[3] *New York Blood Center, Lindsley F Kimball Research Institute, Laboratory of Molecular Parasitology, New York, NY 10065, USA*

**Abstract**

Onchocerciasis or river blindness is a neglected parasitic disease that causes severe dermatitis and visual impairment, predominantly in Africa. The current strategy for controlling and/or eliminating this devastating disease relies on mass administration of a single drug, ivermectin. Neglected tropical disease experts doubt that onchocerciasis can be eliminated through mass drug administration (MDA) with IVM alone. We present in this chapter the rational strategy that was taken to develop a prophylactic vaccine that could accelerate elimination efforts and safeguard the enormous strides made in onchocerciasis control. This vaccine is based on two lead candidate antigens identified by an international partnership, The Onchocerciasis Vaccine for Africa Initiative (TOVA). The protective effects of two lead vaccine candidates in small animal models are presented, as well as the way forward for testing them in the *Onchocerca ochengi* cow infection system. Immune responses against these antigens in humans are reported with putative mechanisms of protective immunity described.

### 21.1 Onchocerciasis Control Programs, their Limitations, and the Need for Additional Supportive Intervention Tools

Onchocerciasis (ONCHO), caused by *Onchocerca volvulus* (*Ov*), is a debilitating eye and skin disease and the world's second-leading infectious cause of blindness in humans; 99% cases are in sub-Saharan Africa (SSA). Current estimates put 120 million people at risk; 20 million are infected, and 1.2 million have vision

*Corresponding author.

*Human and Animal Filariases: Landscape, Challenges, and Control*, First Edition.
Edited by Ronald Kaminsky and Timothy G. Geary.

impairment or blindness [1]. Long the focus of efforts to alleviate morbidity and lost productivity, ONCHO has more recently been targeted for elimination [2, 3]. The three ONCHO control programs aimed at interrupting disease transmission have, since 1989, been based on mass drug administration (MDA) of ivermectin (IVM), a therapy effective at killing microfilariae but not adult worms. Even successes [4–6] now must be weighed against the fact that since 1995, the prevalence of ONCHO has been reduced by only 31% in Africa [7]. The African Programme for Onchocerciasis Control (APOC) called in 2015 for 1.15 billion IVM treatments until 2045 [8], though many neglected tropical disease experts doubt that ONCHO can be eliminated through MDA alone [9], especially given that MDA of IVM cannot easily be used in 11 Central African countries co-endemic for *Loa loa* infections due to the risk of severe adverse events [10, 11]. Moreover, many areas of SSA do not implement ONCHO MDA programs in areas of hypo-endemicity, which could lead to re-introduction to areas undergoing MDA [3]. Of equal concern is the potential emergence of IVM-resistant *Ov*, limiting the long-term effectiveness of MDA [12, 13] and, in time, undermining all the gains achieved by ONCHO control programs. Complicating the resistance issue is that IVM is not administered to children ≤5 years old; and a macrofilaricidal drug, doxycycline, cannot be given to children ≤9 because of the limiting indications for these drugs. These children are not only vulnerable to infection, but they become reservoirs for transmission [14]. For these reasons, APOC called in 2014 for the development and testing of new ONCHO intervention technologies, including a prophylactic vaccine [15]. Importantly, vaccination would become a part of an integrated control strategy that includes existing chemotherapy, new or repurposed macrofilaricidal drugs, and where appropriate, vector control. Furthermore, repositioning vaccines to complement chemotherapy is an innovative, revitalizing concept, one that synergistically supports activities that to date have been mostly focused only on morbidity and/or transmission reduction [16]. In contrast to chemotherapy, which attempts to temporarily cure or reduce infection intensities, morbidity, and transmission, prophylactic vaccines would reduce worm burdens from new infections and induce long-lasting protective host immune responses, capable of being boosted by continued exposure to new infections. The expectation is that an ONCHO vaccine would first be administered to children under the age of five years, prior to their entry into IVM MDA programs. This would thus deliver a key companion technology to ensure the long-term success of MDA, as these vulnerable children may become, if infection is not prevented, reservoirs for transmission [14]. The ONCHO vaccine ultimately could obviate the need for lengthy repeated MDA and, concomitantly, lower the risk of drug failure and reinfection. Moreover, vaccination would be synergistic with later drug therapy, thereby justifying the use of effective vaccines as a long-term solution presently lacking in most control strategies.

The feasibility and proof of concept that antilarval protective immunity to filarial parasites can be induced was first and most consistently demonstrated using irradiation-attenuated infective third-stage larvae (xL3) in several natural hosts (*Onchocerca ochengi (Oo)*/cattle; *Brugia* spp./Rhesus monkeys/cats; *Dirofilaria immitis*/dogs) and other animal models (mice, ferrets, and gerbils) that can be infected with filariae [17–20]. Protection was also shown in a number of these infection model systems using defined recombinant filarial proteins [20–22]. Moreover, two distinctive expressions of anti-L3 protective immunity were described in ONCHO and lymphatic filariasis endemic populations: (i) immunity that impedes the development of a patent infection (microfilaria positive) in putatively immune (PI) individuals [23–27]; and (ii) age-acquired concomitant protective humoral [28] and cellular [29] immunity in infected individuals that prevents most newly acquired L3 infections from developing and results in a stable adult worm burden [30–33].

In this chapter, only the two current and most promising ONCHO vaccine candidates (Ov-103 and Ov-RAL-2) will be covered in detail, as they were chosen to be pursued for clinical development as a prophylactic vaccine against *O. volvulus* L3. Many other promising vaccine candidates against *O. volvulus*, *Wuchereria bancrofti* and *Brugia malayi* (the causative agents of lymphatic filariasis), or *D. immitis* (the causative agent of canine heartworm disease) have been described; however, they will not be the focus of this chapter, as they have been discussed in other publications [22, 34]. The animal models that have been (and can be) used to test the validity of *Ov*-103 and *Ov*-RAL-2 for advancement to preclinical and clinical development will also be described. The research efforts described below are part of The Onchocerciasis Vaccine for Africa Initiative (TOVA, founded in 2015), a collaborative partnership between 14 academic institutions across Africa, Europe, and the United States [35, 36] with a mission to advance at least one ONCHO vaccine into Phase II human trials by 2025 (www.riverblindnessvaccinetova.org).

## 21.2 The Use of the Diffusion Chamber Mouse Model to Support the Development of a Vaccine Against *Onchocerca volvulus* Infective Larvae

Vaccine development against *Ov* has been severely challenged by the absence of suitable animal models. Although chimpanzees and mangabey monkeys are susceptible to infection [37–40], ethical and financial constraints prevent their use in experiments. To overcome this obstacle, a system was developed using diffusion chambers containing *Ov* L3 implanted subcutaneously in animals. This method allows the efficient and complete recovery of all larvae injected into an animal and the unique opportunity to analyze host cells and humoral factors in the parasite microenvironment. Larvae survived at least nine weeks and molted into fourth-stage larvae (L4) in diffusion chambers implanted in rodents, including mice, and in primates, including chimpanzees [41]. The mouse/diffusion chamber system was used to test the hypothesis that protective immunity would develop in mice to the larval stages of *Ov* after vaccination with larvae. The development of protective immunity was determined by a statistically significant reduction in the number of larvae surviving in challenge infections implanted within diffusion chambers. Vaccination of mice with live, dead, or xL3 induced protective immunity, killing approximately 50% of the larvae, whereas immunization with L4 did not [42]. The protective immune response in the vaccinated mice required a Th2-type of CD4 T-cell response, dependent on IL-4 and IL-5 [43, 44]. Protective immunity induced by xL3 was dependent on both IgE and contact of the worms by eosinophils [45].

It was clear from the onset that a vaccine using xL3 would never be practical for use in the field, and that a recombinant antigen-based vaccine was essential. In the first series of studies, antibodies from immune humans, chimpanzees, or rabbits were used to identify *Ov* antigens with vaccine potential. Three of the five antigens identified through this approach, Ov7, OvB8, and Ov64, induced protective immunity only when administered with alum, a Th2 adjuvant. Notably, immunization of mice with a cocktail of the three antigens did not enhance the level of protective immunity, but rather decreased antibody responses compared with monovalent vaccination [46]. Thus, recombinant antigens could induce protective immunity to infection with *Ov* in mice that was Th2-dependent, recapitulating results obtained with xL3.

To overcome variability in the production and validation of recombinant antigens, eight antigens were selected for further analysis based on a set of selection criteria [34, 35]. These antigens were expressed in a single laboratory in both bacterial (*Escherichia coli*) and eukaryotic (*Pichia pastoris*) systems to identify the optimal platform. Mice were immunized with the antigens individually with alum and three of them (*Ov*-103 expressed in *P. pastoris*, *Ov*-RAL-2 expressed in *E. coli*, and *Ov*-CPI2M expressed in either expression system) consistently induced significant protective immunity. Differential cell analysis between diffusion chambers recovered from control and protected mice did not reveal significant differences, and antibody titers to the antigens did not correlate with levels of protective immunity, leaving the mechanism of protective immunity induced by these recombinant antigens unresolved. Immunizing mice with a fusion protein comprised of two or three of these protective antigens, or injection of the three antigens concurrently, did not enhance protective immunity. In all cases, approximately 40% of challenge larvae were killed by the immune response. Antibody titers in mice to the individual antigens did not differ between mice immunized with single antigens, fusion antigens, and concurrent immunization with three antigens, suggesting that these antigens did not compete or antagonize reciprocal immune responses [34].

Immunity to *Ov* induced by xL3 was shown to be Th2-dependent [43–45]. The requirement for Th2 responses was confirmed with several recombinant antigens based on efficacy of alum as the adjuvant [34, 46]. However, there are reports demonstrating that other recombinant antigens that induce immunity to *Ov* require a Th1 response based on the adjuvant [47] or were effective irrespective of adjuvant polarization [48]. To test the hypothesis that vaccine efficacy would be enhanced by the appropriate adjuvant, mice were immunized with recombinant Ov-103 and/or Ov-RAL-2 in the presence of five different adjuvants. Immunizing mice with *Ov*-103 and *Ov*-RAL-2 in conjunction with alum, Advax 2 and MF59 induced significant, yet equivalent, levels of larval killing, whereas these antigens administered with CpG and Advax 1 as adjuvants failed to induce protective immunity. Analysis of spleen cell cytokine responses confirmed that the recombinant vaccine comprised of *Ov*-103 and *Ov*-RAL-2, with all three of the adjuvants, induced a predominately Th2-biased response. IgG1 was the dominant antibody isotype, with antigen-specific IgE absent from these mice. Mice immunized with both *Ov*-103 and *Ov*-RAL-2 in conjunction with each of the three adjuvants displayed enhanced parasite killing compared to single immunizations, with 90% of the worms killed in some cases. Unlike previous studies using multiple-antigen vaccines against *Ov*, antigen-specific antibody titers were significantly increased in mice immunized concurrently with the two antigens; apparently, the antigens interacted synergistically to boost the response to the reciprocal antigen [49]. The requirement for IgG in the killing mechanism induced by the recombinant vaccines was confirmed in studies in which mice that genetically lack IgG were immunized with either *Ov*-103 or *Ov*-RAL-2. These mice developed a robust Th2 immune response but could not kill the larvae in the absence of parasite-specific IgG [50].

Differential cell analysis showed that the types and numbers of cells migrating into the diffusion chamber did not differ between control and immunized mice [49]. It is clear, however, that cells are required for the immune response to kill the worms based on *in vivo* cell exclusion studies, which demonstrated that parasite killing in immunized mice did not occur if cells were prevented from direct contact with the parasites [50]. As an alternative approach to identify the cells involved in controlling the infection after vaccination, chemokine levels in the fluid surrounding the worms in the diffusion chambers were measured. Based on this analysis, it was concluded that neutrophils and eosinophils participate in the protective immune response induced by *Ov*-103, and macrophages and neutrophils participate in immunity induced by *Ov*-RAL-2. Thus, different mechanisms of protective immunity are induced with the two antigens, both dependent on Th2 cytokines and IgG, yet functioning with different effector cells [49].

A new mouse model has recently been developed, in which *Ov* larvae develop in humanized NSG mice. Human stem cells transferred into NSG mice result in mice that lack a murine immune response but have a functional human immune response. As *Ov* will molt and grow in the humanized mice, it would be a useful tool to evaluate human protective immune responses to the infection [51]. Immunized humanized mice would be an interesting tool to use to dissect the mechanism of protective immunity engendered by the human immune response. This information would be critical for the clinical development of a vaccine, providing criteria to assess the efficacy of the vaccine in humans during the immunization process.

## 21.3 Validation of the ONCHO Vaccine in the *Brugia malayi*-Gerbil Infection Model

Because of the limitation of the *Ov* mouse model (which can only determine efficacy against the early larval stages), the *B. malayi* (*Bm*)-gerbil model of subcutaneous infection with L3 was also employed. This provided a valuable complement for evaluating vaccine efficacy against adult and microfilarial worm burdens, the ultimate stages targeted for reduction by a vaccine. Notably, efficacy is much greater when adult worm burdens are determined compared to *Ov* L3 survival in the diffusion chamber model [20, 52]. The two *Bm* homologues (*Bm*-103 and *Bm*-RAL-2) of the

*Ov*-103 and *Ov*-RAL-2 vaccine candidates were expressed and tested individually or after co-administration in separate sites for their protective efficacy [53, 54]. The protective outcomes of co-administered *Bm*-103 and *Bm*-RAL-2 were much improved compared to a single antigen, achieving maximal 61% reduction of worm survival, while the individual antigens induced maximal 39% (*Bm*-103) or 46% (*Bm*-RAL-2) reduction [53]. Importantly, the effect of vaccination on fertility of female worms was assessed by embryogram – i.e. counting embryos, pretzel-stage microfilariae, and stretched microfilariae per female worm – at 120 or 150 dpi, and by counting microfilariae in peripheral blood, verified that co-administered vaccines are better able to reduce the development of all embryonic stages. However, no significant reduction in circulating microfilariae was observed [53], pointing to the need to extend the time point at which to analyze vaccine efficacy on circulating *B. malayi* microfilariae. In the protected gerbils after vaccination using alum-adjuvanted *Bm*-RAL-2, *Bm*-103, or the co-administered vaccines, a strong antigen-specific IgG response to the corresponding antigens was elicited. When these sera samples were used with gerbil peritoneal exudate cells (PECs) as effector cells in antibody-dependent cellular cytotoxicity (ADCC) assays against *Bm* L3 *in vitro*, significant inhibition of motility at 48 hours of culture following the adherence of PECs to the parasite surface was observed [53].

## 21.4 Proof of Principle that an ONCHO Vaccine Can Work in a Natural Environment of Infection

*Oo*, a parasite of cattle, is a sibling species of *Ov* and is transmitted by the same complex of blackfly vectors in West Africa (*Simulium damnosum sensu lato*) as the human parasite [55, 56]. The natural history of *Oo*, including morphology of developmental stages, the prepatent period in the definitive host, and the histological structure of nodules containing adult worms, are remarkably similar between the two species. However, there are some important differences: most *Oo* nodules are intradermal, not subcutaneous as in *Ov*; and *Oo* nodules rarely contain more than a single female worm, which is not true of *Ov*. Cattle with ONCHO are also free of any overt clinical manifestations of disease.

Although originally described in Uganda and Tanzania [57, 58], essentially all epidemiological and experimental research with *Oo* has taken place in Cameroon or has used worms imported from this country for investigations in developed countries. A unique strength of the *Oo* system is that it allows studies on natural protective immunity and the response to vaccination under real-world conditions of exposure to transmission. Thus, it is an ideal system in which to obtain robust data on vaccine efficacy before entering very expensive, late-stage clinical trials in humans. The only site at which detailed epidemiological studies on the transmission of *Oo* have occurred is at the River Vina du Sud, located near the town of Ngaoundéré in the Adamawa Region of northern Cameroon [59, 60]. This is where the first investigation on natural immunity to *Oo* in local cattle took place, which revealed that older animals tended to have lower microfilarial loads, despite a greater number of nodules on average than younger cohorts had, with similar proportions of gravid female worms

[61]. This is consistent with microfilaria-specific concomitant immunity developing naturally in older cattle, as is also seen in human ONCHO with age. Similarly to "putatively immune" humans, a proportion of cattle in Cameroon remain free of infection (or at least have very low parasite loads) despite high levels of natural exposure [62]. However, unlike the situation with humans, it is possible to directly test whether putative immunity develops in exposed cattle. In a key experiment performed close to the *Simulium* breeding site at the Vina du Sud, drug-cured and putatively immune (filaria-negative) cattle from endemic areas were exposed to transmission alongside naïve animals from a non-endemic area. After two years, most of the putatively immune cattle had acquired some nodules and microfilariae but at significantly lower levels than the drug-cured or naïve animals [63]. However, no evidence was found for lower vector attractiveness in putatively immune cattle, indicating that the lower worm burdens observed in these animals were a consequence of immunological control and not differential exposure to infection [63].

The first vaccination experiments performed in the *Oo* cattle system used xL3. Two approaches were employed: vaccination followed by experimental challenge using imported parasites in the UK and a parallel study using the same vaccination protocol but followed by natural exposure near the Vina du Sud for 22 months [63]. Interestingly, in experimental challenge, significant reductions in adult worm burden (males and females) were apparent, whereas under natural exposure, the nodule loads were not significantly lower in vaccinated animals, but there were marked reductions in microfilarial prevalence and density (as well as in the number of gravid female worms). The reasons for these contrasting outcomes are unclear but may relate to differences in immunological control of a "bolus" versus "trickle" challenge following vaccination. The extremely close relationship between *Ov* and *Oo* was subsequently emphasized in an experiment in which vaccination of calves with *Ov* L3 provided almost complete protection against experimental challenge with *Oo* L3 [64]. Conversely, humans residing in regions of northern Cameroon with high cattle densities and consequent transmission of *Oo* appear to be partially protected against *Ov* infection, suggesting that "zooprophylaxis" against the human pathogen can be achieved when *Oo* fails to develop fully in a human host [65].

The xL3 approach would never be practicable as a vaccine strategy for humans; however, the proof of principle afforded by the putative immunity and irradiated vaccine experiments in cattle spurred further development of a recombinant vaccine against ONCHO. Importantly, 18 recombinant *Ov* antigens that were initially evaluated in rodent models of filariasis and against panels of human immune sera were recognized serologically by cattle experimentally infected with *Oo* [66]. Prioritization of the lead vaccine candidates culminated in a major field trial in the *Oo* system using a combination of eight antigens, with each vaccinated calf receiving each antigen separately in either alum or Freund's adjuvant, depending on prior efficacy data from rodents [67]. Calves were protected from blackfly bites during the primary vaccination and boosters, and then exposed continually to infection near the Vina du Sud for 22 months. Surprisingly, at the end of the experiment, all but one vaccinated animal had nodules and the median nodule load was similar to that of the controls. However, only 42% of vaccinated animals had patent infections (microfilariae) compared with 100% of control cattle [67]. In contrast with the results of the xL3 experiment with natural challenge (see above), there was no evidence for reduction in gravid adult female worms in vaccinated animals, suggesting that microfilariae were targeted directly by vaccine-induced immune responses. There were no correlations between antibody levels to a particular antigen and protection against patency with strong statistical support, although there was an inverse trend between the strength of the anti-tropomyosin IgG2 response and patency [67].

It is not known why the eight antigens used in this *Oo* vaccine field trial failed to induce significant protection against adult worm establishment, but it is plausible that some of the immunomodulatory molecules used as vaccine candidates may have interfered with the development of protective immunity. Specifically, it is now known that the abundant larval transcript (ALT) and cysteine proteinase inhibitor (CPI) families of filarial antigens have potent immunomodulatory activity that includes inhibition of antigen processing [68, 69]. To use the proteins as vaccines, their sequences should be modified to ablate amino acid residues that are critical for their immunomodulatory function [34, 54, 70]. Another important consideration with respect to this field trial is that only one of two current lead vaccine candidates, *Ov*-RAL-2, was included, so the potential impact of adding *Ov*-103 is not known. Ongoing vaccine trials of *Ov*-103 and *Ov*-RAL-2 in combination in the *Oo*-cattle system will resolve this question.

## 21.5 Immune Responses in Humans Against the Two *O. volvulus* Lead Vaccine Candidates

To determine if antigen-specific antibody responses and antilarval protective immunity are associated in humans, the presence of anti-*Ov*-103 and anti-*Ov*-RAL-2 cytophilic antibody responses (IgG1 and IgG3) in individuals classified as putatively immune and in infected individuals who developed concomitant immunity with age was analyzed [50]. Notably, 86% of putatively immune individuals and 95% of individuals with concomitant immunity had elevated IgG1 and IgG3 responses to *Ov*-103 and *Ov*-RAL-2. Moreover, it appeared that anti-*Ov*-103- and anti-*Ov*-RAL-2-specific IgG3 responses in infected individuals increased with age. Based on the elevated chemokine levels associated with chemotaxis of neutrophils (KC/CXCL1 and MIP-1α), monocyte/macrophages (MCP-1 and MIP1β), and eosinophils (eotaxin) in protected mice after immunization with *Ov*-103 or *Ov*-RAL-2 [49], the profile of these chemokines was also analyzed in putatively immune and infected humans. Interestingly, both groups contained significantly higher levels of CXCL1 (neutrophils), MCP-1 and MIP-1β (monocyte/macrophages), and IP-10 (an IFN-γ-inducible protein that is a chemoattractant for monocytes and activated T cells) compared to normal human sera [71].

To test whether the elevated anti-*Ov*-103 and anti-*Ov*-RAL-2 cytophilic antibodies present in putatively immune and infected individuals can function in ADCC, *Ov* L3 were cultured in the presence of human-naïve neutrophils or monocytes and monospecific human anti-*Ov*-103 or anti-*Ov*-RAL-2 antibodies. Notably, human monospecific anti-*Ov*-103 antibodies, but not anti-*Ov*-RAL-2 antibodies, significantly inhibited molting of third-stage larvae (L3) *in vitro* by 46% in the presence of naïve human neutrophils, while both anti-*Ov*-103 and anti-*Ov*-RAL-2 antibodies significantly inhibited molting by 70–80% when cultured in the presence of naive human monocytes. Interestingly, inhibition of molting by *Ov*-103 antibodies and monocytes was only partially dependent on contact with cells, while inhibition of molting with *Ov*-RAL-2 antibodies was completely dependent on contact with monocytes. In comparison, significant levels of parasite killing in *Ov*-103 and *Ov*-RAL-2 vaccinated mice only occurred when cells entered the parasite microenvironment [50]. Taken together, these results lead to the conclusion that antibodies to *Ov*-103 and *Ov*-RAL-2 and immune response cells are required for protection in mice, as well as for the development of immunity in humans.

## 21.6 Conclusions and Future Directions

In conclusion, two *Ov* protective vaccine antigens (*Ov*-103 and *Ov*-RAL-2), administered individually or concurrently, with a proven scale-up production pathway and significant efficacy in two small animal models, are ready to move forward into preclinical development. The next evaluative step for the vaccine is being performed in naïve calves against natural infection with *Oo*, a system that mimics immunologically the status of humans living in regions endemic for ONCHO. Furthermore, establishing the immune correlates and mechanisms associated with protective immunity induced by these two ONCHO vaccines in mice, the vaccinated bovine model and protected humans in parallel, will position TOVA to move this promising ONCHO vaccine to a first-in-human trial. This ambitious goal will require a commensurate level of resources for the development and evaluation of optimal formulations, assessment of toxicity in preclinical screens, and cGMP production and scale-up. Nevertheless, it is a critical priority if the ONCHO elimination goals are to be safeguarded and brought to fruition in Africa.

## References


1 GBD (2015). Disease and Injury Incidence and Prevalence Collaborators. 2016. Global, regional, and national incidence, prevalence, and years lived with disability for 310 diseases and injuries, 1990–2015: a systematic analysis for the Global Burden of Disease Study 2015. *Lancet* 388: 1545–1602.

2 Hopkins, A.D. (2016). Neglected tropical diseases in Africa: a new paradigm. *Int. Health* 8 (Suppl 1): i28–i33.

3 Turner, H.C., Walker, M., Churcher, T.S. et al. (2014). Reaching the London Declaration on Neglected Tropical Diseases goals for onchocerciasis: An economic evaluation of increasing the frequency of ivermectin treatment in Africa. *Clin. Infect. Dis.* 59: 923–932.

**4** Diawara, L., Traore, M.O., Badji, A. et al. (2009). Feasibility of onchocerciasis elimination with ivermectin treatment in endemic foci in Africa: first evidence from studies in Mali and Senegal. *PLoS Negl. Trop. Dis.* 3: e000497.

**5** Traore, M.O., Sarr, M.D., Badji, A. et al. (2012). Proof-of-principle of onchocerciasis elimination with ivermectin treatment in endemic foci in Africa: final results of a study in Mali and Senegal. *PLoS Negl. Trop. Dis.* 6: e0001825.

**6** Gustavsen, K., Hopkins, A., and Sauerbrey, M. (2011). Onchocerciasis in the Americas: from arrival to (near) elimination. *Parasites Vectors* 4: 205.

**7** Global Burden of Disease Study (2013). Collaborators. (2015) Global, regional, and national incidence, prevalence, and years lived with disability for 301 acute and chronic diseases and injuries in 188 countries, 1990–2013: a systematic analysis for the Global Burden of Disease Study 2013. *Lancet* 386: 743–800.

**8** Kim, Y.E., Remme, J.H., Steinmann, P. et al. (2015). Control, elimination, and eradication of river blindness: scenarios, timelines, and ivermectin treatment needs in Africa. *PLoS Negl. Trop. Dis.* 9: e0003664.

**9** Keenan, J.D., Hotez, P.J., Amza, A. et al. (2013). Elimination and eradication of neglected tropical diseases with mass drug administrations: a survey of experts. *PLoS Negl. Trop. Dis.* 7: 000e2562.

**10** Kelly-Hope, L.A., Cano, J., Stanton, M.C. et al. (2014). Innovative tools for assessing risks for severe adverse events in areas of overlapping *Loa loa* and other filarial distributions: the application of micro-stratification mapping. *Parasites Vectors* 7: 307.

**11** Molyneux, D.H., Hopkins, A., Bradley, M.H., and Kelly-Hope, L.A. (2014). Multidimensional complexities of filariasis control in an era of large-scale mass drug administration programmes: a can of worms. *Parasites Vectors* 7: 363.

**12** Lustigman, S. and McCarter, J.P. (2007). Ivermectin resistance in *Onchocerca volvulus*: Toward a genetic basis. *PLoS Negl. Trop. Dis.* 1: e000076.

**13** Turner, H.C., Churcher, T.S., Walker, M. et al. (2013). Uncertainty surrounding projections of the long-term impact of ivermectin treatment on human onchocerciasis. *PLoS Negl. Trop. Dis.* 7: e0002169.

**14** Turner, H.C., Walker, M., Lustigman, S. et al. (2015). Human onchocerciasis: Modelling the potential long-term consequences of a vaccination programme. *PLoS Negl. Trop. Dis.* 9: e0003938.

**15** APOC. (2014) Report of the thirty-eigth session os the technical consultative committee (TCC). Ouagadougou, 10 – 14 March 2014.

**16** Lustigman, S., McKerrow, J.H., and Bottazzi, M.E. (2012). Vaccines linked to chemotherapy: a new approach to control helminth infections. In: *Parasitic Helminths: Targets, Screens, Drugs and Vaccines* (ed. C.R. Caffrey), 357–375. Weinheim: Wiley-VCH Verlag GmbH & Co. KGaA.

**17** Denham, D.A. (1980). Vaccination against filarial worms using radiation-attenuated vaccines. *Int. J. Nucl. Med. Biol.* 7: 105–111.

**18** Keiser, P.B. and Nutman, T.B. (2002). Vaccines for filarial infections. In: *In World Class Parasites* (ed. T.R. Klei and T.V. Rajan), 167–178. Norwell MA: Kluwer Academic Press.

**19** Allen, J.E., Adjei, O., Bain, O. et al. (2008). Of mice, cattle, and humans: the immunology and treatment of river blindness. *PLoS Negl.Trop. Dis.* 2: e000217.

**20** Lustigman, S. and Abraham, D. (2009). *Onchocerciasis, in Vaccines for Biodefense and Emerging and Neglected Diseases* (ed. A.D.T. Barrett and L.R. Stanberry), 1379–1400. Amsterdam: Academic Press, Inc – Elsevier Science & Technology.

**21** Lustigman, S., James, E.R., Tawe, W., and Abraham, D. (2002). Towards a recombinant antigen vaccine against *Onchocerca volvulus*. *Trends Parasitol.* 18: 135–141.

**22** Morris, C.P., Evans, H., Larsen, S.E., and Mitre, E. (2013). A comprehensive, model-based review of vaccine and repeat infection trials for filariasis. *Clin. Microbiol. Rev.* 26: 381–421.

**23** Ward, D.J., Nutman, T.B., Zea-Flores, G. et al. (1988). Onchocerciasis and immunity in humans: enhanced T cell responsiveness to parasite antigen in putatively immune individuals. *J. Infect. Dis.* 157: 536–543.

**24** Gallin, M., Edmonds, K., Ellner, J.J. et al. (1988). Cell-mediated immune responses in human infection with *Onchocerca volvulus*. *J. Immunol.* 140: 1999–2007.

**25** Elson, L.H., Guderian, R.H., Araujo, E. et al. (1994). Immunity to onchocerciasis: identification of a putatively immune population in a hyperendemic area of Ecuador. *J. Infect. Dis.* 169: 588–594.

**26** Turaga, P.S., Tierney, T.J., Bennett, K.E. et al. (2000). Immunity to onchocerciasis: cells from putatively immune individuals produce enhanced levels of interleukin-5, gamma interferon, and granulocyte-macrophage colony-stimulating factor in response to *Onchocerca volvulus* larval and male worm antigens. *Infect. Immun.* 68: 1905–1911.

**27** Kazura, J.W. (2000). Resistance to infection with lymphatic-dwelling filarial parasites. In: *Lymphatic Filariasis* (ed. T.B. Nutman), 83–102. London: Imperial College Press.

**28** Kurniawan-Atmadja, A., Sartono, E., Partono, F. et al. (1998). Antibody responses to filarial infective larvae are not dominated by the IgG4 isotype. *Parasite Immunol.* 20: 9–17.

**29** Sartono, E., Kruize, Y.C., Kurniawan, A. et al. (1997). Depression of antigen-specific interleukin-5 and interferon-gamma responses in human lymphatic filariasis as a function of clinical status and age. *J. Infect. Dis.* 175: 1276–1280.

**30** Schulz-Key, H. (1990). Observations on the reproductive biology of *Onchocerca volvulus*. *Acta Leiden.* 59: 27–44.

**31** Duke, B.O.L. and Moore, P.J. (1968). The contributions of different age groups to the transmission of onchocerciasis in a Cameroon forest village. *Trans. R. Soc. Trop. Med. Hyg.* 62: 22–28.

**32** MacDonald, A.J., Turaga, P.S., Harmon-Brown, C. et al. (2002). Differential cytokine and antibody responses to adult and larval stages of *Onchocerca volvulus* consistent with the development of concomitant immunity. *Infect. Immun.* 70: 2796–2804.

**33** Day, K.P., Gregory, W.F., and Maizels, R.M. (1991). Age-specific acquisition of immunity to infective larvae in a bancroftian filariasis endemic area of Papua New Guinea. *Parasite Immunol.* 13: 277–290.

**34** Hess, J.A., Zhan, B., Bonne-Annee, S. et al. (2014). Vaccines to combat river blindness: expression, selection and formulation of vaccines against infection with *Onchocerca volvulus* in a mouse model. *Int. J. Parasitol.* 44: 637–646.

**35** Lustigman, S., Makepeace, B.L., Klei, T.R. et al. (2018). *Onchocerca volvulus*: the road from basic biology to a vaccine. *Trends Parasitol.* 34: 64–79.

**36** Hotez, P.J., Bottazzi, M.E., Zhan, B. et al. (2015). The onchocerciasis vaccine for Africa--TOVA--initiative. *PLoS Negl.Trop. Dis.* 9: e0003422.

**37** Duke, B.O.L. (1980). Observations on *Onchocerca volvulus* in experimentally infected chimpanzees. *Tropenmed. Parasitol.* 31: 41–54.

**38** Eberhard, M.L., Dickerson, J.W., Boyer, A.E. et al. (1991). Experimental *Onchocerca volvulus* infections in mangabey monkeys (*Cercocebus atys*) compared to infections in humans and chimpanzees (*Pan troglodytes*). *Am. J. Trop. Med. Hyg.* 44: 151–160.

**39** Greene, B.M. (1987). Primate model for onchocerciasis research. *Ciba Found Symp.* 127: 236–243.

**40** Prince, A.M., Brotman, B., Johnson, E.H. Jr., et al. (1992). *Onchocerca volvulus*: immunization of chimpanzees with X-irradiated third-stage (L3) larvae. *Exp. Parasitol.* 74: 239–250.

**41** Abraham, D., Lange, A.M., Yutanawiboonchai, W. et al. (1993). Survival and development of larval *Onchocerca volvulus* in diffusion chambers implanted in primate and rodent hosts. *J. Parasitol.* 79: 571–582.

**42** Lange, A.M., Yutanawiboonchai, W., Lok, J.B. et al. (1993). Induction of protective immunity against larval *Onchocerca volvulus* in a mouse model. *Am. J. Trop. Med. Hyg.* 49: 783–788.

**43** Johnson, E.H., Schynder-Candrian, S., Rajan, T.V. et al. (1998). Immune responses to third stage larvae of *Onchocerca volvulus* in interferon-gamma and interleukin-4 knockout mice. *Parasite Immunol.* 20: 319–324.

**44** Lange, A.M., Yutanawiboonchai, W., Scott, P., and Abraham, D. (1994). IL-4- and IL-5-dependent protective immunity to *Onchocerca volvulus* infective larvae in BALB/cBYJ mice. *J. Immunol.* 153: 205–211.

**45** Abraham, D., Leon, O., Schnyder-Candrian, S. et al. (2004). Immunoglobulin E and eosinophil-dependent protective immunity to larval *Onchocerca volvulus* in mice immunized with irradiated larvae. *Infect. Immun.* 72: 810–817.

**46** Abraham, D., Leon, O., Leon, S., and Lustigman, S. (2001). Development of a recombinant antigen vaccine against infection with the filarial worm *Onchocerca volvulus*. *Infect. Immun.* 69: 262–270.

**47** McCarthy, J.S., Wieseman, M., Tropea, J. et al. (2002). *Onchocerca volvulus* glycolytic enzyme fructose-1,6-bisphosphate aldolase as a target for a protective immune response in humans. *Infect. Immun.* 70: 851–858.

**48** MacDonald, A.J., Tawe, W., Leon, O. et al. (2004). Ov-ASP-1, the *Onchocerca volvulus* homologue of the activation associated secreted protein family is immunostimulatory and can induce protective anti-larval immunity. *Parasite Immunol.* 26: 53–62.

**49** Hess, J.A., Zhan, B., Torigian, A.R. et al. (2016). The immunomodulatory role of adjuvants in vaccines formulated with the recombinant antigens Ov-103 and Ov-RAL-2 against *Onchocerca volvulus* in mice. *PLoS Negl.Trop. Dis.* 10: e0004797.

**50** George, J.P., Hess, J.A., Jain, S. et al. (2019). Antibody responses against the vaccine antigens Ov-103 and Ov-RAL-2 are associated with protective immunity to *Onchocerca volvulus* infection in both mice and humans. *PLoS Negl.Trop. Dis.* 13: e0007730.

**51** Patton, J.B., Bennuru, S., Eberhard, M.L. et al. (2018). Development of *Onchocerca volvulus* in humanized NSG mice and detection of parasite biomarkers in urine and serum. *PLoS Negl.Trop. Dis.* 12: e0006977.

**52** Lustigman, S. and Klei, T.R. (2013). Vaccination against filarial nematodes. In: *Parasitic Nematodes, Molecular Biology, Biochemistry and Immunology* (ed. M.W. Kennedy and W. Harnett), 209–244. Oxon and New York: CABI Publishing.

**53** Arumugam, S., Wei, J., Liu, Z. et al. (2016). Vaccination of gerbils with Bm-103 and Bm-RAL-2 concurrently or as a fusion protein confers consistent and improved protection against *Brugia malayi* infection. *PLoS Negl.Trop. Dis.* 10: e0004586.

**54** Arumugam, S., Wei, J., Ward, D. et al. (2014). Vaccination with a genetically modified *Brugia malayi* cysteine protease inhibitor-2 reduces adult parasite numbers and affects the fertility of female worms following a subcutaneous challenge of Mongolian gerbils (*Meriones unguiculatus*) with *B. malayi* infective larvae. *Int. J. Parasitol.* 44: 675–679.

**55** Makepeace, B.L. and Tanya, V.N. (2016). 25 Years of the *Onchocerca ochengi* model. *Trends Parasitol.* 32: 966–978.

**56** Trees, A.J. (1992). *Onchocerca ochengi*: Mimic, model or modulator of *O. volvulus*? *Parasitol. Today* 8: 337–339.

**57** Bwangamoi, O. (1969). Dermatitis in cattle caused by *Onchocerca ochengi* Bwangamoi, 1969, and the effect of the adult filaria on the finished leather. *Bull. Epizoot Dis. Afr.* 17: 435–445.

**58** Bwangamoi, O. (1969). *Onchocerca ochengi* new species, an intradermal parasite of cattle in East Africa. *Bull. Epizoot Dis. Afr.* 17: 321–335.

**59** Wahl, G., Achu-Kwi, M.D., Mbah, D. et al. (1994). Bovine onchocercosis in north Cameroon. *Vet. Parasitol.* 52: 297–311.

**60** Wahl, G., Ekale, D., and Schmitz, A. (1998). *Onchocerca ochengi*: assessment of the *Simulium* vectors in north Cameroon. *Parasitology* 116 (Pt 4): 327–336.

**61** Trees, A.J., Wahl, G., Klager, S., and Renz, A. (1992). Age-related differences in parasitosis may indicate acquired immunity against microfilariae in cattle naturally infected with *Onchocerca ochengi*. *Parasitology* 104 (Pt 2): 247–252.

**62** Achukwi, M.D., Harnett, W., Bradley, J., and Renz, A. (2004). *Onchocerca ochengi* acquisition in zebu Gudali cattle exposed to natural transmission: parasite population dynamics and IgG antibody subclass responses to Ov10/Ov11 recombinant antigens. *Vet. Parasitol.* 122: 35–49.

**63** Tchakoute, V.L., Graham, S.P., Jensen, S.A. et al. (2006). In a bovine model of onchocerciasis, protective immunity exists naturally, is absent in drug-cured hosts, and is induced by vaccination. *Proc. Natl. Acad. Sci. U.S.A.* 103: 5971–5976.

**64** Achukwi, M.D., Harnett, W., Enyong, P., and Renz, A. (2007). Successful vaccination against *Onchocerca ochengi* infestation in cattle using live *Onchocerca volvulus* infective larvae. *Parasite Immunol.* 29: 113–116.

**65** Wahl, G., Enyong, P., Ngosso, A. et al. (1998). *Onchocerca ochengi*: epidemiological evidence of cross-protection against *Onchocerca volvulus* in man. *Parasitology* 116 (Pt 4): 349–362.

**66** Graham, S.P., Wu, Y., Henkle-Duehrsen, K. et al. (1999). Patterns of *Onchocerca volvulus* recombinant antigen recognition in a bovine model of onchocerciasis. *Parasitology* 119: 603–612.

**67** Makepeace, B.L., Jensen, S.A., Laney, S.J. et al. (2009). Immunisation with a multivalent, subunit vaccine reduces patent infection in a natural bovine model of onchocerciasis during intense field exposure. *PLoS Negl.Trop. Dis.* 3: e000544.

**68** Gomez-Escobar, N., Bennett, C., Prieto-Lafuente, L. et al. (2005). Heterologous expression of the filarial nematode alt gene products reveals their potential to inhibit immune function. *BMC Biol.* 3: 8.

**69** Manoury, B., Gregory, W.F., Maizels, R.M., and Watts, C. (2001). Bm-CPI-2, a cystatin homolog secreted by the filarial parasite *Brugia malayi*, inhibits class II MHC-restricted antigen processing. *Curr. Biol.* 11: 447–451.

**70** Babayan, S.A., Luo, H., Gray, N. et al. (2012). Deletion of parasite immune modulatory sequences combined with immune activating signals enhances vaccine mediated protection against filarial nematodes. *PLoS Negl.Trop. Dis.* 6: e0001968.

**71** Lechner, C.J., Gantin, R.G., Seeger, T. et al. (2012). Chemokines and cytokines in patients with an occult *Onchocerca volvulus* infection. *Microbes Infect.* 14: 438–446.

# 22

# Vector Control Approaches to Interrupt Transmission of Human Filarial Parasites

*Lyric Bartholomay**

*University of Wisconsin-Madison, School of Veterinary Medicine, Department of Pathobiological Sciences, 1656 Linden Dr, Madison WI 53706, USA*

**Abstract**

In 1887, Patrick Manson first proposed that a mosquito serves as the "nurse" for the parasites that cause lymphatic filariasis (LF). Thereafter, an age of discovery for vector-borne diseases (VBDs) ensued, and the mosquito became the target of choice for intervention to prevent diseases such as malaria and yellow fever. Lymphatic filariasis, a disease that causes profound suffering but is not acutely lethal, was overlooked in large-scale mosquito and disease control campaigns. With the advent of inexpensive and effective chemotherapeutics to prevent LF transmission, the Global Alliance to Eliminate Lymphatic Filariasis formed and launched the Global Program to Eliminate Lymphatic Filariasis in 2000. Even in this endeavor, mosquitoes as a point of intervention were largely overlooked because mass drug administration (MDA) is so effective. In some endemic countries, LF control by MDA alone, or MDA and a lack of sustained mosquito control, have proven insufficient to eliminate transmission and disease. In this chapter, we primarily review the policy and practice of vector control for LF and present alternative and burgeoning tools that could be useful for endgame LF control. In addition, efforts to control the transmission of *Onchocerca volvulus* by black flies in the genus *Simulium* and *Loa loa* by tabanid flies in the genus *Chrysops* are discussed as examples of campaigns of vector control that were once predominant (onchocerciasis) or have not yet been attempted (loiasis).

## 22.1 Introduction

Manson's watershed discovery of mosquitoes as the intermediate host for *filaria nocturna* (syn. *Wuchereria bancrofti*) in 1877 [1] inspired an age of discovery for mosquito-borne disease to include the pioneering work of Ronald Ross who saw *Plasmodium* parasites developing in the guts of mosquitoes, and Finlay, Lazear, Carter, and Reed, who showed that "loaded" mosquitoes, and not fomites, transmit

*Corresponding author.

*Human and Animal Filariases: Landscape, Challenges, and Control*, First Edition.
Edited by Ronald Kaminsky and Timothy G. Geary.

the yellow fever virus [2, 3]. Thereafter, Boyce concluded that "if the mosquito is destroyed, the life cycle of the parasite is destroyed and the disease must be of necessity cease. This constitutes the fundamental principle of prophylaxis in all the mosquito-borne diseases." At the time, campaigns against the mosquito involved avoidance (e.g. screening houses and sleeping under mosquito nets), mosquito extermination (using natural enemies, kerosene oil as a culicide, wetland drainage to eliminate breeding places, and penalties for harboring larvae in stagnant water), and public education [4].

With the mosquito as a new point of focus for disease control, global campaigns ensued to control malaria and yellow fever – many founded in imperial motives; witness, for example, the extraordinary successes of Major William Gorgas and colleagues in eliminating the threat of yellow fever and malaria in Cuba and Panama [5]. Filariasis control was not represented in those early campaigns. Wilson et al. (2020) provide a comprehensive review of vector control interventions for the elimination of VBD [6]. Of these, they make note of 16 notable, large-scale vector control programs with substantial disease control impacts for mosquito-borne disease from 1900 to present. Only one of those – in the Solomon Islands and Papua New Guinea – relates to LF control.

Indeed, vector control has only rarely, and in focal settings, been the intervention of choice for LF control, perhaps because LF as a disease demanded less attention for intervention. Webber (1979) remarked that "war-time workers surveying the whole of the Pacific found filarial infection in the Solomon Islands one of the highest in the whole region. Nothing was done about it as other disease problems, particularly yaws, malaria, and tuberculosis, were considered priorities" [7]. Yet, the impact of LF on global health is profound for the associated debilitating disease, human suffering, social stigma, and costs to the individual and community. Before interventions, between 1.29 and 1.365 billion people were living at risk of LF in 72 countries. It is estimated that 119–129 million people were infected with LF, and 43 million had clinical disease, with an associated Disability Adjusted Life Years burden of 5.25 million [8, 9]. The total economic burden of LF was estimated to be US$ 5.765 billion annually [10].

The understanding that LF is a VBD inspired the search for vectors of other human filariases, leading to the discovery that *Chrysops* spp. are the vectors for *Loa loa* by Leiper in 1912 [11], followed by the demonstration that *Simulium* spp. are the vectors for *Onchocerca volvulus* by Blacklock in 1926 [12]. Loiasis has generally been regarded as of mild pathological interest [13, 14] and has attracted relatively little attention for therapy or control. In contrast, onchocerciasis was the first human filariasis to attract intensive vector control efforts (see below), which have since been supplanted by chemotherapeutic strategies [15]. Campaigns for onchocerciasis elimination based on both vector control and MDA have largely succeeded in eradicating the parasite from the Americas and have made notable progress in reducing the incidence of infection and blindness in Africa [14–16]. Given that the campaigns of vector control for onchocerciasis have been well reviewed and those for loiasis have been minimal, the major focus of the chapter will be on mosquito abatement for LF control.

## 22.2 Global Health Policies Toward LF Control

The impetus to control LF went public in 1997 based on World Health Assembly resolution WHA50.29 Regarding Elimination of Lymphatic Filariasis as a Public Health Problem [17]. The global community mobilized to form the Global Alliance to Eliminate Lymphatic Filariasis (GAELF) and the Global Programme to Eliminate Lymphatic Filariasis (GPELF) launched in 2000. The fundamental goals of the GPELF were to interrupt transmission of the parasite and to alleviate the suffering of infected individuals and affected communities. Toward this end, the GPELF intended to use MDA with the drug or combinations of drugs of choice according to co-endemicity with *O. volvulus* or *L. loa* and morbidity management and disability prevention (MMDP) to alleviate clinical disease and suffering. The plan included the use of vector control in situations in which MDA alone was not possible or sufficient to interrupt transmission [17], as outlined in Figure 22.1. The GPELF set an ambitious 2020 target date for LF eradication; therefore, it is timely to review the contributions of mosquito control past, present, and future, to LF disease control.

The GPELF plan has evolved and radiated outward and has achieved remarkable coverage and outcomes. Specific policy and guidance related to vector control, in particular, are detailed here. In 2004, the WHO released The Global Strategic Framework for Integrated Vector Management (IVM) [17], which puts forth guidance shaped by fundamental entomological principles for integrated pest management, whereby a pest is surveyed and controlled using a variety of approaches (e.g. environmental management, crop rotation, and chemical insecticides) to a designated, acceptable threshold. The WHO guidance also suggested integration with health sector partners, and integrating vector control for multiple vectors where VBDs are co-endemic in the interest of the many collateral benefits of implementing vector control [18]. The WHO reaffirmed its stance on IVM in a subsequent position statement to encourage member states to strengthen capacity for IVM, defined as "a rational decision-making process for the optimal use of

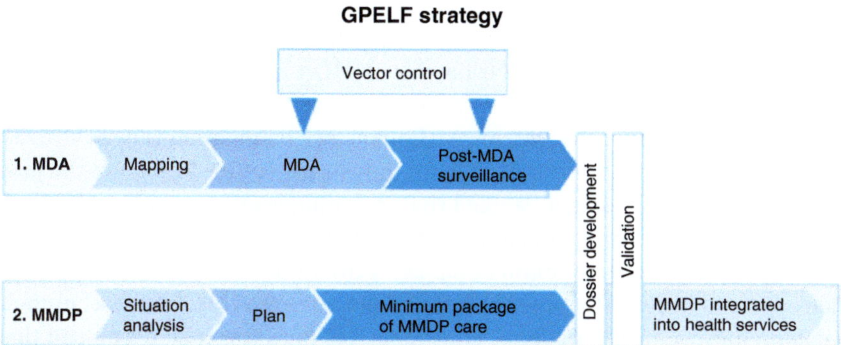


**Figure 22.1** Diagram of the Global Programme to Eliminate Lymphatic Filariasis approach to control LF. Abbreviations: MDA – mass drug administration; MMDP – morbidity management and disability prevention. Source: WHO GPELF Strategy/with permission of WHO.

resources for vector control" [19]. In 2007, WHO published the Global Plan to Combat Neglected Tropical Diseases 2008–2015 and renewed commitment to LF, as a "tool-ready" disease for which infrastructure was in place, and could be scaffolded with "packages" of interventions to benefit "tool-deficient" diseases [20]. In doing so, communities would have access to, for example, MDA to include anthelmintics for control of soil-transmitted helminths and schistosomiasis, bed net distribution and antimalarials for malaria control, and community health information and education [20]. In 2012, the WHO published an entomological handbook for LF elimination programs and provided a new iteration of Figure 22.1 to include vector control for reduction of transmission during MDA and to prevent new infections during post-MDA surveillance [21].

In 2017, the WHO released Guidance for a Global Vector Control Response 2017–2030 [22]. The associated consortium supported the goals of IVM but acknowledged that IVM uptake has been poor because of limited political buy-in and fragmented local and regional infrastructure for vector control at the scale required to tackle multiple VBDs. The new guidance clarifies an approach to achieving both IVM and control for co-endemic VBDs by establishing a foundation of vector control and capacity by strengthening collaboration, engaging communities, enhancing surveillance and evaluation of control efforts, and scaling up efforts to ultimately develop effective and locally adapted vector control [22]. Wilson et al. echo this idea in their review of vector control for elimination of VBD, stating that "there is a need to return to vector control approaches based on a thorough knowledge of the determinants of pathogen transmission, which utilise a range of insecticide and non-insecticide based approaches in a locally tailored manner for more effective and sustainable vector control [6]."

## 22.3 Role of Vector Control in LF Control Programs

As of 2018, 14 GPELF member countries have achieved Stage 5 status, i.e. "validated as having eliminated LF as a public health problem and under surveillance," and 10 more are at Stage 4: "MDA stopped in all endemic districts and under surveillance [23]." These successes are largely attributable to MDA and MMDP. There is good reason to be hopeful that the GPELF will continue to be successful – perhaps on the timescale predicted by the Bill and Melinda Gates Foundation (i.e. 2030) [23]. That said, it is clear that MDA alone will not universally suffice to achieve the goal of LF elimination. Burkhot et al. argued that including vector control in LF elimination strategies would suppress parasite transmission across whole communities, minimize the risk of reintroduction of parasites from microfilaremic patients, and provide collateral benefits for other mosquito-borne disease [24]. Likewise, in their 2009 comprehensive review of vector control for LF, Bockarie et al. concluded that "including vector control would represent an important strategic tool to expedite and sustain the achieved interruption of filariasis transmission" [25]. Irvine et al. modeled the probability of elimination in endemic settings where either *Culex* or *Anopheles* mosquito species are vectors for the parasites and in scenarios where communities with varying levels of microfilaremia received 65% or 80% MDA coverage with or without vector control to reduce bite exposures. With higher endemicity, and either vector type, additional vector control significantly increases the probability of elimination over high-coverage MDA [26].

## 22.4 Arsenal of Mosquito Control Tools for IVM for LF

The tools available to affect mosquito control have arguably changed little since Boyce described "The Campaign Against the Mosquito" in 1909. The key endpoint of both prevention (e.g. screening houses and sleeping under mosquito nets) and mosquito extermination measures (using natural enemies, kerosene oil as culicides, and drainage to get rid of breeding places) is reducing the biting rate and therefore transmission potential from infected mosquitoes. Significant reductions in biting burden and transmission of LF parasites have been achieved for *Culex quinquefasciatus* using larvicides at breeding source sites and for *Anopheles* species using bed nets (see below). It is very likely that sustained mosquito control will also be a keystone to sustained LF control post-MDA. As such, it is critical to appreciate the diversity of mosquito species that transmit LF parasites and to understand the spectrum of control approaches available for the particular mosquito vector species present. As Wilson et al. noted, vector control approaches should be based on a thorough knowledge of the biology of the vector and of pathogen transmission and should utilize a range of tools in a locally tailored manner to optimize disease control potential [6]. Complicating matters, at least 45 mosquito species in the genera *Aedes* or *Ochlerotatus*, *Anopheles*, *Culex*, and *Mansonia* are implicated in the transmission of LF parasites; readers are directed to catalogs of LF mosquito vector species constructed by Raghavan [27], Chow [28], and Nelson [29] to appreciate the diversity of vectors involved, with a note of caution that the nomenclature for many of these species has changed substantially (see Refs [30, 31]). Adding to the complexity of integrating MDA and vector control, some of these species (e.g. *Aedes* vector species in the Pacific Islands) are more effective vectors at lower microfilaremias (see Refs [6, 32]).

The IVM ethos dictates the decision-making process for the optimal use of resources for vector control. The concept stems from Integrated Pest Management (IPM), which was developed primarily for agricultural pests and is summarized as "a decision support system for the selection and use of pest control tactics, singly or harmoniously coordinated into a management strategy, based on cost/benefit analyses that take into account the interests of and impacts on producers, society, and the environment" [33]. In the best-case scenario, when resources permit, IPM and IVM should involve data-driven decision-making for the most cost- and outcome-effective use of control methods, with consideration for product mode of action to pre-empt the development of insecticide resistance. This should be based on surveillance data, to include the target species and life stage, pathogen infection status in adult female mosquitoes, and the presence of insecticide resistance. The WHO provides guidance on the biology, surveillance, and control of key LF vector species in the Handbook of Practical Entomology for National Lymphatic Filariasis Elimination Programmes [21].

The success of LF control in Korea could be considered a natural experiment in IVM alongside MDA. During a 40 year campaign of periodic diethylcarbamazine MDA interventions, residents in endemic communities moved to sleeping indoors and with screened windows, and using mosquito nets and repellents, thereby reducing exposure to mosquito bites [34]. Cheun et al. also suggested that urbanization, industrialization, and use of agricultural pesticides have significantly decreased mosquito populations [34]. There are several additional excellent examples of controlled studies of IVM impacts on LF control. The Vector Control Research Center (VCRC) in Pondicherry, India, was an early adopter of IVM targeting *Cx. quinquefasciatus* as a vector for *W. bancrofti*. The VCRC used sanitation measures for point source reduction, alongside polystyrene bead treatments and predatory fish and measured use of chemical insecticides. Exposure to bites from *Cx. quinquefasciatus* indoors was reduced by 90%. A subsequent return to conventional mosquito control resulted in increased mosquito density [35]. Rueben et al. conducted a comparative study of IVM and MDA and showed dramatic decreases (>90%) in transmission potential and microfilaremia and showed that sustained vector control was essential to sustained LF control [36].

## 22.5 Prevention Measures and Mosquito Extermination Approaches

Here, Boyce's original description of tools for mosquito prevention and extermination are expanded/modernized to include the use of long-lasting insecticidal nets as a physical barrier to mosquito bites and source control reduction for larval breeding habitats, biological control agents for larval control, and chemical insecticides as larvicides and adulticides. The examples below do not represent a comprehensive list of vector control approaches, rather serve to illustrate the differences in vector control approaches applied to different vector species.

### 22.5.1 *Culex* spp. Prevention Measures

Generally speaking, insecticide-treated nets have not been widely used successfully to control *Culex* vectors of LF. Bogh et al. tested the effect of insecticide-treated nets on three LF vectors, *Culex quinquefasciatus*, *Anopheles funestus*, and *An. gambiae*, and saw no change in numbers of indoor or outdoor resting *Cx. quinquefasciatus* [37]. In another study in South India, insecticide-treated curtains placed in eaves and doorways produced an 82% reduction in biting burden from *Cx. quinquefasciatus* [38]. However, Curtis et al. noted "strikingly less [mortality] than that of *Anopheles* with all types of pyrethroid-treated nets and curtains," and a "lack of reduction of *Culex* populations … in communities with widespread use of pyrethroid-treated nets" [39].

### 22.5.2 *Aedes* spp. Prevention Measures

*Aedes* species of LF vectors can be more challenging to target for control as adults because they are outdoor feeding and diurnal. Chambers et al. assessed the potential for insecticide-treated cloth sheets as a preventive measure for *Ae. polynesiensis* control in French Polynesia and observed 98.0% mortality at 24 hour in laboratory conditions. Unfortunately, in field cages, the 24 hr mortality rate was 54.3% compared to 31.2% for controls [40].

### 22.5.3 *Anopheles* spp. Prevention Measures

Bogh (1998) tested the effect of insecticide-treated nets on three LF vectors: *Culex quinquefasciatus*, *An. funestus*, and *An. gambiae*. Nets significantly reduced the vector density for *An. funestus* and *An. gambiae* (99% and 98%, respectively); the authors calculated a 92% reduction in annual transmission potential for *W. bancrofti* [41]. In Uganda, Ashton et al. tested the impact of combining high-coverage, long-lasting insecticidal nets and MDA and showed a marked reduction from 22.3% antigen positivity in 2007 to 6% in >5 year-old children in 2010; however, this study was not designed to test the specific role of these nets in this effect [42]. In East Sepik Province, Papua New Guinea, Reimer et al. tested implementation of bed nets after a 10 year hiatus of LF control. Bed nets reduced the maximum number of bites per person per day from 61.3 to 9.4 bites for 11 months after distribution, the prevalence of infection in *Anopheles* from 1.8% to 0.4%, and the annual transmission potential from up to 325 infective larvae/person/year to 0 [43].

### 22.5.4 Culex Control – Larval Breeding Habitats as Particular Point Source Targets

A small LF focus around Brisbane, Australia, which incidentally was also home to Joseph Bancroft (namesake for *W. bancrofti*), was the subject of a targeted control effort by the Brisbane City Council. Using environmental management (destruction of breeding sites, improving drainage) and biological control with predatory fish to target *Cx. quinquefasciatus* breeding sites, the city was LF free by 1910 [44]. Charleston, South Carolina, USA, appears to have achieved control of filariasis alongside an early Water, Sanitation and Hygiene campaign. The city converted primitive privy vaults to a permanent piped sewerage system that was complete by 1920. This effort eliminated breeding sites preferred by *Cx. quinquefasciatus*, and LF was eliminated as a public health threat by 1930 [45]. It is notable that improved living standards, WASH, and vector control are credited with interrupting LF transmission in Burundi, Cape Verde, Costa Rica, Mauritius, Rwanda, Seychelles, Solomon Islands, Suriname, and Trinidad and Tobago, such that these countries do not need MDA for LF control [46].

*Bacillus sphaericus* and *B. thuringiensis* var *israelensis*, spore-forming, mosquito larva-lethal bacteria, are widely used globally for control of *Culex* vector species. In metropolitan Recife, Brazil, 18 months of targeted treatments of breeding sites for *Cx. quinquefasciatus* took place in 2500 sites and resulted in 60% reduction in mosquito bite burden and a concomitant drop in microfilaremia from 13% to 7.2% [47]. Polystyrene beads have proven a remarkable intervention for heavily polluted point sources of *Cx. quinquefasciatus*. For example, in Makunduchi, Zanzibar, surveyed, larval-infested latrines were treated with polystyrene beads, and the biting burden on individuals dropped from approximately 25 000 bites per year to 440 bites per year; alongside diethylcarbamazine MDA, microfilaremia prevalence dropped from 49.5% to 10.3% [48].

### 22.5.5 *Aedes* spp. Control

Targeting crab hole breeding sites using natural predators (copepods or guppies), or plugging holes, resulted in marked reduction of *Ae. polynesiensis* in Pacific Islands and provided more impact than use of organophosphate sprays or fogs (reviewed in Ref. [24]).

## 22.6 Opportunities for Mosquito Elimination: Vector Control Achieved as a Collateral Benefit from Other Disease Intervention Campaigns

Concerted vector control efforts for diseases other than LF have shown remarkable collateral benefit for reducing mosquito bites and exposure to LF parasites. In the 1970s, the Malaria Eradication Programme in the Solomon Islands implemented household residual spraying of DDT every six months from 1968 to 1976 to control *An. farauti*. In North Choiseul, microfilaremia was monitored in two villages from 1974 to 1977 and revealed a decline from 21.8% to 0% prevalence [7]. More recently, Ashton et al. [42] showed that LLINs for malaria control were protective for *W. bancrofti* infection in Uganda. The WHO, and many authors, strongly suggests leveraging malaria control and LF control when the vector species for the parasite is the same (see, for example, Refs [6, 22, 24, 49, 50]. Stone et al. further note that combined control programs for LF and malaria should factor in high level of vector intervention required for malaria control, as compared to less-intensive but longer term need for control of LF vectors [51].

## 22.7 Challenges to Mosquito Elimination: Insecticide Resistance

Widespread and long-term use of pyrethroid-impregnated bed nets has contributed significantly to insecticide resistance (IR) in malaria vectors that also transmit LF parasites (e.g. *An. gambiae* and *An. funestus*). Across large parts of Africa, pyrethroid resistance is ubiquitous, and incorporation of carbamate and organophosphate insecticides for control is on the rise [52]. The WHO reported resistance to at least three of four classes of insecticide (pyrethroids, organochlorines, carbamates, and organophosphates) in *Anopheles* species from 23 African countries and India [53]. The resistance status of *Anopheles* vector species in Papua New Guinea is being monitored, but thus far, resistance is not evident [54].

## 22.8 Alternative and Emerging Approaches to Vector Control for LF

### 22.8.1 Vector Control Innovations

There is tremendous pressure and motivation to develop novel approaches to mosquito control to pre-empt or overcome hurdles of insecticide resistance and regulatory changes that are diminishing the portfolio of available insecticides. Some of these efforts include (i) developing novel delivery systems (e.g. remote sensing and drones to detect breeding sites and more selectively deliver larvicides [55], attractive toxic sugar baits to attract and kill male and female mosquitoes [56], and eave tubes to concentrate insecticides for control of *An. gambiae* [57]); (ii) developing products with novel modes of action (e.g. spatial repellents to deter blood feeding mosquitoes from homes [58] or molecular mosquitocides [59]); (iii) improving existing products (e.g. synergists for existing [60] and resistance-breaking compounds [61]; and (iv) improving operational practices for integrated insecticide resistance management and product rotation. Modifying populations using genetic means is another potential tool in the arsenal.

### 22.8.2 Genetic/Genomic Interventions

The use of genetically modified mosquitoes as an intervention approach for LF control is not new [62]. Releases of thiotepa-sterilized male *Cx. quinquefasciatus* in the Florida Keys resulted in 96% reduction of the population in five generations (with releases of 18 000 sterile male mosquitoes/day), and more than 99% of egg rafts were sterile [63]. In India, the ICMR/WHO Research Unit on Genetic Control of Mosquitos implemented the same technology for nuisance and vector control of *Cx. quinquefasciatus*. Sterile males were released daily (150 000–300 000 males/day for 5.5 months); despite the remarkable numbers, the release of sterile males resulted in a high level of sterility in the population in only one of two study sites and just for 3 weeks [64]. These experiments were met with the suspicion of exploitation and biowarfare by the general public and a report in the National Herald called the work "neo-imperialism" because of perceptions that the USDA was running experiments in India that could not be done in the United States [65].

Mosquito modification by paratransgenesis is another option for population reduction that involves introducing mosquitoes infected with *Wolbachia* endosymbiotic bacteria that induce sterilizing effects (termed cytoplasmic incompatibility). Laven et al. described successful eradication of *Cx. quinquefasciatus* based on cytoplasmic incompatibility. From March 16 until May 6 1967, 5 000 incompatible males were released daily in a village outside of Rangoon; on May 9 and 10, no viable egg rafts were collected [66]. O'Connor et al. produced an artificially *Wolbachia*-infected strain of *Ae. polynesiensis* (CP) and tested the extent of population suppression in lab and field trials in Motu islands in French Polynesia. After the release of more than 117 000 *Ae. polynesiensis* (CP), the authors observed high rates of insemination of trapped females and a reduction in viable egg production (93% versus 76% at the control and experimental sites, respectively) [67].

The field of mosquito control using molecular tools to alter the mosquito genome is at a watershed moment with the introduction of gene drive and CRISPR-based technologies (see Refs [68, 69]). As a proof of concept for application for LF control in particular, a transposable element transgenic approach for driving production of an antifilarial gene product was developed for *Cx. quinquefasciatus*, with a promoter specifically directing expression to the thoracic musculature – the target site for parasite development [70]. The pipeline of further development of this technology is much further advanced for *Anopheles* species, and *An. gambiae* in particular, in the context of malaria control (see Refs [69, 71]); this could prove useful where *An. gambiae* is both a malaria and a LF parasite vector and could be targeted for population modification. As of publication of this chapter, releases of *Wolbachia*-infected *Ae. aegypti* are underway as part of the Eliminate Dengue Program, and Oxitech is testing sterile, Friendly™ *Ae. aegypti* to suppress populations in Brazil, Florida, the Cayman Islands, Panama, and India. Guidance for testing genetically modified mosquitoes in field settings, with an emphasis on *An. gambiae*, has been proposed [72]. Transgenic approaches to population modification for malaria control may well benefit LF control in the near future.

## 22.9 Vector Control for Onchocerciasis

*Onchocerca volvulus* is transmitted by black flies in the genus *Simulium*; after ingesting microfilariae from an infected human during a blood meal, the parasite undergoes development to the infective L3 stage, which can be introduced into a subsequent human host during a blood meal (see Chapter 2, this volume). Of the many species of *Simulium*, only a subset are competent vectors for *O. volvulus* [73], primarily in the *S. damnosum* complex, of which six sibling species are thought to be the major vectors in Africa [74]. Variation in vector competence among the *S. damnosum* species complex includes physiological or anatomical barriers to parasite development, the extent of attraction to human vs. other hosts, choice of breeding sites, behaviors, etc. Female black flies are hematophagous and lay eggs in selected areas of flowing water, including in rivers and streams, where the larvae subsequently feed and develop over as little as a week before emerging as adults; the common name for onchocerciasis, river blindness, reflects the aquatic stage of the life cycle of the vector. *S. neavei*, the larvae of which associate with freshwater crabs, is also an important vector of this parasite in parts of Africa; its restricted distribution led to the first successful local onchocerciasis elimination program, in Kenya [75].

In the Americas, onchocerciasis was introduced through the slave trade [76]. *S. damnosum* is not present in the Americas [15, 77]. Instead, the primary vector in Guatemala and Mexico was *S. ochraceum*, which is less effective in transmitting the parasite. In South America, vectors included *S. oyapockense*, *S. metallicum*, *S. exiguum*, and *S. guianense*, some of which are as competent as *S. damnosum*. None of these species have a large geographical range, so onchocerciasis in the Americas was found in relatively small and isolated patches. Vector control was not a prominent component of onchocerciasis control programs in the Americas; early efforts included application of the insecticide Paris Green, an arsenical, and clearing of vegetation around *Simulium*-infested streams; neither was notably effective, given the patchy distribution of the vectors and the ease of repopulation of treated areas. Instead, control initially relied on surgical removal of nodules (which was also only modestly effective), and subsequently distribution of ivermectin in MDA campaigns, which has led to the almost complete eradication of the parasite from the Western hemisphere [78].

In contrast, vector control for onchocerciasis in Africa has a long and successful history, which has been well reviewed [15, 79–84]. Although chemotherapy became the mainstay of control efforts following the introduction of ivermectin in the late 1980s, it is important to consider, at least briefly, the role of vector control in achieving significant reductions in the regional prevalence of onchocerciasis and, importantly, in the incidence of parasite-induced blindness.

Recognition that black flies in the genus *Simulium* were vectors for *O. volvulus* led to control that employed insecticides and environmental modification, with notable success in Kenya. Forest clearing near infested rivers and the application of DDT to control populations of *S. neavei* had marked local effect on new infections with *O. volvulus* [75]. Success generated enthusiasm for the implementation of similar efforts in other areas based on the treatment of black fly-infested streams with DDT. Expansion of vector control efforts led to the realization that *S. damnosum* posed greater challenges for control than *S. neavei*. Insecticide applications to suppress adult populations were not sufficiently effective and had broad environmental consequences. As a result, local treatment of waterways with insecticides to kill black fly larvae became the method of choice. Environmental concerns over the actions of DDT led to the selection of alternative insecticides, most notably the organophosphate cholinesterase inhibitor temephos, for operations in areas of West Africa in which *S. damnosum*-mediated transmission was of urgent concern. Although local reductions in vector populations were achieved, the ability of *S. damnosum* to travel considerable distances meant that these reductions were only temporary, leading to the conclusion that vector control would have to be implemented on a much larger geographic scale to achieve sustained disease control. Experience in Kenya around the elimination of *S. neavei* populations revealed that the life span of adult *O. volvulus* was 13–17 years, indicating that the disease would disappear if transmission could be abrogated for more period of time.

This recognition led to initiation of the Onchocerciasis Control Program (OCP) [82] in the mid-1970s to minimize the impact of the parasite in highly endemic areas. Based on transmission dynamics, the extent of savannah environments and the population incidence of infection and eye disease, formal vector control efforts were marshalled in seven countries in West Africa (later expanded to 11) under the auspices of the WHO, the World Bank, and other organizations. Vector control was initially achieved through regular (weekly) aerial spraying of temephos over infested rivers and streams. As resistance to temephos appeared, additional insecticide classes were employed both to minimize the extent of temephos resistance and to forestall the further development of resistance to an extent that could affect the benefits already obtained through OCP operations. Alternative insecticides included pyrethroids, carbamate cholinesterase inhibitors, and a strain of *Bacillus thuringiensis*. This tactic insured that gains made in onchocerciasis control were maintained and expanded over the duration of insecticidal treatment as the primary OCP strategy.

It must be recognized that, before the introduction of ivermectin for use in MDA campaigns, vector control was the only rational and cost-effective method available to reduce the incidence of the disease and the prevalence of parasite-induced blindness. These efforts, as noted, were remarkably effective. Several factors combined to re-orient control efforts from insecticide application to ivermectin MDA. Cost was one; ivermectin could be administered once a year as opposed to the weekly aerial insecticide deliveries, and the drug was donated. Environmental concerns about the routine spraying of broad-spectrum insecticides contributed to the decision as well. The continued success of onchocerciasis control programs is testament to the efficacy of MDA campaigns.

However, there is continuing interest in vector control options in certain circumstances. Modeling suggests that focused, local, short-duration (10 weekly treatments) larviciding campaigns could be a valuable complementary strategy for onchocerciasis in some areas [85]. Additionally, recent experience in Uganda suggests that a combination of vegetation removal through a "slash and clear" operation, coupled with ivermectin MDA, was very successful in reducing fly biting rates, enhancing control efforts [86]. Recruitment of local residents to remove trailing vegetation from streams reduced the number of larvae and the quality of the environment for their development and did not involve the use of insecticides. Modeling suggests that monthly operations of this kind, combined with MDA, could accelerate elimination of onchocerciasis and markedly reduce black fly biting [87]. Integrating vector control with MDA programs may be cost-effective in some situations and deserves additional research investment.

## 22.10 Vector Control for Loiasis

Efforts to control human infections with *L. loa* have been associated almost entirely with efforts to control other filariases, particularly onchocerciasis, because of the rare but very severe adverse events observed in patients with high burdens of *L. loa* mf who are treated with ivermectin or diethylcarbamazine (see Refs [13, 14, 16, 88]). Efforts to extend MDA programs into loiasis regions rely on diagnostic procedures to exclude heavily infected individuals. Because infections with *L. loa* have been presumed to cause few health problems, much less research has been conducted on this parasite than on the filarial species that cause LF and onchocerciasis.

The vectors for *L. loa* are tabanid flies in the genus *Chrysops*; multiple species in this genus are found in Africa, with two being the primary vectors for human loiasis: *C. silacea* and *C. dimidiata* [89, 90]. These species are both anthropophilic and prefer forested, humid habitats near water; eggs are laid in mud overlaid by shallow water. After hatching, larvae burrow into the mud for a prolonged period of development before emerging to feed and mate. Females are hematophagous, and the bites can be painful and may become infected. Once ingested by permissive *Chrysops* spp., *L. loa* mf develop into infective larvae, ending up in the proboscis, from which they enter a host during a subsequent blood meal.

As noted, few attempts were made to control these vectors, which provide significant challenges to a comprehensive program, due in part to the predilection sites for larval development and adult residence. However, considering the barriers to MDA implementation for onchocerciasis and LF control in loiasis regions, a new look at a possible role for vector control in these regions has been recommended [90]. Possible interventions include reducing smoke from wood fires, which attracts adult flies, clearing brush near human habitation, use of repellants, and targeted insecticidal applications; however, more research is needed to quantify the efficacy of these interventions to justify investment in their broader applications in *Loa*-endemic areas. To the extent that onchocerciasis control programs are stalled by the prevalence of heavily infected individuals, these measures may assume sufficient important to warrant implementation.

## 22.11 Conclusion

Vector control for onchocerciasis played an historic and highly significant role in reducing the prevalence of River Blindness and is a testament to the ability of this strategy to ameliorate human illness and suffering. The control of LF across the globe is also a remarkable public health success story and one that is far from over. The original goal for elimination (2020) has passed, and there is much work to be done to advance every endemic country through MDA implementation, reducing infection to the threshold of transmission, and demonstrating sustained reduction over time. Mosquito control has proven effective in reducing bite burdens and thereby transmission potential for major *Culex* and *Anopheles* species vectors of LF parasites and is a useful supplement to MDA where possible and to sustaining transmission below the threshold after MDA is complete. Where possible, there is tremendous collateral benefit to implementing strategies for the dual purpose of malaria and LF control and to integrating control practices to maximize the effectiveness of controls, account for insecticide resistance management, and minimize the environmental impacts of the program. Novel technology development, including new chemical and molecular insecticides, innovative delivery systems, and genetic tools for population modification, will no doubt benefit vector control in LF disease control programs. The extension of novel technologies such as these to *Simulium* and *Chrysops* vector spp. could also have significant impacts on the other major human filariases, although those efforts lag far behind the work already accomplished in mosquitoes.

## References


1 Manson, P. (1878). On the development of *Filaria sanguinis hominis*, and on the mosquito considered as a nurse. *Zool. J. Linn. Soc.* 14 (75): 304–311.

2 Ross, R. (1898). Report on the cultivation of Proteosoma, Labbé, in grey mosquitos. *Indian Med. Gaz.* 33 (12): 448–451.

3 Clements, A.N. and Harbach, R.E. (2017). History of the discovery of the mode of transmission of yellow fever virus. *J. Vector Ecol.* 42 (2): 208–222.

4 Boyce, R. (1909). *Mosquito or Man?* New York, NY: E.P Dutton and Company.

5 Le Prince, J.A.A. and Orenstein, A.J. (1916). *Mosquito Control in Panama; the Eradication of Malaria and Yellow Fever in Cuba and Panama*. New York, London: Putnam.

6 Wilson, A.L., Courtenay, O., Kelly-Hope, L.A. et al. (2020). The importance of vector control for the control and elimination of vector-borne diseases. *PLoS Negl.Trop. Dis.* 14 (1): e0007831.

7 Webber, R.H. (1979). Eradication of *Wuchereria bancrofti* infection through vector control. *Trans. R. Soc. Trop. Med. Hyg.* 73: 722–724.

8 Michael, E., Bundy, D.A., and Grenfell, B.T. (1996). Re-assessing the global prevalence and distribution of lymphatic filariasis. *Parasitology* 112 (Pt 4): 409–428.

9 Mathew, C.G., Bettis, A.A., Chu, B.K. et al. (2020). The health and economic burdens of lymphatic filariasis prior to mass drug administration programs. *Clin. Infect. Dis.* 70 (12): 2561–2567.

10 Redekop, W.K., Lenk, E.J., Luyendijk, M. et al. (2017). The socioeconomic benefit to individuals of achieving the 2020 targets for five preventive chemotherapy neglected tropical diseases. *PLoS Negl.Trop. Dis.* 11 (1): e0005289.

11 Leiper, R. (1913). Report to the advisory committee of the tropical diseases research fund, colonial office London. *Trop. Dis. Bull.* 2: 195–196.

12 Blacklock, D.B. (1927). The insect transmission of *Onchocerca volvulus* (Leuckhart, 1983): the cause of worm nodules in man in Africa. *Br. Med. J.* 1 (3446): 129–133.

13 Boussinesq, M. (2006). Loiasis. *Ann. Trop. Med. Parasitol.* 100 (8): 715–731.

14 Mackenzie, C.D. (2022). Human filarial infections: reflections on the current understanding of their importance, pathobiology and management. In: *Human and Animal Filariases* (ed. R. Kaminsky and T.G. Geary), Chapter 3. Weinheim, Germany: Wiley-VCH.

15 Boatin, B.A. and Richards, F.O. Jr. (2006). Control of onchocerciasis. *Adv. Parasitol.* 61: 349–394.

16 Boatin, B.A., Richards, F.O. Jr., Ramaiah, K.D., and Gyapong, J.O. (2022). Elimination and eradication of human filariases. In: *Human and Animal Filariases* (ed. R. Kaminsky and T.G. Geary), Chapter 12. Weinheim, Germany: Wiley-VCH.

**17** World Health Organization (1997). Ninth plenary meeting, 13 May 1997 A50/VR/9. https://apps.who.int/iris/bitstream/handle/10665/179773/WHA50_R29_eng.pdf (accessed 01 July 2022).

**18** World Health Organization (2004). Global strategic framework for integrated vector management. World Health Organization, Geneva (WHO/CDS/CPE/PVC/2004.10).

**19** World Health Organization (2008). Position statement on integrated vector management. World Health Organization, Geneva (WHO/HTM/NTD/VEM/2008.2).

**20** World Health Organization (2007). *Global Plan to Combat Neglected Tropical Diseases 2008–2015* (ed. L. Savioli and D. Daumerie). Geneva: World Health Organization.

**21** World Health Organization (2013). *Lymphatic Filariasis: A Handbook of Practical Entomology for National Elimination Programmes.* Geneva: World Health Organization.

**22** World Health Organization & UNICEF/UNDP/World Bank/WHO Special Programme for Research and Training in Tropical Diseases (2017). Global vector control response 2017-2030. World Health Organization, Geneva.

**23** WHO (2018). Global programme to eliminate lymphatic filariasis: progress report. *Wkly Epidemiol. Rec.* 94: 457–470.

**24** Burkot, T.R., Durrheim, D.N., Melrose, W.D. et al. (2006). The argument for integrating vector control with multiple drug administration campaigns to ensure elimination of lymphatic filariasis. *Filaria J.* 5: 10.

**25** Bockarie, M.J., Pedersen, E.M., White, G.B., and Michael, E. (2009). Role of vector control in the Global Program to Eliminate Lymphatic Filariasis. *Annu. Rev. Entomol.* 54: 469–487.

**26** Irvine, M.A., Reimer, L.J., Njenga, S.M. et al. (2015). Modelling strategies to break transmission of lymphatic filariasis – aggregation, adherence and vector competence greatly alter elimination. *Parasites Vectors* 8: 547.

**27** Raghavan, N.G.S. (1961). The vectors of human infections by *Wuchereria* species * in endemic areas and their biology. *Bull. World Health Organ.* 24: 177–195.

**28** Chow, C.Y. (1973). Filariasis vectors in the Western Pacific region. *Z. Tropenmed. Parasitol.* 24 (4): 404–418.

**29** Nelson (1978). Mosquito-borne filariasis. In: *Medical Entomology Centenary Symposium Proceedings* (ed. S. Willmott), 15–25. London: Royal Society of Tropical Medicine and Hygiene.

**30** Wilkerson, R.C., Linton, Y.-M., Fonseca, D.M. et al. (2015). Making mosquito taxonomy useful: a stable classification of Tribe Aedini that balances utility with current knowledge of evolutionary relationships. *PLoS One* 10 (7): e0133602.

**31** Harbach, R.E. (2013). The phylogeny and classification of *Anopheles*. In: *Anopheles Mosquitoes – New Insights into Malaria Vectors* (ed. S. Maguin). Rijeka, Croatia: InTech http://dx.doi.org/10.5772/54695.

**32** Pichon, G. (2002). Limitation and facilitation in the vectors and other aspects of the dynamics of filarial transmission: the need for vector control against *Anopheles*-transmitted filariasis. *Ann. Trop. Med. Parasitol.* 96 (Suppl 2): S143–S152.

**33** Kogan, M. (1998). Integrated pest management: historical perspectives and contemporary developments. *Annu. Rev. Entomol.* 43: 243–270.

**34** Cheun, H.-I., Kong, Y., Cho, S.-H. et al. (2009). Successful control of lymphatic filariasis in the Republic of Korea. *Korean J. Parasitol.* 47 (4): 323–335.

**35** Das, P., Manoharan, A., Subramanian, S. et al. (1992). Bancroftian filariasis in Pondicherry, South India – epidemiological impact of recovery of the vector population. *Epidemiol. Infect.* 108 (3): 483–493.

**36** Reuben, R., Rajendran, R., Sunish, I.P. et al. (2001). Annual single-dose diethylcarbamazine plus ivermectin for control of bancroftian filariasis: comparative efficacy with and without vector control. *Ann. Trop. Med. Parasitol.* 95: 361–378.

**37** Bøgh, C., Pedersen, E.M., Mukoko, D.A., and Ouma, J.H. (1998). Permethrin-impregnated bednet effects on resting and feeding behaviour of lymphatic filariasis vector mosquitoes in Kenya. *Med. Vet. Entomol.* 12 (1): 52–59.

**38** Poopathi, S. and Rao, D.R. (1995). Pyrethroid-impregnated hessian curtains for protection against mosquitoes indoors in South India. *Med. Vet. Entomol.* 9: 169–175.

**39** Curtis, C.F., Myamba, J., and Wilkes, T.J. (1996). Comparison of different insecticides and fabrics for anti-mosquito bednets and curtains. *Med. Vet. Entomol.* 10: 1–11.

**40** Chambers, E.W., Bossin, H.C., Ritchie, S.A. et al. (2016). The impact of insecticide-treated cloth targets on the survival of *Stegomyia polynesiensis* (=*Aedes polynesiensis*) under laboratory and semi-field conditions in French Polynesia. *Med. Vet. Entomol.* 30 (3): 247–252.

**41** Pedersen, E.M. and Mukoko, D.A. (2002). Impact of insecticide-treated materials on filaria transmission by the various species of vector mosquito in Africa. *Ann. Trop. Med. Parasitol.* 96 (Suppl 2): S91–S95.

**42** Ashton, R.A., Kyabayinze, D.J., Opio, T. et al. (2011). The impact of mass drug administration and long-lasting insecticidal net distribution on *Wuchereria bancrofti* infection in humans and mosquitoes: an observational study in northern Uganda. *Parasites Vectors* 15: 134.

**43** Reimer, L.J., Thomsen, E.K., Tisch, D.J. et al. (2013). Insecticidal bed nets and filariasis transmission in Papua New Guinea. *N. Engl. J. Med.* 369 (8): 745–753.

**44** Gordon, C.A., Jones, M.K., and McManus, D.P. (2018). The history of Bancroftian lymphatic filariasis in Australasia and Oceania: is there a threat of re-occurrence in mainland Australia? *Trop. Med. Infect. Dis.* 3 (2): 58.

**45** Chernin, E. (1987). The disappearance of Bancroftian filariasis from Charleston, South Carolina. *Am. J. Trop. Med. Hyg.* 37 (1): 111–114.

**46** Rebollo, M.P. and Bockarie, M.J. (2013). Toward the elimination of lymphatic filariasis by 2020: treatment update and impact assessment for the endgame. *Exp. Rev. Anti-Infect. Ther.* 11 (7): 723–731.

**47** Regis, L., Oliveira, C.M.F., Silva-Filha, M.H. et al. (2000). Efficacy of *Bacillus sphaericus* in control of the filariasis vector *Culex quinquefasciatus* in an urban area of Olinda, Brazil. *Trans. R. Soc. Trop. Med. Hyg.* 94 (5): 488–492.

**48** Maxwell, C.A., Curtis, C.F., Haji, H. et al. (1990). Control of Bancroftian filariasis by integrating therapy with vector control using polystyrene beads in wet pit latrines. *Trans. R. Soc. Trop. Med. Hyg.* 84 (5): 709–714.

**49** van den Berg, H., Kelly-Hope, L.A., and Lindsay, S.W. (2013). Malaria and lymphatic filariasis: the case for integrated vector management. *Lancet Infect. Dis.* 13 (1): 89–94.

**50** World Health Organization (2011). Integrated Vector Management to Control Malaria and Lymphatic Filariasis. WHO Position Statement, World Health Organization, Geneva (WHO/HTM/NTD/PCT/2011.2).

**51** Stone, C.M., Lindsay, S.W., and Chitnis, N. (2014). How effective is integrated vector management against malaria and lymphatic filariasis where the diseases are transmitted by the same vector? *PLoS Negl.Trop. Dis.* 8 (12): e0003393.

**52** Ranson, H. and Lissenden, N. (2016). Insecticide resistance in African *Anopheles* mosquitoes: a worsening situation that needs urgent action to maintain malaria control. *Trends Parasitol.* 32 (3): 187–196.

**53** World Health Organization (2018) Global report on insecticide resistance in malaria vectors: 2010–2016. World Health Organization, Geneva.

**54** Koimbu, G., Czeher, C., Katusele, M. et al. (2018). Status of insecticide resistance in Papua New Guinea: an update from nation-wide monitoring of *Anopheles* mosquitoes. *Am. J. Trop. Med. Hyg.* 98 (1): 162–165.

**55** Case, E., Shragai, T., Harrington, L. et al. (2020). Evaluation of unmanned aerial vehicles and neural networks for integrated mosquito management of *Aedes albopictus* (Diptera: Culicidae). *J. Med. Entomol.* 57 (5): 1588–1595.

**56** Qualls, W.A., Müller, G.C., Traore, S.F. et al. (2015). Indoor use of attractive toxic sugar bait (ATSB) to effectively control malaria vectors in Mali. *West Afr. Mal. J.* 14: 301.

**57** Snetselaar, J., Njiru, B.N., Gachie, B. et al. (2017). Eave tubes for malaria control in Africa: prototyping and evaluation against *Anopheles gambiae* s.s. and *Anopheles arabiensis* under semi-field conditions in western Kenya. *Malar. J.* 16: 276.

**58** Achee, N.L., Bangs, M.J., Farlow, R. et al. (2012). Spatial repellents: from discovery and development to evidence-based validation. *Malar. J.* 11: 164.

**59** Airs, P.M. and Bartholomay, L.C. (2018). Molecular and nano-scale alternatives to traditional insecticides for *in situ* control of mosquito vectors. In: *Advances in the Biorational Control of Medical and Veterinary Pests* (ed. E.J. Norris, J.R. Coats, A.D. Gross and J.M. Clark), 75–99. Washington, D.C.: American Chemical Society.

**60** Norris, E.J., Gross, A.D., Bartholomay, L.C., and Coats, J.R. (2019). Plant essential oils synergize various pyrethroid insecticides and antagonize malathion in *Aedes aegypti*. *Med. Vet. Entomol.* 33 (4): 453–466.

**61** Carlier, P.R., Bloomquist, J.R., Totrov, M., and Li, J. (2017). Discovery of species-selective and resistance-breaking anticholinesterase insecticides for the malaria mosquito. *Curr. Med. Chem.* 24 (27): 2946–2958.

**62** World Health Organization (2012) Global plan for insecticide resistance management in malaria vectors. World Health Organization, Geneva.

**63** Patterson, R.S., Weidhass, D.E., Ford, H.R., and Lofgren, C.S. (1970). Suppression and elimination of an island population of *Culex pipiens quinquefasciatus* with sterile males. *Science* 168: 1368–1370.

**64** Yasuno, M., MacDonald, W.W., Curtis, C.F. et al. (1978). A control experiment with chemosterilised *Culex pipiens quinquefasciatus* Wied. in a village near Delhi surrounded by a breeding-free zone. *Jpn. J. Sanit. Zool.* 29 (4): 325–343.

**65** Anonymous (1975). Oh, New Delhi; oh, Geneva. *Nature* 256: 355–357.

**66** Laven, H. (1967). Eradication of *Culex pipiens fatigans* through cytoplasmic incompatibility. *Nature* 216: 383–384.

**67** O'Connor, L., Plichart, C., Sang, A.C. et al. (2012). Open release of male mosquitoes infected with a *Wolbachia* biopesticide: field performance and infection containment. *PLoS Negl.Trop. Dis.* 6 (11): e0001797.

**68** Marshall, J.M., Raban, R.R., Kandul, N.P. et al. (2019). Winning the tug-of-war between effector gene design and pathogen evolution in vector population replacement strategies. *Front. Genet.* 10: 1072.

**69** Carballar-Lejarazú, R. and James, A.A. (2017). Population modification of Anopheline species to control malaria transmission. *Pathog. Global Health* 111 (8): 424–435.

**70** Allen, M.L. and Christensen, B.M. (2004). Flight muscle-specific expression of act88F: GFP in transgenic *Culex quinquefasciatus* Say (Diptera: Culicidae). *Parasitol. Int.* 53 (4): 307–314.

**71** Hammond, A.M. and Galizi, R. (2017). Gene drives to fight malaria: current state and future directions. *Pathog. Global Health* 111: 412–423.

**72** James, S., Collins, F.H., Welkhoff, P.A. et al. (2018). Pathway to deployment of gene drive mosquitoes as a potential biocontrol tool for elimination of malaria in Sub-Saharan Africa: Recommendations of a scientific working group†. *Am. J. Trop. Med. Hyg.* 98 (Suppl): 1–49.

**73** Post, R.J., Mustapha, M., and Krueger, A. (2007). Taxonomy and inventory of the cytospecies and cytotypes of the *Simulium damnosum* complex (Diptera: Simuliidae) in relation to onchocerciasis. *Trop. Med. Int. Health* 12 (11): 1342–1353.

**74** Crosskey, R.W. (1987). A taxa summary for the *Simulium damnosum* complex, with special reference to distribution outside the control areas of West Africa. *Ann. Trop. Med. Parasitol.* 81: 181–192.

**75** McMahon, J.P., Highton, R.B., and Goiny, H. (1958). The eradication of *Simulium neavei* from Kenya. *Bull. World Health Organ.* 19: 75–107.

**76** Gustavsen, K., Hopkins, A., and Sauerbrey, M. (2011). Onchocerciasis in the Americas: from arrival to (near) elimination. *Parasites Vectors* 4: 205.

**77** Shelley, A.J. (1988). Vector aspects of the epidemiology of onchocerciasis in Latin America. *Annu. Rev. Entomol.* 33: 337–366.

**78** Sauerbrey, M., Rakers, L.J., and Richards, F.O. Jr. (2018). Progress toward elimination of onchocerciasis in the Americas. *Int. Health* 10 (suppl10): i71–i78.

**79** McMahon, J.P. (1967). A review of the control of *Simulium* vectors of onchocerciasis. *Bull. World Health Organ.* 37: 415–430.

**80** Walsh, J.F. (1990). Review of vector control prior to the OCP. *Acta Leiden* 59 (1-2): 61–78.

**81** Davies, J.B. (1994). Sixty years of onchocerciasis vector control: a chronological summary with comments on eradication, reinvasion and insecticide resistance. *Annu. Rev. Entomol.* 39: 23–45.

**82** Samba, E.M. (1994). *The Onchocerciasis Control Programme in West Africa: an example of effective public health management*. Geneva: World Health Organization.

**83** Hougard, J.-M., Yaméogo, L., Sékétéli, A. et al. (1997). Twenty-five years of blackfly control in the Onchocerciasis Control Programme in West Africa. *Parasitol. Today* 13: 425–431.

**84** World Health Organization (2002). *Success in Africa: The Onchocerciasis Control Programme in West Africa 1974-2002*. Geneva: World Health Organization.

**85** Routledge, I., Walker, M., Cheke, R.A. et al. (2018). Modelling the impact of larviciding on the population dynamics and biting rates of *Simulium damnosum* (*s.l.*): implications for vector control as a complementary strategy for onchocerciasis elimination in Africa. *Parasites Vectors* 11: 316.

**86** Jacob, B.G., Loum, D., Lakwo, T.L. et al. (2018). Community-directed vector control to supplement mass drug distribution for onchocerciasis elimination in the Madi mid-North focus of Northern Uganda. *PLoS Negl.Trop. Dis.* 12: e0006702.

**87** Smith, M.E., Bilal, S., Lakwo, T.L. et al. (2019). Accelerating river blindness elimination by supplementing MDA with a vegetation "slash and clear" vector control strategy: a data-driven modeling analysis. *Sci. Rep.* 9: 15274.

**88** Specht, S., Kamngo, J., and Geary, T.G. (2022). Antifilarial chemotherapy for humans: current options. In: *Human and Animal Filariases* (ed. R. Kaminsky and T.G. Geary), Chapter 7. Weinheim, Germany: Wiley-VCH.

**89** Kouam, M.K. and Kamgno, J. (2017). *The African Chrysops, in Biological Control of Pest and Vector Insects* (ed. V.D.C. Shields). London: InTech Open https://doi.org/10.5772/67111.

**90** Kelly-Hope, L., Paulo, R., Thomas, B. et al. (2017). *Loa loa* vectors *Chrysops* spp.: perspectives on research, distribution, bionomics, and implications for elimination of lymphatic filariasis and onchocerciasis. *Parasites Vectors* 10 (1): 172.

# 23

# Vector Control Approaches for Canine Filariasis

*Sofija Todorovic[1], Tanja McKay[1,*], and Phillip Kaufman[2]*

[1]Arkansas State University, Department of Biological Sciences, Jonesboro, AR 72467, USA
[2]Texas A&M University, Department of Entomology, College Station, TX 77843-2475, USA

**Abstract**

Mosquitoes are important vectors of *Dirofilaria immitis* and *Dirofilaria repens*, the nematodes that cause canine filariasis. Depending on geographical location, the specific vector species can vary. Species of *Aedes*, *Anopheles*, *Culex*, and *Psorophora* are competent for vectoring *Dirofilaria* spp. However, on a global scale, much remains to be known about which mosquito species are vectors of *Dirofilaria* and research is needed to address vector competency and seasonality of transmission not only between regions, but among species and their willingness to feed on specific hosts, including dogs. This chapter provides an overview of the important mosquito species known to potentially transmit canine filariasis. The mosquito life cycle and *Dirofilaria* interactions are briefly discussed. The majority of the chapter is dedicated to Integrated Mosquito Management (IMM) strategies that can be implemented by mosquito control districts, veterinarians, and pet owners, as well as research-based biological control methods for mosquito vector population suppression.

## 23.1 Introduction

A variety of nematode species vectored by arthropods infect domestic canines (Table 23.1). Tahir and colleagues provide an overview of many filarial nematodes associated with pets [12]. *Onchocerca lupi* first described in 1967 in a wolf in Georgia [13] causes ocular lesions in dogs and cats in Europe [11] and has more recently been more commonly reported in the United States [14, 15]. The black fly *Simulium tribulatum* is reported as the putative vector in southern California [10]. Louse flies and ticks transmit *Acanthocheilonema drancunculoides* [1, 2] and *Ceropithifilaria* spp. can be vectored by ticks [7] in the Mediterranean region. Fleas and lice can vector *Acanthocheilonema reconditum* [3, 4]. Other filarial species in canines include *Brugia malayi* and *Brugia pahangi*, which can cause lymphatic

*Corresponding author.

*Human and Animal Filariases: Landscape, Challenges, and Control*, First Edition.
Edited by Ronald Kaminsky and Timothy G. Geary.

**Table 23.1** Filarial parasites of canines and their arthropod vector associations.

| Filarial species | Vector |
|---|---|
| *Acanthocheilonema dracunculoides* | Louse flies and ticks [1, 2] |
| *Acanthocheilonema reconditum* | Fleas and lice [3, 4] |
| *Brugia malayi* | Mosquitoes [5] |
| *Brugia pahangi* | Mosquitoes [6] |
| *Cercopithifilaria bainae* | Ticks [7] |
| *Cercopithifilaria grassii* | Ticks [7] |
| *Dirofilaria immitis* | Mosquitoes [8] |
| *Dirofilaria repens* | Mosquitoes [9] |
| *Onchocerca lupi* | Black flies [10] |
| *Thelazia californiensis* | Non-biting flies (*Drosophila*) [11] |
| *Thelazia callipaeda* | Non-biting flies (*Drosophila*) [11] |

filariasis in humans [16, 17] with dogs and cats as important reservoirs for human infection [18]. Of all the filarial species known to cause canine infection, *Dirofilaria immitis* and *D. repens* receive the most attention due to the significant prevalence of infections in dogs and the ensuing damage an infection can cause. *D. immitis* has a cosmopolitan distribution [19], with *D. repens* currently distributed only in the old world [20]. This chapter focuses on mosquitoes, the only vectors of *Dirofilaria* spp. The chapter begins with a brief overview of the mosquito life cycle and *Dirofilaria* interactions. It then outlines the various control strategies used to mitigate mosquitoes. Some of these strategies are more related to research-based ideas and have the potential for use. However, the main focus of this chapter is to introduce the reader to the important vector species as well as mosquito control strategies that could be implemented by either mosquito control districts, the veterinarian or the pet owner, each helping to decrease the transmission of *Dirofilaria* spp.

## 23.2 Mosquito Life Cycle and Habitats

Mosquitoes (Diptera: Culicidae) are ubiquitous as they have a short lifespan and reproduce multiple times a year (i.e. multivoltine) [21]. Mosquitoes undergo holometabolous development, also referred to as complete metamorphosis, beginning with the aquatic phase for immature development. Eggs are laid either singly or in groups in different types of habitats tied to individual species preferences. For example, *Culex* and *Anopheles* species oviposit eggs on the surface of stagnant water near marshes, whereas *Aedes* species generally deposit eggs on moist soil or on dry edges of water-holding containers [22]. After hatching, larvae develop through a series of instars, eventually entering the pupal stage. Development from egg to adult varies by mosquito species but typically takes 10 to 14 days depending on environmental conditions, including ambient temperature as well as food source and availability [22, 23].

Several medically important mosquitoes rely on varying amounts of water in natural and artificial containers to complete larval and pupal development [24]. These containers may include natural accumulations of water, such as rock pools [25], tree holes [26, 27], and axils of bromeliads [28, 29] or other phylotelmata. Novel invasive mosquitoes have moved into new geographical areas through the used tire industry. *Aedes (Ochlerotatus) japonicus japonicus* and *Aedes albopictus* were introduced from Japan and temperate regions of Asia, respectively, to the United States through the movement of automobile tires [30–33]. *Ae. j. japonicus* inhabit natural containers, such as rock pools along stream beds [34] that are excellent development sites in the United States, similar to its development habitats in its home range of Japan [25]. *Ae. albopictus* has a cosmopolitan distribution, as the adults can withstand cold temperatures and enter diapause in winter [33]. Artificial containers such as used tires can provide optimal environmental conditions for mosquito development [35, 36].

Adult female mosquitoes of medical and veterinary importance are anautogenous, meaning a blood meal is required for egg production, which allows for the potential to serve as disease-causing pathogen vectors [37]. Mosquito host-seeking for blood feeding is aided by ocular structures and ommatidia dimensions (i.e. cells of compound eyes) [38] in addition to sensory cues, including carbon dioxide, lactic acid, heat, moisture [22, 39], and possibly pheromones emitted from microfilaremic hosts [40]. Spatial preference for feeding also varies by mosquito species. For instance, endophagic behaviors (i.e. feeding inside dwellings) are observed in *Aedes aegypti* and *Aedes vexans*, whereas *Ae. albopictus* and *Anopheles punctipennis* are exophagic (i.e. feed outdoors in urban areas) and *Anopheles quadrimaculatus* exhibits a combination of both [41, 42]. Mosquitoes with sylvatic feeding behaviors, such as *Aedes canadensis* and *Aedes sierrensis*, may contribute to dissemination of disease via stray dogs and wildlife reservoirs inhabiting rural areas [42]. Agricultural operations may also influence mosquito populations from genera such as *Anopheles* and *Psorophora*, which are commonly found in flooded rice fields [8, 43].

Peak blood feeding times are classified as crepuscular (i.e. dawn and dusk), diurnal (i.e. daytime), and nocturnal (i.e. nighttime) [42]. *Ae. aegypti* and *Ae. canadensis* are daytime feeders along with *Ae. albopictus*, which has more aggressive feeding tendencies [42, 44]. Crepuscular feeding activity has been observed with *Ae. vexans*, *An. punctipennis*, and *Ae. trivittatus* [38]. Important nocturnal vector species for *D. immitis* include *Culex pipiens quinquefasciatus*, *Anopheles crucians*, and *An. quadrimaculatus* [42].

## 23.3 Important D. immitis Vectors

Vector-borne diseases such as canine filariasis are complex, as the conditions for disease transmission are multidimensional. Multiple physiological interactions occur among *Dirofilaria* spp. during nematode development in mosquitos that serve as intermediate hosts and vectors, as well as in the definitive and aberrant mammalian hosts. The biological and ecological requisites for each mosquito species are also multifaceted and are influenced by physiological and behavioral traits that may ultimately impact the distribution and dynamics of mosquito populations as well as mosquito vector competency.

Ledesma and Harrington provide an overview of potential vectors of *D. immitis* in the United States [42]. Compiled from published studies between 1939 and 2007, Ledesma and Harrington created a list of 25 mosquito species in the United States that were naturally infected with the *D. immitis* $L_3$ stage [42]. Species of *Aedes*, *Anopheles*, *Culex*, and *Psorophora* were presumed to be competent for vectoring *D. immitis*, with the following eight species being the most important vectors in the United States: *Ae. albopictus*, *Ae. canadensis*, *Ae. trivittatus*, *Ae. vexans*, *An. crucians*, *An. punctipennis*, *An. quadrimaculatus*, and *Cx. p. quinquefasciatus*.

In Canada, the prevalence of heartworm infection is low, with most infections reported in southern Ontario. Northern Ontario has been seeing an increase in infection, which may be attributed to climactic change, with temperatures favorable for *D. immitis* development in mosquitoes or inadvertent detection due to an increase in tick-borne disease screening with combination diagnostic tests [45]. The major mosquito vectors of *D. immitis* in southern Ontario include *Ae. vexans*, *Cx. pipiens*, *Mansonia perturbans*, and *Ae. simulans*, with 22 other species reported as potential vectors [45].

In Europe, studies have been conducted on wild-caught mosquitoes and their role in transmitting *Dirofilaria* spp. (Table 23.2). Although many species have been recorded to be positive for *Dirofilaria* spp., *Cx. pipiens* is likely one of the more important vectors throughout Europe. The establishment of new mosquito species in Europe, such *Ae. albopictus*, *Ae. j. japonicus*, and *Ae.* (*Finlaya*) *koreicus*, is a growing concern [81–83]. The rapid distribution of *Ae. albopictus* throughout Europe is concerning due to the potential for inhabiting non-endemic countries, a threat especially with people and pets traveling frequently between endemic and non-endemic regions [9, 84]. Non-endemic countries with few reports of heartworm infections include the United Kingdom, Germany, the Netherlands, Czech Republic, Slovak Republic, Switzerland, Romania, Poland, and regions of France, Spain, and Italy [84]. As new mosquito species are introduced and become established, climate models predict that regions of these northern European countries could see more *Dirofilaria* infections [9]. As globalization and climate warming continues, there is a need to update mosquito-borne disease data.

Bendas and colleagues [85] discussed the gaps in literature and the estimated heartworm prevalence in South American countries, including Argentina, Brazil, Chile, Guiana, Paraguay, Peru, Suriname, Venezuela, and Mexico. Some reported *Dirofilaria* spp. vectors in the genera *Aedes, Anopheles, Culex, and Wyeomyia* have been compiled in Table 23.2, but there is still limited knowledge regarding mosquito vectors in South America and Mexico. On a global scale, much remains to be known about which mosquito species are vectors of *Dirofilaria* and research is needed to address vector competency and seasonality of transmission in many areas throughout the world. The seasonality of transmission is quite different not

**Table 23.2** A list of naturally collected mosquitoes throughout the world (excluding the United States) found positive for *Dirofilaria immitis* and/or *Dirofilaria repens* infection.

| Species | Country | *Dirofilaria* species[a)] | Year of collection |
|---|---|---|---|
| *Aedes aegypti* | French Polynesia | DI | 2003–2004 [46] |
| | Argentina | DI | 2007–2008 [47] |
| | Mexico | DI | 2007 [48] |
| | Russia | DR | 2013–2017 [49] |
| *Aedes albopictus* | Brazil | DI | 1996–1997 [50] |
| | Taiwan | DI | 1997–1998 [51] |
| | Italy | DI | 2000–2002 [44] |
| | Italy | DR | 2002–2003 [52] |
| | Japan | DI | 1985–1987 [53] |
| | Russia | DI | 2013–2017 [49] |
| | France | DI, DR | 2015 [54] |
| *Aedes annulipes* | Moldova | DR | 2010–2015 [55] |
| *Aedes behningi* | Moldova | DI | 2010–2015 [55] |
| *Aedes cantans* | Moldova | DR | 2010–2015 [55] |
| | Russia | DI, DR | 2013–2017 [49] |
| *Aedes caspius* | Moldova | DR | 2010–2015 [55] |
| | Portugal | DI | 2011–2013 [56] |
| | Hungary | DI, DR | 2013 [57] |
| | Romania | DI, DR | 2014 [58] |
| *Aedes cataphylla* | Russia | DI, DR | 2013–2017 [49] |
| *Aedes cinereus* | Russia | DI, DR | 2013–2017 [49] |
| *Aedes communis* | Russia | DI, DR | 2013–2017 [49] |
| *Aedes detritus* | Portugal | DI | 2011–2013 [56] |
| *Aedes flavescens* | Moldova | DR | 2010–2015 [55] |
| *Aedes geniculatus* | Moldova | DR | 2010–2015 [55] |
| | Russia | DI, DR | 2013–2017 [49] |
| *Aedes intudens* | Russia | DI, DR | 2013–2017 [49] |
| *Aedes notoscriptus* | Australia | DI | 1979 [59] |
| *Aedes polynesiensis* | Samoa | DI | 1978–1980 [60] |
| | French Polynesia | DI | 2003–2004 [46] |
| | American Samoa | DI | 2006 [61] |
| *Aedes riparius* | Moldova | DR | 2010–2015 [55] |
| *Aedes samoanus* | Samoa | DI | 1978–1980 [60] |
| *Aedes scapularis* | Brazil | DI | 1995–1996 [62] |
| | Brazil | DI | 1996–1997 [50] |
| | Mexico | DI | 2007 [48] |

(*continued*)

**Table 23.2** (Continued)

| Species | Country | *Dirofilaria* species[a)] | Year of collection |
|---|---|---|---|
| *Aedes (Ochlerotatus) sollicitans* | Mexico | DI | 2007 [63] |
| | Mexico | DI | 2007 [48] |
| *Aedes sticticus* | Moldova | DR | 2010–2015 [55] |
| *Aedes (Ochlerotatus) taeniorhynchus* | Brazil | DI | 1995–1996 [62] |
| | Brazil | DI | 1996–1997 [50] |
| | Mexico | DI | 2007 [63] |
| | Mexico | DI | 2007 [48] |
| *Aedes vexans* | Korea | DI, DR | 2005 [64] |
| | Turkey | DI | 2008–2009 [65] |
| | Czech Republic | DR | 2009–2011 [66] |
| | Serbia | DI, DR | 2013 [67] |
| | Russia | DI, DR | 2013–2017 [49] |
| | Moldova | DR | 2010–2015 [55] |
| | Slovakia | DR | 2013 [68] |
| | Romania | DI, DR | 2014 [58] |
| *Armigeres subalbatus* | Korea | DR | 2005 [64] |
| *Anopheles albimanus* | Mexico | DI | 2007 [48] |
| *Anopheles algeriensis* | Romania | DI | 2014 [58] |
| *Anopheles annulipes* | Australia | DI | 1979 [59] |
| *Anopheles atroparvus* | Portugal | DI | 2011–2013 [56] |
| *Anopheles claviger* | Belarus | DR | 2015 [69] |
| *Anopheles crucians* | Mexico | DI | 2007 [48] |
| *Anopheles daciae* | Germany | DR | 2011–2013 [70] |
| *Anopheles hyrcanus* | Romania | DI, DR | 2014 [58] |
| *Anopheles maculipennis* | Italy | DI | 2000–2002 [71] |
| | Iran | DI | 2005–2006 [72] |
| | Moldova | DI, DR | 2010–2015 [55] |
| | Portugal | DI | 2011–2013 [56] |
| | Romania | DI, DR | 2014 [58] |
| | Germany | DR | 2016 [73] |
| *Anopheles messeae* | Russia | DI, DR | 2013–2017 [49] |
| *Anopheles plumbeus* | Austria | DR | 2015 [74] |
| *Anopheles pseudopictus* | Moldova | DR | 2010–2015 [55] |
| *Anopheles pseudopunctipennis* | Mexico | DI | 2007 [48] |

(*continued*)

**Table 23.2** (Continued)

| **Species** | **Country** | ***Dirofilaria* species[a)]** | **Year of collection** |
|---|---|---|---|
| *Anopheles sinensis* Group (*An. sinensis s. s., An. pullus, An. kleini, An. belenrae, An. lesteri*)[b)] | Korea | DI, DR | 2005 [64] |
| *Anopheles sineroides* | Korea | DI, DR | 2005 [64] |
| *Anopheles triannulatus* | Brazil | DI | 2010 [75] |
| *Coquillettidia richiardii* | Italy | DI | 2000–2002 [71] |
| | Russia | DI, DR | 2013–2017 [49] |
| | Romania | DI, DR | 2014 [58] |
| *Culex annulirostris* | Australia | DI | 1979 [59] |
| *Culex antennatus* | Egypt | DR | 2012–2013 [76] |
| *Culex coronator* | Mexico | DI | 2007 [48] |
| *Culex declarator* | Brazil | DI | 1995–1996 [62] |
| *Culex interrogator* | Mexico | DI | 2007 [63] |
| *Culex modestus* | Moldova | DR | 2010–2015 [55] |
| | Russia | DI, DR | 2013–2017 [49] |
| | Hungary | DI | 2013 [57] |
| *Culex pipiens* | Italy | DI | 1997–1999 [40] |
| | Italy | DI | 2000–2002 [71] |
| | Spain | DI | 2004–2006 [77] |
| | Korea | DI | 2005 [64] |
| | Argentina | DI | 2007–2008 [47] |
| | Turkey | DI | 2008–2009 [65] |
| | Moldova | DI, DR | 2010–2015 [55] |
| | Germany | DI | 2011–2013 [70] |
| | Spain | DI | 2012–2013 [78] |
| | Egypt | DI, DR | 2012–2013 [76] |
| | Portugal | DI | 2011–2013 [56] |
| | Serbia | DI, DR | 2013 [67] |
| | Hungary | DI | 2013 [57] |
| | Russia | DI, DR | 2013–2017 [49] |
| | Romania | DI, DR | 2014 [58] |
| | Belarus | DI | 2015 [69] |
| *Culex pusillus* | Egypt | DI | 2012–2013 [76] |
| *Culex saltanensis* | Brazil | DI | 1995–1996 [62] |

(*continued*)

**Table 23.2** (Continued)

| Species | Country | *Dirofilaria* species[a)] | Year of collection |
|---|---|---|---|
| *Culex theileri* | Portugal | DI | 2002–2003 [79] |
| | Iran | DI | 2005–2006 [72] |
| | Portugal | DI | 2011–2013 [56] |
| *Culex tritaeniorhynchus* | Japan | DI | 1985–1987 [53] |
| *Culex quinquifasciatus* | Australia | DI | 1979 [59] |
| | Brazil | DI | 1995–1996 [62] |
| | Brazil | DI | 1996–1997 [50] |
| | Taiwan | DI | 2005 [64] |
| | Brazil | DI | 2010 [75] |
| | Mexico | DI | 2007 [48] |
| | Mexico | DI | 2016–2017 [80] |
| *Culiseta annulata* | Moldova | DR | 2010–2015 [55] |
| *Culiseta longiareolata* | Moldova | DR | 2010–2015 [55] |
| *Uranotaenia unguiculata* | Moldova | DR | 2010–2015 [55] |
| *Wyeomyia bourrouli* | Brazil | DI | 1995–1996 [62] |

See Ledesma and Harrington [42] for an overview of potential vectors in the United States.
Note: Published studies were only included if wild mosquitoes were positive for *D. immitis* and *D. repens* larvae, regardless of development stage. Studies in which wild mosquitoes were caught and then fed on infected host blood were excluded.
a) DI = *D. immitis*; DR = *D. repens*.
b) Could not be identified to species.

only between regions, but among species and their willingness to feed on specific hosts, including dogs.

## 23.4 Dirofilaria immitis in Mosquitoes

Susceptible female mosquitoes may acquire blood meals from microfilaremic hosts and ingest heartworm microfilariae [19, 86]. The ingested microfilariae enter the mosquito midgut along with the blood meal before migrating into the cells of the Malpighian tubules where they develop into noninfective $L_1$ larvae [19, 87]. The $L_1$ larvae enter the Malpighian tubule lumen (Figure 23.1a) and subsequently molt into $L_2$ and then to $L_3$ larvae (Figure 23.1b) [19]. The non-yet infective $L_3$ larvae then break through the Malpighian tubules and migrate through the hemocoel to the head and labium (Figure 23.2) [19]. If the parasite burden is too heavy, the mosquito may not survive due to impaired Malpighian tubule function [88]. $L_3$ larvae that successfully migrate to the labium are considered to be infective [19] and exhibit positive thermotaxis to facilitate transmission to a mammalian host [42]. The tip

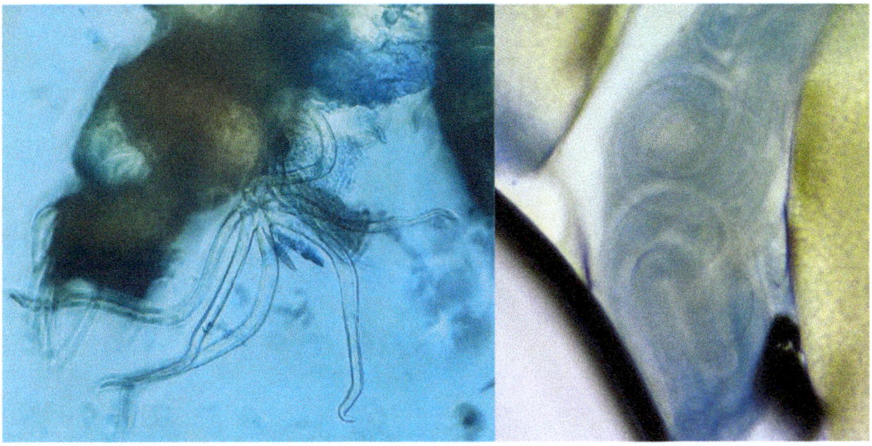

(a) (b)

**Figure 23.1** (a) Malpighian tubules of a mosquito. (b) *D. immitis* $L_3$ larvae in a mosquito Malpighian tubule.

of the mosquito's labium will rupture during a blood meal and infective $L_3$ larvae will enter the host through the punctured epidermis to continue the life cycle in the vertebrate definitive host [19].

With the exceptions of malaria and canine heartworm prophylactics, lymphatic filariasis eradication program measures, and yellow fever vaccination campaigns, no medical interventions have been implemented to prevent mosquito-borne diseases. The commonly available monthly heartworm preventatives for dogs and cats work postinfection by killing $L_3$ and $L_4$ *D. immitis* larvae present in pets at the time of administration to prevent development of adult heartworms [19, 89, 90]. These products have an essential role in preventing heartworm disease in companion animals, but there are concerns regarding owner compliance [91, 92] and lack of efficacy due to resistant strains of *D. immitis* [89, 93, 94]. Therefore, it is vital to incorporate Integrated Mosquito Management (IMM) practices to mitigate mosquitoes through a combination of control strategies.

## 23.5 Integrated Mosquito Management (IMM)

IMM uses an ecosystem-based approach to intervene before mosquitoes become problematic, to be prepared when actions are needed, and in the case of dirofilariasis, minimizing risk of companion animals from adult heartworms. An excellent overview of IMM and its components is provided by the American Mosquito Control Association (AMCA) [95]. The concept of IMM is a long-term strategy to reduce or prevent mosquito abundance, which includes the use of a combination of monitoring and control operations that cannot be accomplished with a single management approach. Given the role of mosquitoes in pathogen transmission to humans, IMM often includes goals of reducing risks of vector-borne diseases to

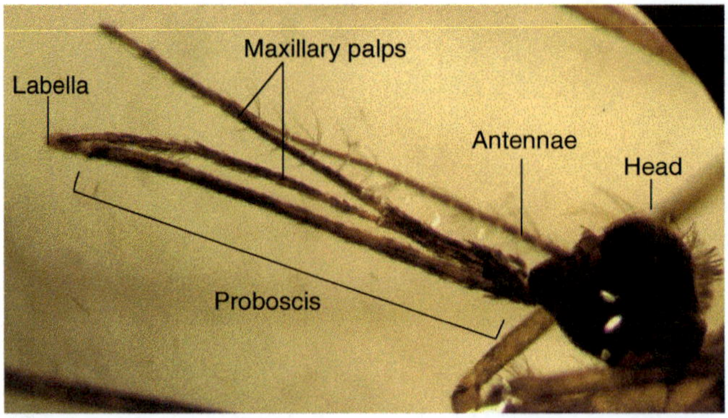


(a)

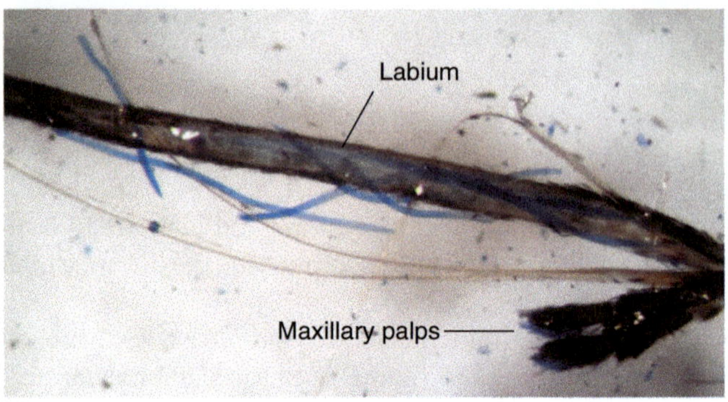


(b)

**Figure 23.2** (a) The head of a female anopheline mosquito with long maxillary palps equal to the length of the proboscis. The proboscis is composed of a labium containing thin stylets, including the mandibles, maxillae, a hypopharynx, and a labrum terminating at the labella. (b) *Aedes aegypti* proboscis dissected to expose *D. immitis* infective $L_3$ larvae (stained with methyl blue).

humans and, by extension, companion animals, while limiting harmful interactions with the environment. It is important to recognize that, due to the varied developmental habitats of mosquitoes, one-size-fits-all solutions are not feasible, and for several important mosquitoes, an area-wide approach, as practiced by mosquito abatement programs, is needed. The major components of IMM (Figure 23.3) are outlined below and include surveillance, physical control, larval source reduction, and adult mosquito control and monitoring for insecticide efficacy and resistance [92]. In this chapter, we focus on the first four components:

1. The first step in all IMM programs is a robust surveillance plan. Mosquito surveillance programs are critical to assess the effectiveness of prevention, avoidance, or suppression tactics. Due to the widespread nature of many mosquito species, these programs are often implemented by mosquito abatement programs executed by local government agencies. Mosquito species identification is an essential component in IMM to recognize the specific invasive and indigenous species that may serve as vectors for nematode transmission or account for

mosquito resistance to insecticides. In community-wide abatement programs, this is done through a wide range of tools not available to or appropriate for the general public. These include carbon dioxide-baited traps, use of sentinel chickens for background virus monitoring, and widespread identification of larger larval developmental sites. For the pet owner, surveillance can best be accomplished by examining containers holding water for developing mosquito larvae and pupae.

2. The next step in a local IMM program includes physical control operations intended to remove all water-holding containers or make them such that they do not hold water (punch holes in bottom) or alter the developmental habitat. Common developmental sites found near residential areas that are often overlooked include rim-less tires, water-filled vases or similar small containers, animal water troughs, and plant holders. Rain gutters should be properly maintained, and debris following home renovation projects should be removed. When immature mosquito populations are found, manual elimination of the immature mosquitoes is required. This should include the emptying of water-holding containers, elimination of debris that may hold water, and, if permanent water-holding containers are present, periodic (2× weekly) scrubbing to remove eggs. Although not technically physical control, the addition of a larvicide is warranted when it is not possible to alter such habitats. Many home improvement and other stores carry such products that are safe to use around pets. If adult mosquitoes become a problem, the next step in IMM is to reduce the impact of heartworm transmission by preventing mosquito bites (see below). On the larger scale, most abatement programs incorporate larval reduction, typically through the use of lower toxicity products that are specifically toxic to mosquito larvae. These products are delivered either by hand or machine to areas known to repeatedly harbor developing mosquitoes.
3. When adult mosquitoes are a concern, reducing their interaction with pets should be attempted. An option is to choose the times of day and amount of time that pets are exposed to mosquitoes outdoors, such as avoiding times when crepuscular vector species feed. This approach is not practical for many animals who spend all of their time outdoors, but consideration of this approach for indoor pets is recommended. Numerous repellent products are available and well studied for human use. Less is known about their use on pets, and it is important to recognize that human-approved products are not necessarily appropriate for pets, and in some cases can make pets sick or even kill them. One challenge with pets is that they will groom topically applied products off themselves, resulting in ingestion of the repellent chemicals. All label instructions should be read and understood before use and, if a product is not labeled for cats or dogs, it should not be used on them. Products are available that have repellency claims for various insects and ticks. Repellency of these products is not universal against all mosquitoes and their effectiveness will vary among individual animals. It must be stressed that use of a repellent product does not replace the use of preventative heartworm medicinal products.

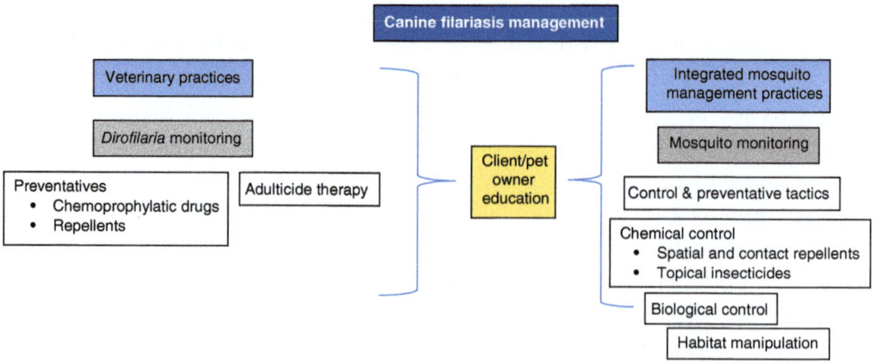


**Figure 23.3** Intervention practices at the veterinary level for canine filariasis with emphasis on Integrated Mosquito Management to mitigate disease transmission.

4. The final step in an IMM program involves using chemicals to kill mosquitoes. Chemical applications can occur in various ways, including spatial or contact repellents for humans, spot-on preventatives for pets, and adulticides or larvicides applied to the environment. Such approaches are most effectively used in mosquito abatement programs, with individual property applications becoming more commonly employed; however, variable efficacy has been reported. Volatile pyrethroids (which also kill mosquitoes), also called spatial repellents, differ from repellents applied to the skin or clothing in that spatial repellents protect a larger space surrounding one or multiple individuals, rather than a body region on a single person. Spatial repellents such as metofluthrin, transfluthrin, and prallethrin, or various botanical extracts, are commercially available as passive emanators, vaporizers, candles, coils, and heated mats for mosquito control in areas with humans and animals [96, 97]. Botanical extracts, including citronella, cedar, lavender, eucalyptus, neem tree oil, peppermint, and lemongrass and lemon eucalyptus oil (para-menthane-3,8-diol), have been shown to deter mosquitoes [97]. The US Environmental Protection Agency (EPA) [98] generated a list of approved spatial and contact repellents for mosquitoes and ticks specific to consumer needs (https://www.epa.gov/insect-repellents/find-repellent-right-you). It should be noted that, as with skin repellents, some spatial repellents should not be used around cats and label directions should be confirmed before use.

Heartworm transmission and survival in vectors can be prevented through delivery of oral (e.g. ivermectin and milbemycin oxime) or topical (e.g. moxidectin and selamectin) anthelmintic drugs in combination with topical insecticides (e.g. imidacloprid–permethrin solution) with repellent effects to prevent blood feeding [99, 100]. However, some mosquito populations have developed resistance to pyrethroid-based insecticides [101], including permethrin, which is commonly used on dogs to control ectoparasites [99]. Additionally, reports of ineffective insecticides have been attributed to mosquito resistance and cross-resistance beyond pyrethroid compounds [99]. A relatively newer non-pyrethroid compound, dinotefuran, is formulated as a topical insecticide (in conjunction with pyriproxyfen and permethrin) to reduce risk of heartworm transmission for dogs because of its low toxicity to animals, fewer reported resistance issues, and efficacy in repelling mosquitoes [99].

Mosquito abatement programs use adulticides and larvicides to suppress mosquito populations on a community-level scale, but mosquitoes may be unaffected by insecticide sprays not due to resistance, but due to applications not delivered to places they rest or oviposit [102, 103]. Many ecological methods for mosquito reduction have been considered due to environmental pollution, insecticide resistance, and public health concerns about chemical insecticide use [104, 105].

Many research-based alternatives to chemical control are discussed below, including using biological control agents against mosquitoes as well as the potential for the use of genetically modified mosquitoes as control agents in population reduction programs. These approaches fall outside of traditional IMM but are worthy of discussion due to their ongoing development and potential use in mosquito abatement.

## 23.6 Biological Control Agents

Entomopathogenic fungi have been investigated as alternatives to chemical insecticides as they are environmentally friendly and minimally toxic to humans and other vertebrates [102] and may reduce the risk of insecticide-resistant mosquito populations [106, 107]. The fungal conidia (i.e. infective spores) attach to the cuticle of adult mosquitoes and produce various proteases that allow penetration to infect the mosquito, followed by release of lethal toxins [107, 108]. Mosquitoes are infected by direct contact with fungal pathogens, which may enable development of entomopathogenic fungi-treated cloths, nets, and sprays to combat mosquito populations [106]. Environments with stagnant water, moisture, moderate temperatures, and sunlight protection are ideal for mosquito development, resting, and possible overwintering [102]. These conditions are also favorable for entomopathogenic fungi, allowing mosquito control in their natural habitats [102]. Common entomopathogenic fungi used as terrestrial mosquito control agents include *Beauveria bassiana* and *Metarhizium anisopliae*, which have effectively reduced *Cx. quinquefasciatus* [102], *Anopheles gambiae*, and *Anopheles stephensi* [106]. Laboratory strains of *Metarhizium brunneum* have been effective against *Ae. aegypti* [108]. In comparison to commercial insecticides with fast-acting mechanisms, pathogenic fungi have a slow-kill effect, resulting in mortality days 3 to 14 posttreatment [102]. The delay in killing mosquitoes may reduce selection for resistance [102, 106]. Further research is necessary to validate entomopathogenic fungi as a biopesticide against specific heartworm vectors. Scholte and colleagues provide a detailed overview of entomopathogenic fungi for mosquito control [109].

The *Aedes* Densonucleosisvirus (AeDNV) is a mosquito-specific virus that is relatively stable in the environment and has potential as a biological control agent [110–112]. This is a less toxic alternative to chemical insecticides as it is not known to infect or replicate in other arthropods or vertebrates [110, 112]. AeDNV infects aquatic mosquito larvae through the anal papillae [110, 111] and disseminates to other tissues with 75% mortality in *Ae. aegypti* [110]. Larvae that do not die following infection continue to develop and eclose as infected adults with reduced lifespan, fecundity, and egg viability and also can vertically transmit AeDNV to progeny [110]. Although entomopathogenic viruses have not been well studied [110, 113], laboratory evidence suggests that AeDNV is effective against species in the *Aedes* and *Culex* genera [110]. In addition, the *Baculovirus* CuniNPV is pathogenic to *Culex nigripalpus* and may be effective as a biological control agent [110, 113]. Viral infections are dose-dependent and maintaining a high viral concentration in the field is difficult, and so more research is necessary to develop persistent infections that significantly reduce wild mosquito populations [110].

Some species of bacteria have been implemented as biological control agents against mosquito vectors. The Gram-positive bacteria, *Bacillus thuringiensis* var. *israelensis* (*Bti*), has been used to selectively target *Ae. aegypti* and *Ae. albopictus* larvae for population control [108, 110]. Spinosad is an environmentally friendly bioinsecticide that kills *Ae. aegypti* and *Anopheles albimanus* larvae and other insect pests in laboratory studies [114]. Although Spinosad is highly toxic to insects, the neurotoxic metabolites that are derived by fermentation of *Saccharopolyspora spinosa* bacteria display low toxicity to mammals, allowing for the potential of this bacterial species for biological control of insects, including mosquitoes [114].

Various aquatic predators of immature mosquitoes, such as fishes, crustaceans, and insects, have potential for mosquito population control. The efficacy of predatory fish such as *Gambusia* for vector control relies on many factors, including their small size, predisposition to ingest large amounts of larvae, affinity for the target insect, and tolerance of high temperatures and pH variations [115]. Haas and Pal [116] give an excellent overview of the variety of fishes that have promise for biological control. Mosquitofish (*Gambusia* spp.) have been used as biological control agents for *Culex tarsalis* [103, 117], *Cx. quinquefasciatus* [102], and various *Aedes* and *Anopheles* spp. [118]. One of the shortcomings in using fish to control mosquitoes is that the mating behaviors of these fish may decline in certain environments (e.g. storm drains), preventing substantial control [102]. As polyphagous predators, mosquitofish also feed on other invertebrates that prey on mosquitoes, which may indirectly benefit mosquito populations by reducing other predator populations [117]. Mosquitoes may also be deterred from laying eggs in areas where predators are prevalent. Gravid female *Culex tarsalis* and *Cx. quinquefasciatus* were deterred from oviposition in the presence of tadpole shrimp (*Triops longicaudatus*) due to water surface agitation [119]. Cyclopoid copepods are also commonly used to control mosquito populations, but Cuthbert and colleagues [120] showed that the freshwater calanoid copepod *Lovenula raynerae* may be more effective, as it is a voracious predator of *Cx. pipiens* larvae, unlike cyclopoid copepods. *Toxorhychites* spp. are mosquitoes that are predatory on container-developing and tree-hole mosquito species, including *Ae. albopictus* and *Aedes triseriatus* [105, 121, 122]. Both adult male and female *Toxorhychites* are non-blood feeding, so they pose no threat of blood-borne pathogen transmission [122].

## 23.7 Future Directions

Many different methods have been investigated to use mosquitoes as biological control agents in attempt to reduce mosquito vector populations. The sterile insect technique (SIT) has been implemented for decades to eradicate agricultural pests, including the tsetse fly *Glossina austeni* and the primary screwworm *Cochliomyia homnivorax* from endemic regions [108]. Using radiation to sterilize male insects to prevent successful reproduction with a wild female has been impractical for mosquito control, because irradiated male mosquitoes generally do not seek mates [108]. Cytoplasmic incompatibility (CI), another control method, utilizes the endosymbiotic bacteria, *Wolbachia*, to suppress mosquito populations. CI results in unviable progeny when *Wolbachia*-infected male mosquitoes, released into the environment, mate with a wild female that is either uninfected or infected with a different strain of *Wolbachia* [108, 123]. The biotechnology company MosquitoMate (www.mosquitomate.com) [124], with EPA approval [125], is using a specific strain of *Wolbachia* to suppress wild *Ae. albopictus* populations in the United States and District of Colombia. Genetically modified (GM) male *Ae. aegypti* mosquitoes were developed by introducing a self-limiting gene that results in premature death of offspring if mated with a wild female mosquito [108]. The biotechnology company Oxitec Ltd. [126] has conducted field trials by releasing GM mosquitoes in Brazil, Panama, Cayman Islands, and the Florida Keys, achieving approximately 90% mosquito suppression. Further research is necessary for these novel technologies, but they show promising results and may reduce transmission of pathogens of human and veterinary importance, potentially including *Dirofilaria* spp., if mosquito vector populations can be successfully suppressed in endemic regions.

Dog heartworm issues are prevalent in many areas of the world. Although many vectors have been properly screened for competence, many others remain to be identified. In other areas of the world, few if any vector incrimination studies have been performed. With a changing climate in many parts of the world, areas once thought to have only seasonal transmission due to vector inactivity during winter months are no longer viewed in that light and this message needs to be communicated to veterinarians and pet owners. Exciting new technologies are becoming available for mosquito abatement; however, these will likely target species of human health interest first. Many important heartworm vectors do not fall in this category, and thus mosquito management and pet owner preventative delivery remain highly important in managing this disease.

## References


1 Nelson, G.S. (1963). *Dipetalonema dracunculoides* (Cobbold, 1870), from the dog in Kenya: with a note on its development in the louse-fly, *Hippobosca longipennis*. *J. Helminthol.* 37 (3): 235–240.

2 Olmeda-García, A.S., Rodríguez-Rodríguez, J.A., and Rojo-Vázquez, F.A. (1993). Experimental transmission of *Dipetalonema dracunculoides* (Cobbold 1870) by *Rhipicephalus sanguineus* (Latreille 1806). *Vet. Parasitol.* 47 (3-4): 339–342.

**3** Nelson, G.S. (1962). *Dipetalonema reconditum* (Grassi, 1889) from the dog with a note on its development in the flea, *Ctenocephalides felis* and the louse, *Heterodoxus spiniger*. *J. Helminthol.* 36 (3): 297–308.

**4** Brianti, E., Gaglio, G., Napoli, E. et al. (2012). New insights into the ecology and biology of *Acanthocheilonema reconditum* (Grassi, 1889) causing canine subcutaneuous filariosis. *Parasitology* 139 (4): 530–536.

**5** Dedkhad, W., Bartholomay, L.C., Christen, B.M. et al. (2018). Effects of cross-mating on susceptibility of synonymous mosquitoes, *Anopheles paraliae* and *Anopheles lesteri* to infection with nocturnally subperoidic *Brugia malayi*. *Acta Trop.* 187: 65–71.

**6** Ewert, A. (1965). Comparative migration of microfilariae and development of *Brugia pahangi* in various mosquitoes. *J. Trop. Med. Hyg.* 14 (2): 254–259.

**7** Otranto, D., Brianti, E., Dantas-Torres, F. et al. (2013). Species diversity of dermal microfilariae of the genus *Cercopithifilaria* infesting dogs in the Mediterranean region. *Parasitology* 140 (1): 99–108.

**8** McKay, T., Bianco, T., and Barnett, S. (2013). Prevalence of *Dirofilaria immitis* (Nematoda: Filarioidea) in mosquitoes from Northeast Arkansas, the United States. *J. Med. Entomol.* 50 (4): 871–878.

**9** Genchi, C., Rinaldi, L., Mortarino, M. et al. (2009). Climate and *Dirofilaria* infection in Europe. *Vet. Parasitol.* 163 (4): 286–292.

**10** Hassan, K.H., Bolcen, S., Kubofcik, J. et al. (2015). Isolation of *Onchocerca lupi* in dogs and black flies, California, USA. *Emerging Infect. Dis.* 21 (5): 789–796.

**11** Otranto, D., Dantas-Torres, F., Brianti, E. et al. (2013). Vector-borne helminths of dogs and humans in Europe. *Parasites Vectors* 6: 16.

**12** Tahir, D., Davoust, B., and Parola, P. (2019). Vector-borne nematode diseases in pets and humans in the Mediterranean Basin: an update. *Vet. World* 12 (10): 1630–1643.

**13** Rodonaja, T.E. (1967). A new nematode, *Oncocerca lupi* n. sp., from *Canis lupus cubanenis*. *Bull. Acad. Sci. Georgian SSR* 45 (3): 715–719.

**14** Otranto, D., Giannelli, A., Trumble, N.S. et al. (2015). Clinical case presentation and review if literature of canine onchoercosis by *Onchocerca lupi* in the United States. *Parasites Vectors* 8: 89.

**15** Cantey, P.T., Weeks, J., Edwards, M. et al. (2016). The emergence of zoonotic *Onchocerca lupi* infection in the United States – A case-series. *Clin. Infect. Dis.* 62 (6): 778–783.

**16** Ambily, V.R., Narayana Pillai, U., Arun, R. et al. (2011). Detection of human filarial parasite *Brugia malayi* in dogs by histochemical staining and molecular techniques. *Vet. Parasitol.* 181: 210–214.

**17** Snowden, K.F. and Hammerberg, B. (1989). The lymphatic pathology of chronic *Brugia pahangi* infection in the dog. *Trans. R. Soc. Trop. Med. Hyg.* 83 (5): 670–678.

**18** Chansiri, K., Tejangkura, T., Kwaosak, P. et al. (2002). PCR based methods for identification of zoonostic *Brugia malayi* microfilariae in domestic cats. *Mol. Cell. Probes* 16 (2): 129–135.

**19** McCall, J.W., Genchi, C., Kramer, L.H. et al. (2008). Heartworm disease in animals and humans. *Adv. Parasitol.* 66: 193–285.

**20** Pampiglione, S., Canestri Trotti, G., and Rivasi, F. (1995). Human dirofilariasis due to *Dirofilaria* (*Nochtiella*) *repens:* a review of world literature. *Parassitologia* 37: 149–193.

**21** Zhong, H.E., Yan, Z., Jones, F., and Brock, C. (2003). Ecological analysis of mosquito light trap collections from west central Florida. *Environ. Entomol.* 4: 807–815.

**22** (AMCA) American Mosquito Control Association. (2018). Mount Laurel, NJ. Web. http://www.mosquito.org (accessed 30 July 2020).

**23** (EPA) United States Environmental Protection Agency. (2017) Mosquito Life Cycle. Washington D.C., USA. Web. https://www.epa.gov/mosquitocontrol/mosquito-life-cycle (accessed 30 July 2020).

**24** Yee, D.A., Allgood, D., and Kneitel, J.M. (2012). Constitutive differences between natural and artificial container mosquito habitats: Vector communities, resources, microorganisms, and habitat parameters. *J. Med. Entomol.* 49 (3): 482–491.

**25** Andreadis, T.G. and Wolfe, R.J. (2010). Evidence for reduction of native mosquitoes with increased expansion of invasive *Ochlerotatus japonicas japonicas* (Diptera: Culicidae) in the Northeastern United States. *J. Med. Entomol.* 47 (1): 43–52.

**26** Wilton, D.P. (1968). Oviposition site selection by the tree-hole mosquito, *Aedes triseriatus* (Say). *J. Med. Entomol.* 5 (2): 189–194.

**27** Yee, D.A., Allgood, D., Kneitel, J.M., and Kuehn, K.A. (2012). Constitutive differences between natural and artificial container mosquito habitats: vector communities, resources, microorgamisms, and habitat parameters. *J. Med. Entomol.* 49 (3): 482–491.

**28** Vajda, E.A., Webb, C.E., Toi, C. et al. (2018). New record of *Wyeomyia mitchellii* (Diptera: Culicidae) on Guam, United States. *J. Med. Entomol.* 55 (2): 447–480.

**29** La Corte, R., Maia, P.C.R., Dolabella, S.S. et al. (2019). Mosquitoes of the Caatinga. III. Larval habitats, frequency, and dynamics of immature and adult stages in a dry Brazilian forest. *J. Med. Entomol.* 56 (1): 120–128.

**30** Peyton, E.L., Campbell, S.R., Candeletti, T.M. et al. (1999). *Aedes (Finlaya) japonicus japonicus* (Theobald), a new introduction into the United States. *J. Am. Mosq. Control Assoc.* 15: 238–241.

**31** Andreadis, T.G., Anderson, J.F., Munstermann, L.E. et al. (2001). Discovery, distribution, and abundance of the newly introduced mosquito *Ochlerotatus japonicas* (Diptera: Culicidae) in Connecticut, USA. *J. Med. Entomol.* 38 (6): 774–779.

**32** Fonseca, D.M., Campbell, S., Crans, W.J. et al. (2001). *Aedes* (*Finlaya*) *japonicas* (Diptera: Culicidae), a newly recognized mosquito in the United States: Analyses of genetic variation in the United States and putative source populations. *J. Med. Entomol.* 38 (2): 135–146.

**33** Fader, J.E. (2016). The importance of interspecific interactions on the present range of the invasive mosquito *Aedes albopictus* (Diptera: Culicidae) and persistence of resident container species in the United States. *J. Med. Entomol.* 38: 135–146.

**34** Gaspar, J.P., McKay, T., and Huss, M.J. (2012). First report of *Aedes japonicas* in natural and artificial habitats in northeastern Arkansas. *J. Am. Mosq. Control Assoc.* 28 (1): 38–42.

**35** Yee, D.A. (2008). Tires as habitats for mosquitoes: a review of studies within the Eastern United States. *J. Med. Entomol.* 45 (4): 581–593.

**36** Lounibos, L.P., Bargielowski, I., Carrasquilla, M.C., and Nishimura, N. (2016). Coexistence of *Aedes aegypti* and *Aedes albopictus* (Diptera: Culicidae) in Peninsular Florida two decades after competitive displacements. *J. Med. Entomol.* 53 (6): 1385–1390.

**37** Shaalan, M.G., Soliman, D.E., Abdou, M.A. et al. (2017). Molecular characterization of vitellogenesis in anautogenous *Culex pipiens pipiens* L. mosquitoes. *Int. J. Mosq. Res.* 4 (2): 5–11.

**38** Land, M.F., Gibson, G., Horwood, J., and Zeil, J. (1999). Fundamental differences in the optical structures of the eyes of nocturnal and diurnal mosquitoes. *J. Comp. Physiol. A* 185: 91–103.

**39** Braack, L., Hunt, R., Koekemoer, L.L. et al. (2015). Biting behavior of African malaria vectors: 1. where do the main vector species bite on the human body? *Parasites Vectors* 8: 2–10.

**40** Capelli, G., di Regalbono, A.F., Simonato, G. et al. (2013). Risk of canine and human exposure to *Dirofilaria immitis* infected mosquitoes in endemic areas of Italy. *Parasites Vectors* 6: 60. https://doi.org/10.1186/1756-3305-6-60.

**41** Kawada, H., Takemura, S., Arikawa, K., and Takagi, M. (2005). Comparative study on nocturnal behavior of *Aedes aegypti* and *Aedes albopictus*. *J. Med. Entomol.* 43 (3): 312–318. https://doi.org/10.1093/jmedent/42.3.312.

**42** Ledesma, N. and Harrington, L. (2011). Mosquito vectors of dog heartworm in the United States: vector status and factors influencing transmission efficiency. *Top. Comp. Anim. Med.* 26 (4): 178–185.

**43** Meisch, M.V., Anderson, A.L., Watson, R.L., and Olson, L. (1982). Mosquito species inhabiting ricefields in five rice growing regions of Arkansas. *J. Am. Mosq. Control Assoc.* 42 (3): 341–346.

**44** Cancrini, G., Frangipane di Regalbono, A., Ricci, I. et al. (2003). *Aedes albopictus* is a natural vector of *Dirofilaria immitis* in Italy. *Vet. Parasitol.* 118 (3-4): 195–202.

**45** McGill, E., Berke, O., Peregrine, A.S., and Weese, J.S. (2019). Epidemiology of canine heartworm (*Dirofilaria immitis*) infection in domestic dogs in Ontario, Canada: geographic distribution, risk factors and effects of climate. *Geospatial Health* 14 (741): 17–24.

**46** Russell, R.C., Webb, C.E., and Davies, N. (2005). *Aedes aegypti* (L.) and *Aedes polynesiensis* Marks (Diptera: Culicidae) in Moorea, French Polynesia: a study of adult population structures and pathogen (*Wuchereria bancrofti* and *Dirofilaria immitis*) infection rates to indicate regional and seasonal epidemiological risk for dengue and filariasis. *J. Med. Entomol.* 42 (6): 1045–1056.

**47** Vezzani, D., Mesplet, M., Eiras, D.F. et al. (2011). PCR detection of *Dirofilaria immitis* in *Aedes aegypti* and *Culex pipiens* from urban temperate Argentina. *Parasitol. Res.* 108 (4): 985–989.

**48** Manrique-Saide, P., Escobedo-Ortegón, J., Bolio-González, M. et al. (2010). Incrimination of the mosquito, *Aedes taeniorhynchus*, as the primary vector of heartworm, *Dirofilaria immitis*, in coastal Yucatan, Mexico. *Med. Vet. Entomol.* 24: 456–460.

**49** Shaikevich, E., Bogacheva, A., and Ganushkina, L. (2019). *Dirofilaria* and *Wolbachia* in mosquitoes (Diptera: Culicidae) in central European Russia and on the Black Sea Coast. *Parasite* 26: https://doi.org/10.1051/parasite/2019002.

**50** Ahid, S.M.M. and Lourenço-de-Oliveria, R. (1999). Mosquitos vetores potenciais de dirofilariose canina na Região Nordeste do Brasil. *Rev. Saúde Pública* 33 (3): 560–565.

**51** Lai, C.H., Tung, K.C., Ooi, H.K., and Wang, J.S. (2008). Susceptibility of mosquitoes in central Taiwan to natural infections of *Dirofilaria immitis*. *Med. Vet. Entomol.* 15 (1): 64–67.

**52** Cancrini, G., Scaramozzino, P., Gabrielli, S. et al. (2007). *Aedes albopictus* and *Culex pipiens* implicated as natural vectors of *Dirofilaria repens* in Central Italy. *J. Med. Entomol.* 44 (6): 1064–1066.

**53** Konishi, E. (1989). *Culex tritaeniorhynchus* and *Aedes albopictus* (Diptera: Culicidae) as natural vectors of *Dirofilaria immitis* (Spriruida: Filariidae) in Miki City, Japan. *J. Med. Entomol.* 26 (4): 294–300.

**54** Tahir, D., Bittar, F., Barré-Cardi, H. et al. (2017). Molecular survey of *Dirofilaria immitis* and *Dirofilaria repens* by new real-time TaqMan® PCR assay in dogs and mosquitoes (Diptera: Culicidae) in Corsica (France). *Vet. Parasitol.* 235: 1–7.

**55** Şuleşco, T., von Thien, H., Toderaş, I. et al. (2016). Circulation of *Dirofilaria repens* and *Dirofilaria immitis* in Moldova. *Parasites Vectors* 9: 627.

**56** Ferreira, C.A.C., Mixao, V.D., Novo, M.T.L.M. et al. (2015). First molecular identification of mosquito vectors of *Dirofilaria immitis* in continental Portugal. *Parasites Vectors* 139 (8): https://doi.org/10.1186/s13071-015-0760-2.

**57** Zittra, C., Kocziha, Z., Pinnyei, S. et al. (2015). Screening blood-fed mosquitoes for the diagnosis of filarioid helminths and avian malaria. *Parasites Vectors* 8: 16.

**58** Tomazatos, A., Cadar, D., Török, E. et al. (2018). Circulation of *Dirofilaria immitis* and *Dirofilaria repens* in the Danube Delta Biosphere Reserve, Romania. *Parasites Vectors* 11: 392.

**59** Russel, R.C. (1985). Report of a field study on mosquito (Diptera: Culicidae) vectors of dog heartworm *Dirofilaria immitis* (Spirurida: Onchocercidae) near Sydney, N. S. W., and the implications for veterinary and public health concern. *Aus. J. Zool.* 33: 461–472.

**60** Samarawickrema, W.A., Kimura, E., Sones, F. et al. (1992). Natural infections of *Dirofilaria immitis* in *Aedes* (*Stegomyia*) *polynesiensis* and *Aedes* (*Finlaya*) *samoanus* and their implication in human health in Samoa. *Trans. R. Soc. Trop. Med. Hyg.* 86: 187–188.

**61** Chambers, E.W., McClintock, S.K., Avery, M.F. et al. (2009). Xenomonitoring if *Wucheria bancrofti* and *Dirofilaria immitis* infections in mosquitoes from American Samoa: trapping considerations and a comparison of polymerase chain reaction assays with dissection. *Am. J. Trop. Med. Hyg.* 80 (5): 774–781.

**62** Labarthe, N., Serrão, M.L., Melo, Y.F. et al. (1998). Potential vectors of *Dirofilaria immitis* (Leidy, 1856) in Itocoatiara, Ocieanic Region of Niterói Municipality, State of Rio de Janeiro, Brazil. *Mem Inst Oswaldo Cruz, Rio de Janeiro* 93 (4): 425–432.

**63** Manrique-Saide, P., Bolio-González, M., Sauri-Arceo, C. et al. (2008). *Ochlerotatus taeniorhynchus*: a probable vector of *Dirofilraia immitis* in coastal areas of Yucatan, Mexico. *J. Med. Entomol.* 45 (1): 169–171.

**64** Lee, S.E., Kim, H.C., Chong, S.T. et al. (2007). Molecular survey of *Dirofilaria immitis* and *Dirofilaria repens* by direct PCR for wild caught mosquitoes in the Republic of Korea. *Vet. Parasitol.* 148: 149–155.

**65** Yildirim, A., Inci, A., Duzlu, O. et al. (2011). *Aedes vexans* and *Culex pipiens* as the potential vectors of *Dirofilaria immitis* in Central Turkey. *Vet. Parasitol.* 178 (1-2): 146–147.

**66** Rudolf, I., Sebesta, O., Mendel, J. et al. (2014). Zoonotic *Dirofilaria repens* (Nematida: Filariodea) in *Aedes vexans* mosquitoes, Czech Republic. *Parasitol. Res.* 113 (12): 4663–4667.

**67** Kurucz, K., Kepner, A., Krtinic, B. et al. (2016). First molecular identification of *Dirofilaria* spp. (Onchoceridae) in mosquitoes from Serbia. *Int. J. Infect. Dis.* 53: 156.

**68** Bocková, E., Iglódyová, A., and Kočišová, A. (2015). Potential mosquito (Diptera: Culicidae) vector of *Dirofilaria repens* and *Dirofilaria immitis* in urban areas of Eastern Slovakia. *Parasitol. Res.* 114: 4487–4492.

**69** Şuleşco, T., Volkova, T., Yashkova, S. et al. (2016). Detection of *Dirofilaria repens* and *Dirofilaria immitis* DNA in mosquitoes from Belarus. *Parasitol. Res.* 115 (9): 3535–3541.

**70** Kronefeld, M., Kampen, H., Sassnau, R., and Werner, D. (2014). Molecular detection of *Dirofilaria immitis*, *Dirofilaria repens* and *Setaria tundra* in mosquitoes from Germany. *Parasites Vectors* 7 (30): https://doi.org/10.1186/1756-3305-7-30.

**71** Cancrini, G., Magi, M., Gabrielli, S. et al. (2006). Natural vectors of Dirofilariasis in rural and urban areas of the Tuscan Region, Central Italy. *J. Med. Entomol.* 43 (3): 547–579.

**72** Azari-Hamidian, S., Yaghoobi-Ershadi, M.R., Javadian, E. et al. (2009). Distribution and ecology of mosquitoes in a focus of dirofilariasis in northwestern Iran, with the first finding of filarial larvae in naturally infected local mosquitoes. *Med. Vet. Entomol.* 23 (2): 111–121.

**73** Heym, E.C., Kampen, H., Krone, O. et al. (2019). Molecular detection of vector-borne pathogens from mosquitoes collected in two zoological gardens in Germany. *Parasitol. Res.* 118: 2097–2105.

**74** Ubleis, S.S., Cuk, C., Nawratil, M. et al. (2018). Xenomonitoring of mosquitoes (Diptera: Culicidae) for the presence of filarioid helminths in Eastern Austria. *Can. J. Infect. Dis. Med. Microbiol.* 2018: https://doi.org/10.1155/2018/9754695.

**75** Ogawa, G.M., Veves da Cruz, E., Nayara Araújo Cunha, P., and Arnha Camargo, L.M. (2013). Canine heartworm disease in Porto Velho: first recor, distribution map and occurrence of positive mosquitoes. *Rev. Bras. Parasitol. Vet. B* 22 (4): 559–564.

**76** Dyab, A.K., Galal, L.A., Mahmoud, A.E., and Mokhtar, Y. (2015). Xenomonitoring of different filarial nematodes using single and multiplex PCR in mosquitoes from Assiut Governorate, Egypt. *Korean J. Parasitol.* 53 (1): 77–83.

**77** Morchón, R., Bargues, M.D., Latorre, J.M. et al. (2008). Haplotype H1 of *Culex pipiens* implicated as natural vector of *Dirofilaria immitis* in an endemic area of Western Spain. *Vector Borne Zoonotic Dis.* 7 (4): https://doi.org/10.1089/vbz.2007.0124.

**78** Bravo-Barriga, D., Parreira, R., Almeida, A.P.G. et al. (2016). *Culex pipiens* as a potential vector for transmission of *Dirofilaria immitis* and other unclassified filarioidea in Southwest Spain. *Vet. Parasitol.* 223: 173–180.

**79** Sanata-Ana, M., Khadem, M., and Capela, R. (2006). Natural infection of *Culex theileri* (Diptera: Culicidae) with *Dirofilaria immitis* (Nematoda: Filarioidae) on Maderia Island, Portugal. *J. Med. Entomol.* 43 (1): 104–106.

**80** Torres-Chable, O.M., Baak-Baak, C.M., Cigarroa-Toledo, N. et al. (2018). Molecular detection of *Dirofilaria immitis* in dogs and mosquitoes in Tabasco, Mexico. *J. Vector Borne Dis.* 55 (2): 151–158.

**81** Knudsen, A.B., Romi, R., and Majori, G. (1996). Occurrence and spread in Italy of *Aedes albopictus*, with implications for its introduction into other parts of Europe. *J. Am. Mosq. Control Assoc.* 12 (12): 177–183.

**82** Kaufman, M.G. and Fonseca, D.M. (2014). Invasion biology of *Aedes japonicus japonicus* (Diptera: Culicidae). *Annu. Rev. Entomol.* 59: 31–49.

**83** Montarsi, F., Ciocchetta, S., Devine, G. et al. (2015). Development of *Dirofilaria immitis* within the mosquito *Aedes (Finlaya) koreicus*, a new invasive species for Europe. *Parasites Vectors* 8: 177.

**84** Genchi, C., Bowman, D., and Drake, J. (2014). Canine heartworm disease (*Dirofilaria immitis*) in Western Europe: survery of veterinary awareness and perceptions. *Parasites Vectors* 7 (206): 2–7.

**85** Bendas, A.J.R., Mendes-De-Almeida, F., Guerrero, J., and Labarthe, N. (2017). Update on *Dirofilaria immitis* epidemiology in South America and Mexico: literature review. *Braz. J. Vet. Res. Anim. Sci.* 54 (4): 319–329.

**86** Brown, H.E., Harrington, L.C., Kaufman, P.E. et al. (2012). Key factors influencing canine heartworm, *Dirofilaria immitis*, in the United States. *Parasites Vectors* 5: 1–9.

**87** Cancrini, G. and Gabrielli, S. (2007). Vectors of Dirofilaria nematodes: biology, behaviour, and host/parasite. In: *Dirofilaria immitis and D. repens in dog and cat and human infections*, 1ste (ed. C. Genchi, R. Rinaldi and C. Guiseppe), 48–58. Via Neova Poggioreale, Naples, Italy: Litografia Vigilante, srl. ISBN: 88-89132-14-0.

**88** Bradley, T.J., Sauerman, D.M. Jr., and Nayar, J.K. (1984). Early cellular responses in the Malpighian tubules of the mosquito *Aedes taeniorhynchus* to infection with *Dirofilaria immitis* (Nematoda). *J. Parasitol.* 70 (1): 82–88.

**89** Bourguinat, C., Keller, K., Bhan, A. et al. (2011). Macrocyclic lactone resistance *in Dirofilaria immitis. Vet. Parasitol.* 181: 388–392.

**90** Knight, D.H. and Lok, J.B. (1998). Seasonality of heartworm infections and implications for chemoprophylaxis. *Clin. Tech. Small Anim. Pract.* 13 (2): 77–82.

**91** (AHS) American Heartworm Society (2013) Heartworm preventative resistance: is it possible? Web. https://www.heartwormsociety.org/newsroom/in-the-news/81-heartworm-preventive-resistance-is-it-possible (accessed 30 July 2020).

**92** Lue, T.W., Pantenburg, D.P., and Crawford, P.M. (2008). Impact of the owner-pet and client-veterinarian bond on the care that pets receive. *JAVMA* 232: 531–540.

**93** Blagburn, B.L., Dillon, A.R., Arther, R.G. et al. (2011). Comparative efficacy of four commercially available heartworm preventative products against the MP3 laboratory strain of *Dirofilaria immitis. Vet. Parasitol.* 176: 189–194.

**94** Blagburn, B.L., Arther, R.G., Dillon, A.L. et al. (2016). Efficacy of four commercially available heartworm preventative products against JYD-34 laboratory strain of *Dirofilaria immitis. Vet. Parasitol.* 191: 1–10.

**95** (AMCA) American Mosquito Control Association. (2017). Best practices for integrated mosquito management: A focused update. Web. https://cdn.ymaws.com/www.mosquito.org/resource/resmgr/docs/Resource_Center/Training_Certification/12.21_amca_guidelines_final_.pdf (accessed 31 July 2020).

**96** Bibbs, C.S. and Kaufman, P.E. (2017). Volatile pyrethroids as potential mosquito abatement tool: a review of pyrethroid-containing spatial repellents. *J. Integr. Pest Manage.* 8 (1): 1–10.

**97** Norris, E.J. and Coats, J.R. (2017). Current and future repellent technologies: the potential of spatial repellents and their place in mosquito-borne disease control. Review. *Int. J. Environ. Res. Public Health* 14 (124): 1–15. https://doi.org/10.3390/ijerph14020124.

**98** (EPA) Environmental Protection Agency (2019) Find the right repellent for you.Web. https://www.epa.gov/insect-repellents/find-repellent-right-you (accessed 31 July 2020).

**99** Franc, M., Genchi, C., Bouhsira, E. et al. (2012). Efficacy of dinotefuran, permethrin, and pyripoxyfen combination spot-on against *Aedes aegypti* mosquitoes on dogs. *Vet. Parasitol.* 189: 333–337.

**100** Hayasaki, M. and Saeki, H. (2009). Inhibition and prevention efficacy against mosquito bloodsucking and *Dirofilaria immitis* infection by administration of topical insecticide. *J. Vet. Med. Sci.* 71 (8): 1049–1052.

**101** Liu, N., Xu, Q., Li, T. et al. (2009). Permethrin resistance and target site insensitivity in the mosquito *Culex quinquefasciatus* in Alabama. *J. Med. Entomol.* 46 (6): 1424–1429.

**102** Popko, D.A., Henke, J.A., Mullens, B.A., and Walton, W.E. (2017). Evaluation of two entomopathogenic fungi for control of *Culex quinquefasciatus* (Diptera: Culicidae) in underground storm drains in Coachella Valley, California, United States. *J. Med. Entomol.* 55 (3): 654–665.

**103** Mains, J.W., Kelly, P.H., Dobson, K.L. et al. (2019). Localized control of *Aedes aegypti* (Diptera: Culicidae) in Miami, FL, via inundative releases of *Wolbachia*-infected male mosquitoes. *J. Med. Entomol.* 56 (5): 1296–1303.

**104** Childs, M.R. (2006). Comparison of Glia topminnow and western mosquitofish as biological control agents of mosquitoes. *West. N. Am. Nat.* 66 (2): 181–190.

**105** Shaalan, E.A. and Canyon, D.V. (2009). Aquatic insect predators and mosquito control. Review. *Trop. Biomed.* 26 (3): 223–261.

**106** Heinig, R.L., Paaijmans, K.P., Hancock, P.A., and Thomas, M.B. (2015). The potential for fungal biopesticides to reduce malaria transmission under diverse environmental conditions. *J. Appl. Ecol.* 52: 1558–1566.

**107** Prado, R., Macedo-Salles, P.A., Duprat, R.C. et al. (2019). Action of *Metarhizium brunneum* (Hypocreales: Clavicipitaceae) against oraganophosphate- and pyrethroid-resistant *Aedes aegypti* (Diptera: Culicidae) and the synergistic effects of phenylthiourea. *J. Med. Entomol.* 57 (2): 454–462.

**108** Benelli, G., Jeffries, C.L., and Walker, T. (2016). Biological control of mosquito vectors: past, present, and future. *Insects* 52 (7): 2–18.

**109** Scholte, E., Knols, B.G.J., Samson, R.A., and Takken, W. (2004). Entomopathegenic fungi for mosquito control: a review. *J. Insect Sci.* 4 (1): 19. doi: 10.1093/jis/4.1.19.

**110** Suchman, E.L., Piper, J., Wise De Valdez, M. et al. (2009). *Aedes aegypti* densonucleosis virus amplifies, spreads, and reduces host populations in laboratory cage studies. *J. Med. Entomol.* 46 (4): 909–918.

**111** Wise De Valdez, M.R., Suchman, E.L., Carlson, J.O., and Black, W.C. (2010). A large scale laboratory cage trial of *Aedes* densonucleosis virus (AeDNV). *J. Med. Entomol.* 47 (3): 392–399.

**112** Liu, P., Xiaocong, L., Gu, J. et al. (2016). Development of non-defective recombinant densovirus vectors for microRNA delivery in the invasive vector mosquito, *Aedes albopictus*. *Sci. Rep.* 6: 20979. https://doi.org/10.1038/srep20979.

**113** Afonso, C.L., Tulman, E.R., Lu, Z. et al. (2001). Genome sequence of baculovirus pathogenic for *Culex nigripalpus*. *J. Virolol.* 75 (22): 11157–11165.

**114** Bond, J.G., Marina, C.F., and Williams, T. (2004). The naturally derived insecticide Spinosad is highly toxic to *Aedes* and *Anopheles* mosquito larvae. *Med. Vet. Entomol.* 18: 50–56.

**115** de Góes Cavalcanti, P., de Paula Júnior, F., Soares Pontes, R.S. et al. (2009). Survival of larvivorous fish used for biological control of *Aedes aegypti* larvae in domestic containers with different chlorine concentrations. *J. Med. Entomol.* 46 (4): 841–844.

**116** Haas, R. and Pal, R. (1984). Mosquito larvivorous fishes. *Bull. Entomol. Soc. Am.* 30 (1): 17–25.

**117** Blaustein, L. and Karban, R. (1990). Indirect effects of the mosquitofish *Gambusia affinis* on the mosquito *Culex tarsalis*. *Limnol. Oceanogr.* 35 (3): 767–771.

**118** Lounibos, L.P., Nishimura, N., and Dewald, L.B. (1992). Predation of *Mansonia* (Diptera: Culicidae) by native mosquitofish in southern Florida. *J. Med. Entomol.* 29 (2): 236–241.

**119** Tietze, N.S. and Mulla, M.S. (1991). Biological control of *Culex* mosquitoes (Diptera: Culicidae) by tadpole shrimp *Triops longicaudatus* (Notostraca: Triopsidae). *J. Med. Entomol.* 28 (1): 24–31.

**120** Cuthbert, R.N., Dalu, T., Wasserman, R.J. et al. (2018). Calanoid copepods: an overlooked tool in the control of disease vector mosquitoes. *J. Med. Entomol.* 55 (6): 1656–1658. https://doi.org/10.1093/jme/tjy132.

**121** Bonnet, D.D. and Mukaida, T. (1957). A copepod predacious on mosquito larvae. *J. Am. Mosq. Control Assoc.* 17 (2): 99–100.

**122** Lounibos, L.P., Escher, R.L., Nishimura, N., and Juliano, S.A. (1997). Long-term dynamics of a predator used for biological control and decoupling from mosquito prey in a subtropical treehole ecosystem. *Oecologia* 111: 189–200.

**123** Lees, R.S., Gilles, J.R., Hendrichs, J. et al. (2015). Back to the future: the sterile insect technique against mosquito disease vectors. *Curr. Opin. Insect Sci.* 10: 156–162.

**124** MosquitoMate, Inc. 2020. Lexington, KY. Web. https://mosquitomate.com/ (accessed 31 July 2020).

**125** (EPA) United States Environmental Protection Agency (2017). *EPA registers the Wolbachia ZAP strain in live male Asian tiger mosquitoes.* Washington D.C., USA Web: https://www.epa.gov/pesticides/epa-registers-wolbachia-zap-strain-live-male-asian-tiger-mosquitoes (accessed 30 July 2020).

**126** Oxitec Ltd. 2020. Milton Park, Abingdon, UK. Web. www.oxitec.com (accessed 31 July 2020).

# 24

# *Wolbachia* Endosymbionts as Treatment Targets for Filarial Diseases

*Marc P. Hübner*[1,2,3,*], *Kenneth Pfarr*[1,3], *and Achim Hoerauf*[1,2,3]

[1] *University Hospital Bonn, Institute for Medical Microbiology, Immunology and Parasitology, Venusberg-Campus 1, Building 63, Bonn 53127, Germany*
[2] *Cluster of Excellence of the University of Bonn, Bonn, Germany*
[3] *German Center for Infection Research (DZIF), Partner site Bonn-Cologne, Bonn, Germany*

**Abstract**

*Wolbachia* bacteria are endosymbionts of many parasitic filarial nematodes. They are present in human pathogenic filarial species causing lymphatic filariasis (*Wuchereria bancrofti*, *Brugia malayi*, and *Brugia timori*), onchocerciasis (*Onchocerca volvulus*), mansonellosis (*Mansonella perstans* and *Mansonella ozzardi*), but not in filariae causing loiasis (*Loa loa*). Furthermore, they are present in a large number of filariae parasitizing animals, including the canine filariae *Dirofilaria immitis* and *Dirofilaria repens*. *Wolbachia* are proposed to provide the filarial nematodes with essential factors like heme, nucleotides, and riboflavin. The filarial species that contain *Wolbachia* depend on the endosymbionts for development, embryogenesis, and survival. Due to this crucial role of *Wolbachia* for filarial transmission and survival, drugs that target the *Wolbachia* endosymbionts provide an alternative treatment option that is being exploited. Over the past decades, it was shown that the anti-wolbachial drug doxycycline is a safe treatment that leads to permanent sterilization of the female adult worms and provides a macrofilaricidal effect, i.e. it kills adult worms, in humans suffering from onchocerciasis and lymphatic filariasis. As microfilariae are slowly cleared over several weeks by anti-wolbachials from the skin or circulation, anti-wolbachial therapies do not trigger inflammatory responses such as those seen after diethylcarbamazine treatment in onchocerciasis and lymphatic filariasis patients. Furthermore, the lack of *Wolbachia* endosymbionts in *L. loa* prevents potential life-threatening serious adverse events that can occur in highly microfilaremic patients treated with diethylcarbamazine or ivermectin. Similarly, doxycycline is included in the recommended combination treatment with ivermectin and melarsomine for canine heartworm disease and was suggested to lessen treatment-associated pathology in comparison to melarsomine treatment alone or in combination with ivermectin.

*Corresponding author.

*Human and Animal Filariases: Landscape, Challenges, and Control*, First Edition.
Edited by Ronald Kaminsky and Timothy G. Geary.

Thus, anti-*Wolbachia* compounds have several advantages compared to drugs currently used for human mass drug administration. Nevertheless, contraindications for doxycycline and its treatment regimen of at least four weeks prevent its broader use for human filariasis. Therefore, novel or improved anti-*Wolbachia* therapies are being discovered and developed, including high-dose rifampicin, moxifloxacin, ABBV-4083 (Flubentylosin), Corallopyronin A, AWZ-1066S, and combination therapies with other antibiotics or direct acting drugs to reduce the treatment time required to clear filarial infection or permanently block the transmission of the disease.

## 24.1 *Wolbachia* Endosymbionts of Filariae

Among the human pathogenic filariae, *Wolbachia* are present in *Brugia malayi*, *B. timori*, and *Wuchereria bancrofti* [1, 2], the causative agents of lymphatic filariasis; in *Onchocerca volvulus* [3–7], the causative agent of onchocerciasis (river blindness); in *Mansonella perstans* [8] and *Mansonella ozzardi* [9]; but are absent from *Loa loa* [10–12] and the closely related nematode *Dracunculus medinensis* [13]. Whether *Mansonella streptocerca* harbors *Wolbachia* endosymbionts is not known.

*Wolbachia* endosymbionts have also been found in filarial nematodes of animals, including filariae parasitizing cattle (*Onchocerca ochengi*, *Onchocerca gutturosa*, and *Onchocerca lienalis*) [3, 14, 15], cats (*Brugia pahangi*) [1, 16], dogs (*Dirofilaria immitis* [16, 17] and *Dirofilaria repens* [18]), rodents (*Litomosoides sigmodontis*) [19], and monkeys (*Dipetolonema gracile*) [20], but are not present in *Onchocerca flexuosa* in deer [21], *Litomosoides yutajensis* in bats [22], or *Acanthocheilonema viteae* in rodents [19].

Phylogenetic analysis showed that the *Wolbachia* present in filarial nematodes belong to the supergroups C, D, and F, with supergroup C containing the *Onchocerca* and *Dirofilaria* species, supergroup D the *Brugia* spp., *W. bancrofti* and *Litomosoides* spp., and supergroup F *Mansonella* spp. [23]. Supergroup J contains *Wolbachia* of *Dipetalonema gracile*. The supergroups A, B, E, F, and I represent *Wolbachia* of arthropods [23].

Intracellular bacteria were initially described in filariae in electron micrographs from *D. immitis*, *B. malayi*, and *O. volvulus* [24, 25]. They were found in large numbers in all examined larval stages, oocytes, and in the lateral chords and hypodermis of both sexes, but never in the male genitals. Based on the morphology, these intracellular bacteria were described as rickettsia-like endobacteria that were transovarially transmitted to the next generation [26]. However, the therapeutic potential of *Wolbachia* in filariae was not immediately recognized, e.g. treatment of Mongolian gerbils (*Meriones unguiculatus*) with tetracycline-protected animals from infection with *B. pahangi*, but the effect was not attributed to the endobacteria [27]. The endobacteria were later identified as close relatives of *Wolbachia* endosymbionts described in arthropods by sequencing the 16S rDNA gene [17]. After identification of the endobacteria as *Wolbachia*, studies using animal filariae confirmed that the endobacteria were vertically transmitted and that tetracycline treatment resulted in loss of *Wolbachia* from *B. pahangi*, *D. immitis*, *O. ochengi*, and *L. sigmodontis* and this correlated with loss of microfilariae from the infected animals and infertility of adult female *L. sigmodontis* [14, 16, 19, 28]. Hoerauf et al. showed that the effect of tetracycline was specific to *Wolbachia* because treatment of animals infected with *A. viteae*, a filarial species without *Wolbachia*, did not result in a loss of microfilariae from the blood nor did the adult female worms become infertile [19]. A first indication that depletion of *Wolbachia* is macrofilaricidal was provided by studies in the *O. ochengi* bovine model in which nodules containing adult worms disappeared after treatment with oxytetracycline, and the loss of nodules was preceded by depletion of *Wolbachia* [29].

The reason(s) for this dramatic effect on filarial nematode embryogenesis/development and survival is not completely understood, but several hypotheses have been proposed based on the annotated genomes of *B. malayi* and their *Wolbachia* endosymbionts (*w*Bm) [30–32]. This analysis revealed that *w*Bm had a highly reduced genome. It encoded complete pathways for the *de novo* synthesis of purines and pyrimidines, but not for the *de novo* synthesis of amino acids, with the exception of *meso*-diaminopimelate, an amino acid required to produce peptidoglycan. The *w*Bm genome also encoded complete pathways for the synthesis of heme, riboflavin, and flavin adenine dinucleotide, all of which are involved in the production of cellular energy, but are not completely encoded in – or are completely lacking from – the nematode genome. Experimental evidence supported the importance of *Wolbachia* heme biosynthesis in nematode biology. Using inhibitors of heme biosynthesis, it was shown that *B. malayi* in culture became immotile and died, an effect that could not be reversed by adding a source of heme to the culture medium [33]. This was supported by microarray results that found a significant upregulation in the expression of mitochondrial genes involved in ATP synthesis, specifically subunits that required heme and riboflavin, after tetracycline-depletion of *Wolbachia* from *L. sigmodontis* infecting *M. unguiculatus* [34]. The same upregulation was not seen in *A. viteae* after four weeks of tetracycline treatment of infected animals [34]. An essential role of *Wolbachia* as a source of riboflavin has also been shown. Treating *B. malayi* nematodes in culture with doxycycline resulted in immobility of adult female worms and a halt in the release of microfilariae. Supplementing the medium with riboflavin (vitamin $B_6$) partially rescued worms treated with doxycycline [35]. Thus, *Wolbachia* appear to be essential to those filarial nematodes harboring the endosymbionts for nucleotides, heme, riboflavin, and flavin adenine dinucleotide and may also be a source of ATP to their filarial hosts during energy intensive stages, e.g. embryogenesis and development.

## 24.2 *Wolbachia* as Targets for Human Filarial Diseases

### 24.2.1 Doxycycline: The First Safe Macrofilaricidal Drug for Human Filariasis

Doxycycline, a second-generation tetracycline antibiotic (Table 24.1), is the first safe macrofilaricidal drug for human filarial disease. Following experimental

**Table 24.1** List of anti-wolbachial compounds and candidates, their effect and *Wolbachia* target.

| Drug (class) | Target in *Wolbachia* | Activity | Comment | Status | References |
|---|---|---|---|---|---|
| Doxycycline (Tetracycline) | Inhibits protein synthesis by binding the 30S ribosomal subunit, thus preventing binding of the aminoacyl-tRNA | Blocks embryogenesis, macrofilaricidal in onchocerciasis and lymphatic filariasis | Extended daily treatments of 3–6 wk are required. Contraindicated in children <8 yr, pregnant and lactating women. | Individual therapy for onchocerciasis, lymphatic filariasis. | [36–39] |
| Minocycline (Tetracycline) | Inhibits protein synthesis by binding the 30S ribosomal subunit, thus preventing binding of the aminoacyl-tRNA | Partial inhibition of embryogenesis | Superiority to doxycycline given at comparable treatment lengths has to be confirmed in clinical trials. | Phase 2 clinical trial with 3-wk treatment has been performed for onchocerciasis. | [40, 41] |
| Rifampicin (Rifamycin) | Binds to prokaryotic DNA-dependent RNA polymerase, precluding subsequent translation | Partial inhibition of embryogenesis (10 mg/kg/d dose) | 10 mg/kg/d were suboptimal for the treatment of human filariasis. High dose (30–35 mg/kg/d) predicted to be 15-fold more potent than doxycycline *in vivo*. | Phase 2 clinical study using 10 mg/kg/d for onchocerciasis has been performed. High dose studies are ongoing. | [42–44] |
| Rifapentine (Rifamycin) | Binds to prokaryotic DNA-dependent RNA polymerase, precluding subsequent translation | Rifapentine plus moxifloxacin combination reduced treatment time required to clear *Wolbachia* in *in vivo* animal experiments | Combination of rifapentine and moxifloxacin is expected to be superior to doxycycline therapy. | Phase 2 clinical trial comparing the combination of rifapentine plus moxifloxacin in onchocerciasis is ongoing. | [45] |

| | | | | | |
|---|---|---|---|---|---|
| Moxifloxacin (Quinolone) | Inhibitor of bacterial DNA gyrase, topoisomerase II and IV, preventing DNA and RNA synthesis | Rifapentine plus moxifloxacin combination reduced treatment time required to clear *Wolbachia* in *in vivo* animal experiments | Combination of rifapentine and moxifloxacin is expected to be superior to doxycycline therapy. | Phase 2 clinical trial comparing the combination of rifapentine plus moxifloxacin in onchocerciasis is ongoing. | [45] |
| ABBV-4083 (Flubentylosin) (Macrolide) | Inhibits protein synthesis by biding to the 50S subunit of the bacterial ribosome | Animal studies suggest inhibition of embryogenesis and clearance of microfilariae | Superiority to doxycycline shown in animal studies | Phase 1 clinical study completed. Phase 2 clinical trial in onchocerciasis patients is ongoing. | [46, 47] |
| AWZ-1066S (Azaquinazoline) | Inhibition of aspects of protein synthesis | Animal studies suggest inhibition of embryogenesis and clearance of microfilariae | Superiority to doxycycline and minocycline shown in animal studies | Phase 1 clinical study is ongoing. | [48] |
| AN11251 (Pleuromutilin) | Inhibits protein synthesis by binding to the 50S subunit of the bacterial ribosome | *Wolbachia* depletion | Experimental animal studies suggest at least equal *Wolbachia* depletion as doxycycline | Preclinical candidate | [49, 112] |
| Corallopyronin A (α-Pyrone) | Noncompetitive inhibitor of bacterial DNA-dependent RNA polymerase | Animal studies have shown inhibition of embryogenesis, clearance of microfilariae and macrofilaricidal efficacy | Animal studies suggest superiority to doxycycline | Preclinical development, phase 1 clinical study scheduled for 2024. | [50, 119] |

confirmation in animal models that doxycycline treatment depletes *Wolbachia* endosymbionts of filariae, inhibits embryogenesis, and leads, over time, to the clearance of microfilariae and adult worms [14, 16, 19, 51], Hoerauf et al. demonstrated in onchocerciasis patients that doxycycline given at 100 mg daily for six weeks depletes *Wolbachia* bacteria from *O. volvulus*, inhibits embryogenesis, leads to long-term amicrofilaridermia, and is macrofilaricidal [52, 53]. Subsequent clinical trials in bancroftian filariasis patients demonstrated that doxycycline is also effective, leading to a 96% *Wolbachia* reduction and a 99% reduction in microfilaremia one year after treatment with 200 mg doxycycline for six weeks and complete amicrofilaremia when ivermectin, in addition, is given as a single dose four months after doxycycline was administered [54]. Following proof of a safe macrofilaricidal treatment with doxycycline, several clinical studies in onchocerciasis and lymphatic filariasis patients were performed to elaborate the optimal treatment duration and dose and to identify the kinetics of efficacy.

With regard to onchocerciasis, it was shown that following a six-week doxycycline regimen (100 mg daily), *Wolbachia* gradually decline from month 2 to 6 following treatment, with inhibition of embryogenesis by six months, and decline of microfilariae by 11 months after treatment start. These effects were maintained for 18 months, but no macrofilaricidal efficacy was observed at that time point [55]. If ivermectin was added to the treatment, microfilariae quickly declined (as expected from the mode of action of ivermectin), but microfilariae loads did not recover, in contrast to what is observed with ivermectin treatment alone. In accordance with the histological findings from onchocercomata removed after doxycycline treatment and *Wolbachia* depletion, adult female worms became infertile. Inhibition of embryogenesis was not improved if doxycycline was given for six weeks at 200 mg/d compared to 100 mg/d; at 20 months after treatment, both regimens led to only 1.2% of onchocercomata with embryogenesis in female adult worms compared to 28.8% in the placebo control, with complete *Wolbachia* absence in 96% of treated female worms [56] (https://doi.org/10.1186/ISRCTN68861628). Shortening the 200 mg/d doxycycline treatment from six to four or two weeks demonstrated that at least four weeks of doxycycline treatment are required in onchocerciasis patients. Six and four weeks of doxycycline treatment reduced microfilarial loads by 18 months after treatment start, with six weeks being, by trend, more efficacious than four weeks, and two weeks resulting in no significant reduction [57]. In accordance, two weeks of 200 mg/d doxycycline therapy failed to deplete *Wolbachia* [57]. Comparison of the six and four week 200 mg/d doxycycline treatment in onchocerciasis patients in a randomized, placebo-controlled trial, in which ivermectin was additionally given as single dose six months after initiation of doxycycline treatment, demonstrated that the six-week therapy was macrofilaricidal, leading to the death of ~60% of the female adult worms at 20 and 27 months after treatment compared to ~50% in the four-week group [58]. The actual macrofilaricidal efficacy may have been even underestimated, as reinfections occur after doxycycline treatment in areas of ongoing transmission [59]. The six-week doxycycline therapy was also more efficacious than the four-week therapy in maintaining the inhibition of embryogenesis and reducing microfilariae loads [58]. Reduction of doxycycline treatment time to five weeks at 100 mg/d was also effective for onchocerciasis, leading to the death of 49% of female adult worms, depletion of *Wolbachia* and complete inhibition of embryogenesis two years after treatment [60]. Importantly, a community trial including 12 936 subjects and a second placebo-controlled study demonstrated that large-scale six-week doxycycline therapy is feasible and safe in areas co-endemic for loiasis [61, 62]. Thus, doxycycline treatment overcomes previous restrictions of MDA treatment in areas of loiasis co-endemicity, where DEC and ivermectin cannot be given in an MDA approach due to the risk of the development of life-threatening serious adverse events in *L. loa* patients with high microfilariae loads [36, 63–65]. The placebo-controlled study in onchocerciasis patients further compared the efficacy of 200 mg doxycycline given daily for six weeks with or without additional ivermectin (150 μg) treatment given at four months. In concordance with the previous studies, combined treatment with ivermectin led to amicrofilaridermia in 89% of the patients, compared to 67% of the doxycycline and 21% of the ivermectin only-treated patients [62]. Both doxycycline treatments completely cleared *Wolbachia* and embryonic stages within the female filariae and mediated a comparable macrofilaricidal effect [62]. Importantly, microfilariae with reduced *Wolbachia* levels from doxycycline-treated patients are retarded in their development into infective stage larvae (L3) in the *Simulium* vector, indicating that the transmission of onchocerciasis may be inhibited by anti-*Wolbachia* therapy [66].

Four years after the 6-week doxycycline treatment in the large-scale trial, patients previously treated with doxycycline had a significantly lower prevalence of microfilaridermia despite ongoing transmission and subsequent one to two rounds of annual ivermectin treatments in this area [67]. Similarly, meta-analytic modeling that included data from the onchocerciasis trials with doxycycline given at 100 and 200 mg for four, five, or six weeks showed that the life span of adult *O. volvulus* is reduced by 70–80%, from 10–15 years to 2–3 years [68]. Importantly, in an area of endemic onchocerciasis with persistent microfilaridermia despite multiple rounds of ivermectin MDA treatments, i.e. suboptimal ivermectin responses, doxycycline therapy (100 mg daily for six weeks) followed by ivermectin treatment at 3 and 12 months was also shown to be efficacious [69]. Comparable to the previous studies, doxycycline had cleared the *Wolbachia* in 95% of the surgically recovered female worms by 20 months after treatment. No microfilariae were detected in any nodule taken from doxycycline-treated patients (compared to 34.1% in patients treated only with ivermectin), 97% of the patients had no detectable microfilaridermia and a significant macrofilaricidal effect was observed [69], demonstrating that doxycycline therapy is also efficacious in suboptimal ivermectin responders. These data indicate that four to six weeks of doxycycline therapy provide long-term inhibition of embryogenesis, thereby leading to amicrofilaridermia over time and slow death of adult filariae. The slow clearance of microfilariae due to inhibition of embryogenesis and a lack of a direct microfilaricidal efficacy in combination with the slow-acting macrofilaricidal efficacy has several advantages, as it prevents serious adverse events that are observed, e.g. during the clearance of *O. volvulus* microfilariae by DEC [70, 71] and is safe to administer in areas co-endemic for loiasis [61, 62].

Following the initial proof that anti-wolbachial therapy with doxycycline is also effective for lymphatic filariasis [54], a randomized, placebo-controlled study was performed for bancroftian filariasis and demonstrated that treatment with 200 mg doxycycline daily for eight weeks almost completely cleared microfilariae 8–14 months following treatment, with only 22% of the treated patients having remaining viable worms. None of the doxycycline patients had serious adverse events [72]. Giving bancroftian filariasis patients doxycycline 200 mg daily for a shorter period of six weeks followed by a single administration of ivermectin (150–200 μg/kg) and albendazole (400 mg) four weeks later also resulted in a significant reduction in *Wolbachia*, microfilarial loads and adult worm survival at 24 months after treatment compared to the placebo control [73]. A further reduction of doxycycline therapy with 200 mg daily for four weeks followed by a single ivermectin dose after four months confirmed that this regimen is also effective, leading to a significant reduction of *Wolbachia*, microfilariae levels and the presence of only 20% viable adult worms at 24 months [74]. New data from another placebo-controlled trial will show that reduction of the doxycycline regimen to four to five weeks of 100 mg/d is equivalent (not inferior) to the 200 mg/d regimens, reaching 86% macrofilaricidal efficacy at 18 months compared to 76% with the previous gold standard (ISRCTN15216778) [75]. Similarly, reducing the doxycycline dose to 100 mg daily for six weeks followed by a single administration of DEC (6 mg/kg) and albendazole (400 mg) after four months in *B. malayi* patients reduced the microfilariae count by 77% after one year in the group receiving doxycycline alone and by 88% in patients receiving doxycycline plus the other drugs [76]. In contrast, microfilariae reductions of 26.7% were observed in patients who only received DEC and albendazole [76]. Importantly, prior treatment with doxycycline also reduced the occurrence of serious adverse events associated with DEC plus albendazole treatment [76], which is expected to be due to the doxycycline-induced depletion of *Wolbachia*, preventing inflammatory responses that otherwise occur when *Wolbachia* are released from dying microfilariae [77]. Further shortening of the doxycycline regimen with 200 mg/d for three weeks or 10 days followed by a single dose of DEC four months later showed that three weeks of doxycycline treatment also leads to the complete ablation of microfilaremia and a significant reduction of worm nests in the scrotums of *W. bancrofti* patients after one year, whereas the 10-day doxycycline-treated group had results comparable to the placebo control [78].

With regard to infections with *M. perstans*, six weeks of 200 mg/d doxycycline treatment was effective in an open-label clinical trial in Ghana, depleting *Wolbachia* from microfilariae by >1 log, and resulting in a complete clearance of microfilariae in the majority of patients two years after treatment start [79].

Thus, the filariae causing lymphatic filariasis, *Brugia* spp. and *W. bancrofti*, are more susceptible to treatment with doxycycline compared to *O. volvulus*, with three to four weeks of doxycycline therapy leading to long-term reduction of microfilaremia and adult worm death. Furthermore, depletion of *Wolbachia* by doxycycline therapy reduced the incidence of serious adverse events that occur with

DEC. Doxycycline-containing regimens are frequently administered by doctors in outpatient clinics, not only in endemic countries with health systems that provide individual care for lymphatic filariasis but also in Europe and the US. Most recently, Brazil and the Bolivarian Republic of Venezuela began using doxycycline treatment as end-game strategy to clear remaining onchocerciasis disease foci [80].

#### 24.2.1.1 Lymphedema Treatment with Doxycycline Is Superior to Simple Hygiene

In addition to the macrofilaricidal efficacy and long-term sterilization achieved by doxycycline therapy in lymphatic filariasis patients, this treatment also ameliorates the disease manifestations of lymphatic filariasis [73]. *W. bancrofti* lymphedema patients treated for six weeks with 200 mg doxycycline daily followed by a single administration of ivermectin and albendazole at four months had significantly lower plasma levels of vascular endothelial growth factor VEGF-C and its soluble receptor sVEGFR-3, significantly lower mean stage of lymphedema at 12 months, and by 24 months significantly smaller dilatation of the supratesticular lymphatic vessels [73]. Furthermore, doxycycline treatment reduced pro-inflammatory cytokines such as TNF that are associated with the presence of *Wolbachia* and microfilariae [81]. Significant reduction in lymph vessel dilatation was also achieved with three weeks of 200 mg/d doxycycline therapy followed by a single dose of DEC four months later [78]. Results from a randomized placebo controlled clinical trial confirmed that patients with mild to moderate lymphedema pathology (lymphedema scores of two to three according to the scoring system established by Dreyer et al. [82]) had significant reductions in lymphedema pathology after one and two years and fewer dermatolymphangioadenitis attacks in this time frame after being administered doxycycline daily at 200 mg for six weeks in addition to standard hygiene morbidity treatment [37]. Importantly, the beneficial effect of doxycycline therapy on lymphedema was independent of the presence of viable adult worms [37]. Nevertheless, not all factors leading to lymphedema are related to *Wolbachia* endosymbionts of filariae, as pro-angiogenic factors such as VEGF-A, VEGF-C, VEGF-D, and angiopoietins were increased in bancroftian lymphedema patients as well as in patients with subclinical microfilaremia compared to endemic normal controls, but doxycycline treatment failed to change their levels 1-year after treatment, indicating that the filariae *per se* also contribute to pathology [83].

These results indicate that, compared to the standard lymphedema treatment including regular cleaning with soap and water, nail and skin care, topical antibiotics or antifungals, exercise and shoes, doxycycline therapy improve or halts the progression of the lymphedema.

In an ongoing multi-center clinical trial in Ghana (ISRCTN14042737), Mali (NCT02927496), Sri Lanka (NCT02929134), India (NCT02929121), and Tanzania (ISRCTN65756724), the beneficial effects on lymphedema using 200 mg doxycycline given for six weeks are being compared to a reduced doxycycline dose of 100 mg given for six weeks and the standard lymphedema treatment in the absence of doxycycline therapy. The goal of this multi-center trial is to provide evidence to support the implementation of doxycycline therapy into the current treatment programs for morbidity control, a task that has not had as great success as MDA, but which is necessary to improve the lives of millions with lymphedema or hydrocele.

### 24.2.2 Recommendations for Doxycycline Therapy for Lymphatic Filariasis and Onchocerciasis

Based on the results described above, it is recommended that, for onchocerciasis, doxycycline is given daily for four weeks at 200 mg or five weeks at 100 mg to achieve permanent sterilization of female adult worms and permanent amicrofilaridermia. For macrofilaricidal efficacy in onchocerciasis, 200 mg doxycycline should be given daily for six weeks. In lymphatic filariasis, doxycycline is macrofilaricidal at 100 mg given daily for four weeks, and reduces lymphedema and hydrocele pathology following six weeks of daily 200 mg doxycycline. For lymphedema patients, it is recommended to repeat the doxycycline therapy yearly or every other year [37].

### 24.2.3 Limitations

Although doxycycline is the first safe macrofilaricidal treatment for lymphatic filariasis and onchocerciasis, limitations are evident. The treatment duration of at least four weeks of daily doxycycline to achieve permanent sterilization in onchocerciasis or macrofilaricidal efficacy in lymphatic filariasis is an option only for individual therapy, not for mass drug administration. Furthermore, doxycycline forms a stable calcium complex in any bone-forming tissue, which may affect bone growth and lead to teeth discoloration, and is therefore contraindicated for use in children below the age of 8 or breast-feeding women. In addition, doxycycline crosses the placenta and has been shown to have toxic effects on the developing fetus and therefore cannot given to pregnant women.

## 24.3 *Wolbachia* as Targets for *D. immitis* Infections of Dogs

It was demonstrated by Bandi et al. in 1999 that after the end of a regimen of 30 days of 30 mg/kg doxycycline in naturally infected dogs, filarial embryogenesis was inhibited compared to control animals [16]. This was associated with the restriction of *Wolbachia* to the caudal end containing the ovary of the adult filariae in the treated animals, whereas *Wolbachia* were present throughout the reproductive tract of the filariae recovered from the control animals. As necropsies were performed immediately after treatment end, *D. immitis* adult worms showed no impaired motility and microfilariae levels were comparable to the counts before treatment started [16], which we now know is due to the fact that inhibition of embryogenesis precedes the clearance of microfilariae and clearance of adult worms by several months. Twice-daily administration of 10 mg/kg doxycycline for 30 days to experimentally infected dogs improved efficacy and led to a gradual decline of microfilaremia that was evident at the end of treatment and continued until necropsy 12–13 months after treatment started [84]. This doxycycline treatment regimen reduced the adult worm recovery in comparison to the control animals and remaining adult filariae were moribund [84]. Furthermore, L3 larvae isolated from mosquitoes that fed on doxycycline-treated dogs 73–77 and 161–164 days after treatment did not establish infections in dogs [84], demonstrating that doxycycline treatment prevents the transmission of the disease long-term.

Experimental combination studies using beagles transplanted with adult *D. immitis* were performed to compare the efficacy of treatments at weeks 0–6, 10–12, 16–18, 22–26, and 28–34 with ivermectin (6 μg/kg/wk orally), doxycycline (10 mg/kg/d orally), and a combination of both treatments [85]. The doxycycline/ivermectin combination therapy cleared microfilaremia by 12 weeks after treatment, whereas most dogs that received doxycycline or ivermectin alone had reduced but not eliminated microfilariae at the end of the study 36 weeks after treatment [85]. The combination therapy completely inhibited embryogenesis and had superior macrofilaricidal efficacy compared to the single drug treatments [85]. In contrast, ivermectin treatment alone cleared microfilaremia within days, but only led to a temporary inhibition of embryogenesis followed by a rebound of microfilaremia. Doxycycline treatment, on the other hand, does not directly affect microfilariae; therefore, the microfilariae clearance due to doxycycline-inhibited embryogenesis occurs more slowly, i.e. according to the half-life of microfilariae. Results from this study were confirmed in a subsequent study which additionally demonstrated that L3 larvae obtained from mosquitoes that fed on doxycycline-treated dogs were not infective [86].

Furthermore, it was shown that combinations with doxycycline can reduce the occurrence of lung and arterial pathology caused by the registered adulticidal drug melarsomine dihydrochloride. The combination of ivermectin (6 μg/kg monthly) with doxycycline (20 mg/kg daily for 30 days) and melarsomine dihydrochloride (2.5 mg/kg intramuscularly at week 12, followed by two injections one month later that were 24 hours apart) as well as the combination treatment of doxycycline and melarsomine led to fewer lung lesions compared to dogs that received melarsomine alone, and the ivermectin and doxycycline combination resulted in fewer severe arterial lesions [87]. This beneficial impact of the combination of doxycycline, ivermectin, and melarsomine was further shown in an analysis from medical reports of dogs that were treated either with 2–3 doses of ivermectin before melarsomine treatment and received either four weeks with 10 mg/kg twice-daily doxycycline or not [88]. No heartworm-caused deaths were reported for the dogs receiving doxycycline before melarsomine treatment and those dogs had fewer respiratory conditions compared to the group that did not receive doxycycline [88]. As an adulticidal treatment for *D. immitis*, the American Heartworm Society recommends treatment with the standard monthly ivermectin (6 μg/kg) treatment, plus one round of doxycycline (10 mg/kg twice-daily) given for four weeks, and treatment with melarsomine dihydrochloride (2.5 mg/kg) intramuscularly at day 60, 90, and 91. This regimen reduces microfilariae loads, leads to adult worm death, and prevents the aggravation of pulmonary damage [89].

To identify novel combination therapies for dirofilariasis, combinations of doxycycline (10 mg/kg/d) for 40 days followed by a 7-day treatment with the saponin acaciaside (10 mg/kg/d) were tested [90]. This combination led to almost complete clearance of microfilariae, superior to the single drug administrations, that was maintained for four months until the end of the study [90]. In addition, given the greater potency of minocycline versus doxycycline in animal models of filariasis [91], and the increasing costs of doxycycline compared to minocycline, minocycline was also considered for heartworm disease [40, 92]. The comparison of 28-day treatment of doxycycline or minocycline given at 10 or 5 mg/kg twice-daily in addition to a six-month treatment with ivermectin/pyrantel and melarsomine dihydrochloride (2.5 mg/kg) intramuscularly at days 60, 90, and 91 showed that all dogs treated with 10 mg/kg doxycycline were *Wolbachia* negative four weeks after doxycycline treatment. In comparison, 2/8 dogs treated with 5 mg/kg doxycycline and 10 mg/kg minocycline and 3/8 dogs treated with 5 mg/kg minocycline remained *Wolbachia* positive [92]. Gastrointestinal problems were more common in dogs treated with 10 mg/kg doxycycline or minocycline compared to the 5 mg/kg doxycycline or minocycline groups [92]. Based on those observations, it is recommended to use the treatment regimen of the American Heartworm Society with ivermectin in combination with four weeks of doxycycline (10 mg/kg twice-daily) and the three-dose melarsomine treatment and, in the case of gastrointestinal side effects, reduce the doxycycline dose to 5 mg/kg [92].

## 24.4 Identification of Anti-wolbachials with an Improved Profile

Based on the essential role of *Wolbachia* endosymbionts for filarial development, transmission, and survival, but the limitations of doxycycline, more potent anti-wolbachials are required. A recently developed high throughput *in vitro* screening system using a *Wolbachia* insect cell line allowed screening for compounds with potent anti-wolbachial activity [93, 94]. Rodent models infected with diverse filarial nematodes [95–98] were subsequently used for *in vivo* pre-clinical screening. Based on those results, several novel treatments are currently in clinical trials for filariasis or are under consideration for clinical trials.

### 24.4.1 Clinical Trials

#### 24.4.1.1 Rifampicin

In the search for antibiotics with improved efficacy against *Wolbachia* endosymbionts of filariae and that overcome some of the limitations of doxycycline, rifampicin is of interest (Table 24.1), as it can be administered to children. Rifampicin was initially tested in the *L. sigmodontis* mouse model and shown to deplete *Wolbachia* endosymbionts and reduce filarial development and adult worm survival [99]. A subsequent human clinical study tested two and four weeks of 10 mg/kg/d rifampicin in onchocerciasis patients in Ghana and compared its efficacy against six-week doxycycline treatment (100 mg/d) as well as untreated patients [42]. Eighteen months after the four-week rifampicin treatment, filarial embryogenesis was reduced and viable female *O. volvulus* filariae contained significantly fewer *Wolbachia* endosymbionts compared to untreated patients [42]. Nevertheless, complete depletion of the *Wolbachia* endosymbionts was not achieved 18 months after the two and four weeks of 10 mg/kg/d rifampicin treatment, and 78–100% of the male and female adult worms were viable [42]. Compared to the six-week doxycycline regimen, 10 mg/kg/d rifampicin was inferior in *Wolbachia* depletion [42]. The reason for this inferior performance may lie in suboptimal dosing. In subsequent preclinical dose-escalation studies in animal models using *B. malayi* and *O. ochengi*, PK-PD analysis showed that 10 mg/kg/d rifampicin, which is the bioequivalent dose of the 600 mg/d dose used in humans, is suboptimal and a bioequivalent human high dose of 30–35 mg/kg/d (approximately 1.5–2 g/d in humans) achieved >90% *Wolbachia* reduction within 7 and 14 days in those models, respectively [43]. Thus, high-dose rifampicin has the potential to require only a one to two-week duration of treatment for human filariasis [43]. Initial clinical studies using 20 and 35 mg/kg rifampicin for tuberculosis treatment were shown not to cause increased toxicity in comparison to the standard 10 mg/kg dose [100–102]. In addition, shortened treatment durations of one to two weeks are not expected to lead to the development of *Mycobacterium tuberculosis* resistance to rifampicin [103]. Based on these findings, additional phase 2 clinical studies using high-dose rifampicin treatment are ongoing (ISRCTN38954299).

#### 24.4.1.2 Minocycline

Minocycline (Table 24.1) was shown in several animal models of filariasis to be active against *Wolbachia* endosymbionts and to be more quickly acting than doxycycline [45, 46, 91, 96]. A subsequent human field trial compared the efficacy of three weeks 200 mg/d minocycline with three and four weeks of doxycycline at 200 mg/d in onchocerciasis patients in Ghana [41]. Six months after treatment, *Wolbachia* were absent in 72.7% of female adult worms from minocycline-treated patients, 64.1% of three-week doxycycline-treated patients and 98.8% of four-week doxycycline-treated patients [41]. Thus, in line with the animal models, *Wolbachia* depletion was, by trend, superior after three-week minocycline treatment in comparison to doxycycline given for the same duration, but inferior to the four-week doxycycline gold standard treatment regimen [41]. Future studies in a fully randomized clinical trial are required to confirm the superiority of three-week minocycline therapy in comparison to three weeks of doxycycline treatment, but a short-term regimen of two weeks or less for onchocerciasis seems not to be achievable with a minocycline monotherapy.

#### 24.4.1.3 Combinations of Anti-wolbachials

In addition to the identification of novel or repurposed drugs with an improved efficacy against the *Wolbachia* endosymbionts of filariae, several experimental and clinical studies indicated that combinations of anti-*Wolbachia* compounds with albendazole or additional anti-*Wolbachia* compounds may allow shorter treatment regimens.

#### 24.4.1.4 Anti-wolbachials Plus Albendazole

Albendazole is an anthelmintic benzimidazole used for mass drug administration with DEC or ivermectin for the treatment of lymphatic filariasis and onchocerciasis, respectively. Given as a monotherapy semiannually, albendazole was reported to reduce peripheral blood microfilaremia and *W. bancrofti* antigenemia in lymphatic filariasis patients [104] and prolonged treatment for 3–10 days was shown to partially inhibit embryogenesis and reduced microfilariae levels [105, 106]. However, a recent Cochrane study analyzing 13 randomized clinical trials with lymphatic filariasis patients identified little to no impact of albendazole treatment alone or in combination with the microfilaricidal drugs DEC or ivermectin on microfilariae prevalence or adult worm clearance [107]. In accordance, a single dose of 800 mg albendazole in combination with 200 μg/kg ivermectin did not improve the sterilization or killing of adult *O. volvulus* worms or lead to prolonged clearance of microfilariae compared to ivermectin alone, irrespective of a semiannual or annual treatment [108].

In the *B. malayi* gerbil model, albendazole has a synergistic effect on *Wolbachia* depletion as well as microfilariae clearance in combination with antibiotics of the tetracycline (minocycline) and rifamycin (rifampicin) classes, although albendazole on its own has no anti-*Wolbachia* efficacy [109]. Interestingly, the synergistic efficacy of anti-*Wolbachia* candidates and albendazole was more prominent in female adult worms, which led to the conclusion that the combination acts on the germline *Wolbachia* population [109]. Albendazole, as a β-tubulin inhibitor, may interfere with the dividing *Wolbachia* bacteria and enhance the efficacy of anti-wolbachials [109]. These preclinical results indicate that combination therapies of anti-*Wolbachia* compounds with improved potency and albendazole may allow regimens of seven or fewer treatment days [109].

An initial pilot study testing three-week 200 mg/d doxycycline with albendazole given at 800 mg/d for the final three days of therapy demonstrated that the combination improves *Wolbachia* depletion and provides an inhibitory effect on embryogenesis compared to three-week doxycycline monotherapy, although it was inferior to monotherapy with doxycycline given for four weeks with 200 mg/d [41].

#### 24.4.1.5 Rifapentine Plus Moxifloxacin

Preclinical studies in the *L. sigmodontis* mouse model indicated that combinations of different antibiotics can improve efficacy against *Wolbachia* endosymbionts and reduce the treatment time required to deplete them [45]. Matrix testing of combinations of rifamycins (rifapentine, rifampicin), gyrase inhibitors (sparfloxacin, ciprofloxacin, moxifloxacin) as well as tetracyclines and derivates (doxycycline, tigecycline, minocycline, methacycline) identified the combination of moxifloxacin and rifapentine (Table 24.1) as most effective in clearing *Wolbachia* endosymbionts of *L. sigmodontis*, allowing oral treatments as short as four days in this model [45]. Based on those results, a phase two clinical trial in onchocerciasis patients in Ghana started in 2018, evaluating a combination of 900 mg/d rifapentine plus 400 mg/d moxifloxacin given for 14 or 7 days compared to patients receiving the standard anti-*Wolbachia* therapy of 200 mg doxycycline for four weeks or untreated controls (https://doi.org/10.1186/ISRCTN43697583).

#### 24.4.1.6 Rifampicin Plus Doxycycline

Based on the efficacy of rifampicin against *Wolbachia* endosymbionts in the *L. sigmodontis* model [99] and onchocerciasis patients [42], a combination of two weeks doxycycline (200 mg/d) and 10 mg/kg/d rifampicin was compared to four weeks of 200 mg/d doxycycline in bancroftian filariasis patients [110]. Both treatments reduced *Wolbachia* within microfilariae and reduced microfilariae levels in the patients significantly. However, whereas four weeks of doxycycline mediated complete clearance of the adult worm burden, the two-week combination was less effective (50% reduction) [109]. Two phase two clinical studies subsequently tested whether prolonged administration of rifampicin at 10 mg/kg/d in combination with doxycycline for three weeks provides a shortened treatment regimen for onchocerciasis and lymphatic filariasis compared to doxycycline monotherapy. The onchocerciasis study included six treatment arms, with doxycycline given at 200 mg/d and 100 mg/d for six weeks, rifampicin given at 10 mg/kg/d for six weeks, a combination of doxycycline (200 mg/d) and rifampicin (10 mg/kg/d) for three weeks and placebo controls (https://doi.org/10.1186/ISRCTN68861628). While 100 and 200 mg/kg doxycycline given for six weeks both inhibited embryogenesis in 98.8% of female worms within recovered onchocercomata at 20 months after treatment, the combination with rifampicin and doxycycline given for three weeks inhibited embryogenesis in 87.3% of the female worms. Rifampicin treatment alone and the placebo group had inhibited embryogenesis in 64.8% and 71.2% of worms, respectively [56]. The corresponding study with lymphatic filariasis patients included seven treatment arms using four weeks of doxycycline at 200 mg/d, four and five weeks of doxycycline at 100 mg/d, 10 days, two weeks and three weeks of a combination of doxycycline (200 mg/d) plus rifampicin (10 mg/kg/d) (https://doi.org/10.1186/ISRCTN15216778). Results of this study showed that three weeks of the combination with doxycycline and rifampicin led to a significantly reduced number of patients positive for microfilariae at 12 months compared to the placebo control [75]. As mentioned above for this study, four weeks of 100 mg/d doxycycline therapy is comparable to five-week treatment with 100 mg/d doxycycline with regard to macrofilaricidal efficacy at 18 months, leading to the recommendation to use 100 mg/d doxycycline for four weeks for the treatment of lymphatic filariasis [75]. As mentioned above, PK-PD analysis suggests that 10 mg/kg/d rifampicin is suboptimal and future studies should assess combinations with rifampicin high dose of 30–35 mg/kg/d [43].

In conclusion, no combination therapy has been proven, in humans, to be equal to- or better than the current "gold standard" doxycycline monotherapy in filariasis patients. However, data from animal models suggest that monotherapy or therapy with high-dose rifampicin with and without doxycycline or the combination of moxifloxacin and rifapentine have the potential to allow shorter treatment regimens.

#### 24.4.1.7 ABBV-4083 (Flubentylosin)

ABBV-4083 is an analog of the veterinary macrolide antibiotic Tylosin A with an improved pharmacokinetic profile and efficacy (Table 24.1) against *Wolbachia* endosymbionts of filariae. Whereas Tylosin A has limited oral bioavailability, this characteristic was vastly improved with ABBV-4083 [46, 111]. Compared to doxycycline, ABBV-4083 was superior in activity against *Wolbachia* in all filariasis animal models tested (*B. malayi*, *L. sigmodontis*, and *O. ochengi*) [46, 111]. Treatment regimens of one to two weeks resulted in >90% *Wolbachia* depletion in these animal models and blocked filarial embryogenesis and microfilariae release, surpassing the efficacy of three to four weeks therapy with doxycycline or minocycline [46, 111]. Additional experiments in the *L. sigmodontis* rodent model demonstrated that ABBV-4083 induces *Wolbachia* depletion as soon as three days after treatment begins and *Wolbachia* depletion continues in the following weeks after treatment ends [47]. Furthermore, it was shown in the *L. sigmodontis* model that up to four missed ABBV-4083 treatments can be subsequently completed without impairing efficacy [47]. Beneficial preclinical safety assessments [46, 111] and completed phase 1 clinical studies support the progression of ABBV-4083 to phase 2 clinical studies (https://www.dndi.org/diseases-projects/portfolio/abbv-4083/). Thus, ABBV-4083 represents a next-generation macrofilaricidal oral drug candidate for the treatment of human filarial diseases, which may allow treatment regimens of 14 days or less. It has completed clinical phase 1 evaluation and phase 2 trials in onchocerciasis patients are ongoing.

### 24.4.2 Pre-clinical Candidates

#### 24.4.2.1 AWZ1066S

The azaquinazoline AWZ1066S (Table 24.1) is a highly selective anti-*Wolbachia* candidate with improved oral efficacy in preclinical animal models of filariasis compared to human bioequivalent doses of doxycycline and minocycline [48]. AWZ1066S provides maximal clearance of *Wolbachia* within one day of *in vitro* drug exposure, markedly reducing the time to *Wolbachia* clearance compared to other anti-*Wolbachia* drugs such as doxycycline, moxifloxacin, minocycline, and rifampicin [48]. *In vivo*, seven-day treatment with AWZ1066S achieved a *Wolbachia* reduction of ≥98% in the *B. malayi* SCID mouse and *L. sigmodontis* gerbil models and inhibited embryogenesis [48]. Microfilariae gradually declined in AWZ1066S treated *L. sigmodontis*-infected jirds beginning six weeks after starting treatment with sustained amicrofilaremia from 14 weeks after treatment start. This suggests that clearance of microfilariae is due to the inhibition of embryogenesis rather than direct microfilaricidal efficacy of AWZ1066S [48]. The potent and highly selective anti-wolbachial efficacy of AWZ1066, its promising preclinical safety assessment [48], and human PK simulations predicting >90% *Wolbachia* reduction in >90% of patients using an oral seven-day regimen with 10 mg/kg [48] are in line with the target product profile criteria for novel drug candidates for human filariasis. AWZ1066S is currently being assessed for its safety in clinical phase 1 studies.

#### 24.4.2.2 AN11251

Pleuromutilins are inhibitors of the ribosomal protein synthesis complex in Gram-positive bacteria. The boron-pleuromutilin AN11251 (Table 24.1) was identified using *in vitro* assays in which it exhibited very high potency against *Wolbachia* [49]. AN11251 had good oral bioavailability and achieved *Wolbachia* depletion of >99% in the *L. sigmodontis* mouse model in a 10–14-day treatment regimen, which was superior to the bioequivalent human dose of doxycycline [49, 112]. Preliminary *in vitro* and *in vivo* safety assessment support further evaluation of AN11251 as a preclinical anti-*Wolbachia* candidate for human filarial diseases.

#### 24.4.2.3 Corallopyronin A

Corallopyronin A is an α-pyrone ring-containing natural product from *Corallococcus coralloides* [113] (Table 24.1). It is highly effective against Gram-positive bacteria but has low efficacy against Gram-negative bacteria [114]. It is a non-competitive inhibitor of bacterial DNA-dependent RNA polymerase and is effective against rifampicin-resistant *Staphylococcus aureus* [114, 115]. Corallopyronin A, and the related myxopyronins [114, 116], targets the switch region rather than the active site like rifamycins [117, 118]. Although this compound is not typically active against Gram-negative bacteria, *Wolbachia* (with a rudimentary cell wall and lipopolysaccharide-free outer membrane) are susceptible to Corallopyronin A *in vitro* and *in vivo* [50]. *Wolbachia* were depleted by more than four logs from *L. sigmodontis* in mice compared to worms from untreated animals. In accordance with this result, Corallopyronin A is the only anti-wolbachial drug candidate reported so far in the *L. sigmodontis* model that has demonstrated a robust and reproducible reduction of adult worm burdens, alone and in combination [119]. All *in vitro* and *in vivo* safety and toxicity tests indicate that Corallopyronin A is safe and non-toxic [120], thus supporting its development to human trials (phase 1 clinical studies scheduled for 2024) by the German Center for Infection Research (DZIF, www.dzif.de).

## 24.5 Conclusion

*Wolbachia* endosymbionts of filariae are excellent targets for novel drugs, as therapy aimed at them can overcome several issues that may be present in direct acting drug candidates. Thus, *Wolbachia*-targeting drugs should be safe to administer in areas co-endemic for *L. loa*, which does not harbor these endosymbionts. Similarly, the slow decline of microfilariae levels provided by anti-wolbachials is expected to prevent serious adverse events (blindness, severe dermatitis) that are manifested following DEC treatment in onchocerciasis patients. Most importantly, anti-wolbachials have been shown to cause permanent sterility of female adult filariae and lead to a slow macrofilaricidal effect. Doxycycline is the first safe and well tolerated macrofilaricide for human use and also improves lymphedema pathology. Moreover, doxycycline therapy ameliorates pathology induced by melarsomine therapy during canine dirofilariasis and is now a part of the American Heartworm Association recommended treatment regimen. However, doxycycline therapy has limitations, including the need for extended treatment regimens and, in the case of human filariasis, the exclusion of children, and pregnant and breast-feeding women from treatment. Thus, further research is required to identify novel treatment regimens that overcome these limitations and allow shortened duration of therapy. Novel, more quickly-acting anti-wolbachials have been recently identified and combination studies using different anti-wolbachials or anti-wolbachials with albendazole have shown the feasibility of macrofilaricidal treatment regimens as short as 7–10 days. These novel treatments are urgently required to meet the Sustainable Development Goals and the WHO goal to eliminate lymphatic filariasis and onchocerciasis by 2030.

## References


**1** Taylor, M.J., Bilo, K., Cross, H.F. et al. (1999). 16S rDNA phylogeny and ultrastructural characterization of *Wolbachia* intracellular bacteria of the filarial nematodes *Brugia malayi*, *B. pahangi*, and *Wuchereria bancrofti*. *Exp. Parasitol.* 91: 356–361.

**2** Fischer, P., Wibowo, H., Pischke, S. et al. (2002). PCR-based detection and identification of the filarial parasite *Brugia timori* from Alor Island, Indonesia. *Ann. Trop. Med. Parasitol.* 96: 809–821.

**3** Henkle-Duhrsen, K., Eckelt, V.H., Wildenburg, G. et al. (1998). Gene structure, activity and localization of a catalase from intracellular bacteria in *Onchocerca volvulus*. *Mol. Biochem. Parasitol.* 96: 69–81.

**4** Tamarozzi, F., Halliday, A., Gentil, K. et al. (2011). Onchocerciasis: the role of *Wolbachia* bacterial endosymbionts in parasite biology, disease pathogenesis, and treatment. *Clin. Microbiol. Rev.* 24: 459–468.

**5** Taylor, M.J., Bandi, C., and Hoerauf, A. (2005). *Wolbachia* bacterial endosymbionts of filarial nematodes. *Adv. Parasitol.* 60: 245–284.

**6** Pfarr, K. and Hoerauf, A. (2005). The annotated genome of *Wolbachia* from the filarial nematode *Brugia malayi*: what it means for progress in antifilarial medicine. *PLoS Med.* 2: e110.

**7** Taylor, M.J. and Hoerauf, A. (1999). *Wolbachia* bacteria of filarial nematodes. *Parasitol. Today* 15: 437–442.

**8** Keiser, P.B., Coulibaly, Y., Kubofcik, J. et al. (2008). Molecular identification of *Wolbachia* from the filarial nematode *Mansonella perstans*. *Mol. Biochem. Parasitol.* 160: 123–128.

**9** Casiraghi, M., Favia, G., Cancrini, G. et al. (2001). Molecular identification of *Wolbachia* from the filarial nematode *Mansonella ozzardi*. *Parasitol. Res.* 87: 417–420.

**10** McGarry, H.F., Pfarr, K., Egerton, G. et al. (2003). Evidence against *Wolbachia* symbiosis in *Loa loa*. *Filaria J.* 2: 9.

**11** Desjardins, C.A., Cerqueira, G.C., Goldberg, J.M. et al. (2013). Genomics of *Loa loa*, a *Wolbachia*-free filarial parasite of humans. *Nat. Genet.* 45: 495–500.

**12** Büttner, D.W., Wanji, S., Bazzocchi, C. et al. (2003). Obligatory symbiotic *Wolbachia* endobacteria are absent from *Loa loa*. *Filaria J.* 2: 10.

**13** Foster, J.M., Landmann, F., Ford, L. et al. (2014). Absence of *Wolbachia* endobacteria in the human parasitic nematode *Dracunculus medinensis* and two related *Dracunculus* species infecting wildlife. *Parasites Vectors* 7: 140.

**14** Langworthy, N.G., Renz, A., Mackenstedt, U. et al. (2000). Macrofilaricidal activity of tetracycline against the filarial nematode *Onchocerca ochengi*: elimination of *Wolbachia* precedes worm death and suggests a dependent relationship. *Proc. Biol. Sci.* 267: 1063–1069.

**15** Townson, S., Hutton, D., Siemienska, J. et al. (2000). Antibiotics and *Wolbachia* in filarial nematodes: antifilarial activity of rifampicin, oxytetracycline and chloramphenicol against *Onchocerca gutturosa*, *Onchocerca lienalis* and *Brugia pahangi*. *Ann. Trop. Med. Parasitol.* 94: 801–816.

**16** Bandi, C., McCall, J.W., Genchi, C. et al. (1999). Effects of tetracycline on the filarial worms *Brugia pahangi* and *Dirofilaria immitis* and their bacterial endosymbionts *Wolbachia*. *Int. J. Parasitol.* 29: 357–364.

**17** Sironi, M., Bandi, C., Sacchi, L. et al. (1995). Molecular evidence for a close relative of the arthropod endosymbiont *Wolbachia* in a filarial worm. *Mol. Biochem. Parasitol.* 74: 223–227.

**18** Grandi, G., Morchon, R., Kramer, L. et al. (2008). *Wolbachia* in *Dirofilaria repens*, an agent causing human subcutaneous dirofilariasis. *J. Parasitol.* 94: 1421–1423.

**19** Hoerauf, A., Nissen-Pahle, K., Schmetz, C. et al. (1999). Tetracycline therapy targets intracellular bacteria in the filarial nematode *Litomosoides sigmodontis* and results in filarial infertility. *J. Clin. Invest.* 103: 11–18.

**20** Casiraghi, M., Bordenstein, S.R., Baldo, L. et al. (2005). Phylogeny of *Wolbachia* pipientis based on gltA, groEL and ftsZ gene sequences: clustering of arthropod and nematode symbionts in the F supergroup, and evidence for further diversity in the *Wolbachia* tree. *Microbiology* 151: 4015–4022.

**21** Brattig, N.W., Buttne, D.W., and Hoerauf, A. (2001). Neutrophil accumulation around *Onchocerca* worms and chemotaxis of neutrophils are dependent on *Wolbachia* endobacteria. *Microbes Infect.* 3: 439–446.

**22** Guerrero, R., Bain, O., Attout, T., and Martin, C. (2006). The infective larva of *Litomosoides yutajensis* Guerrero et al., 2003 (Nematoda: Onchocercidae), a *Wolbachia*-free filaria from bat. *Parasite* 13: 127–130.

**23** Ferri, E., Bain, O., Barbuto, M. et al. (2011). New insights into the evolution of *Wolbachia* infections in filarial nematodes inferred from a large range of screened species. *PLoS One* 6: e20843.

**24** McLaren, D.J., Worms, M.J., Laurence, B.R., and Simpson, M.G. (1975). Micro-organisms in filarial larvae (Nematoda). *Trans. R. Soc. Trop. Med. Hyg.* 69: 509–514.

**25** McLaren, D.J. (1972). Ultrastructural studies on microfilariae (Nematoda: Filarioidea). *Parasitology* 65: 317–332.

**26** Kozek, W.J. and Marroquin, H.F. (1977). Intracytoplasmic bacteria in *Onchocerca volvulus*. *Am. J. Trop. Med. Hyg.* 26: 663–678.

**27** Bosshardt, S.C., McCall, J.W., Coleman, S.U. et al. (1993). Prophylactic activity of tetracycline against *Brugia pahangi* infection in jirds (*Meriones unguiculatus*). *J. Parasitol.* 79: 775–777.

**28** Genchi, C., Sacchi, L., Bandi, C., and Venco, L. (1998). Preliminary results on the effect of tetracycline on the embryogenesis and symbiotic bacteria (*Wolbachia*) of *Dirofilaria immitis*. An update and discussion. *Parassitologia* 40: 247–249.

**29** Gilbert, J., Nfon, C.K., Makepeace, B.L. et al. (2005). Antibiotic chemotherapy of onchocerciasis: in a bovine model, killing of adult parasites requires a sustained depletion of endosymbiotic bacteria (*Wolbachia* species). *J. Infect. Dis.* 192: 1483–1493.

**30** Ghedin, E., Wang, S., Foster, J.M., and Slatko, B.E. (2004). First sequenced genome of a parasitic nematode. *Trends Parasitol.* 20: 151–153.

**31** Ghedin, E., Wang, S., Spiro, D. et al. (2007). Draft genome of the filarial nematode parasite *Brugia malayi*. *Science* 317: 1756–1760.

**32** Foster, J., Ganatra, M., Kamal, I. et al. (2005). The *Wolbachia* genome of *Brugia malayi*: endosymbiont evolution within a human pathogenic nematode. *PLoS Biol.* 3: e121.

**33** Wu, B., Novelli, J., Foster, J. et al. (2009). The heme biosynthetic pathway of the obligate *Wolbachia* endosymbiont of *Brugia malayi* as a potential anti-filarial drug target. *PLoS Negl. Trop. Dis.* 3: e475.

**34** Strubing, U., Lucius, R., Hoerauf, A., and Pfarr, K.M. (2010). Mitochondrial genes for heme-dependent respiratory chain complexes are up-regulated after depletion of *Wolbachia* from filarial nematodes. *Int. J. Parasitol.* 40: 1193–1202.

**35** Li, Z. and Carlow, C.K. (2012). Characterization of transcription factors that regulate the type IV secretion system and riboflavin biosynthesis in *Wolbachia* of *Brugia malayi*. *PLoS One* 7: e51597.

**36** Hoerauf, A., Pfarr, K., Mand, S. et al. (2011). Filariasis in Africa--treatment challenges and prospects. *Clin. Microbiol. Infect.* 17: 977–985.

**37** Mand, S., Debrah, A.Y., Klarmann, U. et al. (2012). Doxycycline improves filarial lymphedema independent of active filarial infection: a randomized controlled trial. *Clin. Infect. Dis.* 55: 621–630.

**38** Chopra, I. and Roberts, M. (2001). Tetracycline antibiotics: mode of action, applications, molecular biology, and epidemiology of bacterial resistance. *Microbiol. Mol. Biol. Rev.* 65: 232–260.

**39** Taylor, M.J., Hoerauf, A., and Bockarie, M. (2010). Lymphatic filariasis and onchocerciasis. *Lancet* 376: 1175–1185.

**40** Papich, M.G. (2017). Considerations for using minocycline vs doxycycline for treatment of canine heartworm disease. *Parasites Vectors* 10: 493.

**41** Klarmann-Schulz, U., Specht, S., Debrah, A.Y. et al. (2017). Comparison of doxycycline, minocycline, doxycycline plus albendazole and albendazole alone in their efficacy against onchocerciasis in a randomized, open-label, pilot trial. *PLoS Negl. Trop. Dis.* 11: e0005156.

**42** Specht, S., Mand, S., Marfo-Debrekyei, Y. et al. (2008). Efficacy of 2- and 4-week rifampicin treatment on the *Wolbachia* of *Onchocerca volvulus*. *Parasitol. Res.* 103: 1303–1309.

**43** Aljayyoussi, G., Tyrer, H.E., Ford, L. et al. (2017). Short-course, high-dose rifampicin achieves *Wolbachia* depletion predictive of curative outcomes in preclinical models of lymphatic filariasis and onchocerciasis. *Sci. Rep.* 7: 210.

**44** Campbell, E.A., Korzheva, N., Mustaev, A. et al. (2001). Structural mechanism for rifampicin inhibition of bacterial RNApolymerase. *Cell* 104: 901–912.

**45** Specht, S., Pfarr, K.M., Arriens, S. et al. (2018). Combinations of registered drugs reduce treatment times required to deplete *Wolbachia* in the *Litomosoides sigmodontis* mouse model. *PLoS Negl. Trop. Dis.* 12: e0006116.

**46** Taylor, M.J., von Geldern, T.W., Ford, L. et al. (2019). Preclinical development of an oral anti-*Wolbachia* macrolide drug for the treatment of lymphatic filariasis and onchocerciasis. *Sci. Transl. Med.* 11: eaau2086.

**47** Hübner, M.P., Koschel, M., Struever, D. et al. (2019). In vivo kinetics of *Wolbachia* depletion by ABBV-4083 in *L. sigmodontis* adult worms and microfilariae. *PLoS Negl. Trop. Dis.* 13: e0007636.

**48** Hong, W.D., Benayoud, F., Nixon, G.L. et al. (2019). AWZ1066S, a highly specific anti-*Wolbachia* drug candidate for a short-course treatment of filariasis. *Proc. Natl. Acad. Sci. U.S.A.* 116: 1414–1419.

**49** Jacobs, R.T., Lunde, C.S., Freund, Y.R. et al. (2019). Boron-pleuromutilins as anti-*Wolbachia* agents withpotential for treatment of onchocerciasis and lymphatic filariasis. *J. Med. Chem.* 62: 2521–25240.

**50** Schiefer, A., Schmitz, A., Schaberle, T.F. et al. (2012). Corallopyronin A specifically targets and depletes essential obligate *Wolbachia* endobacteria from filarial nematodes in vivo. *J. Infect. Dis.* 206: 249–257.

**51** Hoerauf, A., Volkmann, L., Nissen-Paehle, K. et al. (2000). Targeting of *Wolbachia* endobacteria in *Litomosoides sigmodontis*: comparison of tetracyclines with chloramphenicol, macrolides and ciprofloxacin. *Trop. Med. Int. Health* 5: 275–279.

**52** Hoerauf, A., Volkmann, L., Hamelmann, C. et al. (2000). Endosymbiotic bacteria in worms as targets for a novel chemotherapy in filariasis. *Lancet* 355: 1242–1243.

**53** Hoerauf, A., Mand, S., Adjei, O. et al. (2001). Depletion of *Wolbachia* endobacteria in *Onchocerca volvulus* by doxycycline and microfilaridermia after ivermectin treatment. *Lancet* 357: 1415–1416.

**54** Hoerauf, A., Mand, S., Fischer, K. et al. (2003). Doxycycline as a novel strategy against bancroftian filariasis-depletion of *Wolbachia* endosymbionts from *Wuchereria bancrofti* and stop of microfilaria production. *Med. Microbiol. Immunol.* 192: 211–216.

**55** Hoerauf, A., Mand, S., Volkmann, L. et al. (2003). Doxycycline in the treatment of human onchocerciasis: kinetics of *Wolbachia* endobacteria reduction and of inhibition of embryogenesis in female *Onchocerca* worms. *Microbes Infect.* 5: 261–273.

**56** Batsa-Debrah, L., Specht, S., Klarmann Schulz, U. et al. (2017). Doxycycline for the treatment of onchocerciasis: a daily dose of 100 mg for 6 weeks shows reduction of fertile female *Onchocerca volvulus* equivalent to 200 mg/d. *Am. J. Trop. Med. Hyg.* 97: 165.

**57** Debrah, A.Y., Mand, S., Marfo-Debrekyei, Y. et al. (2006). Assessment of microfilarial loads in the skin of onchocerciasis patients after treatment with different regimens of doxycycline plus ivermectin. *Filaria J.* 5: 1.

**58** Hoerauf, A., Specht, S., Buttner, M. et al. (2008). *Wolbachia* endobacteria depletion by doxycycline as antifilarial therapy has macrofilaricidal activity in onchocerciasis: a randomized placebo-controlled study. *Med. Microbiol. Immunol.* 197: 295–311.

**59** Specht, S., Hoerauf, A., Adjei, O. et al. (2009). Newly acquired *Onchocerca volvulus* filariae after doxycycline treatment. *Parasitol. Res.* 106: 23–31.

**60** Hoerauf, A., Specht, S., Marfo-Debrekyei, Y. et al. (2009). Efficacy of 5-week doxycycline treatment on adult *Onchocerca volvulus*. *Parasitol. Res.* 104: 437–447.

**61** Wanji, S., Tendongfor, N., Nji, T. et al. (2009). Community-directed delivery of doxycycline for the treatment of onchocerciasis in areas of co-endemicity with loiasis in Cameroon. *Parasites Vectors* 2: 39.

**62** Turner, J.D., Tendongfor, N., Esum, M. et al. (2010). Macrofilaricidal activity after doxycycline only treatment of *Onchocerca volvulus* in an area of *Loa loa* co-endemicity: a randomized controlled trial. *PLoS Negl. Trop. Dis.* 4: e660.

**63** Kamgno, J., Pion, S.D., Mackenzie, C.D. et al. (2009). *Loa loa* microfilarial periodicity in ivermectin-treated patients: comparison between those developing and those free of serious adverse events. *Am. J. Trop. Med. Hyg.* 81: 1056–1061.

**64** Gardon, J., Gardon-Wendel, N., Demanga, N. et al. (1997). Serious reactions after mass treatment of onchocerciasis with ivermectin in an area endemic for *Loa loa* infection. *Lancet* 350: 18–22.

**65** Boussinesq, M. (2012). Loiasis: new epidemiologic insights and proposed treatment strategy. *J. Travel Med.* 19: 140–143.

**66** Albers, A., Esum, M.E., Tendongfor, N. et al. (2012). Retarded *Onchocerca volvulus* L1 to L3 larval development in the *Simulium damnosum* vector after anti-wolbachial treatment of the human host. *Parasites Vectors* 5: 12.

**67** Tamarozzi, F., Tendongfor, N., Enyong, P.A. et al. (2012). Long term impact of large scale community-directed delivery of doxycycline for the treatment of onchocerciasis. *Parasites Vectors* 5: 53.

**68** Walker, M., Specht, S., Churcher, T.S. et al. (2014). Therapeutic efficacy and macrofilaricidal activity of doxycycline for the treatment of river blindness. *Clin. Infect. Dis.* 60: 1199–1207.

**69** Debrah, A.Y., Specht, S., Klarmann-Schulz, U. et al. (2015). Doxycycline leads to sterility and enhanced killing of female *Onchocerca volvulus* worms in an area with persistent microfilaridermia after repeated ivermectin treatment: a randomized, placebo-controlled, double-blind trial. *Clin. Infect. Dis.* 61: 517–526.

**70** Saint Andre, A., Blackwell, N.M., Hall, L.R. et al. (2002). The role of endosymbiotic *Wolbachia* bacteria in the pathogenesis of river blindness. *Science* 295: 1892–1895.

**71** Rivas-Alcala, A.R., Greene, B.M., Taylor, H.R. et al. (1981). Chemotherapy of onchocerciasis: a controlled comparison of mebendazole, levamisole, and diethylcarbamazine. *Lancet* 2 (8245): 485–490.

**72** Taylor, M.J., Makunde, W.H., McGarry, H.F. et al. (2005). Macrofilaricidal activity after doxycycline treatment of *Wuchereria bancrofti*: a double-blind, randomised placebo-controlled trial. *Lancet* 365: 2116–2121.

**73** Debrah, A.Y., Mand, S., Specht, S. et al. (2006). Doxycycline reduces plasma VEGF-C/sVEGFR-3 and improves pathology in lymphatic filariasis. *PLoS Pathog.* 2: e92.

**74** Debrah, A.Y., Mand, S., Marfo-Debrekyei, Y. et al. (2007). Macrofilaricidal effect of 4 weeks of treatment with doxycycline on *Wuchereria bancrofti*. *Trop. Med. Int. Health* 12: 1433–1441.

**75** Klarmann, U., Debrah, A.Y., Mand, S. et al. (2012). Shortening the timeframe and dosage of antiwolbachia therapy: doxycycline alone versus doxycycline plus rifampicin in their efficacy against lymphatic filariasis; a randomized, doubleblind, placebo-controlled trial. *Am. J. Trop. Med. Hyg.* 87: 157.

**76** Supali, T., Djuardi, Y., Pfarr, K.M. et al. (2008). Doxycycline treatment of *Brugia malayi*-infected persons reduces microfilaremia and adverse reactions after diethylcarbamazine and albendazole treatment. *Clin. Infect. Dis.* 46: 1385–1393.

**77** Cross, H.F., Haarbrink, M., Egerton, G. et al. (2001). Severe reactions to filarial chemotherapy and release of *Wolbachia* endosymbionts into blood. *Lancet* 358: 1873–1875.

**78** Mand, S., Pfarr, K., Sahoo, P.K. et al. (2009). Macrofilaricidal activity and amelioration of lymphatic pathology in bancroftian filariasis after 3 weeks of doxycycline followed by single-dose diethylcarbamazine. *Am. J. Trop. Med. Hyg.* 81: 702–711.

**79** Batsa Debrah, L., Phillips, R.O., Pfarr, K. et al. (2019). The efficacy of doxycycline treatment on *Mansonella perstans* infection: an open-label, randomized trial in Ghana. *Am. J. Trop. Med. Hyg.* 101: 84–92.

**80** WHO (2019). Progress in eliminating onchocerciasis in the WHO Region of the Americas: doxycycline treatment as an end-game strategy. *Wkly. Epidemiol. Rep.* 37: 415–419.

**81** Turner, J.D., Mand, S., Debrah, A.Y. et al. (2006). A randomized, double-blind clinical trial of a 3-week course of doxycycline plus albendazole and ivermectin for the treatment of *Wuchereria bancrofti* infection. *Clin. Infect. Dis.* 42: 1081–1089.

**82** Dreyer, G., Medeiros, Z., Netto, M.J. et al. (1999). Acute attacks in the extremities of persons living in an area endemic for bancroftian filariasis: differentiation of two syndromes. *Trans. R. Soc. Trop. Med. Hyg.* 93: 413–417.

**83** Bennuru, S., Maldarelli, G., Kumaraswami, V. et al. (2010). Elevated levels of plasma angiogenic factors are associated with human lymphatic filarial infections. *Am. J. Trop. Med. Hyg.* 83: 884–890.

**84** McCall, J.W., Kramer, L., Genchi, C. et al. (2014). Effects of doxycycline on heartworm embryogenesis, transmission, circulating microfilaria, and adult worms in microfilaremic dogs. *Vet. Parasitol.* 206: 5–13.

**85** Bazzocchi, C., Mortarino, M., Grandi, G. et al. (2008). Combined ivermectin and doxycycline treatment has microfilaricidal and adulticidal activity against *Dirofilaria immitis* in experimentally infected dogs. *Int. J. Parasitol.* 38: 1401–1410.

**86** McCall, J.W., Genchi, C., Kramer, L. et al. (2008). Heartworm and *Wolbachia*: therapeutic implications. *Vet. Parasitol.* 158: 204–214.

**87** Kramer, L., Grandi, G., Passeri, B. et al. (2011). Evaluation of lung pathology in *Dirofilaria immitis*-experimentally infected dogs treated with doxycycline or a combination of doxycycline and ivermectin before administration of melarsomine dihydrochloride. *Vet. Parasitol.* 176: 357–360.

**88** Nelson, C.T., Myrick, E.S., and Nelson, T.A. (2017). Clinical benefits of incorporating doxycycline into a canine heartworm treatment protocol. *Parasites Vectors* 10: 515.

**89** Serrano-Parreno, B., Carreton, E., Caro-Vadillo, A. et al. (2017). Pulmonary hypertension in dogs with heartworm before and after the adulticide protocol recommended by the American Heartworm Society. *Vet. Parasitol.* 236: 34–37.

**90** Datta, S., Maitra, S., Gayen, P., and Sinha Babu, S.P. (2009). Improved efficacy of tetracycline by acaciasides on *Dirofilaria immitis*. *Parasitol. Res.* 105: 697–702.

**91** Sharma, R., Al Jayoussi, G., Tyrer, H.E. et al. (2016). Minocycline as a re-purposed anti-*Wolbachia* macrofilaricide: superiority compared with doxycycline regimens in a murine infection model of human lymphatic filariasis. *Sci. Rep.* 6: 23458.

**92** Savadelis, M.D., Day, K.M., Bradner, J.L. et al. (2018). Efficacy and side effects of doxycycline versus minocycline in the three-dose melarsomine canine adulticidal heartworm treatment protocol. *Parasites Vectors* 11: 671.

**93** Clare, R.H., Bardelle, C., Harper, P. et al. (2019). Industrial scale high-throughput screening delivers multiple fast acting macrofilaricides. *Nat. Commun.* 10: 11.

**94** Johnston, K.L., Cook, D.A.N., Berry, N.G. et al. (2017). Identification and prioritization of novel anti-Wolbachia chemotypes from screening a 10,000-compound diversity library. *Sci. Adv.* 3: eaao1551.

**95** Pionnier, N.P., Sjoberg, H., Chunda, V.C. et al. (2019). Mouse models of *Loa loa*. *Nat. Commun.* 10: 1429.

**96** Halliday, A., Guimaraes, A.F., Tyrer, H.E. et al. (2014). A murine macrofilaricide pre-clinical screening model for onchocerciasis and lymphatic filariasis. *Parasites Vectors* 7: 472.

**97** Bakowski, M.A. and McNamara, C.W. (2019). Advances in antiwolbachial drug discovery for treatment of parasitic filarial worm infections. *Trop. Med. Infect. Dis.* 4: 108.

**98** Hübner, M.P., Torrero, M.N., McCall, J.W., and Mitre, E. (2009). *Litomosoides sigmodontis*: a simple method to infect mice with L3 larvae obtained from the pleural space of recently infected jirds (*Meriones unguiculatus*). *Exp. Parasitol.* 123: 95–98.

**99** Volkmann, L., Fischer, K., Taylor, M., and Hoerauf, A. (2003). Antibiotic therapy in murine filariasis (*Litomosoides sigmodontis*): comparative effects of doxycycline and rifampicin on *Wolbachia* and filarial viability. *Trop. Med. Int. Health* 8: 392–401.

**100** Boeree, M.J., Diacon, A.H., Dawson, R. et al. (2015). A dose-ranging trial to optimize the dose of rifampin in the treatment of tuberculosis. *Am. J. Respir. Crit. Care Med.* 191: 1058–1065.

**101** Boeree, M.J., Heinrich, N., Aarnoutse, R. et al. (2017). High-dose rifampicin, moxifloxacin, and SQ109 for treating tuberculosis: a multi-arm, multi-stage randomised controlled trial. *Lancet Infect. Dis.* 17: 39–49.

**102** Velasquez, G.E., Brooks, M.B., Coit, J.M. et al. (2018). Efficacy and safety of high-dose rifampin in pulmonary tuberculosis. A randomized controlled trial. *Am. J. Respir. Crit. Care Med.* 198: 657–666.

**103** Gruneberg, R.N., Emmerson, A.M., and Cremer, A.W. (1985). Rifampicin for non-tuberculous infections? *Chemotherapy* 31: 324–328.

**104** Pion, S.D.S., Chesnais, C.B., Weil, G.J. et al. (2017). Effect of 3 years of biannual mass drug administration with albendazole on lymphatic filariasis and soil-transmitted helminth infections: a community-based study in Republic of the Congo. *Lancet Infect. Dis.* 17: 763–769.

**105** Awadzi, K., Hero, M., Opoku, O. et al. (1991). The chemotherapy of onchocerciasis. XV. Studies with albendazole. *Trop. Med. Parasitol.* 42: 356–360.

**106** Cline, B.L., Hernandez, J.L., Mather, F.J. et al. (1992). Albendazole in the treatment of onchocerciasis: double-blind clinical trial in Venezuela. *Am. J. Trop. Med. Hyg.* 47: 512–520.

**107** Macfarlane, C.L., Budhathoki, S.S., Johnson, S. et al. (2019). Albendazole alone or in combination with microfilaricidal drugs for lymphatic filariasis. *Cochrane Database Syst. Rev.* 1: Cd003753.

**108** Batsa Debrah, L., Klarmann-Schulz, U., Osei-Mensah, J. et al. (2020). Comparison of repeated doses of ivermectin versus ivermectin plus albendazole for treatment of onchocerciasis – a randomized open-label clinical trial. *Clin. Infect. Dis.* 71: 933–943.

**109** Turner, J.D., Sharma, R., Al Jayoussi, G. et al. (2017). Albendazole and antibiotics synergize to deliver short-course anti-*Wolbachia* curative treatments in preclinical models of filariasis. *Proc. Natl. Acad. Sci. U.S.A.* 114: E9712–E9721.

**110** Debrah, A.Y., Mand, S., Marfo-Debrekyei, Y. et al. (2011). Macrofilaricidal activity in *Wuchereria bancrofti* after 2 weeks treatment with a combination of rifampicin plus doxycycline. *J. Parasitol. Res.* 2011: 201617.

**111** von Geldern, T.W., Morton, H.E., Clark, R.F. et al. (2019). Discovery of ABBV-4083, a novel analog of Tylosin A that has potent anti-*Wolbachia* and anti-filarial activity. *PLoS Negl. Trop. Dis.* 13: e0007159.

**112** Ehrens, A., Lunde, C.S., Jacobs, R.T. et al. (2020). In vivo efficacy of the boron-pleuromutilin AN11251 against *Wolbachia* of the rodent filarial nematode *Litomosoides sigmodontis*. *PLoS Negl. Trop. Dis.* 14: e0007957.

**113** Schaberle, T.F., Schmitz, A., Zocher, G. et al. (2015). Insights into structure-activity relationships of bacterial RNA polymerase inhibiting corallopyronin derivatives. *J. Nat. Prod.* 78: 2505–2509.

**114** Irschik, H., Jansen, R., Hofle, G. et al. (1985). The corallopyronins, new inhibitors of bacterial RNA synthesis from *Myxobacteria*. *J. Antibiot.* 38: 145–152.

**115** O'Neill, A., Oliva, B., Storey, C. et al. (2000). RNA polymerase inhibitors with activity against rifampin-resistant mutants of *Staphylococcus aureus*. *Antimicrob. Agents Chemother.* 44: 3163–3166.

**116** Rentsch, A. and Kalesse, M. (2012). The total synthesis of corallopyronin A and myxopyronin B. *Angew. Chem. Int. Ed.* 51: 11381–11384.

**117** Belogurov, G.A., Vassylyeva, M.N., Sevostyanova, A. et al. (2009). Transcription inactivation through local refolding of the RNA polymerase structure. *Nature* 457: 332–335.

**118** Mukhopadhyay, J., Das, K., Ismail, S. et al. (2008). The RNA polymerase "switch region" is a target for inhibitors. *Cell* 135: 295–307.

**119** Schiefer, A., Hübner, M.P., Krome, A. et al. (2020). Corallopyronin A for short-course anti-wolbachial, macrofilaricidal treatment of filarial infections. *PLoS Negl. Trop. Dis.* 14: e0008930.

**120** Pfarr, K., Schiefer, A., Krome, A. et al. (2018). Corallopyronin A: an effective antiwolbachial compound for the treatment of filarial infections and antibiotic for antimicrobial resistant STIs. *Am. J. Trop. Med. Hyg.* 99: 667–668.

# Index



*Human and Animal Filariases: Landscape, Challenges, and Control*, First Edition.
Edited by Ronald Kaminsky and Timothy G. Geary.

## *p*